AF844187

P91-1525

TN
871
.W44
1990

Developments in Petroleum Science, 28

well cementing

EDITED BY

ERIK B. NELSON

Dowell Schlumberger, Z. I. de Molina la Chazotte
42000 St. Etienne, FRANCE

SHELL WESTERN E&P INC.
WOODCREEK LIBRARY
P. O. BOX 4423
HOUSTON, TX 77210

ELSEVIER

Amsterdam — Oxford — New York — Tokyo 1990

ELSEVIER SCIENCE PUBLISHERS B.V.
Sara Burgerhartstraat 25
P.O. Box 211, 1000 AE Amsterdam, The Netherlands

Distributors for the United States and Canada:

ELSEVIER SCIENCE PUBLISHING COMPANY INC.
655, Avenue of the Americas
New York, NY 10010, U.S.A.

ISBN 0-444-88751-2 (Vol. 28)

© Schlumberger Educational Services, 5000 Gulf Freeway, Houston, Texas 77023

All rights reserved. No part of this publication may be reproduced, stored in a retrieval system or transmitted in any form or by any means, electronic, mechanical, photocopying, recording or otherwise, without the prior written permission of the publisher, Elsevier Science Publishers B.V./ Physical Sciences & Engineering Division, P.O. Box 330, 1000 AH Amsterdam, The Netherlands.

Special regulations for readers in the USA – This publication has been registered with the Copyright Clearance Center Inc. (CCC), Salem, Massachusetts. Information can be obtained from the CCC about conditions under which photocopies of parts of this publication may be made in the USA. All other copyright questions, including photocopying outside of the USA, should be referred to the publisher.

No responsibility is assumed by the publisher for any injury and/or damage to persons or property as a matter of products liability, negligence or otherwise, or from any use or operation of any methods, products, instructions or ideas contained in the material herein.

Printed in The Netherlands

Contents

ACKNOWLEDGMENTS

Following the success of *Reservoir Stimulation* (edited by M. J. Economides and K. G. Nolte), Schlumberger Educational Services (SES) decided to produce a companion work concerning well cementing technology. In early 1988, I was invited to organize the project and serve as the editor. In light of the high standards set by previous cementing texts, I accepted the task (my first foray into such territory) with not a little trepidation. It is my sincere hope that the industry will find the result, *Well Cementing*, to be a worthy addition to the petroleum literature. During the two-year gestation period of *Well Cementing*, I have become deeply indebted to many people and organizations without whose generous assistance this project could never have been completed.

The SES production team was headed by Bill Diggons. His positive attitude and patience were very much appreciated. The production manager, Martha Dutton, shepherded this project through many difficulties. Her dedication and perseverance far exceeded the call of duty. Our proofreader, Judith Barton, was involved through the duration of the project, from the initial manuscript drafts to the final layout. Her meticulous attention to grammar, composition, and style greatly improved the readability of each chapter. To give the textbook a consistent "look," artists Martha Dutton, Patti McKee, Mike Mitchell, and Doug Slovak were obliged to redraw virtually all of the graphic material submitted by the authors. In many cases they worked miracles, transforming very rough drawings into clear and coherent illustrations. Layout and typesetting were performed by Publishing Resource Group, headed by Kathy Rubin, and assisted by Susan Price. The references were diligently researched by Rana Rottenberg. I would also like to thank Brigitte Barthelemy, Pat Hoffman, Chris Jones, Sharon Jurek, and Norma McCombs for their fine efforts.

This textbook has benefitted substantially from the technical assistance of many people who reviewed the material and suggested corrections and changes. I wish to express gratitude to the following who gave so generously of their time—Robert Beirute (Amoco), George Birch (Dowell Schlumberger), Simon Bittleston (Schlumberger Cambridge Research), Gary Briggs (Shell), D. G. Calvert (Mobil), Robert Cooper (Dowell Schlumberger), K. M. Cowan (Shell), Michael J. Economides (Mining University, Leoben, Austria), W. H. Grant (Chevron), Tom Griffin (Dowell Schlumberger), Jacques Jutten (Dowell Schlumberger), S. R. Keller (Exxon), Johnny Love (LaFarge Cement), Geoff Maitland (Schlumberger Cambridge Research), Gilles Michel (Dowell Schlumberger), Larry K. Moran (Conoco), Anthony Pearson (Schlumberger Cambridge Research), Phil Rae (Dowell Schlumberger), Michel Richebourg (Dowell Schlumberger), Robert C. Smith (Amoco), and Terry R. Smith (Shell).

I am also most grateful to many publishing companies and organizations, especially the Society of Petroleum Engineers and the American Petroleum Institute, for the permission to reproduce tables and figures from their publications.

Finally, special thanks go to Chris Hall who, being a veteran of multi-author textbook production, provided much valuable advice and moral support.

Erik B. Nelson
Saint-Etienne, France
16 March 1990

Preface

Robert C. Smith

Amoco Production Company

OBJECTIVES OF PRIMARY CEMENTING

Primary cementing is the process of placing cement in the annulus between the casing and the formations exposed to the wellbore. Since its inception in 1903, the major objective of primary cementing has always been to provide zonal isolation in the wellbore of oil, gas, and water wells (Smith, 1984; Smith, 1987), e.g., to exclude fluids such as water or gas in one zone from oil in another zone. To achieve this objective, a hydraulic seal must be obtained between the casing and the cement, and between the cement and the formations, while at the same time preventing fluid channels in the cement sheath (Fig. 1). This requirement makes primary cementing the most important operation performed on a well. Without complete zonal isolation in the wellbore, the well may never reach its full producing potential. Remedial work required to repair a faulty cementing job may do irreparable harm to the producing formation. In addition to the possibility of lost reserves and lower producing rates, start-up of production (revenue) is delayed. Other problems may arise, such as not being able to confine stimulation treatments to the producing zone, or confining secondary and tertiary fields to the pay zone.

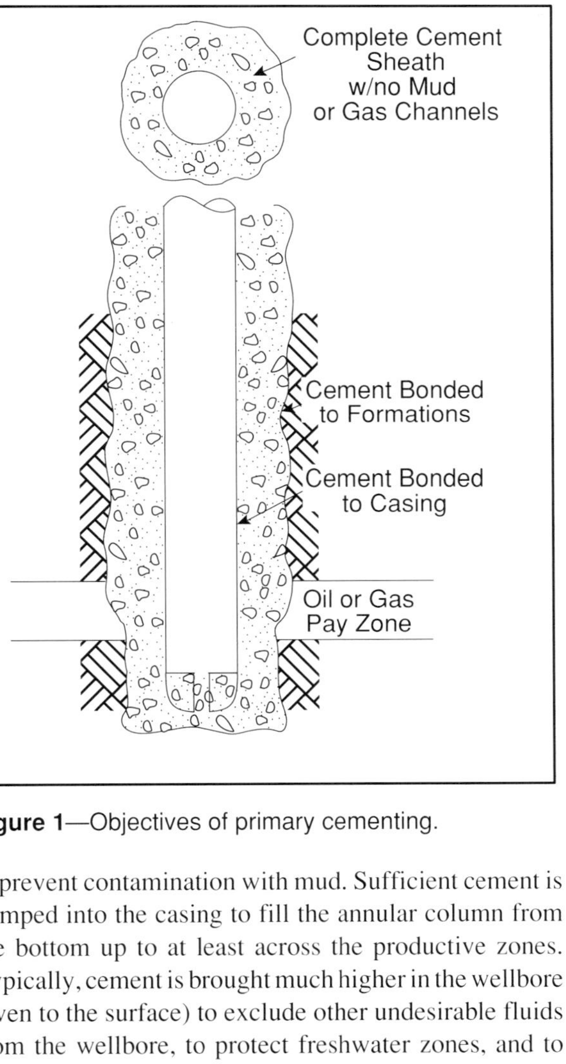

Figure 1—Objectives of primary cementing.

THE BASIC CEMENTING PROCESS

The basic process for accomplishing a primary cementing job uses the two-plug method for pumping and displacement. This method was first used in 1910 in shallow wells in California (Smith, 1987). After drilling the well to the desired depth, the drillpipe is removed and a larger string of casing is run into the well until it reaches the bottom of the well. At this time, the drilling mud used to remove formation cuttings during drilling the well is still in the wellbore. This mud must be removed and replaced with hardened cement. The process to accomplish this is the two-plug cementing method (Fig. 2). Two plugs are used to isolate the cement as it is pumped down the casing to prevent contamination with mud. Sufficient cement is pumped into the casing to fill the annular column from the bottom up to at least across the productive zones. Typically, cement is brought much higher in the wellbore (even to the surface) to exclude other undesirable fluids from the wellbore, to protect freshwater zones, and to protect the casing from corrosion. The cementing process is completed when a pressure increase at the surface indicates the top plug has reached the landing collar, or float collar, and displacement with mud or water is termi-

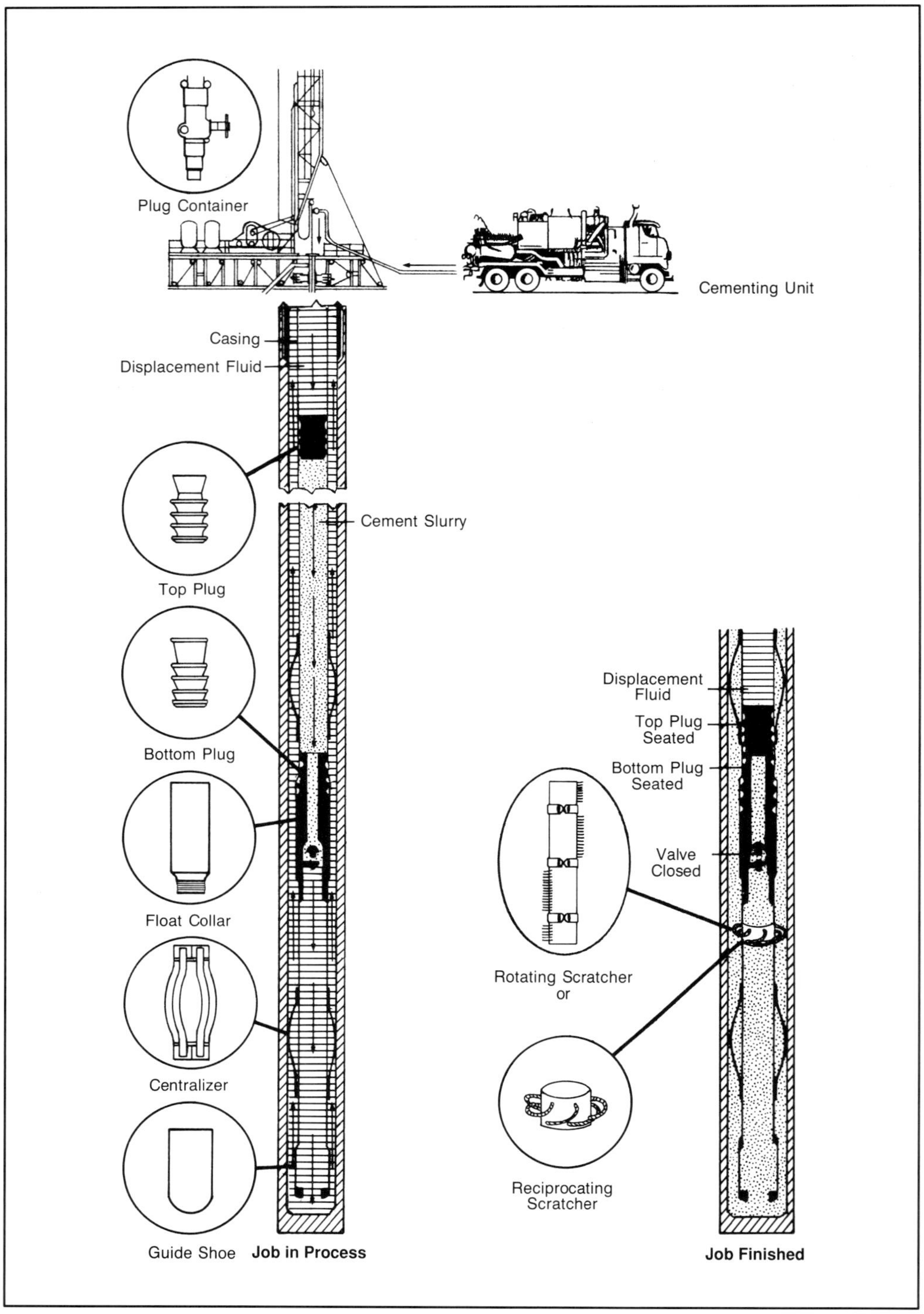

Figure 2—Typical primary cementing job.

nated. The well is left shut in for a time to allow the cement to harden before beginning completion work or drilling out to a deeper horizon.

Although wells are drilled deeper today (30,000 ft or more), technology has advanced, and cementing practices have changed, the basic two-plug cementing method described above is still used today. The advances that have been made since then have been aimed at engineering the job for the application, and doing it at the lowest cost. Let's examine some of the major technological advances that have been made down through history, and how some cementing practices have changed.

TECHNOLOGICAL ADVANCES

Available Cements

During the early days, only one or two cements were available for cementing. As wells became deeper, more flexibility in cement performance was required than could be achieved with available cements. It was with the advent of the API Standardization Committee in 1937 that more and better cements were developed (Smith, 1987). Today, eight API classes of cements are available, each with distinct characteristics (API, 1984).

Cement Additives

Cement additives have played an important role in the advancement of cementing technology. To properly use the available cements, additives were developed to control the major cement properties, i.e., thickening time, consistency, fluid-loss rate, free water, setting time, etc. Consequently, a wide variety of cement additives is now available to alter cement properties to meet most well conditions. For example, calcium lignosulfonates and other retarders maintain the cement in a slurry form to allow long pumping times for great depths and at high bottomhole temperatures.

Fluid-Loss Control

Perhaps one of the most notable developments among all the additives is the one that controls the fluid-loss rate of the cement and maintains the proper water-to-cement ratio. These additives made their debut in the early 1950s in response to deeper drilling below 10,000 to 12,000 ft. For a cement to be pumpable, excess water above that required for proper hydration is required. Some or all of this excess water can be easily squeezed from the slurry, if the cement encounters a permeable formation in the wellbore during the cement job. The loss of only a portion of this water can significantly alter the cement properties. Thickening time, for example, is decreased with water loss. At the deeper depths where longer pump times are required, thickening times must be predictable. Any change in the water ratio downhole can drastically reduce the thickening time, such that the job is terminated prematurely. If a high portion of the excess water is squeezed from the slurry, the cement may experience what many call a "flash set." At this point, the cement is no longer pumpable and the job is terminated prematurely. Fluid-loss additives tie up the excess water, and prevent it from being squeezed from the slurry (Shell and Wynne, 1958). Usually, when a job is terminated prematurely, remedial work is required.

Reduction in WOC Time

In the early 1960s, a significant development occurred in cement design which has allowed tremendous savings in rig costs to be realized. This was made possible by reducing the time for the cement to harden, the waiting-on-cement (WOC) time. During the early days, WOC time averaged 10 days and in some instances up to 28 days before operations could be resumed. As late as 1961, the WOC time still averaged about 24 hours. The cost of rig days was considerable. In 1961, a technique for reducing this time to as little as eight hours surfaced (Bearden and Lane, 1961). The tensile strength of cement required to support pipe and allow drillout operations to resume was determined to be only 8 psi. To achieve this strength at the earliest possible time required proper use of accelerators to obtain early strength development. The projected savings to an industry that drilled 45,000 wells per year was 30,000 rig days per year based on cutting the WOC time from 24 hours to 8 hours. In the peak years of the 1980s when the industry drilled over 80,000 wells per year, the rig-day savings was even more dramatic.

Density-Altering Additives

The density of neat cement, i.e., water and cement, varies from 14.8 to 16.4 lb/gal depending on the API Class of cement used. In many cases of high bottomhole formation pressures, this density is too low to control the well fluids. In other cases, lower density cements are required to prevent lost circulation during the cement job. Many additives have been developed to control and meet density requirements. The groupings are shown in Fig. 3 for the most common additives (Smith, 1984). The heavy

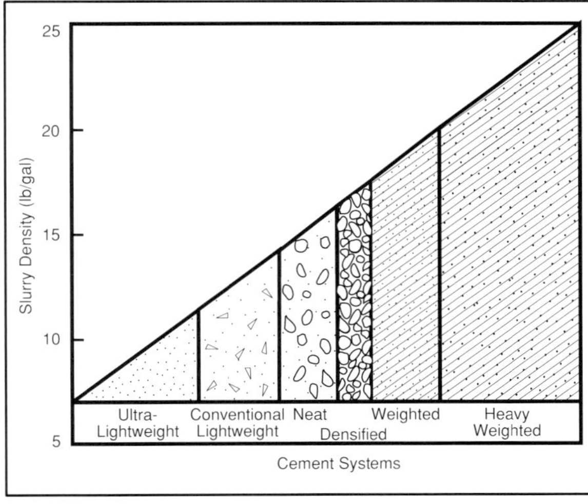

Figure 3—Density-altering additives vs. slurry density within which they are used.

materials add weight to the slurry to achieve higher densities. To lower the density, other additives either allow large quantities of lightweight water to be added to the cement, or they are low specific gravity materials, or they impart a combination of these effects.

Testing Equipment

One of the most outstanding developments of mechanical testing devices for cement slurry design was the high-temperature, high-pressure thickening time tester developed in 1939 by R. F. Farris (retired, Amoco Production Company) (Smith, 1987). This device allowed a more accurate determination of the thickening time of cement slurries under a simulated downhole environment of temperature and pressure. This device continues to be the standard for the industry 50 years later, and is part of the API Specification 10 for well cements.

Flow After Cementing

Perhaps the most important development for deeper high-pressure gas wells has been the control of flow after cementing. Without proper slurry design, natural gas can invade and flow through the cement matrix during the WOC time. This gas must be prevented from invading the cement. Failure to prevent gas migration can cause such problems as high annular pressures at the surface, blowouts, poor zonal isolation, loss of gas to nonproductive zones, poor stimulation, low producing rates, etc. All of these are costly to correct. It is generally acknowledged in the industry that the mechanism that allows gas invasion into the cement matrix is the gel-strength development of the slurry as it changes from a liquid to a solid. In this condition, the cement loses its ability to transmit hydrostatic pressure, and gas invasion may occur. Other mechanisms include excessive fluid loss, bridging, and the formation of microannuli.

There are several successful methods (Cheung and Beirute, 1985; Garcia and Clark, 1976; Webster and Eikerts, 1979; Bannister et al., 1983; Tinsley et al.; 1980; Griffin et al., 1979) to control gas migration as shown in Fig. 4, each with its advantages. Usually a combination of methods works best. In selecting optimum methods for controlling gas migration, many well conditions must be considered: formation pressure, permeability, gas flow rate, bottomhole temperature, wellbore geometry, well deviation, height of the cement column, and formation fracture pressure.

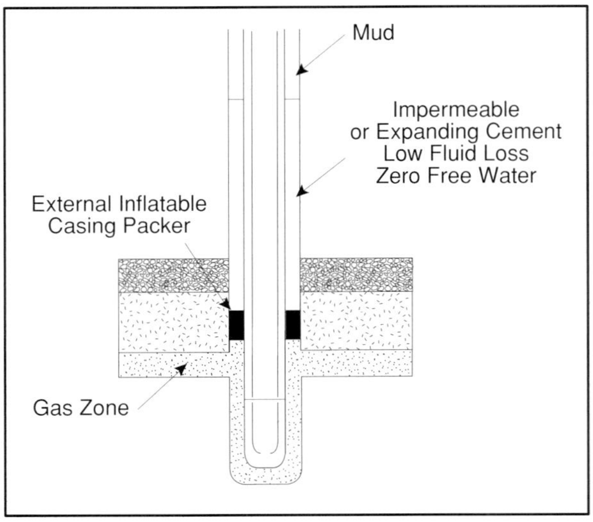

Figure 4—Methods of preventing flow after cementing.

WELL PREPARATION AND HOLE CONDITIONING

Uppermost in all planning and drilling decisions must be that the wellbore be cementable. The ideal cementable wellbore (Smith, 1984; Shryock and Smith, 1980) and its requirements are shown in Fig. P–5. The drillers must keep these requirements foremost in all plans. It is im-

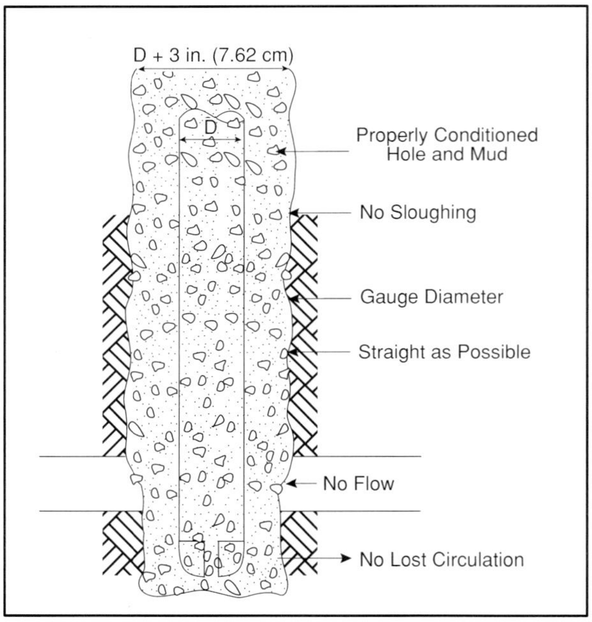

Figure 5—Ideal cementable wellbore requirements.

perative that the cementable wellbore not be sacrificed in the efforts to reduce drilling days and mud costs. The cost of repairing a faulty cement job can far exceed savings in drilling costs.

Mud displacement efficiency during the cementing job can be enhanced by properly conditioning the mud (Clark and Carter, 1973; Haut and Crook, 1980). This is one phase of the entire operation that should not be rushed—up to 24 hours may be required to properly condition the mud and wellbore after the casing is on the bottom. At best, a cement slurry can only follow the path of the drilling mud circulating ahead of it in the annulus. Therefore, the time required to properly condition the mud and the hole will be very well spent. Centralization of the casing, as well as pipe movement during mud conditioning and cementing, also improves the chances for a successful cement job. Beneficial results are obtained with either pipe reciprocation or rotation, or both simultaneously.

JOB EXECUTION AND MONITORING

Currently, technology is expanding rapidly in the area of job execution. This is a process that has gained momentum over the past 10 years. During this time, equipment and techniques have been developed to properly monitor all of the many parameters of a cement job (Smith, 1982; Beirute, 1984; Smith, 1984). In turn, this allows timely decisions to make changes during execution to improve job success. Recorded data normally include pump rate in, annulus rate out, wellhead pressure (at the cementing head), density of fluids pumped in and those returning (using radioactivity devices or equivalent), cumulative displacement volume, cumulative return volume, and hook load during pipe reciprocation (Smith, 1984). To enable the job supervisor to make timely decisions, a central monitoring point, such as a monitoring van or portable electronic data recorder, is useful (Smith, 1984).

OTHER ADVANCES

In a short preface, it is impossible to cover all of the important technological developments that have occurred over the years. A discussion of these advances would fill a complete volume. Suffice it to say that in my opinion, adequate technology is available to successfully cement, on the first attempt, over 90% of the wells drilled. This technology is available in the other major areas of consideration not discussed above, such as slurry design (Smith, 1987; Suman and Ellis, 1977; API Task Group, 1977; Venditto and George, 1984; API, 1984), blending of bulk materials (Pace et al., 1984; Gerke et al., 1985), slurry mixing, casing hardware, and quality control (Clark and Carter, 1973). Each area requires special attention and offers many challenges.

REFERENCES

API Task Group: "Better Temperature Readings Promise Better Cement Jobs," Drilling (Aug. 1977).

API, API Specifications for Materials and Testing for Well Cements, Second Edition, API Spec. 10, Dallas (1984).

Bannister, C. E., Shuster, G. E., Wooldridge, L. A., Jones, M. J., and Birch, A. G.: "Critical Design Parameters to Prevent Gas Invasion During Cementing Operations," paper SPE 11982, 1983.

Bearden, W. G. and Lane, R. D.: "You Can Engineer Cementing Operations to Eliminate Wasteful WOC Time," Oil and Gas J. (July 3, 1961), p. 104.

Beirute, R. M.: "The Phenomenon of Free Fall During Primary Cementing," paper SPE 13045, 1984.

Cheung, P. R. and Beirute, R. M.: "Gas Flow in Cements," JPT (June 1985) 1041–1048.

Clark, C. R. and Carter, L. G.: "Mud Displacement With Cement Slurries," JPT (July 1973) 775–783.

Garcia, J. A. and Clark, C. R.: "An Investigation of Annular Gas Flow Following Cementing Operations," paper SPE 5701, 1976.

Gerke, R. R., Simon, J. M., Logan, J. L. and Sabins, F. L.: "A Study of Bulk Cement Handling and Testing Procedures," paper SPE 14196, 1985.

Griffin, T. J., Spangle, L. B., and Nelson, E. B.: "New Expanding Cement Promotes Better Bonding," Oil and Gas Journal (June 25, 1979) 143–151.

Haut, R. C. and Crook, R. J., Jr.: "Primary Cementing: Optimized for Maximum Mud Displacement," World Oil (Nov. 1980).

Pace, R. S., McElfresh, P. M., Cobb, J. A., Smith C. L. and Olsberg, M. A.: "Improved Bulk Blending Techniques for Accurate and Uniform Cement Blends," paper SPE 13041, 1984.

Shell, F. J. and Wynne, R. A.: "Application of Low-Water Loss Cement Slurries," API Paper No. 875–12–1, Spring Meeting of Rocky Mtn. District, Denver, CO, 21–23 April, 1958.

Shryock, S. H. and Smith, D. K.: "Geothermal Cementing—The State-of-the-Art," Halliburton Services Brochure C–1274 (1980).

Smith, D. K.: Cementing, Monograph Series, SPE, Dallas (1987).

Smith, R. C.: "Successful Primary Cementing Can Be a Reality," JPT (Nov. 1984) 1851–1858.

Smith, R. C.: "Successful Primary Cementing Checklist," Oil and Gas J. (Nov. 1, 1982).

Suman, G. O., Jr. and Ellis, R. C.: "Cementing Handbook," World Oil (1977).

Tinsley, J. M., Miller, E. C., and Sutton, D. L.: "Study of Factors Causing Annular Gas Flow Following Primary Cementing," JPT (Aug. 1980) 1427–1437.

Venditto, J. J. and George, C. R.: "Better Wellbore Temperature Data Equal Better Cement Job," World Oil (Feb. 1984)

Webster, W. W. and Eikerts, J. V.: "Flow After Cementing—A Field Study and Laboratory Model," paper SPE 8259, 1979.

Introduction

Erik B. Nelson

Dowell Schlumberger

Well cementing technology is an amalgam of many interdependent scientific and engineering disciplines, including chemistry, geology, physics, and petroleum, mechanical, and electrical engineering. Each is essential to achieve the primary goal of well cementing—zonal isolation. By preparing this textbook, the authors have aspired to produce a comprehensive and up-to-date reference concerning the application of these disciplines toward cementing a well.

Well Cementing is organized generally in four principal sections. The first section (comprised only of Chapter 1) applies reservoir engineering concepts to illustrate how the quality of the hydraulic seal provided by the cement sheath can affect well performance. The second section (Chapters 2 through 11) presents information which must be considered during the design phase of a cementing treatment. Various aspects of cement job exeScution are covered in the third section (Chapters 12 through 15). The fourth section (Chapter 16) addresses cement job evaluation.

In the Preface, Robert C. Smith states that "primary cementing is the most important operation performed on a well." Indeed, from operational experience, few would dispute that no other event has a greater impact on the production potential of a well. Yet it is interesting to note that very little work has been published regarding the quantification of zonal isolation from a reservoir engineering point of view. In Chapter 1, common reservoir engineering concepts are used to derive a theoretical Index of Zonal Isolation (IZI), which can be used to calculate the maximum tolerable cement sheath permeability (matrix and interfacial). The IZI concept is subsequently applied to typical wellbore scenarios, and the results further underscore the critical importance of cement sheath integrity.

Chapter 2 is concerned with the central unifying theme of this textbook—Portland cement. The physical and chemical properties, and the performance of this remarkable material, are crucial to every facet of well cementing technology. This chapter presents (in a well cementing context) a review of the manufacture, chemical composition, hydration chemistry, and classification of Portland cements.

Well cementing exposes Portland cement to conditions far different from those anticipated by its inventor. Cement systems must be designed to be pumped under conditions ranging from below freezing in permafrost zones to greater than 1,000° F (538°C) in some thermal recovery wells. After placement, the cement systems must preserve their integrity and provide zonal isolation during the life of the well. It has only been possible to accommodate such a wide range of conditions through the development of additives which modify the available Portland cements for individual well requirements. The impressive array of cement additives used in the well cementing industry is discussed in Chapter 3. The chemical nature of the various classes of additives is described, and typical performance data are provided. In addition, building upon the material presented in Chapter 2, the mechanisms by which the additives operate are also explained.

The rheology of well cement systems is discussed in Chapter 4. A review of the relevant rheological models and concepts is presented, followed by a discussion specific to particle-laden fluids. The rheological behavior of a cement slurry must be optimized to effectively remove drilling mud from the annulus. The appropriate cement slurry design is a function of many parameters, including the wellbore geometry, casing hardware, formation integrity, drilling mud characteristics, presence of spacers and washes, and mixing conditions. A large amount of theoretical and experimental work concerning mud removal has been performed since 1940, yet this subject remains controversial today. Chapter 5 is a review of the work performed to date, contrasting the opposing viewpoints, and distilling some mud removal guidelines with which the majority of workers in this field would agree.

The interactions between cement systems and the formations with which they come into contact are important topics. Such interactions encompass three principal effects—fluid loss, formation damage, and lost circulation. It is generally acknowledged that an inappropriate level of fluid-loss control is often responsible for primary and remedial cementing failures. In addition, invasion of cement filtrate into the formation may be damaging to production. Chapter 6 is a discussion of static and dynamic fluid-loss processes, the deposition of cement filter cakes on formation surfaces, and the influence of a previously deposited mudcake on the fluid-loss process. Another section of Chapter 6 is a review of methods for preventing or correcting lost circulation. Since lost circulation is best attacked before the cementing process is initiated, the treatment of this problem during drilling is also presented.

As well cementing technology has advanced, many problems have been encountered for which special cement systems have been developed. Cement technologies specific to such problems as slurry fallback, lost circulation, microannuli, salt formations, permafrost, and corrosive well environments are presented in Chapter 7. The compositions of the cement systems (several of which do not involve Portland cement) are explained, and typical performance data are provided.

Annular gas migration has been a topic of intense interest and controversy for many years, and a thorough review is presented in Chapter 8. This complex phenomenon may occur during drilling or well completion procedures, and has long been recognized as one of the most troublesome problems of the petroleum industry. The causes and consequences of gas migration are discussed, and theoretical and experimental models are described. In addition, methods to predict and solve gas migration problems are discussed.

The physical and chemical behavior of well cements changes significantly at high temperatures and pressures; consequently, special guidelines must be followed to design cement systems which will provide adequate casing protection and zonal isolation throughout the life of so-called "thermal wells." In addition, the presence of corrosive zones and weak formations must frequently be considered. Thermal cementing encompasses three principal types of wells—deep oil and gas wells, geothermal wells, and thermal recovery (steamflood and fireflood) wells. In Chapter 9, each scenario is discussed separately, because the cement system design parameters can differ significantly. The chemistry of thermal cements is also presented, and data are provided to illustrate the long-term performance of typical systems.

The proper mixing and placement of well cements rely upon the application of electrical and mechanical technology. Chapter 10 focuses on cementing equipment and casing hardware. In line with the trend toward deeper wells and more severe working environments, this technology has become increasingly sophisticated, and the equipment has become more flexible in application and more reliable in operation. First, an extensive discussion is presented concerning the various types of equipment for bulk handling, storage, cement mixing, and pumping. In addition, the special considerations for onshore and offshore cementing, as well as cementing in remote locations, are discussed. The second section of this chapter is a discussion on the wide variety of casing hardware (float equipment, cementing plugs, stage tools, centralizers, scratchers, etc.), and explains its operation. This discussion is supported by an extensive series of illustrations.

Chapters 2 through 10 contain information the engineer must consider when designing a cement system, or choosing the proper equipment for the cementing treatment. Sophisticated computer programs are available to perform most job design tasks; nevertheless, this has not diminished the need for simple engineering common sense. The methodology by which an engineer may systematically develop an optimum cement job design is discussed in Chapter 11. An example of the job design procedure is also presented.

Chapter 12 is a presentation of primary cementing techniques. This chapter provides an explanation of the relevant primary cementing terminology, the classification of casing strings, and the special problems associated with the cementation of each type of string. The cementing of large-diameter casings, stage cementing, and liner cementing are also covered.

Chapter 13 is devoted to remedial cementing techniques—squeeze cementing and plug cementing. The theoretical basis for squeeze cementing is explained, followed by a discussion of placement techniques, including low- and high-pressure squeezes, Bradenhead squeezes, and hesitation squeezes. Next, information concerning the design and preparation of cement slurries is provided. Finally, the application of squeeze cementing techniques to solve various problems, common misconceptions concerning squeeze cementing, and the evaluation of a squeeze job are discussed. In the section devoted to plug cementing, the reasons for performing such jobs, placement techniques, job design considerations, and job evaluation are covered.

Foamed cement is a system in which nitrogen or air, as a density-reducing medium, is incorporated into the slurry to obtain a low-density cement with physical properties far superior to those made by conventional methods. In recent years, as the technology for preparing such systems in the field has improved, foamed cement has become commonplace. Chapter 14 is a discussion of all aspects of foamed cement technology. First, the thermodynamic and physico-chemical bases for foamed cements are explained, followed by a discussion of foam rheology. Second, the design of a foamed cement treatment is described, including laboratory testing, pre-job planning, and engineering. Third, the execution of a foamed cement job is covered, together with safety considerations, the configuration of field equipment, and the mixing procedure. Finally, the field applications for which foamed cement is appropriate are described, including some case histories.

Chapter 15 is a discussion of horizontal well cementing. At present, most horizontal holes can be completed without cementing. However, when cementing is necessary, such jobs are among the most critical. This chapter is a review of the classification of horizontal wells, reservoir engineering justification for horizontal drainholes, reservoir scenarios for which horizontal wells are appropriate, and completion procedures. Mud removal can be extremely problematic in horizontal wellbores. This chapter presents the experimental work which has been performed to model the problem in the laboratory, and to determine the optimum techniques for achieving proper cement placement. Guidelines are presented regarding mud properties, casing movement and centralization, use of preflushes and spacer fluids, and cement slurry properties.

After a well has been cemented, the results are often evaluated to check whether the objectives have been reached. Chapter 16 is a comprehensive presentation of the techniques presently available to perform such evaluations. These include hydraulic testing, nondestructive methods such as temperature, nuclear or noise logging, and acoustic cement logging. The theoretical basis of each technique is discussed, the measuring devices are described, and the interpretation of the results is explained. The interpretation discussion is supported by many illustrations.

Three appendices are included in this textbook to supplement the material covered in the chapters. Appendix A is a digest of rheological equations commonly used in well cementing, presented in a tabular format. Appendix B is a discussion of laboratory cement testing procedures, and the equipment commonly used to perform such tests. Appendix C is a presentation of common cementing calculations for slurry design, primary and remedial cementing, and foamed cementing. Most of these calculations are performed today by computer; nevertheless, this material has been included for the reader's reference.

As stated earlier, this text has been written to provide the reader with up-to-date technical information concerning well cementing. Since work to produce this book began in March 1988, virtually all aspects of cementing technology have continued to advance at a rapid pace; consequently, we were obliged to continually revise and update most chapters until press time. While this has been somewhat exasperating for the authors, it is a strong indication of the industry's continuing commitment to the improvement of well cementing technology.

We have attempted to present the material in a logical and easily understandable form, and to reduce the aura of mystery which seems to be associated with many aspects of this technology. It is our fervent hope that this book will be a useful addition to the reader's reference library.

1 Implications of Cementing on Well Performance

Michael J. Economides*

Dowell Schlumberger

1–1 INTRODUCTION

Zonal isolation is surely the most important function of the cement sheath. As will be shown in this introductory chapter, zonal isolation is so critical that no shortchanging in the quality of the cement and the cement/casing or cement/formation bonds can ever be justified. Flow of fluids *along* the cement sheath is invariably an undesirable occurrence. For a producing well, this is manifested either by the loss of reservoir fluids through crossflow along the cement sheath, or by the influx of underground fluids from other formations into the active layer. For an injector, the injected fluids may escape into unintended layers through the cement sheath. During hydraulic fracturing, escape of fluids through an imperfect cement sheath may result in either undesirable fracture-height migration or screenout of the intended fracture in the targeted formation because of the fracturing fluid loss. In all cases, the direction of the flow of fluids into or out of the active layer is opposite to the direction of the pressure gradient and proportional to its value.

While flow of any fluid along and through the cement sheath is undesirable, upward gas flow or "gas migration" through and along the cement sheath has received particular attention. As early as 1963, Guyvoronsky and Farukshin identified the possibility of gas percolation through the matrix of a gelling cement slurry, and measured permeabilities up to 300 md. Several investigators studied the gas migration phenomenon and methods for its minimization (Carter and Slagle, 1970; Levine et al., 1980; Parcevaux et al., 1985; Stewart and Schouten, 1988). A comprehensive review of the subject is presented in Chapter 8.

Portland cement systems of normal density (≈16.0 lb/gal or 1.93 g/cm^3) usually exhibit extremely low matrix permeability, if allowed to set undisturbed. The literature quotes values in the microdarcy range. However, gas migration can open additional flow paths, in the form of interconnected porosity through the setting cement. The resulting set cement suffers from an unnaturally high permeability, because of this earlier disruption, and may not provide a competent seal. Flow paths may also take the form of discrete conductive channels (microannuli) at the pipe/cement or cement/formation interfaces. These paths, and their effective width, then correspond to a certain permeability that far outweighs the intrinsic permeability of the undisturbed set cement. As can be seen in Section 1–2, even a seemingly small microannulus width results in a very large effective permeability through the cement sheath.

The adhesion of the hardened cement to the pipe and the shear stress required to detach it, thus creating a microannulus, should be of primary concern during hydraulic fracturing. Surprisingly, only a cursory treatment of the subject is found in the literature. An outline of the issue is presented in Section 1–4.

1–2 ZONAL ISOLATION

While, as mentioned earlier, zonal isolation is the most important function of cementing, the necessary amount of zonal isolation is not often quantified. A simple way to attempt this is to compare the producing rate of the active layer into the well with the contributions of an overlying or underlying formation through the cement sheath.

Figure 1–1 is a representation of a typical completion configuration. In the middle is a perforated interval with two other potentially producing intervals (one above and one below) separated by some "impermeable" layers, of thickness $(\Delta L)_1$ and $(\Delta L)_2$, respectively.

For simplicity, let us consider steady-state flow into the well from the producing layer. The equation describing this rate for a radial oil reservoir is easily derived from Darcy's law, and is given below in oilfield units.

* Now at Mining University, Leoben, Austria

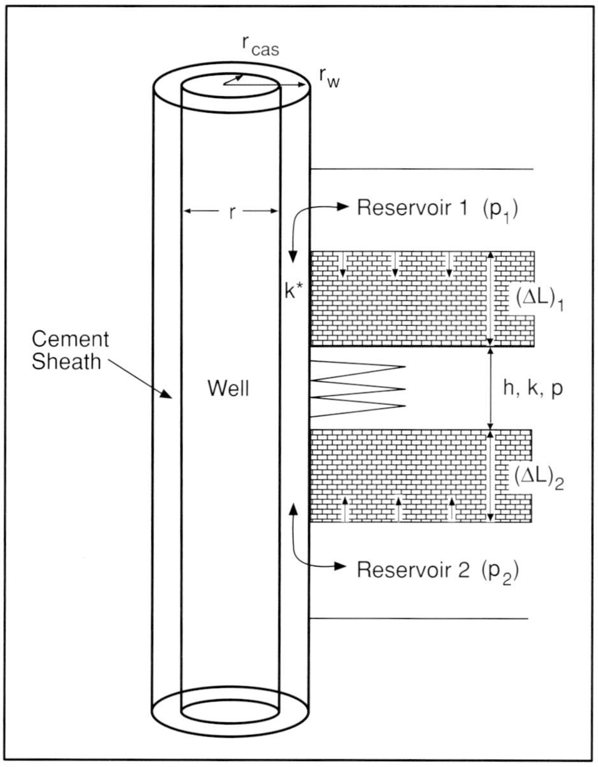

Figure 1–1—Typical well completion configuration.

$$q = \frac{kh\,(p_e - p_{wf})}{141.2\,B\mu\,[\ln\frac{r_e}{r_w} + s]}, \tag{1–1a}$$

where:

q = flow rate (stb/D),
k = permeability (md),
h = thickness (ft),
p_e = reservoir pressure (psi),
p_{wf} = flowing bottom hole pressure (psi),
μ = viscosity (cp),
s = skin factor, and
B = formation volume factor.

For a gas well, the analogous equation is

$$q = \frac{kh\,(p_e^2 - p_{wf}^{\ 2})}{1424\,\mu\,ZT\,[ln\frac{r_e}{r_w} + s]}, \tag{1–1b}$$

where:

q = flow rate (Mscf/D),
Z = gas deviation factor, and
T = reservoir temperature (°R).

Crossflow from the adjoining formations into the producing layer is likely to occur, because a pressure gradient is formed between them. The rate of flow is proportional to the vertical permeability.

For flow into the producing layer from another formation, the largest vertical pressure gradient would be at the cement sheath, which must have at least as low a permeability as the barrier layers. From the geometry shown in Fig. 1–1, the area of flow through the cement sheath is equal to

$$A = \pi\,(r_w^{\ 2} - r_{cas}^{\ 2}). \tag{1–2}$$

Darcy's law can be applied along the cement annulus. Thus, from the generalized expression

$$q = \frac{kA}{\Delta L\mu}\,\Delta P, \tag{1–3}$$

and replacing A as given by Eq. 1–2, an expression giving the flow rate (in oilfield units) through the cement sheath can be obtained.

$$q_{cem} = \frac{k^*\pi\,(r_w^{\ 2} - r_{cas}^{\ 2})\,(p_e - p_{wf})}{141.2\,B\mu\,(\Delta L)_1}. \tag{1–4}$$

Equation 1–4 provides the oil flow rate that can be either through the cement sheath "matrix" permeability, through a microannulus formed within the sheath, or through a microannulus formed between the cement and casing or the cement and the formation. The permeability k^* is an equivalent permeability value and it can be related to the width of the microannulus, as will be shown later in the chapter.

In Eq. 1–4, if the pressure in the adjoining layer is equal to the initial pressure of the producing formation, then p_i becomes p_e. For new wells, this is a reasonable assumption and it will be used here for simplicity. Analogous expressions to Eq. 1–4 can be readily derived for the flow of gas or water. In the case of gas, the expression is

$$q_{cem} = \frac{k^*\pi\,(r_w^{\ 2} - r_{cas}^{\ 2})\,(p_i^2 - p_{wf}^{\ 2})}{1424\,\mu ZT\,(\Delta L)_1}, \tag{1–5}$$

where

q = flow rate (Mscf/D),
Z = gas deviation factor, and
T = reservoir temperature (°R).

As can be seen, the relationship is between rate and pressure squared, which one should expect in the case of gas. An even more appropriate expression is between rate and the real-gas pseudopressure function. This calculation can be readily available in most instances. Equation 1–4 is applicable for the flow of water if the B and μ used are those for water instead of oil.

Using Eq. 1–4, the oil flow rate through the cement sheath can be estimated for various values of equivalent permeability. Table 1–1 contains some typical values

r_w	= 0.406 ft (8¾-in. OD)
r_{cas}	= 0.328 ft (7⅞-in. OD)
p_i	= 3000 psi
B	= 1.1 resbbl/stb
μ	= 1 cp
$(\Delta L)_1$	= 20 ft
p_{wf}	= 1000 psi

Table 1–1—Well and reservoir data for oil flow along cement sheath.

from reservoir and well data. The distance between the target reservoir and an adjoining formation, ΔL_1, is taken as equal to 20 ft. Figure 1–2 is a graph of the steady-state oil flow rate for a range of k^*, using the data in Table 1–1. Figure 1–3 is an analogous example for a gas well, using the data in Table 1–2 and Eq. 1–5. The relationship between these equivalent permeability values and the size of the channel that may cause them will be discussed in the next subsection. As can be seen from Figs. 1–2 and 1–3, the flow rates can be substantial.

1–2.1 Index of Zonal Isolation (IZI)

Dividing Eq. 1–1a by Eq. 1–4, the ratio of the flow rate into the well from the *intended* formation to the flow rate through the cement is defined here as the *Index of Zonal Isolation* (IZI) and is given by 1–6.

$$IZI = \frac{q}{q_{cem}} = \frac{kh\,\Delta L}{\pi k^* \left(r_w^2 - r_{cas}^2\right)\left(\ln \frac{r_e}{r_w} + s\right)}, \qquad (1\text{–}6)$$

Interestingly, all variables that distinguish Eq. 1–1a (for oil and water) and Eq. 1–1b (for gas) are the same as those evident in Eq. 1–4 (for oil and water) and Eq. 1–5 (for gas). Thus, the IZI expression as given by Eq. 1–6 is valid for any fluid. The expression given by Eq. 1–6 assumes that the initial reservoir pressures are essentially equal in the two formations. If the pressures are not equal, then the pressure gradients should remain in the respective top and bottom of the right-hand side of Eq. 1–6.

Equation 1–6 can provide the quantification of zonal isolation. It can be used either to calculate the required cement equivalent permeability to provide a desired flow-rate ratio or, for a given cement permeability, what would be the flow rate through the cement sheath from

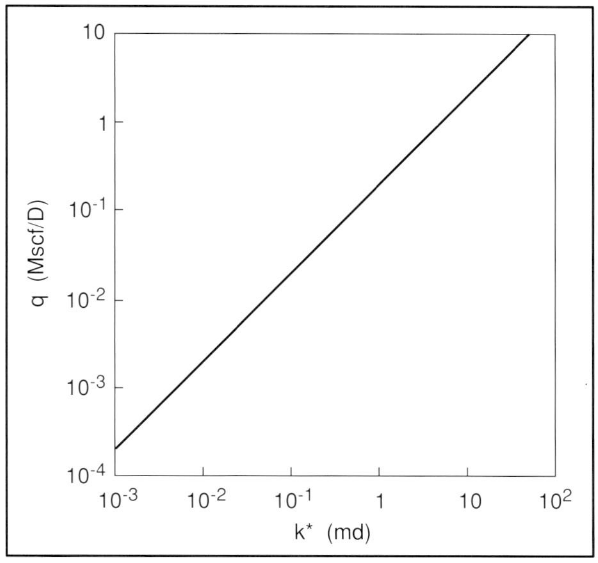

Figure 1–3—Gas flow rate along cement sheath for a range of cement equivalent permeabilities.

r_w	= 0.406 ft (8¾-in. OD)
r_{cas}	= 0.328 ft (7⅞-in. OD)
p_i	= 3000 psi
p_{wf}	= 1000 psi
μ	= 0.025 cp
Z	= 0.95
T	= 640° R
$(\Delta L)_1$	= 20 ft

Table 1–2—Well and reservoir data for gas flow along cement sheath.

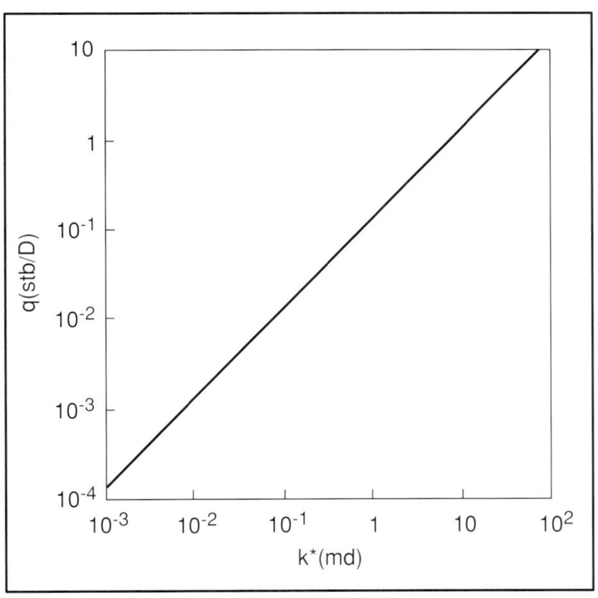

Figure 1–2—Well and reservoir data for gas flow along cement sheath.

adjoining layers. As discussed earlier, the cement permeability k^* is an equivalent permeability value, resulting either from the presence of a microannulus or from an unnaturally high cement-matrix permeability. The latter may be precipitated by the disruptive effects of fluid invasion as the cement changes from liquid to solid. The permeability for the flow through a slot is given by the well known

$$k = \frac{w^2}{12}, \qquad (1\text{–}7)$$

where 12 is a geometric factor. In oilfield units the relationship is

$$k = 5.4 \times 10^{10}\, w^2 \qquad (1\text{–}8)$$

where k is in md and w in inches. The constant is equal to 8.4×10^{13} if w is in meters. The relationship implied by Eq. 1–8 is significant. While a large matrix permeability within the cement sheath is unlikely (of the magnitudes shown in Figs. 1–2 and 1–3), a large equivalent permeability can result from a relatively small microannulus width.

Equation 1–6 can be used also as an evaluation tool to detect flow through the sheath. If a vertical interference or a multilayer test is done and the reservoir is well defined, then crossflow through the adjoining low-permeability layers may be calculated (Ehlig-Economides and Ayoub, 1986). As a result, the ideal flow rate from the targeted interval can be calculated.

Deviations from this value can be attributed to flow through an imperfect cement sheath and, using Eq. 1–6, the permeability of the cement can be extracted. The net flow rate through the perforated interval is

$$q = q_{res} + \Sigma q_{cf} + \Sigma q_{cem}, \qquad (1\text{–}9)$$

where:

q_{res} = lateral reservoir flow rate,

Σq_{cf} = crossflow contributions through the barrier, and

q_{cem} = contributions through the cement sheath.

Figure 1–4 is a graph for an example well using an 80-acre spacing, a skin effect equal to 5, and r_w equal to 0.406 ft. The group $kh\Delta L$ is graphed on the abscissa while the cement permeability k^* is graphed on the left ordinate. On the right ordinate is the equivalent path width squared that would result in similar flow rate. Two curves are offered: one for 50 and another for 100 of the q/q_{cem} ratio (IZI). As can be seen, the cement permeability requirements and the need for more zonal isolation become more critical for lower permeability producing formations that are separated by thin barriers. In both cases,

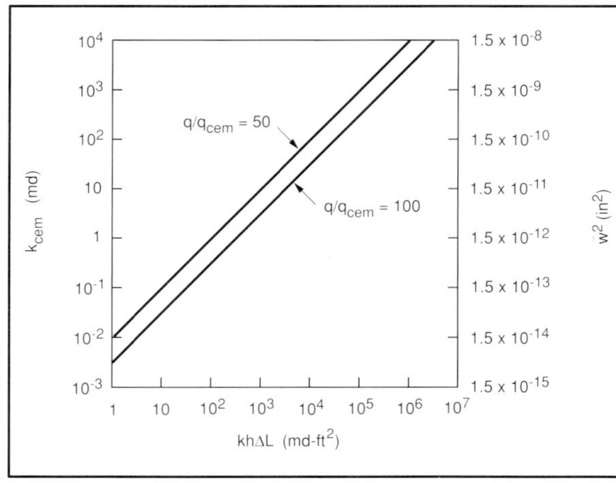

Figure 1–4—Example of the IZI concept.

the product $kh\Delta L$ becomes small, requiring a small cement permeability. This would not be a problem if only the innate matrix permeability of the cement sheath is considered. For most cements, this permeability is less than 0.01 md.

However, the presence of a *continuous* microannulus can totally reverse and severely aggravate the situation. The width squared of the microannulus is graphed on the right ordinate of Fig. 1–4. As can be seen, for a typical reservoir ($k = 4$ md, $h = 50$ ft, $\Delta L = 50$ ft, resulting in $kh\,\Delta L = 10^4$) for a $q/q_{cem} = 50$, the microannulus width must be less than 4.5×10^{-5} in. (1.1 μm), which corresponds to an equivalent permeability of 120 md. It is important to point out that such a microannulus width is two orders of magnitude smaller than the average diameter of a cement grain, is well within most casing roughness tolerances, and would probably not be detectable by bond logging. In addition, downhole pressure changes of a few psi would be sufficient to cause casing diameter fluctuations within this realm. Such microannuli would probably not be continuous; nevertheless, these calculations clearly demonstrate the extreme importance of obtaining an intimate bond between the cement sheath and casing and formation surfaces.

The quantified IZI then becomes an important variable to control. For tight reservoirs, if only absolute contributions or losses from or into adjoining formations are of concern, then a low IZI can be tolerated. However, it should be remembered, especially in the case where influx of foreign fluids such as gases, water or oil of different physical properties is evident, the minimum tolerable IZI may be very high and contingent on the production facilities at the wellhead. In such cases, even more stringent requirements in the IZI may be necessary in tight, thinly separated formations as implied in Eq. 1–6.

1–3 CEMENT-TO-PIPE BOND AND HYDRAULIC FRACTURING

Unfortunately, and surprisingly, this is an area of research that has not received its due attention. Handin (1965) attempted to characterize the "strength" of oil well cements at downhole pressure/temperature conditions. He characterized the compressive strength of cements and determined the ultimate strength at failure. He concluded that "oil-well cements become very ductile even under low effective confining pressures." Because of the magnitude of the ultimate compressive strengths at normal system densities, these cements have mechanical constitutive properties similar to sedimentary rocks under similar confining conditions.

However, hydraulic fracturing is a tensile failure mechanism and a cement sheath is potentially subjected to two phenomena: fracture propagation *within* the cement sheath and/or the dislodging of the cement sheath from the pipe by overcoming the cement-to-pipe bond. In either case, the net result is the creation of an annulus (fracture within the cement or between the cement and the pipe).

For the fracture-height migration within the cement, there is currently ongoing research to characterize this phenomenon. In general, it would be desirable if the fracture height within the cement is at the most equal or, preferably, less than the fracture height within the fractured interval. If the fracture height within the cement is larger than the reservoir fracture height, undesirable communication will ensue. The quantity ΔL in Eq. 1–6 will be effectively reduced substantially.

Of particular interest is the shear bond strength which is the adhesion strength between cement and pipe. Parcevaux and Sault (1984) showed that there is no apparent correlation between the cement compressive strength and the shear bond strength. Furthermore, they determined that the shear bond strength ranges from 1,000 psi ($\approx$ 7 MPa) for standard cement to 1,800 psi $\approx$ 12 MPa) for cements containing bond-enhancing agents (BA), as shown in Fig. 1–5. These values would imply that for many reservoirs where the tensile strength of the rock is larger than 1,000 psi, the adhesion between cement and pipe will fail first, resulting in the occurrence of a microannulus along the pipe. This has major implications both for the loss of fracturing fluids during the stimulation treatment as well as the migration of reservoir fluids following the treatment. In such a situation, remedial cementing would be indicated. The cement shear bond is outlined in more detail in Chapter 8.

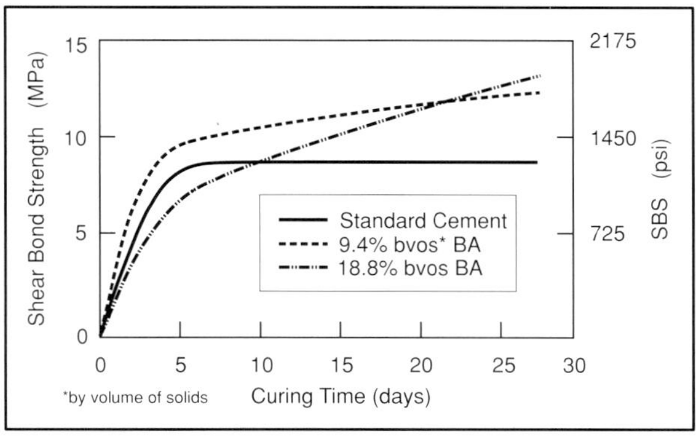

Figure 1–5—Cement shear bond strength development at 20°C.

1–5 CONCLUSION

The above discussion demonstrates that the ability of a well to achieve its production potential is influenced most by the degree of zonal isolation achieved during the completion. The quality of the cement sheath is in turn the most important factor influencing zonal isolation. Therefore, the cementation of a well should be of critical importance to every operator. The chapters to follow discuss the many interdependent facets which the engineer must consider to design, execute, and evaluate a successful cement job.

1–6 ACKNOWLEDGMENT

The author wishes to thank Phil Rae for valuable suggestions and insights on this subject.

1–7 REFERENCES

Bannister, C. E., Shuster, G. E., Wooldridge, L. A., and Jones, M. J.: "Critical Design Parameters to Prevent Gas Invasion During Cementing Operations," paper SPE 11982, 1983.

Carter, L. G. and Slagle, K. A.: "A Study of Completion Practices to Minimize Gas Communications," paper SPE 3164, 1970.

Cheung, P. R. and Beirute, R. M.: "Gas Flow in Cements," *JPT* (June 1985) 1041–1048.

Ehlig-Economides, C. A. and Ayoub, J. A.: "Vertical Interference Testing Across a Low Permeability Zone," *SPEFE* (Oct. 1986) 497–510.

Garcia, J. A. and Clark, C. R.: "An Investigation of Annular Gas Flow Following Cementing Operations," paper SPE 5701, 1976.

Guyvoronsky, A. A. and Farukshin, L. K.: "Hydrostatic Pressure of Cement Slurry," *Neftyanik* (1963) No. 10, 3–32 (translated from Russian).

Handin, J.: "Strength of Oil Well Cements at Downhole Pressure-Temperature Conditions," *SPEJ* (Dec. 1965) 341–347.

Lee, S. T., Chien, M. C. H., and Culham, W. G.: "Vertical Single-Well Pulse Testing of a Three-Layer Stratified Reservoir," paper SPE 13429, 1984.

Levine, D. C., Thomas, E. W., Bezner, H. P., and Tolle, G. C.: "How to Prevent Annular Gas Flow Following Cementing Operations," *World Oil* (Oct. 1980) 85–94.

Parcevaux, P., Piot, B., and Vercaemer, C.: "Annular Gas Flow: A Hazard-Free Solution," *Pet. Inform.* (July 1985) 34–38.

Parcevaux, P. A. and Sault, P. H.: "Cement Shrinkage and Elasticity: A New Approach for a Good Zonal Isolation," paper SPE 13176, 1984.

Parcevaux, P.: "Mechanisms of Gas Channeling During Primary Cementation: Methods for Prevention and Repair," *Chemishe Produckte in des Erdongewinnung,* Clausthal Technical U., Clausthal-Zellerfeld, (Sept. 6, 1984).

Stewart, R. B. and Schouten, F. C.: "Gas Invasion and Migration in Cemented Annuli: Causes and Cures," *SPEDE* (March 1988) 77–82.

1–8 NOMENCLATURE

B = formation volume factor

h = formation thickness

k = effective formation permeability

p = reservoir pressure

p_i = initial reservoir pressure

q = surface flow rate

r = radial distance

r_{cas} = casing diameter

r_w = wellbore radius

s = wellbore skin factor

t = time

Greek Symbols

μ = viscosity

ϕ = porosity, fraction of bulk volume

Subscripts

i = initial condition

wf = flowing wellbore condition

2

Chemistry and Characterization of Portland Cement

Michel Michaux, Erik B. Nelson, and Benoit Vidick

Dowell Schlumberger

2–1 INTRODUCTION

Portland cement is by far the most important binding material in terms of quantity produced; indeed, it is possible that it may be the most ubiquitous manufactured material. Portland cement is used in nearly all well cementing operations. The conditions to which Portland cements are exposed in a well differ significantly from those encountered at ambient conditions during construction operations; as a result, special Portland cements are manufactured for use as well cements. Certain other cements, used to a far lesser extent for the solution of special well problems, are discussed in Chapters 7 and 9.

Portland cement is the most common example of a *hydraulic cement.* Such cements set and develop compressive strength as a result of hydration, which involves chemical reactions between water and the compounds present in the cement, *not* upon a drying-out process. The setting and hardening occur not only if the cement/water mixture is left to stand in air, but also if it is placed in water. The development of strength is predictable, uniform and relatively rapid. The set cement also has low permeability, and is nearly insoluble in water; therefore, exposure to water does not destroy the hardened material. Such attributes are essential for a cement to achieve and maintain zonal isolation.

In this chapter, fundamental information is presented regarding the manufacture, hydration and classification of Portland cements. In addition, the effects of various chemical and physical parameters upon performance are discussed. Several excellent textbooks were relied upon heavily to produce this overview of cement technology: Taylor (1964); Lea (1971); Ghosh (1983); and Barnes (1983).

2–2 CHEMICAL NOTATION

A special chemical notation established by cement chemists is frequently used in this chapter. The chemical formulas of many cement compounds can be expressed as a sum of oxides; for example, tricalcium silicate, Ca_3SiO_5, can be written as $3CaO \cdot SiO_2$. Abbreviations are given for the oxides most frequently encountered, such as C for CaO and S for SiO_2. Thus Ca_3SiO_5 becomes C_3S. A list of abbreviations is given below.

$C = CaO$	$F = Fe_2O_3$	$N = Na_2O$	$P = P_2O_5$
$A = Al_2O_3$	$M = MgO$	$K = K_2O$	$f = FeO$
$S = SiO_2$	$H = H_2O$	$L = Li_2O$	$T = TiO_2$

Others are sometimes used, such as $\bar{S} = SO_3$ and $\bar{C} = CO_2$. This convention of using a shortened notation was adopted as a simple method for describing compounds whose complete molecular formulas occupy much space.

2–3 MANUFACTURING OF PORTLAND CEMENT

Portland cement consists principally of four compounds: tricalcium silicate (C_3S), dicalcium silicate (C_2S), tricalcium aluminate (C_3A) and tetracalcium aluminoferrite (C_4AF). These compounds are formed in a kiln by a series of reactions at temperatures as high as 1500°C between lime, silica, alumina and iron oxide.

In the manufacturing process selected raw materials are ground to a fine powder, and proportioned in such a way that the resulting mixture has a desired chemical composition. After blending, the raw material mixture is fed into a kiln and converted to cement clinker. The clinker is cooled, a small amount of gypsum (3% to 5%) is added, and the mixture is pulverized. The pulverized product is finished Portland cement.

2–3.1 Raw Materials

Two types of raw materials are needed to prepare a mixture that will produce Portland cement: "calcareous" materials which contain lime, and "argillaceous" materials which contain alumina, silica and iron oxide. Depending upon the location of the cement plant, a great variety of natural and artificial raw materials is employed.

The most important calcareous materials are sedimentary and metamorphic limestones, coral, shell deposits and "cement rock," which naturally has a composition similar to Portland cement. Artificial calcareous materials include precipitated calcium carbonate and other alkali wastes from various industrial processes.

Natural argillaceous materials frequently used as raw materials include clays, shales, marls, mudstones, slate, schist, volcanic ashes and alluvial silt. Blast furnace slag from steelworks and fly ash from coal-fired power plants are the most important artificial sources.

When selecting the raw materials, it is important to consider impurities which can have significant effects on the properties of the finished cement. These include magnesia (M), fluorine compounds, phosphates, lead oxide, zinc oxide and alkalis. After clinkering in the kiln, such impurities are often in solid solution within the principal cement phases, resulting in a change of reactivity. Excess magnesia (>5%) can cause a disruptive delayed expansion of the set cement, a condition known as "unsoundness." The presence of more than 0.1% fluorine in the raw materials, usually as calcium fluoride, results in a significant decrease in cement strength. Phosphates can have a beneficial effect on strength at a level of 0.20% to 0.25%; however, they have a deleterious effect at concentrations exceeding 0.5%. Lead and zinc oxides have a deleterious effect upon cement properties. The effect of alkalis is variable. The total alkali content, expressed as sodium oxide (N), generally should not exceed 0.6%, because of adverse reactions with certain types of siliceous aggregates.

2–3.2 Raw Material Preparation

Before calcination in the kiln, the raw materials must first be pulverized to a fine powder, and uniformly blended in a way such that the bulk composition corresponds to that required to manufacture a particular type of Portland cement. Although each cement plant has its own specific method, there are two general processes in use today: the dry process and the wet process. In the dry process, grinding and blending are done with dry materials. In the wet process, the grinding and blending operations use a watery slurry.

A schematic diagram of the dry process is shown in Fig. 2–1. The raw materials are crushed, dried in rotary driers, proportioned to obtain the correct bulk composition, and then ground in tube mills consisting of rotating steel cylinders containing steel balls or other grinding media. The ground material passes through a pneumatic size classifier, in which the air velocity is sufficient to carry ground material of the required fineness. Coarser particles are thrown out by centrifugal action. The ground material is stored in several silos. The chemical composition varies from silo to silo; therefore, another opportunity exists to reblend and "fine tune" the mixture which will go to the kiln.

The wet process is illustrated in Fig. 2–2.The raw materials are initially proportioned in the dry state. Water is added, and further size reduction occurs in a grinding mill. Size classification is performed by pumping the resulting slurry past a vibrating screen. Coarser material is returned to the mill for regrinding. The slurry is stored in basins equipped with rotating arms and compressed air agitation to keep the mixture homogeneous. The chemical composition of the slurries varies slightly from basin to basin. Thus final adjustments of composition can be performed by blending the slurries from various basins.

For many years, the wet process was preferred because more accurate control of the raw mix was possible; however, significantly more fuel was required for the kiln to evaporate the water. The increased cost of fuel in recent years has forced a return to the dry process, and the

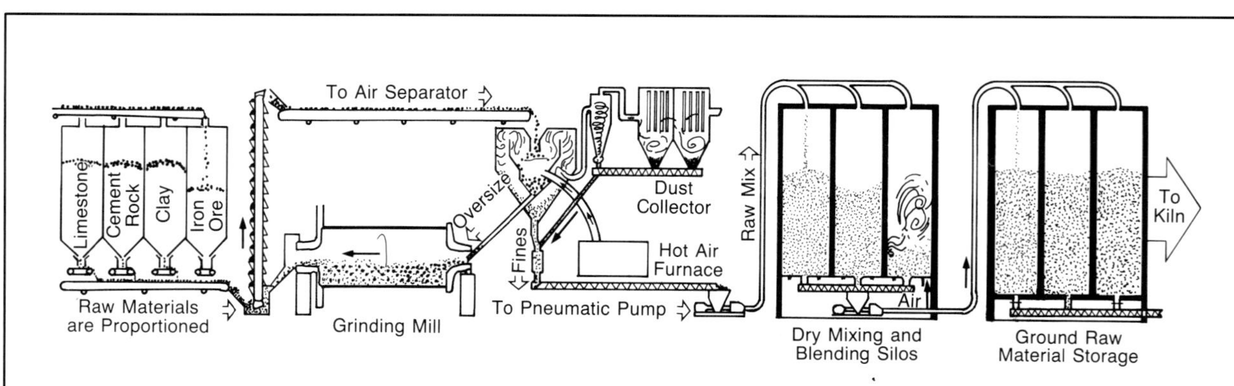

Figure 2–1—Schematic flow diagram of the Dry Process (from Portland Cement Association, 1969).

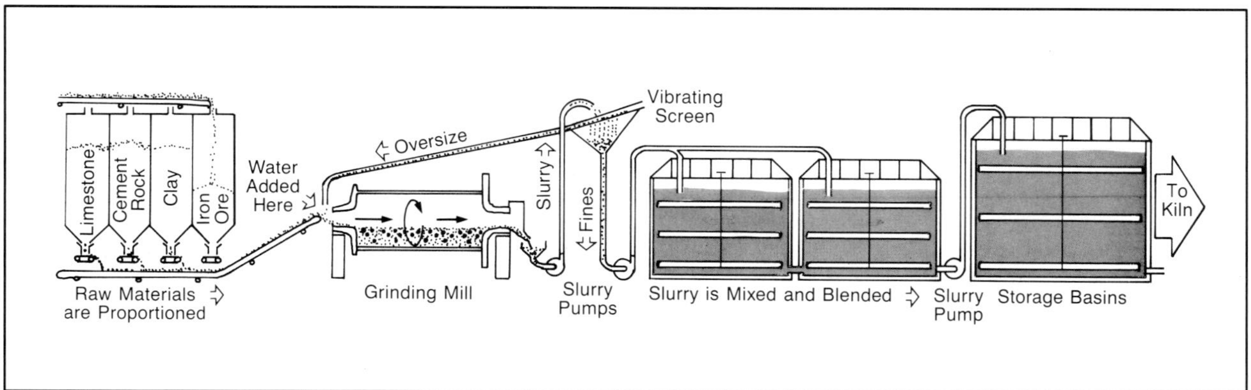

Figure 2–2—Schematic flow diagram of the Wet Process (from Portland Cement Association, 1969).

technology has been developed to obtain improved control of raw material composition.

2–3.3 Heat Treatment

Having achieved the appropriate degree of size reduction, classification and blending of the raw materials, heat treatment is performed in a rotary kiln which is usually preceded by a preheater. This step is shown in Fig. 2–3. The kiln is slightly inclined and rotates at 1 to 4 RPM; as a result, the solid material passes through the kiln as it rotates. Depending upon the cement plant, the fuel can be oil, gas or pulverized coal.

A complex series of reactions takes place in the kiln, whereby the raw materials are converted to "clinker." There are six temperature zones in a kiln, and the temperature ranges and reaction profiles are shown in Table 2–1. Evaporation of free water occurs in Zone I. Water removal occurs very quickly if the dry process has been used; however, up to one-half the length of the kiln can be devoted to drying with a wet-process system. During preheating (Zone II), dehydroxylation of the clay minerals occurs. In Zones III and IV, several important reactions occur. Dehydroxylation of clay minerals is completed, and the products crystallize. Calcium carbonate decomposes to free lime, releasing large quantities of carbon dioxide. The production of various calcium aluminates and ferrites also begins. The sintering zone, Zone V, occupies a small portion of the kiln; however, most of the principal cement phases are produced at this stage. At this point, part of the reaction mixture liquefies. At the maximum temperature in the sintering zone, also known as the "clinkering temperature," the formation of C_2S and C_3S

Zone	Temperature Range (°C)	Reaction Profile
I	Up to 200	Evaporation
II	200 to 800	Preheating
III	800 to 1100	Decarbonation
IV	1100 to 1300	Exothermic Reactions
V	1300 to 1500 to 1300	Sintering
VI	1300 to 1000	Cooling

Table 2–1—Reaction zones in rotary cement kiln.

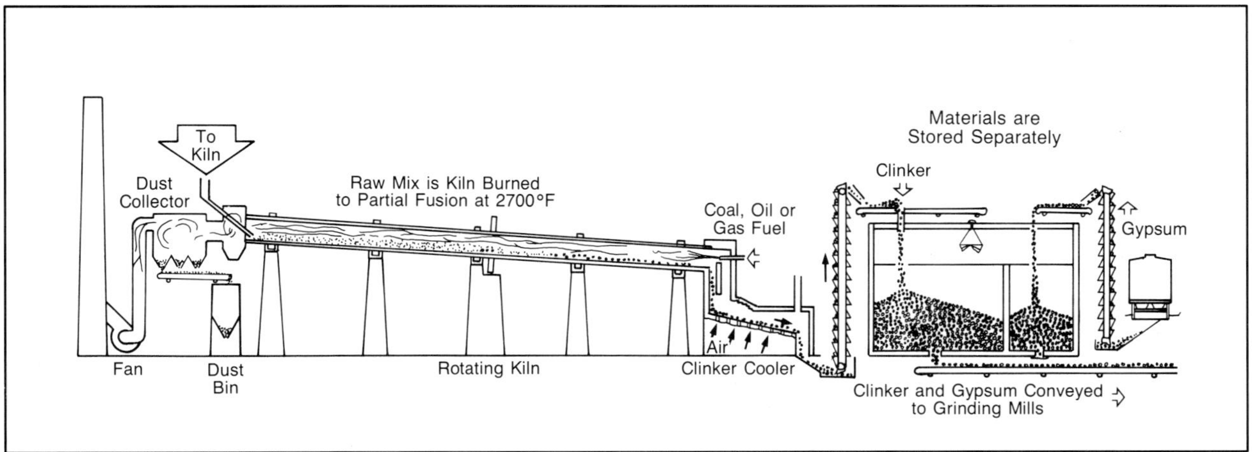

Figure 2–3—Schematic flow diagram of the burning process (from Portland Cement Association, 1969).

is completed. The uncombined lime, alumina and iron oxide are contained in the liquid phase. During the cooling phase (Zone VI), the C_3A and C_4AF crystallize as the liquid phase disappears.

2-3.4 Cooling

The quality of the clinker and the finished cement is very dependent upon the rate of cooling. The best clinker is obtained by cooling slowly to about 2,282°F (1250°C) followed by rapid cooling, usually 32° to 36°F/min (18–20°C/min).

When the cooling rate is slow, 7° to 9°F/min (4° to 5°C/min), the C_3A and C_4AF develop a high degree of crystallinity, the C_3S and C_2S crystals become highly ordered and the free MgO forms crystals (mineral name: periclase). This results in a cement which is less hydraulically active. Early compressive strength is high, but longer term strength is low. Because of the formation of periclase, cements which have cooled slowly tend to demonstrate a higher degree of unsoundness.

When the cooling rate is fast, the liquid phase which formed during Zone V in the kiln solidifies to a glass. The C_3A and C_4AF remain trapped in the glassy phase, and the crystallinity of the C_2S and C_3S is less ordered. The free MgO also remains in the glassy phase; as a result, it is less active and the resulting cement is less apt to demonstrate unsoundness. Early compressive strength is lower, but longer term strength is higher.

The general behavior described above is based upon general observations of cement behavior at ambient conditions. As of this writing, it is unclear whether the cooling method is relevant to the behavior of Portland cements at the higher temperatures and pressures encountered during well cementing operations.

Figure 2–4 is a microscope photograph of a typical Portland cement clinker. The various clinker phases have distinct crystal habits, and each is identified in the figure.

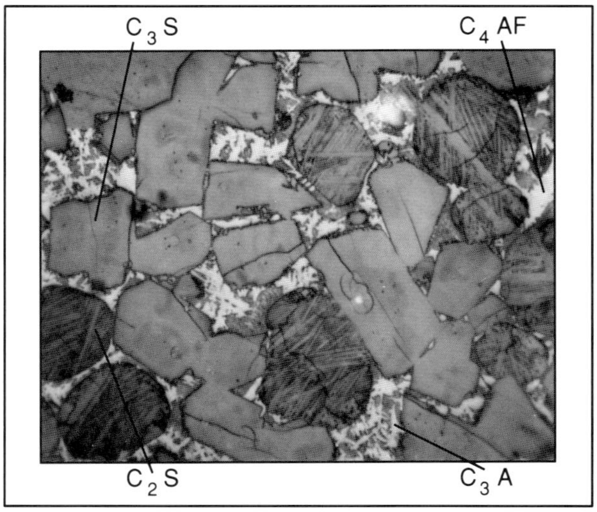

Figure 2–4—Thin-section microscopic view of Portland cement clinker (photograph supplied by LaFarge-Coppée).

2-3.5 Grinding

As shown in Fig. 2–5, the finished cement is produced by grinding the clinker with gypsum ($C\overline{S}H_2$) which, for reasons which will be explained later, prevents a phenomenon known as "flash set." Most cement is produced in tubular mills partly filled with hard steel balls and, depending upon the type of cement being manufactured, the clinker is ground to a given particle-size distribution. The particle size of the cement grains varies from 1–100 μm.

The ball milling process is inherently inefficient, with 97–99% of the energy input being converted to heat. Consequently, it is necessary to cool the mill. If the cement reaches an excessively high temperature, too much of the gypsum can dehydrate to form calcium sulfate hemihydrate ($C\overline{S}H_{1/2}$) or soluble anhydrite ($C\overline{S}$). Such

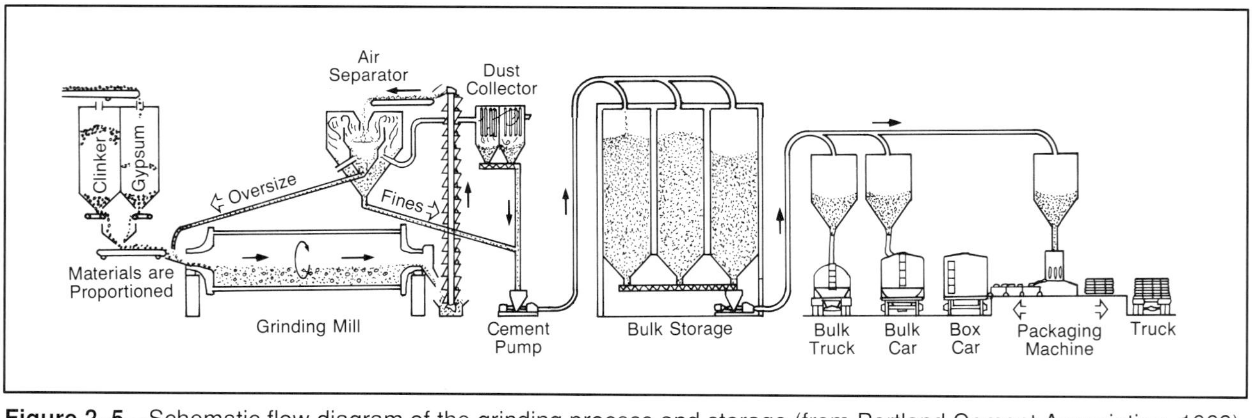

Figure 2–5—Schematic flow diagram of the grinding process and storage (from Portland Cement Association, 1969).

compounds, while still able to prevent the flash set, can cause another phenomenon called "false set," which will also be discussed later in this chapter.

2–3.6 Storage

After the finished cement emerges from the grinder, it is stored in large airtight silos. For reasons which are explained later, it is important to protect the product from humidity and carbon dioxide. Frequently, there are several silos for a particular type of cement. In such cases, cement from different silos can be blended to maintain a more consistent product.

2–4 HYDRATION OF THE CLINKER PHASES

The compounds present in Portland cement are anhydrous. When brought into contact with water, they are attacked or decomposed forming hydrated compounds. Supersaturated and unstable solutions are formed, gradually depositing their excess solids. Since the solubilities of the original anhydrous compounds are much higher than those of the hydration products, complete hydration should ultimately occur.

Research concerning cement hydration has largely consisted of studying the behavior of individual cement components in an aqueous environment, and relating the findings to the behavior of the multicomponent system—Portland cement. The principal components of Portland cement (C_3S, C_2S, C_3A and C_4AF) display different hydration kinetics and form different hydration products. This chapter follows the same path, first presenting the contributions of the individual phases in this section, and finally discussing their combined performance in Portland cement in the following section.

2–4.1 Hydration of the Silicate Phases

The silicate phases in Portland cement are the most abundant, often comprising more than 80% of the total material. C_3S is the principal constituent, with a concentration as high as 70%. The quantity of C_2S normally does not exceed 20%.

As shown in the idealized chemical equations below, the hydration products for both phases are calcium silicate hydrate and calcium hydroxide (also known as portlandite).

$$2C_3S + 6H \rightarrow C_3S_2H_3 + 3CH \qquad (2\text{–}1)$$

$$2C_2S + 4H \rightarrow C_3S_2H_3 + CH \qquad (2\text{–}2)$$

The calcium silicate hydrate does not have the exact composition of $C_3S_2H_3$; instead, the C:S and H:S ratios are variable depending upon such factors as the calcium concentration in the aqueous phase (Barret et al., 1980a and 1980b), temperature (Odler and Skalny, 1973), the presence of additives (Odler and Skalny, 1971) and aging (Barnes, 1983). The material is quasi-amorphous, and thus is commonly called "C-S-H gel." C-S-H gel comprises roughly 70% of fully hydrated Portland cement at ambient conditions, and is considered as the principal binder of hardened cement. By contrast, the calcium hydroxide is highly crystalline, and occurs as hexagonal plates. Its concentration in hardened cement is usually between 15% to 20%.

After a brisk but brief initial hydration when added to water, the silicate phases experience a period of low reactivity, called the "induction period." Therefore, they do not significantly influence the rheology of the cement slurry. Substantial hydration eventually resumes and, as shown in Fig. 2–6, the hydration rate of C_3S exceeds that of C_2S by a wide margin. Because of its abundance, and the massive formation of C-S-H gel, the hydration of C_3S is largely responsible for the beginning of the set and early strength development. The hydration of C_2S is significant only in terms of the final strength of the hardened cement.

The mechanism of C_2S hydration is very similar to that of C_3S; therefore, only C_3S is considered in this chapter. The hydration of C_3S is considered to be an model for the hydration behavior of Portland cement.

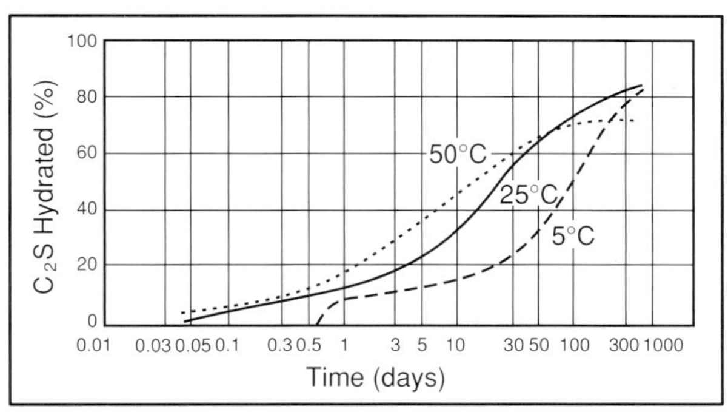

Figure 2–6a—Hydration of C_2S vs time.

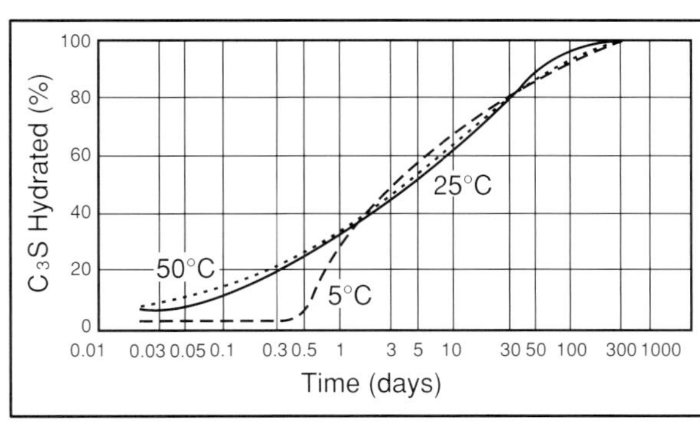

Figure 2–6b—Hydration of C_3S vs time.

The hydration of C_3S is an exothermic process; therefore, the hydration rate can be followed by conduction calorimetry. From the thermogram given in Fig. 2–7, five hydration stages are arbitrarily defined.

I. Preinduction Period
II. Induction Period
III. Acceleration Period
IV. Deceleration Period
V. Diffusion Period

2–4.1.1 Preinduction Period

The duration of the preinduction period is only a few minutes, during and immediately following mixing. A large exotherm is observed at this time, resulting from the wetting of the powder and the rapidity of the initial hydration. From a physical standpoint, an initial layer of C-S-H gel is formed over the anhydrous C_3S surfaces. A generally accepted chemical mechanism, proposed by Barret (1986), is based upon a dissolution/precipitation model.

When C_3S comes into contact with water, a surface protonation occurs leading to the transformation of O^{2-} and SiO_4^{4-} ions in the first layer of the crystal lattice into OH^- and $H_3SiO_4^-$ ions. This almost instantaneous reaction is immediately followed by the congruent dissolution of the protonated surface, according the following equation.

$$2Ca_3SiO_5 + 8H_2O \rightarrow 6\,Ca^{2+} + 10\,OH^- + 2H_3SiO_4^- \quad (2\text{–}3)$$

The solution becomes supersaturated very quickly with respect to C-S-H gel, and C-S-H gel precipitation occurs (Barret and Bertandrie, 1986 and Ménétrier, 1977).

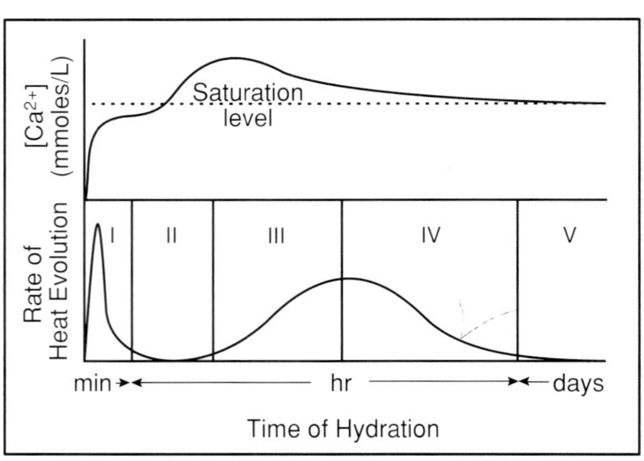

Figure 2–7—Schematic representation of changes taking place in C_3S-water system.

$$2Ca^{2+} + 2\,OH^- + 2H_3SiO_4^- \rightarrow Ca_2(OH)_2H_4Si_2O_7 + H_2O \quad (2\text{–}4)$$

Equation 2–4 assumes that the initial C-S-H gel has a C:S ratio of about 1.0 (Ménétrier, 1977). In addition, the silicate anions in the C-S-H gel are, at short hydration times, dimeric (Michaux et al., 1983). The precipitation of C-S-H gel takes place at the C_3S/solution interface, where the ionic concentrations are the highest; consequently, a thin layer is deposited on the C_3S surface.

Addition of Eqs. 2–3 and 2–4 produces the following.

$$2Ca_3SiO_5 + 7H_2O \rightarrow Ca_2(OH)_2H_4Si_2O_7 + 4Ca^{2+} + 8\,OH^- \quad (2\text{–}5)$$

During the preinduction period, critical supersaturation with respect to calcium hydroxide is not reached; therefore, as indicated in Equation 2–5, the concentration of lime increases as further hydration continues.

2–4.1.2 Induction Period

As explained earlier, relatively little hydration activity is observed during the induction period. The rate of heat liberation dramatically falls. Additional C-S-H gel is slowly precipitated, and the Ca^{2+} and OH^- concentrations continue to rise. When critical supersaturation is finally reached, precipitation of calcium hydroxide begins to occur. A recommencement of significant hydration is observed, thus signaling the end of the induction period. At ambient temperatures, the duration of the induction period is a few hours.

The termination mechanism of the induction period is still a subject of debate among cement chemists. Many theories have been proposed; however, they are often more complementary than contradictory. Generally speaking, they fall into one of two broader theories: the protective layer theory and the delayed nucleation theory.

According to the protective layer theory (Powers, 1961 and de Jong et al., 1967), the permeability of the initially precipitated C-S-H gel is very low; consequently, further hydration is inhibited, and an induction period occurs. Within this theory, two termination mechanisms have been proposed. According to Powers (1961), Double et al. (1978), and Thomas and Powers (1981), osmotic force is developed within the C-S-H gel layer as hydration continues. The gel layer eventually bursts, resulting in a large release of silicates into the solution and a massive formation of C-S-H gel. The other mechanism, proposed by de Jong et al. (1967), holds that the C-S-H gel layer undergoes a morphological change, resulting in increased permeability. Consequently, water more easily penetrates the layer, and hydration accelerates.

The protective layer theory treats the precipitation of calcium hydroxide as merely a consequence of the increased hydration rate. According to the delayed nucleation theory, the calcium hydroxide precipitation acts as a trigger for the acceleration of hydration. Within this theory, a number of diverse mechanisms have been proposed regarding the induction period. Skalny and Young (1980) and Tadros et al. (1976) considered that the induction period is one of slow C_3S dissolution. Ca^{2+} and OH^- ions pass into the solution, and the degree of supersaturation with respect to lime continues to increase; thus, further C_3S hydration is retarded because of the high Ca^{2+} concentration in the interfacial region. Eventually, sufficient supersaturation (~1.5 to 2.0 times the saturation value) accumulates to form stable $Ca(OH)_2$ nuclei and precipitation commences, thus ending the induction period. Fierens and Verhaegen (1976) did not agree; instead, they proposed a mechanism involving rapid chemisorption of water onto preferential sites on the C_3S surface. The hydration products nucleate onto the active sites, and accelerated hydration commences when the nuclei reach a critical size.

2–4.1.3 Acceleration and Deceleration Periods

At the end of the induction period, only a small percentage of the C_3S has hydrated. The acceleration and deceleration periods, also collectively known as the "setting period," represent the interval of most rapid hydration. During the acceleration period, solid $Ca(OH)_2$ crystallizes from solution and C-S-H gel deposits into the available water-filled space. The hydrates intergrow, a cohesive network is formed and the system begins to develop strength.

The porosity of the system decreases as a consequence of the deposition of hydrates. Eventually, the transportation of ionic species and water through the network of C-S-H gel is hindered, and the hydration rate decelerates. At ambient conditions, these events occur within several days.

2–4.1.4 Diffusion Period

Hydration continues at a slow pace owing to the ever-decreasing system porosity, the network of hydrated products becomes more and more dense, and strength increases. There is no evidence of major structural changes; however, polymerization of the silicate anions of C-S-H gel has been observed (Dent-Glasser et al., 1978). The duration of the diffusion period is indefinite at ambient conditions. Portlandite crystals continue to grow and engulf the hydrating C_3S grains; as a result, total hydration is never attained (see Fig. 2–8).

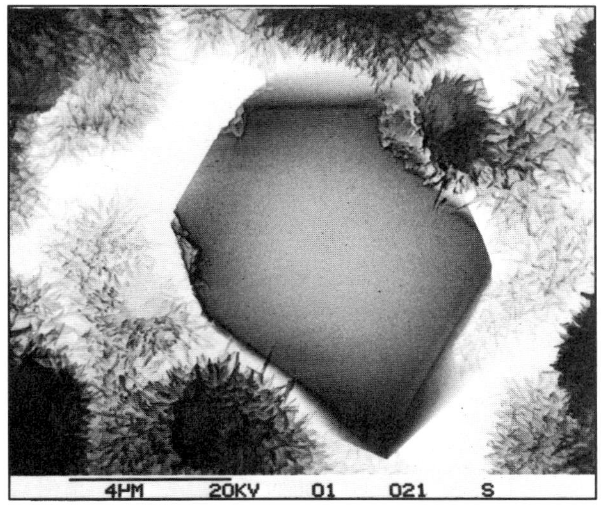

Figure 2–8—Photograph of precipitated $Ca(OH)_2$ in C-S-H gel matrix.

2–4.2 Hydration of the Aluminate Phases

The aluminate phases, especially C_3A, are the most reactive at short hydration times. Although their abundance is considerably lower than the silicates, they have a significant influence upon the rheology of the cement slurry and early strength development of the set cement. C_3A hydration is emphasized in this section. The hydration of C_4AF is very similar to that of C_3A, but much slower (Ramachandran and Beaudoin, 1980).

As with C_3S, the first hydration step of C_3A is an interfacial reaction between the surface of the anhydrous solid and water. This irreversible reaction leads to the hydroxylation of the superficial anions AlO_2^- and O_2^- into $[Al(OH)_4]^-$ and OH^- anions (Bertrandie and Barret, 1986), resulting in a congruent dissolution of the protonated surface.

$$Ca_3Al_2O_6 + 6\,H_2O \rightarrow 3Ca^{2+} + 2[Al(OH)_4]^- + 4OH^- \quad (2\text{–}6)$$

The solution quickly becomes supersaturated with respect to some calcium aluminate hydrates, leading to their precipitation.

$$6Ca^{2+} + 4[Al\,(OH)_4]^- + 8OH^- + 15H_2O \rightarrow Ca_2[Al\,(OH)_5]_2 \cdot 3H_2O + 2[Ca_2\,Al(OH)_7 \cdot 6H_2O] \quad (2\text{–}7)$$

By adding Eqs. 2–6 and 2–7, the following equation is obtained using cement chemistry notation.

$$2C_3A + 27H \rightarrow C_2AH_8 + C_4AH_{19} \quad (2\text{–}8)$$

The calcium aluminate hydrates in Eq. 2–8 are metastable, and occur as hexagonal crystals. They eventually convert to the more stable cubic form, C_3AH_6, as shown below. At ambient conditions, this reaction occurs within several days.

$$C_2AH_8 + C_4AH_{19} \rightarrow 2C_3AH_6 + 15H \quad (2\text{–}9)$$

Unlike the calcium silicate hydrates, the calcium aluminate hydrates are not amorphous, and do not form a protective layer at the C_3A surfaces; consequently, as shown in Fig. 2–9, no induction period is observed, and the hydration goes to completion very rapidly. If such uncontrolled hydration is allowed to occur in a Portland cement slurry, severe rheological difficulties are experienced.

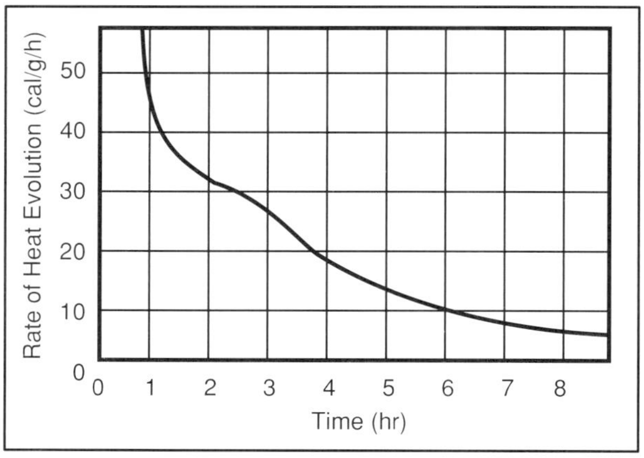

Figure 2–9—Thermogram of C_3A hydration (25°C).

C_3A hydration is controlled by the addition of 3 to 5% gypsum to the cement clinker before grinding, as described earlier in this chapter. Upon contact with water, part of the gypsum dissolves. The calcium and sulfate ions released in solution react with the aluminate and hydroxyl ions released by the C_3A to form a calcium trisulfoaluminate hydrate, known as the mineral ettringite.

$$6Ca^{2+} + 2[Al(OH)_4]^- + 3SO_4^{2-} + 4OH^- + 26H_2O \rightarrow Ca_6[Al(OH)_6]_2(SO_4)_3 \cdot 26H_2O$$

or, the global reaction can be written as

$$C_3A + 3C\bar{S}H_2 + 26H \rightarrow C_3A \cdot 3C\bar{S} \cdot 32H \quad (2\text{–}10)$$

As shown in Fig. 2–10, ettringite occurs as needle-shaped crystals which precipitate onto the C_3A surfaces, hindering further rapid hydration. Thus, as shown in Fig. 2–11, an "induction period" is artificially created. During this period, the gypsum is gradually consumed and ettringite continues to precipitate. The retardation of C_3A hydration ceases and rapid hydration resumes, when the supply of gypsum is exhausted. The sulfate ion concentration sharply drops. Ettringite becomes unstable, and converts to a platy calcium monosulfoaluminate hydrate.

Figure 2–10—Photograph of ettringite crystals (photograph courtesy of Dr. Herbert Pöllmann, Univ. of Erlangen).

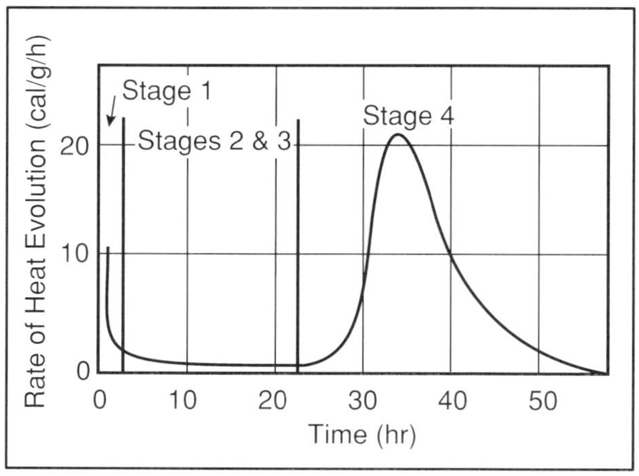

Figure 2–11—Thermogram of C_3A hydration with gypsum (25°C).

$$C_3A \cdot 3C\bar{S} \cdot 32H + 2C_3A + 4H \rightarrow 3C_3A \cdot C\bar{S} \cdot 12H \quad (2\text{–}11)$$

Any remaining unhydrated C_3A forms calcium aluminate hydrate as shown in Eq. 2–8 (Bensted, 1976).

2–5 HYDRATION OF PORTLAND CEMENTS —THE MULTICOMPONENT SYSTEM

The hydration of Portland cement is a sequence of overlapping chemical reactions between clinker components, calcium sulfate and water, leading to continuous cement slurry thickening and hardening. Although the hydration of C_3S is often used as a model for the hydration of Portland cement, it must be kept in mind that many additional parameters are involved.

From a chemical point of view, Portland cement hydration is a complex dissolution/precipitation process in which, unlike the hydration of the individual pure phases, the various hydration reactions proceed simultaneously at differing rates. The phases also influence each other. For example, the hydration of C_3A is modified by the presence of hydrating C_3S, because the production of calcium hydroxide reinforces the retarding action of gypsum. None of the clinker minerals is pure. Depending upon the composition of the raw materials, each contains alien oxides in solid solution which alter their reactivity.

The hydration products are also impure. The C-S-H gel incorporates significant amounts of aluminum, iron and sulfur, while the ettringite and monosulfoaluminate phases contain silicon. The calcium hydroxide also contains small quantities of foreign ions, chiefly silicate.

A typical schematic thermogram of Portland cement hydration is shown in Fig. 2–12. It can roughly be described as the addition of the thermograms for C_3S and C_3A, adjusted for relative concentration.

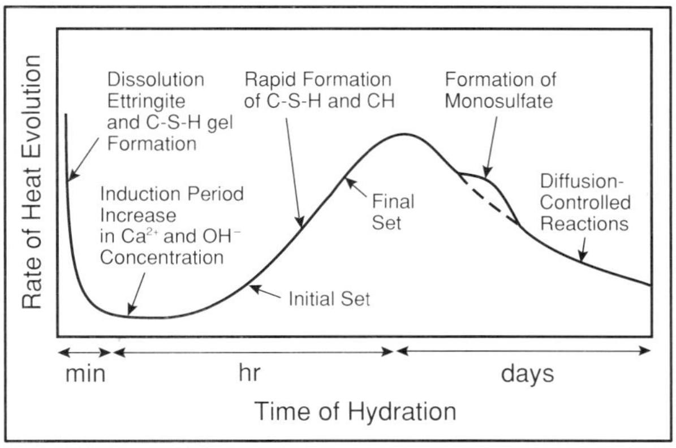

Figure 2–12—Schematic representation of Portland cement hydration.

2–5.1 Volume Changes During Setting

When Portland cements react with water, the system cement plus water undergoes a net volume diminution. This is an absolute volume decrease, and occurs because the absolute density of the hydrated material is greater than that of the initial reactants. Table 2–2 shows the change of absolute volume with time for a number of Portland cements.

Despite the decrease in absolute volume, the external dimensions of the set cement, or the bulk volume remain the same or slightly increase. To accomplish this, the internal porosity of the system increases.

In the confined environment of a wellbore, the decrease in absolute volume can affect the transmission of hydrostatic pressure to the formation, and can affect the cement's ability to prevent annular fluid migration. This subject is thoroughly discussed in Chapter 8.

	No.	1 day	7 days	28 days	100 days
Portland cement	1	2.8	4.8	6.0	6.9
Portland cement	2	1.7	4.4	—	6.3
Portland cement without gypsum	3	2.7	8.0	8.6	8.7
	4	2.6	6.3	7.5	7.6

Table 2–2—Percentage absolute volume diminution of Portland cements (from Lea, 1971).

2–5.2 Effect of Temperature

Temperature is one of the major factors affecting the hydration of Portland cement. The hydration rate of the cement and the nature, stability and morphology of the hydration products are strongly dependent upon this parameter.

Elevated hydration temperatures accelerate the hydration of cement. As illustrated by the calorimetry curves in Fig. 2–13, the duration of the induction and setting periods is shortened, and the rate of hydration during the setting period is much higher. However, upon extended curing, the degree of hydration and the ultimate strength are often reduced. This is most probably related to the formation of a dense layer of C-S-H gel around the C_3S surfaces, hindering their complete hydration (Bentur et al., 1979).

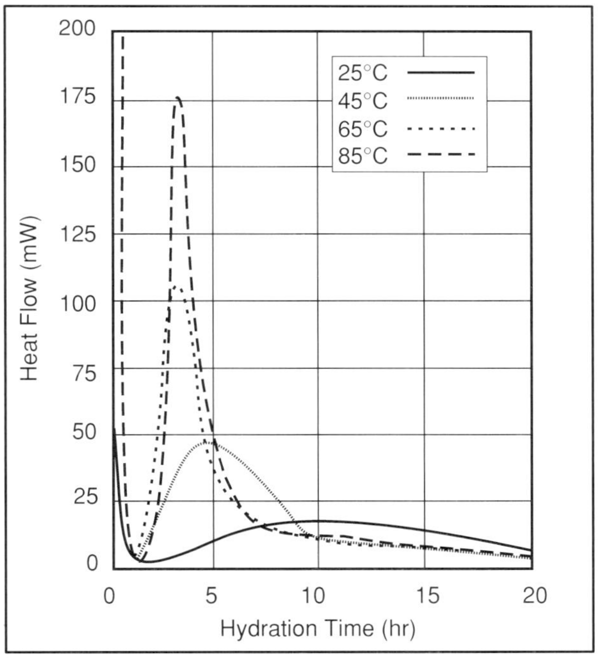

Figure 2–13—Effect of temperature upon hydration kinetics of Class G Portland cement.

Up to 104°F (40°C), the hydration products are the same as those which occur at ambient conditions. Certain changes occur in the microstructure and morphology of C-S-H gel at higher temperatures: the material becomes more fibrous and individualized, and a higher degree of silicate polymerization is observed. At curing temperatures exceeding 230°F (110°C), C-S-H gel is no longer stable, and crystalline calcium silicate hydrates are eventually formed. This subject is thoroughly discussed in Chapter 9.

The conversion of the hexagonal aluminate hydrates to the cubic form (Eq. 2–9) is strongly accelerated by temperature. Above 176°F (80°C) C_3AH_6 is directly formed.

The behavior of the calcium sulfoaluminates is also dependent upon curing temperature. Above 140°F (60°C) ettringite is no longer stable, and decomposes to calcium monosulfoaluminate and gypsum (Lea, 1970; Barvinok et al., 1976).

$$C_3A \cdot 3C\overline{S} \cdot 32H \rightarrow C_3A \cdot C\overline{S} \cdot 12H + 2C\overline{S} + 20H \qquad (2\text{–}12)$$

However, other researchers have recorded higher stability limits for ettringite, up to 230°F (110°C) (Lach and Bures, 1974). The calcium monosulfoaluminate is reported to be stable up to 374°F (190°C) (Satava and Veprek, 1975).

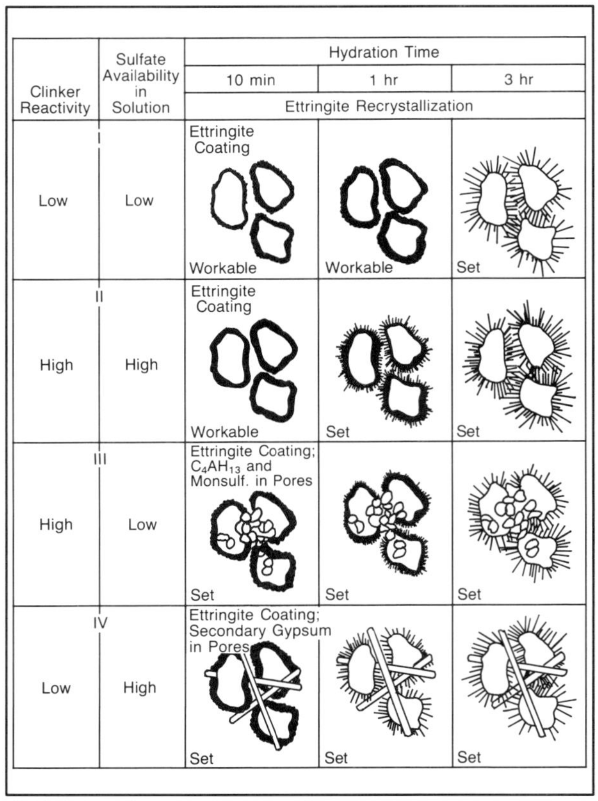

Figure 2–14—Schematic diagram of structure development in the setting of Portland cement in relation to the reactivity of the clinker and to sulfate availability (from Ghosh, 1983).

2–5.3 Flash Set and False Set

When Portland cement clinker is ground alone (i.e., without gypsum) and mixed with water the C_3A rapidly reacts, the temperature markedly increases, and an irreversible stiffening occurs followed quickly by a pseudo-set. This phenomenon is called a "flash set," or sometimes a "quick set." As discussed earlier during the discussion of aluminate hydration, the uncontrolled C_3A hydration can be prevented by the addition of gypsum to the system. This is why gypsum is ground in with the clinker during the manufacture of Portland cement. For optimum cement performance, the quantity of gypsum must be balanced according to the reactivity of the clinker (Fig. 2–14).

It is important to point out that a flash set can still occur if the quantity of gypsum in the cement is insufficient with respect to the reactivity of the clinker. Unfortunately, no simple rule exists to determine the optimum gypsum content, as this depends upon a variety of parameters, including cement particle size distribution, the alkalis and the aluminate phase content (Lerch, 1946; Ost, 1974).

Because of the heat generated during the grinding process at the cement mill, the calcium sulfate in Portland cement is dehydrated to a variable extent. In some cases, calcium sulfate hemihydrate ($C\overline{S}H_{1/2}$) and/or soluble anhydrite ($C\overline{S}$) are the only forms of calcium sulfate present. At ambient temperature, the solubilities of $C\overline{S}H_{1/2}$ and $C\overline{S}$ are approximately twice that of gypsum; therefore, upon hydration, the aqueous phase of the cement slurry quickly becomes supersaturated with respect to gypsum. To relieve this condition, so-called "secondary gypsum" is precipitated. A marked stiffening or gelation of the cement slurry, known as "false set," is observed.

False sets are reversible upon vigorous slurry agitation; however, such agitation would not be possible during most well cementing operations, particularly if the slurry is mixed continuously. The addition of a dispersant can be useful for reducing the rheological impact of false sets with cements known to have such inclinations (Chapter 3).

2–5.4 Effects of Aging

The performance of Portland cement can be affected significantly by exposure to the atmosphere and/or high temperatures during storage in sacks or silos. The principal effects upon well cements include the following (Silk, 1986).

- Increased Thickening Time
- Decreased Compressive Strength
- Decreased Heat of Hydration
- Increased Slurry Viscosity

The effects are principally due to carbonation of the calcium silicate hydrate phases, and partial hydration of the free CaO. The rate at which these processes occur is directly related to the relative humidity of the storage environment. The effects of limited cement exposure to air during transport operations have been shown to be less severe (Cobb and Pace, 1985).

When Portland cement is stored in hot regions, the temperature in the silo can be sufficiently high to result in the dehydration of gypsum (Locher et al., 1980). Such cements would be more apt to exhibit the false-set phenomenon. Thus, when designing cement systems for a particular job, it is always prudent to perform the laboratory tests with samples of the cement to be used at the wellsite.

If sufficient potassium sulfate is present as an impurity in the cement, a reaction with gypsum can occur resulting in the formation of syngenite.

$$2CaSO_4 \cdot 2H_2O + K_2SO_4 \rightarrow$$

$$\underset{\text{syngenite}}{CaK_2(SO_4)_2 \cdot H_2O} + CaSO_4 \cdot \tfrac{1}{2}H_2O + 2.5H_2O \qquad (2\text{–}13)$$

The water liberated during this reaction can prehydrate the aluminate phases. When the cement is eventually hydrated in water, an imbalance exists between the aluminates and sulfates, often leading to a false set.

2–5.5 Influence of Alkalis

The principal alkaline elements found in Portland cement are sodium and potassium. They have been shown to affect setting and strength development; thus, the amounts of these substances are usually held below 1% (expressed as oxides).

The effects of alkalis upon strength development are unpredictable, and dependent upon a large number of significant parameters. Alkalis have been shown to improve compressive strength (Sudakas et al., 1978), and to be deleterious (Chernikh et al., 1963). Jawed and Skalny (1978) demonstrated a positive effect upon early strength, but a negative effect upon long-term strength.

2–5.6 Influence of Particle-Size Distribution

The particle size distribution (sometimes called fineness) is an important parameter with respect to cement reactivity and slurry rheology. The fineness of cement is usually determined by turbidimetry (Wagner method) or by measuring the air permeability of a small layer of lightly compacted cement (Blaine method) (Appendix B). With the assumption that the cement particles are spherical, such information is used to calculate a theoretical surface area; however, this method underestimates the true surface area (Vidick et al., 1987), as measured by the BET gas-adsorption method (Table 2–3).

Sample	Surface Area (m^2/g) Blaine	BET
A	0.2	0.8
B	0.3	0.5
C	0.4	1.2

Table 2–3—Surface area of anhydrous Class G cements as measured by two techniques (from Vidick, 1987).

The water-to-cement ratio required to wet the cement particles and prepare a pumpable slurry is directly related to the surface area (Sprung et al., 1985). Thus, for consistency of performance, the fineness is controlled by the cement manufacturer.

The development of compressive strength is often correlated with the cement's surface area (Frigione and Marra, 1976; Bakchoutov et al., 1980). Generally, the results indicate that cements with narrow particle-size distributions tend to develop higher compressive strength. Regourd et al. (1978) showed that the rate of hydration is accelerated by high surface area, but that it is difficult to separate the effects of fineness from those of chemical composition. Hunt (1986) and Hunt and Elspass (1986), working with a selection of well cements, found a good correlation between the Blaine fineness and thickening time (Fig. 2–15).

2–5.7 Sulfate Resistance

Downhole brines commonly contain magnesium and sodium sulfates, and detrimental effects can result when such solutions react with certain cement hydration products. Magnesium and sodium sulfates react with precipitated calcium hydroxide to form magnesium and sodium hydroxides, and calcium sulfate. The calcium sulfate can

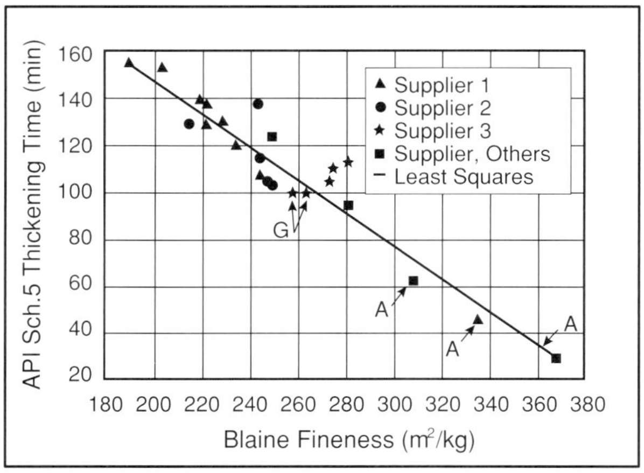

Figure 2–15—Linear regression of thickening time and Blaine fineness from Class A and G cements (from Hunt, 1986).

in turn react with the aluminates to form secondary ettringite.

$$Ca(OH)_2 + MgSO_4 + 2H_2O \rightarrow CaSO_4 \cdot 2H_2O + Mg(OH)_2 \quad (2\text{–}14)$$

Swelling occurs due to the replacement of $Ca(OH)_2$ by $Mg(OH)_2$.

$$Ca(OH)_2 + Na_2SO_4 + 2H_2O \rightarrow CaSO_4 \cdot 2H_2O + 2NaOH \quad (2\text{–}15)$$

An increase in cement porosity occurs, because NaOH is much more soluble than $Ca(OH)_2$.

$$3CaO \cdot Al_2O_3 \cdot 6H_2O + 3(CaSO_4 \cdot 2H_2O) + 20H_2O \rightarrow 3CaO \cdot Al_2O_3 \cdot 3CaSO_4 \cdot 32H_2O$$

or

$$C_3AH_6 + 3C\overline{S}H_2 + 20H \rightarrow C_3A \cdot 3C\overline{S} \cdot 32H \quad (2\text{–}16)$$

When ettringite forms *after* the cement has developed strength, an expansion occurs. As discussed in Chapter 7, a limited amount of expansion can be beneficial in terms of bonding; however, uncontrolled cement expansion leads to loss of compressive strength, cracking and damage to tubulars.

Portland cements with low C_3A contents are less susceptible to sulfate attack (American Petroleum Institute, 1955) after setting. In addition, because the solubility of magnesium and sodium sulfate is low above 140°F (60°C), sulfate attack is not normally a serious problem at that temperature or higher (Suman and Ellis, 1977). In any event, as discussed in Chapter 3, sulfate attack can be substantially reduced by the addition of "pozzolanic materials" such as fly ash to the cement system.

2–6 CLASSIFICATION OF PORTLAND CEMENTS

Portland cements are manufactured to meet certain chemical and physical standards which depend upon their application. To promote consistency of performance among cement manufacturers, classification systems and specifications have been established by various user groups. The best known systems are those of the American Society for Testing and Materials (ASTM) and the American Petroleum Institute (API).

2–6.1 Classification Criteria

The principal chemical criterion for classifying Portland cements is the relative distribution of the main clinker phases, known as the "potential phase composition." Despite vigorous research over the last 100 years, a reliable direct method for determining the concentrations of clinker phases in Portland cement has yet to surface. This goal is elusive because of the phases' chemical similarity. Methods such as petrographic microscopy, X-ray diffraction, and various physical and chemical separation techniques are qualitative to semiquantitative at best (Taylor, 1964; Aldridge, 1982). The most widely accepted method of expressing the relative amounts of the principal clinker phases relies upon a series of calculations based upon the oxide composition of the cement. This method, first introduced by Bogue (1929), is based upon various phase equilibria relationships between the cement components. Bogue's method suffers from various limitations, but remains a yardstick by which cements are classified. The Bogue equations are listed in Table 2–4. Limits on the amounts of alkalis, free CaO, MgO and SO_3, insoluble residue and the loss on ignition are also specified for some classes of Portland cements.

Physical parameters which appear in specifications include the fineness of the cement, and the performance of the cement according to standardized tests. The performance tests include measurements of thickening time, compressive strength, expansion and free water. The reader is referred to Appendix B for a complete description of the test methods and equipment.

2–6.2 API Classification System

Specifications for well cements were established by the API, because the conditions to which Portland cement is exposed in wells can differ radically from those experienced in construction applications. There are currently eight classes of API Portland cements, designated A through H. They are arranged according to the depths to which they are placed, and the temperatures and pressures to which they are exposed.

When the ratio of percentages of aluminum oxide to ferric oxide is 0.64 or more, the percentages of tricalcium silicate, dicalcium silicate, tricalcium aluminate, and tetracalcium aluminoferrite shall be calculated from the chemical analysis as follows:

Tricalcium silicate = $(4.071 \times \% CaO) - (7.600 \times \% SiO_2) - (6.718 \times \% Al_2O_3) - (1.430 \times \% Fe_2O_3) - (2.852 \times \% SO_3)$

Dicalcium silicate = $(2.867 \times \% SiO_2) - (0.7544 \times \% C_3S)$

Tricalcium aluminate = $(2.650 \times \% Al_2O_3) - (1.692 \times \% Fe_2O_3)$

Tetracalcium aluminoferrite = $3.043 \times \% Fe_2O_3$

When the alumina-ferric oxide ratio is less than 0.64, a calcium aluminoferrite solid solution (expressed as $ss(C_4AF + C_2F)$) is formed. Contents of this solid solution and of tricalcium silicate shall be calculated by the following formulas:

$ss(C_4AF + C_2F) = (2.100 \times \% Al_2O_3) + (1.702 \times \% Fe_2O_3)$

Tricalcium silicate = $(4.071 \times \% CaO) - (7.600 \times \% SiO_2) - (4.479 \times \% Al_2O_3) - (2.859 \times \% Fe_2O_3) - (2.852 \times \% SO_3)$.

No tricalcium aluminate will be present in cements of this composition. Dicalcium silicate shall be calculated as previously shown.

In the calculation of C_3A, the values of Al_2O_3 and Fe_2O_3 determined to the nearest 0.01% shall be used. In the calculation of other compounds, the oxides determined to the nearest 0.1% shall be used. All values calculated as described above shall be reported to the nearest 1%.

Table 2–4—Bogue equations for calculating potential phase composition (from ASTM Method C 114).

Within some classes, cements with varying degrees of sulfate resistance (as determined by C_3A content) are sanctioned: ordinary (O), moderate sulfate resistance (MSR) and high sulfate resistance (HSR). The chemical and physical specifications are listed in Tables 2–5 and 2–6, respectively. Table 2–7 lists typical compositions and surface-area ranges for certain API cements. Below is a general description of each API class, with its ASTM equivalent when appropriate.

Class A: Intended for use from surface to a depth of 6,000 ft (1,830 m), when special properties are not required. Available only in Ordinary type, Class A is similar to ASTM Type I cement.

Class B: Intended for use from surface to a depth of 6,000 ft (1,830 m), when conditions require moderate to high sulfate resistance. Class B is similar to ASTM Type II, and has a lower C_3A content than Class A.

Class C: Intended for use from surface to a depth of 6,000 ft (1,830 m), when conditions require high early strength. Class C is available in all three degrees of sulfate resistance, and is roughly equivalent to ASTM Type III. To achieve high early strength, the C_3S content and the surface area are relatively high.

Classes D, E and F are also known as "retarded cements," intended for use in deeper wells. The retardation is accomplished by significantly reducing the amount of faster-hydrating phases (C_3S and C_3A), and increasing the particle size of the cement grains. Since these classes were first manufactured, the technology of chemical retarders has significantly improved; consequently, they are rarely found today.

Class D: Intended for use at depths from 6,000 ft (1,830 m) to 10,000 ft (3,050 m), under conditions of moderately high temperatures and pressures. It is available in MSR and HSR types.

Class E: Intended for use from 10,000 ft (3,050 m) to 14,000 ft (4,270 m) depth, under conditions of high temperatures and pressures. It is available in MSR and HSR types.

Class F: Intended for use from 10,000 ft (3,050 m) to 16,000 ft (4,880 m) depth, under conditions of extremely high temperatures and pressures. It is available in MSR and HSR types.

Classes G and H were developed in response to the improved technology in slurry acceleration and retardation by chemical means. The manufacturer is prohibited from adding special chemicals, such as glycols or acetates, to the clinker. Such chemicals improve the efficiency of grinding, but have been shown to interfere with various cement additives. Classes G and H are by far the most commonly used well cements today.

Class G: Intended for use as a basic well cement from
Class H: surface to 8,000 ft (2,440 m) depth as manufactured, or can be used with accelerators and retarders to cover a wide range of well depths and temperatures. No additions other than calcium sulfate or water, or both, shall be interground or blended with the clinker during manufacture of Class G and H well cements. They are available in MSR and HSR types.

The chemical compositions of Classes G and H are essentially identical. The principal difference is the surface area. Class H is significantly coarser than Class G, as evidenced by their different water requirements.

REFERENCES

Aldridge, L.P.: "Accuracy and Precision of Phase Analysis in Portland Cement by Bogue, Microscopic and X-ray Diffraction Methods," *Cement and Concrete Res.* (1982) **12**, 381–398.

American Petroleum Institute: "Report of Cooperative Tests on Sulfate Resistance of Cement and Additives," API Mid-Continent Dist. Study Committee on Cementing Practices and Testing of Oil Well Cements, 1955.

Bakchoutov, V. S., Al-Vardi, K. H., Pin-Khouan, T. and Nikolaeva, M. K.: "Study of the Grain Composition of Oil-Well Cements," *Proc.*, Seventh Intl. Cong. Chem. Cement, Paris (1980) **5**, 203.

Barnes, P.: *Structure and Performance of Cements*, Applied Science Publishers Ltd., London (1983).

Barret, P. and Bertrandie, D.: "Fundamental Hydration Kinetic Features of the Major Cement Constituents: Tricalcium Silicate (Ca_3SiO_5) and Beta-Dicalcium Silicate (βCa_2SiO_4)," *J. Chim. Phys.* (1980) **83**, 765–775.

Barret, P., Bertrandie, D., and Ménétrier, D.: "Comparative Study of C-S-H Formation From Supersaturated Solutions and C_3S Solution Mixtures," *Proc.*, Seventh Intl. Cong. Chem. Cement, Paris, (1980) 2, II/261– II/266.

Barret, P., Ménétrier, D., Bertrandie, D., and Regourd, M.: "Thermodynamic and Kinetic Aspects of C_3S Passage in Solution and C-S-H Formation from Supersaturated Solutions and C_3S Solution Mixtures," *Proc.*, Seventh Intl. Cong. Chem. Cement, Paris (1980) **2**, II/279-II/284.

Barret, P.: "Hydration Mechanism of Calcium Silicates (C_3S, C_2S) and Cement Compounds, Through the General Concepts of the Reactivity of Solids," *Proc.*, Eighth Intl. Cong. Chem. Cement, Paris(1986) **3**, 86–92.

Barvinok, M. S., Komokhov, P. S., and Bondareva, N. F.: "Effect of Temperature and Additives on the Early Hardening Stage," *Proc.*, Sixth Intl. Congr. Chem Cement, Paris (1976) **2**, 151–155.

Bensted, J.: "Fase Ferritica Uno Studio Spettroscopio All'Infrarosso," *Il Cemento* (1976) **73**, 45–51.

Bentur, A., Berger, R. L., Kung, J. H., Milestone, N. B., and Young, J. F.: "Structural Properties of Calcium Silicate Pastes—Pt. 2 : Effect of Curing Temperature," *J. Amer. Ceramic Soc.* (1979) **62**, 362–366.

Bertrandie, D. and Barret, P.: "Initial Interfacial Steps in Hydration of Calcium Aluminates as Cement Compounds," *Proc.*, Eighth Intl. Cong. Chem. Cement, Paris (1986) **3**, 79–85.

Bogue, R. H.: "Calculation of the Compounds in Portland Cement," *Ind. Eng. Chem. Anal. Ed.* (1929) **1**, 192–197.

	Cement Class					
	A	B	C	D,E,F	G	H
Ordinary Type (O)						
Magnesium oxide (MgO), maximum, %	6.0		6.0			
Sulfur trioxide (SO_3), maximum, %	3.5		4.5			
Loss on ignition, maximum, %	3.0		3.0			
Insoluble residue, maximum, %	0.75		0.75			
Tricalcium aluminate ($3CaO \cdot Al_2O_3$), maximum, %			15			
Moderate Sulfate-Resistant Type (MSR)						
Magnesium oxide (MgO), maximum, %		6.0	6.0	6.0	6.0	6.0
Sulfur trioxide (SO_3), maximum, %		3.0	3.5	3.0	3.0	3.0
Loss on ignition, maximum, %		3.0	3.0	3.0	3.0	3.0
Insoluble residue, maximum, %		0.75	0.75	0.75	0.75	0.75
Tricalcium silicate ($3CaO \cdot SiO_2$), maximum, %					58	58
Tricalcium silicate ($3CaO \cdot SiO_2$), minimum, %					48	48
Tricalcium aluminate ($3CaO \cdot Al_2O_3$), maximum, %		8	8	8	8	8
Total alkali content expressed as sodium oxide (Na_2O) equivalent, maximum, %					0.75	0.75
High Sulfate-Resistant Type (HSR)						
Magnesium oxide (MgO), maximum, %		6.0	6.0	6.0	6.0	6.0
Sulfur trioxide (SO_3), maximum, %		3.0	3.5	3.0	3.0	3.0
Loss on ignition, maximum, %		3.0	3.0	3.0	3.0	3.0
Insoluble residue, maximum, %		0.75	0.75	0.75	0.75	0.75
Tricalcium silicate ($3CaO \cdot SiO_2$), maximum, %					65	65
Tricalcium silicate ($3CaO \cdot SiO_2$), minimum, %					48	48
Tricalcium aluminate ($3CaO \cdot Al_2O_3$), maximum, %		3	3	3	3	3
Tetracalcium aluminoferrite ($4CaO \cdot Al_2O_3 \cdot Fe_2O_3$) plus twice the tricalcium aluminate ($3CaO \cdot Al_2O_3$), maximum, %		24	24	24	24	24
Total alkali content expressed as sodium oxide (Na_2O) equivalent, maximum, %					0.75	0.75

Table 2–5—Chemical requirements for API Portland cements (from API Spec 10: Materials and Testing for Well Cements).

Well Cement Class				A	B	C	D	E	F	G	H
Water, % by weight of well cement				46	46	56	38	38	38	44	38
Soundness (autoclave expansion), maximum, %				0.80	0.80	0.80	0.80	0.80	0.80	0.80	0.80
Fineness* (specific surface), minimum, m²/kg				150	160	220	—	—	—	—	—
Free-water content, maximum, mL				—	—	—	—	—	—	3.5**	3.5**
	Schedule Number	Curing Temp F° (°C)	Curing Pressure. psi (kPa)	Minimum Compressive Strength, psi (MPa)							
Compressive Strength Test, 8-Hour Curing Time	—	100 (38)	Atmos.	250 (1.7)	200 (1.4)	300 (2.1)	— —	— —	— —	300 (2.1)	300 (2.1)
	—	140 (60)	Atmos.	— —	— —	— —	— —	— —	— —	1500 (10.3)	1500 (10.3)
	6S	230 (110)	3000 (20,700)	— —	— —	— —	500 (3.5)	— —	— —	— —	— —
	8S	290 (143)	3000 (20,700)	— —	— —	— —	— —	500 (3.5)	— —	— —	— —
	9S	320 (160)	3000 (20,700)	— —	— —	— —	— —	— —	500 (3.5)	— —	— —
Compressive Strength Test, 12-Hour Curing Time	8S	290 (143)	3000 (20,700).	— —	— —	— —	— —	— —	— —	— —	— —
	Schedule Number	Curing Temp. F° (°C)	Curing Pressure. psi (kPa)	Minimum Compressive Strength, psi (MPa)							
Compressive Strength Test, 24-Hour Curing Time	—	100 (38)	Atmos.	1800 (12.4)	1500 (10.3)	2000 (13.8)	— —	— —	— —	— —	— —
	4S	170 (77)	3000 (20,700)	— —	— —	— —	1000 (6.9)	1000 (6.9)	— —	— —	— —
	6S	230 (110)	3000 (20,700)	— —	— —	— —	2000 (13.8)	— —	1000 (6.9)	— —	— —
	8S	290 (143)	3000 (20,700)	— —	— —	— —	— —	2000 (13.8)	— —	— —	— —
	9S	320 (160)	3000 (20,700)	— —	— —	— —	— —	— —	1000 (6.9)	— —	— —
	10S	350 (177)	3000 (20,700)	— —	— —	— —	— —	— —	— —	— —	— —
	Specification Test Schedule Number	Maximum Consistency 15- to 30-min. Stirring Period, B_c+		Minimum Thickening Time, min.***							
Pressure Temperature Thickening Time Test	1	30		90	90	90	—	—	—	—	—
	4	30		90	90	90	90	—	—	—	—
	5	30		—	—	—	—	—	—	90	90
	5	30		—	—	—	—	—	—	120 max++	120 max.++
	6	30		—	—	—	100	100	100	—	—
	8	30		—	—	—	—	154	—	—	—
	9	30		—	—	—	—	—	190	—	—

*Determined by Wagner turbidmeter apparatus described in ASTM C 115: *Fineness of Portland Cement by the Turbidmeter.*

**Based on 250-mL volume, percentage equivalent of 3.5 mL is 1.4%.

+Bearden units of slurry consistency (Bc).

Bc—Bearden units of consistency obtained on a pressurived consistometer as defined in Section 8 of API Spec 10 and calibrated as per the same section.

ABc—Bearden units of consistency obtained on an atmosphere pressure consistometer as defined in Section 9 of API Spec 10 and calibrated as per the same section.

The relationship between Bc and ABc is approximately Bc x 0.69 = ABc. This relationship is valid only for units of consistency less than 30 Bc.

***Thickening time requirements are based on 75 percentile values of the total cementing times observbed in the casing survey, plus a 25% safety factor.

++Maximum thickening time requirement for Schedule 5 is 120 minutes.

Table 2–6—Physical requirements for API Portland cements (parenthetical values are in metric units) (from API Spec. 10: Materials and Testing for Well Cements).

API Class	ASTM Type	Typical Potential Phase Composition (%)				Typical Fineness (cm^2/g)
		C_3S	β-C_2S	C_3A	C_4AF	
A	I	45	27	11	8	1600
B	II	44	31	5	13	1600
C	III	53	19	11	9	2200
D		28	49	4	12	1500
E		38	43	4	9	1500
G	(II)	50	30	5	12	1800
H	(II)	50	30	5	12	1600

Table 2–7—Typical composition and fineness of API cements (from Nelson, 1983).

Chernikh, V. F. et al.: *Tsement* (1963) **5**.

Cobb, J. A. and Pace, R. S.: "Elements Affecting the Thickening Time of a Cement Blend," paper SPE 14195, 1985.

de Jong, J. G. M., Stein, H. N., and Stevels, J. M.: "Hydration of Tricalcium Silicate," *J. Appl. Chem.* (1967) **17**, 246–250.

Dent-Glasser, L. S., Lachowski, E. E., Mohan, K., Taylor, H.F.W.: "A Multi-Method Study of C_3S Hydration," *Cement and Concrete Res.* (1978) **8,** 733–739.

Double, D. D., Hellawall, A., and Perry, S. J.: "The Hydration of Portland Cement," *Proc.*, Royal Soc. of London (1978) Ser. A **359,** 435–451.

Fierens, P. and Verhaegen, J. P.: "Effect of Water on Pure and Doped Tricalcium Silicate Using the Techniques of Adsorboluminescence," *Cement and Concrete Res.* (1975) **5,** 233–238.

Fierens, P. and Verhaegen, J. P.: "Hydration of Tricalcium Silicate in Paste—Kinetics of Calcium Ion Dissolution in the Aqueous Phase," *Cement and Concrete Res.* (1976) **6,** 337–342.

Fierens, P. and Verhaegen, J. P.: "Induction Period of Hydration of Tricalcium Silicate," *Cement and Concrete Res.* (1976) **6**, 287–292.

Fierens, P. and Verhaegen, J. P.: "Microcathodoluminescence of Tricalcium Silicate," *Il Cemento* (1976) **73**, 39–44.

Fierens, P. and Verhaegen, J. P.: "Nucleophilic Properties of the Surface of Tricalcium Silicate," *Cement and Concrete Res.* (1976) **6,** 103–11.

Fierens, P. and Verhaegen, J. P.: "Thermoluminescence Applied to the Kinetics of the Chemisorption of Water by Tricalcium Silicate," *Silicates Ind.* (1974) **39,** 125–130.

Frigione, G. and Marra, S.: "Relationship Between Particle Size Distribution and Compressive Strength in Portland Cement," *Cement and Concrete Res.* (1976) **6,** 113–127.

Ghosh, S. N., ed: *Advances in Cement Technology*, Pergamon Press Ltd., Oxford (1983).

Hunt, L. P. and Elspass, C. W.: "Particle-Size Properties of Oil-Well Cements," *Cement and Concrete Res.* (1986) **16**, 805–812.

Hunt, L. P.: "Prediction of Thickening Time of Well Cements from Blaine Air Permeability," *Cement and Concrete Res.* (1986) **16,** 190–198.

Jawed, I. and Skalny, J.: "Alkalis in Cement: A Review—Pt. 2: Effects of Alkalis on the Hydration and Performance of Portland Cement," *Cement and. Concrete Res.* (1978) **8**, 37-51.

Lach, V. and Bures, J.: "Phase Composition and Microstructure of Cement Paste Hydrated at Elevated Temperatures," *Proc.*, Sixth Intl. Cong. Chem Cement, Paris (1974) **2**, 129–135.

Lea, F. M.: *The Chemistry of Cement and Concrete*, Chemical Publishing Co., Inc., New York (1971).

Lerch, W.: Portland Cement Res. *Laboratory Bull.* (1946) **12**.

Locher, F. W., Richartz, W., and Sprung, S.: "Setting of Cement. Part II. Effect of Adding Calcium Sulfate," *Zement-Kalk-Gips* (1980) **33,** 271–277.

Ménétrier, D.: DSc thesis, Universite de Dijon, Dijon, France (1977).

Michaux, M., Ménétrier, D., and Barret, P.: *Comptes Rendus Acad. Sci.* (1983) Series 2, **296**, 1043–1046.

Michaux, M.: "Contribution à L'Etude de la Constitution de L'Hydrosilicate de Calcium et au Mecanisme de sa Formation par Hydratation du Silicate Tricalcique en Présence ou Non D'Additifs," DSc thesis, Université de Dijon, Dijon, France (1984).

Nelson, E. B.: "Portland Cements Characterized, Evaluated," *Oil and Gas J.* (Feb. 1983) 73–77.

Odler, I. and Skalny, J.: "Hydration of Tricalcium Silicate at Elevated Temperatures," *J. Appl. Chem.* Biotechnol. (1973) **23**, 661–667.

Odler, I. and Skalny, J.: "Influence of Calcium Chloride on Paste Hydration of Tricalcium Silicate," *J. Amer Ceramic Soc.* (1971) **54**, 362–364.

Ost, B. W.: "Optimum Sulfate Content of Portland Cements," *Amer. Ceramic Soc. Bull.* (1974) **53**, No. 8, 579–580.

Portland Cements, Portland Cement Association, Skokie, IL, (1969).

Powers, T. C.: "Some Physical Aspects of Hydration of Portland Cement," *J. Res. Dev. Lab. Portland Cement Assoc.* (1961) **3**, 47–56.

Ramachandran, V. S. and Beaudoin, J. J.: "Hydration of C_4AF + Gypsum: Study of Various Factors," *Proc.*, Seventh Intl. Cong. Chem. Cement, Paris (1980) **2,** II/25–II/30.

Regourd, M., Hornain, H., and Mortureux, B.: *Ciments, Bétons, Platres et Chaux* (March 1978) 712 .

Satava, V. and Veprek, O.: *J. Amer Ceramic Soc.* (1975) **58,** 857.

Silk, I.M.: "Exposure to Moisture Alters Well Cement," *Pet. Eng. Intl.* (1986) **58**, 45–49.

Skalny, J. and Young, J. F.: "Mechanisms of Portland Cement Hydration," *Proc.*, Seventh Int. Cong. Chem. Cement, Paris (1980) 1, 1–52.

Sprung, S., Kuhlmann, K. and Ellerbrock, H. G.: "Particle Size Distribution and Properties of Cement Part II: Water Demand of Portland Cement," *Zement-Kalk-Gips* (1985) **11**, 275.

Sudakas, L. G., Zozulya, R. A., Kokurkina, A. V., and Sorokina, V. A.: "Alkalies, Microstructure and Activity of Industrial-Grade Cement Clinkers," *Tsemen*t (1978) **12**, 11–12.

Suman, G. O. and Ellis, R. C.: *Cementing Oil and Gas Wells ... Including Casing Handling Procedures*, World Oil, Houston, 1977.

Tadros, M. E., Skalny, J. and Kalyoncu, R. S.: "Early Hydration of Tricalcium Silicate," *J. Amer. Ceramic Soc.* (1976) **59**, 344–347.

Taylor, H. F. W., ed: *The Chemistry of Cements*, Academic Press Inc. Ltd., London (1964).

Thomas, N. L. and Double, D. D.: "Calcium and Silicon Concentrations in Solution During the Early Hydration of Portland Cement and Tricalcium Silicate, "*Cement and Concrete Res.* (1981) **11**, 675–687.

Vidick, B., Oberste-Padtberg, R., Laurent, J. P., and Rondelez, F.: "Selective Surface Determination of the Silicate Phases in Portland Cement Powders Using Alkyltrichlorosilane," *Cement and Concrete Res.* (1987) **17**, 624.

3

Cement Additives and Mechanisms of Action

Erik B. Nelson, Jean-François Baret and Michel Michaux

Dowell Schlumberger

3–1 INTRODUCTION

In well cementing, Portland cement systems are routinely designed for temperatures ranging from below freezing in permafrost zones to 700°F (350°C) in thermal recovery and geothermal wells. Well cements encounter the pressure range from near ambient in shallow wells to more than 30,000 psi (200 MPa) in deep wells. In addition to severe temperatures and pressures, well cements must often be designed to contend with weak or porous formations, corrosive fluids, and overpressured formation fluids. It has been possible to accommodate such a wide range of conditions only through the development of cement additives. Additives modify the behavior of the cement system, ideally allowing successful slurry placement between the casing and the formation, rapid compressive strength development, and adequate zonal isolation during the lifetime of the well.

Today, over 100 additives for well cements are available, many of which can be supplied in solid or liquid forms. Eight categories of additives are generally recognized.

1. *Accelerators*: chemicals which reduce the setting time of a cement system, and increase the rate of compressive strength development.
2. *Retarders*: chemicals which extend the setting time of a cement system.
3. *Extenders*: materials which lower the density of a cement system, and/or reduce the quantity of cement per unit volume of set product.
4. *Weighting Agents*: materials which increase the density of a cement system.
5. *Dispersants*: chemicals which reduce the viscosity of a cement slurry.
6. *Fluid-Loss Control Agents*: materials which control the loss of the aqueous phase of a cement system to the formation.
7. *Lost Circulation Control Agents*: materials which control the loss of cement slurry to weak or vugular formations.
8. *Specialty Additives*: miscellaneous additives, e.g., antifoam agents, fibers, etc.

In this chapter, each of the above categories is discussed individually. The physical and chemical phenomena with which the additives must contend, as well as examples of additives and proposed mechanisms of action, are discussed in detail. A thorough review of Chapter 2 is recommended before reading this chapter.

3–2 VARIABILITY OF ADDITIVE RESPONSE

Typical performance data for many additives are presented throughout this chapter. It is important for the reader to understand that this information is presented solely to illustrate general trends, and should not be used for design purposes. Most additives are strongly influenced by the chemical and physical properties of the cement, which are highly variable even within a given API classification. Consequently, a wide spectrum of results can be obtained with the same slurry design. The important cement parameters include the following:

- particle size distribution,
- distribution of silicate and aluminate phases,
- reactivity of hydrating phases,
- gypsum/hemihydrate ratio, and total sulfate content,
- free alkali content, and
- chemical nature, quantity, and specific surface area of initial hydration products.

Other important parameters include temperature, pressure, additive concentration, mixing energy, mixing order and water-to-cement ratio.

Figure 3–1 is a graphic illustration of the variability of additive response to cements. The figure compares the

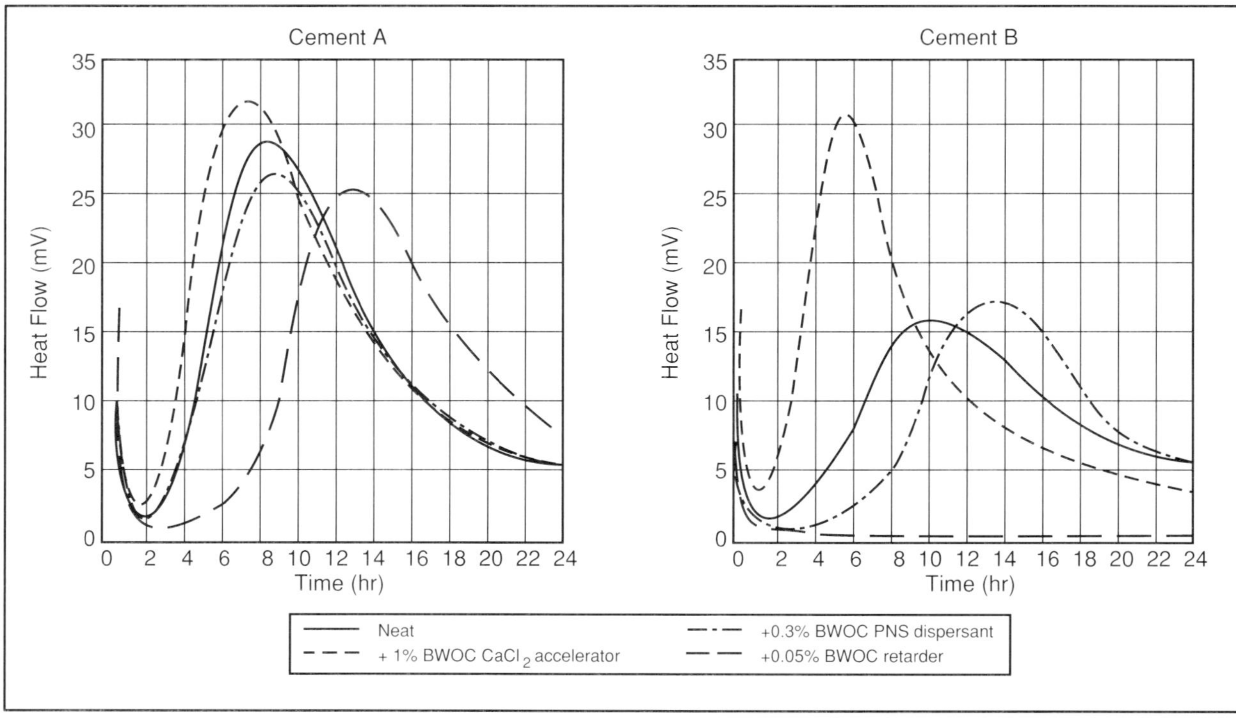

Figure 3–1—Calorimetric behavior of Cements A and B in the presence of different additives.

hydration behavior of two API Class G cements. Conduction calorimetry curves were generated for the neat slurries, and for three additional slurries containing an accelerator, a retarder or a dispersant. Scrutiny of the curves reveals significant differences in hydration behavior.

Because of the complexity of the cement hydration process, and the large number of parameters involved, the only practical method for cement slurry design (and avoiding unpleasant surprises at the wellsite) is thorough laboratory testing before the job. It is essential that the tests be performed with a representative sample of the cement to be used during the cement job.

3–3 ACCELERATORS

Accelerators are added to cement slurries to shorten the setting time (Stages I and II of the hydration scheme described in Chapter 2) and/or to accelerate the hardening process (Stages III and IV). They are often used to offset the set delay caused by certain other additives, such as dispersants and fluid-loss control agents (Odler et al., 1978).

3–3.1 Examples

Many inorganic salts are accelerators of Portland cement. Among these, the chlorides are the best known; however, an accelerating action is also reported for many other salts including carbonates, silicates (especially sodium silicate), aluminates, nitrates, nitrites, sulfates, thiosulfates, and alkaline bases such as sodium, potassium and ammonium hydroxides.

Among the chlorides, the accelerating action becomes stronger by passing from monovalent to bivalent and trivalent chlorides, and as the radius of the accompanying cation increases (Skalny and Maycock, 1975). Edwards and Angstadt (1966) suggested that cations and anions may be ranked according to their efficiency as accelerators for Portland cement.

$$Ca^{2+} > Mg^{2+} > Li^{+} > Na^{+} > H_2O$$

$$OH^{-} > Cl^{-} > Br^{-} > NO_3^{-} > SO_4^{2-} = H_2O$$

Calcium chloride is undoubtedly the most efficient and economical of all accelerators. Regardless of concentration, it always acts as an accelerator (Table 3–1). It is normally added at concentrations between 2% to 4% by weight of cement (BWOC). Results are unpredictable at concentrations exceeding 6% BWOC, and premature setting may occur.

Sodium chloride affects the thickening time and compressive strength development of Portland cement in different ways, depending upon its concentration and the curing temperature (Fig. 3–2). NaCl acts as an accelerator at concentrations up to 10% by weight of mix water

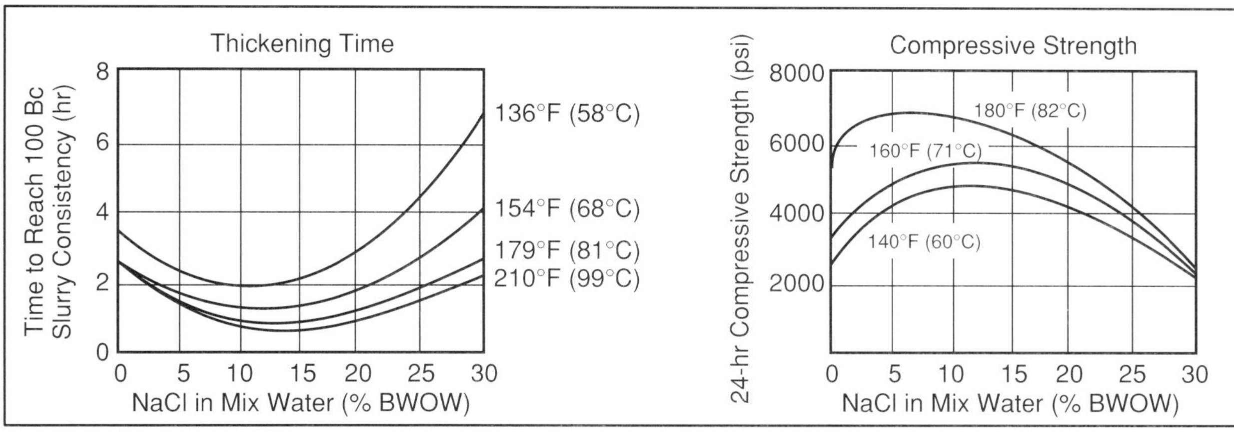

Figure 3–2—Effect of sodium chloride on thickening time and compressive strength/development.

Thickening Time of Neat Cement Slurries Accelerated by Flake Calcium Chloride			
	Thickening Time (hr:min)		
$CaCl_2$ (% BWOC)	91°F	103°F	113°F
0	4:00	3:30	2:32
2	1:17	1:11	1:01
4	1:15	1:02	0:59

Compressive Strength Development for Accelerated Cement Slurries									
	Compressive Strength (psi) at Temperature and Time Indicated								
$CaCl_2$	60°F			80°F			100°F		
%	6 hr	12 hr	24 hr	6 hr	12 hr	24 hr	6 hr	12 hr	24 hr
0	Not Set	60	415	45	370	1260	370	840	1780
2	125	480	1510	410	1020	2510	1110	2370	3950
4	125	650	1570	545	1245	2890	1320	2560	4450

Table 3–1—Effects of calcium chloride upon the performance of Portland cement systems.

(BWOW). Between 10% to 18% (BWOW) NaCl is essentially neutral, and thickening times are similar to those obtained with fresh water. The addition of NaCl concentrations above 18% BWOW causes retardation. Sodium chloride is not a very efficient accelerator, and should be used only when calcium chloride is not available at the wellsite.

Seawater is used extensively for mixing cement slurries on offshore locations. It contains up to 25 g/L NaCl, resulting in acceleration. The presence of magnesium (about 1.5 g/L) also must be taken into account (Chapter 7).

Sodium silicate is normally used as a cement extender; however, it also has an accelerating effect. Sodium silicate reacts with Ca^{2+} ions in the aqueous phase of the cement slurry to form additional nuclei of C-S-H gel, thus hastening the end of the induction period.

Organic accelerators exist, including calcium formate ($Ca(HCOO)_2$), oxalic acid ($H_2C_2O_4$) and triethanolamine (TEA: $N(C_2H_4OH)_3$) (Singh and Agha, 1983; Pauri et al., 1986; Ramachandran, 1973; 1976). The latter is an accelerator of the aluminate phases, and a retarder of the silicate phases. TEA is not normally used alone, but in combination with other additives to counteract excessive retardation caused by some dispersants. To the authors' knowledge, such organic accelerators have not yet been used in well cementing.

3–3.2 Calcium Chloride—Mechanisms of Action

Calcium chloride is by far the most common accelerator for Portland cement. The mechanisms by which it operates are complex, and still not completely understood. Several hypotheses have been described in the literature, and are summarized below.

3–3.2.1 Effects on the Hydration of Principal Portland Cement Phases

It is sometimes proposed that the acceleration of set is the result of an increase in hydration rate of the aluminate phases/gypsum system (Bensted, 1978; Traetteberg and Grattan-Bellew, 1975). Chloride ions enhance the formation of ettringite until the gypsum is consumed (Ténoutasse, 1978). If free C_3A remains, calcium monochloroaluminate ($C_3A \cdot CaCl_2 \cdot 10H_2O$) forms. The more rapid set of the cement slurry is also attributed to the crystalline shape of ettringite, which occurs as very fine needles (Bensted, 1978; Young et al., 1973).

By contrast, Stein (1961) and Edwards and Angstadt (1966) concluded that accelerators do not promote the hydration of the C_3A, but predominantly accelerate the hydration of C_3S. This accelerating action of calcium chloride is confirmed by studying the hydration of the pure silicate phase, C_3S (Odler and Skalny, 1971) and C_2S (Collepardi and Massidda, 1973).

3–3.2.2 Change in C-S-H Structure

The hydration of Portland cement is often seen as being controlled by the diffusion of water and ionic species through the initial protective C-S-H gel coating (Chapter 2). Therefore, the rate of hydration should depend strongly on the permeability of the coating. A morphological change of the C-S-H gel to a more open flocculated structure would enhance diffusion and accelerate hydration. Such a process has been confirmed in studies with pure C_3S (Odler and Skalny, 1971; Traetteberg et al., 1974; Ben-Dor and Perez, 1976). The C-S-H gel has a higher C/S ratio, and a crumpled foil morphology rather than the usual spicular one. In the presence of calcium chloride, C-S-H gel has a higher specific surface (Collepardi and Marchese, 1972) and a higher degree of silicate anion polymerization (Hirljac et al., 1983). A change in the pore-size distribution of hydrated C_3S (Skalny et al., 1971; Young et al., 1973) and C_2S (Odler and Skalny, 1971) has also been evidenced. The morphology of calcium hydroxide (portlandite) is also affected by the presence of chloride ions (Berger and McGregor, 1972).

3–3.2.3 Diffusion of Chloride Ions

Kondo et al. (1977) determined the diffusion rate of anions and cations of alkaline and alkaline-earth chlorides through a set Portland cement plate. They concluded that the diffusion coefficient of the chloride ion is much higher than that of the cation accompanying it. Since the chloride ions diffuse into the C-S-H gel layer more quickly than the cations, a counterdiffusion of hydroxyl ions occurs to maintain the electrical balance. Therefore, the precipitation of portlandite, ending the induction period, takes place earlier. These authors have also established that only a small amount of chloride ions is incorporated into the C-S-H lattice, but may be chemisorbed onto the C-S-H surface.

Singh and Ojha (1981) believed that calcium chloride accelerates C_3S hydration because chloride ions have a smaller ionic size, and a greater tendency to diffuse into the C-S-H membrane than hydroxyl ions. Therefore, an increase in the internal pressure takes place more quickly, causing an early bursting of the C-S-H membrane, and an acceleration of hydration.

3–3.2.4 Change in Aqueous Phase Composition

Michaux et al. (1989) showed that the presence of calcium chloride strongly modifies the distribution of ionic species in the aqueous phase of well cement slurries. Because of the introduction of chloride ions which do not participate in the formation of hydration products during the induction period, a decrease of hydroxyl and sulfate concentrations and an increase of calcium concentration are observed. Kurczyk and Schwiete (1960) proposed that the accelerating action of calcium chloride is related to a decrease of alkalinity in the aqueous phase, enhancing the dissolution rate of lime.

Stadelmann and Wieker (1985) investigated the influence of a large number of inorganic salts on the hydration of C_3S. They showed C_3S hydration to be accelerated by increasing the solubility of calcium hydroxide in the aqueous phase, e.g., with $CaCl_2$. Conversely, retardation was observed when the solubility of calcium hydroxide decreased, e.g., with a high NaCl concentration.

Wu and Young (1984) demonstrated that the addition of calcium salts affects the dissolution rate of C_3S. When the concentration of calcium in the aqueous phase was monitored with time, the maximum was always reached earlier in the presence of chloride ions. Thus, precipitation of calcium hydroxide (and the end of the induction period) occurred earlier.

In conclusion, it is apparent that many factors are involved simultaneously in the acceleration of Portland cement by calcium chloride. Physical and chemical phenomena are involved. The presence of chloride ions alters the structure and increases the permeability of the C-S-H gel layer. In addition, calcium chloride significantly alters the distribution of ionic species in the aqueous phase, resulting in a faster hydration rate.

3–3.3 Secondary Effects of Calcium Chloride

In addition to acceleration of the initial set, several other effects are observed when calcium chloride is present in a Portland cement system. Some effects are not beneficial; as a result, calcium chloride should be used judiciously depending upon well conditions. A summary of the more important secondary effects is given below.

3–3.3.1 Heat of Hydration

The presence of $CaCl_2$ increases the rate of heat generation during the first hours after slurry mixing. If the wellbore is thermally insulated to a sufficient degree, the temperature of the cement, casing, and surrounding formation can increase by as much as 50° to 60°F (27° to 33°C) after slurry placement. An auto-acceleration of hydration results.

More importantly, increased casing expansion occurs because of the temperature rise. Since steel casing and cement do not have the same coefficient of thermal expansion, the casing may shrink away from the cement when the hydration heat eventually dissipates. This results in a so-called "thermal microannulus," and zonal isolation is compromised (Pilkington, 1988). Additional research must be performed to better quantify this effect, and to determine the most susceptible wellbore environments.

3–3.3.2 Slurry Rheology

According to Collepardi (1971), calcium chloride increases the yield point of a cement slurry, but initially does not affect the plastic viscosity. After a 30-minute hydration at ambient conditions, the plastic viscosity begins to increase. Slurries containing calcium chloride also tend to have a higher degree of thixotropy; as a result, particle sedimentation is seldom a problem.

3–3.3.3 Compressive Strength Development

Calcium chloride significantly increases the rate of compressive strength development during the first few days after placement. The magnitude of this effect depends upon the curing temperature and the $CaCl_2$ concentration (Table 3–1).

3–3.3.4 Shrinkage

Calcium chloride has been shown to increase volumetric shrinkage by 10% to 50% in concretes (Shideler, 1952). This is due mainly to the higher degree of hydration, and changes in hydration products (Collepardi and Massida, 1973). Such data cannot be directly translated to well cements, because the service conditions are very different. To the authors' knowledge, a thorough investigation of the dimensional stability of calcium chloride-accelerated well cements has not been performed. The magnitude of the shrinkage effect with concretes suggests that such a study is overdue.

3–3.3.5 Permeability

Initially, the permeability of set cement containing calcium chloride is reduced. This is due to the higher volume of hydration products present compared to an additive-free cement. At later ages, when the degree of hydration is similar for both systems, the set cement containing $CaCl_2$ is more permeable (Gouda, 1973).

3–3.3.6 Sulfate Resistance

Since the ultimate permeability of calcium chloride-accelerated systems is higher, the resistance to aggressive sulfate solutions is reduced (Shideler, 1952; Gouda, 1973). However, as discussed in Chapter 2, the C_3A content of the cement is the principal controlling factor.

3–4 RETARDERS

Like acceleration, the mechanism of set retardation of Portland cement is still a matter of controversy. Several theories have been proposed, but none is able to fully explain the retardation process by itself. Two principal factors must be considered: the chemical nature of the retarder, and the cement phase (silicate or aluminate) upon which the retarder acts. Four principal theories have been proposed, and are summarized below.

1. *Adsorption Theory*: retardation is due to the adsorption of the retarder onto the surface of the hydration products, thereby inhibiting contact with water.
2. *Precipitation Theory*: the retarder reacts with calcium and/or hydroxyl ions in the aqueous phase, forming an insoluble and impermeable layer around the cement grains.
3. *Nucleation Theory*: the retarder adsorbs on the nuclei of hydration products, poisoning their future growth.
4. *Complexation Theory*: calcium ions are chelated by the retarder, preventing the formation of nuclei.

It is probable that all of the above effects are involved to some extent in the retardation process. Despite the uncertainty regarding the mechanisms of retardation, the chemical technology is very well developed. The major chemical classes of retarders, as well as proposed mechanisms of action, are discussed individually below.

3–4.1 Lignosulfonates

The most commonly used retarders for well cements are the sodium and calcium salts of lignosulfonic acids (Fig. 3–3). Lignosulfonates are polymers derived from wood pulp; therefore, they are usually unrefined and contain various amounts of saccharide compounds. The average molecular weight varies from about 20,000 to 30,000. Since purified lignosulfonates lose much of their retarding power, the set-retarding action of these additives is often attributed to the presence of low-molecular-weight carbohydrates (Chatterji, 1967; Milestone, 1976; 1979), such as pentoses (xylose and arabinose), hexoses (mannose, glucose, fructose, rhamnose and galactose), and by aldonic acids (especially xylonic and gluconic acids).

Lignosulfonate retarders are effective with all Portland cements, and are generally added in concentrations ranging from 0.1% to 1.5% BWOC (Fig. 3–4). Depending upon their carbohydrate content and chemical structure (e.g., molecular weight distribution, degree of sul-

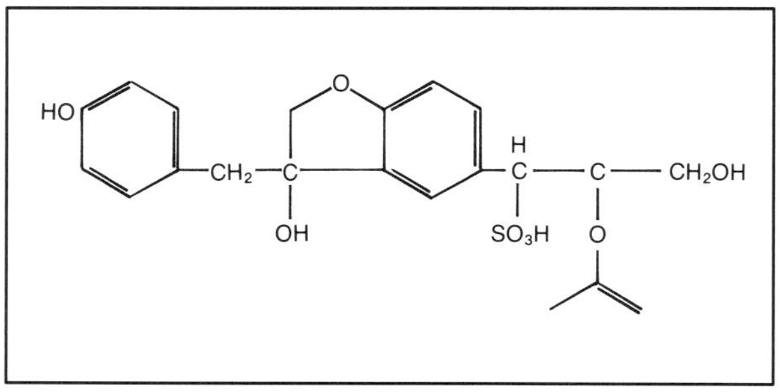

Figure 3–3—Basic lignosulfonate chemical structure.

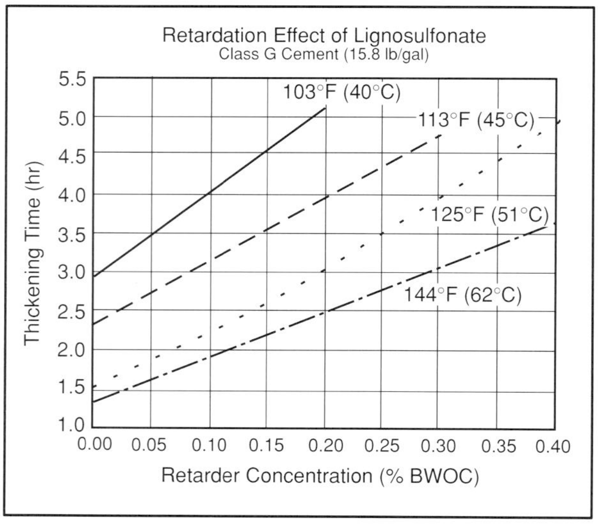

Figure 3–4—Retardation effect of lignosulfonate.

fonation, etc.), and the nature of the cement, they are effective to about 250°F (122°C) bottom-hole circulating temperature (BHCT). The effective temperature range of lignosulfonates can be extended to as high as 600°F (315°C), when blended with sodium borate (Section 3–4.6).

It is now well-established that lignosulfonate retarders predominantly affect the kinetics of C_3S hydration; however, their effects upon C_3A hydration are not insignificant (Stein, 1961; Angstadt and Hurley, 1963). The retardation mechanism of the lignosulfonates is generally thought to be a combination of the adsorption and nucleation theories.

Ramachandran (1972) has shown that the sulfonate and hydroxyl groups adsorb onto the C-S-H gel layer. Because of the very high specific surface area of C-S-H gel, the lignosulfonate can be considered to be incorporated into the hydrate structure, with a consequential change of morphology to a more impermeable structure (Ciach and Swenson, 1971). A waterproofing action of the adsorbed lignosulfonate, preventing further significant hydration, also was proposed (Jennings et al., 1986).

Some of the lignosulfonate remains in the aqueous phase. It may be in a free state and/or linked to calcium ions through electrostatic interactions. It has been shown that at low lignosulfonate concentrations, the crystal growth (and probably the nucleation) of calcium hydroxide is inhibited (Jawed et al., 1979). Although the same experiment has not yet been performed with C-S-H gel, a similar result would be expected. A significant change in the size and morphology of the calcium hydroxide crystals was also observed when C_3S was hydrated in the presence of lignosulfonates (Berger and McGregor, 1972). These results suggest that if the nucleation and crystal growth of hydration products are hindered by the presence of additives, the hydration rate of C_3S will be similarly affected.

Lignosulfonate retarders perform best with low-C_3A cements. When C_3A is hydrated in the presence of organic additives such as lignosulfonates, the solution concentration of the additives quickly falls. The hydration products of C_3A initially have a much stronger adsorptive effect than those of C_3S (Blank et al., 1963; Rossington and Runk, 1968). In a Portland cement system, C_3A hydration can prevent a significant quantity of lignosulfonate from reaching the surfaces of C_3S hydration products; as a result, the efficiency of the additive is reduced (Young, 1969).

3–4.2 Hydroxycarboxylic Acids

Hydroxycarboxylic acids contain hydroxyl and carboxyl groups in their molecular structures (Fig. 3–5). Gluconate and glucoheptonate salts are the most widely-used materials in this category. They have a powerful retarding action, and can easily cause overretardation at bottom hole circulating temperatures less than 200°F (93°C). As shown in Fig. 3–6, these materials are efficient to temperatures approaching 300°F (150°C).

Another hydroxycarboxylic acid with a strong retarding effect is citric acid. Citric acid also is effective as a cement dispersant (Section 3–5), and is normally used at concentrations between 0.1% to 0.3% BWOC.

The retarding action of hydroxycarboxylic acids and their salts is generally attributed to the presence of alpha- or beta-hydroxycarboxylic groups (HO-C-CO_2H and HO-C-C-CO_2H, respectively) which are capable of strongly chelating a metal cation, such as calcium (Double, 1983). Highly stable five-or six-membered rings are formed, which partially adsorb onto the hydrated cement surface, and poison nucleation sites of hydration products. Similarly to lignosulfonates, hydroxycarboxylic acids act more efficiently with low-C_3A cements.

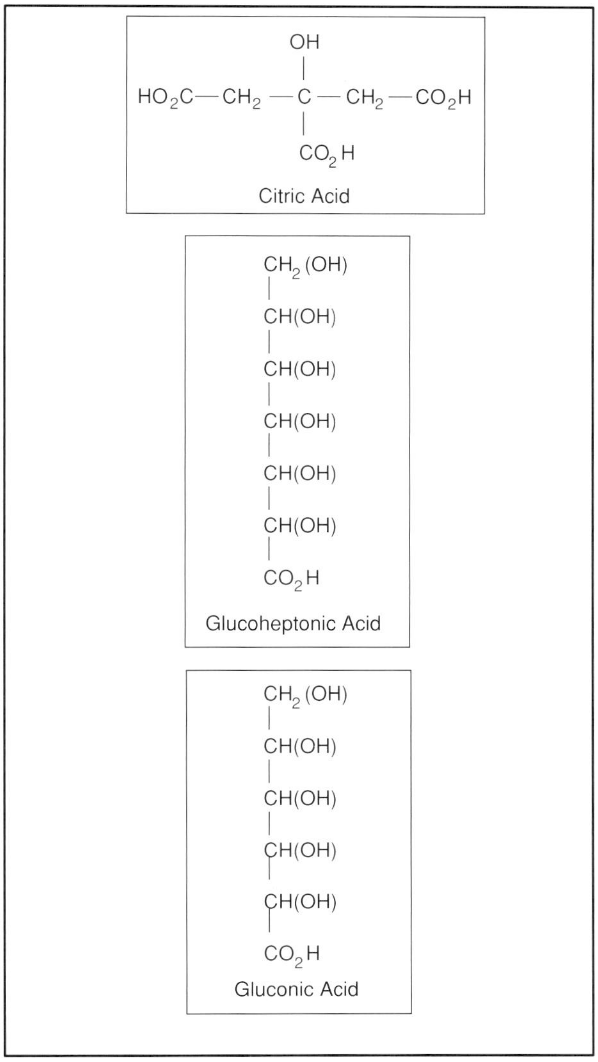

Figure 3–5—Molecular structures of hydroxycarboxylic acid retarders.

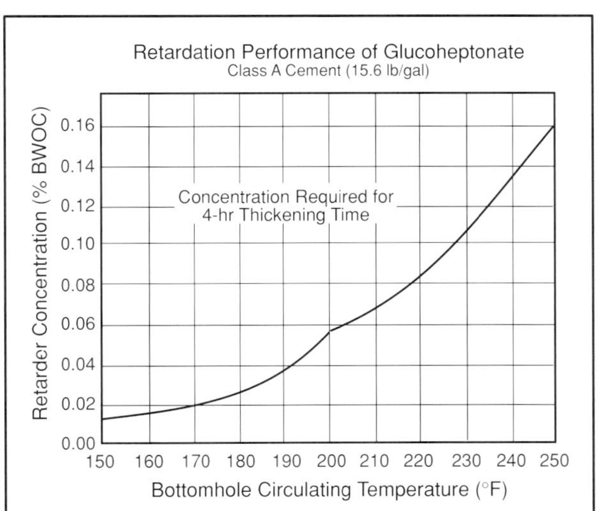

Figure 3–6—Retardation performance of glucoheptonate.

3–4.3 Saccharide Compounds

Saccharide compounds (so-called sugars, Fig. 3–7) are known as excellent retarders of Portland cement. The best retarders in this category are those containing a five-membered ring, such as sucrose and raffinose (Bruere, 1966; Previte, 1971; Thomas and Birchall, 1983). Such compounds are not commonly used in well cementing, because the degree of retardation is very sensitive to small variations in concentration.

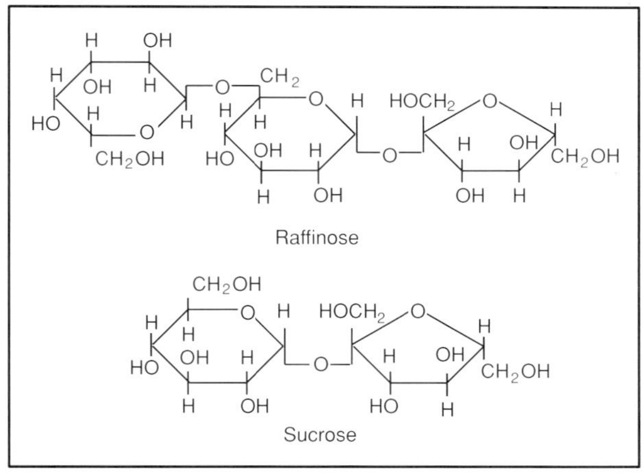

Figure 3–7—Structures of saccharide retarders.

The retarding action of saccharide compounds has been investigated thoroughly, and has been shown to be dependent upon the compounds' susceptibility to degradation by alkaline hydrolysis. The sugars are converted to saccharinic acids containing alpha-hydroxycarbonyl groups (HO-C-C=O), which adsorb strongly onto C-S-H gel surfaces (Taplin, 1960). Inhibition of hydration is thought to occur when the nucleation sites of the C-S-H gel are poisoned by the adsorbed sugar acid anions (Milestone, 1979).

3–4.4 Cellulose Derivatives

Cellulose polymers are polysaccharides derived from wood or other vegetals, and are stable to the alkaline conditions of cement slurries. Set retardation is probably the result of adsorption of the polymer onto the hydrated cement surface. The active sites are the ethylene oxide links and carboxyl groups.

The most common cellulosic retarder is carboxymethylhydroxyethylcellulose (CMHEC) (Shell and Wynn, 1958). Its molecular structure is shown in Fig. 3–36. CMHEC is an effective retarder at temperatures up to about 250°F (120°C) (Rust and Wood, 1966). Typical performance data are presented in Fig. 3–8.

A number of secondary effects are observed with CMHEC. It is often used as a fluid-loss control agent

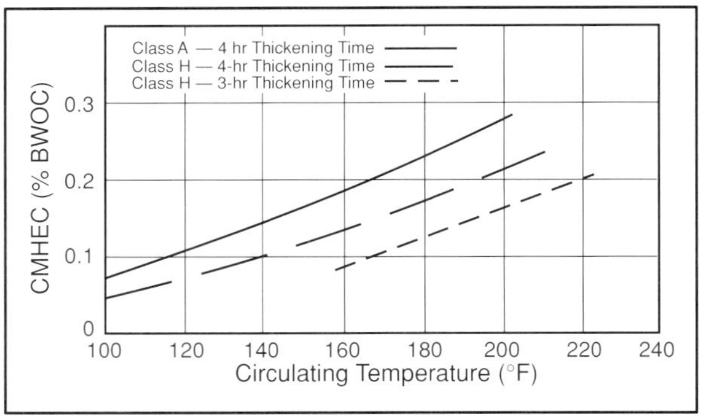

Figure 3–8—Typical thickening times obtained with CMHEC (using Class A and Class H cements).

(Section 3–8). In addition, CMHEC significantly increases the viscosity of the slurry.

3–4.5 Organophosphonates

Alkylene phosphonic acids and their salts have been recently identified as set-retarding additives for well cements. Such materials have excellent hydrolytic stability and, depending upon the molecular backbone, are effective to circulating temperatures as high as 400°F (204°C) (Nelson, 1984; Sutton et al., 1985, Nelson, 1987). Phosphomethylated compounds containing quaternary ammonium groups also are efficient (Crump and Wilson, 1984). Organophosphonates are advantageous for well cementing applications because of their apparent insensitivity to subtle variations in cement composition and tendency to lower the viscosity of high-density cement slurries. Very little is known concerning the mechanism of action; however, it is probable that the phosphonate groups (Fig. 3–9) adsorb onto the hydrated cement surface much like the other types of retarders.

Performance data for an organophosphonate presently used in the field is shown in Figure 3–10.

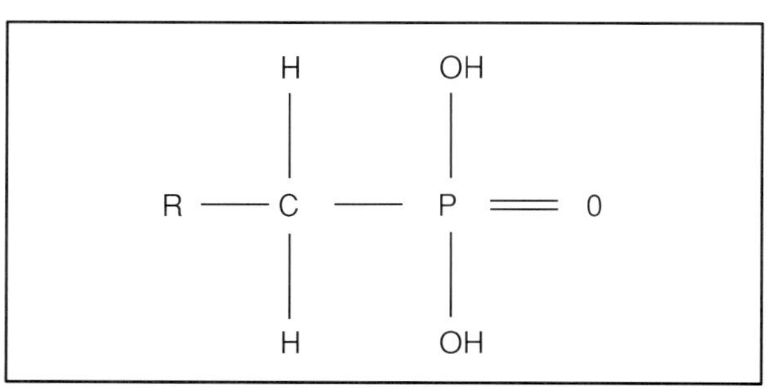

Figure 3–9—Alkylene phosphonate structure.

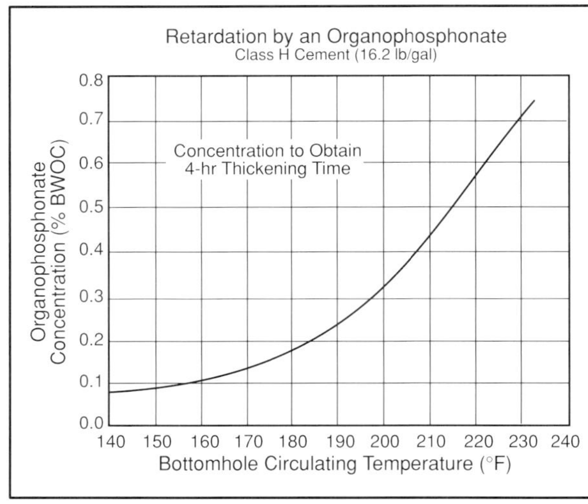

Figure 3–10—Retardation performance of organophosphonate.

3–4.6 Inorganic Compounds

Many inorganic compounds retard the hydration of Portland cement. The major classes of materials are listed below.

- *Acids and Salts Thereof:* boric, phosphoric, hydrofluoric and chromic
- *Sodium Chloride:* concentrations > 20% BWOW (Section 3–2)
- *Oxides*: zinc and lead

In well cementing, zinc oxide (ZnO) is sometimes used for retarding thixotropic cements, because it does not affect the slurry rheology (Chapter 7), nor does it affect the hydration of the C_3A-gypsum system (Ramachandran, 1986). The retardation effect of ZnO is attributed to the precipitation of zinc hydroxide onto the cement grains (Arliguie and Grandet, 1985). $Zn(OH)_2$ has a low solubility ($K_s = 1.8 \cdot 10^{-14}$), and is deposited as a colloidal gel; consequently, the layer has low permeability. The retardation effect ends when the gelatinous zinc hydroxide eventually transforms to crystalline calcium hydroxyzincate.

$$2Zn(OH)_2 + 2OH^- + Ca^{2+} + 2H_2O \rightarrow CaZn_2(OH)_6 \cdot 2H_2O \qquad (3\text{–}1)$$

Sodium tetraborate decahydrate (borax: $Na_2B_4O_7 \cdot 10H_2O$) is commonly used as a "retarder aid." It has the ability to extend the effective temperature range of most lignosulfonate retarders to as high as 600°F (315°C);

however, it can be detrimental to the effectiveness of cellulosic and polyamine fluid-loss additives.

3–5 EXTENDERS

Cement extenders are routinely used to accomplish one or both of the following.

Reduce Slurry Density—A reduction of slurry density reduces the hydrostatic pressure during cementing. This helps to prevent induced lost circulation because of the breakdown of weak formations. In addition, the number of stages required to cement a well may be reduced.

Increase Slurry Yield—Extenders reduce the amount of cement required to produce a given volume of set product. This results in a greater economy. Extenders can be classified into one of three categories, depending upon the mechanism of density reduction/yield increase. Often more than one type of extender is used in the same slurry.

Water Extenders—Extenders such as clays and various water viscosifying agents allow the addition of excess water to achieve slurry extension. Such extenders maintain a homogeneous slurry, and prevent the development of excessive free water.

Low-Density Aggregates—The densities of the materials in this varied category are lower than that of Portland cement (3.15 g/cm^3). Thus, the density of the slurry is reduced when significant quantities of such extenders are present.

Gaseous Extenders—Nitrogen or air can be used to prepare foamed cements with exceptionally low densities, yet sufficient compressive strength. The preparation and placement of such cement systems are complex, and a thorough treatment is given in Chapter 14.

A list of the common extenders with general information regarding their performance characteristics appears in Table 3–2.

Extender	Range of Slurry Densities Obtainable (lb/gal) 6 11 16	Performance Features and Other Benefits
Bentonite	11.5–15	Assists fluid-loss control.
Fly Ashes	13.1–14.1	Resist corrosive fluids.
Sodium Silicates	11.1–14.5	Only low percentages required. Ideal for seawater mixing.
Microspheres	8.5–15	Good compressive strength, thermal stability, and insulating properties.
Foamed Cement	6–15	Excellent strength and low permeability.

Table 3–2—Summary of extenders.

3–5.1 Clays

The term "clay" refers to a material composed chiefly of one or more "clay minerals." Clay minerals are essentially hydrous aluminum silicates of the phyllosilicate group (Hurlbut, 1971), where the silica tetrahedra are arranged in sheets. Such minerals have a platy or flaky habit and one prominent cleavage. In some, magnesium or iron substitutes in part for aluminum, and alkalis or alkaline earths may also be present as essential components.

The most frequently used clay-base extender is bentonite, also known as "gel," which contains at least 85% of the clay mineral smectite (also called montmorillonite). It is obtained primarily from mines in Wyoming and South Dakota. Smectite, $NaAl_2(AlSi_3O_{10})(OH)_2$, is composed of two flat sheets of silica tetrahedra sandwiching one sheet of alumina octahedra. Bentonite has the unusual property of expanding several times its original volume when placed in water, resulting in higher fluid viscosity, gel strength, and solids suspending ability.

Bentonite is added in concentrations up to 20% BWOC. Above 6%, the addition of a dispersant is usually necessary to reduce the slurry viscosity and gel strength. The API recommends that 5.3% additional water (BWOC) be added for each 1% bentonite for all API classes of cement; however, testing is necessary to determine the optimum water content with a particular cement. As shown in Table 3–3, the slurry density decreases and the yield increases quickly with bentonite concentration; however, as shown in Fig. 3–11, there is a price to be paid in terms of compressive strength. Cement permeability also increases with bentonite concentration; therefore, such cements are less resistant to sulfate waters and corrosive fluids. High concentrations of

% Bentonite	Class G — 44% Water		
	Water (gal/sk)	Slurry Density (lb/gal)	Yield (ft^3/sk)
0	4.97	15.8	1.14
2	6.17	15.0	1.31
4	7.36	14.4	1.48
6	8.56	13.9	1.65
8	9.76	13.5	1.82
10	10.95	13.1	1.99
12	12.15	12.7	2.16
16	14.55	12.3	2.51
20	16.94	11.9	2.85

Table 3–3—Effect of bentonite upon cement slurry properties.

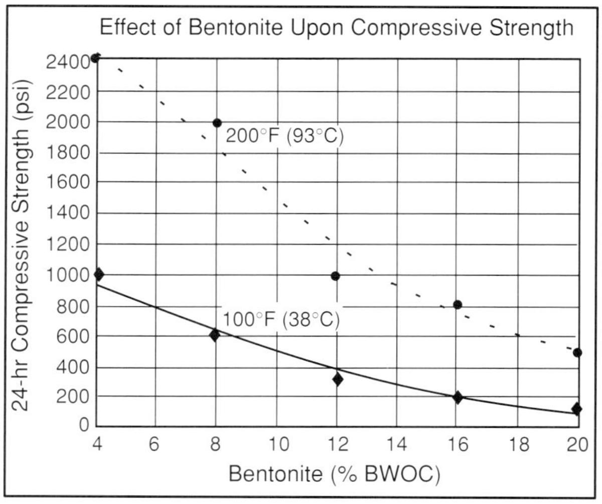

Figure 3–11—Effect of bentonite upon compressive strength.

bentonite tend to improve fluid-loss control. In addition, bentonite is an effective extender at elevated temperatures (Chapter 9).

The presence of high concentrations of Ca^{2+} ion in the aqueous phase of a cement slurry inhibits the hydration of bentonite; therefore, the extending properties of bentonite can be greatly enhanced if the material is allowed to completely hydrate in the mix water prior to slurry mixing. A slurry containing 2% prehydrated bentonite BWOC is equivalent to one containing 8% dry-blended bentonite (Table 3–4). Complete hydration of a good quality bentonite (no beneficiating agents added) occurs in about 30 min. The thickening time of prehydrated bentonite slurries is generally the same as that for dry-blended slurries of the same density. It should also be noted that prehydrating the bentonite does not appreciably change the final compressive strength.

Bentonite can be prehydrated in sea water or light brine, but the salt inhibits the hydration, and the slurry yield is reduced. Bentonite is not effective as an extender in highly saline cement slurries. Under such circumstances another clay mineral, attapulgite, is frequently used (Smith and Calvert, 1974). Attapulgite, $(Mg,Al)_5Si_8O_{22}(OH)_4 \cdot 4H_2O$, is also known as "salt-gel," and occurs as fibrous needles which provide viscosity by association when they become dispersed in water. Unlike bentonite, no improvement in fluid-loss control is obtained when attapulgite is present in the slurry.

% Pre-hydrated Bentonite	% Dry-Blended Bentonite	Fresh Water (gal/sk)	Slurry Density (lb/gal) Prehy-drated	Dry Blend	Slurry Yield (ft3/sk) Prehy-drated	Dry Blend
0	0	5.2	—	15.6	—	1.18
0.5	2	6.4	14.8	14.8	1.34	1.35
1.0	4	7.6	14.1	14.2	1.50	1.52
1.5	6	8.8	13.5	13.7	1.66	1.69
2.0	8	10.0	13.1	13.3	1.83	1.86
2.5	10	11.2	12.7	12.9	1.99	2.03
3.0	12	12.4	12.4	12.6	2.16	2.20
4.0	16	14.8	11.9	12.2	2.48	2.55
5.0	20	17.2	11.5	11.8	2.81	2.89

Table 3–4—Comparison of prehydrated and dry-blended bentonite slurry properties.

3–5.2 Sodium Silicates

Silicate extenders react with lime in the cement or with calcium chloride to form a calcium silicate gel. The gel structure provides sufficient viscosity to allow the use of large quantities of mix water without excessive free-water separation. This is a totally distinct process from that exhibited by the clay extenders, which absorb water. Sodium silicates are most frequently used, and are available in solid or liquid form. A major advantage of the silicates is their efficiency, which facilitates storage and handling. However, because of their tendency to accelerate, they tend to reduce the effectiveness of other additives, retarders and fluid-loss agents in particular.

The solid sodium silicate, Na_2SiO_3 (sodium metasilicate), is normally dry blended with the cement. If it is added to fresh mix water prior to slurry preparation, a gel may not form unless calcium chloride is also added. The recommended concentration of Na_2SiO_3 ranges from 0.2% to 3.0% BWOC. These concentrations provide a slurry density range of from 14.5 to 11.0 lb/gal (1.75 to 1.35 g/cm^3). The typical properties and performance of sodium metasilicate-extended cement systems is shown in Table 3–5.

The liquid sodium silicate, $Na_2O \cdot (3\text{–}5)SiO_2$ (also called water glass), is added to the mix water prior to slurry mixing. If calcium chloride is to be included in the slurry, it must be added to the mix water before the sodium silicate to obtain sufficient extending properties. Other materials can be added at any time. The normal concentration range is 0.2 to 0.6 gal/sk. Typical performance data are presented in Table 3–6.

3–5.3 Pozzolans

Pozzolans are perhaps the most important group of cement extenders, and are defined in accordance with ASTM designation C-219-55 as follows:

> *"A silicous or siliceous and aluminous material, which in itself possesses little or no cementitious value, but will, in finely divided form and in the presence of moisture, chemically react with calcium hydroxide at ordinary temperatures to form compounds possessing cementitious properties."*

Thus, pozzolans not only extend Portland cement sys-

Sodium Metasilicate (% BWOC)	Slurry Density (lb/gal)	Slurry Yield (ft³/sk)	Water gal/sk	Water %	Compressive Strength, 24 hr (psi) 120°F	Compressive Strength, 24 hr (psi) 140°F	Thickening Time (hr:min) 103°F	Thickening Time (hr:min) 113°F	Thickening Time (hr:min) 125°F	Thickening Time (hr:min) 140°F
0	15.8	1.15	4.97	44	4770	5310	+4:05	3:10	2:35	—
0.15	14.5	1.38	6.77	60	1746	2248	3:20	2:37	2:10	—
1.0	14.5	1.38	6.77	60	1896	2175	2:40	1:34	—	—
0.25	14.0	1.51	7.68	68	1420	1510	—	—	—	—
1.0	14.0	1.51	7.68	68	1640	1723	—	—	—	—
0.5	13.5	1.66	8.81	78	946	1278	—	3:30	2:10	—
2.0	13.5	1.66	8.81	78	1327	1420	1:53	1:28	—	—
0.5	13.0	1.84	10.17	90	750	927	—	—	—	—
2.0	13.0	1.84	10.17	90	120	1080	—	—	—	—
0.75	12.5	2.05	11.75	104	382	625	+5:00	+5:00	+5:00	+5:00
2.0	12.5	2.05	11.75	104	633	653	+5:00	1:43	—	—
1.0	12.0	2.32	13.78	122	265	380	—	—	—	—
2.0	12.0	2.32	13.78	122	420	510	—	—	—	—
1.5	11.5	2.69	16.6	147	147	230	+5:00	+5:00	+5:00	+5:00
3.0	11.5	2.69	16.6	147	271	289	+5:00	1:27	—	—
2.0	11.0	3.20	20.34	180	102	175	—	—	—	—
3.0	11.0	3.20	20.34	180	145	205	—	—	—	—

Table 3–5—Typical Class G + sodium metasilicate data.

Slurry Density (lb/gal)	Slurry Density (g/cm³)	Liquid Silicate Concentration (gal/sk)	Thickening Time at BHCT (hr:min) 103°F (39°C)	113°F (45°C)	175°F (79°C)
14.2	1.70	0.20	2:20	1:40	—
13.6	1.63	0.30	3:00	2:00	—
13.0	1.56	0.36	3:40	2:20	—
12.5	1.50	0.42	4:00+	2:30	1:50
12.0	1.44	0.50	4:00+	4:00+	3:10
11.5	1.38	0.60	4:00+	4:00+	3:50

Slurry Density (lb/gal)	Slurry Density (g/cm³)	Compressive Strength at BHST (24 hr (psi)) 95°F (35°C)	110°F (43°C)	140°F (60°C)	170°F (77°C)	200°F (93°C)
14.2	1.70	2200	2550	2300	2100	2000
13.6	1.63	1150	—	1450	—	1350
13.0	1.56	900	—	1050	—	1050
12.5	1.50	850	850	850	850	850
12.0	1.44	500	—	500	—	500
11.5	1.38	250	350	300	300	300

Table 3–6—Effect of liquid sodium silicate upon cement slurry performance.*

*API Class G cement

tems, but also react and contribute to the compressive strength of the set product. There are two types of pozzolans: (1) natural pozzolans, which include volcanic ashes and diatomaceous earth, and (2) artificial pozzolans such as certain fly ashes.

When one 94-lb sack of cement hydrates, about 20 to 23 lb of free $Ca(OH)_2$ is liberated. By itself, $Ca(OH)_2$ contributes nothing to the strength of the set cement and is fairly soluble; thus, it can be eventually dissolved and removed by water contacting the cement. This contributes to a weakening of the cement. When a pozzolan is present, the silica combines with the free $Ca(OH)_2$ to form a stable cementitious compound (secondary C-S-H) which is very durable.

The water permeability of set pozzolan/cement systems is usually less than 0.001 md, if the system is not extended by the addition of a large amount of water. The low permeability of the set cement, as well as the decrease of free $Ca(OH)_2$ content, resists the encroachment of sulfate water and other corrosive fluids. Should corrosive waters nevertheless enter the set pozzolanic cement, damage is further prevented by another mechanism. An ion exchange process occurs because of the presence of zeolites in the pozzolan, and the alkalis are rendered less harmful.

There are two notation systems commonly used for mixing pozzolan cements. The first is a volume ratio based upon bulk volume. A 1:1 ratio indicates one cubic foot of pozzolan and one cubic foot of cement. The first figure indicates the volume of pozzolan, and the second indicates the volume of cement. This system is used primarily with very light pozzolans.

The second mixing system is the most widely used. It is based on the "equivalent sack." A sack of Portland cement has an absolute volume of 3.59 gal. In other words, one sack of cement when mixed with water will increase the volume of the mix by 3.59 gal. An equivalent sack is that weight of pozzolan that also has an absolute volume of 3.59 gallons. Thus, different pozzolans have different equivalent sack weights. The ratio for mixtures based upon equivalent sacks is designated as 25:75, 50:50, 75:25 or whatever ratio is desired. The term 25:75 indicates ¼ equivalent sack of pozzolan and ¾ sack of Portland cement.

The weights of other additives (except salt) are calculated as a percentage by weight of the "sack" of pozzolan/cement blend. Salt is always calculated as a percentage of the mix water.

As an example, an equivalent sack of one typical fly ash is 74 lb. A 50:50 blend with this pozzolan would require 37 lb of fly ash and 47 lb of Portland cement. Thus, 84 lb of this blend would displace 3.59 gal. Additive concentrations would then be calculated as a percentage of an 84-lb sack, not the usual 94-lb sack of Portland cement.

3–5.3.1 Diatomaceous Earth

Diatomaceous earth is composed of the siliceous skeletons of diatoms deposited from either fresh- or sea-water. The main constituent of diatomaceous earth is opal, an amorphous form of hydrous silica containing up to 10% water. For use as a pozzolanic extender, diatomaceous earth is ground to a fineness approaching that of Portland cement; consequently, the material has a large surface area and a high water demand.

Diatomaceous earth imparts slurry properties similar to those of bentonite slurries; however, it does not increase the slurry viscosity to such a high degree. In addition, because of its pozzolanic activity, set cements containing diatomaceous earth are stronger than their bentonitic counterparts. The principal disadvantage of diatomaceous earth is its cost. Typical slurry properties and performance of diatomaceous earth slurries are shown in Table 3–7.

3–5.3.2 Fly Ashes

Fly ash is the residue from power plants which burn pulverized coal (Davis et al., 1937). The ash is carried forward in the gases as fused particles which solidify into a roughly spherical shape. The ash is very finely divided, with a surface area roughly approximating that of Portland cements. The major constituent of fly ash is a glass chiefly composed of silica and alumina with some iron oxide, lime, alkalies and magnesia. Quartz, mullite, hematite and magnetite, as well as some combustible matter, are also found. The composition and properties of fly ash can vary widely depending upon the source of the coal and the efficiency of the power plant; accordingly, the specific gravities of fly ashes can vary from about 2.0 to 2.7 (Lea, 1971).

According to ASTM specifications, three types of fly ash are recognized: Types N, F and C. As shown in Table 3–8, the distinction is made on chemical grounds. Type F

	Mineral Admixture Class		
	N	F	C
Silicon dioxide (SiO_2 plus aluminum oxide (Al_2O_3) plus iron oxide (Fe_2O_3), min., %	70	70	50
Sulfur trioxide (SO_3), max., %	4	5	5
Moisture content, max., %	3	3	3
Loss on ignition, max., %	10	12	6

Table 3–8—Chemical requirements for fly ashes.

Diatomaceous Earth (%)	Water (gal/sk)	Slurry Weight (lb/gal)	Slurry Volume (ft^3/sk)
0	5.2	15.6	1.18
10	10.2	13.2	1.92
20	13.5	12.4	2.42
30	18.2	11.7	3.12
40	25.6	11.0	4.19

Diatomaceous Earth (%)	Compressive Strength of API Class A Cement (psi)							
	After Curing 24 hr at Temp. and Press. of				After Curing 72 hr at Temp. and Press. of			
	80°F ambient	95°F 800 psi	110°F 1600 psi	140°F 3000 psi	80°F ambient	95°F 800 psi	110°F 1600 psi	140°F 3000 psi
0	1360	1560	2005	2620	2890	3565	4275	4325
10	110	360	520	750	440	660	945	1125
20	70	190	270	710	240	345	645	1000
40	15	30	50	260	70	150	220	630

Table 3–7—Effect of diatomaceous earth on API classes A and H cements.

fly ashes are most frequently used in well cementing. They are normally produced from burning anthracite or bituminous coals. Type C fly ashes, made from lignite or subbituminous coals, are less siliceous, and some contain more than 10% lime; as a result, many of them are themselves cementitious and thus do not fit the strict definition of a pozzolanic material.

Normally, 2% bentonite is used in Type F fly ash/Portland cement systems to improve the slurry properties and prevent the development of free water. In Table 3–9, slurry data for different ratios of Type F fly ash and cement are presented.

The use of Type C fly ashes as extenders for well cements is relatively new. Because of the significant amount of lime in such fly ashes, the rheological effects must be carefully monitored. In addition, Type C ashes are highly individual depending upon the source, and special slurry preparation guidelines are required for each.

Some Type C fly ashes are sufficiently cementitious to be used as the principal component of a well cement. Such systems have been developed for application in shallow wells having circulating temperatures up to 120°F (49°C). Compressive strength development is often more rapid than that observed with conventional Portland cement systems.

3.5.3.3 Commercial Lightweight Cements

Commercial oil-well cements, such as Trinity Lite-Wate (Trademark of General Portland Cement Company) and TXI Lightweight (Trademark of Texas Industries) are special formulations composed of interground Portland cement clinker and lightweight siliceous aggregates; consequently, some pozzolanic activity occurs. They are convenient and time-saving for the service company. The particle-size distribution of such cements is very fine, and the normal slurry density range is from 11.9 to 13.7 lb/gal (1.43 to 1.64 g/cm^3).

3–5.3.4 Silica

Two forms of finely divided silica are used in well cements: α-quartz and condensed silica fume. Silica as α-quartz is used most frequently for the prevention of strength retrogression when Portland cement systems are placed in thermal wells (Chapter 9). Two particle sizes are routinely used: "silica sand," with an average particle size of about 100 μm, and "silica flour," with an average particle size of about 15 μm. Due primarily to cost, these materials are rarely used for slurry extension alone.

Condensed silica fume (also called microsilica) is a byproduct of the production of silicon, ferrosilicon and other silicon alloys. The individual particles are glassy, amorphous microspheres. The mean particle size is usually between 0.1μm and 0.2 μm about 50 to 100 times finer than Portland cement or fly ash; consequently, the surface area is extremely high (15,000 to 25,000 m^2/kg).

Condensed silica fume is highly reactive and, because of its fineness and purity, is the most effective pozzolanic material currently available (Parker, 1985). The high degree of pozzolanic activity has allowed the introduction of low-density cement systems with a higher rate of compressive strength development (Carathers and Crook, 1987). The high surface area of condensed silica fume increases the water demand to prepare a pumpable slurry; therefore, slurries with densities as low as 11.0 lb/gal (1.32 g/cm^3) can be prepared which have little or no free water. The normal concentration of this material is about 15% BWOC; however, up to 28% BWOC is possible.

The fineness of condensed silica fume also promotes improved fluid-loss control, perhaps by reducing the permeability of the initial cement filter cake. For this reason, it is also used for the prevention of annular fluid migration (Chapter 8). In addition, it is being introduced as a source of silica in thermal cement systems (Chapter 9).

Ratio*		Weight of Components (lb)		Minimum Water Requirement			Maximum Water Requirement		
Fly Ash	Class H	Fly Ash	Class H	Water (gal/sk)	Slurry Density (lb/gal)	Slurry Volume (ft^3/sk)	Water (gal/sk)**	Slurry Density (lb/gal)	Slurry Volume (ft^3/sk)**
25	75	18.5	70.5	5.24	15.1	1.19	5.64	14.7	1.25
35	65	25.9	61.5	5.17	15.0	1.18	5.73	14.6	1.26
50	50	37.0	47.0	5.00	14.7	1.16	5.80	14.2	1.27
65	35	48.1	32.9	4.85	14.5	1.14	5.89	13.8	1.28
75	25	55.5	23.5	4.75	14.3	1.12	5.96	13.5	1.29

* All systems contain 2% bentonite by weight of fly ash/cement blend.
** Based on the weight of an equivalent sack of the specific blend.

Table 3–9—Properties of fly ash/Class H cement systems.

3–5.4 Lightweight Particles

Lightweight particle extenders reduce the density of the slurry because of their low density with respect to the cement particles. They include expanded perlite, powdered coal, gilsonite, and either glass or ceramic microspheres. As a general rule, extenders in this category are inert within the cement matrix.

3–5.4.1 Expanded Perlite

Perlite is a crushed volcanic glass which expands when heated to the point of incipient fusion (Lea, 1971). The expanded perlite product generally has a bulk density of 7.75 lb/ft^3, which allows the preparation of competent cement slurries with densities as low as 12.0 lb/gal (1.44 g/cm^3). A small quantity of bentonite (2% to 4% BWOC) is added to prevent the segregation of the perlite particles from the slurry.

Expanded perlite contains open and closed pores and matrix. Under hydrostatic pressure, the open pores fill with water, and some of the closed pores are crushed; as a result, the perlite becomes heavier. Therefore, to prepare an expanded perlite slurry which will have a given density downhole, it is necessary to mix a lower density slurry at the surface. At 3,000 psi, the specific gravity of expanded perlite is 2.40. Table 3–10 shows some typical slurry designs, and illustrates the differences in slurry density observed at atmospheric pressure and at 3,000 psi.

3–5.4.2 Gilsonite

Gilsonite is a naturally occurring asphaltite mineral, found primarily in deposits located in Colorado and Utah. The specific gravity of gilsonite is 1.07. The water requirement for gilsonite is low, about 2 gal/ft^3; thus, it is possible to prepare low-density cement systems which develop relatively high compressive strength (Slagle and Carter, 1959). Up to 50 lb of gilsonite can be used per sack of Portland cement, to obtain slurry densities as low as 12.0 lb/gal (1.44 g/cm^3); however, mixing difficulties may be experienced at such high concentrations. Bentonite is often included in such slurries.

Gilsonite is a black, angular solid, with a wide particle size range (up to 0.6 cm), and is often used to prevent lost circulation (Chapter 6). Gilsonite has a melting point of 385°F (196°C). Some softening occurs above 240°F (116°C), and particles may tend to fuse. As a result, the use of gilsonite is not recommended in wells with bottom hole static temperatures above 300°F (149°C).

3–5.4.3 Powdered Coal

As an extender, the performance of powdered coal is very similar to that of gilsonite. Its specific gravity is slightly higher (1.30). Like gilsonite, it is coarsely ground and often used as a material to prevent lost circulation. Unlike gilsonite, the melting point of powdered coal is 1,000°F (538°C), which allows the use of powdered coal in thermal well environments.

Between 12.5 and 25 lb of powdered coal are normally added per sack of cement, and slurries with densities as low as 11.9 lb/gal (1.43 g/cm^3) can be prepared. Bentonite is also often incorporated in powdered coal slurries. Table 3–11 illustrates typical slurry designs for powdered coal systems.

3–5.4.4 Microspheres

Extending cement slurries with microspheres is a relatively recent development. Microspheres are small gas-filled beads with specific gravities normally between 0.4 and 0.6. Such low specific gravities allow the preparation of high strength/low permeability cements with densities as low as 8.5 lb/gal (1.02 g/cm^3). Two types of microspheres are available: glass and ceramic.

The original application of microspheres was for the primary cementing of conductor and surface pipes, where washouts and low fracturing pressures are common. However, they are used much more extensively today, and in many cases microsphere cements have eliminated the need for multistage cementing. A significant limitation of microspheres is their inability to withstand high hydrostatic pressure; thus, they cannot be used in deep wells. Microsphere cement systems require special care in design and mixing, and the procedures are briefly described below.

A wide selection of glass microspheres is available for reducing slurry density (Smith et al., 1980). They are generally classified according to the maximum hydrostatic pressure they can withstand. The average particle size is similar to that of cement. The particle-size distribution may vary over a range of from 20 to 200 μm with walls 0.5 to 2.0 μm thick. Most grades of glass microspheres withstand pressures up to 5,000 psi; however, special grades with thicker walls and higher specific gravity will survive to 10,000 psi. Glass microspheres are significantly more expensive than their ceramic counterparts; thus, their use is relatively infrequent.

Ceramic microspheres are derived from fly ashes; thus, the composition of the shell is aluminosilicate. The

Slurry Properties at Various Pressures									
Cement/ Perlite Ratio (by volume) (sk:ft³)	Bentonite (%)	Mix Water (gal/sk)	Atmospheric			3000 psi			Compressive Strength (24 hr, 100°F, 3000 psi)
			Slurry Density (lb/gal)	Slurry Density (lb/ft³)	Slurry Volume (ft³/sk)	Slurry Density (lb/gal)	Slurry Density (lb/ft³)	Slurry Volume (ft³/sk)	
1:½	2	6.5	13.80	103.2	1.52	14.85	111.1	1.41	
	2	7.0	13.58	101.6	1.58	14.57	109.0	1.47	2800
	2	7.5	13.36	99.9	1.65	14.29	106.9	1.54	
	2	8.0	13.16	98.4	1.72	14.02	104.9	1.61	2200
	2	8.5	12.98	97.1	1.78	13.75	102.8	1.67	
1:1	2	9.0	12.26	91.7	2.00	13.71	102.5	1.79	1950
	2	9.5	12.15	90.9	2.07	13.55	101.3	1.86	
	2	10.0	12.02	89.9	2.14	13.37	100.0	1.93	1500
	2	10.5	11.91	89.1	2.20	13.20	98.7	1.99	
	2	11.0	11.81	88.3	2.27	13.04	97.5	2.06	1050
1:1½	2	10.5	11.50	86.0	2.36	13.31	99.6	2.04	
	2	11.0	11.41	85.3	2.43	13.16	98.4	2.11	1125
	2	11.5	11.31	84.6	2.49	13.00	97.2	2.17	
	2	12.0	11.23	84.0	2.56	12.86	96.2	2.24	1050
	2	12.5	11.17	83.6	2.63	12.71	95.6	2.31	890
	4	11.5	11.38	85.1	2.50	13.04	97.5	2.18	1170
	4	12.0	11.29	84.4	2.57	12.91	96.6	2.25	1000
	4	12.5	11.21	83.8	2.64	12.77	95.5	2.32	860
	4	13.0	11.15	83.4	2.70	12.65	94.6	2.38	740
	4	13.5	11.09	82.9	2.77	12.53	93.7	2.45	650
	4	14.0	11.03	82.5	2.84	12.43	93.0	2.52	600
1:2	2	12.0	10.92	81.7	2.72	12.98	97.1	2.29	1300
	2	12.5	10.86	81.2	2.78	12.82	95.9	2.35	
	2	13.0	10.80	80.8	2.85	12.71	95.1	2.42	1025
	2	13.5	10.75	80.4	2.92	12.60	94.2	2.49	
	2	14.0	10.69	80.0	2.98	12.49	93.4	2.55	775
	2	14.5	10.63	79.5	3.04	12.39	92.7	2.61	
	4	13.0	10.85	81.1	2.86	12.76	95.4	2.43	1000
	4	13.5	10.79	80.7	2.93	12.64	94.5	2.50	870
	4	14.0	10.73	80.3	2.99	12.53	93.7	2.56	760
	4	14.5	10.69	80.0	3.06	12.43	93.0	2.63	670
	4	15.0	10.65	79.7	3.13	12.33	92.2	2.70	590
	4	15.5	10.60	79.3	3.19	12.22	91.4	2.76	520

Data are based on the use of Class A cement

Table 3–10—Properties of cement systems containing expanded perlite + bentonite.

composition of the gas inside is a mixture of CO_2 and N_2. The microspheres are heavier than their glass counterparts with a specific gravity of 0.7 and a bulk density of 25 lb/ft³; thus, a higher concentration is necessary to achieve low slurry densities (Harms and Sutton, 1981).

As mentioned earlier, hollow microspheres are susceptible to breakage and collapse when exposed to high hydrostatic pressure; as a result, the density of the slurry increases. This increase can be predicted and, as shown in Fig. 3–12, can be taken into account in the design calculations. The use of ceramic microspheres is not recommended when bottom hole pressures exceed 4,500 psi.

It is important to ensure that the microspheres do not separate from the cement particles during the course of the blending process. The microspheres must be thoroughly dry-blended with the cement and not premixed in the water. Any variation in the ratio of microspheres to cement will result in erratic densities during mixing.

Microspheres are compatible with any class of cement. Figure 3–13 illustrates the amount of microspheres required to achieve slurry densities between 8.5 and 15.0 lb/gal (1.02 and 1.80 g/cm³). Mix water requirements are shown in Fig. 3–14, and slurry yields in Fig. 3–15. The relationship between the density of ceramic microsphere system density and compressive strength is illustrated in Table 3–12.

Bentonite (%)	Powdered Coal (lb/sk)	Water (gal/sk)	Slurry Density (lb/gal)	Slurry Volume (lb/ft^3)	Bentonite (%)	Powdered Coal (lb/sk)	Water (gal/sk)	Slurry Density (lb/gal)	Slurry Volume (ft^3/sk)
0	0	5.20	15.6	1.18	6	0	8.78	13.7	1.69
	5	5.40	15.2	1.26		5	8.98	13.5	1.77
	10	5.60	14.9	1.35		10	9.18	13.3	1.86
	12.5	5.70	14.7	1.40		12.5	9.28	13.3	1.91
	15	5.80	14.6	1.44		15	9.38	13.2	1.95
	20	6.00	14.3	1.53		20	9.58	13.0	2.04
	25	6.20	14.1	1.62		25	9.78	12.9	2.13
	30	6.40	14.0	1.71		30	9.98	12.8	2.22
	40	6.80	13.5	1.88		40	10.38	12.6	2.39
	50	7.20	13.2	2.06		50	10.78	12.4	2.57
2	0	6.39	14.8	1.35	8	0	9.98	13.3	1.86
	5	6.59	14.5	1.43		5	10.18	13.1	1.95
	10	6.79	14.3	1.52		10	10.38	13.0	2.04
	12.5	6.89	14.1	1.57		12.5	10.48	12.9	2.08
	15	6.99	14.0	1.61		15	10.58	12.9	2.12
	20	7.19	13.8	1.70		20	10.78	12.8	2.21
	25	7.39	13.6	1.79		25	10.98	12.7	2.30
	30	7.59	13.5	1.88		30	11.18	12.6	2.39
	40	7.99	13.2	2.05		40	11.58	12.4	2.57
	50	8.39	12.9	2.23		50	11.98	12.2	2.74
4	0	7.59	14.2	1.52	12	0	12.37	12.6	2.20
	5	7.78	14.0	1.60		5	12.57	12.5	2.29
	10	7.98	13.7	1.69		10	12.77	12.4	2.38
	12.5	8.08	13.6	1.74		12.5	12.87	12.4	2.42
	15	8.18	13.6	1.78		15	12.97	12.4	2.47
	20	8.38	13.4	1.87		20	12.17	12.3	2.56
	25	8.58	13.3	1.96		25	13.37	12.2	2.64
	30	8.78	13.2	2.03		30	13.57	12.1	2.73
	40	9.18	12.9	2.22		40	13.98	12.0	2.91
	50	9.58	12.7	2.40		50	14.38	11.9	3.09

Table 3–11—Physical slurry properties of Class A cement with powdered coal and bentonite.

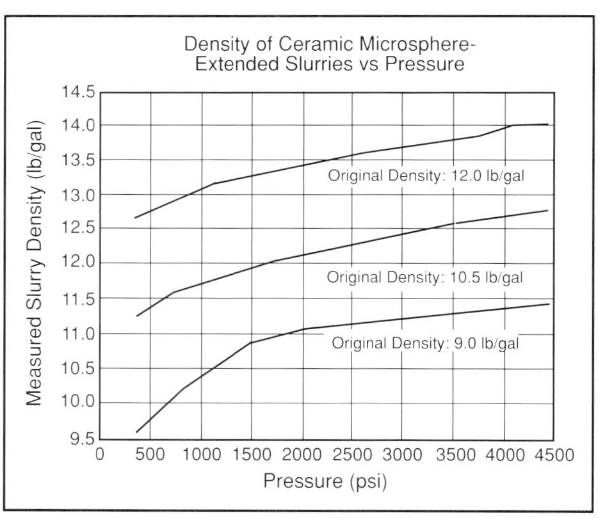

Figure 3–12—Density of ceramic microsphere-extended slurries vs pressure.

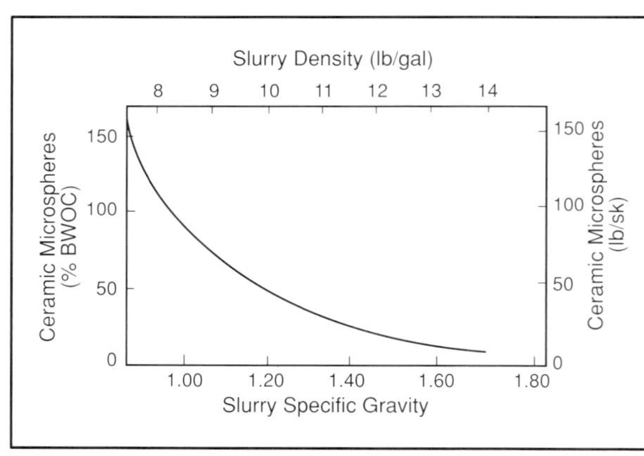

Figure 3–13—Microsphere concentration requirements.

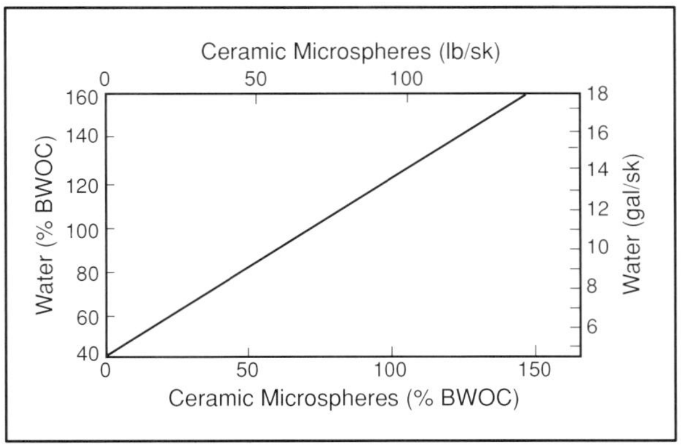

Figure 3–14—Water requirements for ceramic microsphere cement systems.

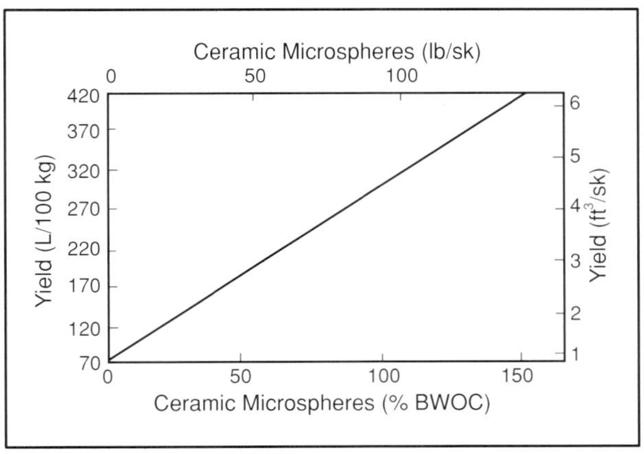

Figure 3–15—Yield of ceramic microsphere systems.

Curing Pressure (psi)	Compressive Strength Data (psi)						
	Slurry Mixing Densities (lb/gal)						
	8.5	9	9.5	10	10.5	11	11.5
0	55	100	160	250	270	—	420
800	115	115	125	250	250	450	470
2000	—	—	175	315	355	420	480
3000	215	—	250	295	295	435	640

All slurries were cured 24 hr at 80°F.

Table 3–12—Compressive strength data for ceramic microsphere slurries mixed with Class G cement, 1% calcium chloride, and 0.4% PNS dispersant.

3–5.5 Nitrogen

Foamed cement is a system in which nitrogen, as the density-reducing medium, is incorporated directly into the slurry to obtain a low-density cement. The system requires the use of specially formulated base cement slurries to create a homogeneous system with high compressive strength and low permeability. Nitrogen allows the preparation of competent cement systems with densities as low as 7.0 lb/gal (0.84 g/cm^3).

The design, preparation and placement of foamed cements are sufficiently complex to warrant a separate chapter devoted entirely to the subject. The reader is referred to Chapter 14 for a complete discussion of this important technology.

3–6 WEIGHTING AGENTS

High pore pressures, unstable wellbores and deformable/plastic formations are controlled by high hydrostatic pressures. Under such conditions, mud densities in excess of 18.0 lb/gal (2.16 g/cm^3) are common. To maintain control of such wells, cement slurries of equal or higher density are also necessary.

One method of increasing the cement slurry density is simply to reduce the amount of mix water. To maintain pumpability, the addition of a dispersant is required. The principal disadvantage of "reduced water slurries" is the difficulty of simultaneously achieving adequate fluid-loss control, acceptable slurry rheology, and no solids settling. Without excellent fluid-loss control, the risk of slurry bridging is higher. If solids settling occurs, the compressive strength and bonding will not be uniform across the cemented interval. The maximum slurry density attainable by this method is 18.0 lb/gal (2.16 g/cm^3).

When higher slurry densities are required, materials with a high specific gravity are added. To be acceptable as a weighting agent, such materials must meet several criteria.

- The particle-size distribution of the material must be compatible with the cement. Large particles tend to settle out of the slurry, while small particles tend to increase slurry viscosity.
- The water requirement must be low.
- The material must be inert with respect to cement hydration, and compatible with other cement additives.

The most common weighting agents for cement slurries are ilmenite, hematite and barite. A summary of their physical properties appears in Table 3–13. The concentrations of each material normally required to achieve a given slurry density are plotted in Fig. 3–16.

Material	Specific Gravity	Absolute Volume (gal/lb)	Color	Additional Water Requirement (gal/lb)
Ilmenite	4.45	0.027	Black	0.00
Hematite	4.95	0.024	Red	0.0023
Barite	4.33	0.028	White	0.024

Table 3–13—Physical properties of weighting agents for cement slurries.

3–6.1 Ilmenite

Ilmenite ($FeTiO_3$), a black granular material, has a specific gravity of 4.45. It has little effect upon cement slurry thickening time and compressive strength development. As currently supplied, the particle size distribution of ilmenite is rather coarse; therefore, the slurry viscosity must be carefully adjusted to prevent sedimentation. Slurry densities in excess of 20.0 lb/gal (2.4 g/cm^3) are easily attainable with ilmenite.

3–6.2 Hematite

With a specific gravity of 4.95, hematite (Fe_2O_3) is a very efficient weighting agent. The material occurs as red crystalline granules. Unlike ilmenite, it is currently supplied with a fine particle-size distribution. At high hematite concentrations, addition of a dispersant is often necessary to prevent excessive slurry viscosity. Hematite is routinely used to prepare cement slurries with densities up to 19.0 lb/gal (2.28 g/cm^3); however, slurries with densities as high as 22 lb/gal (2.64 g/cm^3) can be prepared.

3–6.3 Barite

Barite ($BaSO_4$), a white powdery material, is readily available at most oil field locations; however, it is not an efficient weighting agent compared to ilmenite or hematite. Although it has a high specific gravity (4.33), additional water is required to wet its particles, and its effectiveness as a densifier is significantly diminished. The additional water also decreases the compressive strength of the set cement. Nevertheless, slurries with densities up to 19.0 lb/gal (2.28 g/cm^3) can be prepared with barite.

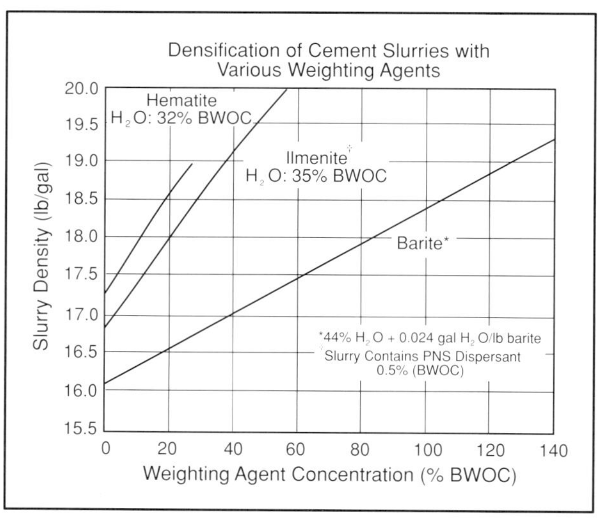

Figure 3–16—Densification of cement slurries with various weighting agents.

3–7 DISPERSANTS

Well cement slurries are highly concentrated suspensions of solid particles in water. The solids content can be as high as 70%. The rheology of such suspensions is related to the supporting liquid rheology, the solid volume fraction (volume of particles/total volume) and to interparticle interactions. In a cement slurry, the interstitial fluid is an aqueous solution of many ionic species and organic additives. Therefore, the rheology can differ greatly from that of water. The solids content of the slurry is a direct function of the slurry density. Particle interactions depend primarily on the surface charge distribution. Cement dispersants, also known in the construction industry as "superplasticizers," adjust the particle surface charges to obtain the desired rheological properties of the slurry.

This section discusses the electrical properties of cement grains in an aqueous medium, the relationship between the Bingham viscoplastic behavior of the slurry and interparticle attractions, and the types of chemicals which are effective cement dispersants. Finally, the effects of dispersants on slurry rheology and homogeneity are discussed.

3–7.1 Surface Ionization of Cement Particles in an Aqueous Medium

As discussed in Chapter 2, the hydrolysis of C-S-H leads to a charged surface.

$$-Si-OH + OH^- \rightleftharpoons -Si-O^- + H_2O \quad (3\text{–}2)$$

The free calcium ions in the solution react with the negatively charged groups on the grain surfaces. One calcium ion may bind two Si - O$^-$ groups which may be, as shown in Fig. 3–17, either on the same grain or bridging two grains (Thomas and Double, 1981). The bridging occurs because of the large cement surface area, and competition for calcium ions between adsorption sites. A portion

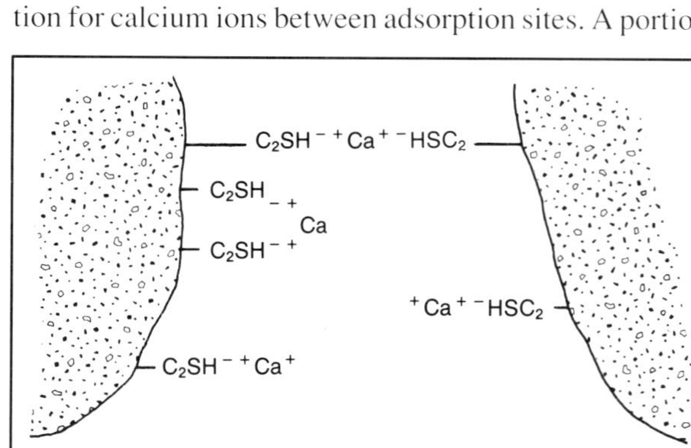

Figure 3–17—Cement grain interactions.

of a cement grain may be positively charged, owing to calcium adsorption, while another part is negatively charged. As a result, interactions occur between oppositely charged patches. Were it not for bridging, the cement grains would be covered uniformly by positive charges, leading to spontaneous dispersion.

3–7.2 Viscoplasticity of Cement Slurries and Mechanism of Dispersion

When cement powder and water are mixed, a structure is formed throughout the slurry which prevents flow below a given shear stress threshold: the yield value. This is the result of the previously-described electrostatic interactions between particles. At low shear stresses, below the yield value, the slurry behaves as a solid. It may undertake some finite deformations, be compressed or eventually creep, but it does not flow. Above the yield value it behaves as a liquid with, in the Bingham model, a well-defined plastic viscosity (Wilkinson, 1960). The reader is referred to Chapter 4 for a complete presentation concerning cement slurry rheology.

As can be seen in Fig. 3–18 (Baret, 1988), the experimental shear-stress/shear-rate curves are approximately linear. The slope of the line is the "plastic viscosity," and its ordinate at the origin is the "yield value." However, the "apparent viscosity," i.e., the shear-stress/shear-rate ratio, is not a constant. Instead, it decreases with increasing shear stress. This plasticity results from the breaking of the electrostatic structure under shear. Once the yield value is exceeded, the slurry no longer behaves as a singular unit; instead, it is broken into pieces, and aggregates of particles move among one another. These aggregates contain entrapped interstitial water; as a result, the effective volume of the dispersed phase is larger than that of the cement grains.

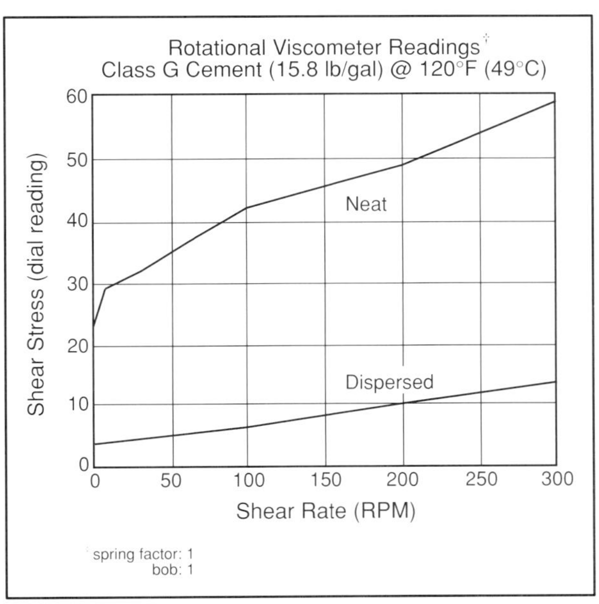

Figure 3–18—Rheological data for a neat and a dispersed cement slurry.

The volume of the dispersed phase is the key factor which determines the rheology of the dispersion. For example, in the first-order analysis leading to Einstein's relation (Einstein, 1926)

$$\mu = \mu_0 (1 + 2.5\phi_s) \qquad (3\text{–}3)$$

the viscosity of a dispersion (μ), made with a base fluid of viscosity (μ_0), depends only on the volume fraction (ϕ_s) occupied by the dispersed phase. In more sophisticated models (Petrie, 1976) for concentrated dispersions, the volume fraction of the dispersed phase remains the determining parameter. Thus, large cement particle aggregates correspond to high slurry viscosity.

It is seen in Fig. 3–18 that aggregate disruption can be achieved either by shearing or by adding a dispersant. Both actions release a portion of the entrapped water in the aggregates; hence, the effective volume of the dispersed phase is decreased, and the slurry viscosity falls. The viscosity reaches a minimum when all aggregates are destroyed (Figure 3–19), resulting in a dispersion of individual particles (Shaw, 1980).

Figure 3–19—Dispersion vs flocculation.

As discussed earlier, when cement is slurried in water, positively charged and negatively charged patches exist on the cement grain surfaces. These patches interact with one another to create a continuous structural network. At high solids concentrations, this network must be broken if the slurry is to be pumpable. When certain polyanions are added to the slurry, they adsorb onto the positively charged sites, and thus suppress particle interactions. Obviously, polycations could do the same by interacting with the negatively charged surface sites, but in so doing they would compete with calcium adsorption and thus impair the cement hydration process.

A hydrolyzed silanol or aluminol group on a cement grain surface ($-Si-O^- + Ca^+$) bears a negative charge which may adsorb onto a calcium ion. As shown in Fig.

3–20, a polyanion molecule may adsorb there and bring several negative charges. The amount adsorbed varies with the concentration of dispersant, as shown by the adsorption isotherm shown in Fig. 3–21. The cement particles become uniformly negatively charged. This effect may be observed by measuring the zeta potential, a function of the particle charge, of a dilute cement suspension. Figure 3–21 also shows that for polynaphthalene sulfonate, the surface charge levels off when adsorption reaches a plateau (Daimon and Roy, 1978; Michaux and Defossé, 1986; Andersen, 1986). The charged particles repel each other; as a result, flocculation is defeated and the slurry is dispersed.

In the case of nonionic polymers, and to some extent also with polyelectrolytes, particle repulsion can be ensured by a mechanism other than the electrostatic repulsion. Entropic and enthalpic contributions may forbid polymer chain entanglement, thus preventing close contact between two particles covered by an adsorbed polymer layer (Derham et al., 1974; Hunter, 1987) (Figure 3–22).

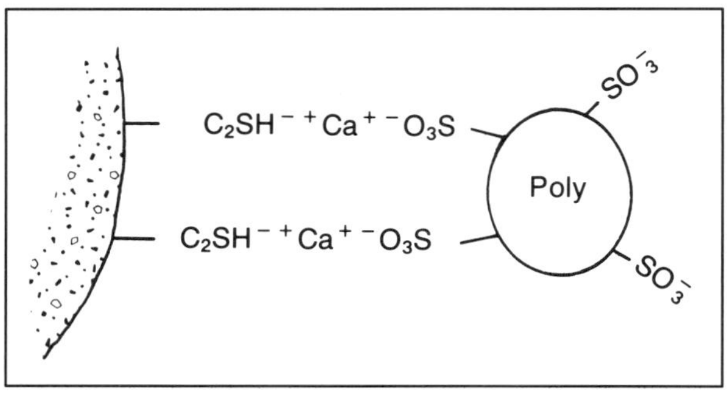

Figure 3–20—Polyanion adsorption on cement particle surface.

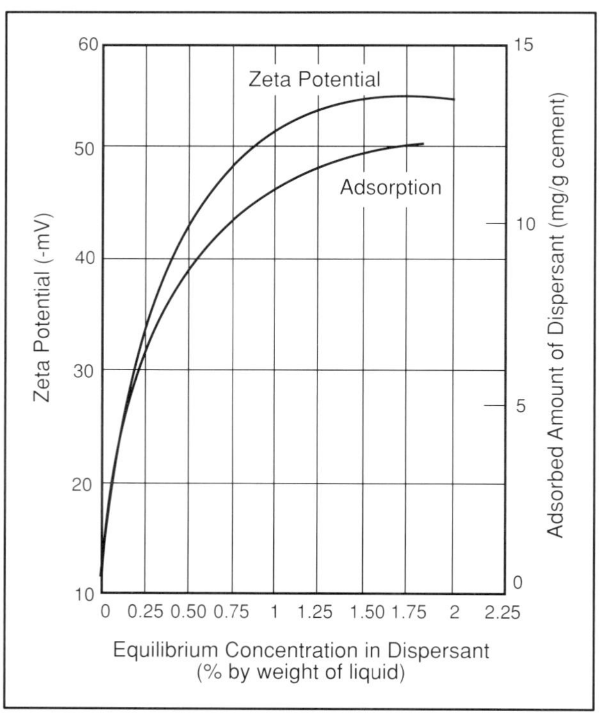

Figure 3–21—Zeta potential and adsorption isotherm for a diluted cement suspension (77°F, 25°C).

3–7.3 Chemical Composition of Cement Dispersants

Sulfonates are the most common cement dispersants. The preferred materials generally have 5 to 50 sulfonate groups attached to a highly branched polymer backbone. Branched polymers are more desirable, because the range of concentration for which they may bridge two particles is much narrower (Ruehrwein and Ward, 1952; Goodwin, 1982) (Figure 3–23). However, some linear polymers, as well as small organic molecules carrying several anionic groups, are also effective.

Polymelamine sulfonate (PMS) is used most frequently in the construction industry (Malhotra and Malanka, 1979), and to a limited extent in well cementing. Mela-

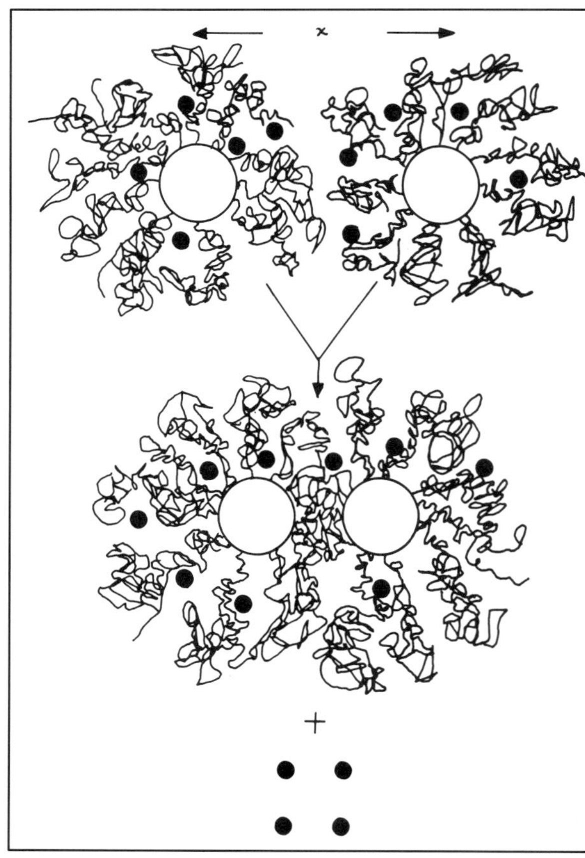

Figure 3–22—Schematic representation of steric stabilization of a cement dispersion by an adsorbed polymer. The bottom configuration corresponds to a higher free energy.

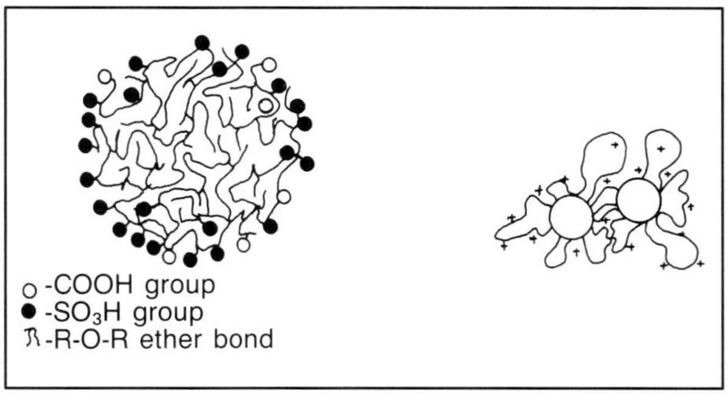

Figure 3–23—Schematic representation of a branched polymer (lignosulfonate) in water, and of particle bridging induced at low concentration of linear polymer.

mine reacts with formaldehyde to form trimethylol melamine, which is in turn sulfonated with bisulfite and condensed to form a polymer. The product is available commercially in solid form or as a water solution (20% and 40%). As shown in Fig. 3–24, about 0.4% PMS (BWOC) is typically required to achieve proper dispersion. This product is effective only at temperatures less than 185°F (85°C) because of limited chemical stability. The structure of the base unit is shown in Fig. 3–25.

Polynaphthalene sulfonate (PNS or NSFC) is a condensation product of β-naphthalene sulfonate and formaldehyde (Tucker, 1932), with high variability in the degree of branching and the molecular weight (Rixom, 1974;

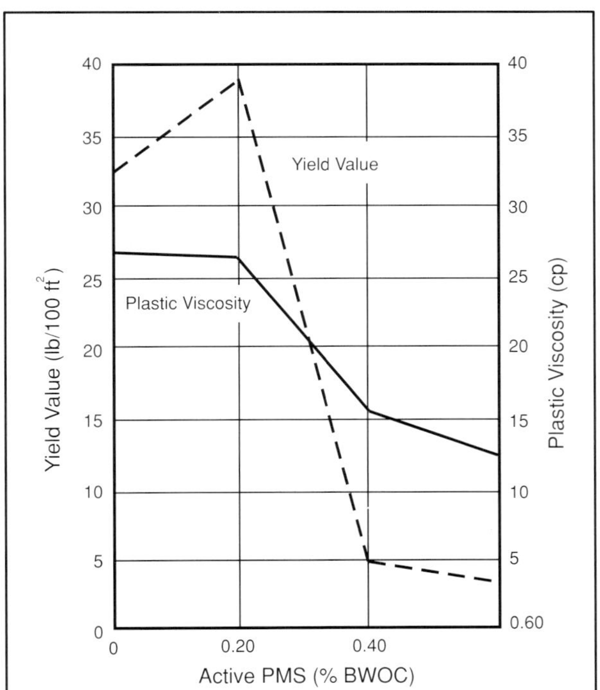

Figure 3–24—Yield value and plastic viscosity of a Class G slurry at 120°F (49°C).

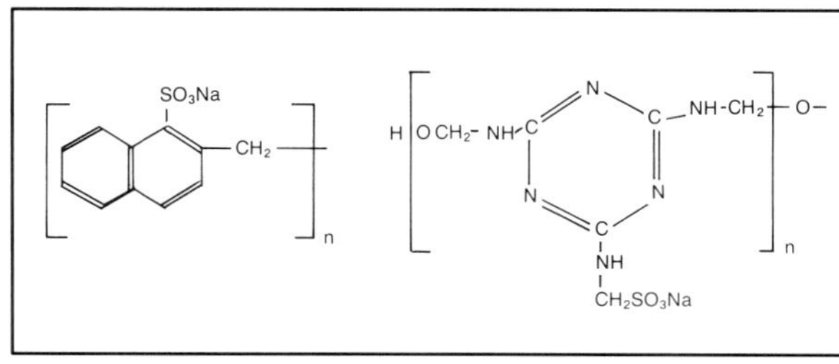

Figure 3–25—Polynaphthalene sulfonate and polymelamine sulfonate repeating units.

Costa et al., 1982). The repeating unit has the structure shown in Fig. 3–25 (Rixom, 1978). The commercial material is supplied as a powder or a 40% aqueous solution. For fresh water slurries, 0.5% to 1.5% active BWOC is normally required for effective slurry dispersion; however, as shown in Fig. 3–26, concentrations as high as 4% BWOC may be necessary for slurries containing NaCl (Michaux and Oberste-Padtberg, 1986). The dispersive ability of PNS is highly variable depending upon the cement. Fig. 3–27 (Michaux et al., 1986), a plot of the yield values for several cements vs the concentration of dispersant, demonstrates the complexity of the PNS molecular interactions with the cement grain surface. PNS is by far the most common dispersant for well cements.

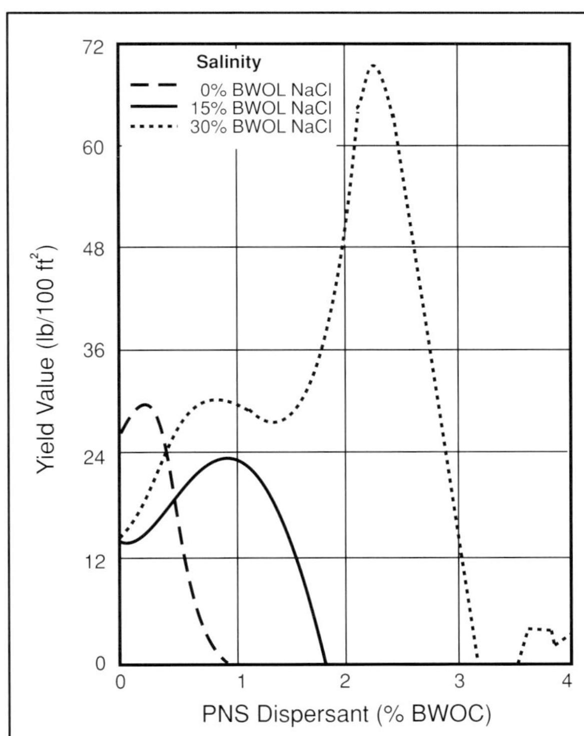

Figure 3–26—Influence of NaCL concentration on dispersing ability of PNS (15.8 lb/gal Class G slurry, 77°F, 25°C).

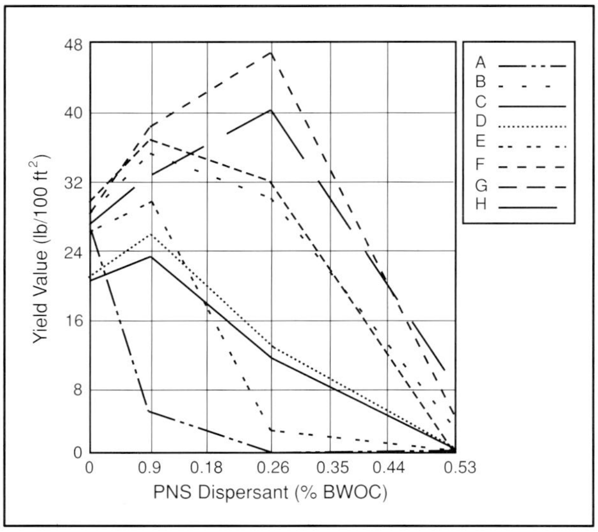

Figure 3–27—Yield value vs PNS concentration for different API Class G cements (77°F, 25°C).

Lignosulfonates are most frequently used as dispersants in drilling mud formulations (Lummus and Azar, 1986), but are also effective in cement slurries (Detroit, 1980). However, since they act simultaneously as retarders, they cannot be used at lower temperatures. Other lignin derivatives such as lignin carboxylic acids (Every and Jacob, 1978) are more effective as cement dispersants than the lignin sulfonic acids, but they also retard the set. Lignin derivatives are obtained from byproducts of the paper industry. They are inexpensive, and tend to be ill-defined chemically. The commercial products are predominantly sodium or calcium salts, with sugar contents between 1% and 30%. It is also important to note that the performance of some lignosulfonates is very sensitive to cement quality, and gelation difficulties are possible.

Polystyrene sulfonates are effective cement dispersants; however, they are rarely used for this purpose because of cost (Biagini, 1982). Polyacrylates (MacWilliams and Wirt, 1978) and copolymers such as sulfonated styrene-indene (Begou, 1978) or styrene-maleic anhydride (MacWilliams and Wirt, 1978) also have good fluidizing properties if they are used in conjunction with inorganic compounds, such as alkali metal or ammonium salts of carbonates, bicarbonates, oxalates, silicates, aluminates and borates.

Hydroxylated polysaccharides of low molecular weight, formed by hydrolysis of starch, cellulose or hemicellulose (Rixom, 1978), and other non-ionic polymers such as cellulose derivatives, ethylene oxide polymers, polyvinyl alcohol and polyglycol (Burge, 1978) have dispersive properties. However, set retardation is a side effect.

Non polymeric chemicals such as hydroxycarboxylic acids can have strong dispersing properties. As discussed earlier, they are all powerful retarders (Double, 1983). A typical example is citric acid (Messenger, 1978), which is often used in salt cement systems.

3–7.4 Rheology of Dispersed Slurries

In Figs. 3–18 and 3–27 it has been seen that with sufficient dispersant, a cement slurry has a zero yield value and behaves as a Newtonian fluid. It is interesting to observe how the yield value varies with dispersant concentration. Results with PNS (Michaux and Defossé, 1986) are displayed in Fig. 3–28. The yield value first begins to increase with dispersant concentration, and then decreases steeply to zero. At low dispersant concentrations, there is an excess of positively charged sites. The maximum yield value reflects the point of maximum particle interaction, when an exact balance exists between negative and positive surface sites. At a higher dispersant concentration, the grain surfaces are completely covered by negative charges; consequently, the yield value is zero because of electrostatic repulsion (Kondo et al., 1978).

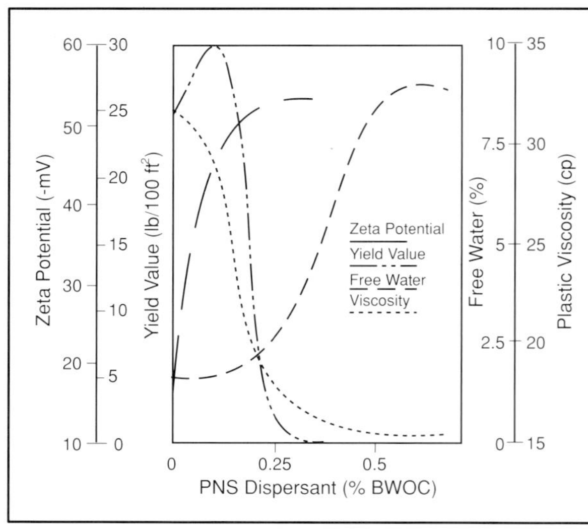

Figure 3–28—Yield value, plastic viscosity, zeta potential, and free water for a cement slurry at 85°C.

The effect of dispersants upon cement slurry viscosity is often different from that observed with the yield value. Although the electrostatic interactions between cement particles increase initially with dispersant concentration, the size of the particle aggregates immediately begins to decrease. Consequently, the volume of immobilized water decreases and, as shown in Fig. 3–28, the slurry viscosity also decreases continuously with dispersant concentration.

3–7.5 Particle Settling and Free Water

As a side effect of dispersant addition, the slurry may show sedimentation, a slurry density gradient from the top to the bottom of a container, and/or free water, a layer of non particle-laden fluid on top of the slurry. It is possible for free water to occur, and a homogeneous slurry to exist below. It is also possible for sedimentation to occur without the formation of a separate water layer.

Free Water: When cement particles in a suspension are not completely dispersed, they interact through electrostatic forces. A flocculated structure forms which supports the weight of a given particle. If the annulus in the well is sufficiently narrow, the weight of the particles is transmitted to the walls, and the slurry is self-supporting. Such cases are rare; consequently, the weight of the cement particles is transmitted to the bottom by the gel lattice, and structural deformation occurs. Water is squeezed out of the lower portion of the slurry, and is accommodated in the higher, less-stressed layers. The ability of the upper layers to accommodate the additional water is limited; thus, a layer of water may form at the top of the slurry (Fig. 3–29).

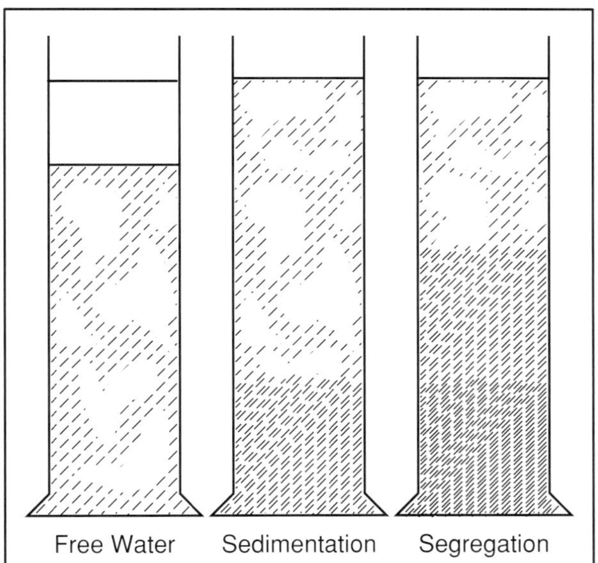

Figure 3–29—Three different cement slurry settling processes.

Sedimentation: As described in the previous sections, dispersants suppress interactions between cement particles by neutralizing positively charged sites. When the process is complete, the particles repel each other through double-layer interactions. The range of action of these forces is very short because of the high ionic content of the medium. Therefore, the repulsive forces allow smooth packing of the particles. In a fully dispersed slurry, the particles are free to move and, in particular, free to fall in the gravity field and collect at the container bottom. In reality, this ideal situation never occurs; instead, a density gradient is established. Three explanations to this may be proposed, which all incorporate the concept of particle polydispersity: small and large particles do not behave identically.

1. Smaller particles have not settled yet.
2. Smaller particles are prevented from settling by Brownian motion.
3. The flocculated gel exists, but is not sufficiently strong to support the larger particles.

3–7.6 Prevention of Free Water and Slurry Sedimentation

Nonhomogeneous cement columns are not acceptable, particularly when the wellbore is highly deviated or horizontal (Chapter 15). Sufficient mechanical strength of set cement and proper zonal isolation are jeopardized under such circumstances. Careful study of Fig. 3–28, a plot of free water and yield value vs. dispersant concentration, reveals a narrow range (between 0.2% and 0.3% BWOC) within which the slurry is sufficiently fluid and yet stable. In a field environment, control of additive concentration within such a narrow range is difficult. Therefore, "anti-settling agents" are often added to broaden the concentration range within which low yield values and low free water can be obtained (Fig. 3–30). Anti-settling agents are materials which restore some of the yield value, but at a level compatible with the pumping conditions and the friction pressure the well formation can bear. Examples of such materials are discussed below.

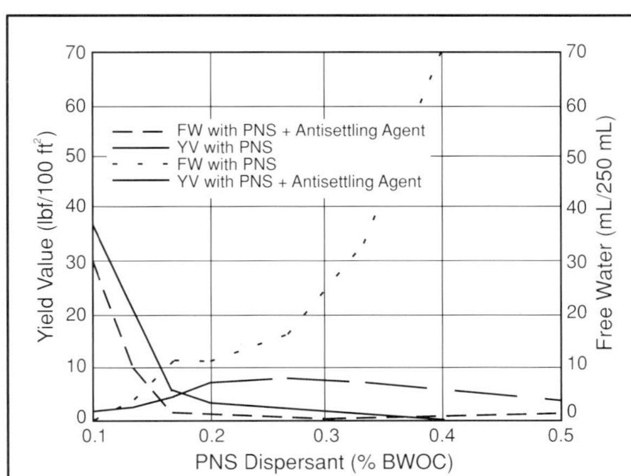

Figure 3–30—Yield value and free water behavior of Class G cement slurries with and without anti-settling agent (15.8 lb/gal, 185°F, 85°C).

Bentonite may be used to reduce slurry settling (Morgan and Dumbauld, 1954). As discussed in Section 3–5, bentonite has the ability to absorb large quantities of water; as a result, slurry homogeneity is preserved.

Various *hydrosoluble polymers* reduce sedimentation by increasing the viscosity of the interstitial water. The most commonly used materials are cellulosic derivatives, such as hydroxyethylcellulose.

Sea water and silicates can improve slurry stability (Childs et al., 1984). In addition, metallic salts such as $NiCl_2$ and $MgCl_2$, build weak but extensive hydroxide structure throughout the slurry volume (Defossé, 1985; Kar, 1986). As shown in Fig. 3–31, such structure building substantially reduces free water.

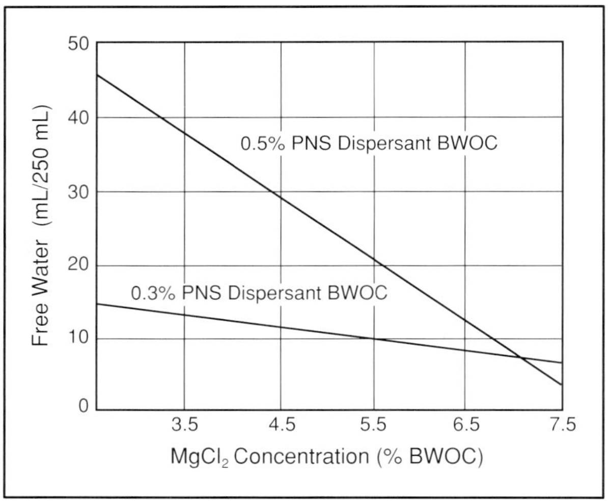

Figure 3–31—Free water development of 15.8 lb/gal Class G slurries with two PNS dispersant concentrations (185°F, 85°C).

The efficiency of anti-settling additives can be evaluated by measuring the density gradient in a column of set cement. A test slurry is placed in a cylinder and allowed to set. Wafers of the set cement are extracted from the top, middle and bottom of the column. The weight difference between the wafers gives an indication of the degree of slurry sedimentation. Figure 3–32 illustrates typical results for two 15.8-lb/gal (1.9 g/cm³) slurries.

3–8 FLUID-LOSS CONTROL AGENTS

When a cement slurry is placed across a permeable formation under pressure, a filtration process occurs. The aqueous phase of the slurry escapes into the formation, leaving the cement particles behind. Such a process is commonly known as "fluid loss," and is described in detail in Chapter 6.

If fluid loss is not controlled, several serious consequences may result which can lead to job failure. As the volume of the aqueous phase decreases, the slurry density increases; as a result, the performance of the slurry (rheology, thickening time, etc.) diverges from the original design. If sufficient fluid is lost to the formation, the slurry becomes unpumpable.

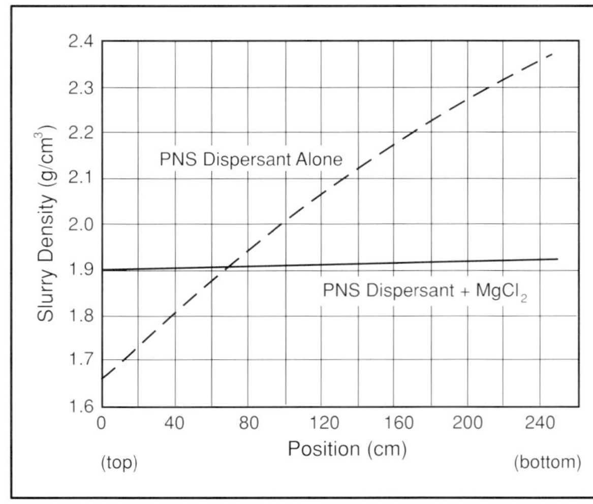

Figure 3–32—Comparison of density gradients in set cement columns (15.8 lb/gal, 185°F, 85°C).

The API fluid-loss rate of a neat cement slurry (Appendix B) generally exceeds 1,500 mL/30 min. As discussed in Chapter 6, an API fluid-loss rate less than 50 mL/30 min is often required to maintain adequate slurry performance. To accomplish such a reduction in the fluid-loss rate, materials known as "fluid-loss control agents" are included in the slurry design.

At present, the exact mechanisms by which fluid-loss control agents operate are not completely understood; however, several processes are known to occur. Once fluid-loss commences across a formation, a filter cake of cement solids is deposited on the formation surface. Fluid-loss agents decrease the filtration rate by reducing the permeability of filter cake, and/or by increasing the viscosity of the aqueous phase.

Two principal classes of fluid-loss additives exist: finely divided particulate materials and water-soluble polymers. The chemical and physical nature of each type of material, as well as mechanistic hypotheses, are discussed in this section.

3–8.1 Particulate Materials

The first fluid-loss control agent for cement slurries was bentonite (Cutforth, 1949). Because of the small size of its platelets (Section 3–3), bentonite can enter the filter cake and lodge between the cement particles. As a result, the permeability of the filter cake decreases. In addition, particulate systems such as carbonate powder, asphaltenes, thermoplastic resins, etc., are used to control fluid loss.

As described in Chapter 7, latex cements demonstrate excellent fluid-loss control. Latices are emulsion polymers, usually supplied as milky suspensions of very small spherical polymer particles (generally between 200 to 500 nm in diameter). Most latex dispersions contain about 50% solids. Like bentonite, such small particles can physically plug small pores in the cement filter cake.

The most common latices for well cements are those of vinylidene chloride (Eberhard and Park, 1958), polyvinyl acetate (Woodard and Merkle, 1962) and, more recently, styrene-butadiene (Parcevaux et al., 1985). The first two materials are limited to temperatures below 122°F (50°C). Styrene-butadiene latex has been applied at temperatures up to 350°F (176°C). Figure 3–33 is a plot of fluid-loss rate vs styrene-butadiene latex concentration for various cement slurries.

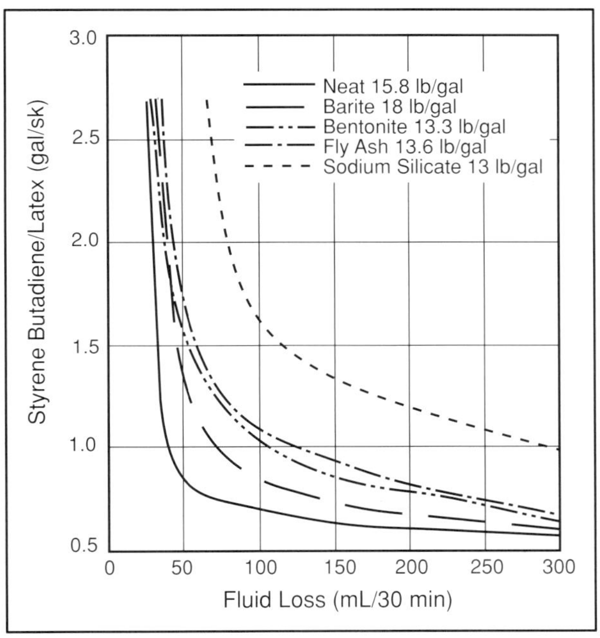

Figure 3–33—Fluid-loss behavior of latex-modified cement slurries at 185°F (85°C).

3–8.2 Water-Soluble Polymers

Water-soluble polymers received much attention as fluid-loss agents in the early 1940s, when they were first used in drilling fluids. Today, such materials are used extensively as fluid-loss control agents for well cement slurries. In general terms, they operate by simultaneously increasing the viscosity of the aqueous phase and decreasing the filter-cake permeability.

The viscosity of a polymer solution is dependent upon the concentration and the molecular weight. For example, as seen in Fig. 3–34, a 2% solution of low-molecular-weight hydroxyethylcellulose (HEC) may have a viscosity of 500 cP, but the viscosity of an equally concentrated solution of high-molecular-weight HEC can be as high as 50,000 cP (Aqualon, 1987). Such high viscosity would certainly decrease the filtration rate; however, this strategy alone cannot be relied upon to provide fluid-loss control, because slurry mixing would be impossible.

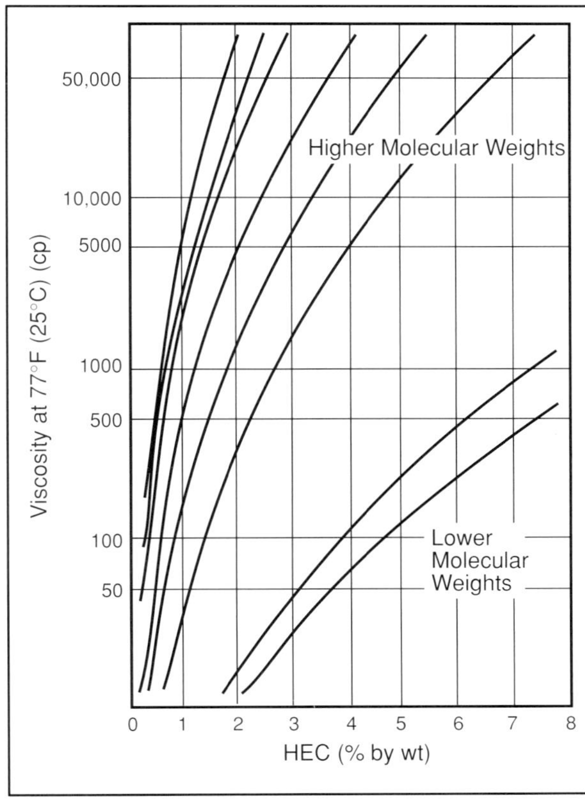

Figure 3–34—Concentration and molecular weight effect on viscosity of aqueous solutions of hydroxyethylcellulose (HEC).

Reduction of filter-cake permeability is the more important parameter with regard to fluid-loss control. When a slurry contains sufficient fluid-loss control agent to provide an API fluid-loss rate of 25 mL/30 min, the resulting filter cake is approximately 1,000 times less permeable than that obtained with a neat slurry (Binkley et al., 1957; Desbrières, 1988); whereas, the interstitial water viscosity increases, at most, five times (Table 3–14).

The size of the pores in the cement filter cake can be evaluated by mercury porosimetry. The typical size distribution is shown in Fig. 3–35, which shows the median diameter to be 1 µm. The typical radius of gyration of a polymer molecule is less than 1,000 Å (0.1 µm); therefore, only clusters of molecules would be sufficiently

Additive	Filter-Cake Permeability (md)	Viscosity of the Filtrate (cp)	Efficiency Ratio	Fluid-Loss Volume (mL/30 min)
None.	5100	1	1	1600
A—0.35%	924	2.24	0.280	450
A—0.60%	140	4.48	0.077	173
A—0.80%	6.1	3.70	0.018	45
A—1.00%	4.9	3.32	0.017	20
B—0.30%	770	3.10	0.217	300
B—0.80%	5.1	4.80	0.014	26
B—1.30%	1.3	2.30	0.011	12
C—0.08 GPS	1825	1.01	0.596	240
C—0.20 GPS	21	1.05	0.058	43
C—0.40 GPS	15	2.05	0.038	14

Table 3–14—Efficiency of different polymers in decreasing cake permeability and increasing filtrate viscosity at 25°C (80°F) (from Desbrières, 1988).

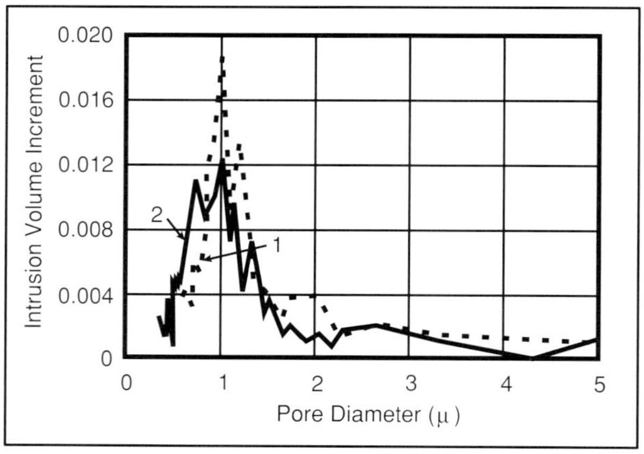

Figure 3–35—Pore diameters of two Class G cement filter cakes (15.8 lb/gal with 0.5% PNS BWOC, no fluid-loss additive).

large to obstruct a pore in the filter cake. Water-soluble polymers can form weakly bonded colloidal aggregates in solution, which are sufficiently stable to become wedged in the filter-cake constrictions (Christian et al., 1976). Such polymers may also adsorb onto the cement grain surfaces, and thus reduce the size of the pores. More likely, a superposition of these two phenomena, adsorption plus aggregation, is the true mechanism of action of polymeric fluid-loss agents.

Cement slurries containing water-soluble polymers must be well dispersed to obtain optimum fluid-loss control. Sulfonated aromatic polymers or salt are almost always added in conjunction with these materials. As described in Section 5, dispersants improve the packing of cement grains (and perhaps the polymer aggregates) in the filter cake. Thus, as shown in Table 3–15, dispersants reduce the permeability of the cement filter cake and can provide some degree of fluid-loss control on their own (Smith, 1987). However, one must bear in mind that overdispersion and sedimentation of the slurry may artificially improve the results of the API fluid-loss test (Appendix B).

Cement: API Classes A and G
API Fluid–Loss Test
Screen: 325 mesh
Pressure: 1000 psi
Temperature 80°F

PNS Dispersant (%)	Fluid Loss (mL/30 min) at a Water Ratio (gal/sk) of 3.78	4.24	4.75	5.2
0.50	490	504	580	690
0.75	310	368	476	530
1.00	174	208	222	286
1.25	118	130	146	224
1.50	72	80	92	—
1.75	50	54	64	—
2.00	36	40	48	—

Table 3–15—API fluid loss of densified cement slurries (from Smith, 1987).

Several classes of water-soluble polymers have been identified as useful fluid-loss control agents. The chemical properties and performance of each are discussed separately in the following sections.

3–8.2.1 Cellulose Derivatives

The first polymer used as a fluid-loss additive was a protein (i.e., a polypeptide) extracted from soy beans (Alcorn and Bond, 1944). Shortly thereafter ethylenediaminecarboxymethylcellulose (Lea and Fisher, 1949) and other cellulose derivatives were introduced (Lea, 1949; Cutforth, 1949). In the late 1950s, carboxymethylhydroxyethylcellulose (CMHEC) was introduced as a fluid-loss additive for cement slurries, and is still widely used today (Shell and Wynn, 1958; Greminger, 1958). The basic unit structure of CMHEC is shown in Fig. 3–36.

More recently (Chatterji and Brake, 1982; Chatterji et al., 1984), the performance of CMHEC has been improved by adjusting the degree of substitution (DS) from 0.1 to 0.7 (carboxymethyl) and the mole ratio of ethylene oxide to anhydroglucose (MS) from about 0.7 to about 2.5 (Fig. 3–36). According to Chatterji, et al., (1984) the performance of CMHEC in salt slurries can be improved by the addition of a hydroxycarboxylic acid such as tartaric acid.

The most common cellulosic fluid-loss control agent is hydroxyethylcellulose (HEC), with a DS range between 0.25 and 2.5 (Hook, 1969). The basic structural unit is shown in Figure 3–37. Various molecular weights of the polymer are used, depending upon the density of

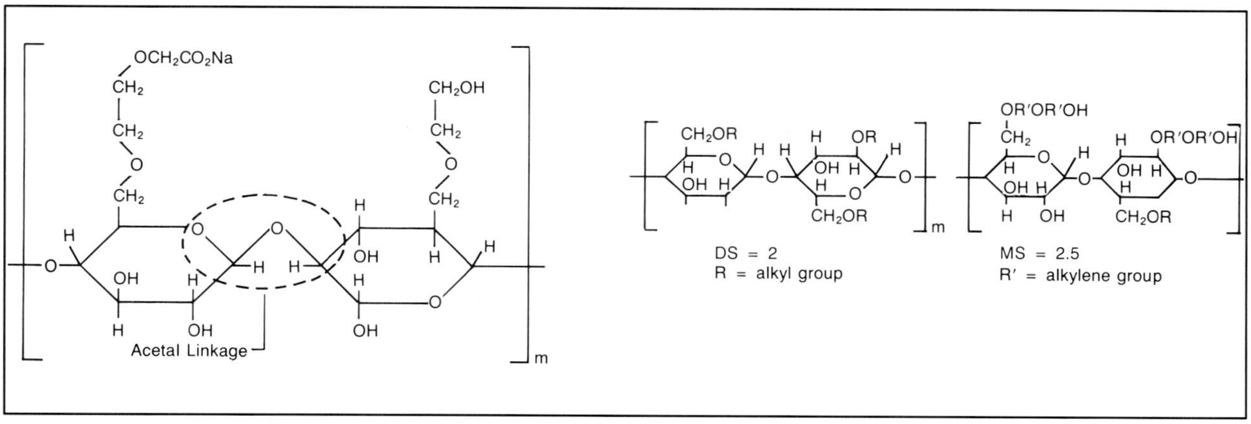

Figure 3–36—CMHEC molecular structure and illustration of DS and MS concepts.

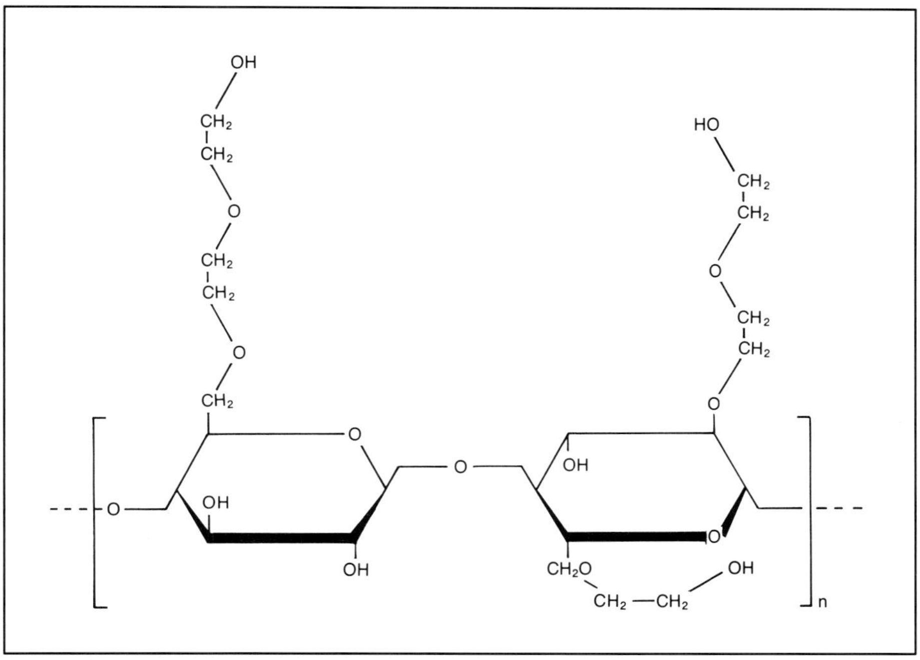

Figure 3–37—Idealized structure of hydroxyethylcellulose (HEC).

the cement slurry. For normal-density slurries an HEC of medium molecular weight (2% solution viscosity: 40 cP) is used. The typical fluid-loss control performance of this material is shown in Figure 3–38. A higher-molecular weight HEC is used for lower-density slurries (2% solution viscosity: 180 cP), and the typical performance in bentonite-extended slurries is shown in Figure 3–39.

HEC, as well as hydroxypropylcellulose (HPC), with a DS range of about 0.9 to 2.8, and a MS range of about 1.0 to 6.0, are disclosed as fluid-loss control additives when used in conjunction with high molecular weight xanthan gum (MW 2,000,000) (Baker and Harrison, 1984).

All cellulosic fluid-loss additives share certain disadvantages. They are effective water viscosifiers; as a result, they can increase the difficulty of slurry mixing, and ultimately cause undesirable viscosification of the cement slurry. At temperatures less than about 150°F (65°C), cellulosic fluid-loss additives are efficient retarders; thus, care must be taken to avoid overretardation of the slurry. Also, as shown in Figs. 3–38 and 3–39, the efficiency of the cellulose polymers decreases with increasing temperature. Cellulosic fluid-loss control agents are not normally used at circulating temperatures above 200°F (93°C).

3–8.2.2 Non-Ionic Synthetic Polymers

Polyvinylpyrrolidone (PVP) may be used simply with naphthalenesulfonate-formaldehyde condensate dispersants (Boncan and Gandy, 1986). It is also known to improve fluid-loss control when added with CMHEC (Hale, 1981) or HEC (Chatterji and Brake, 1982; Chatterji et al., 1984).

Complex mixtures containing polyvinylpyrrolidone, maleic anhydride-N-vinylpyrrolidone copolymer and poly(arylvinylbenzyl) ammonium chloride, i.e., a polycation (Wahl, 1964), have been reported as effective fluid-loss control additives. In addition, N-vinylpyrrolidone can be copolymerized with styrenesulfonate to form a product with satisfying fluid-loss control properties (Newlove et al., 1984; Sedillo et al., 1987).

Poly(vinyl alcohol) (PVAL) is frequently used as a fluid-loss control additive (Harrison, 1968; Carpenter, 1986). This material is particularly advantageous for low-temperature applications, at 100°F (38°C) and below, because it has no retarding effect and is compatible with accelerators such as calcium chloride. The fluid-loss control behavior of PVAL is shown in Fig. 3–40. It is important to note the sharp threshold effect associated with this additive: within a very short concentration range, the fluid-loss rate falls from 500 mL/30 min to 20mL/30min.

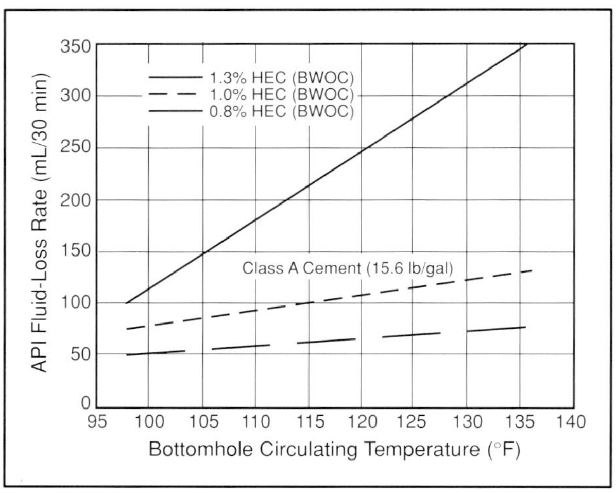

Figure 3–38—Typical fluid-loss control performance of hydroxyethylcellulose in normal-density slurries.

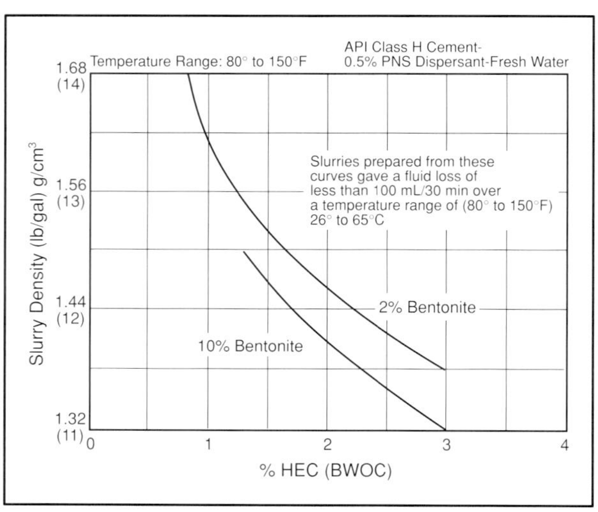

Figure 3–39—Typical fluid-loss control performance for HEC in low-density slurries.

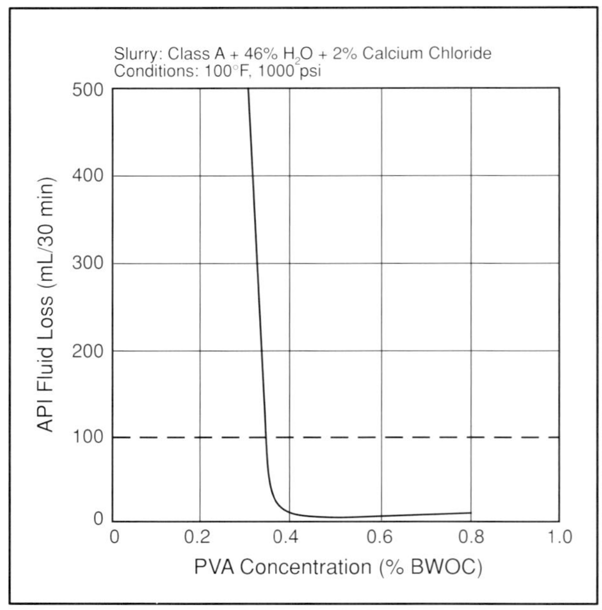

Figure 3–40—API fluid loss vs concentration of poly(vinyl alcohol).

3–8.2.3 Anionic Synthetic Polymers

The largest group of anionic polymer fluid-loss additives is composed of co-or terpolymers derived from acrylamide (AAm). Polyacrylamide is nonionic and is not used by itself in cement slurries. Partially hydrolyzed polyacrylamide containing various proportions of acrylic acid (AA) or acrylate units, is often added to drilling muds; however, because of the strong interaction between the carboxylate groups and cement grain surfaces, often resulting in retardation or flocculation, it is difficult to use in well cement slurries. Nevertheless, some applications have been reported using a material with a low AA/AAm ratio, about 0.1 (McKenzie and McElfresh, 1982).

The copolymers of acrylamide most often described in the patent literature contain a sulfonate monomer: 2-acrylamido-2-methylpropanesulfonic acid (AMPS). The structural formula is shown in Fig. 3–41. AMPS has been copolymerized with the following materials to produce fluid-loss control agents.

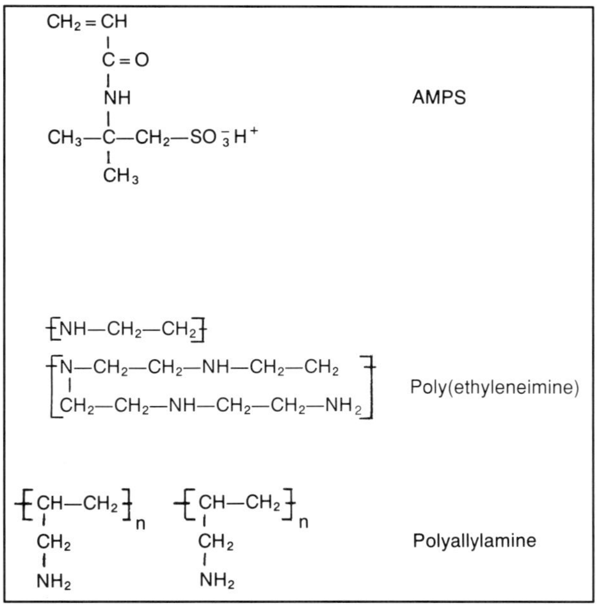

Figure 3–41—2-acrylamido-2-methyl propane sulfonic acid (AMPS) structure, poly(ethylene imine) repeating unit and branching, and polyallyamine structure.

- Acrylamide (AAm) (Presinski et al., 1977; Boncan and Gandy, 1986)
- N,N-dimethylacrylamide (NNDMA) (Rao, 1986; Brothers, 1987; George and Gerke, 1985; Fry et al., 1987).

Terpolymers of AMPS are also used, as described below.

- AMPS + AAm + itaconic acid (IA) (Savoly et al., 1987)
- AMPS + AA + N-methyl-N-vinyl acetamide (NMVA) (Defossé, 1985)
- AAm + vinyl sulfonate + NMVA (Hille et al., 1987)
- AA(AAm) + NMVA + AMPS (Hille et al., 1987)

AMPS may be also part of a copolymer or a terpolymer, grafted to a lignin backbone, associated with acrylonitrile, NNDMA or AA. These complex polymers are claimed to be efficient in salt slurries (Fry et al., 1987).

Figure 3–42 illustrates the typical concentrations of the terpolymer AMPS/AA/NMVA which provide an API fluid-loss rate of about 100 mL/30 min at various temperatures. Data are presented for two Class G cements, which also contain a PNS dispersant.

Sulfonated poly(vinyl aromatics) such as sulfonated polystyrene (SPS) (Martin, 1966; Newlove et al., 1984; Sedillo et al., 1987) and sulfonated polyvinyltoluene (SPVT) (Wahl et al., 1963) have been identified as useful fluid-loss control agents. A blend of SPVT, PNS and a sulfonated copolymer of styrene and maleic anhydride is effective in salt cement systems (Nelson, 1986). The fluid-loss control performance of this material in a salt-saturated cement slurry is shown in Fig. 3–43.

3–6.6 Cationic Polymers

Poly(ethyleneimine), shown in Fig. 3–41, is an example of a polyalkylene polyamine which has been widely used as fluid-loss additive (Gibson and Kucera, 1970; Scott

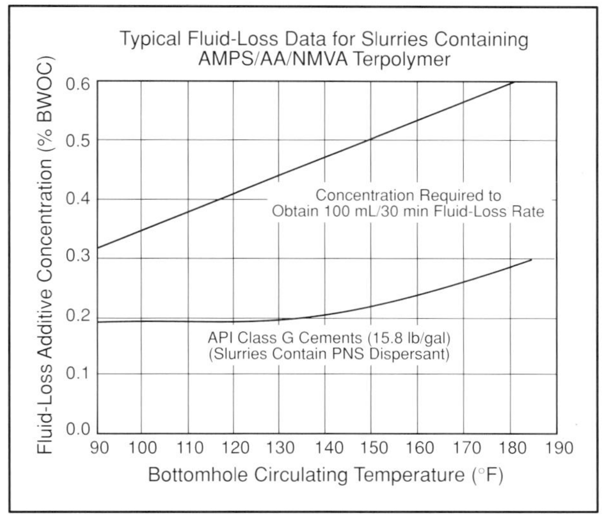

Figure 3–42—Typical fluid-loss data for slurries containing AMPS/AA/NMVA terpolymer.

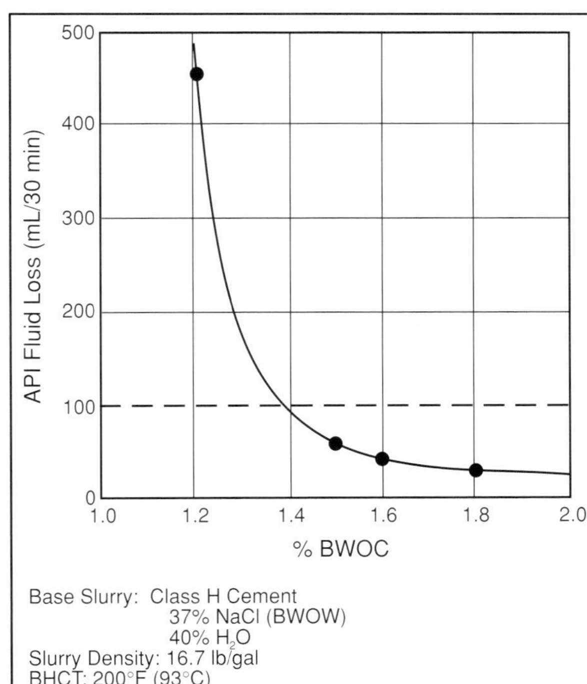

Figure 3–43—Fluid-loss control performance of blend of sulfonated poly(vinylaromatics) in salt-saturated cement slurries.

et al., 1970; McKenzie, 1984). The molecular weight range within which poly(ethyleneimine) is effective is from 10,000 to 1,000,000. Its structure is likely to be highly branched; therefore, all three types of amine groups (primary, secondary and tertiary) should be present in the chain.

The dispersant PNS must be present with poly(ethyleneimine) to obtain significant fluid-loss control. An insoluble association is made between the two polymers to create particles which provide fluid-loss control. As shown in Figure 3–44, fluid-loss control improves as the molecular weight of the poly(ethyleneimine) increases.

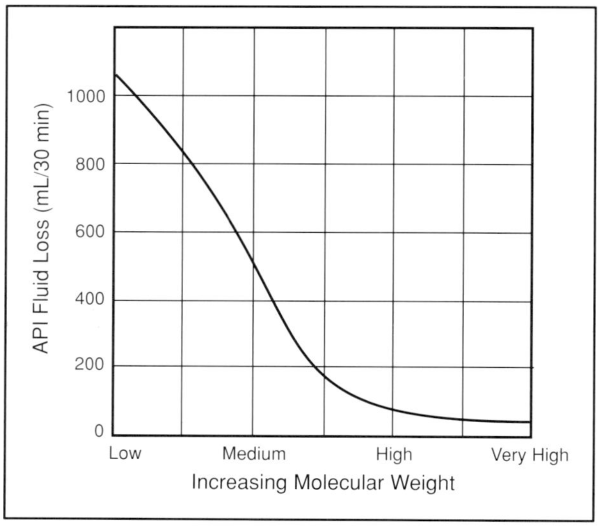

Figure 3–44—Influence of polyamine molecular weight on fluid-loss control.

The principal advantage of poly(ethyleneimine) as a fluid-loss control agent is its effectiveness at high temperatures. As shown in Table 3–16, poly(ethyleneimine) provides excellent fluid-loss control at circulating temperatures as high as 436°F (225°C). A notable disadvantage of poly(ethyleneimine) is its tendency to promote slurry sedimentation (Section 3–5). Although the sedimentation is preventable, slurry design can be very difficult.

Polyallylamine has been reported by Roark, et al., (1986; 1987) as an effective fluid-loss control agent. Instead of being part of the chain backbone, the amine group is pendant (Fig. 3–41). This material can also be slightly crosslinked to decrease slurry sedimentation. Table 3–17 shows the fluid-loss control performance of polyallylamine at two molecular weights.

Various quaternary ammonium or sulfonium monomers can be copolymerized with various materials to obtain effective fluid-loss control agents. Several are described below.

FLA (% BWOC)	PNS (% BWOC)	Ilmenite (lb/sk)	Slurry Density (lb/gal)	Temp. (°F)	Fluid Loss (mL/30 min)
0.1	0.5	—	16.2	290	20
0.1	0.5	—	16.2	315	30
0.13	0.5	—	16.2	337	18
0.15	1.0	—	16.8	299	8
0.15	1.5	—	19.0	380	34
0.15	1.5	—	20.0	370	40
0.18	1.0	5	17.4	342	30
0.18	1.0	30	18.2	370	90
0.18	1.0	25	18.0	400	78
0.2	1.2	95	19.2	436	16
0.25	1.5	70	19.0	380	10
0.25	1.5	70	19.0	380	11

Note: Fluid-loss tests were run with a differential pressure of 500 psi (750 psi with 250-psi backpressure).

Table 3–16—Typical fluid-loss data with polyethyleneimine fluid-loss additive (FLA).

Molecular Weight	API Fluid Loss (mL/30 min)
10,000	121
150,000	142

Table 3–17—Comparison of two molecular weights of polyallylamine polymers added in the concentration of 2% BWOC, with 0.66% of lignosulfonate; the fluid-loss tests were performed at 150°F using Class G cement (from Roark et al., 1987).

- Alkyl ammonium chloride or sulfonium chloride (Wahl and Dever, 1963).
- Dimethyl-diallyl ammonium chloride (DM-DAAC) (Reese et al., 1985; 1986).
- Methacrylamidopropyltrimethyl ammonium chloride (MAPTAC) (Peiffer, et al., 1986; 1987)

The alkyl ammonium and sulfonium chloride is co-polymerized with vinylbenzene to obtain poly(aryl-vinylbenzyl)alkyl ammonium or sulfonium chlorides. DM-DAAC is copolymerized with acrylic acid (AA) or methacrylic acid. MAPTAC is copolymerized with styrene sulfonate (SS) or acrylamide (AAm). Such materials are ampholytic polymers bearing negative and positive charges at a high pH (such as the aqueous phase of a Portland cement slurry).

3-9 LOST CIRCULATION PREVENTION AGENTS

The loss of circulation during a primary cementing job is a serious problem which usually results in having to perform remedial cementing. Circulation losses tend to occur in vuggy or cavernous formations, and particularly in highly fractured incompetent zones, which break down at relatively low hydrostatic pressures.

Usually, the operator will have experienced some circulation difficulties during drilling; thus, measures can be taken to prevent their occurrence during cementing. A thorough discussion of the causes of and solutions for lost circulation is presented in Chapter 6; however, in this chapter, it is appropriate to briefly mention the common cement additives used for the prevention of lost circulation.

3–9.1 Bridging Materials

Many lost-circulation problems are controlled by the addition of materials which physically bridge over fractures, and block weak zones. Such materials increase the resistance of the zone to pressure parting. As a general rule, they are chemically inert with respect to Portland cement hydration.

Granular materials such as gilsonite and granular coal are excellent bridging agents. As discussed in Section 3–5, they are also used extensively as cement extenders. They are added in concentrations similar to those specified in Section 3–5. Other granular materials used less often include ground walnut or pecan shells, coarse bentonite, and even corn cobs.

Another important bridging agent is cellophane flakes. As the cement slurry encounters the lost-circulation zone, the flakes form a mat at the face of the fracture. The thickness of the flakes is usually 0.02 to 0.06 mm, and the planar dimensions are less than 1 cm on each side. The normal concentration of cellophane flakes is between 0.125–0.500 lb/sk.

3–9.2 Thixotropic Cements

When the vugular or cavernous zones are so large that bridging agents are ineffective, thixotropic cements are often indicated. When such slurries enter the formation, they are no longer subjected to shear; as a result, they gel and become self-supporting. Eventually, the lost-circulation zone is plugged. The chemical nature of such systems is thoroughly presented in Chapter 7.

3–10 MISCELLANEOUS CEMENT ADDITIVES

There are a number of materials added to cement slurries which do not fit into any general category. These include antifoam agents, fibrous additives to improve cement durability, radioactive tracing agents and mud decontaminants.

3–10.1 Antifoam Agents

Many cement additives can cause the slurry to foam during mixing. Excessive slurry foaming can have several undesirable consequences. Slurry gelation can result, and cavitation in the mixing system can occur with loss of hydraulic pressure. In addition, air entrainment can indirectly result in higher-than-desired slurry densities. During slurry mixing, a densitometer is used to help field personnel proportion the ingredients (Chapter 10). If air is present in the slurry at the surface, the density of the system "cement + water + air" is measured. Since the air becomes compressed downhole, the densitometer underestimates the true downhole slurry density. Antifoam agents are usually added to the mix water or dry blended with the cement to prevent such problems.

Antifoam agents produce a shift in surface tension and/or alter the dispersibility of solids so that the conditions required to produce a foam are no longer present. In general, antifoams must have the following characteristics to be effective.

- Insoluble in the foaming system.
- A lower surface tension than the foaming system (Lichtman and Gammon, 1979).

The antifoam functions largely by spreading on the surface of the foam or entering the foam. Since the film formed by the spread of antifoam on the surface of a foaming liquid does not support foam, the foam situation is alleviated.

In well cementing, two classes of antifoam agents are commonly used: polyglycol ethers and silicones. Very small concentrations are necessary to achieve adequate foam prevention, usually less than 0.1% by weight of mix water.

Poly(propylene glycol) is most frequently used because of its lower cost, and is effective in most situations; however, it must be present in the system before mixing. Field experience has shown that post addition of poly(propylene glycol) is inefficient, and in some cases foam stabilization can result.

The silicones are highly effective antifoam agents. They are suspensions of finely divided particles of silica dispersed in polydimethylsiloxane or similar silicones. Oil-in-water emulsions at 10% to 30% activity also exist. Unlike the polyglycol ethers, the silicones will defeat a foam regardless of when they are added to the system.

3–10.2 Strengthening Agents

Fibrous materials are available which, when added to well cements in concentrations between 0.15% and 0.5% BWOC, increase the cement's resistance to the stresses associated with perforation, drill collars, etc. (Carter et al., 1968). Such materials transmit localized stresses more evenly throughout the cement matrix. Nylon fibers, with fiber lengths varying up to 1 in., are most commonly used.

Another material which dramatically improves the impact resistance and flexural strength of well cements is particulated rubber (Hook, 1971). This material is usually added in concentrations up to 5% BWOC. Latex-modified cements also exhibit improved flexural strength (Chapter 7).

3–10.3 Radioactive Tracing Agents

Cement slurries can be made radioactive to more easily determine their location behind casing. Radioactive tracers were at one time used to determine the fill-up or top of the cement column; however, temperature surveys and cement bond logs have largely assumed this function. Radioactive slurries still find occasional use in remedial cementing when it is desired to locate the slurry after placement. A base radiation log is run prior to the cement job to measure the natural formation radioactivity. After the job is completed, another radiation log is generated, and the location of the remedial slurry is determined by comparison with the base log (Chapter 16).

The most common radioactive agents for well cementing are ${}_{53}I^{131}$ (half-life: 8.1 days) and ${}_{77}Ir^{192}$ (half-life: 74 days). The iodine is generally available as a liquid. Sand or glass beads tagged with iridium 192 are often available in areas where tracers are used with hydraulic fracturing fluids.

3–10.4 Mud Decontaminants

Certain chemicals in drilling fluids, such as tannins, lignins, starches, celluloses and various chemically-treated lignosulfonates, can severely retard a Portland cement slurry. To minimize such effects should the cement slurry and the mud become intermixed, chemicals such as paraformaldehyde or blends of paraformaldehyde and sodium chromate are effective (Beach and Goins, 1957).

3–11 SUMMARY

Table 3–18 summarizes the major categories of well cement additives, their principal benefits, chemical compositions, and mechanisms of action.

REFERENCES

Alcorn, I. W. and Bond, D. C.: "Cementing Earth Bores," U.S. Patent No. 2,469,353 (1944).

Andersen, P. J.: "The Effect of Superplasticizers and Air-Entraining Agents on the Zeta Potential of Cement Particles," *Cement and Concrete Res.* (1986) **16**, 931–940.

Angstadt, R. L. and Hurley, F. R.: "Hydration of the Alite Phase in Portland Cement," *Nature* (1963) **197**, 688.

Aqualon: Customer Leaflet No. 33,007–F3, 1987.

Arliguie, G. and Grandet, J.: "Etude par Calorimetrie de L'Hydratation du Ciment Portland en Presence de Zinc," *Cement and Concrete Res.* (1985) **15**, 825–832.

Baker, W. S. and Harrison, J. J.: "Cement Composition and Method of Cement Casing in a Well," U.S. Patent No. 4,462,836 (1984).

Baret, J. F.: "Dispersants and Antisettling Agents for Oilwell Cement Slurries," *R. Soc. Chem.* (1988) **,67**, 57–61.

Beach, H. J. and Goins, W. C. Jr.: "A Method of Protecting Cements Against the Harmful Effects of Mud Decontamination," *Trans.*, AIME (1957) **210**, 148–152.

Begou, P.: "Products With a Fluidifying Action for Mineral Pastes and Binders," U.S. Patent No. 4,071,493 (1978).

Ben-Dor, L. and Perez, D.: "Influence of Admixtures on Strength Development of Portland Cement and on the Microstructure of Tricalcium Silicate," *J. Mater. Sci.* (1976) **11**, 239–245.

Bensted, J.: "Effect of Accelerator Additives on the Early Hydration of Portland Cement," *Il Cemento* (1978) **1**, 13–20.

Berger, R. L. and McGregor, J. D.: "Influence of Admixtures on the Morphology of Calcium Hydroxide Formed During Tricalcium Silicate Hydration," *Cement and Concrete Res.* (1972) **2**, 43–55.

Biagini, S., Ferrari, G., Maniscalco, V., Casolaro, M., Tanzi, M. C., Rusconi, L.: "Sulfonated Polystyrene as Superplasticizer," *Il Cemento* (1982) **4**, 345–354.

Binkley, G. W., Dumbauld, G. K., and Collins, R. E.: "Factors Affecting the Rate of Deposition of Cement in Unfractured Perforations During Squeeze-Cementing Operations," paper SPE 891–G, 1957.

Blank, B., Rossington, D. R., and Weinland, L. A.: "Adsorption of Admixtures on Portland Cement," *J. Amer. Ceramic Soc.* (1963) **46**, 395–399.

Boncan, V. G. and Gandy, R.: "Well Cementing Method Using an AM/AMPS Fluid-Loss Additive Blend," U.S. Patent No. 4,632,186 (1986).

Brothers, L. E.: "Method of Reducing Fluid-Loss in Cement Compositions Containing Substantial Salt Concentrations," U.S. Patent No. 4,640,942 (1987).

Bruere, G. M.: "Bleeding of Cement Pastes Containing Paraffin Wax Emulsions and Clays," *Cement and Concrete Res.* (1974) **4**, 557–566.

Bruere, G. M.: "Set-Retarding Effects of Sugars in Portland Cement Pastes," *Nature* (1966) **212**, 502–503.

Burge, T.: "Additive for Mortar and Concrete," U.S. Patent No. 4,069,062 (1978).

Carathers, K. and Crook, R.: "Surface Pipe Cement Gives High Early Strength With New Cement Additive," *Proc.*, Southwestern Petroleum Short Course, Lubbock, TX (1987) 12–19.

Carpenter, R. B.: "Matrix Control Cementing Slurry," U.S. Patent No. 4,569,395 (1986).

Carter, L. G. et al.: "Resilient Cement Decreases Perforating Damage," presented at the API Mid-Continent Dist. Div. of Production Spring Meeting, Amarillo, TX (1968).

Additive Category	Benefit	Chemical Composition	Mechanism of Action
accelerator	–shorter thickening time –higher early compressive strength	$CaCl_2$ NaCl	increased permeability of C-S-H gel layer[1]
		sodium silicates	formation of C-S-H gel nuclei by reaction with Ca^{2+} ions
retarder[2]	longer thickening time	lignosulfonates hydroxycarboxylic acids cellulose derivatives organophosphonates certain inorganic compounds	adsorption onto C-S-H gel layer, reducing permeability prevention of nucleation and growth of hydration products chelation of calcium ions precipitation of impermeable solids on C-S-H gel layer
extender	–lower slurry density –higher slurry yield	bentonite	absorption of water
		sodium silicates	formation of C-S-H gel + absorption of water
		pozzolans gilsonite powdered coal microspheres	lower density than cement
		nitrogen	foamed cement
weighting agent	higher slurry density	barite ($BaSO_4$) hematite (Fe_2O_3) ilmenite ($FeTiO_3$)	higher density than cement
dispersant	lower slurry viscosity	polynaphthalene sulfonate polymelamine sulfonate lignosulfonates polystyrene sulfonate hydroxylated polysaccharides hydroxycarboxylic acids	induce electrostatic repulsion of cement grains
fluid-loss additive	reduced slurry dehydration	cellulosic polymers	increased viscosity of aqueous phase of slurry
		polyamines sulfonated aromatic polymers polyvinylpyrrolidone polyvinylalcohol AMPS copolymers or terpolymers	reduced permeability of cement filter cake
		bentonite latices	particle bridging of cement filter cake
lost-circulation control agent	prevent loss of slurry to formation	gilsonite granular coal cellophane flakes nut shells	bridging effect across formation
		gypsum certain soluble sulfate salts bentonite crosslinked cellulosic polymers	induce thixotropic behavior of slurry[3]
Miscellaneous			
antifoam agent	reduced air entrainment aid for slurry mixing	polyglycol ethers silicones	insoluble in foaming system lower surface tension than foaming system
strengthening agent	increase shock resistance and/or flexural strength of set cement	nylon fibers ground rubber	transmit localized stresses more evenly throughout cement matrix
radioactive tracing agent	easier determination of location behind casing	$_{53}I^{131}$ $_{77}Ir^{192}$	emission of radioactivity

[1] proposed theoretical mechanism
[2] More than one mechanism may apply for certain classes of retarders. See text for clarification.
[3] discussed in Chapter 7

Table 3–18—Summary of additives and mechanisms of action.

Chatterji, J. and Brake, B. G.: "Water-Loss Reducing Additives for Salt Water Cement Slurries," U.K. Patent No. 2,080,812A (1982).

Chatterji, J., Brake, B. G., and Tinsley, J. M.: "Liquid Water-Loss Reducing Additives for Cement Slurries," U.S. Patent No. 4,466,837 (1984).

Chatterji, S.: "Electron-Optical and X-ray Diffraction Investigation of the Effects of Lignosulphonates on the Hydration of C_3A," *Indian Concrete J.*, (1967) **41**, 151–160.

Childs, J. D., Brothers, L. E., and Taylor, M. J.: "Method of Preparing a Lightweight Cement Composition from Sea Water," U.S. Patent No. 4,450,009 (1984).

Christian, W. W., Chatterji, J. and Ostroot, G. W.: "Gas Leakage in Primary Cementing—A Field Study and Laboratory Investigation," *JPT* (Nov. 1976) 1361–1369.

Ciach, T. D. and Swenson, E. G.: "Morphology and Microstructure of Hydrating Portland Cement and Its Constituents—Pt. 2.: Changes in Hydration of Calcium Silicates Alone and in the Presence of Triethanolamine and Calcium Lignosulfonate, Both With and Without Gypsum," *Cement and Concrete Res.* (1971) **1**, 159–176.

Collepardi, M. and Marchese, B.: "Morphology and Surface Properties of Hydrated Tricalcium Silicate Pastes," *Cement and Concrete Res.* (1972) **2**, 57–65.

Collepardi, M. and Massidda, L.: "Hydration of Beta Dicalcium Silicate Alone and in the Presence of $CaCl_2$ or C_2H_5OH," *J. Amer. Ceramic Soc.* (1973) **56**, 181–183.

Collepardi, M.: "Il Comportamento Reologico delle Paste Cementizie," *Il Cemento* (1971) **68**, 99–106.

Costa, U., Massazza, F., and Barril, A.: "Adsorption of Superplasticizers on C_3S: Changes in Zeta Potential and Rheology of Pastes," *Il Cemento* (1982) *4*, 323–336.

Crump, D. K. and Wilson, D. A.: "Set-Retarding Additives for Cement From Aminomethylenephosphonic Acid Derivatives," U.S. Patent No. 4,468,252 (1984).

Cutforth, H. G.: "Low Water-Loss Cement Slurry and Method of Cementing a Well Therewith," U.S. Patent No. 2,598,675 (1949).

Daimon, M. and Roy, D. M.: "Rheological Properties of Cement Mixes—Pt. I: Methods, Preliminary Experiments, and Adsorption Studies," *Cement and Concrete Res.* (1978) **8**, 753–764.

Daimon, M. and Roy, D. M.: "Rheological Properties of Cement Mixes: II. Zeta Potential and Preliminary Viscosity Studies," *Cement and Concrete. Res.* (1979) **9**, 103–109.

Davis, R. E., Carlson, R. W., Kelly, J. W., and Davis, H. E.: "Properties of Cements and Concretes Containing Fly Ash," *J. Amer. Concrete Inst.* (1937) **33**, 577–612.

Defossé, C: "Cement Slurry Compositions for Cementing Oil Wells, Adapted to Control Free Water and Corresponding Cementing Process," French Patent No. 2,540,097 (1985).

Defossé, C.: "Fluid-Loss Additive for Cement," French Patent No. 2,540,098 (1985).

Derham, K. W., Goldsbrough, J., and Gordon, M.: "Pulse-Induced Critical Scattering (PICS) from Polymer Solutions," *J. Pure and Appl. Chem.* (1974) **38**, 97–115.

Desbriéres, J.: "Influence of Polymeric Additives on Cement Filter-Cake Permeability," *R. Soc. Chem.* (1988) **67**, 62–67.

Detroit, W. J.: "Lignosulfonate Derivatives," U.S. Patent No. 4,219,471 (1980).

Double, D. D.: "New Developments in Understanding the Chemistry of Cement Hydration," *Trans.* Royal Soc. London (1983) Ser. A 310, 53–66.

Eberhard, J. F. and Park, A.: "Portland Cement-Vinylidene Chloride Polymer Composition, Method of Making, and Method of Using," U.S. Patent No. 2,819,239 (1958).

Edwards, G. C. and Angstadt, R. L.: "The Effect of Some Soluble Inorganic Admixtures on the Early Hydration of Portland Cement," *J. Appl. Chem.* (1966) **16**, 166–168.

Einstein, A.: *Investigation on the Theory of Brownian Movement*, Methuen, New York (1926); Dover, New York (1956).

Every, R. L. and Jacob, J. T.: "Production of Raw Mix Cement Slurries Having Reduced Water Content," U.S. Patent No. 4,115,139 (1978).

Fry, S. E., Childs, J. D., Brothers, L. E. and Lindsey, D. W.: "Method of Reducing Fluid Loss in Cement Compositions Which May Contain Substantial Salt Concentrations," U.S. Patent No. 4,676,317 (1987).

Fry, S. E., Childs, J. D., Brothers, L. E., and Lindsey, D. W.: "Method of Reducing Fluid Loss in Cement Compositions Which May Contain Substantial Salt Concentrations," U.S. Patent No. 4,703,801 (1987).

George, C. R. and Gerke, R. R.: "Oilfield Cements," European Patent Application No. 0,163,459 (1985).

Gibson, D. L. and Kucera, C. H.: "Low-Water-Loss Aqueous Cement Slurry and Method of Use," U.S. Patent No. 3,491,049 (1970).

Colloidal Dispersion, J. W. Goodwin (ed.), Royal Society of Chemistry, London (1982).

Gouda, V. K., Mourad, W. E., and Mikhail, R. S.: "Additives to Cement Pastes: Simultaneous Effects on Pore Structure and Corrosion of Steel Reinforcement," *J. Colloid and Interface Sci.* (1973) **43**, No. 2, 293–302.

Greminger, G. K.: "Hydraulic Cement Compositions for Wells," U.S. Patent No. 2,844,480 (1958).

Hale, B. W.: "Well-Cementing Method Using Low Fluid-Loss Cement Slurry," U.S. Patent No. 4,258,790 (1981).

Harms, W. M and Sutton, D. L.: "Ultralow Density Cementing Compositions," paper SPE 9597, 1981.

Harrison, H. T.: "Aqueous Cementing Composition Adaptable to High Turbulent Flow and Method of Cementing a Well Using Same," U.S. Patent No. 3,409,080 (1968).

Hille, M., Friede, W., Wittkus, H., Engelhardt, F., and Riegel, U.: "Cement Slurries for Deep Holes, With a Copolymer Content for Reducing the Water Loss," Canadian Patent No. 1,228,373 (1987).

Hirljac, J., Wu, Z. Q. and Young, J. F.: "Silicate Polymerization During the Hydration of Alite," *Cement and Concrete. Res.* (1983) **13**, 877–886.

Hook, F. E.: "Aqueous Cement Slurry; and Method of Use," U.S. Patent No. 3,483,007 (1969).

Hook, F. E.: Unpublished Data, 1971.

Hunter, R. J.: *Foundation of Colloid Science*, Clarendon, Oxford (1987).

Hurlbut, C. S.: *Dana's Manual of Mineralogy*, John Wiley and Sons, New York (1971).

Jawed, I., Klemm, W. A. and Skalny, J.: "Hydration of Cement-Lignosulfonate-Alkali Carbonate System," *J. Amer. Ceramic Soc.* (1979) **62**, 461–464.

Jennings, H. M., Taleb, H., Frohnsdorff, G., and Clifton, J. R.: "Interpretation of the Effects of Retarding Admixtures on Pastes of C_3S, C_3A Plus Gypsum, and Portland Cement," *Proc.*, Eighth Intl. Cong. Chem. Cement, Rio de Janeiro (1986) 239–243.

Kar, K. K. and Burba, J. L.: "Metallo-Organo Aluminates as Lubricant Additives," U.S. Patent No. 4,610,797 (1986).

Kondo, R., Daimon, M, Sakai, E., and Ushiyama, H.: "Influence of Inorganic Salts on the Hydration of Tricalcium Silicate," *J. Appl. Chem. Biotechnol.* (1977) **27**, 191–197.

Kondo, R., Daimon, M., and Sakai, E.: "Interaction Between Cement and Organic Polyelectrolytes," *Il Cemento* (1978) **3**, 225.

Kurczyk, H. G. and Schwiete, H. E.: "Elektronenmikroskopische un Thermochemische Untersuchungen ber die Hydratation der Calciumsilikate $3CaO \cdot SiO_2$ und $\beta \cdot 2CaO \cdot SiO_2$ und den Einflub von Calciumchlorid und Gips auf den Hydrationsvorgang," *Tonind.-Zeit. und Keramische Rundsch.* (1960) **24**, 585–597.

Lea, P. J. and Fisher, H. B.: "Low Water-Loss Cement and Process of Making," U.S. Patent No. 2,614,634 (1949).

Lea, P.J.: "Low-Water-Loss Cement Slurry," U.S. Patent No. 2,614,998 (1949).

Lea, F. M.: *The Chemistry of Cement and Concrete*, Chemical Publishing Co., Inc., New York (1971).

Lichtman, I. and Gammon, T.: "Defoamers," *Kirk-Otmer Encyclopedia of Chemical Technology*, M. Grayson and D. Eckroth (eds.), John Wiley and Sons, New York (1979) **7**, 430–448.

Lummus, J. L. and Azar, J. J.: *Drilling Fluids Optimization*, PennWell Publishing Co., Tulsa, OK (1986).

MacWilliams, D. C. and Wirt, J. R.: "Aqueous Suspensions," U.S. Patent No. 4,126,480 (1978).

Malhotra, V. M. and Malanka, D.: *Superplasticizers in Concrete*, American Concrete Inst., Detroit (1979).

Martin, R. C.: "Cement Composition," U.S. Patent No. 3,234,154 (1966).

McKenzie, L. F. and McElfresh, P. M.: "Acrylamide-Acrylic Acid Copolymers for Cement Fluid-Loss Control," paper SPE 10623, 1982.

McKenzie, L. F.: "Polyamine Fluid-Loss Additive for Oilwell Cements," U.S. Patent No. 4,482,383 (1984).

Messenger, J. V.: "Cementing Against Evaporites," U.S. Patent No. 4,089,376 (1978).

Michaux, M. and Defoss , C.: "Oilwell Cement Slurries—Pt. 1:. Microstructural Approach of Their Rheology," *Cement and Concrete Res.* (1986) **16**, 23–30.

Michaux, M. and Oberste-Padtberg, R.: "The Effect of Mix-Water Salinity on Cement Slurries Containing Polynaphthalene Sulfonate," *Proc.*, Eighth Intl. Cong. Chem. Cement, Rio de Janeiro (1986) **4**, 239.

Michaux, M., Oberste-Padtberg, R., and Defosse, C.: "Oilwell Cement Slurries—Pt. 2: Adsorption Behaviour of Dispersants," *Cement and Concrete Res.* (1986) **16**, 921–930.

Michaux, M., Fletcher, P., and Vidick, B.: "Evolution at Early Hydration Times of the Chemical Composition of Liquid Phase of Oilwell Cement Pastes With and Without Additives," *Cement and Concrete Res.* (1989) **19**, 443–456.

Milestone, N. B.: "The Effect of Lignosulphonate Fractions on the Hydration of Tricalcium Aluminate," *Cement and Concrete Res.* (1976) **6**, 89–102.

Milestone, N. B.: "Hydration of Tricalcium Silicate in the Presence of Lignosulphonates, Glucose, and Sodium Gluconate," *J. Amer. Ceramic Soc.* (1979) **62**, 321–324.

Morgan, B. E. and Dumbauld, G. K.: "Bentonite Cement Proving Successful in Permanent-Type Squeeze Operations," *World Oil* (Nov. 1954) 220.

Nelson, E. B.: "Well Treating Composition," U.K. Patent Application GB 2,157,279A, 1984.

Nelson, E. B.: "Sulfonated Poly (Vinyl Aromatics) as Fluid-Loss Additives for Salt Cement Slurries," U.S. Patent No. 4,601,758 (1986).

Nelson, E. B.: "Well Treating Process and Composition," Canadian Patent No. 1,216,742 (1987).

Newlove, J. C., Portnoy, R. C., Schulz, D. N., and Kitano, K.: "Fluid Loss Control in Oilfield Cements," U.S. Patent No. 4,480,693 (1984).

Odler, I. and Skalny, J.: "Influence of Calcium Chloride on Paste Hydration of Tricalcium Silicate," *J. Amer. Ceramic Soc.* (1971) **54**, 362–363.

Odler, I. and Skalny, J.: "Pore Structure of Hydrated Calcium Silicate—Pt. 2: Influence of Calcium Chloride on the Pore Structure of β-Dicalcium Silicate," *J. Colloid and Interface Sci.* (1971) **36**, No. 3, 293–297.

Odler, I., Duckstein, U., and Becker, T.: "On the Combined Effect of Water Soluble Lignosulfonates and Carbonates on Portland Cement and Clinker Pastes—Pt. 1: Physical Properties," *Cement and Concrete Res.* (1978) **8**, 469–479.

Parcevaux, P. A. et al.: "Cement Compositions for Cementing Wells, Allowing Pressure Gas-Channeling in the Cemented Annulus to be Controlled," U.S. Patent No. 4,537,918 (1985).

Parker, D. G.: "Microsilica Concrete—Pt .1: The Material," Concrete Soc (UK) Current Practice Sheet No. 104 (Oct. 1985).

Pauri, M., Monosi, S., Moriconi, G., and Collepardi, M.: "Effect of Triethanolamine on the Tricalcium Silicate Hydration," *Proc.*, Eighth Intl. Cong. Chem. Cement, Rio de Janeiro (1986) **3**, 125–129.

Peiffer, D. G., Lundberg, R. D., Sedillo, L., and Newlove, J. C.: "Fluid-Loss Control in Oilfield Cements," U.S. Patent No. 4,626,285 (1986).

Peiffer, D. G., Lundberg, R. D., Sedillo, L. and Newlove, J. C.: "Fluid Loss Control in Oilfield Cements," U.S. Patent No. 4,683,952 (1987).

Petrie, E. M.: "Effect of Surfactant on the Viscosity of Portland Cement-Water Dispersions," *Ind. Eng. Chem. Prod. Res. Dev.* (1976) **15**, No. 4, 242–249.

Pilkington, P. E.: "Pressure Needed to Reduce Microannulus Effect on CBL," *Oil and Gas J.* (May 30, 1988) 68–74.

Presinski, L. J., Cook, M. M., and Adams, S. L.: "Low Fluid Loss Cementing Compositions Containing Hydrolyzed Acrylamide/2-Acrylamido-2-Methylpropane Sulfonic Acid Derivative Copolymers and Their Use," U.S. Patent No. 4,015,991 (1977).

Previte, R. W.: "Some Insights on the Mechanism of Saccharide Set Retardation of Portland Cement," *Cement and Concrete Res.* (1971) **1**, 301–316.

Ramachandran, V. S.: "Interaction of Calcium Lignosulfonate With Tricalcium Silicate, Hydrated Tricalcium Silicate, and Calcium Hydroxide," *Cement and Concrete Res.* (1972) **2**, 179–194.

Ramachandran, V. S.: "Action of Triethanolamine on the Hydration of Tricalcium Aluminate," *Cement and Concrete Res.* (1973) **3**, 41–54.

Ramachandran, V. S.: *Calcium Chloride in Concrete*, Applied Science Publishers, Englewood, NJ (1976) 216.

Ramachandran, V. S.: "Hydration of Cement—Role of Triethanolamine," *Cement and Concrete Res.* (1976) **6**, 623–632.

Ramachandran, V. S.: "Admixture and Addition Interactions in the Cement-Water System," *Il Cemento* (1986) **1**, 13–38.

Rao, S. P. and Burkhalter, J. F.: "Oilfield Cementing Methods and Compositions," European Patent Appl. No. 0,192,447 (1986).

Reese, D. W., Pace, R. S., and McKenzie, L. F.: "Nonretarding Fluid-Loss Additives for Well Cementing Compositions," U.S. Patent No. 4,610,306 (1986).

Rhehrwein, R. A. and Ward, D. W.: "Mechanism of Clay Aggregation by Polyelectrolytes," *Soil Sci.* (1952) **73**, 485–492.

Rixom, M. R.: "Development of an Admixture to Produce Flowing or Self-Compacting Concrete," *Precast Concrete* (1974) **5**, 633–637.

Rixom, M. R.: *Chemical Admixtures for Concrete*, John Wiley and Sons, New York (1978).

Roark, D. N., Nugent, A. Jr., and Bandlish, B. K.: "Fluid-Loss Control and Compositions for Use Therein," European Patent Appl. No. 0,201,355 (1986).

Roark, D. N., Nugent, A. Jr., and Bandlish, B. K.: "Fluid-Loss Control in Well Cement Slurries," U.S. Patent No. 4,657,948 (1987).

Roark, D. N., Nugent, A. Jr., and Bandlish, B. K.: "Fluid-Loss Control in Well Cement Slurries," U.S. Patent No. 4,698,380 (1987).

Roark, D. N., Nugent, A. Jr., and Bandlish, B. K.: "Fluid-Loss Control In Well Cement Slurries," U.S. Patent No. 4,706,755 (1987).

Rossington, D. R. and Runk, E. J.: "Adsorption of Admixtures on Portland Cement Hydration Products," *J. Amer. Ceramic Soc.* (1968) **51**, 46–50.

Rust, C. F. and Wood, W. D.: "Laboratory Evaluations and Field Testing of Silica-CMHEC-Cement Mixtures," *JPT* (Nov. 1966) 25–29.

Savoly, A., Villa, J. A., Garvey, C. M., and Resnick, A. L.: "Fluid Loss Agents for Oilwell Cementing Compositions," U.S. Patent No. 4,674,574 (1987).

Scott, L. J., Gibson, D. L., Hook, F. E., and Kucera, C. E.: "Cementing Composition and Well Cementing Process Employing Same," U.S. Patent No. 3,511,314 (1970).

Sedillo, L. P., Newlove, J. C., and Portnoy, R. C.: "Fluid-Loss Control in Oilfield Cements," U.S. Patent No. 4,659,750 (1987).

Shaw, D. J.: *Introduction to Colloid and Surface Chemistry*, Butterworths, London (1980).

Shell, F. J. and Wynne, R. A.: "Application of Low-Water-Loss Cement Slurries," paper API 875–12–I, 1958.

Shideler, J. J.: "Calcium Chloride in Concrete," *J. Amer. Concrete Inst.* (1952) **48**, 537–559.

Singh, N. B. and Abha, K.: "Effect of Calcium Formate on the Hydration of Tricalcium Silicate," *Cement and Concrete Res.* (1983) **13**, 619–625.

Singh, N. B. and Ojha, P. N.: "Effect of Calcium Chloride on the Hydration of Tricalcium Silicate, *J. Mater. Sci.* (1981) **16**, 2675–2681.

Skalny, J. and Maycock, J. N.: "Mechanisms of Acceleration by Calcium Chloride: A Review," *J. Testing and Evaluation* (1975) **3**, No. 4, 303–311.

Skalny, J., Odler, I., and Hagymassy, J. Jr.: "Pore Structure of Hydrated Calcium Silicates. 1. Influence of Calcium Chloride on the Pore Structure of Hydrated Tricalcium Silicate," *J. Colloid and Interface Sci.* (1971) **35**, No. 3, 434–440.

Slagle, K. A. and Carter, L. G.: "Gilsonite—A Unique Additive for Oilwell Cements," *Drill. and Prod. Prac.*, API (1959) 318–328.

Smith, D.K.: *Cementing*, Henry A. Doherty Series, SPE, Richardson, TX (1987).

Smith, R. C. and Calvert, D. G.: "The Use of Sea Water in Well Cementing," paper SPE 5030, 1974.

Smith, R. C., Powers, C. A., and Dobkins, T. A.: "A New Ultralightweight Cement With Super Strength," *JPT* (Aug. 1980) 1438–1444.

Stadelmann, C. and Wieker, W.: "On the Influence of Inorganic Salts on the Hydration of Tricalcium Silicate," *Il Cemento* (1985) **4**, 203–210.

Stein, H. N.: "Influence of Some Additives on the Hydration Reactions of Portland Cement—Pt. I: Nonionic Organic Additives," *J. Appl. Chem.* (1961) **11**, 474–482.

Stein, H. N.: "Influence of Some Additives on the Hydration Reactions of Portland Cement—Pt. 2: Electrolytes," *J. Appl. Chem.* (1961) **11**, 474–482.

Sutton, D. L., Sabins, F. L., and Love, R.: "Well Cementing Methods and Compositions," European Patent Appl. No. 853,069,268 (1985).

Taplin, J. H.: "Discussion of Some Chemical Additions and Admixtures in Cement Paste and Concrete," *Proc.*, Fourth Intl. Cong. Chem. Cement, Washington, D.C. (1960) **2**, 924.

Tenoutasse, N.: "The Hydration Mechanism of C_3A and C_3S in the Presence of Calcium Chloride and Calcium Sulphate," *Proc.*, Fifth Intl. Cong. Chem. Cement, Paris (1978) Pt. 2, 372–378.

Thomas, N. L. and Birchall, J. D.: "The Retarding Action of Sugars on Cement Hydration," *Cement and Concrete Res.* (1983) **13**, 830–842.

Thomas, N. L. and Double, D. D.: "Calcium and Silicon Concentrations in Solution During the Early Hydration of Portland Cement and Tricalcium Silicate," *Cement and Concrete Res.* (1981) **11**, 675–687.

Traetteberg, A. and Grattan-Bellew, P. E.: "Hydration of 3CaO $\cdot Al_2O_3$ and 3CaO$\cdot Al_2O_3$ + Gypsum With and Without $CaCl_2$," *J. Amer. Ceramic Soc.* (1975) **58**, 221–227.

Traetteberg, A., Ramachandran, V. S., and Grattan-Bellew, P. E.: "A Study of the Microstructure of Tricalcium Silicate in the Presence of Calcium Chloride," *Cement and Concrete Res.* (1974) **4**, 203–221.

Tucker, G. R.: "Concrete and Hydraulic Cement," U.S. Patent No. 2,141,569 (1938).

Wahl., W. W. and Dever, C. D.: "Water-Loss Control of Aqueous Cement Slurries by Addition of Quaternary Ammonium Polymers or Sulfonium Polymers," U.S. Patent No. 3,094,501 (1963).

Wahl, W. W. and Dever, C. D.: "Hydraulic Cement Composition Containing a Mixture of Polymeric Additaments and Method of Cementing a Well Therewith," U.S. Patent No. 3,140,269 (1964).

Wahl, W. W., Dever, C. D., and Ryan, R. F.: "Low Water-Loss Cement Composition," U.S. Patent No. 3,086,588 (1963).

Wilkinson, W. L.: Non-Newtonian Fluids, Pergamon Press, New York (1960).

Woodard, G. W. and Merkle, G. H.: "Composition of Hydraulic Cement and Polyvinyl Acetate and Use Thereof," U.S. Patent No. 3,158, 520 (1952).

Wu, Z. Q. and Young. J. F.: "Formation of Calcium Hydroxide from Aqueous Suspensions of Tricalcium Silicate," *J. Amer. Ceramic Soc.* (1984) **67**, 48–51.

Young, J. F.: "Influence of Tricalcium Aluminate on the Hydration of Calcium Silicates," *J. Amer. Ceramic Soc.* (1969) **52**, 44–46.

Young, J. F., Berger, R. L., Lawrence, F. V. Jr.: "Studies on the Hydration of Tricalcium Silicate Pastes—Pt. 3: Influence of Admixtures on Hydration and Strength Development," *Cement and Concrete Res.* (1973) **3**, 689–700.

4

Rheology of Well Cement Slurries

Dominique Guillot

Dowell Schlumberger

4–1 INTRODUCTION

A proper understanding of cement slurry rheology is important to design, execute and evaluate a primary cementation. An adequate rheological characterization of cement slurries is necessary for many reasons, including—

- evaluation of slurry mixability and pumpability,
- determination of the pressure-vs-depth relationship during and after placement,
- calculation of the return rate when free fall is occurring,
- prediction of the temperature profile when placing cement in the hole, and
- design of the displacement rate required to achieve optimum mud removal.

Despite a great amount of research performed during the past 50 years, a complete characterization of the rheology of cement slurries has yet to be achieved. This is due to the complexity of cement slurry rheological behavior, which depends on many different factors such as—

- water-to-cement ratio,
- specific surface of the powder, and more precisely the size and the shape of cement grains,
- chemical composition of the cement and the relative distribution of the components at the surface of the grains,
- presence of additives, and
- mixing and testing procedures.

The influence of these factors on cement slurry properties is described elsewhere (Chapters 2, 3, and 5, and Appendix B). This chapter concentrates on the rheological characterization and flow behavior of cement slurries in a wellbore.

4–2 SOME RHEOLOGICAL PRINCIPLES

4–2.1 Terminology

Rheology is concerned with the flow and deformation of materials in response to applied stresses. The equations which describe the flow of any fluid are the equations of conservation of mass, momentum, and energy. They cannot be solved without assuming one or more constitutive equations which relate the deformation of the fluid (strain) to the imposed forces (stress). One such equation relates the *shear-stress* tensor τ to the *shear-rate* tensor $\dot{\gamma}$. The form of this equation for cements is the restrictive meaning given to "rheology" in the following developments.

Since the tensorial notation may not be familiar to some readers, it is worthwhile taking the example of simple shear flow for which both tensors (shear stress and shear rate) have only one nonzero component. A fluid is considered that is contained between two parallel plates, one of them moving with a velocity V (Fig. 4–1). The shear stress τ represents the force per unit area which causes the fluid to flow. In this case, a force balance shows shear stress to be uniform throughout the fluid and equal to the force per unit area necessary to move one of the plates at velocity V, while maintaining the other one in a fixed position. The field unit of stress is lbf/100 ft^2, while the SI unit is the pascal (Pa or N m^{-2}) with 1 lbf/100

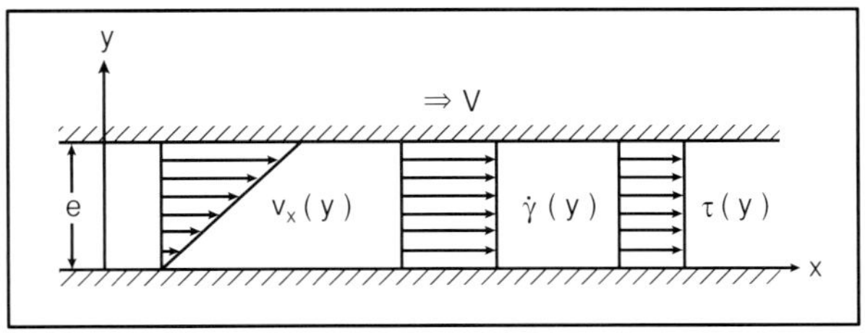

Figure 4–1—Flow between parallel plates (upper plate is moving at velocity V).

ft^2 = 0.4788 Pa. The shear rate or rate of strain $\dot{\gamma}$ is here equivalent to the velocity gradient, since

$$\frac{d}{dy}\left(\frac{dx}{dt}\right) = \frac{d}{dt}\left(\frac{dx}{dy}\right) = \frac{d\gamma}{dt}$$

where γ is the strain.

It is also uniform in this particular case and, hence, equal to the moving plate velocity V divided by the distance between the plates e. Shear rates are expressed in reciprocal seconds (s^{-1}). The force necessary to move one of the plates at a given velocity V is determined by a fluid property called its viscosity, which is defined as the ratio of the shear stress to the shear rate. Viscosities are commonly expressed in centipoises (cp), but the corresponding SI unit is Pa s with 1 cp = 1 mPa s.[1]

For flow situations more complex than the one just described, the shear-rate tensor can have several components that are nonzero. The apparent viscosity is then a scalar quantity that relates certain elements of the shear-stress tensor to those of the rate of strain tensor. When considering shearing flows of time-independent incompressible fluids, the viscosity is either a constant or depends only on a quantity called the second invariant of the shear-rate tensor. For such complex flows, the magnitude of this tensor (i.e., the square root of one-half of its second invariant) is defined as the shear rate (Bird et al., 1979).

Most fluids exhibit a shear-rate-dependent viscosity which is nontrivial to characterize. For fluids such as cement slurries, the viscosity is not only a function of the shear rate currently being applied, but also of the past shear history. They exhibit a time-dependent behavior which is even more difficult to characterize. However, for practical oilfield purposes, cement slurries are (almost) invariably represented by time-independent models.

4–2.2 Time-Independent Rheological Models

It is worthwhile to present a few examples of rheological models most widely used to describe the rheological behavior of cement slurries. These rheological models are a mathematical expression for the shear stress or the viscosity as a function of the shear rate.

Newtonian Model

In this model, the shear stress is proportional to the rate of shear; therefore, the viscosity is a constant (η) which is usually expressed in cp.

[1] Unless indicated otherwise, all equations in this chapter are expressed in SI units.

$$\eta = \frac{\tau}{\dot{\gamma}} = constant \qquad (4\text{–}1)$$

The rheogram (stress-rate vs strain-rate curve) of the fluid is a straight line of slope η passing through the origin (Fig. 4–2). To characterize the behavior of such fluids, laboratory work is minimal because, in principle, a single measurement of shear stress at one shear rate is all that is necessary. Typical Newtonian fluids used in cementing operations are water, some chemical washes, gasoline, and light oil.

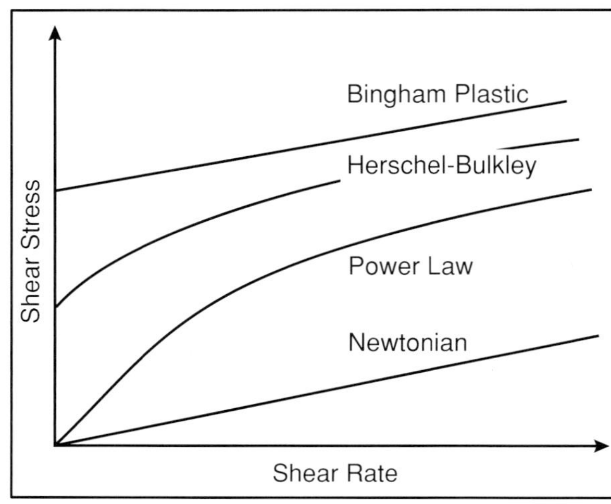

Figure 4–2—Examples of flow curves used in the petroleum industry.

Non-Newtonian Models

Most cement slurries exhibit a much more complicated non-Newtonian behavior. Generally their viscosity is a function of the shear rate, and also of the shear history as discussed later. A distinction is usually made between shear thinning fluids for which the viscosity decreases with the rate of shear, and shear thickening fluids for which the reverse is true. Generally speaking, cement slurries fall in the first category, and the most popular models describing the rheological properties of cement slurries are the *power law model* and the *Bingham plastic model*.

The equation for the power law model can be written as

$$\tau = k \times \dot{\gamma}^{n} \qquad (4\text{–}2)$$

where n, called the *Power Law Index*, is a dimensionless parameter which quantifies the degree of non-Newtonian behavior of the fluid (for shear thinning fluids, $n < 1$). The quantity k, expressed in lbf s^n/ft^2 (1 lbf s^n/ft^2 = 47.88 Pa s^n), is called the *Consistency Index* because it is proportional to the apparent viscosity of a power law fluid.

The power law relationship is represented by the curved line through the origin in Fig. 4–2. The corresponding apparent viscosity decreases with the rate of shear, from infinity at zero shear rate to zero at infinite shear rate. This is not physically sound without restriction, because there must be a limiting finite viscosity at high shear rates for any type of fluid; nevertheless, the power law model has been found to represent the behavior of many different types of fluids, including cement slurries, within a limited shear-rate range.

The Bingham plastic model is represented by the equation

$$\tau = \tau_y + \mu_p \times \dot{\gamma} \ , \qquad (4\text{–}3)$$

if $\tau \geq \tau_y$.

$$\dot{\gamma} = 0 \ , \qquad (4\text{–}4)$$

if $\tau \leq \tau_y$.

It is the simplest model describing the behavior of a special kind of fluid which does not flow unless submitted to a minimum stress, called the *yield stress* (τ_y)—a phenomenon which is very common in concentrated suspensions such as cement slurries. Yield stresses are expressed in the usual unit for stress, i.e., lbf/100 ft^2 (1 lbf/100 ft^2 = 0.4788 Pa). Above the yield stress, the Bingham plastic model assumes that the shear stress is linearly related to the shear rate (Fig. 4–2). In this case, the corresponding apparent viscosity decreases from infinity at zero shear rate to the *plastic viscosity* (μ_p) at infinite shear rate. Plastic viscosities are expressed in cp. This model suffers from serious limitations which will be discussed in detail later. Several other more realistic models used to describe the rheological properties of cement slurries include the Casson (1959), Vocadlo (Parzonka and Vocadlo, 1968)[2], and Herschel-Bulkley (1926) models which are described by Eqs. 4–5, 4–6, and 4–7, respectively.

$$\sqrt{\tau} = \sqrt{\tau_y} + \sqrt{\mu_p} \times \sqrt{\dot{\gamma}} \qquad (4\text{–}5)$$

$$\tau = \left[\tau_y^{\,1/n} + k^{\,1/n} \times \sqrt{\dot{\gamma}}\right]^n \qquad (4\text{–}6)$$

$$\tau = \tau_y + k \times \dot{\gamma}^n \qquad (4\text{–}7)$$

[2] This model is sometimes improperly attributed to Robertson and Stiff (1976).

All these models combine the concept of a yield stress τ_y with shear thinning behavior, represented by a variety of power law relationships. In these cases the rheogram is curved, but possesses a finite intercept (Fig. 4–2). Like the Bingham model, the Casson model has the advantage of possessing only two parameters; however, it is less flexible than the three-parameter models which reduce to the Bingham plastic model as n tends toward 1.

4–2.3 Time-Dependent Rheological Behavior

The rheological properties of cement slurries can be not only shear-rate dependent, but also time dependent. This can occur for two reasons. First, there are physical interactions between the cement particles in suspension which result in a loose structure whose nature determines the rheology. This structure is very sensitive to the way in which the fluid is deformed. For such materials, an equilibrium structure and a corresponding shear stress can be associated with any particular shear rate. However, the equilibrium can only be reached if the shear rate is applied for a sufficient length of time. Prior to reaching equilibrium, the structure progressively builds up or breaks down, depending on whether the previously applied shear rate was higher or lower than the current rate. This is associated with an increase or a decrease of the shear stress until an asymptotic value is reached (Fig. 4–3). This time-dependent phenomenon is called *thixotropy*. In thixotropic fluids, the process is frequently assumed to be reversible. However, this is seldom the case with cement slurries, because there is a second source of time dependency—continuous chemical reactions which modify slurry properties with time in an irreversible manner. Nevertheless, the situation is simplified somewhat during the induction period (Chapter 2), particularly for retarded cement slurries, where any time dependence is dominated by thixotropic effects.

4–2.4 Shear-Rate Ranges Encountered in a Wellbore

As explained above, the rheological behavior of cement slurries is extremely complex, and the simple models given in Section 4–2.2 are only able to describe their behavior under limited ranges of flow conditions. Therefore, before attempting to characterize and model the rheological properties of a cement slurry, it is absolutely essential to have an idea of the rate of *strain* to which it is submitted while being placed in the wellbore.

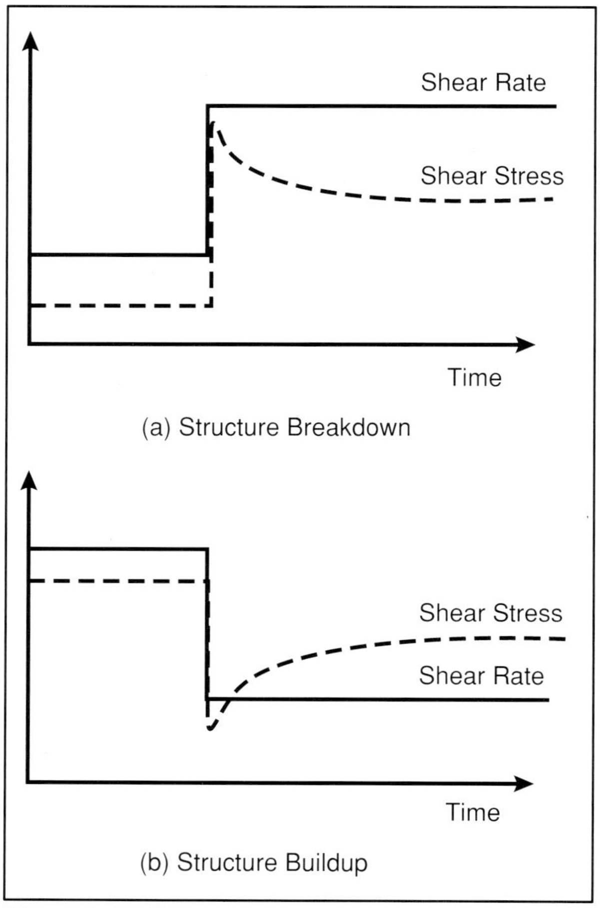

Figure 4–3—Time-dependent response of a thixotropic fluid to a step change in shear rate.

For example, the flow of a cement slurry between two concentric pipes of radii R_o and $R_i < R_o$ is considered. It is assumed that the fluid is incompressible and inelastic. Provided the flow is laminar[3], steady, and isothermal, the z component of the equation of motion along the axis of symmetry reduces to (Bird et al., 1960)

$$\frac{1}{r}\frac{d}{dr}(r\tau_{rz}) = -\frac{dP^*}{dz} \qquad (4\text{–}8)$$

where

P^* = total pressure, given by $P^* = p + \rho g_z z$,

r = radial distance from the symmetry axis such that $R_i < r < R_o$,

p = pressure due to friction,

ρ = fluid density, and

g_z = z component of gravity.

It can be integrated for any kind of fluid.

$$\tau_{rz} = -\frac{1}{2}\frac{dP^*}{dz}\left[r - \frac{(\lambda R_o)^2}{r}\right] , \qquad (4\text{–}9)$$

where

λR_o is the radial position at which $\tau_{rz} = 0$.

Since

$$\tau_{rz} = \eta(\dot{\gamma}_{rz}) \cdot \dot{\gamma}_{rz} = \eta\left(\frac{dv}{dr}\right) \cdot \frac{dv}{dr} , \qquad (4\text{–}10)$$

then

$$\eta\left(\frac{dv}{dr}\right) \cdot \frac{dv}{dr} = -\frac{1}{2}\frac{dP^*}{dz}\left(r - \frac{(\lambda R_o)^2}{r}\right) . \qquad (4\text{–}11)$$

This general expression is used for various flow situations relevant to the wellbore geometry.

4–2.4.1 Laminar Flow in a Pipe

For the particular case of a pipe of radius R, $\lambda = 0$, and using Eq. 4–9, the shear-stress profile varies linearly from zero along the symmetry axis to a maximum value at the wall τ_w.

$$\tau = \tau_{rz} = -\frac{r}{2}\frac{dP^*}{dz} = \frac{r}{R}\tau_w . \qquad (4\text{–}12)$$

Equation 4–11 reduces to

$$\eta\left(\frac{dv}{dr}\right) \cdot \frac{dv}{dr} = \frac{1}{2}\frac{dP^*}{dz}\left(r - \frac{(\lambda R_o)^2}{r}\right) . \qquad (4\text{–}13)$$

Integrating from radius r to the wall ($r = R$), and assuming the velocity at the wall to be zero, gives a general expression for the velocity at a distance r from the pipe axis.

$$v(r) = -2\frac{dP^*}{dz}\int_{\tau(r)}^{\tau_w}\frac{\tau}{\eta\left(\frac{dv}{dr}\right)}d\tau = -2\frac{dP^*}{dz}\int_{\tau(r)}^{\tau_w}\dot{\gamma}d\tau$$

$$(4\text{–}14)$$

The volumetric flow rate Q or the average velocity V (i.e., the volumetric flow rate per unit cross-sectional area) can be derived from the velocity profile through an integration by parts and rearranged to give

$$\frac{4Q}{\pi R^3} = \frac{4V}{R} = \frac{4}{\tau_w^3}\int_0^{\tau_w}\frac{\tau^3}{\eta\left(\frac{dv}{dr}\right)}d\tau = \frac{4}{\tau_w^3}\int_0^{\tau_w}\tau^2\dot{\gamma}(\tau)d\tau .$$

$$(4\text{–}15)$$

[3] Laminar flow is discussed in detail in Section 4–6. For the time being, the fluid particles are assumed to flow along streamlines which are parallel to the main direction of flow.

A particularly useful form of Eq. 4–15 gives the expression for the shear rate at the wall $\dot{\gamma}_w$

$$\dot{\gamma}_w = \frac{3n' + 1}{4n'} \times \frac{4V}{R} , \qquad (4\text{–}16)$$

where

$$n' = \frac{d \log (\tau_w)}{d \log (4V/R)} . \qquad (4\text{–}17)$$

4–2.4.2 Laminar Flow in a Narrow Concentric Annulus

In the case of axial annular flow, there is no general expression for the velocity profile and the volume flux. However, for most cementing applications, the annular gap (R_o–R_i) is sufficiently small compared to the wellbore radius R_o that one can assume the annulus to be a rectangular slot with a width and thickness of $w = \pi(R_o + R_i)$, and $e = (R_o - R_i)$, respectively (Section 4–6.4). Expressions for the shear-stress profile, velocity profile and volume flux can be easily derived in the same way as for a pipe with r now being the distance from the plane of symmetry of the slot.

$$\tau = \tau_{rz} = -\frac{r}{2}\frac{dP^*}{dz} = \frac{r}{2}\tau_w \qquad (4\text{–}18)$$

$$v(r) = -\frac{dP^*}{dz}\int_{\tau(r)}^{\tau_w} \frac{\tau}{\eta\left(\frac{dv}{dr}\right)} d\tau = -\frac{dP^*}{dz}\int_{\tau(r)}^{\tau_w} \dot{\gamma} d\tau \qquad (4\text{–}19)$$

$$\frac{6Q}{we^2} = \frac{6V}{e} = \frac{3}{\tau_w^2}\int_0^{\tau_w} \frac{\tau^2}{\eta\left(\frac{dv}{dr}\right)} d\tau = \frac{3}{\tau_w^2}\int_0^{\tau_w} \tau\dot{\gamma}(\tau) d\tau \qquad (4\text{–}20)$$

$$\dot{\gamma}_w = \frac{2n' + 1}{3n'} \times \frac{6V}{e} \qquad (4\text{–}21)$$

$$n' = \frac{d\log(\tau_w)}{d\log(6V/e)} \qquad (4\text{–}22)$$

For fluids exhibiting a yield stress τ_y, the lower limit of the integral in Eqs. 4–15 and 4–20 should be replaced by τ_y. The same modification applies to Eqs. 4–14 and 4–19, if $\tau(r) \leq \tau_y$.

4–2.4.3 Shear-Rate/Shear-Stress Range in a Pipe or Narrow Concentric Annulus

As can be seen from Eqs. 4–12 and 4–18, the shear-stress profiles in pipes and narrow annuli are well defined, whatever the rheological properties of the fluid; however, they are dependent upon the friction pressure (Eq. 4–9), a quantity which is usually unknown.

On the other hand, the shear rate varies from zero at the pipe axis or on the plane of symmetry of the annulus, to a maximum value $\dot{\gamma}_w$ at the wall, with a radial variation which depends on the non-Newtonian behavior of the fluid, characterized by the value of n' (Eqs. 4–16 and 4–17 for pipes, and 4–21 and 4–22 for narrow annuli). It is only for Newtonian fluids ($n' = 1$) and for power law fluids ($n' = n$ = constant), that this parameter is constant (independent of V or $\dot{\gamma}_w$). In such cases, the value of the shear rate at the wall can be derived from the average velocity and the dimensions of the flow path. The shear rate at the wall for Newtonian fluids, which is

$$\dot{\gamma}_{Nw} = \frac{4V}{R} \qquad (4\text{–}23)$$

for pipes, and

$$\dot{\gamma}_{Nw} = \frac{6V}{(R_o - R_i)} \qquad (4\text{–}24)$$

for narrow concentric annuli, represents a lower limit for the shear rate at the wall for non-Newtonian fluids, provided they are shear thinning (i.e., $n' < 1$, which is the case of most cement slurries).

$$\dot{\gamma}_{Nw} \leq \dot{\gamma}_w \qquad (4\text{–}25)$$

In fact, experience shows that for most cement slurries, n′ is usually greater than 0.1, e.g.,

$$\dot{\gamma}_w \leq 3.25 \times \dot{\gamma}_{Nw} \qquad (4\text{–}26)$$

in pipes, and

$$\dot{\gamma}_w \leq 4 \times \dot{\gamma}_{Nw} \qquad (4\text{–}27)$$

in narrow annuli.

Thus, the shear rate at the wall $\dot{\gamma}_w$ for non-Newtonian fluids is not very well defined unless the precise rheology of the fluid is known. It is always worthwhile to calculate the value which a Newtonian fluid would experience in a given application. Some typical figures for $\dot{\gamma}_{Nw}$ are given in Table 4–1.

As can be expected from Eqs. 4–16 and 4–21, the Newtonian shear rate at the wall is extremely sensitive to the pipe diameter or annular size and, therefore, may vary significantly from one case to another. Generally speaking, the variations in the true shear rate at the wall due to variations in hole geometry may be greater than those due to variations in n' (i.e., in the non-Newtonian behavior of the fluids).

As stated earlier, the shear rate is not uniform across the gap in either of these geometries. Therefore, theoretically speaking, solving Eqs. 4–15 and 4–20 requires a knowledge of the shear-stress/shear-rate relationship in the range from the shear rate at the wall down to zero shear rate. In fact, these equations are such that volume fluxes depend mainly on the local shear-stress/shear-rate relationship in a region just below τ_w or $\dot{\gamma}_w$. This is also broadly the case for velocity profiles.

When dealing with time-dependent fluids, the problem is relatively more complex. Not only is the shear rate nonuniform in these two geometries, but also the time during which a given shear rate is applied needs to be considered. Thus, for example, in perfect laminar flow, fluid particles flowing at different radial positions relative to the pipe axis or within an annulus experience widely different shear histories. A particle on or near the pipe axis experiences a low shear rate for a relatively short time, while a particle near the wall sees a high shear rate for a relatively long time.

4–3 EQUIPMENT AND EXPERIMENTAL PROCEDURES

4–3.1 Coaxial Cylinder Viscometers

This geometry is the basis for the standard API specifications for the rheological evaluation of oilfield fluids.

4–3.1.1 Principle and Flow Equations

The test material is confined between two concentric cylinders of radii R_2 and R_1 ($R_2 > R_1$), one of which is rotated at a velocity Ω. It will be assumed for the time being that

	Shear Rate (s⁻¹) Flow Rates			
Pipe Diameter (in.)	1 159	2 318	5 795	10 (BPM) 1590 (L/min)
2	206	412	1030	2060
5	13.2	26.4	65.9	132
10	1.65	3.29	8.23	16.5
	Shear Rate (s⁻¹) Flow Rates			
Annular Size (ID–OD) (in.)	1 159	2 318	5 795	10 (BPM) 1590 (L/min)
4–5½	116	231	578	1160
7–8½	70.8	142	354	708
9⅝–12¼	16.4	32.8	81.9	164
13⅜–17½	4.60	9.19	23.0	46.0

Table 4–1—Newtonian shear rates for various pipe diameters, annular geometries, and flow rates.

fluid elements are moving in concentric circles around the common axis (Fig. 4–4). In steady state, a momentum balance shows that the shear stress τ at any radius r is given by (Whorlow, 1980, p. 116)

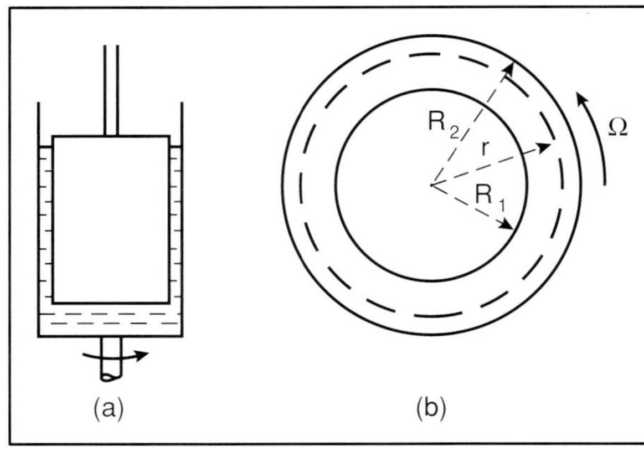

Figure 4–4—Schematic representation of a coaxial cylinder viscometer, (a) vertical section (b) horizontal section (after Whorlow, 1980).

$$\tau = \frac{T}{2\pi r^2} \qquad (4\text{–}28)$$

where T is the torque acting per unit length on a cylindrical surface of any radius r. In practice, T is measured from the torque acting on the static cylinder of length L. This expression shows that the shear stress decreases from a maximum value $\tau_1 = T/2\pi R_1^2$ at the inner cylinder surface to $\tau_2 = T/2\pi R_2^2$ at the outer cylinder surface. Shear stress (and therefore the shear rate) will be uniform only if the radius ratio $s = R_2/R_1$ is close to unity. It is important to point out that the more shear thinning the fluid, the more drastic must be the condition on the radius ratio, because the shear-rate range corresponding to a given shear-stress range is increasingly wider.

The governing flow equation in a coaxial cylinder viscometer is (Whorlow, 1980)

$$\Omega = \frac{1}{2}\int_{\tau_2}^{\tau_1} \frac{\dot{\gamma}}{\tau}\, d\tau \qquad (4\text{–}29)$$

Since both limits of the integral are functions of the torque, there is no general analytical expression for the shear rate and the viscosity of a non-Newtonian fluid flowing in such a geometry. Therefore, the shear-rate profile cannot be determined a priori, because it depends on the precise non-Newtonian behavior of the fluid, as well as on the rotational speed and the dimensions of the geometry. To use such equipment to measure the flow curve for a non-Newtonian fluid, it is necessary to either assume a specific rheological model to use in conjunction with Eq. 4–29, or to make R_2/R_1 sufficiently close to unity that the variations of shear stress across the gap are negligible.

In many ways, the situation is similar to that described for pipe flow or annular flow, but a major difference exists between these geometries. In pipes and annuli, the minimum shear rate is always zero. In coaxial cylinder viscometers, it is always nonzero, except under specific circumstances such as when the fluid exhibits a yield stress. In this case, if the rotational speed is sufficiently low such that

$$\tau_2 \leq \tau_y \leq \tau_1 \ , \tag{4–30}$$

i.e., if on a cylindrical surface of radius r ($R_1 \leq r \leq R_2$) the shear stress is smaller than the yield stress of the fluid, then the effective annular gap is reduced. Since the rate of shear is zero from R_2 to r_y, this parameter is defined by

$$r_y = \sqrt{\frac{T}{2\pi\tau_y}} \ . \tag{4–31}$$

Equation 4–29 then becomes

$$\Omega = \frac{1}{2}\int_{\tau_y}^{\tau_1} \frac{\dot{\gamma}}{\tau} d\tau \ . \tag{4–32}$$

When the condition of Eq. 4–30 is satisfied, the flow regime is sometimes called *plug flow*, because part of the velocity profile is flat and the material between R_2 and r_y moves as a plug.

4–3.1.2 Validity of Equations for Coaxial Cylinder Viscometers

End Effects

In the equations developed in Section 4–3.1.1, the torque per unit length of any cylindrical surface of radius r was assumed to be known. However, since coaxial cylinder viscometers have a finite length, the shear flow in the annular gap which determines the measured torque is not homogeneous. The flow pattern is significantly modified close to the top and the bottom of the gap. In addition, the fluid which may be present and which is sheared above and below the inner cylinder also contributes to the measured torque. Very often, end effects of this kind are assumed to be proportional to the undisturbed stress, and an extra cylinder length or a torque correction factor allows them to be taken into account. This factor is usually measured for Newtonian fluids, and applied to all fluids without regard to which rheological model is most appropriate. A more reliable procedure consists of performing the measurements with different levels of fluid in the gap. For each rotational speed, the measured torque is a linear function of the fluid height in the gap, and the slope is the torque per unit length. Since this procedure is quite cumbersome, some geometries have been specifically designed to minimize end effects (Fig. 4–5).

Annular Gap Size

The flow equations in Section 4–3.1.1 also assume the fluid to be homogeneous in the annular gap. Since cement slurries are concentrated suspensions, they can only be considered homogeneous if the annular gap size is at least 10 times the size of the largest particles. In view of the particle-size distribution of oil-well cement powder, the gap size should be approximately 1 mm. Strictly, what should be considered is the size of particle aggregates, a quantity which is much more difficult to determine. In the absence of quantitative information, rheological measurements should be performed with different gap sizes. If the experimental data are dependent upon the gap size, the homogeneity of the fluid is questionable.

Departure From Circular Streamlines

Above a given rotational velocity (depending upon the fluid characteristics), the particles no longer move in concentric circles about the axis of rotation of the equipment, and the flow becomes too complex to permit the rheological characterization of the fluid. For cement slurries, this may only be a problem in equipment where the inner cylinder rotates. In such cases, the rotational velocity should be smaller than a critical value which, for Newtonian fluids, is given by Taylor (1923) as

$$\Omega < 41.3 \times \frac{R_2^{1/2}}{R_1(R_2 - R_1)^{3/2}} \times \frac{\eta}{\rho} \ . \tag{4–33}$$

For non-Newtonian fluids, an estimate of the critical velocity can be obtained using Eq. 4–33, but with an apparent viscosity corresponding to the appropriate shear rate. This procedure can lead to large errors if the fluid possesses elastic as well as viscous characteristics (Bird et al., 1979), but such effects are unlikely to be significant for most cement slurries.

4–3.1.3 Flow of Model Fluids in Coaxial Cylinder Viscometers

When a rheological model is assumed for the fluid to be characterized, a simple analytical expression can sometimes be determined for the torque as a function of the rotational speed.

For a Newtonian fluid, the flow equation is

$$\frac{T}{2\pi R_1^2} = \eta \times \frac{2s^2\Omega}{s^2 - 1} \tag{4–34}$$

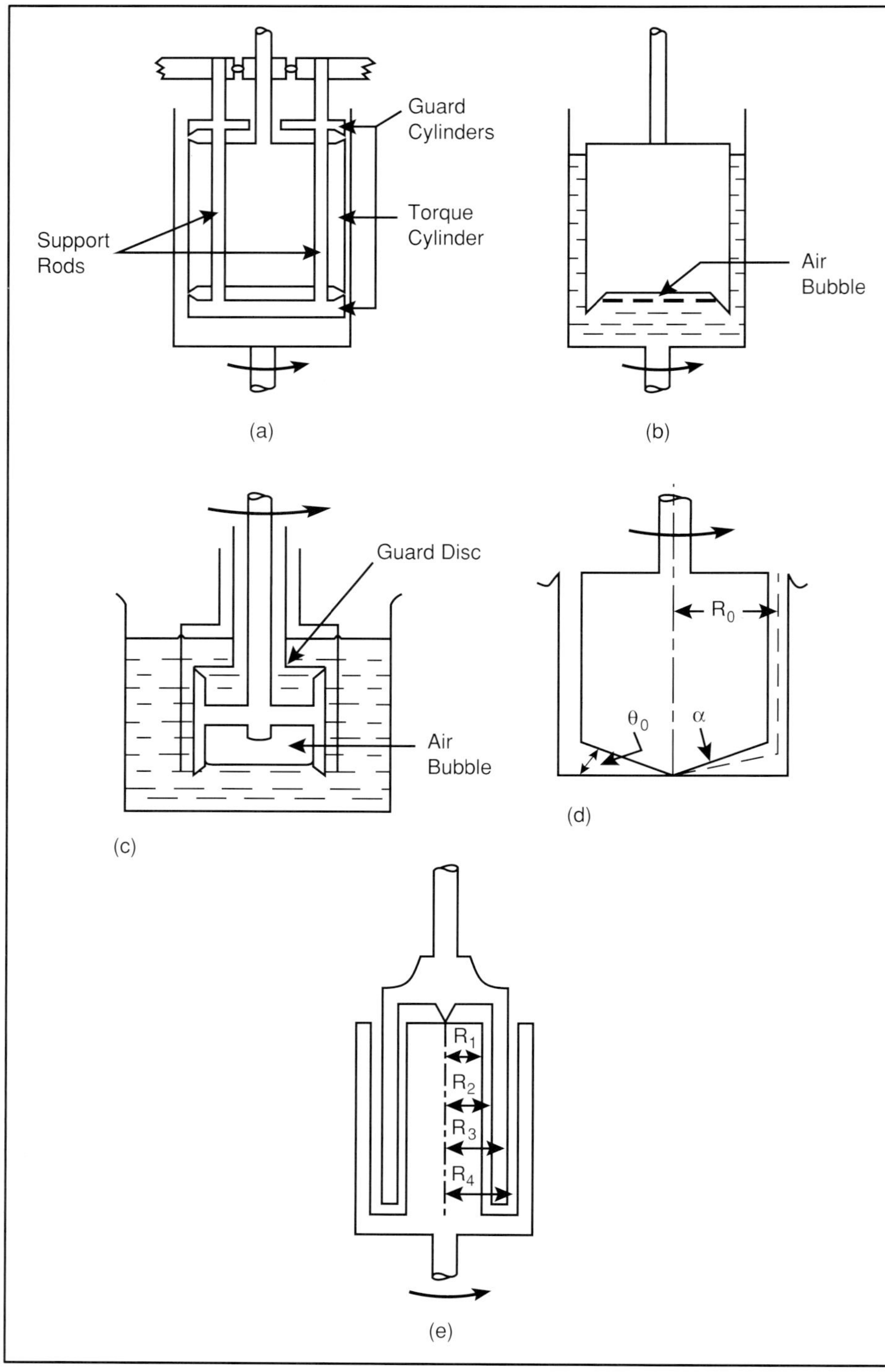

Figure 4–5—Methods for eliminating end effects. (a) guard cylinders, (b) trapped air bubble, (c) Ferranti portable viscometer, (d) Mooney-Ewart viscometer, (e) Moore-Davies double viscometer (after Whorlow, 1980).

and the shear rate at the inner and outer surfaces are, respectively,

$$\dot{\gamma}_1 = \frac{2s^2\Omega}{s^2 - 1} \quad (4\text{–}35)$$

and

$$\dot{\gamma}_2 = \frac{2\Omega}{s^2 - 1} \ , \quad (4\text{–}36)$$

where

$s = R_2/R_1$.

For a power law fluid, the corresponding equations are

$$\frac{T}{2\pi R_1^2} = k \times \left(\frac{2s^{2/n}\Omega}{n(s^{2/n} - 1)}\right)^n \ , \quad (4\text{–}37)$$

$$\dot{\gamma}_1 = \frac{2s^{2/n}\Omega}{n(s^{2/n} - 1)} \ , \quad (4\text{–}38)$$

and

$$\dot{\gamma}_2 = \frac{2\Omega}{n(s^{2/n} - 1)} \ . \quad (4\text{–}39)$$

For a Bingham plastic fluid, different equations apply depending on the torque value. If $T > 2\pi R_2^2\tau_y$, then all the fluid in the gap is in laminar shear flow and the governing equations are

$$\frac{T}{2\pi R_1^2} = \frac{2s^2}{s^2 - 1} \times \left[\mu_p\Omega + \tau_y \ln(s)\right] \ , \quad (4\text{–}40)$$

$$\dot{\gamma}_1 = \frac{2s^2\Omega}{s^2 - 1} + \left[\frac{2s^2\ln(s)}{s^2 - 1} - 1\right] \times \frac{\tau_y}{\mu_p} \ , \quad (4\text{–}41)$$

and

$$\dot{\gamma}_2 = \frac{2\Omega}{s^2 - 1} + \left[\frac{2\ln(s)}{s^2 - 1} - 1\right] \times \frac{\tau_y}{\mu_p} \ . \quad (4\text{–}42)$$

If $2\pi R_1^2\tau_y < T < 2\pi R_2^2\tau_y$, part of the fluid is in plug flow and expressions for $\dot{\gamma}_1$ and $\dot{\gamma}_2$ are implicit.

$$\Omega = \frac{T}{2\pi R_1^2\mu_p} \times \left[1 - \frac{2\pi R_1^2\tau_y}{T}\right] - \frac{\tau_y}{2\mu_p} \times \ln\left[\frac{T}{2\pi R_1^2\tau_y}\right] \quad (4\text{–}43)$$

If $T < 2\pi R_1^2\tau_y$, then none of the fluid can flow and

$$\Omega = 0 \ . \quad (4\text{–}44)$$

Thus, for a Newtonian fluid, there is a linear relationship between the torque and the rotational speed. For a power law fluid, there is a power law relationship between the two for all cylinder sizes. For Bingham plastic fluids, as for all fluids exhibiting a yield stress, the equations are more complex. In the absence of a plug flow region, there is a linear relationship between the torque and the rotational speed, with an apparent intercept equal to

$$T = \tau_y \frac{4\pi R_1^2 R_2^2}{R_2^2 - R_1^2} \ln(R_2 / R_1) \ .$$

Below a given torque value $T = 2\pi R_2^2\tau_y$, the relationship becomes independent of the outer radius R_2, and nonlinear with an intercept $T = 2\pi R_1^2\tau_y$ for $\Omega = 0$ (Fig. 4–6).

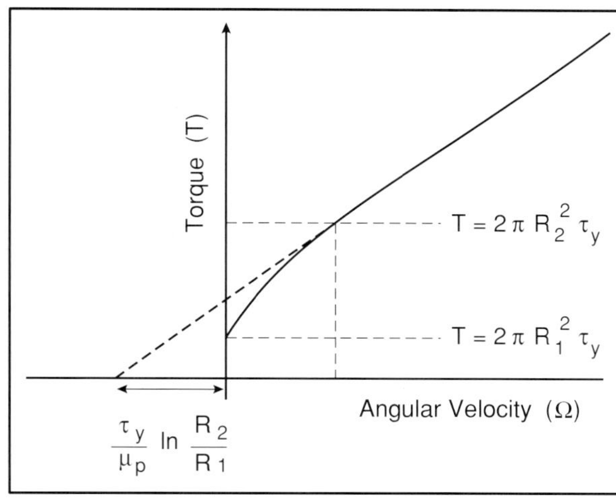

Figure 4–6—Torque/angular velocity graph for a Bingham plastic fluid in a coaxial cylinder viscometer.

Therefore, deriving the rheological parameters for the Newtonian and the power law models from a series of torque/rotational speed measurements is straightforward. However, this is not the case for the Bingham plastic model, and for fluids exhibiting a yield stress in general. Indeed, the flow behavior is described by Eqs. 4–40 and 4–43, whose limit of validity depends on one of the parameters which it is desired to measure—the yield stress. This problem is usually overlooked, and all data are fitted according to the linear equation (Eq. 4–40).

4–3.1.4 Narrow Gap Approximation

When the radius ratio of the cylinders is close to one, the shear stress and the shear rate can be considered as uniform in the annular gap, and given by

$$\tau_a = \frac{T}{2\pi R_a^2} \ , \quad (4\text{–}45)$$

and

$$\dot{\gamma}_a = \frac{R_a\,\Omega}{R_2 - R_1}, \qquad (4\text{–}46)$$

where

$$R_a = \frac{R_2 + R_1}{2}. \qquad (4\text{–}47)$$

Therefore, values for the shear stress and the shear rate can be derived directly from the torques and the rotational speeds. The errors resulting from using this approximation can easily be determined. For power law fluids,

$$\frac{k_{narrow}}{k} = \frac{(s+1)^2}{4} \times \left[\frac{4}{n} \times \frac{s-1}{s+1} \times \frac{s^{2/n}}{s^{2/n}-1}\right]^n. \qquad (4\text{–}48)$$

For Bingham plastic fluids,

$$\frac{\mu_{p-narrow}}{\mu_p} = \frac{(s+1)^4}{16 s^2} \qquad (4\text{–}49)$$

and

$$\frac{\tau_{y-narrow}}{\tau_y} = \frac{8s^2\,ln(s)}{(s-1)(s+1)^3}. \qquad (4\text{–}50)$$

When this approximation holds it presents a major advantage, because calculating the integral in Eq. 4–29 or 4–32 would no longer be necessary. Shear-stress and shear-rate values can be derived directly from the characteristics of the geometry, and from the torque/rotational speed values.

4–3.1.5 More General Analyses

For situations where the narrow gap approximation does not hold, several methods have been developed to calculate the shear stress and the corresponding shear-rate values in the gap, without assuming a rheological model (Whorlow, 1980). Solutions have been obtained in the form of a series, but all require the determination of at least the first-order derivative of the experimental curve (Ω, T). Therefore, these methods can only be applied with caution because they suppose that—

- the linearity of the torque measuring device is excellent,
- the spacing of the (Ω,T) data in a given shear-rate range is sufficiently close for accurate definition of the slope,
- the reproducibility of the results is excellent, and
- the torque at a given rotational speed is time independent.

Unfortunately, these conditions are almost never met simultaneously when characterizing cement slurries.

4–3.1.6 Standard Oilfield Equipment and Procedures

Equipment

The standard equipment used to characterize the rheological properties of cement slurries and other oilfield fluids (drilling muds, spacers, fracturing fluids, etc.) is a coaxial cylinder viscometer, the main features of which were defined by Savins and Roper in 1954. The fluid, contained in a large cup, is sheared between an outer sleeve (the rotor) and an inner cylinder (the bob), which is attached to a torque measuring device (Fig. 4–7). The characteristics of the geometry are

R_2 = 0.725 in. (1.842 cm),

R_1 = 0.679 in.(1.725 cm), and

L = 1.5 in. (3.8 cm).

Depending upon the particular model, the outer sleeve can be rotated at two (600 and 300 RPM), six (600, 300, 200, 100, 6, and 3 RPM), or more (previous values plus possibly 60, 30, 20, 10, 6, 3, 2, and 1 RPM) rotational speeds. This covers a shear-rate range from at least 5 s^{-1} to 1,022 s^{-1} (these values are calculated using the Newtonian shear-rate formula at the inner cylinder surface). The six-speed models are the most commonly used in the oil industry. The torque is measured from the deflection of a torsional spring indicated on a scale reading in degrees. The standard torsional spring has a nominal range from zero to 0.117 N-m, which corresponds to a shear-stress range from 0 to 153 Pa (calculated at the inner cylinder surface). Most manufacturers provide other springs with stiffnesses of one-fifth, one-half, two, or five times that of the standard spring.

Before discussing the experimental procedures in detail, and the equations which are used to treat the data, it is worthwhile to mention that, when well maintained, the accuracy of the torque measuring device of most standard oilfield equipment is reasonable. Once calibrated, a typical error to be expected for a shear stress of 5 Pa (i.e., a reading of 10 degrees with the standard spring) is of the order of $\pm$15%. Nevertheless, this figure is much higher if the bearing spring supporting the inner cylinder shaft is damaged, and it is not unusual to encounter equipment for which the relative error is of the order of $\pm$50% at such low shear stresses (Fig. 4–8). This creates problems

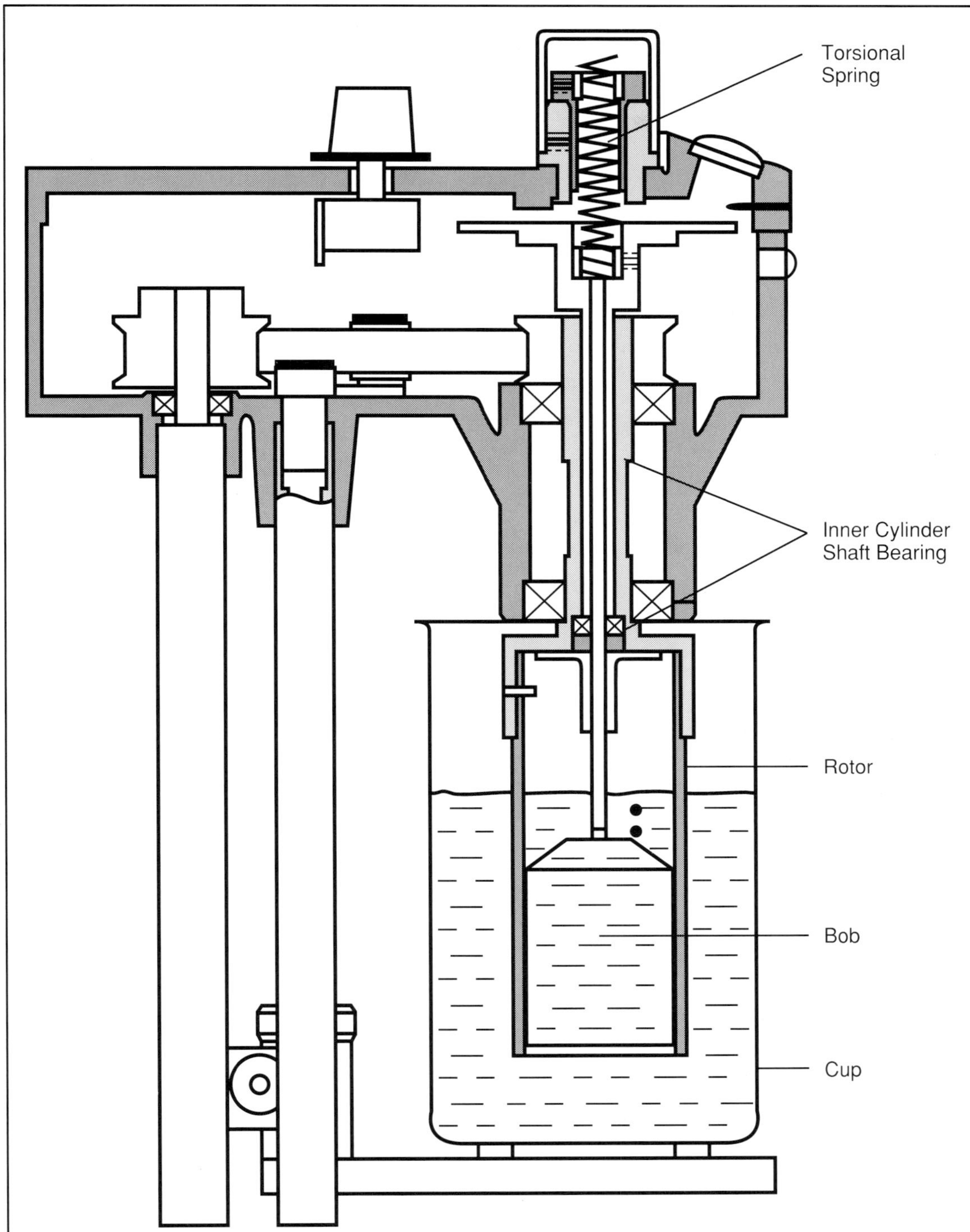

Figure 4–7—Schematic diagram of a couette-type coaxial cylinder viscometer (drawing courtesy EG&G Chandler Engineering).

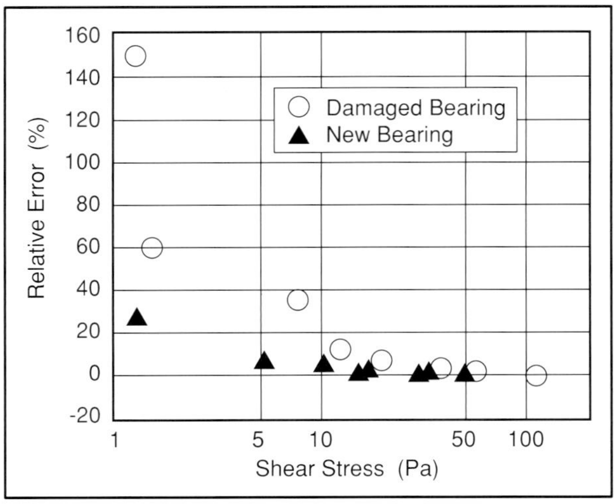

Figure 4–8—Relative error of shear-stress measurements using standard oilfield equipment (test performed with a Newtonian oil using the standard API procedure).

when trying to characterize the rheology of low-viscosity fluids, such as dispersed slurries.

Experimental Procedure

The experimental procedure (described in API Spec 10 [1988]) consists of shearing the fluid at the highest rotational speed for one minute before recording the corresponding torque reading. The rotational speed is then decreased step by step to the minimum shear rate, and the corresponding torque readings are recorded after 20 s of rotation at each rotational velocity.

The top rotational speed recommended in API Spec 10 has been reduced from 600 to 300 RPM (1,022 s^{-1} to 511 s^{-1}) in view of a comparative study performed among several laboratories. The repeatability of results was found to be greatly improved by limiting the maximum rotational speed to 300 RPM (Figs. 4–9 and 4–10) (Beirute, 1986). Unfortunately, this new procedure is not yet applied by all users. This creates confusion, because the measurements are often dependent on the procedure.

Since API Spec 10 now recommends against the use of the 600-RPM speed, the standard two-speed equipment should no longer be used. The six-speed models also suffer from a severe limitation. Since the 6- and 3-RPM readings are not very accurate, or are affected by slippage at the wall (Section 4–4.1.3), the user is left with three useful readings at 100, 200, and 300 RPM. These rotational speeds correspond to a fairly narrow shear-rate range (170 s^{-1} to 511 s^{-1}). Therefore, when the maximum shear rate experienced by a cement slurry while being placed in the wellbore is likely to be lower than 170 s^{-1}, the use of equipment allowing measurements between 6 and 100 RPM (10 s^{-1} and 170 s^{-1}) is strongly recommended.

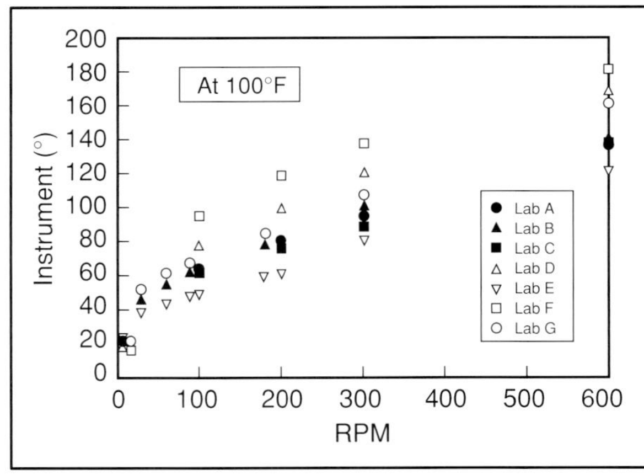

Figure 4–9—Poor repeatability of rheological data measured by several laboratories using the same cement materials, mixing method, and test procedure involving 600-RPM reading (Class H cement + 38% water BWOC) (after Beirute, 1986).

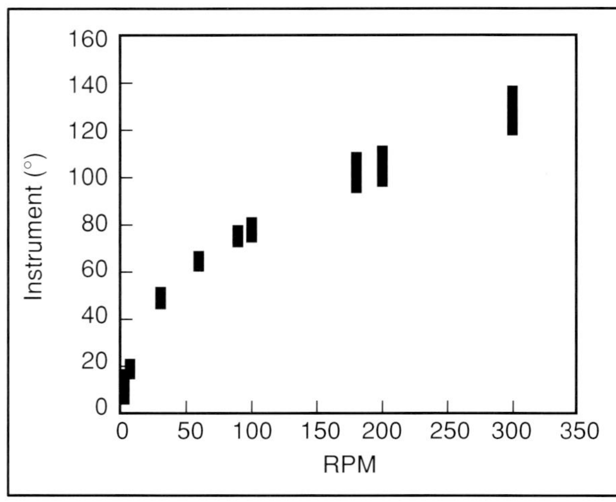

Figure 4–10—Improvement of repeatability of rheological data as a result of limiting maximum rotational speed to 300 RPM (compare with Fig. 4–9) (Class H cement + 38% water BWOC) (after Beirute, 1986).

Data Analysis

Earlier, it was stressed that the formula giving the shear rate at the inner cylinder surface for a Newtonian fluid (Eq. 4–35) is valid only for a Newtonian fluid. Therefore, the recommended API procedure (which consists of converting rotational speeds to Newtonian shear rates at the inner cylindrical surface) is often not correct. It leads to an overestimation of the Consistency Index for power law fluids and of the yield stress for Bingham plastic fluids (not taking into account the plug flow region). The expressions are given by Eqs. 4–51 and 4–52, respectively.

$$\frac{k_{API}}{k} = \left[\frac{s^{2/n}}{ns^2} \times \frac{s^2 - 1}{s^{2/n} - 1}\right]^n \qquad (4\text{–}51)$$

$$\frac{\tau_{y-API}}{\tau_y} = \frac{2s^2}{s^2 - 1} \ln(s) \qquad (4\text{–}52)$$

The corresponding errors for the standard geometry used in the oil industry ($s = 1.068$) range from 0.0% to 6.7% for the Consistency Index of power law fluids, when the Power Law Index varies from zero to one. For Bingham plastic fluids, the error is zero for the plastic viscosity and 6.7% for the yield stress. One may consider these errors as being negligible for practical purposes; however, since there is a risk that the same approach may be used with other geometries exhibiting a much higher radius ratio, a better recommendation is to use the exact equations (Eqs. 4–37 to 4–40) which are no more complicated.

As mentioned earlier, another possibility when using the standard oilfield geometry is to adopt the narrow gap approximation (Eqs. 4–45 to 4–47 in Section 4–3.1.4), which gives the following.

$$R_a = 0.70 \text{ in. } (1.78 \text{ cm})$$

$$\dot{\gamma}_a\,(\text{s}^{-1}) = 15.2 \times \Omega\,(\text{rad s}^{-1}), \text{ or}$$
$$\dot{\gamma}_a(\text{s}^{-1}) = 1.60 \times \Omega\,(\text{RPM})$$

$$\tau_a\,(\text{Pa}) = 0.477 \times \theta$$
(reading with standard spring 1)

$$\tau_a\,(\text{lbf/100 ft}^2) = 0.996 \times \theta$$
(reading with standard spring 1)

With the standard oilfield geometry, this leads to an overestimation (Eqs. 4–48 to 4–50) of 0.2% for the plastic viscosity, and an underestimation of 0.8% for the yield stress. For power law fluids, the errors are of the same order of magnitude, i.e., negligible. It can be shown that this is true for other rheological models that are used to describe the behavior of cement slurries (Casson, Vocadlo, Herschel-Bulkley, etc.). Therefore, as suggested by Mannheimer (1982), the expressions for the shear rate and the shear stress recommended in API Spec 10 could advantageously be replaced by the expressions derived from the narrow gap approximation for the standard oilfield geometry.

4–3.2 Pipe and Slit Viscometers

4–3.2.1 Principle and Flow Equations

Pipe or slit viscometers can seem attractive for characterizing the rheological properties of cement slurries, because the shear history in such equipment matches that which the test fluid experiences in a cylindrical string or a narrow annulus. The fluid is usually pumped in the flow geometry, and the corresponding friction pressure drop across the device is measured. From the flow equations developed earlier (Eqs. 4–15 and 4–20), one can see that when the fluid flows in a pipe or a slit, it is not necessary to determine the true rheogram for the fluid (i.e., the shear-rate/shear-stress relationship). The Newtonian shear rate ($\dot{\gamma}_{Nw}$) vs shear stress (τ_w) relationship at the wall is independent of the pipe or slit size and, therefore, can be used to predict the flow-rate/friction-pressure relationship in laminar flow for any size, provided this is performed over the same Newtonian shear-rate range. However, this is not always possible to achieve for cementing applications; generally speaking, one must have access to the true shear-rate/shear-stress relationship. Two procedures can be used depending on whether or not a rheological model is assumed for the fluid to be characterized. If no model is assumed, the Newtonian shear rate at the wall must be converted to the true shear rate at the wall using Eqs. 4–16, 4–17, 4–21, and 4–22. This necessitates calculating the derivative of the (Q, dp/dz) flow curve. If a rheological model is assumed for $\dot{\gamma}(\tau)$, Eqs. 4–15 and 4–20 can be integrated sometimes analytically or alternatively using numerical procedures.

4–3.2.2 Validity of Pipe and Slit Viscometer Equations

In the equations just developed, it has been assumed that the flow is fully established; in other words, the flow is not affected by the proximity of the entrance or the outlet of the geometry. Since pipe or slit viscometers are not very often used to characterize cement slurries, the reader is referred to the texts by Walters (1975) and Whorlow (1980) for further details concerning these end effects. The validity of the equations is also limited to laminar flow, which is discussed in detail later in Sections 4–6.2 and 4–6.4.

4–3.2.3 Fluid Flow in Pipe or Slit Viscometers

In this section, specific rheological models are inserted into the equations of Sections 4–1.4.1 and 4–1.4.2 to give explicit relationships between the frictional pressure drop and the fluid flow rate (Walters, 1975; Whorlow, 1980). For a Newtonian fluid, the pipe-flow equation is

$$\frac{dp}{dz} = \frac{128\eta Q}{\pi D^4}\ , \qquad (4\text{–}53)$$

where

Q is the volumetric flow rate = $\pi R^2 V$.

For a power law fluid, the corresponding equation is

$$\frac{dp}{dz} = \frac{4k}{d^{3n+1}} \times \left[\frac{3n+1}{4n} \cdot \frac{8Q}{\pi}\right]^n . \qquad (4\text{–}54)$$

Notice that in the oil industry, reference is often made to a *Pipe Consistency Index* k' which is defined as

$$k' = \left[\frac{3n+1}{4n}\right]^n \times k\ . \qquad (4\text{–}55)$$

For Bingham plastic fluids, the flow equation is implicit in flow rate.

$$\xi = \frac{32Q}{\pi D^3} \times \frac{\mu_p}{\tau_y} = \frac{1}{\psi} \times \left[1 + \frac{4}{3}\psi + \frac{1}{3}\psi^4\right]\ , \qquad (4\text{–}56)$$

where

ψ = (τ_y/τ_w) is the inverse of a dimensionless shear stress and

ξ is a dimensionless shear rate which Eqs. 4–23 and 4–24 show to be $\dot{\gamma}_{Nw} \times (\mu_p/\tau_y)$.

The corresponding equations for a slit of width w and thickness e are the following.

$$\frac{dp}{dz} = \frac{12\eta Q}{we^3} \qquad (4\text{–}57)$$

$$\frac{dp}{dz} = \frac{2k}{e^{2n+1}} \times \left[\frac{2n+1}{n} \times \frac{2Q}{w}\right]^n \qquad (4\text{–}58)$$

$$\xi = \frac{6Q}{we^3} \times \frac{\mu_p}{\tau_y} = \frac{1}{\psi} \times \left[1 - \frac{3}{2}\psi + \frac{1}{2}\psi^3\right] \qquad (4\text{–}59)$$

Thus, for Newtonian fluids, there is a linear relationship between flow rate and friction pressure. For power law fluids, there is a power law relationship between the two fluids. For Bingham plastic fluids, the relationship is nonlinear, with an intercept proportional to τ_y (Fig. 4–11). The last term of Eqs. 4–56 and 4–59 can sometimes be neglected, and the equations are then explicit in flow rate.

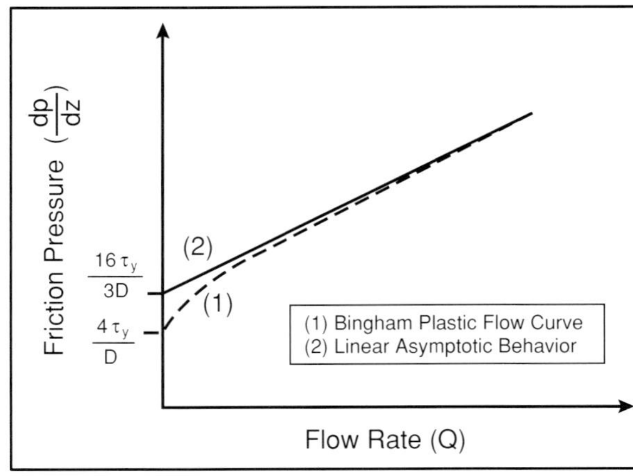

Figure 4–11—Flow curve of a Bingham plastic fluid in a pipe.

$$\frac{dp}{dz} = \frac{128\mu_p Q}{\pi D^4} + \frac{16}{3} \times \frac{\tau_y}{D} \qquad (4\text{–}60)$$

$$\frac{dp}{dz} = \frac{12\mu_p Q}{we^3} + \frac{3\tau_y}{e} \qquad (4\text{–}61)$$

This can be done provided the dimensionless shear rate ξ is sufficiently large. For example, if

$$\frac{32Q}{\pi D^3} \times \frac{\mu_p}{\tau_y} > 2.95$$

for pipe flow, or

$$\frac{6Q}{we^3} \times \frac{\mu_p}{\tau_y} > 6.45$$

for annular flow, calculating the friction pressures from Eqs. 4–60 and 4–61 will induce a relative error of less than 0.1%.

Equations describing the laminar flow of Bingham plastic fluids in pipes and annuli are often expressed in terms of other dimensionless parameters (i.e., the Hedstrom number He and the Bingham Reynolds number Re_{BG}). From the definitions of these parameters, which are given in Appendix A together with the corresponding flow equations, one can see that the dimensionless shear rate ξ is such that $\xi = 8\,Re_{BG}/He$ in pipes and $\xi = 12\,Re_{BG}/He$ in annuli. Therefore, when compared to the Bingham Reynolds number, the higher the Hedstrom number the less Newtonian is the behavior of the fluid.

4–3.3 Other Viscometers

A number of other rheological techniques are available to characterize the rheological properties of cement slurries under flowing conditions or at rest. To characterize their non-Newtonian flow behavior, rotational viscometers (like coaxial cylinder viscometers) can be used with different fixtures such as cone-and-plate or plate-and-plate geometries. The basic principle is always the same. The test fluid is sheared between two surfaces—one of them is fixed, and the other one is either rotated at a constant velocity or at a constant torque. The flow pattern is such that shear rate and shear stress can be derived in a simple way from the rotational speed and torque. Notice that where the torque is imposed the equipment is effectively a constant stress system, because the shear stress is often proportional to the torque.

Other techniques using the same flow geometries, or different methods such as vanes (Section 4–5), are more specifically dedicated to the characterization of viscoelastic material. They can be used to study the rheological properties of cement slurries at rest. The basic aim of these experiments is the measurement of the stress/strain ratio. Such techniques include transient methods such as stress relaxation and creep, or sinusoidal methods such as dynamic experiments where stress and strain vary with time. The amplitude of the deformation can be low if one is interested in the viscoelastic properties of the material, or high if the objective is to characterize the yield strength of the material.

An extensive discussion of the above techniques is beyond the scope of this chapter. For additional information, the reader is referred to Walters (1975) and Whorlow (1980).

4–4 DATA ANALYSIS AND RHEOLOGICAL MODELS

4–4.1 Coaxial Cylinder Viscometer

4–4.1.1 Examples

Some typical data obtained at ambient temperature using the standard oilfield equipment and procedure are shown in Figs. 4–12 and 4–13. The higher readings correspond to a neat Class G cement slurry mixed at 15.8 lb/gal (1.90 g/cm^3), and the lower readings to the same formulation to which 0.1 gal/sk of a lignosulfonate dispersant has been added. For both cases, the line corresponds to a fit of the five highest readings (excluding the 3- and 6-RPM readings at 5 and 10 s^{-1}) to the full Bingham plastic equation (Eq. 4–40). The rheological parameters are reported in Table 4–2.

The behavior of the dispersed formulation follows the Bingham plastic model almost perfectly. This is remarkable because for low shear rates (5 to 10 s^{-1}), the fitted curve is based on an extrapolation of the data obtained at higher shear rates (50 to 500 s^{-1}). On the other hand, the formulation which does not contain additives (with the

Fluid	Correlation Coefficient	Plastic Viscosity (mPa s)	Yield Stress (Pa)
Neat Slurry	0.983	27	14
Dispersed Slurry	0.999	25	2.4

Table 4–2—Rheological parameters for Class G cement slurries with and without a dispersant.

exception of an antifoam) exhibits significantly different behavior. Above 50 s^{-1}, the Bingham plastic model gives a reasonable description of the properties up to 500 s^{-1}. However, the experimental data show a definitive curvature toward the shear rate axis on the linear graph even at high shear rates. This means that extrapolation using this model is likely to overestimate the shear stress for any particular shear rate above 500 s^{-1}. The Bingham plastic model also significantly overestimates the experimental shear stresses at low shear rates. However, the 3- and 6-RPM readings (5 and 10 s^{-1}) are affected by apparent slippage at the wall (as will be explained later in Section

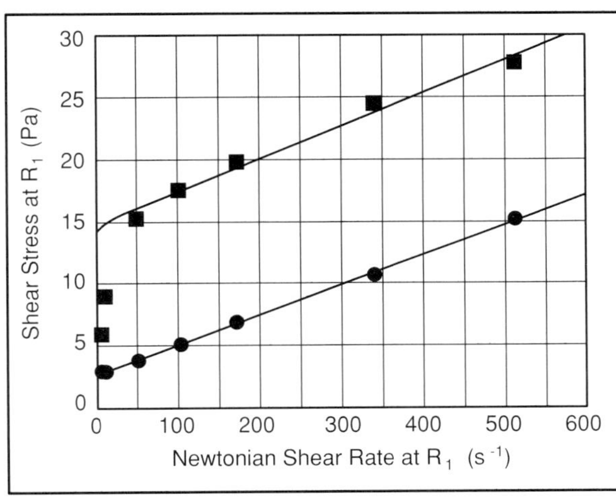

Figure 4–12—Flow curve of two cement slurries in a standard coaxial cylinder viscometer—linear scale.

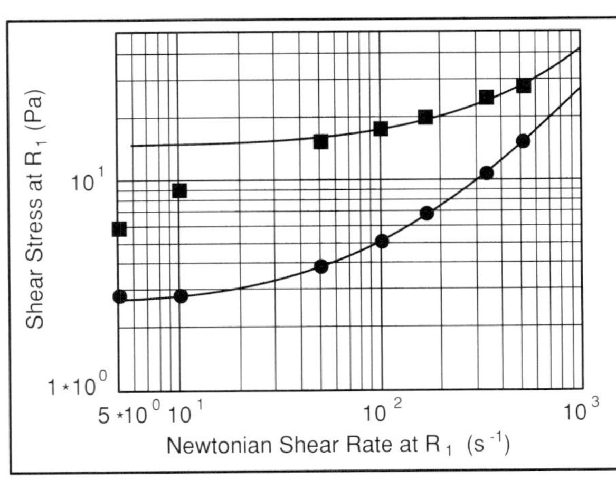

Figure 4–13—Flow curve of two cement slurries in a standard coaxial cylinder viscometer—log-log scale.

4–4.1.3) and should not be considered. Notice that the 30-RPM (50 s^{-1}) reading for the neat formulation does not satisfy the condition for Eq. 4–40 to be applicable. This means that according to the plastic-viscosity and yield-stress values obtained, plug flow is still present at this rotational speed.

It is also worthwhile to mention that the common practice of using only two high-rotational-speed readings to determine the rheological parameters of a given model can often be misleading. In the case of the dispersed formulation, good results are obtained because the fluid behaved according to the Bingham plastic model throughout the investigated shear-rate range. For the neat formulation, using only the 300- and the 200-RPM readings would lead to a plastic viscosity of 20 mPa s and a yield stress of 18 Pa. Since the actual rheogram is curved toward the shear-rate axis, a higher yield stress and a lower plastic viscosity are obtained when fitting only the high-shear data to a Bingham plastic model. Therefore, this procedure tends to give a better description of the shear-stress/shear-rate relationship at high shear rates, but it also overestimates shear stresses at low shear rates to a larger extent than the global fit procedure.

4–4.1.2 End Effects

With standard oilfield equipment, the end correction factor recommended by manufacturers is 1.064. It is in fact hidden in the spring calibration constant, which is 1.064 times lower than the nominal constant. This value is in agreement with measurements performed on Newtonian oils by Mannheimer (1988) and by the author. However, the author has found that end effects can account for up to 16% of the measured torque when testing cement slurries (Fig. 4–14), indicating that with the current standard procedure shear stresses can be overestimated by up to 10%.

Unfortunately, today there is no clear understanding of how end effects vary with the non-Newtonian behavior of the fluids; therefore, no simple procedure can be proposed to take them into account in a systematic way. Nevertheless, when trying to compare results obtained with different instruments, one must be aware that end effects can account for differences in measured shear stresses.

4–4.1.3 Slippage at the Wall

As explained earlier, once converted to shear-stress/shear-rate data, the torque/angular velocity relationship for a given fluid should be independent of the annular gap size. Several authors (Tattersall, 1973; Mannheimer, 1983 and 1988; Lapasin et al., 1983; Denis and Guillot, 1987; Haimoni, 1987) have shown that this is not always the case with cement slurries, in particular at low shear rates (Fig. 4–15). The correct interpretation of this effect is not trivial. One of the possible reasons for such a dependency is the fact that the fluid is not homogeneous throughout the gap. In particular, close to the rheometer walls, it is plausible that the concentration of cement particles is smaller than that of the bulk of the fluid. Another explanation which has already been mentioned is the presence of particle aggregates in the annular gap, the size of which may not be negligible when compared to the gap size. Mannheimer (1983; 1988) and others have attempted to analyze this phenomenon in terms of a slip velocity V_s (i.e., the velocity of the test fluid at the wall is

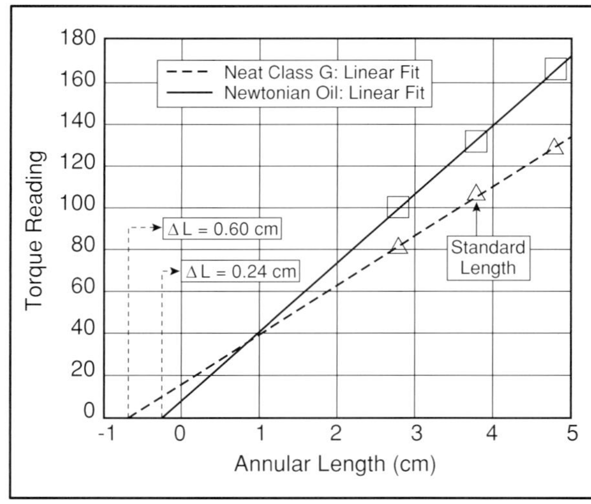

Figure 4–14—Graphical determination of end effects with a modified coaxial cylinder viscometer (ΔL is the length that should be added to the inner cylinder length *L* to account for end effects).

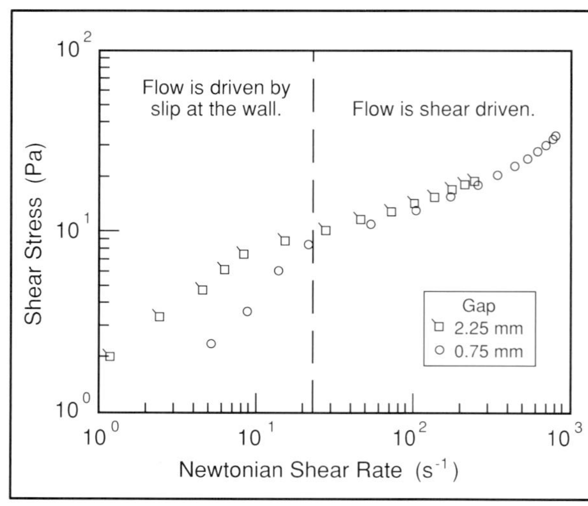

Figure 4–15—Flow curves of a neat Class G cement slurry in a coaxial cylinder viscometer with two different annular gaps (after Denis et al., 1987).

assumed to be nonzero). Such an assumption implies that Eq. 4–29 is no longer valid, and should be replaced by

$$\Omega = \frac{1}{2}\int_{\tau_2}^{\tau_1} \frac{\dot{\gamma}}{\tau}\, d\tau + \frac{V_s(R_1)}{R_1} + \frac{V_s(R_2)}{R_2} \quad . \qquad (4\text{–}62)$$

Assuming the slip velocity depends only on the shear stress at the wall for a constant shear stress at the outer cylinder surface (Mooney, 1931),

$$\lim_{R_1 \to R_2} \Omega = 2\,\frac{V_s(\tau_2)}{R_2} \quad . \qquad (4\text{–}63)$$

Therefore, the effect of wall slip could be accounted for by performing experiments with different inner cylinder radii. This analysis, which has been simplified by Mannheimer (1982) for narrow annular gaps, has not been conclusively validated. As can be seen in Fig. 4–16, the percentage of the flow due to slip does not vary consistently with shear stress. In a first series of tests, Mannheimer (1982) found the effect of slip velocity to be negligible above a given shear stress. Later, using different cements, conflicting results were obtained. The coaxial cylinder viscometer data, corrected for wall slip, were shown not to agree with laminar friction-pressure data in large-diameter pipes (Mannheimer, 1988).

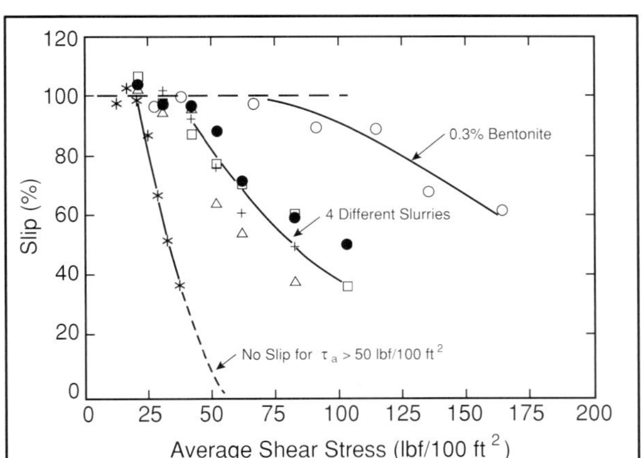

Figure 4–16—Effect of shear stress on percent slip measured with a concentric cylinder viscometer (slurry contains 38% water BWOC) (after Mannheimer, 1988).

Another approach to wall slip consists of trying to minimize the phenomenon, using grooved cylindrical surfaces. However, the reliability of the procedure with oil-well cement slurries is questionable, because the measured shear stresses depend on the depth of the serrations (Haimoni, 1987).

Haimoni (1987) tried to combine these two approaches (i.e., varying the gap size and the surface roughness of the cylinders) while making measurements on the same material. Although he was not able to propose a method to account for apparent slippage at the wall, he concluded that this phenomenon seems to have negligible consequences on the measurements performed in a coaxial cylinder viscometer once plug flow is eliminated.

Using data affected by slippage at the wall, if not detected, can lead to completely erroneous conclusions on the behavior of the test fluid at low shear rates. For example, if one fits the data of the neat cement formulation presented in Fig. 4–12 to a power law model, quite good results are obtained in the whole shear-rate range as shown on a linear graph in Fig. 4–17, and it could be concluded that the fluid exhibits no measurable yield stress. However, rerunning the test with a wider gap would show that data at 5 and 10 s^{-1} are affected by slippage at the wall and, therefore, should not be used for characterizing the rheological properties of the fluid.

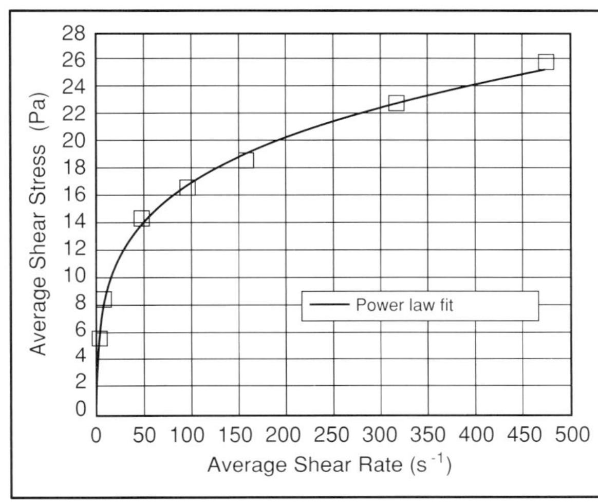

Figure 4–17—Power law fit to the rheological data of the neat cement formulation presented in Fig. 4–12.

Thus, in the absence of a proven method of allowing for wall slippage, coaxial cylinder viscometer data which are affected by this phenomenon should not be used when trying to determine rheological parameters. These data points can often be detected on a log-log plot of the torque vs rotational speed, which usually shows a drastic change in curvature (Fig. 4–13). Very often the experimental data falling below this breaking point are affected by slippage at the wall. This assumption can be checked by rerunning the test with a different gap size. Experimental data which do not satisfy the condition for Eq. 4–40 to be valid should also be discarded.

4-4.1.4 Particle Migration

Particle migration due to gravitational or centrifugal forces may also affect the rheological measurements. For the results to be meaningful, the test fluid should not segregate during the measurement. Before measuring the rheological properties of a cement slurry, it is essential to ensure that particle segregation does not occur under static conditions (leading to free water and sedimentation). Unfortunately, this does not necessarily mean that it will not occur under dynamic conditions because

- the apparent viscosity of the fluid usually decreases with shear, and
- under dynamic conditions, the centrifugal forces can be greater than the gravitational forces.

Sedimentation

Sedimentation can occur in standard oilfield equipment, but the design is such that measurements are not too strongly affected unless the problem is extremely severe. First, the dead volume of fluid above the inner cylinder ensures that, if sedimentation is occurring, the concentration of cement particles in the gap does not decrease instantaneously as would be the case if it were not present. Second, when going from a high rotational speed to a low speed, or vice versa, vertical movement of the fluid in the gap is likely to occur and renew the fluid in the gap from the reservoir of fluid in the cup. Third, it seems also that even at a constant rotational speed, the test fluid is sometimes submitted to a strong pumping circulation of fluid through the gap.

When using other systems (such as closed cup systems as shown in Fig. 4–5b) great care should be taken during all steps of the testing procedure to ensure that the experimental results are not biased by cement particle settling. The phenomenon may even occur in consistometer cups, where cement slurries are conditioned prior to measuring their rheological properties. Therefore, the test slurry should be carefully homogenized prior to taking a sample for the rheological test. In addition, one should verify that the measured torques at a given rotational speed are stable. If they continuously decrease, particle sedimentation is likely to occur (although it may sometimes be difficult to differentiate this from thixotropy). The measured torque may first decrease and then increase, because a bank of cement particles accumulating at the bottom of the cup enters the annular gap. This explains why closed cup geometries should be used with care for characterizing the rheological properties of cement slurries.

Centrifugation

If one considers a cement particle flowing at one-half the rotational speed of the rotor in standard oilfield equipment, it is submitted to the following centrifugal acceleration.

$$T = \frac{\Omega^2}{4} \times R_a \qquad (4\text{–}64)$$

At 600 RPM, this is about 18 m s^{-2} (i.e., almost twice the gravitational acceleration). Therefore, if cement particles settle under gravity, they are even more likely to migrate in the rheometer because of the centrifugal forces. This can occur not only in the annular gap, but also in the dead volume of fluid above the inner cylinder. The migration of cement particles in this portion of the flow geometry is even promoted by the deformation of the free surface of the fluid due also to centrifugal forces. Once centrifuged at high rotational speeds, the particles seem to migrate in the annular gap, and to irreversibly affect the readings taken at lower speeds. This problem can be solved by suppressing the dead volume of fluid above the inner cylinder (i.e., by positioning the cup at a lower level than the standard level) (Fig. 4–18). Unfortunately, this solution is not universal because it may create some problems with cement formulations exhibiting a settling tendency. Not all cement formulations show such behavior, and the best way to detect it is to run a speed hysteresis cycle. When the ramp-down readings are much higher than the ramp-up readings, centrifugation can be suspected to have affected the results. The lower readings should be preferred to characterize the properties of the test fluid.

4–4.2 Pipe Viscometer

Pipe viscometers have also been used to characterize the rheological properties of cement slurries, but their use

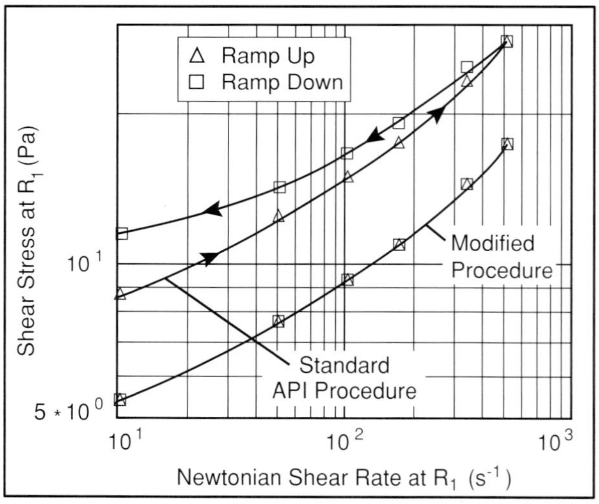

Figure 4–18—Speed hysteresis cycles performed on a neat Class G cement slurry, using the API standard procedure and a modified procedure.

has been usually limited to a laboratory environment, because they are quite cumbersome and the results obtained can be inconsistent. Bannister (1980) and Mannheimer (1988) observed that flow curves of cement slurries in small-diameter pipes are diameter dependent (Fig. 4–19). Experimental results have also been published (Fig. 4–20) showing that the diameter dependency can be negligible for large-diameter pipes and above a minimum shear rate or minimum shear stress (Denis and Guillot, 1987). Unfortunately, these diameters are so large that the corresponding equipment cannot be used routinely to characterize the rheological properties of cement slurries. Therefore, several authors have attempted to cope with the behavior observed in small-diameter pipes.

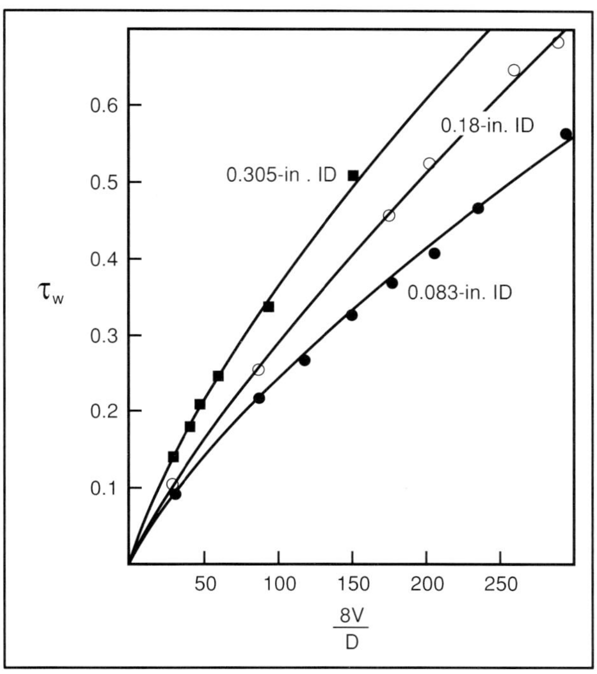

Figure 4–19—Rheological measurements using a pipe-flow rheometer (slurry: Class H + 0.36% hydroxyethylcellulose + 40% water BWOC)—80°F. The flow curves are pipe diameter dependent (after Bannister, 1980).

4–4.2.1 Slippage at the Wall

An analysis in terms of wall slippage, similar to the one performed for coaxial cylinder viscometers, can also be performed for pipe viscometers. If the velocity of the fluid is assumed to be V_s at the pipe wall, Eq. 4–15 becomes (Oldroyd, 1949)

$$\frac{4V}{R} = \frac{4}{\tau_w^3}\int_0^{\tau_w} \tau^2 \dot{\gamma}\, d\tau + \frac{4V_s}{R} \quad . \qquad (4–65)$$

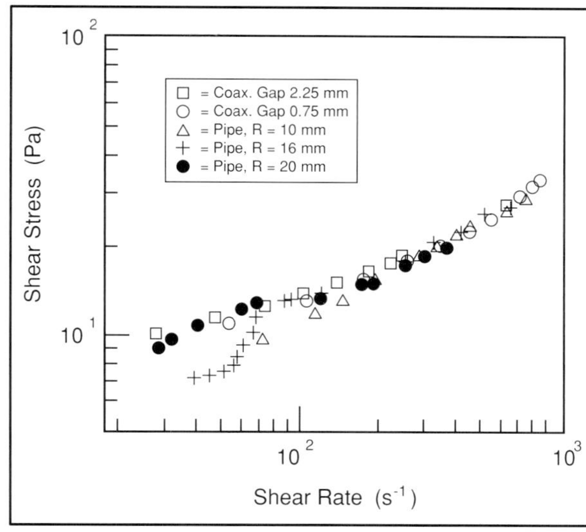

Figure 4–20—Pipe- and coaxial-flow results for a neat Class G cement slurry (shear rates are corrected for non-Newtonian effects). Above 200 s^{-1} there is good agreement between the different data sets.

If V_s is assumed to be only shear-stress dependent, Eq. 4–65 can be differentiated for a constant value of shear stress at the wall to obtain the expression for the slip velocity.

$$V_S(\tau_w) = \left[\frac{d(4\,V/R)}{dR^{-1}}\right]_{\tau_w = constant} \quad . \qquad (4–66)$$

Thus, the effect of wall slip can in principle be accounted for by performing flow experiments in pipes of different diameters. As mentioned above, such an analysis can only be performed if the slip velocity depends simply on the shear stress at the wall. Mannheimer (1988) showed that this is not necessarily the case, and that the slip velocity can also be affected by the surface roughness of the pipe. This may lead to meaningless conclusions, e.g., that slippage at the wall accounts for more than 100% of the flow! When experimental precautions were taken to ensure that the surface roughness of the pipes used was the same, suitable results were obtained by Mannheimer (1988), but he gave no experimental evidence that pipe viscometer data corrected for apparent slippage at the wall can be used to predict laminar friction pressures in field-size pipes or annuli.

Bannister (1980) used a different approach to analyze pipe viscometer data. The procedure in fact only applies provided the flow curves for different pipe diameters can be described by a power law relationship with the same Power Law Index n', and a Consistency Index k'_R that is pipe-radius dependent.

$$\tau_w = k'_R \times \left[\frac{4V}{R}\right]^{n'} \qquad (4–67)$$

It is then straightforward to show that the Power Law Index n of the fluid is n', and that the apparent slip velocity is given by

$$V_s = C_s \times \left[\frac{1}{\tau_w}\right]^{1/n'} , \qquad (4\text{–}68)$$

where C_s is a constant. The Pipe Consistency Index of the fluid k' can be derived from the following relationship.

$$\left[\frac{1}{k'_R}\right]^{1/n'} = \left[\frac{1}{k'}\right]^{1/n'} + \frac{4C_s}{R} \qquad (4\text{–}69)$$

Using this procedure, Bannister (1980) was able to predict the friction pressure in a large-diameter pipe (1.815 in. ID) from friction-pressure measurements obtained with a laboratory-scale, pipe-flow loop (0.083 in. < ID < 0.305 in.) for a specific cement slurry formulation (Table 4–3).

	PRESSURE DROP (PSI)		
Pump Rate (BPM)	Fann 35[(1)] Reading	Pipe/Flow Rheometer	Field Evaluation
0.5	16	25	24
1.0	24	36	37
1.5	32	45	43
1.75	36	49	48

(1) Rheological data analyzed using Bingham Plastic Model.

Table 4–3—Calculated pressure drops for a Class H cement slurry (38% water, 0.1% retarder, 0.1% prehydrated bentonite) flowing through 1.815-in. ID pipe (98°F) (after Bannister, 1980).

4–4.3 Comparison Between Different Equipment

When trying to characterize the rheological behavior of materials as complex as cement slurries, it is essential to ensure that the measurements are not equipment dependent. It has already been mentioned that there are very good reasons for believing that this is not true. Thus, several authors have compared the rheological measurements performed with different types of equipment, usually a coaxial cylinder viscometer and a pipe viscometer. For such a comparison to be significant, it must be performed within a shear-rate range common to both apparatuses.

Denis and Guillot (1987) showed that reasonable agreement between a pipe viscometer and a specific coaxial cylinder viscometer can be obtained with some cement slurry formulations, provided the rheological data are not affected by slippage at the wall (Fig. 4–20). However, when cement slurries are characterized with the standard oilfield viscometer, the results have quite often been found to be significantly different from those obtained with pipe viscometers, even when using large-diameter pipes to minimize the effects of apparent slippage at the wall (Bannister, 1980; Mannheimer, 1983; Denis and Guillot, 1987). This is not surprising when one considers the number of problems which can be encountered with oilfield equipment.

In an attempt to solve this problem, Shah and Sutton (1989) tried to obtain a statistical correlation between the measurements performed with a standard oilfield viscometer and a pipe viscometer. They used a modified coaxial cylinder viscometer to allow for vertical circulation of the slurry in the annular gap, the circulation being stopped while a measurement was taken at a given rotational speed. For a wide variety of cement slurry formulations, they compared the rheological parameters obtained by fitting the experimental data obtained with their modified viscometer $[(\mu_p)_c, (\tau_y)_c]$ and a pipe-flow loop $[(\mu_p)_p, (\tau_y)_p]$ to a Bingham plastic model. They found the following correlation for the plastic viscosities when expressed in cp (Fig. 4–21)

$$(\mu_p)_p = 0.962 \times \left[(\mu_p)_c\right]^{0.9815} , \qquad (4\text{–}70)$$

indicating that the plastic viscosities obtained with the pipe viscometer were of the order of 10% lower than those obtained with the coaxial cylinder viscometer. For the yield stresses, those obtained from the pipe-flow data were overestimated by a factor 1.333, and those obtained from the coaxial cylinder viscometer by a factor 1.067, because in both cases the shear rate at the wall was assumed to be the Newtonian value which is not the case for a Bingham plastic fluid. Therefore, once the yield stresses are corrected, the correlation of Shah and Sutton (1989) (Fig. 4–22) becomes

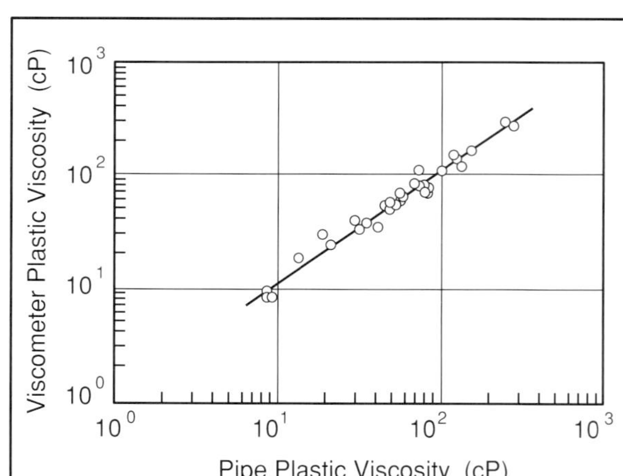

Figure 4–21—Plastic-viscosity relationship between standard coaxial cylinder and pipe viscometers (after Shah et al., 1989).

$$(\tau_y)_p = 1.273 \times (\tau_y)_c = 1.61, \qquad (4\text{–}71)$$

where yield stresses are expressed in lbf/100 ft^2. This indicates that the yield stresses obtained with the pipe viscometer were between 0% and 27% higher than those obtained with the coaxial cylinder viscometer. This empirical procedure is quite useful, but it suffers from one limitation—the cement slurries were assumed to be described by a Bingham plastic model, which is not necessarily the case as will be shown below.

4–4.4 Which is the Best Rheological Model?

The power law and Bingham plastic models are most widely used to describe the rheological properties of cement slurries. Both can describe the shear-stress/shear-rate relationship for a given cement slurry quite well within a limited shear-rate range. However, when attempting to describe the behavior of cement slurries over a wide shear-rate range, the situation is different.

The power law model suffers from limitations, because—

- most cement slurries exhibit a yield stress, and the power law model does not include such a parameter; and
- the viscosity of any fluid at high shear rates should tend toward a nonzero value, which again is not taken into account in the power law model.

Thus, the power law model underestimates the shear stresses at both low and high shear rates.

The Bingham plastic model does not have such drawbacks. It includes both a yield stress τ_y and a limiting viscosity μp at infinite shear rates. Nevertheless, not all cement slurries are very well described by the Bingham plastic model. When plotted on a linear graph (shear stress vs shear rate), some rheological data show a definite curvature toward the shear-rate axis (Fig. 4–12). When this is the case, the Bingham plastic model behaves in a manner opposite to the power law model, i.e., an overestimation of the shear stresses occurs at both low and high shear rates. The low shear behavior is a difficult problem to solve, because the data at low shear rates can be affected by slippage at the wall. However, the overestimation of the shear stress at high shear rates may also be a problem, specifically for predicting friction pressures in pipes and annuli outside the shear-rate range investigated with a coaxial cylinder viscometer (Guillot and Denis, 1988). Several models have been used in an attempt to solve this problem, such as the Casson, Vocadlo, or Herschel-Bulkley models. Most have been found to better fit the rheological behavior of cement slurry formulations. A comparison of Fig. 4–23 and 4–12 shows that, for this specific example, the Herschel-Bulkley model describes the rheological behavior better than the Bingham plastic model when the data are not affected by slip at the wall (i.e., above 40 s^{-1}). However, the use of these models is now fairly limited for several reasons.

- It is not yet clear whether (and by how much) the raw data obtained with a coaxial cylinder viscometer are affected by end effects, slippage at the wall, and particle migration.
- Most cement slurries are characterized with a six-speed standard oilfield rotational viscometer where,

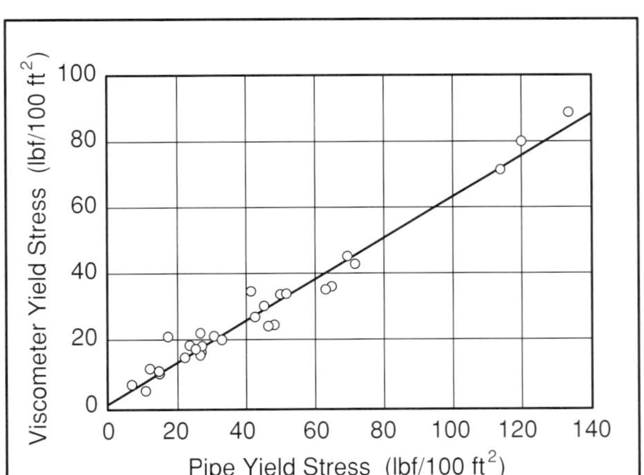

Figure 4–22—Yield-stress relationship between standard coaxial cylinder and pipe viscometers (after Shah et al., 1989).

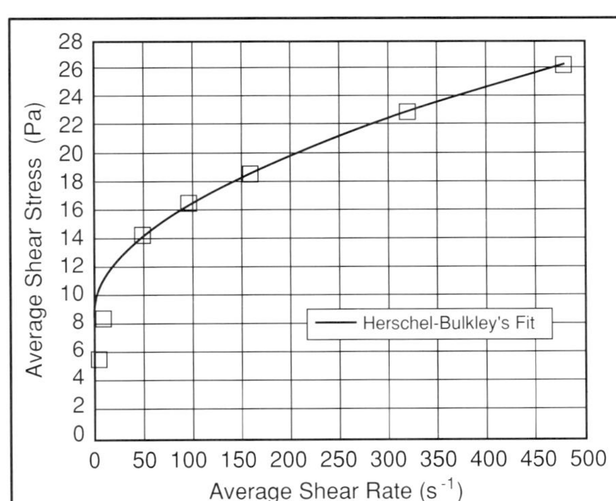

Figure 4–23—Herschel-Bulkley fit to the rheological data of the neat cement formulation presented in Fig. 4–12.

as mentioned earlier, often only three readings are useful for fitting the data to a model.

4–4.5 Temperature and Pressure Dependence

The pressure and temperature dependence of the rheological properties of cement slurries is not well understood, because the standard oilfield equipment allows measurements to be performed only at atmospheric pressure, and at temperatures below 80° to 90°C. Limited studies at higher temperatures suggest that cement slurry stability, which is already a concern below 80 to 90°C, is even more problematic at higher temperatures.

Very little work has been devoted to the pressure dependence of the rheological properties of cement slurries. Besides the lack of equipment, the principal reason is that cement slurries are water-based; in view of the low compressibility and viscosity-pressure dependence of water, the effect of pressure on their flow properties has usually been considered to be negligible. This is most probably the case for most systems, except those exhibiting a high solid-to-liquid ratio. For such formulations, the higher compressibility of the liquid phase when compared to the solid phase is likely to give a significant viscosity increase with increasing pressure, through an increase of the solid-to-liquid ratio. The viscosity of solid suspensions increases roughly exponentially with the solid volume fraction, tending toward infinity as close packing is approached. Hence, it becomes increasingly sensitive to pressure as the solid content increases.

On the other hand, temperature can have a drastic effect on the cement slurry rheology, but the extent of this effect is highly dependent on the cement brand and the additives in the formulation. The differences in temperature dependence are shown in Figs. 4–24 and 4–25. The first formulation contains a hydrosoluble polymer (hydroxyethylcellulose) which viscosifies the interstitial water and contributes significantly to the slurry viscosity. Since the polymer solution viscosity itself is temperature sensitive, the plastic viscosity of the slurry follows the same continuous downward trend, while the yield stress remains almost constant. The behavior of the second system (containing a dispersant and latex) is much more complicated. The plastic viscosity of the slurry first decreases by a factor of two between 25° and 45°C, and then increases more slowly from 45° to 85°C. Meanwhile, the yield stress increases slowly but continuously throughout the temperature range investigated.

These two examples illustrate the fact that there is currently little hope of finding a general model to describe the temperature dependence of the cement slurry rheology. What can probably be done is to define some typical behavior which could be described by the same model, but these studies are at a research level today.

Most cement placement simulators used to design primary cementing jobs, being isothermal, employ a single figure which is measured at the estimated BHCT or at the

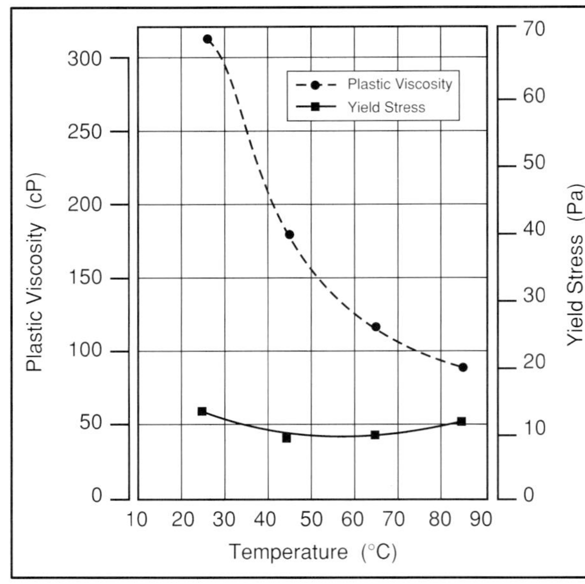

Figure 4–24—Temperature dependence of the Bingham plastic parameters of a cement formulation containing a cellulose derivative.

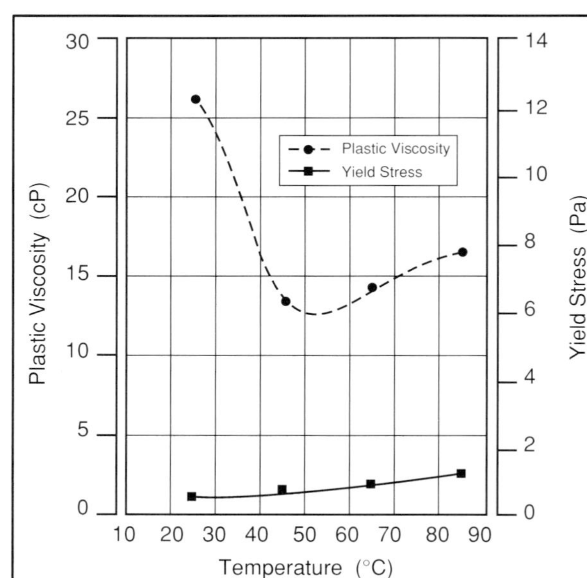

Figure 4–25—Temperature dependence of the Bingham plastic parameters of a cement formulation containing a dispersant and a latex.

maximum temperature allowed by the equipment (i.e., 80° to 90°C).

4–5 TIME-DEPENDENT RHEOLOGICAL BEHAVIOR OF CEMENT SLURRIES

In the oil industry, little attention has been paid to the complete characterization of the thixotropic behavior of cement slurries. The high shear imposed at the beginning of the standard test procedure is intended to break down the structure the fluid may have built up prior to the test. However, this assumes that 60 s at the maximum shear rate is sufficient time to enable the structure to reach an equilibrium, which may well not be the case. In a similar way, when running the speed down, the fluid is sheared for 20 s at each step before the reading is taken. Depending on whether the aim is to characterize a structure which has been previously broken at high shear, or the equilibrium structure at each shear rate, the duration of the step may either be too long or too short. Thus, the current procedure is not adapted to thixotropic cement slurries, nor is it suited to detect whether or not a given slurry exhibits thixotropic properties. This situation could perhaps be improved by adopting a different procedure which would consist, for example, of increasing the rotational speed first and then decreasing it; this cycle would be repeated until an equilibrium is reached. The extent of the hysteresis in the measured shear stress would at least give some measure of the extent of the thixotropic nature of a given slurry.

For the time being, the word "thixotropy" in the oil industry is commonly associated with the ability of a given fluid to build up a structure upon standing. This structure is usually characterized by its "gel strength," which is the minimum shear stress required to shear a fluid at a measurable flow rate. Following the standard procedure defined by the API for drilling muds, gel strengths of cement slurries are usually evaluated by measuring the peak value of the shear stress upon sudden application of a shear rate of 5.11 s^{-1} after a given rest period. Unfortunately, the results obtained with this experimental method are questionable for two main reasons.

- It has already been mentioned that the low shear behavior of cement slurries is very often affected by slippage at the wall. This is even more so for thixotropic systems, because the majority of the experimental results show that the higher the yield stress of the fluid the larger the shear-rate range affected by slippage at the wall.
- The results obtained may vary from one piece of equipment to another, depending on the inertia of the fixture and on the stiffness of the measuring device.

Very little can be done on the standard oilfield equipment regarding the second point, and one must be aware that even in the absence of slippage at the wall (e.g., with drilling muds), these gel-strength values can be underestimated (Speers et al., 1987). Other devices have been developed to better characterize the gel-strength development of cement slurries (Sabins et al.,1980). However, in most cases, the stress distribution in these devices is not known, and what is actually measured is a "consistency" which is difficult to correlate with the true material gel strength.

The technique which looks the most promising today for characterizing the gel strength of at least highly thixotropic cement slurries is the shear vane method. The standard coaxial cylinder geometry is replaced by a vane (Fig. 4–26). Provided the vane is rotated at a sufficiently low speed, the sheared surface is cylindrical, and the maximum torque recorded can be used to calculate the gel strength of the material. The advantage of this method, which is commonly used in soil mechanics, is that it is not affected by slippage at the wall because the shear surface is within the material itself.

The structure buildup of a given cement slurry can also be followed through oscillatory dynamic tests, measuring the evolution of the storage (elastic) and loss (viscous) moduli vs time (Hannant and Keating, 1985; Chow

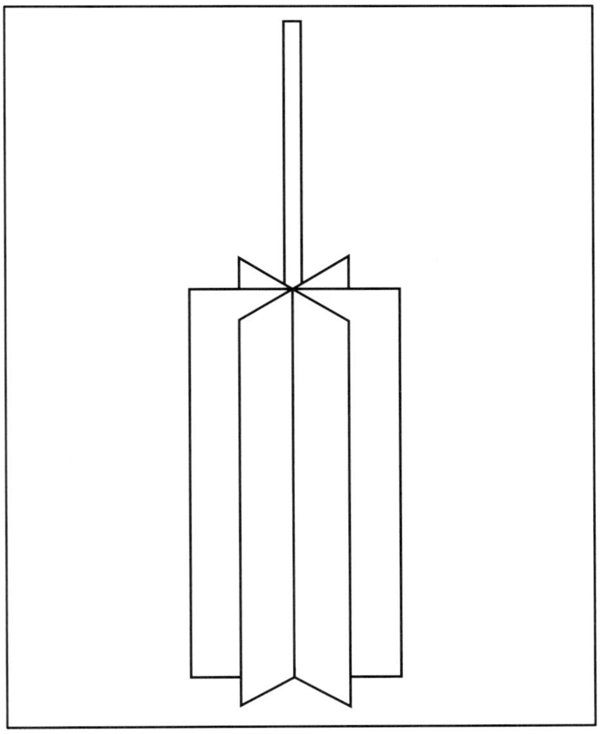

Figure 4–26—Schematic of a six-blade vane geometry.

et al., 1988), but these techniques do not give direct access to the gel strength.

A very important point which needs to be stressed at this stage, and which is frequently forgotten, is that most cement slurries exhibit a structural change not only upon standing but also under the condition of constant shear rate and temperature. For example, the evolution of shear stress as a function of time for a given cement formulation in standard oilfield equipment at 511 s^{-1} is shown in Fig. 4–27. It appears that this time-dependent behavior is not only shear-history dependent, a problem which has been addressed at the beginning of this subsection, but also that it is due to the on-going chemical reactions in the material. Once again, this effect is rarely investigated. Therefore, in the absence of further information, one must conclude that the properties which have been presented so far are only representative of the material at a given age and rate of mixing.

4–6 FLOW BEHAVIOR OF CEMENT SLURRIES IN THE WELLBORE ENVIRONMENT

In this section, some of the consequences of the rheological behavior of cement slurries (described so far for their flow within the wellbore) are investigated.

4–6.1 Pipe Laminar Flow

The equations for the velocity profile and for the volume flux for laminar flow in pipes have already been developed. Solutions were given for the volume flux of the two commonly used model fluids. They are summarized in Appendix A. In the same table are also reported the corresponding equations for the velocity profiles.

It is to be noticed that the velocity profiles for power law fluids depend only on the Power Law Index. The lower the Power Law Index the flatter the velocity profile, whatever the flow rate or the pipe diameter, provided the flow regime remains laminar (Fig. 4–28). For Bingham plastic fluids, two equations are necessary to describe a velocity profile because part of it, around the pipe axis, is flat, while the rest of it is a parabola. Velocity profiles also depend on a single parameter—the dimensionless shear stress ψ (= τ_y/τ_w). Another parameter which could be used is the dimensionless shear rate or $\xi = \dot{\gamma}_{Nw} \times (\mu_p/\tau_y)$, but the equations then become implicit. Thus, the normalized velocity profiles for such fluids are flow-rate dependent. Given the pipe diameter or the annular gap, the smaller the average velocity and the plastic viscosity-to-yield stress ratio (μ_p/τ_y), the flatter the velocity profile (Fig. 4–29). Notice that the dimensionless shear stress ψ also represents the fraction of the pipe diameter where the profile is totally flat. This is why this parameter is sometimes called the *plug-to-pipe ratio.*

4–6.2 Pipe Turbulent Flow

Regardless of the type of fluid, once a critical flow rate in a given pipe is exceeded, streamlines are no longer parallel to the main direction of flow. Fluid particles become subject to random fluctuations in velocity both in amplitude and direction. In fact, velocity fluctuations are not completely random. Near the wall, fluctuations in the axial direction are greater than those in the radial direction, and both approach zero at the wall. Such flow instability

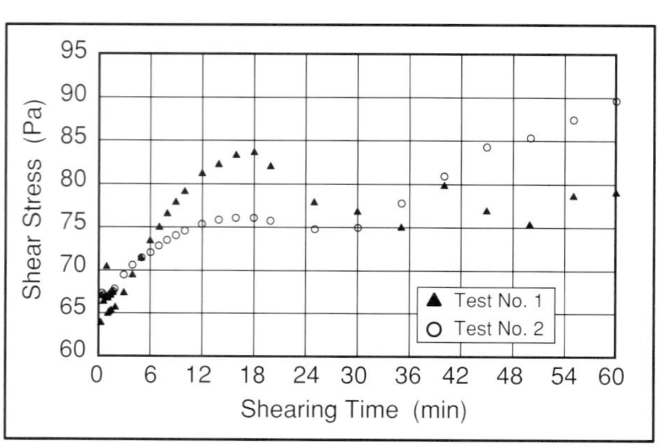

Figure 4–27—Shear stress against shearing time (results obtained using a standard oilfield coaxial viscometer at a shear rate of 511 s^{-1}).

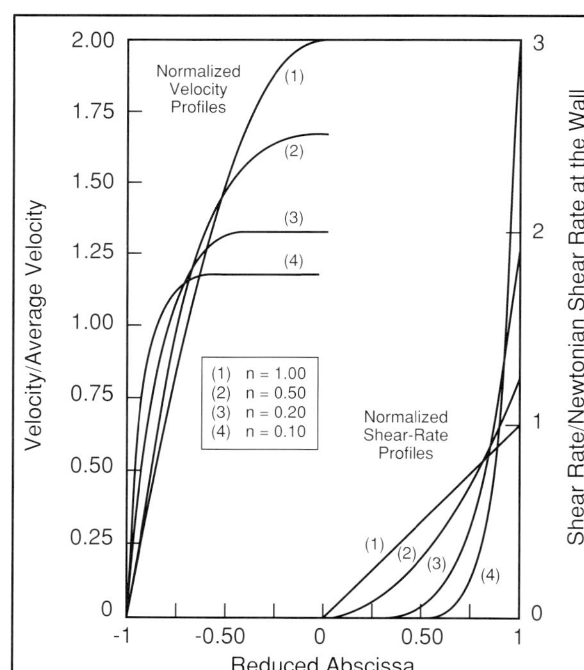

Figure 4–28—Normalized velocity- and shear-rate profiles for a power law fluid flowing in a pipe (n = Power Law Index).

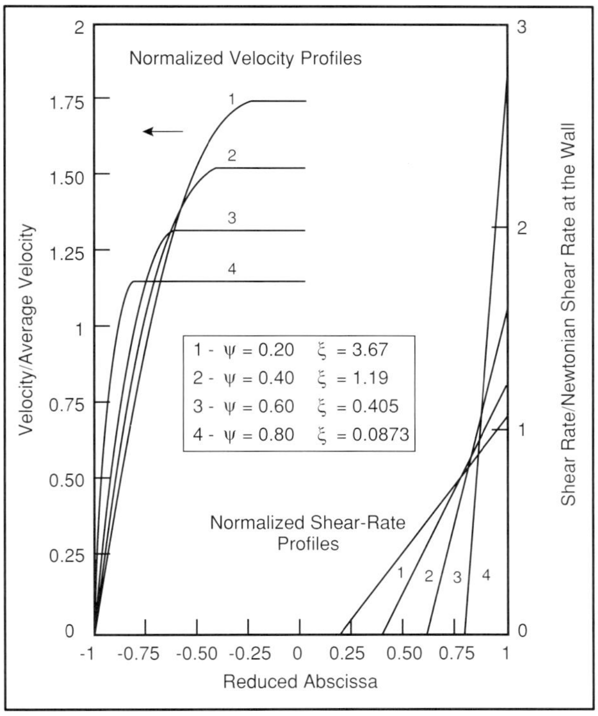

Figure 4–29—Normalized velocity- and shear-rate profiles for a Bingham plastic fluid flowing in a pipe (ψ = dimensionless shear stress, ξ = dimensionless shear rate).

starts for a given value of a dimensionless parameter, the *Reynolds number* (*Re*) which, for Newtonian fluids, is defined by

$$Re = \frac{\rho V D}{\eta} \qquad (4\text{–}72)$$

Departure from laminar flow occurs as the Reynolds number increases beyond a value of 2,100. A transition regime which is not very well characterized exists up to Re = 3,000. Above this value, flow becomes turbulent. The resistance to flow at the pipe wall is then expressed as

$$\frac{1}{\sqrt{fr}} = A \log\left[Re\sqrt{fr}\right] + C \qquad (4\text{–}73)$$

where *fr*, the *Fanning friction factor*, is defined by

$$fr = \frac{2\tau_w}{\rho V^2} \ . \qquad (4\text{–}74)$$

In Eq. 4–73, which was first proposed by von Karman in 1930 (Schlichting, 1979), parameters *A* and *C* depend on the roughness of the pipe. For turbulent flow in smooth pipes, A = 4.0 and C = –0.4.

With these definitions it should be noticed that, in laminar flow

$$fr = \frac{16}{Re} \ . \qquad (4\text{–}75)$$

In the transition regime, the friction-factor/Reynolds number relationship is not uniquely defined, but for most engineering applications, a linear interpolation is made on a log-log scale between the laminar value of *fr* at a Reynolds number of 2,100 and its value at a Reynolds number of 3,000 (Fig. 4–30).

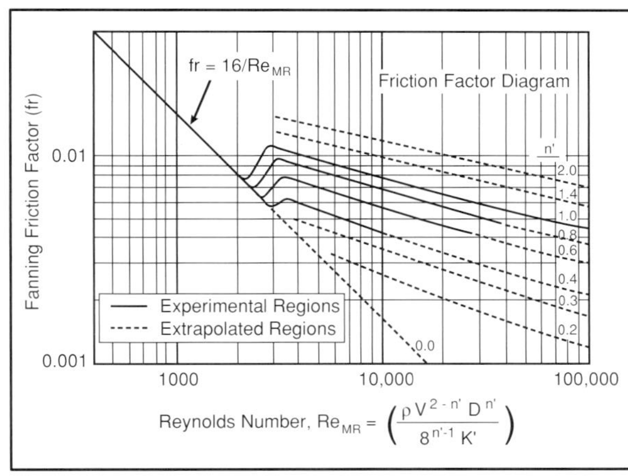

Figure 4–30—Relationship between Fanning friction factor and the generalized Reynolds number. Note that, for a given Reynolds number, *fr* is strongly dependent on the value of n′ (from Dodge and Metzner, 1959).

Similar equations have been developed for non-Newtonian fluids. The main problem here is to determine which viscosity should be used in the expression for the Reynolds number, because it is shear-rate dependent.

For Bingham plastic fluids, the simplest method (Hedstrom, 1952) consists of assuming that once turbulent flow is reached, the fluid behaves like a Newtonian one with a viscosity equal to its plastic viscosity (the procedure is described in API Spec 10). This indicates that the relevant Reynolds number in turbulent flow is

$$Re_{BG} = \frac{\rho V D}{\mu_p} \ . \qquad (4\text{–}76)$$

Equation 4–73 is then used to calculate friction pressures for a given flow rate (Fig. 4–30). This assumption has been established empirically for smooth pipes by several authors working with different types of fluids (Govier and Aziz, 1972). Unfortunately, it does not seem to hold for all cement slurries. Guillot and Denis (1988) showed that this procedure can lead to a considerable overestimation of friction factors (Fig. 4–31).

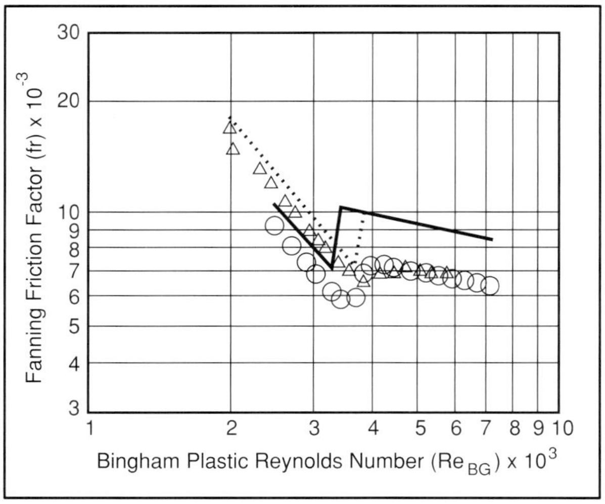

Figure 4–31—Fanning friction factor/Reynolds number graph for a given cement formulation. Circles and triangles are experimental data for 16- and 20-mm pipe, respectively. The continuous (16-mm) and the dotted (20-mm) lines were calculated following API procedures for Bingham plastic fluids (i.e., in turbulent flow fluids are assumed to behave like Newtonian fluids with a viscosity μ_p) (after Guillot and Denis, 1988).

Other methods for calculating turbulent friction pressures of Bingham plastic fluids in pipes have been developed (Govier and Aziz, 1972), but their validity has not been fully established for cement slurries. In addition, all of these procedures assume that the Bingham plastic model describes reasonably well the rheological properties of the fluid considered. Unfortunately, as explained earlier, this is not always the case.

A more general approach, which does not suffer from this limitation, is very often preferable. Dodge and Metzner (1959) proposed to generalize Eq. 4–73 to describe the turbulent flow of nonelastic non-Newtonian fluids in smooth pipes (Fig. 4–30).

$$\frac{1}{\sqrt{fr}} = A_n{}' \times \log\left[Re_{MR}\, fr^{1-n'/2}\right] + C_n{}' \qquad (4\text{–}77)$$

where $A_{n'}$ and $C_{n'}$ are a function of n' only. The generalized Reynolds number, Re_{MR}, is defined by Metzner and Reed (1955) as

$$Re_{MR} = \frac{\rho V^{2-n'} D^{n'}}{8^{n'-1} k'} \ . \qquad (4\text{–}78)$$

The local power law parameters n' and k' are defined by

$$n' = \frac{d \log (\tau_w)}{d \log (8V_L/D)} \qquad (4\text{–}79)$$

and

$$k' = \frac{\tau_w}{(8V_L/D)^{n'}} \ . \qquad (4\text{–}80)$$

V_L is the average velocity for the same shear stress at the wall τ_w, if the flow is laminar. Notice that for power law fluids,

$$n' = n \qquad (4\text{–}81)$$

and

$$k' = \left[\frac{3n+1}{4n}\right]^n k \ . \qquad (4\text{–}82)$$

These equations where first developed for power law fluids (i.e., for $n' = n$ = constant), but Dodge and Metzner (1959) extended their application to other nonelastic non-Newtonian fluids. This is justified by the fact that, in turbulent flow, only the shear in very close proximity to the wall contributes significantly to the flow rate. Dodge and Metzner (1959) gave experimental evidence that this is correct. For the non-Newtonian fluids they tested, with n' values from 0.36 to 1.0, and Re_{MR} values from 2,900 to 35,000, they empirically found that, for smooth pipes

$$A_n{}' = \frac{4.0}{(n')^{0.75}} \quad \text{and}$$

$$C_n{}' = \frac{-0.40}{(n')^{1.2}} \ .$$

Dodge and Metzner (1959) found their method gave a remarkable prediction of friction pressures for the fluids with which they were working (Fig. 4–32). Very good results were also obtained by Guillot and Denis (1988) with cement slurries whose rheological properties were described by a three-parameter model (Fig. 4–33).

Notice that Eq. 4–77 is implicit in the friction factor even for power law fluids. For most engineering applications, it can be replaced by an explicit expression which is given in Appendix A (Tables A-3 and A-4). For non-power law fluids, even when using this explicit expression, the equation remains implicit in the friction factor and should be solved numerically. For Bingham plastic fluids, an explicit expression for the Reynolds number can be determined, provided the dimensionless shear stress is sufficiently small. This leads to simpler expressions for the flow equations, as shown in Appendix A (Table A-6).

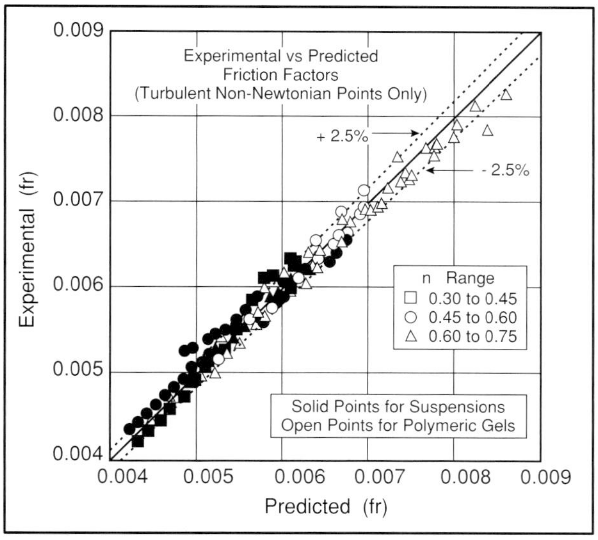

Figure 4–32—Comparison of experimental friction factors with those predicted (after Dodge and Metzner, 1959).

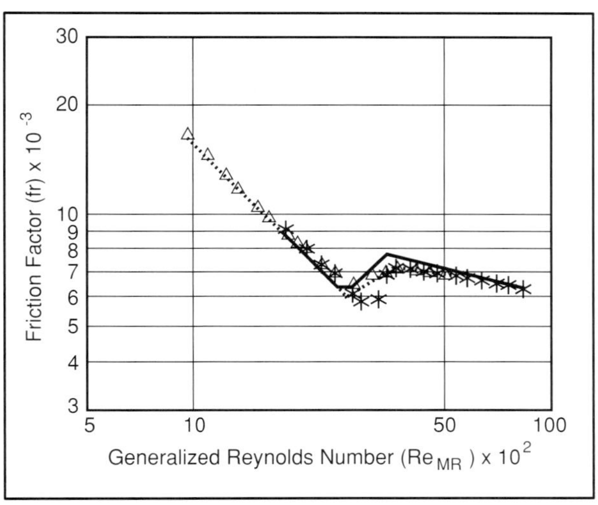

Figure 4–33—Fanning friction factor/generalized Reynolds number graph for a given cement formulation. Symbols correspond to raw data. Lines correspond to calculated values according to the Dodge and Metzner equation, the fluid being described by a three-parameter model.

4–6.3 Transition From Laminar Flow to Turbulent Flow in Pipes

The question of the transition in pipes from laminar flow to turbulent flow of cement slurries is still open today. Most experimental results show that if the fluid is less Newtonian, the critical Reynolds numbers Re_1 corresponding to the end of the purely laminar-flow regime and Re_2 to the beginning of the fully turbulent-flow regime will be higher. Several theories have been developed to account for this variation (Ryan and Johnson, 1959; Hanks, 1963), most of them being specific to a given rheological model. Since there is very little evidence that one of these models better applies to cement slurries, it is reasonable to follow the same generalized approach as for friction pressures in turbulent flow. The critical values shown on the (fr, Re_{MR}) diagram (Fig. 4–30) correspond roughly to the following variation of the critical Reynolds numbers.

$$Re_1 = 3250 - 1150 \times n' \quad (4\text{–}83)$$

$$Re_2 = 4150 - 1150 \times n' \quad (4\text{–}84)$$

As in the case of the friction-factor/Reynolds number equation in turbulent flow, this equation is implicit for nonpower law fluids, and has to be solved numerically for the critical fluid velocity V_L.

4–6.4 Laminar Flow in Concentric Annuli

Equations describing the flow in narrow concentric annuli are given in Appendix A. Qualitatively speaking, the results are the same as for pipe flow. Examples of velocity profiles for power law fluids and Bingham plastic fluids are given in Figs. 4–34 and 4–35, respectively.

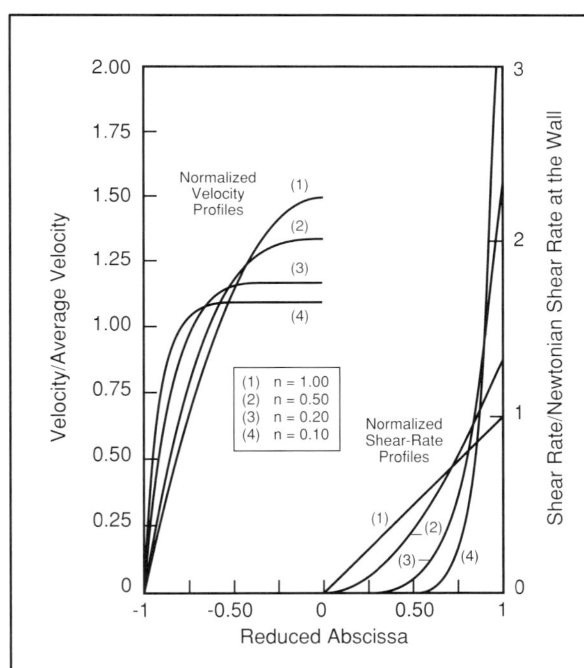

Figure 4–34—Normalized velocity- and shear-rate profiles for a power law fluid flowing in a slot or narrow annulus (ψ = dimensionless shear stress, ξ = dimensionless shear rate).

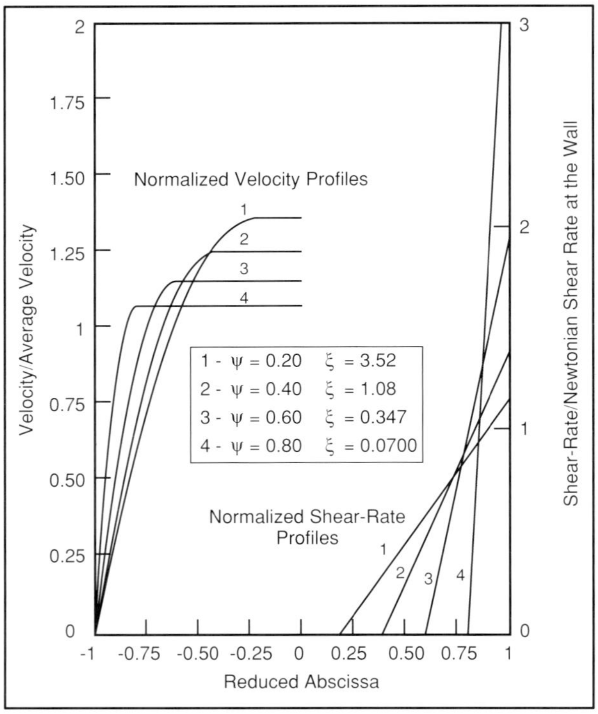

Figure 4–35—Normalized velocity- and shear-rate profiles for a Bingham plastic fluid flowing in a slot or narrow annulus (ψ = dimensionless shear stress, ξ = dimensionless shear rate).

For large concentric annuli, the flow equations were first developed by Fredrickson and Bird (1958) for power law and Bingham plastic fluids. An improved formulation for power law fluids has since been obtained by Hanks and Larsen (1979). For Bingham plastic fluids, these equations are given below.

$$\frac{32Q}{\pi D^3} = \frac{\tau_y}{\mu_p} \times \frac{1}{\psi} \times \left[(1-\alpha^4) - 2\lambda(\lambda-\psi)(1-\alpha^2) - \frac{4}{3}(1-\alpha^3)\psi + \frac{1}{3}(2\lambda-\psi)^3\psi\right] . \quad (4\text{–}85)$$

Here λ is the largest normalized distance from the pipe axis where the shear stress is equal to the yield stress of the fluid, the value of which is defined by the following implicit equation.

$$2\lambda(\lambda-\psi)\ln\left[\frac{\lambda-\psi}{\alpha\lambda}\right] - 1 + (\psi+\alpha)^2 + 2\psi(1-\lambda) = 0 , \quad (4\text{–}86)$$

where α is the radius ratio.

For power law fluids, the flow is described by

$$\frac{32Q}{\pi D^3} = \frac{4n}{3n+1} \times \left[\frac{\tau_w}{k}\right]^{1/n} \left[(1-\lambda^2)^{1+1/n} - \alpha^{1-1/n}(\lambda^2-\alpha^2)^{1+1/n}\right] , \quad (4\text{–}87)$$

where λ is the normalized distance from the pipe axis where the shear stress is zero or where the velocity reaches its maximum; its value is given by the solution of

$$\int_{\alpha}^{\lambda}\left[\frac{\lambda^2}{x} - x\right]^{1/n} dx - \int_{\lambda}^{1}\left[x - \frac{\lambda^2}{x}\right]^{1/n} dx = 0 \cdot \quad (4\text{–}88)$$

For both rheological models, the flow equations are implicit, and they can only be solved numerically. Since the narrow gap equations are much simpler to solve, the question that needs to be addressed is, "What are the errors associated with this approximation?" This really depends on the application. If one is trying to determine the flow rate corresponding to a given friction pressure this approximation is not very accurate, especially for large gap sizes, as shown in Fig. 4–36 for different Power Law Indices. Similar errors are obtained with Bingham plastic fluids.

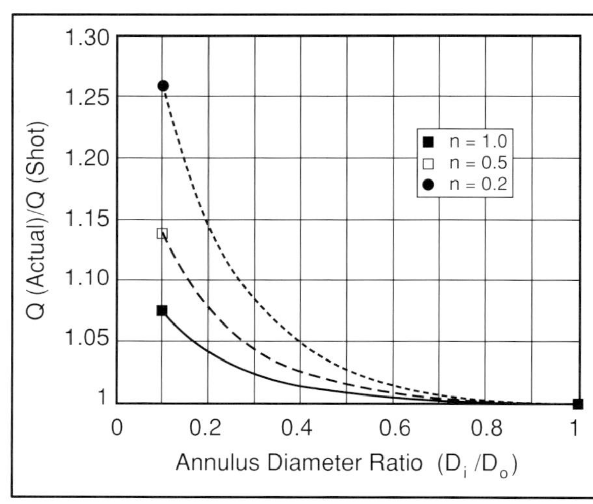

Figure 4–36—Comparison of flow rates at the same friction pressures, calculated using Eqs. 4–85 and 4–86 (or the slot approximation for different Power Law Indices).

On the other hand, when trying to do the reverse calculation (i.e., determine the friction pressure corresponding to a given flow rate), even for an annulus diameter ratio as low as 0.3 the corresponding error is lower than 2.5% both for power law and Bingham plastic fluids. This is likely to be true for any generalized non-Newtonian model, provided that the fluid is shear thinning. Therefore, it is reasonable to conclude that the narrow gap approximation is a good engineering approximation to determine laminar friction pressure of cement slurries in annuli because—

- in most circumstances, annuli are relatively narrow during cementing operations,
- for the diameter ratio in question, this approximation provides an upper limit for the friction pressures, and
- in practice, friction pressures are often negligible for large-diameter ratios.

4–6.5 Turbulent Flow in Concentric Annuli

The question which naturally arises for turbulent flow in concentric annuli is which length scale should be used in the definition of the Reynolds number. Different proposals have been made, such as $(D_o - D_i)/2$, $(D_o - D_i)$, $\sqrt{2/3}\,(D_o - D_i)$, $(2/3)(D_o - D_i)$, or even more complex expressions. Since there is little theoretical justification for using one instead of the other, the oil industry usually adopts the simplest form $(D_o - D_i)$, which in fact corresponds to the hydraulic diameter of the annulus. Therefore, the Reynolds number expression for a Newtonian fluid becomes

$$R_e = \frac{\rho V(D_o - D_i)}{\eta} \quad . \qquad (4\text{–}89)$$

When the definition of the friction factor remains the same (Eq. 4–74), the laminar flow equation for a Newtonian fluid flowing in a narrow concentric annulus is given by

$$fr = \frac{24}{Re} \quad . \qquad (4\text{–}90)$$

For this expression to remain valid for non-Newtonian fluids, following Metzner and Reed (1955), one can define the generalized Reynolds number as

$$Re_{AN} = \frac{\rho\, V^{2-n'}(D_o - D_i)^{n'}}{12^{n'-1}\, k'} \qquad (4\text{–}91)$$

and the local power law parameters n' and k' by

$$n' = \frac{d \log\ \tau_w}{d \log\,[12V_L/(D_o - D_i)]} \qquad (4\text{–}92)$$

and

$$k' = \frac{\tau_w}{\left[12V_L/(D_o - D_i)\right]^{n'}} \quad . \qquad (4\text{–}93)$$

V_L is the average velocity for the same shear stress at the wall τ_w, if the flow is laminar.

For power law fluids,

$$n' = n \qquad (4\text{–}94)$$

and

$$k' = \left[\frac{2n+1}{3n}\right]^n k \quad . \qquad (4\text{–}95)$$

Again, the main interest of these definitions is that Eq. 4–90 represents the true laminar flow equation for any non-Newtonian fluid flowing in a narrow concentric annulus.

It has already been mentioned that the definition of the Reynolds number was quite arbitrary and, therefore, it is not obvious that Eqs 4–73 and 4–77 can be used to calculate turbulent friction pressures in annuli. For Newtonian fluids, it seems that turbulent friction factors lie between the curve defined by Eq. 4–73 for low-diameter ratios D_i/D_o, and the curve corresponding to

$$\frac{1}{\sqrt{fr}} = A \times \log\left[(2/3)Re\sqrt{fr}\right] + C \qquad (4\text{–}96)$$

for high-diameter ratios (i.e., for narrow annuli) (Jones and Leung, 1981). Therefore, for the sake of simplicity, the narrow gap approximation (Eq. 4–96) can be used for all diameter ratios because, as in the case of laminar flow, it gives an upper limit for the friction factor whatever the diameter ratio is. For non-Newtonian fluids, it appears reasonable to follow the same approach and to replace Eq. 4–77 by

$$\frac{1}{\sqrt{fr}} = A_n' \times log\,[(2/3)\, Re_{AN} fr^{1-n'/2}] + C_n' \quad . \qquad (4\text{–}97)$$

This equation is different from the one which is recommended in API Spec 10 ([Eq. 4–77] with the hydraulic diameter replacing the pipe diameter in the expression for the Reynolds number). However, as in laminar flow, this approximation leads to an underestimation of the friction pressures in turbulent flow for Newtonian fluids, and is likely to do so for non-Newtonian fluids as well. Nevertheless there are good reasons for preferring Eq. 4–97 to Eq. 4–77, there is currently a lack of data on cement slurries to fully support the validity of Eq. 4–97.

4–6.6 Transition From Laminar Flow to Turbulent Flow in Annuli

In the oil industry, it is usually assumed that the transition from laminar flow to turbulent flow occurs at the same critical values of the Reynolds number in pipes and annuli, the Reynolds number being defined according to Eq. 4–91 in the latter case. However, most of the theoretical and experimental literature shows that, for annuli, the pipe values should be increased as a function of the annular diameter ratio. In particular, for Newtonian fluids flowing in narrow annuli, the critical value is approximately 2,800 for Re_1 and 3,600 for Re_2, significantly higher than the corresponding pipe values for the Reynolds number defined by Eq. 4–89. Hanks (1963) developed a theory for the flow of Bingham plastic fluids in rectangular slots and annuli, indicating that critical Reynolds numbers for narrow annuli are higher than for pipes. So although there are few published experimental data on cement slurries to validate theoretical critical Reynolds number values in annuli, one may assume provisionally that the current industry practice leads to anunderestimation of the critical flow rate for turbulent flow onset of about 15% to 20%.

4–6.7 Time-Smoothed Velocity Profiles in Pipe or Annular Turbulent Flow

To describe time-smoothed velocity profiles in turbulent flow, a distinction is usually made between three zones—a viscous sublayer close to the walls where viscous effects are dominant, the turbulent core itself away from the wall where purely viscous effects are negligible, and a transition zone in between. Each of these zones is characterized by a given range of dimensionless distance from the wall y^+, which for power law fluids is expressed by

$$y^+ = \frac{y^n \, v_f \, \rho}{k} , \qquad (4\text{–}98)$$

where

v_f = friction velocity, given by

$$v_f = \sqrt{\frac{\tau_w}{\rho}} ,$$

is a measure of the turbulent eddying. Semi-empirical formulas have been developed to describe (in each zone) the time-smoothed velocity profiles for non-Newtonian fluids flowing in pipes or annuli, and the reader is referred to the texts by Schlichting (1979) and Govier and Aziz (1972) for further details. The important conclusions for cementing applications include the following—

- for a given fluid flowing in turbulent flow, the higher the Reynolds number the flatter the time-smoothed velocity profiles;
- time-smoothed velocity profiles for power law fluids are much flatter in turbulent flow than in laminar flow; and
- for Bingham plastic fluids, the ratio of the maximum time-smoothed velocity to the average velocity increases in laminar flow up to the lower limit of the laminar transition range, and then decreases as the Reynolds number increases.

4–6.8 Flow in Eccentric Annuli

The effect of pipe eccentricity on the flow of wellbore fluids in annuli is seldom taken into account today in numerical simulators used to design or evaluate cementing operations. Nevertheless, as discussed in Chapter 5, pipe eccentricity plays a predominant role in the mud-circulation and mud-displacement processes.

The effect of eccentricity on velocity profiles and pressure gradients of non-Newtonian fluids flowing in annuli has been the subject of several publications (McLean et al., 1967; Mitsuishi and Aoyagi, 1973; Iyoho and Azar, 1981; Luo and Peden, 1987). Since there is no simple analytical solution to such a difficult problem, especially for fluids exhibiting a yield stress, several simplified approaches have been adopted. It is only recently, however, that full numerical solutions for the flow of Bingham plastic fluids in eccentric annuli have been developed (Walton and Bittleston, 1990). Going into the details of these models goes beyond the scope of this chapter and, since most of them have not been fully validated, the author has chosen to adopt a simple model to present the qualitative effect of casing eccentricity on circulation efficiency. This model has been used by several authors in a more or less similar manner and for different purposes—notably for mud removal (McLean et al., 1967) and for cuttings transport (Iyoho and Azar, 1981). The eccentric annular geometry is considered as being equivalent to a series of independent rectangular slots of varying heights (Fig. 4–37)[4] The model is referred to as the *basic slot model*. For a fixed pressure gradient, the contribution of each angular sector to the flow rate is determined using the equations given in Appendix A. The reverse problem of calculating the friction pressure knowing the flow rate is then solved numerically. Thus, this model is based on a narrow annulus approximation where the annular gap is assumed to vary slowly with azimuthal position; therefore, results will be presented only for a high-diameter ratio (i.e., $D_i/D_o = 0.8$).

Notice that in the following developments, eccentricity ε is defined as the distance between the axis of the cylinders divided by the average annular gap; however, following the common practice in the oil industry, the *pipe standoff STO*, defined in API Spec 10, where $STO = (1 - \varepsilon) \times 100$, will be used.

[4] For fluids exhibiting a yield stress, this approximation introduces errors which lead to an incorrect description of the plug flow on the wide side of the annulus (Walton and Bittleston, 1990).

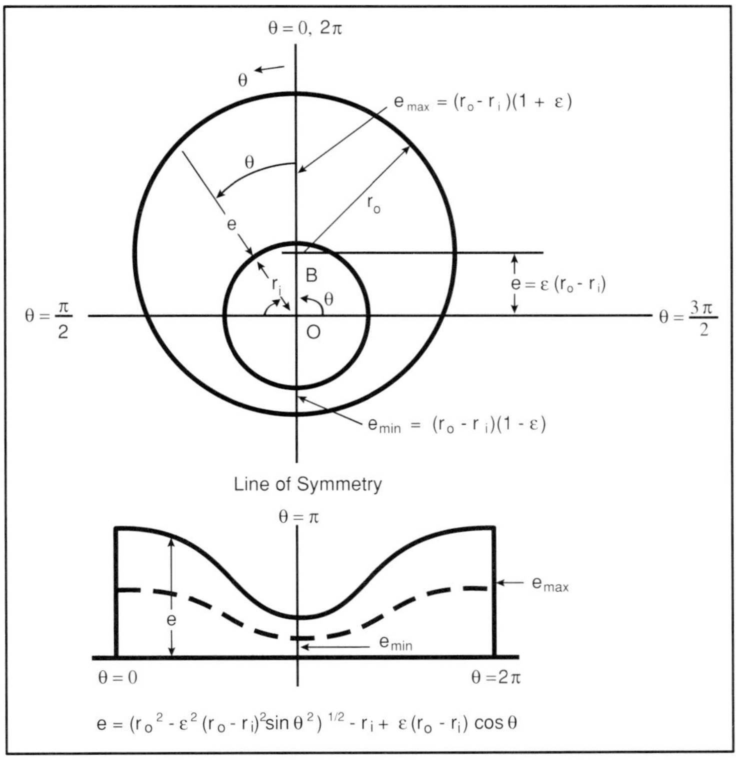

Figure 4–37—Profile of the slot equivalent to the eccentric annulus (after Iyoho and Azar, 1981).

The major effect of eccentricity is to distort the velocity distribution around the annulus, the flow favoring the widest part of the annulus as opposed to the narrowest part (Fig. 4–38). As will be discussed later, since both the velocity and the annular gap vary azimuthally around the annulus, some parameters must now be defined locally. For example, the local Reynolds number for a given annular gap e can be defined by—

$$Re(e) = \frac{\rho\, v(e)^{2-n}\,(2e)^{n'}}{12^{n'-1}\,k'}, \qquad (4\text{–}99)$$

where $v(e)$ is the average velocity along the local annular gap e.

First, the situations are considered where the fluid is in laminar flow all around the annulus, i.e., all local Reynolds numbers are smaller than the critical Reynolds number Re_1. For power law fluids, one can show that the velocity distribution is a function of the annular diameter ratio, the standoff, and the Power Law Index. Since only narrow annuli are considered, the velocity profiles will depend only on two parameters, i.e., STO and n. To get an idea of the distortion of the velocity distribution due to eccentricity, it is worthwhile to calculate the ratio of the average velocity along the widest and the narrowest annular gap (V_{max} and V_{min}, respectively) to the average velocity through the total section area (V). These two parameters are plotted in Fig. 4–39 as a function of API standoff for three different Power Law Indices. As can be seen, as standoff decreases, the average velocity on the widest side first increases and then levels off, while on the narrowest side it decreases quickly toward negligible values. It is also noticeable that when the Power Law Index is low, the distortion of the velocity distribution is worse.

For fluids exhibiting a yield stress, the reduction of the velocity on the narrow side can be even more drastic because the shear stress may locally be lower than the yield stress of the fluid, which implies that the local velocity of the fluid is zero. For Bingham plastic fluids in particular, the dimensionless parameters relevant to the velocity distribution are the same as for a power law fluid, except that the Power Law Index is replaced by the dimensionless shear rate ξ. The effect of ξ on the velocity distribution is similar to that of the Power Law Index (i.e., the lower the value of ξ, the worse the distortion of the velocity distribution). A critical value for the dimensionless shear rate ξ for the fluid to start to flow on the narrow side of the annulus can be defined as discussed in Chapter 5. Pipe eccentricity also affects the friction-pressure/flow-rate relationship. A typical example of the reduc-

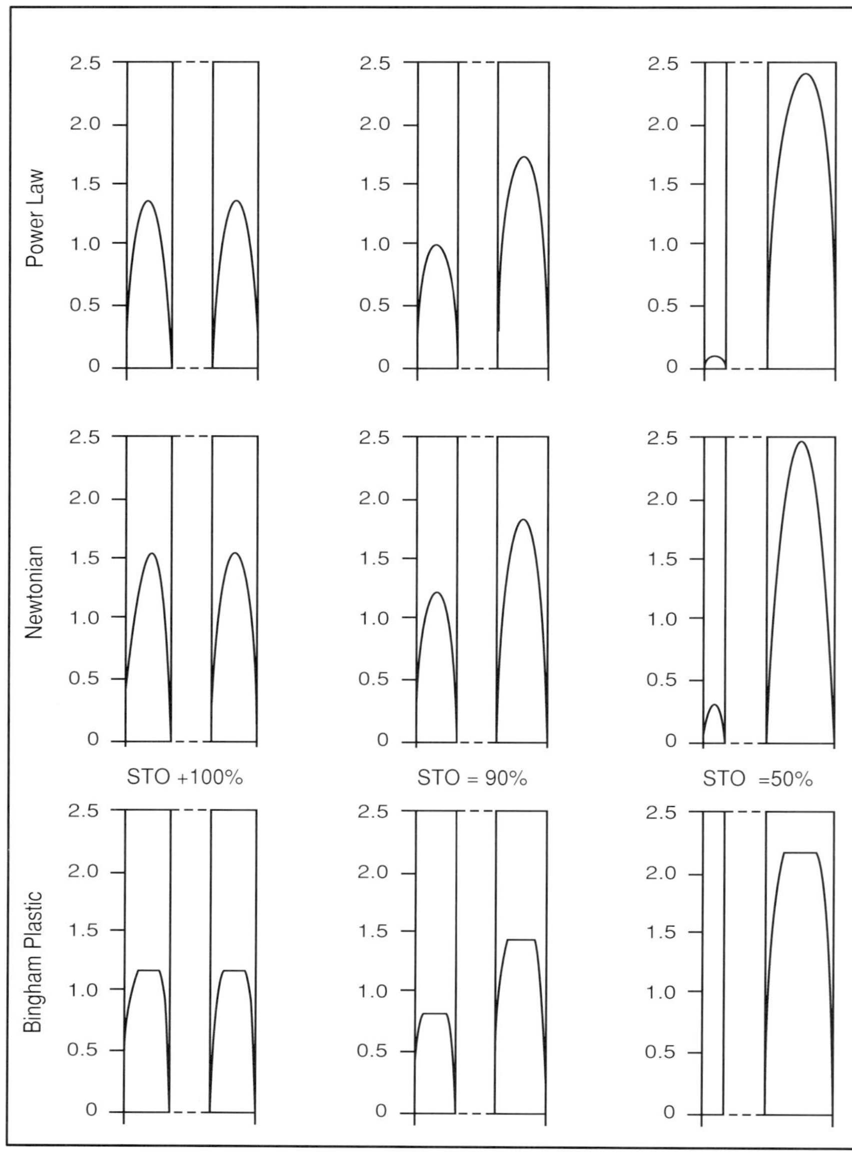

Figure 4–38—Typical example of velocity profile on the narrow and wide sides of eccentric annuli for model fluids.

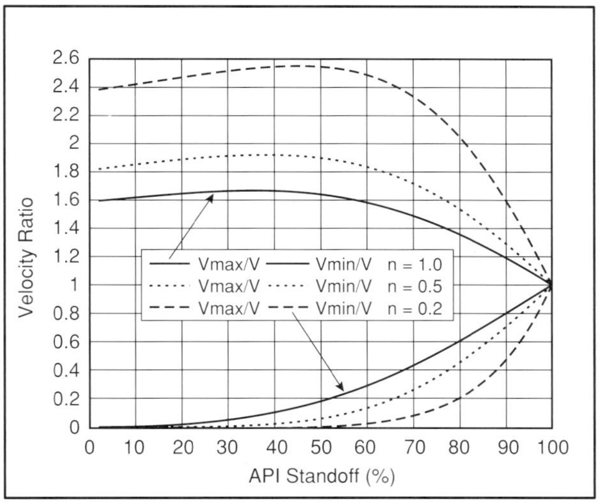

Figure 4–39—Ratio of average velocities along the wide and narrow side of the annulus to the total average velocity for different Power Law Indices. Values are calculated using the basic slot model in laminar flow.

tion of friction pressures due to eccentricity is shown in Fig. 4–40, where the ratios of friction pressures calculated for eccentric annuli to those calculated for the corresponding concentric annulus are plotted against standoff for a power law fluid with a Power Law Index of 0.5. For narrow annuli, the relative reduction of friction pressure depends on the Power Law Index for power law fluids, and on the dimensionless shear rate for Bingham plastic fluids. Theoretically, it can vary between 1.0 and slightly less than 0.4 for shear thinning fluids.

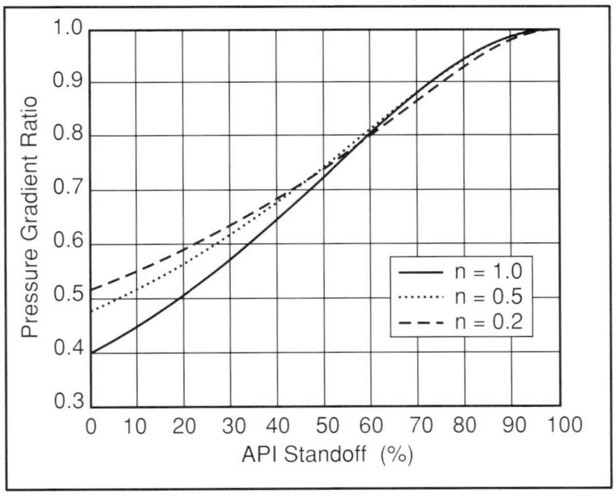

Figure 4–40—Eccentric annulus/concentric annulus friction-pressure ratios (D_i/D_o = 0.8) for different Power Law Indices calculated using the basic slot model—laminar flow.

As mentioned earlier, the uneven velocity distribution in eccentric annuli also has a consequence on transition to turbulent flow. Because the Reynolds number value depends on the local average velocity and on the annular gap, parameters that both vary around the annulus, turbulence is likely to appear first in the sector with the maximum separation between the walls and then to extend all around the annulus as flow increases. The consequence of this is that the laminar flow and turbulent flow regimes can coexist in a given eccentric annulus. Using the basic slot model, one can define the following—

- an average critical Reynolds number Re_{1ecc} at which purely laminar flow ends on the wide side of the annulus, and which therefore represents the maximum average Reynolds number for the fluid to be in purely laminar flow all around the annulus, and
- an average critical Reynolds number Re_{2ecc} at which the fluid begins to be in full turbulent flow on the narrow side of the annulus, and which therefore represents the minimum average Reynolds number for the fluid to be in full turbulent flow all around the annulus.

These two parameters, normalized to the corresponding values for a concentric annulus, Re_1 and Re_2, are plotted in Figs. 4–41 and 4–42 for power law fluids with three different Power Law Indices. These curves, which are typical of any nonelastic shear thinning fluid, show that as standoff decreases there is a wider and wider average Reynolds number range in which both flow regimes coexist, the flow regime starting to be turbulent on the wide side of the annulus earlier than expected from concentric flow calculations, and remaining laminar on the narrow side of the annulus later than expected from concentric flow calculations. Notice also that Re_{1ecc} is dependent on the Power Law Index but Re_{2ecc} is not.

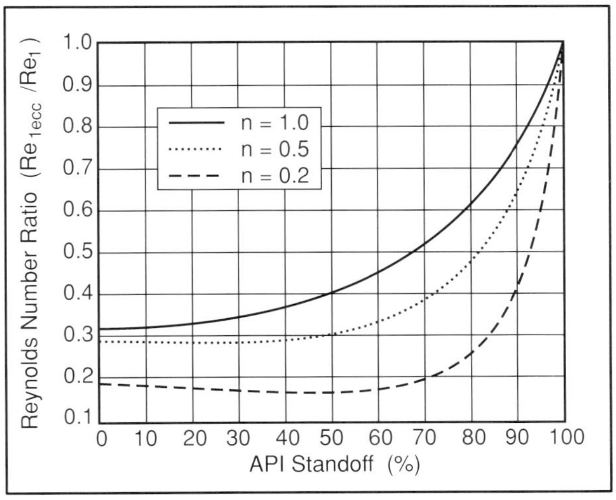

Figure 4–41—Maximum normalized average Reynolds numbers for different power law fluids to be in laminar flow around an eccentric annulus (D_i/D_o = 0.8).

When a given fluid is in turbulent flow all around the annulus, the velocity distribution is less distorted, and the friction pressures less reduced by eccentricity than in laminar flow, as shown by Figs. 4–43 and 4–44.

As stated earlier, although the results presented above have not been fully validated quantitatively, their trends give at least a qualitative idea of the effect of pipe standoff on the flow of wellbore fluids in the annulus. One can conclude that the effect of eccentricity should be taken into account when—

- friction pressures play a significant role, for example U-tubing prediction in relatively small annuli, or friction pressures for slim holes, and

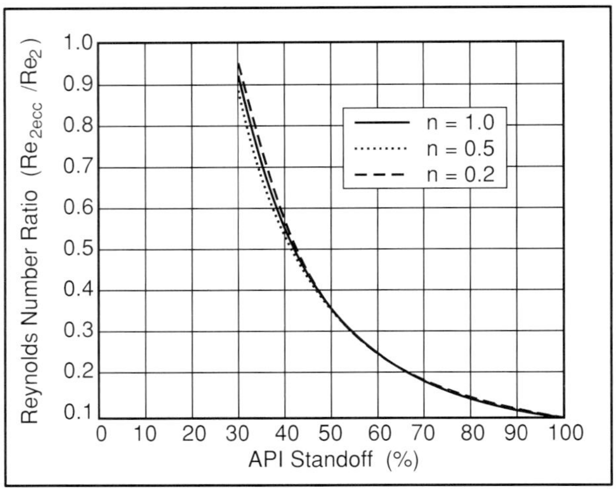

Figure 4–42—Minimum normalized average Reynolds numbers for different power law fluids to be in turbulent flow around an eccentric annulus ($D_i/D_o = 0.8$).

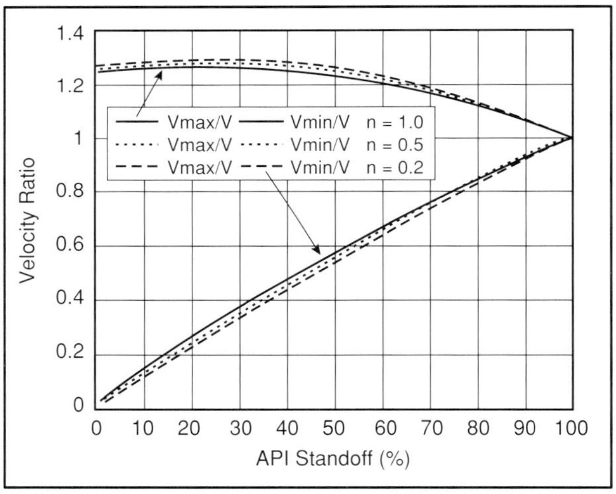

Figure 4–43—Ratio of average velocities on the wide and narrow side of the annulus to the total average velocity for different Power Law Indices. Values are calculated using the basic slot model in turbulent flow.

- velocity distribution plays a significant role, e.g., mud circulation.

4–7 CONCLUSIONS

The accurate and reliable rheological characterization of oilwell cement slurries still presents a problem for the industry. These fluids exhibit a fairly complex rheological behavior which depends not only on fluid composition and on the mixing procedure, as explained in other chapters, but also on shear history, temperature, and on the particular testing procedures used.

It has been shown that the coaxial cylinder viscometer, which is commonly used for measuring the rheological properties of cement slurries, can suffer (unless extreme care is taken) from severe limitations as a result of particle migration, end effects, or slippage at the rheometer wall. Similar problems, if not worse, have been encountered in attempting to use pipe-flow viscometers, and it seems that coaxial cylinder and vane rheometers still remain the most useful instruments for characterizing cement slurries. It is clear that research studies in this area are still needed to further optimize the testing equipment and procedures. Nevertheless, the current standard procedures used in the industry could be improved by running speed hysteresis loops which would allow the user to detect any time-dependent effects during the measurement, whether due to the thixotropic properties of the fluid or to the migration of particles.

Equations describing the flow of cement slurries in pipes and annuli have focused on two widely used rheological models—the power law and the Bingham

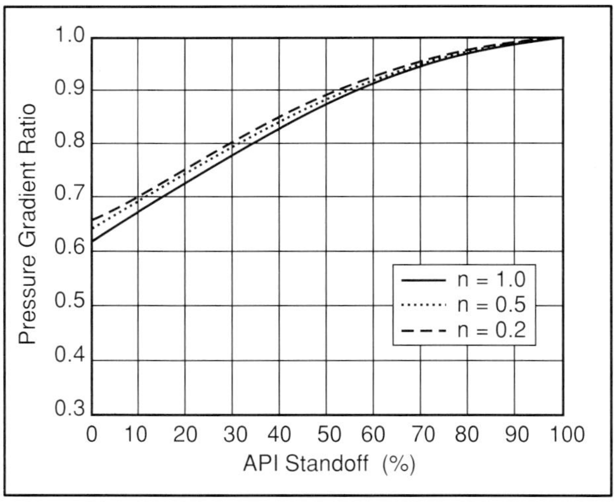

Figure 4–44—Eccentric annulus/concentric annulus friction-pressure ratios ($D_i/D_o = 0.8$) for different Power Law Indices calculated using the basic slot model—turbulent flow.

plastic models. Their limitations have also been discussed, and some more realistic alternatives presented. Because of the lack of experimental data, the validity of many of these equations (at least in turbulent flow) is not yet fully established. However, further progress in this domain depends on a better and more complete characterization of the rheological properties of cement slurries to enable unambiguous experimental evaluation of the fluid mechanics.

NOMENCLATURE

A, A_n'	—	Constants in the friction-factor/ Reynolds number relationship
C, C_n'	—	Constants in the friction-factor/ Reynolds number relationship
C_s	$N^{1/n}\ m^{1-1/n}\ s^{-1}$	Slip coefficient in a pipe
D	m	Inner diameter of a pipe
D_o, D_i	m	Outer and inner diameter of an annulus
e	m	Thickness of a rectangular slot or local annular gap
fr	—	Fanning friction factor
g_z	$m\ s^{-2}$	Component of the gravity acceleration in the main direction of the flow
He	—	Hedstrom number
k	Pa sn	Consistency Index of a power law fluid, or constant in other rheological models
k'	Pa sn′	Local Consistency Index
k'_R	Pa sn	Diameter-dependent Local Consistency Index
L	m	Length of a pipe, an annulus, or a coaxial cylinder viscometer geometry
n	—	Power Law Index of a power law fluid, or constant in other rheological models
n'	—	Local Power Law Index
P^*	Pa	Total pressure
p	Pa	Frictional pressure
Q	$m^3\ s^{-1}$	Volumetric flow rate
r	m	Distance from pipe axis or from the plane of symmetry of a rectangular slot
r_y	m	Shortest distance from rotational axis of a coaxial cylinder viscometer where the shear stress is zero
R	m	Inner radius of a pipe
R_a	m	Mean radius of a coaxial cylinder viscometer
R_o, R_i	m	Outer and inner radius of an annulus
R_1, R_2	m	Inner and outer radius of a coaxial cylinder viscometer
Re	—	Reynolds number
Re_{BG}	—	Bingham plastic Reynolds number
Re_{MR}	—	Metzner and Reed Reynolds number for a pipe
Re_{AN}	—	Generalized Reynolds number for a narrow annulus
Re_1, Re_2	—	Critical Reynolds number for the upper limit of the.laminar flow regime and the lower limit of the turbulent flow regime, respectively
Re_{1ecc}	—	Critical Reynolds number for the upper limit of the laminar flow regime on the wide side of an eccentric annulus
Re_{2ecc}	—	Critical Reynolds number for the lower limit of the turbulent flow regime on the narrow side of an eccentric annulus
s	—	Coaxial cylinder viscometer radius ratio R_2/R_1
STO	%	API standoff
T	N-m	Measured or imposed torque on a rotational viscometer
v	$m\ s^{-1}$	Velocity of a fluid particle
v_f	—	Friction velocity
V	$m\ s^{-1}$	Volumetric flow rate per unit of section area
V_L	$m\ s^{-1}$	Volumetric flow rate per unit of section area corresponding to a given shear stress at the wall assuming flow regime is laminar
w	m	Width of a rectangular slot
y^+	—	Dimensionless distance from a wall
z	m	Axial coordinate in the main direction of flow

α	—	Annulus diameter ratio D_i/D_o
ε	—	Eccentricity of an annulus
$\dot{\gamma}$	s^{-1}	Shear rate
$\dot{\gamma}_a$	s^{-1}	Average shear rate in a coaxial cylinder viscometer
$\dot{\gamma}_w$	s^{-1}	Shear rate at the wall of a pipe or of a narrow concentric annulus
$\dot{\gamma}_{Nw}$	s^{-1}	Newtonian shear rate at the wall of a pipe or of a narrow concentric annulus
$\dot{\gamma}_1, \dot{\gamma}_2$	s^{-1}	Shear rate at inner and outer cylinder surface of a coaxial cylinder viscometer
η	Pa s	Shear rate dependent viscosity or viscosity of a Newtonian fluid
λ	—	Normalized minimum distance from the axis of symmetry a concentric annulus annulus where shear stress is nil
μ_p	Pa s	Plastic viscosity of a Bingham plastic fluid
ρ	kg m^{-3}	Fluid density
τ	Pa	Shear stress
τ_a	Pa	Average shear stress in a coaxial cylinder viscometer
τ_w	Pa	Shear stress at the wall of a pipe or of a narrow concentric annulus
τ_y	Pa	Fluid yield stress
τ_1, τ_2	Pa	Shear stress at inner and outer cylinder surface of a coaxial cylinder viscometer
Ω	rad s^{-1}	Rotational velocity
ξ	—	Dimensionless shear rate
ψ	—	Dimensionless shear stress

REFERENCES

American Petroleum Institute: "API Spec 10," *Specifications for Materials and Testing for Well Cements*, fourth edition, API, Dallas (1988).

Bannister, C. E.: "Rheological Evaluation of Cement Slurries: Methods and Models," paper SPE 9284, 1980.

Beirute, R. M.: "API Revises Procedures to Measure Cement Slurry Rheology," *Oil & Gas J.* (Sept. 22, 1986) 36–38.

Bird, R. B., Stewart, W. E., and Lightfoot, E. N.: *Transport Phenomena*, John Wiley & Sons, New York (1960).

Bird, R. B., Armstrong, R. C., and Hassager, O.: *Dynamics of Polymeric Liquids*, second edition, Wiley, New York (1979).

Casson, N.: "Flow Equation for Pigment-Oil Suspensions of the Printing Ink Type," *Rheology of Dispersed Systems* (C.C. Mill, ed.), Pergamon Press, Oxford (1959) 84–104.

Chow, T. W., McIntire, L. V., Kunze, K. R., and Cooke, C. E.: "The Rheological Properties of Cement Slurries: Effect of Vibration, Hydration Conditions and Additives," *SPEPE* (Nov. 1988) 543–550.

Denis, J. H., and Guillot, D. J.: "Prediction of Cement Slurry Laminar Pressure Drops by Rotational Viscometry," paper SPE 16137, 1987.

Dodge, D. W. and Metzner, A. B.: "Turbulent Flow of non-Newtonian Systems," *AIChE J.* (June 1959), **5**, No. 2, 189–204.

Fredrickson, A. G. and Bird, R. B.: "Non-Newtonian Flow in Annuli," *Ind. & Eng. Chem.* (March 1958) **50**, No. 3, 347–352.

Govier, G. W. and Aziz, K.: *The Flow of Complex Mixtures in Pipes*, R. E. Krieger Publishing Co., Malabar, FL (1972).

Guillot, D. J. and Denis, J. H.: "Prediction of Laminar and Turbulent Friction Pressures of Cement Slurries in Pipes and Centered Annuli," paper SPE 18377, 1988.

Haimoni, A. M.: "Rheology of a Specific Oilwell Cement," Thesis, Surrey U., England (Dec. 1987).

Hanks, R. W.: "The Laminar-Turbulent Transition for Fluids with a Yield Stress," AIChE J. (1963) **9**, No. 3, 306–309.

Hanks, R. W. and Larsen, K. M.: "The Flow of Power Law non-Newtonian Fluids in Concentric Annuli," *Ind. & Eng. Chem. Fundamentals*, (1979) **18**, No. 1, 33–35.

Hannant, D. J. and Keating, J.: "Equipment for Assessing the Development of Structure in Fresh Cement Pastes by the Measurement of Shear Modulus," *Cem. & Concrete Res.* (1985) **15**, 605–612.

Hedstrom, B. O. A.: "Flow of Plastics Materials in Pipes," *Ind. & Eng. Chem.* (March 1952) **44**, No. 3, 651–656.

Herschel, W. H. and Bulkley, R.: "Konsistenzmessungen von gummi-benzöllösungen," *Kolloid-Z* (1926) **39**, 291.

Iyoho, A. W. and Azar, J. J.: "An Accurate Slot Flow Model for non-Newtonian Fluid Flow Through Eccentric Annuli," *SPEJ* (Oct. 1981) 565–572.

Jones, O. C., Jr. and Leung, J. C. M.: "An Improvement in the Calculation of Turbulent Friction in Smooth Concentric Annuli," *J. Fluids Eng.* (1981) **103**, 615–623.

Lapasin, R., Papo, A., and Rajgelj, S.: "Flow Behavior of Fresh Cement Pastes. A Comparison of Different Rheological Instruments and Techniques," *Cem. & Concrete Res.* (1983) **13**, 349–356.

Luo, Y. and Peden, J. M.: "Flow of Drilling Fluids Through Eccentric Annuli," paper SPE 16692, 1987.

Mannheimer, R. J.: "Rheological Evaluation of Cement Slurries," Report No. SwRI 6836, prepared for API (May 1982).

Mannheimer, R. J.: "Effect of Slip on Flow Properties of Cement Slurries Can Flaw Resistance Calculations," *Oil & Gas J.* (Dec. 5, 1983) 144–147.

Mannheimer, R. J.: "Laminar and Turbulent Flow of Cement Slurries in Large Diameter Pipe," Final Report No. SwRI-8983, prepared for API (September 1988).

McLean, R. H., Manry, C. W., and Whitaker, W. W.: "Displacement Mechanics in Primary Cementing," *JPT* (Feb. 1967) 251–260.

Metzner, A. D. and Reed, J. C.: "Flow of non-Newtonian Fluids—Correlation of the Laminar and Turbulent Flow Regions," *AIChE J.* (1955) **1**, 434–440.

Mitsuishi, N. and Aoyagi, Y.: "Non-Newtonian Fluid Flow in an Eccentric Annulus," *J. Chem. Eng. Japan* (1973) **6**, No. 5, 402–408.

Mooney, M.: *J. Rheology* (1931) **2**, 210.

Oldroyd, J. G.: *J. Colloid Sci.* (1949) **4**, 333.

Parzonka, W. and Vocadlo, J.: "Méthode de la caractérisitique du comportement rhéologique des substances viscoplastiques d'aprés les mesures au viscosim tre de Couette (modéle nouveau trois paramétres)," *Rheologica Acta* (1968) **7**, 260–265.

Robertson, R. E. and Stiff, H. A. Jr.: "An Improved Mathematical Model for Relating Shear Stress to Shear Rate in Drilling Fluids and Cement Slurries," *SPEJ.* (Feb. 1976) 31–36.

Ryan, N. W. and Johnson, M. M.: "Transition from Laminar to Turbulent Flow in Pipes," *AIChE J.* (Dec. 1959) **5**, No. 4, 433–435.

Sabins, F. L., Tinsley, J. M., and Sutton, D. L.: "Transition Time of Cement Slurries Between the Fluid and Set State," paper SPE 9285, 1980.

Savins, J. G. and Roper, W. F.: "A Direct-Indicating Viscometer for Drilling Fluids," *Drill. and Prod. Prac.*, API (1954) **7**.

Schlichting, H.: *Boundary Layer Theory*, McGraw-Hill Book Co., New York (1979).

Shah, S. N. and Sutton, D. L.: "New Friction Correlation for Cements From Pipe and Rotational Viscometer Data," paper SPE 19539, 1989.

Speers, R. A., Holme, K. R., Tung, M. A., and Williamson, W. T.: "Drilling Fluid Shear Stress Overshoot Behavior," *Rheol. Acta* (1987) **26**, No. 5, 447–452.

Tattersall, G. H.: "Present Problems Associated With the Study of Cement Paste Rheology," *Fresh Concrete: Important Properties and Their Measurement, Proc.* RILEM Seminar, Leeds U. (March 1973) **1**, 2.3–1 to 2.3–18.

Taylor, G. I.: "Stability of a Viscous Liquid Contained Between Two Rotating Cylinders," *Trans. Royal. Soc. London* (1923), Ser. A223, 289–293.

Uner, D., Ozgen, C., and Tosun, I.: "Flow of a Power Law Fluid in an Eccentric Annulus," *SPEDE* (Sept. 1989) 269–272.

Walters, K.: *Rheometry*, Chapman & Hall, London (1975).

Walton, I. C. and Bittleston, S. H.: "The Flow of a Bingham Plastic Fluid in a Narrow Eccentric Annulus," *J. Fluid Mech.* (1990).

Whorlow, R. W.: *Rheological Techniques*, Ellis Horwood Ltd., Chichester (1980).

5

Mud Removal

Dominique Guillot, Hugo Hendriks,
Françoise Callet, and Benoit Vidick

Dowell Schlumberger

5-1 INTRODUCTION

The main objective of a primary cement job is to provide complete and permanent isolation of the permeable zones located behind the casing. To meet this objective, the drilling mud and the preflush (if any) must be fully removed from the annulus, and the annular space must be completely filled with cement slurry. Once in place, the cement must harden and develop the necessary mechanical properties to maintain a hydraulic seal throughout the life of the well. Therefore, good mud removal and proper slurry placement are essential to obtain well isolation.

Incomplete mud displacement can leave a continuous mud channel across the zones of interest, thereby favoring interzonal communication. Bonding and cement seal durability are also related to the efficiency of the displacement process. This is why mud displacement has been a topic of interest for such a long time in the well cementing community.

Research concerning the cement placement process began in the 1930s. Some key factors influencing primary cement job failures were identified, and solutions were proposed as early as 1940. Using a large-scale simulator, Jones and Berdine (1940) showed that poor zonal isolation could be attributed to channeling of the cement slurry through the mud, a phenomenon which they found to be "promoted" by casing eccentricity. The presence of a residual mud cake at the cement/formation interface was also identified as a cause of poor mud displacement. To minimize cement channeling, Jones and Berdine (1940) proposed to centralize the casing. They also found effective ways to remove the mud cake, including fluid jets, scrapers or scratchers, casing reciprocation, and possibly pumping acid ahead of the cement slurry.

In spite of this early work, mud displacement remains a subject of much current experimental and theoretical work. This is partly due to the increasing complexity of the problem (deeper wells, deviated wells, etc.). However, the major difficulty arises from the fact that both the experimental and theoretical approaches suffer from severe limitations. At first glance, a theoretical approach seems quite attractive, because there are major drawbacks associated with the experimental devices.

- One of the key parameters in the mud removal process, the length-to-annular-gap ratio, is limited. In the laboratory, one is typically limited to a ratio of no more than 500, while in the field this parameter is on the order of 10^4. This prevents the observation of axial deformation of the interface between two fluids, because of the eccentricity, on a long length scale. For example, the effect of gravity in eccentric annuli cannot be fully investigated for this reason. One may argue that, in theory, the length-to-annular-gap ratio could be maintained with an experimental device (reducing the annular gap to a very small value); unfortunately, this is extremely difficult because dynamic similarity requires that all the dimensionless parameters (of which there are at least six) be matched to the corresponding field values.
- Secondly, in view of the number of parameters involved, an experimental investigation of the displacement efficiency over the complete dimensionless parameter space, for the displacement of one history-independent fluid by a second, would represent an enormous amount of work.

Great care should taken in attempting to extrapolate experimental results outside of the parameter ranges within which they have been obtained. It is also important to mention that some of the key parameters (such as rheology) are sometimes very difficult to measure (Chapter 4). In addition, in most of the published experimental studies where cement was used as the displacing fluid, very little information is available regarding its compatibility or incompatibility with the displaced mud. It is now well known that this may strongly affect the results.

The theoretical approach also has its limitations. The complete modeling of the displacement process presents a formidable task, even for the most sophisticated computers. For example, one must contend with unsteady mass and momentum transfer between non-Newtonian fluids of different properties in an asymmetric geometry. Until recently (Walton and Bittleston, 1990), calculation of the velocity field for a single non-Newtonian fluid flowing in an eccentric annulus was limited by computational power and the availability of numerical techniques. In addition, some of the key parameters are rarely taken into account (e.g., fluid thixotropy, static and dynamic filter-cake deposition and erosion, and chemical interactions between fluids). Usually, the fluids are assumed to be separated by a sharp interface with no interfacial mixing. It is worthwhile to mention that attempts to model interfacial instabilities due to density or viscosity differences are still at an early stage. They are often limited to two dimensions, and valid only for Newtonian fluids.

It is obvious that further progress in the area of mud displacement can only be achieved by a combination of experimental and theoretical studies. This approach is currently being followed by several research groups. Unfortunately, this was rarely the case in the past, and it is not surprising that there is no consensus today regarding the optimum design of a primary cement job for successful mud removal and cement placement. In this chapter, an attempt is made to distinguish the areas of consensus and those of controversy. Field data supporting successful cementing practices are cited. However, it is difficult to quantify the success of a job bearing in mind that—

- controlling and monitoring an actual cement job is not as easy as with an experimental model,
- the properties of fluids as mixed in the field are not often measured, and they can be quite different from laboratory-mixed systems (Section 5–5), and
- totally different evaluation techniques are used, and interpretation is sometimes not straightforward (Chapter 16).

This chapter is organized as follows. First, the problem of mud conditioning and circulation is addressed. This is a key point, because the success of a primary cement job may depend on the suitability of the well for cementing. Next, a discussion of mud displacement is presented. The ideal case of mud displacement, occurring in a concentric annulus between two impermeable walls (the so-called "mobile" mud), is discussed first. Second, the effects of eccentricity are considered. Third, the difficult problem of removing the "immobile" mud, which can easily be bypassed by the displacing fluid, is discussed. Fourth, it will be shown how casing movement can help to overcome some of the previously described problems, and contribute to the success of critical primary cementing jobs. Fifth, the problem of the interactions between the mud and cement slurry, which often necessitates the use of spacers and washes, is discussed. Sixth, the influence of density fluctuations and mixing energy on cement slurry properties is presented. The chapter concludes with qualitative recommendations for achieving successful mud removal and cement placement.

5–2 DISPLACEMENT EFFICIENCY

The most commonly used parameter for defining the ability of a given fluid to displace another is the *displacement efficiency* (Eff*). Consider an annulus of volume Vol and length L which is filled with Fluid No. 1 (the fluid to be displaced) flowing at a given volumetric rate Q (Fig. 5–1). At time $t = 0$, Fluid No. 2 (the displacing fluid) suddenly replaces Fluid No. 1 at the inlet of the annulus ($Z = 0$). At any time $t > 0$, the displacement efficiency is defined as the fraction of annular volume occupied by Fluid No. 2. In other words, for case (d) in Fig. 5–1, the displacement efficiency would be the shaded area divided by the area between $Z = 0$ and $Z = L$. The natural time scale which allows one to define a nondimensional time t^* is the ratio of the flow rate Q to the annular volume Vol.

$$t^* = t \times \frac{Q}{Vol} \qquad (5\text{–}1)$$

This parameter is equivalent to the number of annular volumes pumped. Notice that with these definitions, the displacement efficiency is equal to t^* [case (b)] until Fluid No. 2 appears for the first time at the outlet of the annulus [case (c)]. This is defined as the *breakthrough time* t_b. Afterward [case (d)], the displacement efficiency asymptotes to a constant value which may be less than one, indicating that the annulus still contains Fluid No. 1 (Fig. 5–2).

The displacement efficiency concept can sometimes be misleading, especially for high eccentricities. Under such circumstances, a large channel of bypassed mud in the narrow part of the annulus would correspond to a small proportion of the total flow area. This situation is illustrated in Fig. 5–3.

5–3 WELL PREPARATION

5–3.1 Borehole

As stated by C. W. Sauer (1987) in his review on the state of the art of mud displacement, "it should not be believed that the cement job should go all right regardless of what

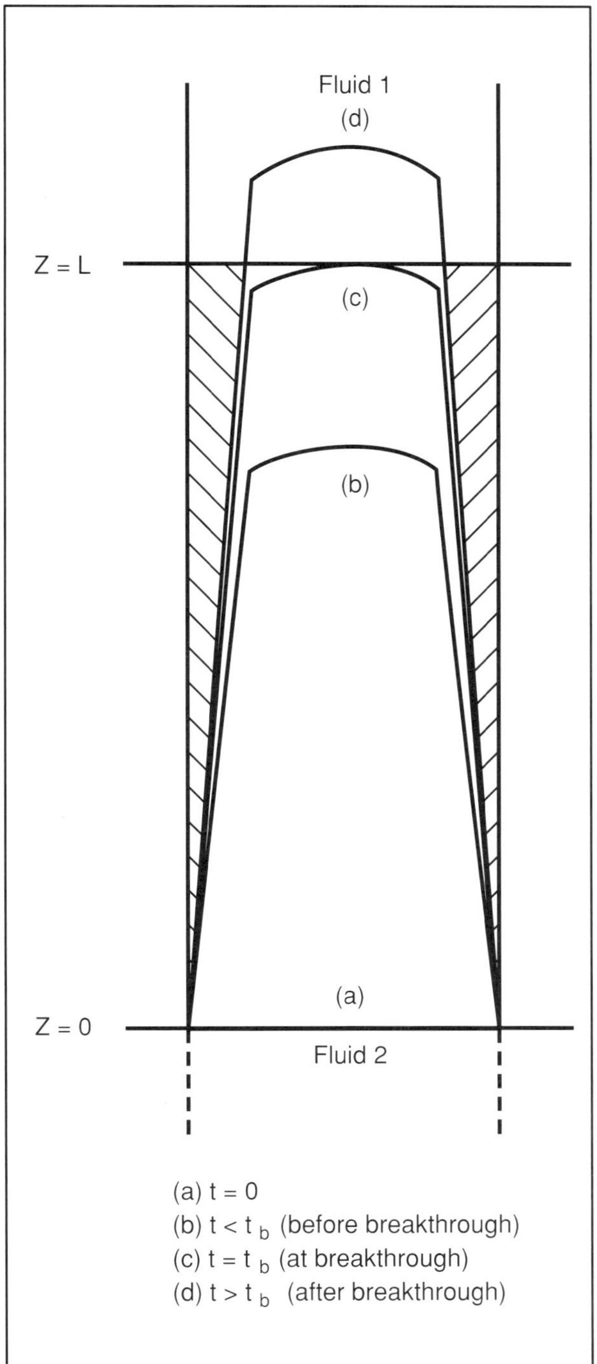

Figure 5–1—Schematic of interface profiles at different times during the displacement of Fluid No. 1 by Fluid No. 2.

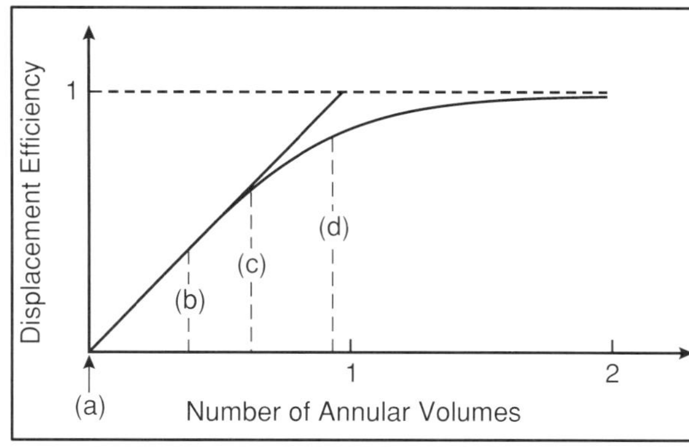

Figure 5–2—Schematic of a displacement efficiency curve.

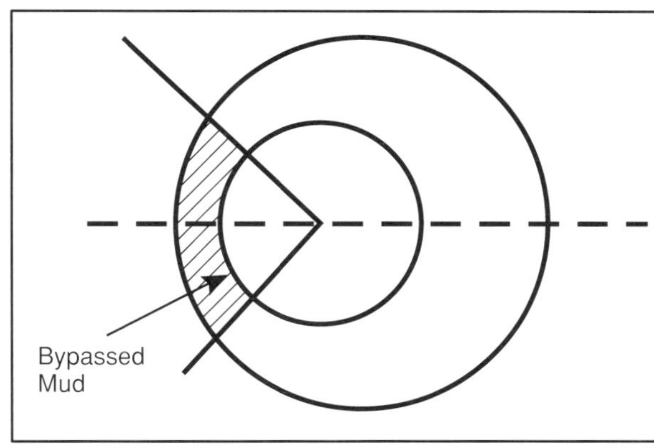

Figure 5–3—Schematic diagram of bypassed mud in an eccentric annulus.

else is or has taken place during the drilling to casing point." A poorly drilled hole may have several washed out zones which are difficult to clean out, regardless of the displacement rate. Furthermore, these washed out pockets tend to hold gelled and/or dehydrated mud which may be dragged out by the cement slurry, contaminating the cement column above. Crooked holes make casing centralization difficult; consequently, the removal of the mud from the narrow side of the annulus is problematic. Poorly treated mud could induce washouts or thick filter cakes which would be difficult if not impossible to remove. While good drilling practices do not guarantee a successful cement job, they may prevent a failure. Although it is understandable that the objective for the drilling engineer is to drill the well safely, and as fast and economically as possible, this should be accomplished bearing in mind that one of the ultimate goals during drilling is to provide the optimum wellbore for successful cementing—

- a well with controlled subsurface pressures,
- a smooth hole with a minimum number of doglegs,
- an in-gauge hole,
- a stabilized borehole,
- a hole cleaned from cuttings, and
- a correctly treated mud that will give thin, dynamic filter cakes in front of permeable zones.

Unfortunately, such an ideal situation cannot always be achieved. Therefore, cement placement techniques must often be designed to minimize the influence of poor well preparation.

5-3.2 Mud Conditioning

Drilling muds have properties which are designed to facilitate drilling operations and provide proper cuttings transport, but are not necessarily conducive to efficient mud displacement. Therefore, it may be necessary to *condition* the mud, i.e., to modify its properties. Prior to placing cement in the wellbore, two mud characteristics can be changed—density and rheology. Anticipating the best conditions for displacement, it is desirable to reduce the mud density to the minimum wellbore density limit. However, the mud density is usually maintained close to the wellbore pressure limit during drilling. Reducing the mud's gel strength, yield stress, and plastic viscosity is recognized as being very beneficial, because the driving forces necessary to displace the mud are reduced, and its mobility is increased. Of course, this can only be done when the cuttings have been cleaned from the borehole. Care must also be taken to prevent the settling of weighting agents. This may represent a major constraint for highly deviated holes (Keller et al., 1983; Chapter 15).

The mud rheology can be modified by adding water (which also reduces the density) or dispersants to the mud at the surface. It is necessary to circulate the mud until its rheological properties are within the desired range. This necessitates circulation for at least one hole volume, and ideally should be done before removing the drillpipe. Otherwise, unconditioned mud may have sufficient time to gel during the pseudostatic period (while removing drillpipe, logging, and running casing).

The running of casing should be performed carefully to avoid fracturing the formations. The equivalent flow rate in the annulus (Q_{ann}) as a function of the speed (V_{run}) at which the casing is run is given by the following equation.

$$Q_{ann} = V_{run} \times A_{pipe} \ , \tag{5–2}$$

where A_{pipe} = surface area of pipe.

A quick calculation shows that these rates are not negligible. For example, a 7-in. pipe run at 3 ft/s (1 m/s) gives rise to an annular rate of 8.6 BPM. Since the casing is not run continuously, the velocity is not constant, and inertial forces also contribute to the annular pressure. Mathematical models for calculating the associated pressure surges can be found in the literature (Mitchell, 1988).

Mud circulation is also necessary after the casing is in place. Unfortunately, it is very common to condition the mud only at this stage. Circulation is beneficial in the following ways—

- ensures that the hole is cleaned from cuttings,
- ensures that gas flow is not occurring,
- homogenizes the mud after treatment on the surface,
- reduces mud yield stress and plastic viscosity because most drilling muds are thixotropic, and
- erodes the gelled and/or dehydrated mud that is trapped in washouts, on the narrow side of an eccentric annulus, and at the walls of permeable formations.

Cuttings or gelled or dehydrated mud, which are eroded while running the casing, can lead to an excessive pressure buildup when circulation is resumed. Therefore, it is often desirable to circulate the annulus at intermediate intervals before the bottom of the hole is reached.

These qualitative recommendations are not very helpful for the completion engineer who must design the mud circulation phases before removing the drillpipe, and after the casing is in place. For this purpose, it is possible to use mathematical models to predict and/or measure mud circulation.

5-3.2.1 Modeling of Mud Circulation

Imagine instantaneously marking all the particles which enter the system through the inlet to the annulus at time $t=0$. At later times $t>0$, the position of all these particles can be tracked from a knowledge of the velocity field. As the marked particles move around the annulus, their position shows the boundary between the fluid inside the system at $t=0$ and which still remains in the system, and the fluid which has entered the system only after $t=0$. In effect, they show the displacement of the fluid by itself.

The *circulation efficiency* at any time t is the volume of "new" fluid in the annulus divided by the total annular volume. Therefore, it is essentially the same as displacement efficiency defined previously when applied to the case of a single circulating fluid. As before, the efficiency is equal to the number of annular volumes pumped until time $= t_b$, and thereafter levels off and tends to an asymptotic value not greater than one.

The isothermal flow of an incompressible and inelastic fluid between two pipes of diameters D_o and $D_i < D_o$ is discussed in the following sections. The notations used in this chapter are the same as those defined in Chapter 4, where the basic flow equations are also presented.

5-3.2.2 Laminar Flow in a Concentric Annulus

In laminar flow, circulation efficiency can be calculated by tracking marked particles. This is done by solving the streamline equations. For a concentric annulus, it is directly derived from the nonzero velocity component described in Chapter 4. The data given in this section were calculated using the rectangular slot approximation, the validity of which was also discussed in Chapter 4. Circulation efficiency is plotted vs the number of annular volumes in Figs. 5–4 and 5–5. For power law and Bingham plastic fluids, the curve depends upon a single dimensionless parameter—the Power Law Index n in the former case, and a dimensionless shear stress ψ for the latter which is given by

$$\psi = \frac{\tau_y}{\left[\frac{D_o - D_i}{4} \times \frac{dp}{dz}\right]}, \tag{5–3}$$

where

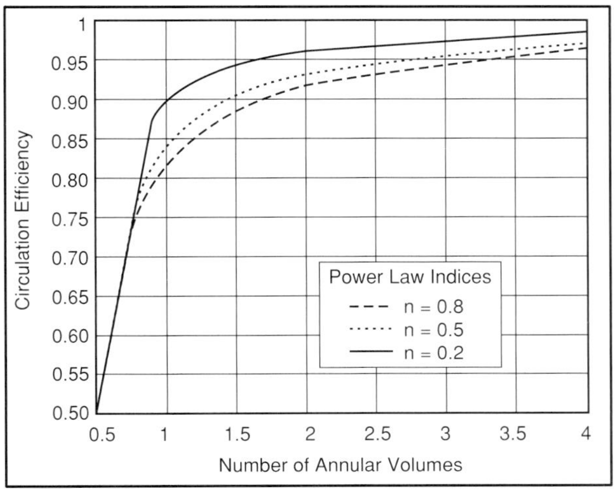

Figure 5–4—Circulation efficiency for a power law fluid in a narrow concentric annulus.

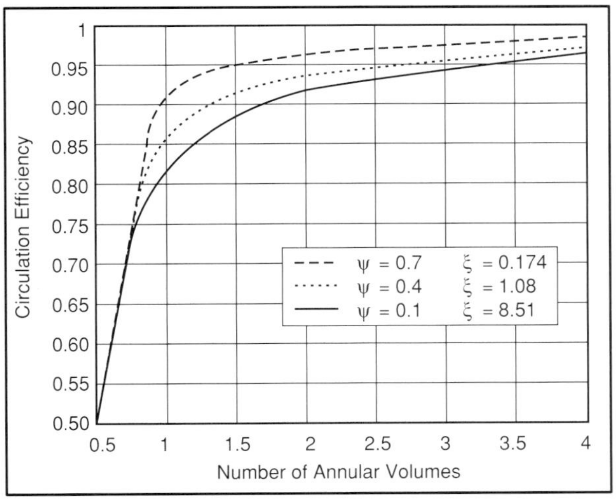

Figure 5–5—Circulation efficiency for a Bingham plastic fluid in a narrow concentric annulus.

(dp/dz) = frictional pressure drop, and

τ_y = yield stress of the fluid.

Since the annulus is supposed to be narrow and concentric, the dimensionless shear stress ψ is also equal to the ratio of the fluid yield stress τ_y to the shear stress at the wall τ_w.

$$\psi = \frac{\tau_y}{\tau_w} \tag{5–4}$$

Notice that the breakthrough time corresponds to the ratio of the average velocity to the maximum velocity. This value, which is equal to two-thirds for Newtonian fluids, is higher for shear thinning fluids as explained in Chapter 4. After breakthrough, the efficiency approaches 100%, a value that can be theoretically reached at infinite time, because the velocity of the fluid particles at the annular walls is assumed to be zero (no slip at the wall).

The figures show that the more shear thinning the fluid (i.e., the smaller the Power Law Index or the larger the dimensionless shear stress), the more efficient the circulation. Circulation efficiency of 100% at the breakthrough time would be obtained with an entirely flat velocity profile (which is equivalent to a Power Law Index of zero, or a dimensionless shear stress of one).

It is also important to point out that circulation efficiency does not depend on the flow conditions (flow rate) for power law fluids, while it does for Bingham plastic fluids. As explained in Chapter 4, the governing parameter for Bingham plastic fluids (the dimensionless shear stress) is a decreasing function of another parameter (the dimensionless shear rate ξ) given by Eq. 4-59 in Chapter 4.

$$\xi = \frac{12V}{D_o - D_i} \times \frac{\mu_p}{\tau_y}, \tag{5–5}$$

where

V = volumetric flow rate per unit of surface area, and

μ_p and τ_y = plastic viscosity and yield stress of the Bingham plastic fluid, respectively.

Thus, everything else being equal, the higher the average velocity, the smaller the annular gap, or the higher the μ_p/τ_y ratio, the worse the circulation efficiency for laminar flow in a concentric annulus.

5–3.2.3 Turbulent Flow in a Concentric Annulus

The flatter average velocity profiles which result from turbulent flow (Chapter 4) generally give much higher circulation efficiencies than those for laminar flow. However, the calculated circulation efficiencies are much more complicated, and to develop this problem in detail is beyond the scope of this chapter. The interested reader is invited to read the texts of Schlichting (1979) and Nauman and Buffham (1983).

5-3.2.4 Influence of String Eccentricity

The effect of eccentricity on circulation efficiency is discussed in this section, but similar arguments could be developed with regard to the effects of asymmetric flow geometry due to oval holes. The qualitative effect of casing eccentricity on velocity profiles and pressure gradients was presented in Chapter 4 using the basic slot model. It was shown that when the inner pipe of an annulus is not centered, the velocity distribution around the annulus is distorted, the flow favoring the wider side. This may lead to unusual situations where the flow regime can be laminar on the narrow side of the annulus and turbulent on the wide side, because the local Reynolds number varies azimuthally around the annulus (see Eq. 4-99 for the definition of the local Reynolds number).

When the flow is laminar around the annulus, the effect of eccentricity on circulation efficiency can be derived from the calculated velocity profiles using the basic slot model briefly presented in Chapter 4 (Iyoho and Azar, 1981). The validity of this model is limited to narrow annuli, and the results for a Newtonian fluid are plotted in Fig. 5–6 (assuming a diameter ratio $D_i/D_o = 0.8$). In this simple case, the circulation efficiency depends only on the pipe standoff, provided the annular diameter ratio D_i/D_o is close to unity.

For shear thinning fluids, the situation is more complex. When standoff decreases, the distortion of the velocity profile is such that the breakthrough time decreases and the circulation efficiency deteriorates (Chapter 4). In eccentric annuli, such fluids have a more uneven velocity distribution, and the effect of eccentricity on the circulation efficiency is even more pronounced. The breakthrough time t_b and the rate of increase of circulation efficiency after breakthrough are reduced to a greater extent with decreasing standoff.

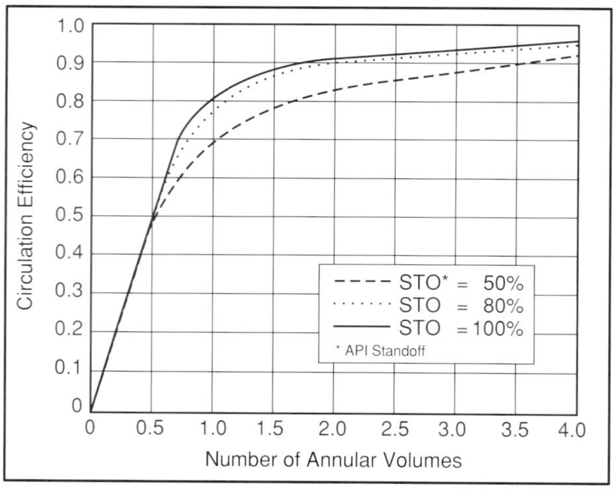

Figure 5–6—Circulation efficiency for Newtonian fluid in an eccentric annulus calculated using the basic slot model ($D_i/D_o = 0.8$).

For power law fluids, provided the annulus diameter ratio is sufficiently close to unity, the circulation efficiency depends upon the pipe standoff and the Power Law Index n. Typical examples of circulation efficiency curves for a Power Law Index of 0.5 are shown in Fig. 5–7.

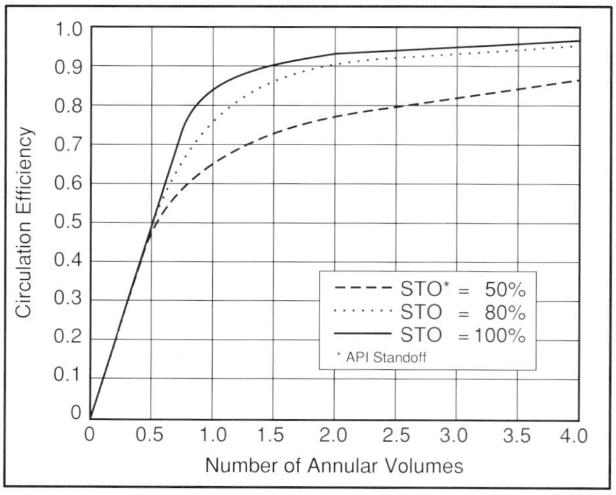

Figure 5–7—Circulation efficiency for power law fluid flowing in an eccentric annulus calculated using the basic slot model ($D_i/D_o = 0.8$ and $n = 0.5$).

For Bingham plastic fluids, the circulation efficiency is dependent upon the pipe standoff, and either the dimensionless shear stress ψ, or the dimensionless shear rate ξ. The latter is preferable for eccentric annuli, because it is constant for a given flow rate regardless of standoff. This is not the case for the former, the friction-pressure/flow-rate relationship being standoff dependent (Chapter 4). For different standoffs, the circulation efficiency of a Bingham plastic fluid flowing at a rate such that $\xi = 0.174$ is shown in Fig. 5–8.

A comparison of Figs. 5–7 and 5–8 to Fig. 5–6 confirms, as expected, that shear thinning fluids are much more affected by pipe eccentricity than Newtonian fluids. The sensitivity of the velocity distribution to fluid rheology, through the Power Law Index or the dimensionless shear rate, has the following consequence. For standoffs typically lower than 80% to 90% (compare Fig. 5–4 to Fig. 5–7, and Fig. 5–5 to Fig. 5–8) and a given number of annular volumes, the more shear thinning the

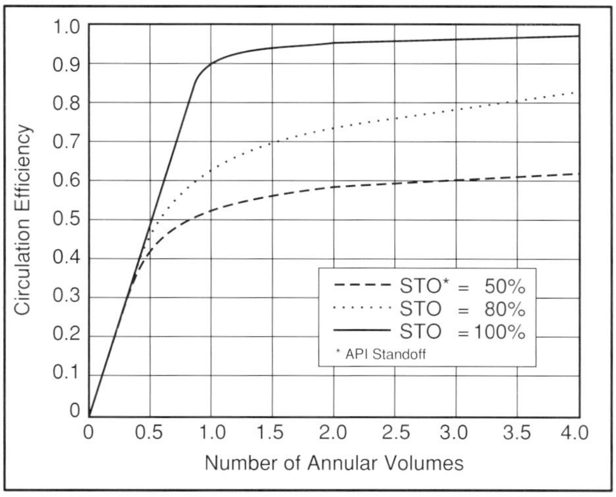

Figure 5–8—Circulation efficiency for a Bingham plastic fluid in an eccentric annulus calculated using the basic slot model ($D_i/D_o = 0.8$ and $\xi = 0.174$).

fluid, the worse the circulation efficiency. Therefore, in this standoff range, the circulation efficiency of power law fluids will increase with the Power Law Index. For Bingham plastic fluids, the higher the dimensionless shear rate ξ, the better the circulation efficiency below 80% to 90% standoff (Fig. 5–9). Thus, for such standoffs, the circulation efficiency can be improved by increasing the flow rate or increasing the μ_p/τ_y ratio. As mentioned earlier, the opposite conclusion was reached for concentric annuli (Figs. 5–5 and 5–9). Since a perfectly concentric annulus never exists in the field, recommendations for improving circulation efficiency in eccentric annuli rather than concentric annuli should be adopted.

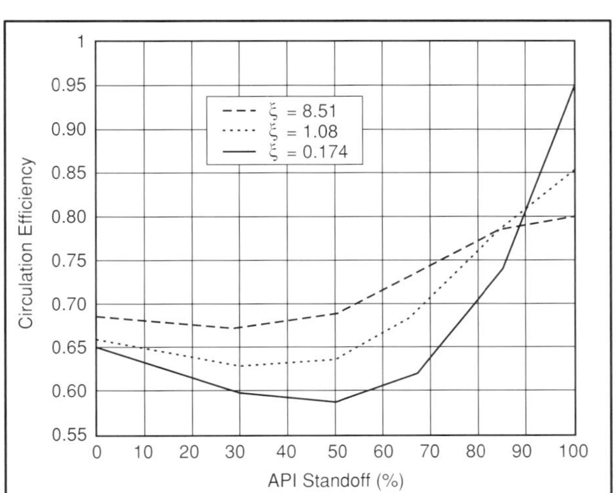

Figure 5–9—Circulation efficiency for a Bingham plastic fluid flowing in different eccentric annuli for a number of annular volumes pumped equal to 1.

Among shear thinning fluids, those exhibiting a yield stress represent a specific case. When the flow rate is sufficiently low, such fluids are stationary in the narrowest part of the annulus because of the uneven distribution of the shear stress around the annulus. The basic slot model shows that this will occur if the shear stress at the wall $\tau_w(e)$ calculated for a local annular gap e is such that

$$\tau_w(e) = \frac{e}{2} \times \frac{dp}{dz} < \tau_y \qquad (5\text{–}6)$$

This situation is not desirable during mud circulation, because the circulation efficiency would asymptote toward a value smaller than one (Fig. 5–8). For this not to be the case, it is necessary to have all of the fluid in movement around the annulus. This can be achieved if the minimum shear stress at the wall (i.e., the shear stress at the wall at the minimum annular gap $[STO(D_o - D_i)/2]$) is greater than the yield stress of the fluid.

$$\frac{dp}{dz} > \frac{4\tau_y}{STO\,(D_o - D_i)}, \text{ or } \psi < STO. \qquad (5\text{–}7)$$

For a Bingham plastic fluid, the corresponding minimum dimensionless shear rate can also be determined using the basic slot model. What this minimum value should be when the flow regime is laminar all around the annulus is shown in Fig. 5–10. However, this is often not the case, especially at low standoff (Chapter 4). For example, consider a mud exhibiting a plastic viscosity of 20 cp (20 mPa s) and a yield stress of 10 lbf/100 ft^2 (4.79 Pa), flowing in a 12$^1/_2$-in.-OD, 9$^5/_8$-in.-ID annulus. The following shows that the minimum flow rate necessary for the fluid to flow around the annulus can be largely overestimated at low standoff if flow is assumed to be truly laminar.

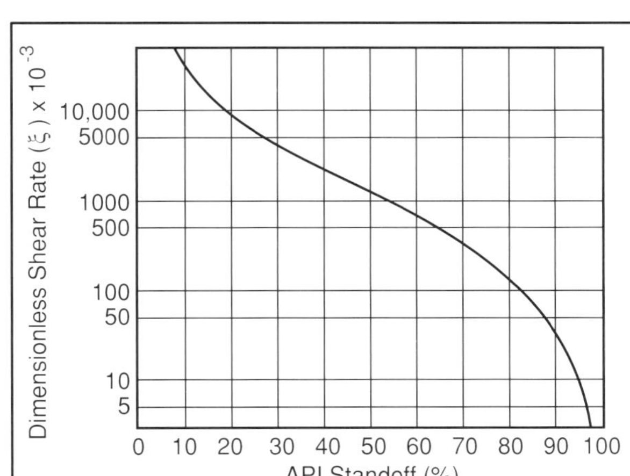

Figure 5–10—Minimum dimensionless shear rate for a Bingham plastic fluid to flow around an eccentric and narrow annulus (flow is assumed to be fully laminar).

STO (%)	Minimum Flow Rate (BPM) (1)	(2)
80	2.0	2.0
60	11	11
40	38	19
20	>100	33

(1) Flow is assumed to be laminar around the annulus.
(2) Variations of flow regime around the annulus are taken into account (Chapter 4).

The effect of pipe standoff on the transition from laminar to turbulent flow is discussed in Chapter 4. It is shown that, once a given fluid is in turbulent flow around the annulus, there is much less distortion of the velocity distribution when compared to a laminar-flow situation. The velocity field is also not sensitive to the shear thinning behavior of the fluid. Therefore, although to the best of the authors' knowledge there is no published model for predicting the circulation efficiency under such conditions, it is bound to be much better in truly turbulent flow than in truly laminar flow.

5-3.2.5 Effect of Gelled Mud and Mud Cake on the Circulation Process

The theoretical results presented thus far have not considered the effects of gelled mud and/or dehydrated mud on the circulation process. Such modifications may not only take place when the mud is static, but also while the mud is being circulated, because wellbore ovality, irregularities (washouts), and casing eccentricity can induce zones where the local velocity of the fluid is zero. This mud is commonly called the *immobile mud.*

When allowed to remain static, most drilling muds develop a structure which is usually characterized by its gel strength. This parameter represents the minimum shear stress value τ_{gel} necessary to induce flow. Drilling muds are designed to exhibit such thixotropic properties, because they must be able to suspend cuttings and the weighting agent when circulation is stopped. Unfortunately, the mud gel strength is partially responsible for the wellhead pressure peak when circulation is resumed. In addition, it can strongly affect the efficiency of the circulation process, especially when the pipe is not centered. Once the mud has been allowed to gel, the force required to overcome it is no longer equal to the yield stress, but to the gel strength. Thus, for a fluid exhibiting a gel strength τ_{gel}, the minimum friction pressure to achieve flow on the narrow side of an eccentered annulus is

$$\frac{dp}{dz} > \frac{4\,\tau_{gel}}{STO\,(D_o - D_i)} \text{ or } \psi < \frac{\tau_y}{\tau_{gel}} \quad . \qquad (5\text{–}8)$$

Many data have been published concerning the gel-strength development of muds as a function of time, but the interpretation of the experimental results obtained by the standard oil industry procedure (Chapter 4) is questionable. For day-to-day applications, the situation is even worse, because the standard practice for measuring the mud gel strength consists of a one-time reading after a maximum of 10 minutes at rest. Ten minutes is far from being representative of the long static periods that muds can experience prior to being circulated (several hours or even days). This lack of valuable information regarding mud properties prevents the development of more comprehensive circulation models. However, the knowledge of mud gel strength is insufficient to determine the circulation efficiency of a mud which has been allowed to gel for a given period. The kinetics of the gel-structure breakdown as a function of shear history must be described to determine the erosion of the gelled mud by the flowing mud. Without such information, one can only determine whether or not mud is flowing on the narrow side of the annulus, not at which velocity it is flowing. Unfortunately, very little is known about this subject.

Nevertheless, one can attempt to characterize mud gel strength, as a function of time, using the standard industry procedure. The minimum friction pressure can be determined using Eq. 5–8, and the corresponding flow rate calculated as if no static gelled mud exists in the annulus.

The presence of a mud cake at the wall of permeable formations is another factor which affects the circulation process. When mud is not flowing across a permeable zone, it is subjected to static filtration. Without sufficient fluid-loss control, an excessively thick filter cake can grow and reduce the size of the annulus. Mud cakes as thick as $^1/_2$ in. (1.2 cm) have been measured by a caliper with poorly treated muds (Table 5–1) (Cowthral, 1982). This partially dehydrated material is difficult to mobilize when circulation is resumed, because both its density and viscosity (especially at low shear rates) are much higher than those of the original mud (Tables 5–2 and 5–3) (Haut and Crook, 1979). Predicting how much mud cake will be eroded when flow is resumed is difficult, because most mud cakes are compressible, and their characteristics vary as a function of distance from the formation. The loose cake furthest from the wall can most probably be eroded by the flow, but removal of the hard cake against the formation is much more difficult.

There is a possible synergism between mud filtration and pipe eccentricity, which would be detrimental to the circulation process. Since the erosion of the deposited filter cake is an increasing function of the shear stress at the formation wall, the mud-cake thickness during circulation is likely to be larger at the narrow side of the annulus.

Cockfield Sand Reference	Mud With Asphalt Additives			Mud Without Asphalt Additives		
	Well No. 13 (in.)	Well No. 22 (in.)	Well No. 23 (in.)	Well No. 10 (in.)	Well No. 14 (in.)	Well No. 15 (in.)
IA	10½	9⅞	9⅝	8⅝	9⅜	9½
IB	10½	9⅞	9⅝	8⅝	9⅜	9½
IC	—	10½	9⅞	—	—	—
IIAU	—	10	9⅝	8⅝	—	—
IIAL	—	10	9⅜	8⅝	9⅜	—
IIB	11¼	9⅞	—	8⅝	9⅜	9⅜
IIC	10¼	9⅞	9⅜	8⅝	8⅜	9⅜
IIIA	10	9½	9⅜	8⅝	8⅞	—
IIIB	10¼	9½	9⅛	8⅝	8⅞	9½
IVAB	10¼	9⅞	9⅛	8⅝	8⅜	9⅛
IVC	10¼	10	9⅛	8½	8⅞	8⅞
VA	—	9⅞	8⅞	8⅝	—	—
VC	10½	9⅞	8⅞	8⅞	8⅞	9⅜

Table 5–1—Comparison of hole sizes measured under similar conditions with two different drilling muds (treated and nontreated) with a 9⅞-in. bit size (after Cowthral, 1982).

	Apparent Viscosity									
	10 s^{-1}		50 s^{-1}		100 s^{-1}		500 s^{-1}		1000 s^{-1}	
Temperature (°F) [°C]	No Water Removed	5% Water Removed	No Water Removed	5% Water Removed	No Water Removed	5% Water Removed	No Water Removed	5% Water Removed	No Water Removed	5% Water Removed
63 [17]	0.048	—	0.08	1.0	0.080	0.70	0.065	0.36	0.070	0.27
151 [66]	0.500	—	0.08	1.6	0.034	0.73	0.021	0.20	0.027	0.16
199 [93]	1.500	—	0.29	2.0	0.130	0.90	0.027	0.18	0.025	0.14
250 [121]	2.100	—	0.78	3.5	0.300	1.70	0.048	0.27	0.039	0.19
300 [149]	7.300	—	4.00	10.0	0.900	7.00	0.120	1.50	0.060	0.62

Table 5–2—Rheology percent of a drilling fluid as a function of temperature and of water removed.

Mud Density		API Fluid Loss	Filter Cake Thickness	Filter Cake Density	
(lb/gal)	(kg/m³)	(cc/30 min)	(mm)	(lb/gal)	(kg/m³)
14.0	1680	20.6	9.5	25.5	3060
14.0	1680	16.6	9.5	25.3	3040
13.9	1670	19.3	10.3	25.2	3030
14.9	1790	21.0	9.5	28.9	3470
14.8	1780	19.2	11.1	29.3	3520
14.8	1780	21.2	8.7	28.3	3390
17.0	2040	3.6	1.6	33.8	4050
17.0	2040	28.0	14.3	34.3	4110
17.0	2040	3.4	6.4	32.6	3910

Table 5–3—Resulting densities of noncirculatable drilling fluid.

The resultant nonuniform thickness of the cake would favor an uneven distribution of the flow path around the annulus, and would further reduce the velocity of the fluid on the narrow side. In extreme cases, the fluid could stop flowing, be subjected to static filtration, and be very difficult to mobilize at a later time.

From the preceding discussion it is clear that a complete understanding of the effects of mud gelation and mud filtration has not yet been achieved. Nevertheless, a qualitative analysis shows that both may have a detrimental effect on the circulation efficiency, in particular when the pipe is not centered.

5–3.2.6 Effect of Casing Movement

Whenever the conditions are such that the majority of the mud in the hole cannot be restored to circulation, a possible solution is to reciprocate and/or rotate the pipe. The effect of pipe movement on the mud circulation process as such has not yet been fully investigated, but the benefits of this technique on mud displacement as a whole, an issue that will be addressed in Section 5–4.3, are not questionable. Both movements are thought to be helpful in mobilizing the slowly moving or even static mud present on the narrow side of an eccentric annulus. Numerical models have been developed to study the influence of casing movement on the flow pattern for simple non-Newtonian fluids (power law or Bingham plastic fluids flowing in laminar flow (Speers et al., 1987). Circulation efficiencies derived from these models show that casing movement can indeed partially counteract the detrimental effect of pipe eccentricity (Fig. 5–11), but it must be stressed that these models do not account for lateral motion of the casing—a likely occurrence during reciprocation as pointed out by McLean et al. (1967).

When used in combination with scratchers, scrapers, or cable wipers, casing movement was also shown to mechanically erode the filter cake, and considerably improve the displacement process (Section 5–4.3). There is currently a lack of quantification for the effects of these mechanical devices on mud circulation; however, there is no doubt that they contribute significantly to the efficiency of the process.

5–3.2.7 Measuring Mud Circulation Efficiency in the Field

The theoretical models mentioned above are still under development, because other relevant parameters such as temperature profile, mud thixotropy, mud cake, etc., are not currently taken into account. Some of the recommendations, e.g., circulation at a rate such that flow regime is turbulent in the annulus, may not be applicable because

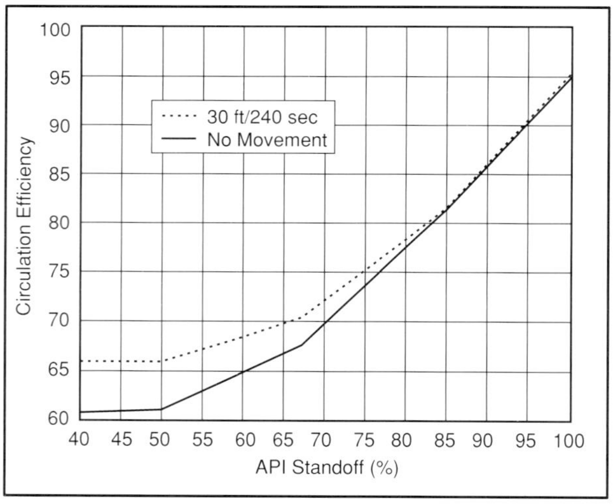

Figure 5–11—Effect of casing reciprocation on circulation efficiency.

of the constraints imposed by the formations, fluids, and field equipment.

- The borehole pressure should be maintained between the pore and fracture pressures.
- It may be desirable to keep the annular velocity of the mud below a certain limit (e.g., the maximum value encountered during drilling) to maintain the stability of the hole.

Although these models could certainly be used as guidelines to design both the rate and the time during which a well should be circulated, they are not. Designing the circulation period before cementing begins usually relies on rules of thumb such as "circulate bottoms-up," which appears to be insufficient in most circumstances. A more suitable approach consists of attempting to measure the mud circulation efficiency (Smith, 1984). This is performed by monitoring the volume of mud which is actually circulating (the "circulatable mud") with a "fluid caliper" or tracer. For this purpose, a small volume of mud is tagged with a tracer and injected at the wellhead. The time necessary for this mud to return to the surface gives an indication of the volume of mud being circulated. This volume is then compared to the hole volume determined from caliper measurements. An illustration of this concept is shown in Fig. 5–12.

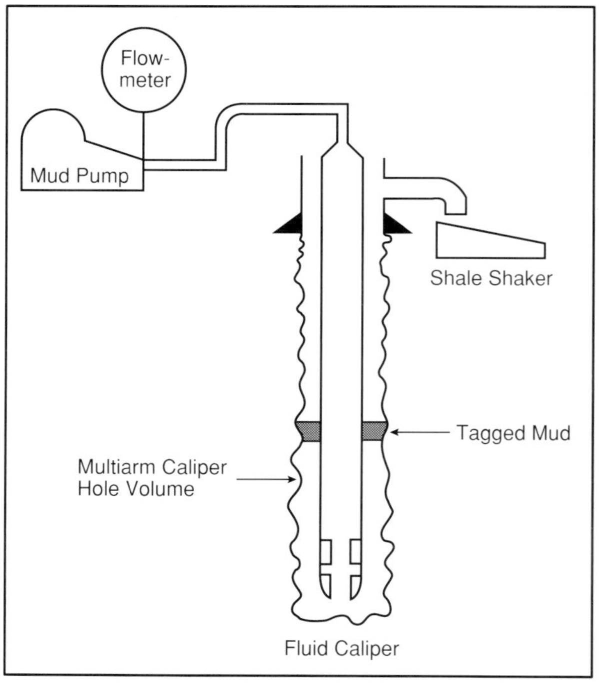

Figure 5–12—Schematic of a well showing the fluid caliper concept used to determine mud circulation efficiency (from Smith, 1989).

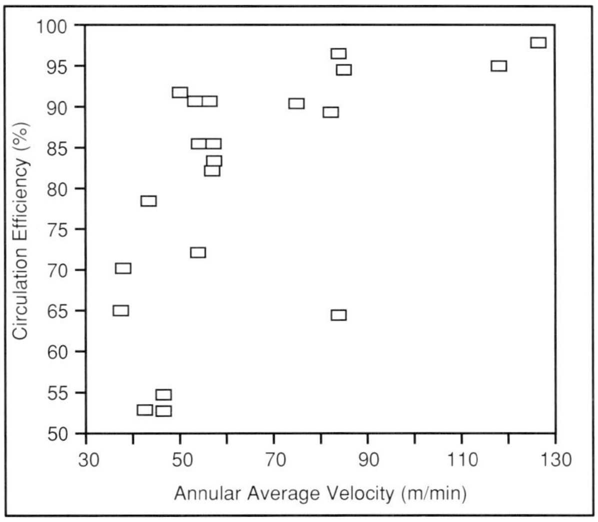

Figure 5–13—Effect of annular velocity on circulation efficiency (from Smith, 1989).

Tracers have included inert particles (oats, rice hulls, and dyes) and reactive materials (carbide pills or radioactive isotopes). Most of these techniques can only provide qualitative results, because it is often not clear what is being measured—time of first appearance or time of maximum concentration of the tracer. A quantitative analysis would require continuous monitoring of the tracer concentration at the return line, and also an interpretation scheme to infer an average circulation velocity from the measurements. Nevertheless, these simple techniques appear to be quite useful. For example, using carbide pills, Smith (1989) advocated designing the flow rate and the circulation time on the assumption that 95% of the calipered hole volume were circulating. His measurements (Fig. 5–13) led him to recommend circulation velocities in excess of 250 ft/min (1.37 m/s). Such a velocity is quite high by oilfield standards, but confirms the necessity to circulate muds at high annular velocities to optimize the efficiency of the process.

5–3.3 Mud Circulation—Conclusions

Ensuring that a large percentage of the mud is actually in circulation is a key to the success of a primary cement job. In view of the complexity of the problem, there is no doubt that sufficient time should be devoted to the design, execution, and evaluation of the mud conditioning and circulation phases prior to cementing. The following qualitative guidelines can be distilled from the preceding discussion.

- The rheological properties of the mud (namely the μ_p/τ_y ratio), mud gel strength, and pipe standoff should be such that the mud is in movement completely around the annulus at an achievable flow rate. This can be done by improving pipe standoff, increasing the μ_p/τ_y ratio, decreasing the mud gel strength, or increasing the flow rate.
- If the above criteria cannot be met, reciprocation or rotation of the pipe should be performed during mud circulation.
- When available, circulation models should be used to better optimize the above parameters in view of improving the circulation efficiency.
- As a rule of thumb, the volume of mud to be circulated should represent at least one full hole volume; however, circulation models can be used to obtain a better estimate of the required mud circulation time.
- Whenever predictions are doubtful, "fluid calipers" should be used to qualitatively measure the efficiency of the circulation process. Circulation should be maintained until 90% of the calipered hole volume are being circulated.

5–4 MUD DISPLACEMENT

Despite what has just been said about velocity profiles and mud circulation efficiencies, it should not be assumed (as is sometimes the case) that the interface profile between two fluids can be derived directly from the velocity profile of one of the two fluids. Mud displacement is much more complicated than mud circulation. In addition to the parameters mentioned earlier, mud displacement is dependent upon the relative properties of the fluids involved (density and rheology), their relative flow regimes, and their eventual interaction when mixed together. To simplify the problem, the displacement of the so-called "mobile mud" (i.e., the displacement of nongelled mud between impermeable walls) is discussed first. Next, the removal of the "immobile" mud, in which case both gelation and filtration are taken into account, is addressed. The last part of the section is dedicated to the influence of pipe standoff and pipe movement on mud displacement.

5–4.1 Displacement of the "Mobile" Mud in Concentric Annuli

One of the first parameters found to have an influence on mud displacement efficiency was the flow regime of the displacing fluid. From pilot-scale studies, Howard and Clark (1948) concluded that when the Reynolds number of the cement slurry was low only 60% of the "circulatable" mud were displaced, whereas 90% to 95% could be displaced when the cement slurry was in the upper laminar or turbulent-flow regime. This issue has subsequently been raised by several authors, but there is still no consensus today concerning the best displacement regime for optimum mud removal. As is demonstrated below, the choice of the proper displacement regime cannot be made outside of the general context of the primary cement job. Hole and pipe sizes, fluid densities, fluid rheological properties, and operational constraints must be taken into account to design a cement job for optimum mud removal. Therefore, to adapt the placement technique to the displacing fluid properties (density and rheology) or vice versa, it is necessary to first understand the action mechanisms of each flow regime.

5-4.1.1 Laminar-Flow Displacement

Modeling of Laminar-Flow Displacement

Modeling the displacement of a fluid by another is a much more difficult problem compared to circulation, because several additional parameters must be taken into account. Everything else being equal, and at least at low flow rates, the displacement of a dense fluid by a lighter one leads to an unstable phenomenon known as *buoyant plume*. Conversely, when the displacing fluid is heavier than the displaced fluid, buoyancy forces tend to flatten the interface and promote efficient displacement.

Differences in rheological properties are also likely to play a role in this process. Everything else being equal, the laminar-flow displacement of a "thin" fluid by a "thick" one will always be more efficient than the reverse situation, which is known to give rise to an unstable interface (Hooper and Grimshaw, 1985).

The above statements are purely qualitative, and do not consider the combined effect of density and rheology. This problem is far from being fully understood at this time. The quantitative conditions ensuring the stability of an interface between two fluids of different properties are still the subject of theoretical studies. However, some partial answers concerning the conditions favoring a flattening of the interface and efficient displacement have been developed.

Consider a displacement occurring in a given annulus. The mass and momentum balances for each fluid are

$$\nabla . v = 0 \qquad (5\text{–}9)$$

and

$$\rho \left[\frac{\partial v}{\partial t} + (v . \nabla) v \right] = -\nabla p + \nabla . \underline{\underline{\tau}} + \rho g \ , \qquad (5\text{–}10)$$

where

v = velocity of fluid,

ρ = density of fluid,

p = pressure in fluid,

g = gravitational acceleration, and

τ = deviatoric part of the stress tensor.

The annulus is assumed to be narrow (i.e., equivalent to a rectangular slot) and concentric. In addition, the interfacial mixing resulting from molecular diffusion is neglected, a hypothesis which seems reasonable because the thickness of the diffusive interface is much smaller than the annular gap.

$$\sqrt{\frac{\kappa L}{V_{ave}}} << (R_o - R_i) \ , \qquad (5\text{–}11)$$

where

κ = diffusivity, and

L = annular length.

For such a displacement, both fluids have only two velocity components—v_x in the direction perpendicular to the plane of symmetry of the slot, and v_z in the main direction of flow. Thus, Eqs. 5–9 and 5–10 become

$$\frac{\partial v_x}{\partial x} + \frac{\partial v_z}{\partial z} = 0 \ , \qquad (5\text{–}12)$$

$$\rho \left[\frac{\partial v_x}{\partial t} + v_x \frac{\partial}{\partial x} v_x + v_z \frac{\partial}{\partial z} v_x \right] = + \frac{\partial}{\partial x} \tau_{xx} + \frac{\partial}{\partial z} \tau_{zx} - \frac{\partial p}{\partial x} + \rho g_x \qquad (5\text{–}13)$$

$$0 = -\frac{\partial p}{\partial y} + \rho g_y \ , \qquad (5\text{–}14)$$

and

$$\rho \left[\frac{\partial v_z}{\partial t} + v_x \frac{\partial}{\partial x} v_z + v_z \frac{\partial}{\partial z} v_z \right] = + \frac{\partial}{\partial x} \tau_{xz} + \frac{\partial}{\partial z} \tau_{zz} - \frac{\partial p}{\partial z} + \rho g_z \ , \qquad (5\text{–}15)$$

with

$$\underline{\underline{\tau}} = \eta \underline{\underline{\dot{\gamma}}} \ . \qquad (5\text{–}16)$$

The velocity components of the shear-rate tensor,

$$\underline{\underline{\dot{\gamma}}} = \begin{bmatrix} 2\frac{\partial}{\partial x} v_x & \left(\frac{\partial}{\partial z} v_x + \frac{\partial}{\partial x} v_z \right) \\ \left(\frac{\partial}{\partial x} v_z + \frac{\partial}{\partial z} v_x \right) & 2\frac{\partial}{\partial z} v_z \end{bmatrix} \ , \qquad (5\text{–}17)$$

and the viscosity are a function of the square root of one-half its second invariant,

$$\sqrt{I_2/2} = \sqrt{2 \left[\left(\frac{\partial}{\partial x} v_x \right)^2 + \left(\frac{\partial}{\partial z} v_z \right)^2 \right] + \left(\frac{\partial}{\partial x} v_z + \frac{\partial}{\partial z} v_x \right)^2} \ . \qquad (5\text{–}18)$$

Since diffusion at the interface has been neglected, the interface is assumed to be stable (i.e., the particles in the interface remain there, and move with the local fluid velocity). If the equation of the interface is given by

$$z = z_I(x,t) \quad , \tag{5–19}$$

then

$$\frac{D(z - z_I)}{Dt} = 0 \tag{5–20}$$

in the interface, where the subscript I refers to values in the interface. The boundary conditions necessary to solve this problem are the no-slip condition at the walls of the annulus, continuity of velocity and stress at the interface between the two fluids, and profile of the interface at time $t = 0$.

Solving this problem requires finding the solution of two sets of simultaneous, nonlinear, partial differential equations, which is a formidable task for even the most sophisticated computers. Progress can only be made by making simplifying assumptions. Very few papers have been published on the subject. One of the most complete pieces of work is by Flumerfelt (1973). He developed an approximate solution for the displacement of a power law fluid by another in laminar flow. Beirute and Flumerfelt (1977) extended this work to fluids following a more general non-Newtonian law—the Robertson and Stiff model. The main assumptions of these analyses are given below.

- Displacement takes place in a narrow rectangular slot.
- For both fluids the flow regime is laminar, and the Reynolds number of each of the fluids Re_i is small when compared to the length-to-gap ratio of the annulus (i.e., $Re_i \ll (L/R_o - R_i)$).
- Displacement occurs under stable conditions (i.e., the interface is supposed to be smooth, a condition which is not quantitatively defined in the paper).
- The fluids are miscible (i.e., surface tension is neglected).
- Molecular diffusion at the interface is negligible.
- The horizontal velocities are negligible.

Beirute and Flumerfelt (1977) recognized that the mass balance was not correct when all of the above assumptions were made. They corrected for the mass balance by multiplying by a correction factor. The correct mass balance would have been obtained if more care had been taken with the interface equation.

In addition to the dimensionless time, the approximate solution depends on five dimensionless parameters for power law fluids (seven for Robertson and Stiff fluids) which are defined as—

a density ratio:

$$K_2 = \frac{\rho_2}{\rho_1} \, , \tag{5–21}$$

an effective viscosity ratio:

$$K_3 = \left[\frac{2k_2}{\rho_1 g\,(R_o - R_i)}\right]^{1/n_2} \times \left[\frac{\rho_1 g\,(R_o - R_i)}{2k_1}\right]^{1/n_1} \, , \tag{5–22}$$

a dimensionless flow rate:

$$K_4 = \frac{2V}{R_o - R_i} \times \left[\frac{2k_1}{\rho_1 g\,(R_o - R_i)}\right]^{1/n_1} \, , \text{ and} \tag{5–23}$$

the Power Law Indices for each fluid: n_1 and n_2.

For Robertson and Stiff fluids, the dimensionless yield stresses for the displaced and displacing fluids (K_5 and K_6, respectively) are

$$K_5 = \left[\frac{2\tau_{y1}}{\rho_1 g\,(R_o - R_i)}\right]^{1/n_1} \, , \text{ and} \tag{5–24}$$

$$K_6 = \frac{1}{K_3} \times \left[\frac{2\tau_{y2}}{\rho_1 g\,(R_o - R_i)}\right]^{1/n_2} \, , \tag{5–25}$$

where

R_o = outer annular radius,
R_i = inner annular radius,
V = average displacement velocity,
g = gravity, and
ρ = density of a fluid with rheological properties defined by $\tau = \left[\tau_y^{1/n} + k^{1/n}\dot{\gamma}\right]^n$.

The subscripts 1 and 2 refer to the displaced and displacing fluids, respectively.

Another parameter, a dimensionless pressure drop $K_1 = ((dp/dz)/\rho_1 g)$, is eliminated while developing the flow equations.

The principal conclusions of Beirute and Flumerfelt (1977) are given below.

- The density ratio (K_2) plays a predominant role, provided the dimensionless flow rate (K_4) is not too high. A K_2 greater than one flattens the profile of the interface, and minimizes channeling (Fig. 5–14).
- Although displacement efficiencies were found to increase with increasing effective viscosity ratios (K_3), the sensitivity to this parameter was found less important than that to the density ratio (K_2) (Fig. 5–15).
- Power Law Indices do not seem to be important in the case of Robertson and Stiff fluids. For power law fluids, better displacement efficiencies are obtained when the Power Law Index of the displacing fluid is lower than that of the displaced fluid.

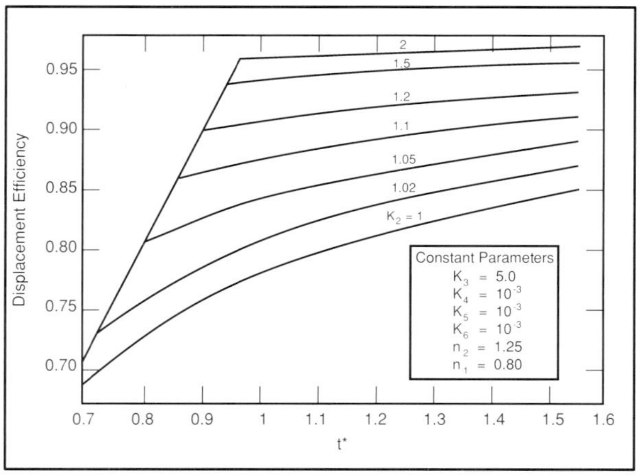

Figure 5–14—Effect of K_2 and t^* on displacement efficiency (after Beirute and Flumerfelt, 1977).

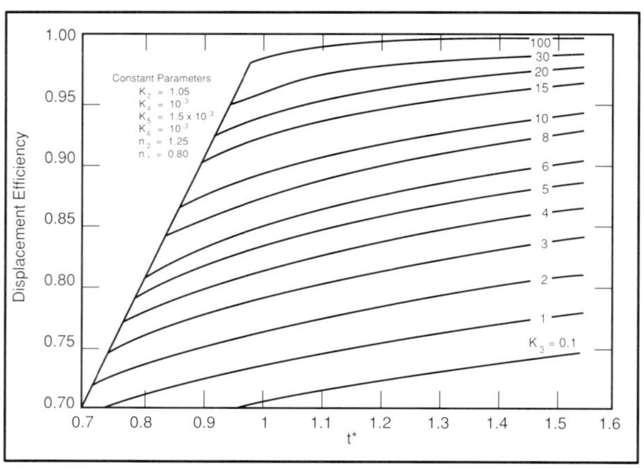

Figure 5–15—Effect of K_3 and t^* on displacement efficiency (after Beirute and Flumerfelt, 1977).

- In the case of Robertson and Stiff fluids, yield stresses are shown to be quite critical. High displacement efficiencies resulted when the dimensionless yield stress values (K_6) of the displacing fluid were higher than those of the displaced fluid (K_5).
- Generally speaking, displacement efficiencies were found to decrease with increasing dimensionless flow rates (K_4).

When looking carefully at the data presented in both papers, it appears that the last conclusion is somehow too drastic. For example, taking the case of power law fluids, Flumerfelt (1973) presented data where the Power Law Indices were varied while keeping the other dimensionless parameters constant (Table 5–4). Consequently, the ratio of apparent viscosities of the two fluids (calculated at the wall for both fluids as if they were pumped independently) changed greatly. Also, depending on the relative value of the Power Law Indices, this ratio could increase or decrease with increasing flow rate (Table 5–5), leading to an improvement or a deterioration of the displacement efficiency. The same argument applies to the effect of yield stresses for Robertson and Stiff fluids.

These theoretical developments regarding the displacement process in a concentric annulus have emphasized the role played by gravitational and viscous forces. With the density ratio and the low-shear-viscosity ratio higher than one (i.e., Power Law Indices ratio lower than one, and a yield-stress ratio higher than one for fluids exhibiting a yield stress), low flow rates contribute to a flattening of the interface profile, and lead to efficient displacements. With increasing flow rate, the influence of the density ratio seems to decrease, and the viscous properties continue to have an impact. Everything else being equal, the more viscous the displacing fluid, the better the results. However, care should be taken to compare apparent fluid viscosities in a shear-rate range representative of the flow conditions encountered.

Experimental Validation

Most of the experimental studies performed on laminar-flow displacement in concentric annuli (Childers, 1968; Zuiderwijk, 1974) are in qualitative agreement with the theoretical study presented above, as far as the relative importance of fluid properties (density and rheology)

Characteristic Groups: $K_2 = 1.0$, $K_3 = 1.0$, $n_1 = 1.0$, $n_2 = 1.0$			
K^4	**$t^* = 1.0$**	**$t^* = 1.2$**	**$t^* = 1.5$**
1×10^{-4}	0.808	0.844	0.879
10^{-3}	0.808	0.844	0.879
10^{-2}	0.808	0.844	0.879
Characteristic Groups: $K_2 = 1.0$, $K_3 = 1.0$, $n_1 = 1.0$, $n_2 = 0.6$			
K^4	**$t^* = 1.0$**	**$t^* = 1.2$**	**$t^* = 1.5$**
1×10^{-4}	0.937	0.982	0.991
10^{-3}	0.926	0.966	0.980
10^{-2}	0.901	0.937	0.958
Characteristic Groups: $K_2 = 1.0$, $K_3 = 1.0$, $n_1 = 0.6$, $n_2 = 1.0$			
K^4	**$t^* = 1.0$**	**$t^* = 1.2$**	**$t^* = 1.5$**
1×10^{-4}	0.467	0.487	0.511
10^{-3}	0.568	0.592	0.621
10^{-2}	0.672	0.702	0.734

Table 5–4—Effect of K_4 on displacement efficiency.

K_4	**$n_1 = 1.0$ $n_2 = 1.0$**	**$n_1 = 1.0$ $n_2 = 0.6$**	**$n_1 = 0.6$ $n_2 = 1.0$**
10^{-4}	1.0	0.0346	28.9
10^{-3}	1.0	0.0868	11.5
10^{-2}	1.0	0.218	4.58

Table 5–5—Ratio of apparent viscosity of displaced fluid (2) to that of displacing fluid (1) when pumped individually. Conditions are the same as those of the displacements described in Table 5–4, i.e., $K_2 = K_3 = 1$.

and flow rates on mud displacement in concentric annuli is concerned.

One of the most extensive studies is that of Zuiderwijk (1974), who performed more that 200 tests. Muds were displaced by five annular volumes of cement, and fairly high displacement efficiencies (>80%) were observed. All fluids were assumed to follow the power law model. Zuiderwijk (1974) concluded that annular velocity was a key parameter in the displacement process. The results showed that, at low velocities, a density ratio greater than one enhanced the displacement efficiency, and gravitational forces appeared to be less important for velocities higher than 1 ft/sec (0.3 m/s). Depending on the prevailing conditions, efficient displacement was obtained at both low and high displacement rates. At low velocities, good results were obtained with cement slurries having a higher viscosity than the mud (Power Law Index ratio $n_c/n_m > 1$). Well-treated muds (i.e., muds with Power Law Indices close to unity) were also found to be easily removed from washout sections, when displaced by a very thin cement slurry at high velocities (in which case $n_c/n_m > 1$), and the efficiency of the process was attributed to the turbulent eddies in the displacing fluid (this point is discussed later). For velocities ranging from 0.5 to 1.5 ft/sec (0.15 to 0.46 m/s), and values of n_c/n_m on the order of one, the displacement efficiency was found to be almost constant.

The slow flow technique, which was developed in the 1960s to overcome some of the practical limitations of turbulent flow displacements, also relies on experimental observations which are partially in agreement with the Beirute and Flumerfelt (1977) studies. In 1965, Parker et al. published the results of an experimental study which showed that good mud displacement could be obtained at low flow rates, provided the displacing fluid (cement slurries in this case) is at least 2 lb/gal (0.24 g/cm^3) heavier than the mud, and the initial gel strength and viscosity of the mud are lower than those of the displacing fluid. They also observed that, under these conditions, the displacement efficiency deteriorated with increasing flow rate.

Excellent results were obtained in the presence of washouts when the annular velocity was less than 90 ft/min (27.4 m/min). The efficiency of the process was attributed to the action of a coagulated mass at the cement/mud interface, which provided a piston-like displacement even in very large washouts and irregularities. At higher flow rates, the cement slurry was observed to break this gel; consequently, poor displacement efficiencies resulted. However, the limited amount of results presented did not allow the clear definition of an optimum displacement velocity range (Fig. 5–16).

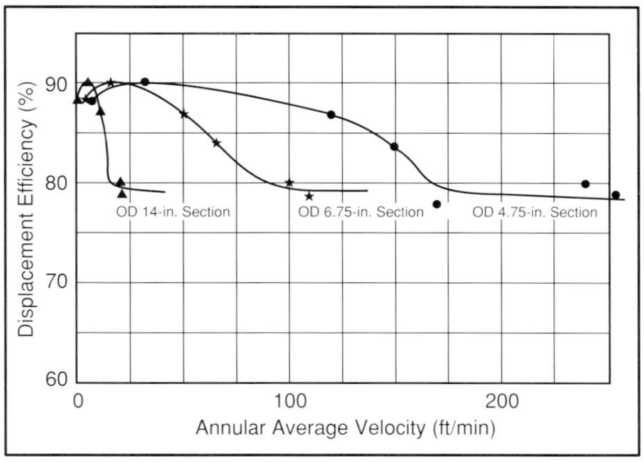

Figure 5–16—Displacement efficiency as a function of annular velocity in various sections of an annulus (ID = 2^3/$_8$-in.) (from Parker et al., 1965).

This technique was later refined. The combined effect of the density and gel-strength differential on mud-displacement efficiency was evaluated and, in the most common case where the mud density was lower than that of the cement, the minimum gel strength required for 100% mud displacement could be calculated from the following empirical equation (Fig. 5–17).

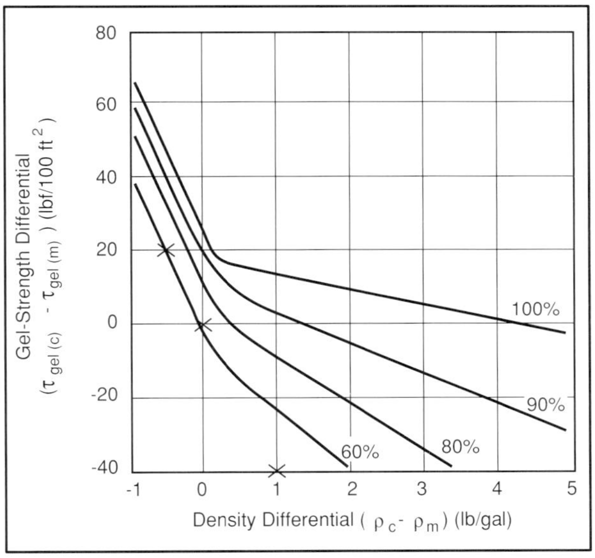

Figure 5–17—Mud displacement efficiency as a function of density and gel-strength differential between cement slurry and mud.

$$\tau_{gel\,(c)} = \tau_{gel\,(m)} + [15 - 2.5\,(\rho_c - \rho_m)]\,, \qquad (5\text{–}26)$$

where

$\tau_{gel\,(c)}$ = cement gel strength (lb/100 ft^2),

$\tau_{gel\,(m)}$ = mud gel strength (lb/100 ft^2),

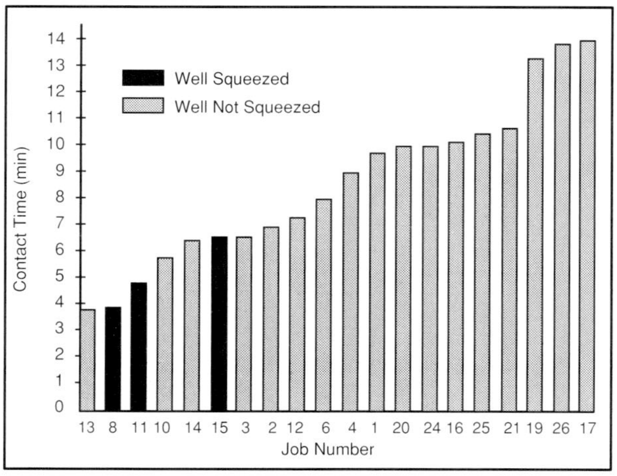

Figure 5–18—Contact time on jobs where turbulent flow was achieved based on field rheology data (from Brice and Holmes, 1964).

ρ_c = cement slurry density (lb/gal), and

ρ_m = mud density (lb/gal).

Unfortunately, these experimental results just provide a qualitative indication concerning the efficiency of the displacement process. The measurements performed do not allow one to quantitatively validate the efficiencies predicted by the numerical displacement models.

5–4.1.2 Turbulent-Flow Displacement

As mentioned earlier, Howard and Clark (1948) obtained good displacement efficiencies when displacing muds with cement slurries in the upper laminar and turbulent-flow regimes. Turbulent flow became commonplace in the 1960s with the introduction of cement formulations which allowed turbulence at achievable pump rates. In 1964, Brice and Holmes published the results of a survey concerning 46 cement jobs performed in southwest Louisiana, an area which was notorious for primary cementing failures. They reiterated the need for turbulent flow, and suggested that the annular space should be in contact with the turbulent displacing fluid for a sufficient time. However, it is difficult to define an optimum contact time from their data (Fig. 5–18).

Today some claim that a four-minute contact time is sufficient, while others claim that 10 minutes are necessary. From the Brice and Holmes data (1964), eight minutes appears to be a reasonable guideline; however, such an assumption would ignore other important factors such as casing movement and centralization. The turbulent-flow-displacement technique has since gained wide acceptance, because it has greatly increased the success ratio of cement jobs in some areas. However, the basic fundamentals underlying this practice are not well understood.

Good mud displacement can often be obtained with a low-viscosity, unweighted fluid such as water, diesel oil, or a chemical wash. Although the stresses generated by such fluids (in the absence of buoyancy forces or with detrimental buoyancy forces) are extremely low, even at a very high Reynolds number, their ability to displace mud supports the following mechanism—turbulent eddies in the displacing fluid cause a drag/erosion/dilution process at the mud/displacing fluid interface (Clark and Carter, 1973; Zuiderwijk, 1974). For weighted turbulent fluids (spacers and scavenger slurries) which are more viscous, the intensity of turbulence is smaller, but the turbulent stresses are bound to be much higher, and the erosion of the mud may be enhanced by the presence of solid particles. For different annular sizes and rates, the Reynolds numbers and friction pressures for water, oil, chemical washes, weighted spacers, and scavengers are shown in Table 5–6.

Generally speaking, gravitational forces are not important when displacing in turbulent flow. This may be attributed to the fact that, in the absence of density differences, the interface between two fluids is already relatively "flat." Therefore, when the displacing fluid is heavier, gravitational forces cannot greatly contribute to improved results. On the other hand, good displacement efficiencies can be obtained with chemical washes which are up to 4 lb/gal (0.48 g/cm^3) lighter than the mud (Graham, 1972). Such results imply that unstable density differences can be countered by the turbulence of a wash.

The suggested mechanisms underlying the efficiency of the turbulent-flow-displacement technique imply that the phenomena are not instantaneous; therefore, the concept of the minimum contact time is probably valid. However, in the absence of a theoretical model concerning this parameter, and the difficulty of deriving it from laboratory experiments (length scale being too short), actual field experience should dictate an absolute value when available.

For the turbulent-flow-displacement technique to be successful, several criteria must be met .

- The displacing fluid must be sufficiently thin for the critical pumping rate to be achievable with field equipment. This implies that the viscosity of the displacing fluid should be much lower than that of the mud, at least under the specific flow conditions. If turbulence is not attained, the displacing fluid may channel through the mud (the so-called "viscous fingering phenomenon").
- The displacing fluid must exhibit excellent fluid-loss properties, especially when its solid-to-liquid ratio is high; otherwise, losses of the base fluid (water or oil)

Newtonian Fluid No. 1 : ρ = 1000 kg/m^3, η = 1 cp — Flow Rates					
Annular Size ID OD in.	**(bbl/min) (L/min)**	**2 318**	**5 795**	**10 1590**	**15 2380**
4 to 5½	Re	28,000	69,000	140,000	210,000
	dp/dz*	8.85	44.3	152	313
7 to 8½	Re	17,000	42,900	85,700	129,000
	dp/dz*	3.78	18.7	63.4	131
9 ⅝ to 12¼	Re	12,100	30,400	60,700	91,100
	dp/dz*	0.389	1.91	6.44	13.2
Turbulent Flow Spacer No. 2 : ρ = 1319 kg/m^3, η = 1, k = 0.38 × 10^{-3} lbf s^n ft^{-2} — Flow Rates					
Annular Size ID OD in.	**(bbl/min) (L/min)**	**2 318**	**5 795**	**10 1590**	**15 2380**
4 to 5½	Re	2030	5060	10,100	15,200
	dp/dz*	19.5	118	384	771
7 to 8½	Re	1240	3100	6210	9340
	dp/dz*	12.0	52.0	167	332
9⅝ to 12¼	Re	879	2220	4,400	6,600
	dp/dz*	1.58	4.2	17.3	34.5
Turbulent Flow Spacer No. 3 : ρ = 1557 kg/m^3, η = 0.49, k = 7.1 × 10^{-3} lbf s^n ft^{-2} — Flow Rates					
Annular Size ID OD in.	**(bbl/min) (L/min)**	**2 318**	**5 795**	**10 1590**	**15 2380**
4 to 5½	Re	1780	7090	20,700	37,200
	dp/dz*	26.3	82.8	233	436
7 to 8½	Re	849	3380	9640	17,800
	dp/dz*	20.7	40.3	112	205
9⅝ to 12¼	Re	285	1140	3240	5980
	dp/dz*	5.77	9.04	15.1	27.9

*Frictional pressure drops *dp/dz* are expressed in psi/1000 ft.

Table 5–6—Reynolds numbers and friction pressure drops calculated for three displacement fluids and three different annuli.

may increase the viscosity, and raise the critical pumping rate for turbulent flow beyond the capabilities of the field equipment.

- The chemical and physical properties of the displacing fluid must be carefully designed (Section 5–4). It is of utmost importance for the displacing fluid to be fully compatible with the mud. In addition, a weighted displacing fluid must be able to suspend the solids required to achieve the designed density—on the surface and under downhole conditions during placement.

Turbulent-flow displacement is usually accepted as being the most efficient technique for achieving good mud removal; however, there are certain well conditions which can make this technique impractical or impossible.

- For an unweighted displacing fluid, the volume necessary to achieve a given contact time may be such that pore pressure cannot be controlled.
- For weighted displacing fluids, the critical pumping rate for turbulent flow may exceed the capabilities of the available equipment, a reduction in flow rate may occur when the displacing fluid rounds the shoe because of U-tubing, or the required volume of fluid could be cost prohibitive.
- Weak formations with low fracture gradients may not be able to withstand the pressures associated with high displacement rates.

5-4.1.3 Removal of Mobile Mud in Concentric Annuli—Conclusions

With regard to the displacement of mobile mud in concentric annuli, it is clear that the relative importance of flow rate and fluid properties is not yet fully understood. Nevertheless, a review of the literature published on the subject allows one to draw some qualitative conclusions.

- Efficient displacements can be obtained both at low and high flow rates, depending on the prevailing conditions.
- In all circumstances, reducing the mud density and rheology will always result in improved efficiency.
- In addition to the flow rate, density and rheology play an important role in the displacement process.
- At high flow rates, very good results can be obtained with thin displacement fluids pumped in at least the upper range of the laminar-flow regime, or even better in the turbulent-flow regime. Under these conditions, the density ratio is no longer a critical variable for the displacement process.
- At low displacement rates, a density ratio greater than one greatly favors the displacement process. A yield stress ratio greater than one also contributes to the displacement efficiency.
- At intermediate flow rates, the situation is less clear. Under such circumstances, it is recommended to maintain a density ratio greater than one, and an "apparent viscosity ratio" less than one (under the prevailing flow conditions). Such conditions favor the flattest possible interface profile and displacement efficiency.

5-4.1.4 Removal of Mobile Mud in Eccentric Annuli

5-4.1.4.1 Theoretical Modeling

Some of the principal causes of cementing difficulties are those which prevent efficient mud circulation. Casing eccentricity and geometrical asymmetry (e.g., in dual completions) are typical examples. As shown earlier, variations in the flow-path dimensions in the cross-sectional area perpendicular to the mainstream favor channeling of the mud through itself, and that of the displacing fluid through the mud. The phenomenon was identified by Jones and Berdine (1940) and Howard and Clark (1948).

In 1967, an in-depth effort to understand the role of casing eccentricity on mud removal was published by McLean et al. First, a model was developed describing the flow of a single Bingham plastic fluid in an eccentric annulus. Next, the model was extended to the displacement problem in the absence of gravitational forces. The results suggested that, in laminar flow, displacements in eccentric annuli are more effectively optimized by increasing the yield-stress ratio rather than the plastic-viscosity ratio. The reasons for this are twofold..

1. Once the yield-stress ratio is higher than a critical value equal to (2 – STO) / STO, the shear stresses generated by the cement are sufficiently high for the mud to flow in the narrowest part of the annulus. Under creeping flow conditions (i.e., near-zero flow rate), the velocity of the mud in the narrowest part of the annulus is equal to the average velocity of the displacing fluid; in effect, there is 100% displacement efficiency. As the flow rate increased, the displacement efficiency decreased (Fig. 5–19).
2. Increasing the plastic viscosity ratio improves the displacement efficiency only at very high flow rates (Fig. 5–20)

While Point No. 2 may be understandable, Point No. 1 is quite paradoxical. Under creeping flow conditions, the cement slurry velocity in the narrow side of the annulus should be nil, because the yield stress of the cement slurry is much higher than that of the mud. This is effectively what McLean et al. (1967) observed; however, this did not prevent the mud being driven by the cement from the narrow side to the wide side. Although Point No. 2 is understandable, the authors presented very little data to support it. The limited number of tests they performed with density differences tends to show that gravitational forces do play a role on mud displacement in an eccentric annulus.

McLean et al. (1967) also investigated displacement at high flow rates in the extended transition from laminar to turbulent flow, but using extremely severe conditions

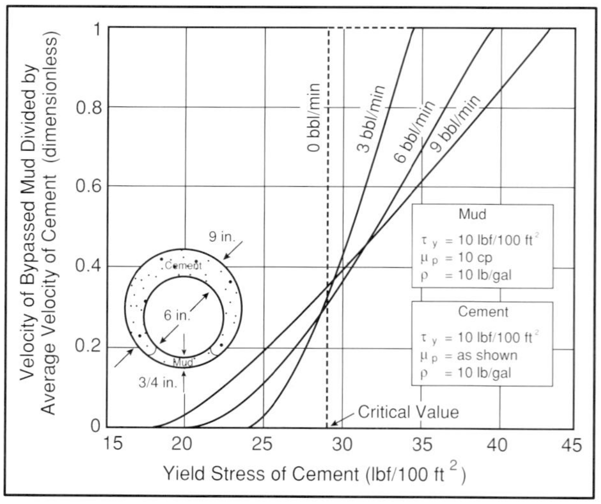

Figure 5–19—Effect of cement yield stress on displacement of a mud from an eccentric annulus (STO = 50%) (after McLean et al., 1967).

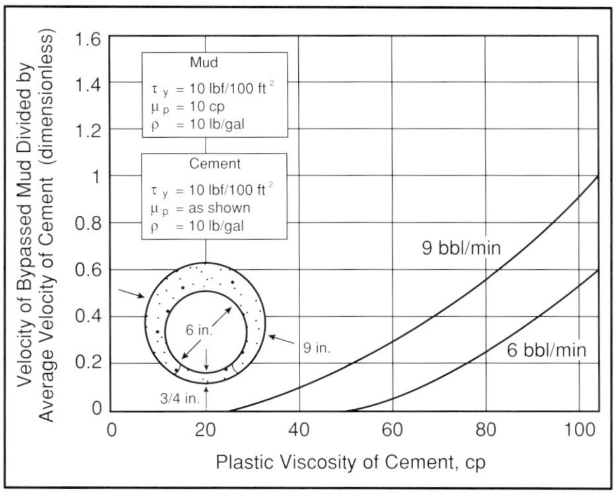

Figure 5–20—Effect of cement plastic viscosity on displacement of a mud from an eccentric annulus (STO = 50%) (after McLean et al., 1967).

(with the inner pipe lying against the outer pipe). Under these circumstances, when displacing a mud or a mock cement exhibiting a yield stress, better results were obtained with a "thick" displacing fluid in laminar flow than a "thin" fluid in partial turbulent flow. These experiments were performed at the same flow rate and, in both cases, the displacement efficiency increased with the flow rate. On the basis of their theoretical work and limited experimental studies (they intentionally did not allow the muds to gel), McLean et al. (1967) suggested that viscous displacing fluids are preferred to thin fluids. While thin displacing fluids extend the area of turbulent flow, the drag and pressure gradient are reduced.

Other authors have attempted to model the effect of eccentricity on the displacement process. On the basis of the same annular geometry used by McLean et al. (1967), Graham (1972) reemphasized the relative viscosity concept in the absence of gravitational forces, and used a mobility concept defined as the ratio of the flow rate to the friction pressure.

A mobility ratio greater than one was shown to be desirable for optimizing the displacement. Since different fluids exhibit different changes in mobility with changing flow conditions, optimum results could be obtained at either high or low flow rates. Even under the best conditions, the velocity of the interface was always greater in the wide part of the annulus than in the narrow side (Fig. 5–21). To overcome the resulting difference in the level of the interface, and ensure that cement would reach the target level on the narrow side, Graham (1972) applied the knowledge of the interfacial velocity around the annulus, and proposed to pump an excess volume of cement slurry.

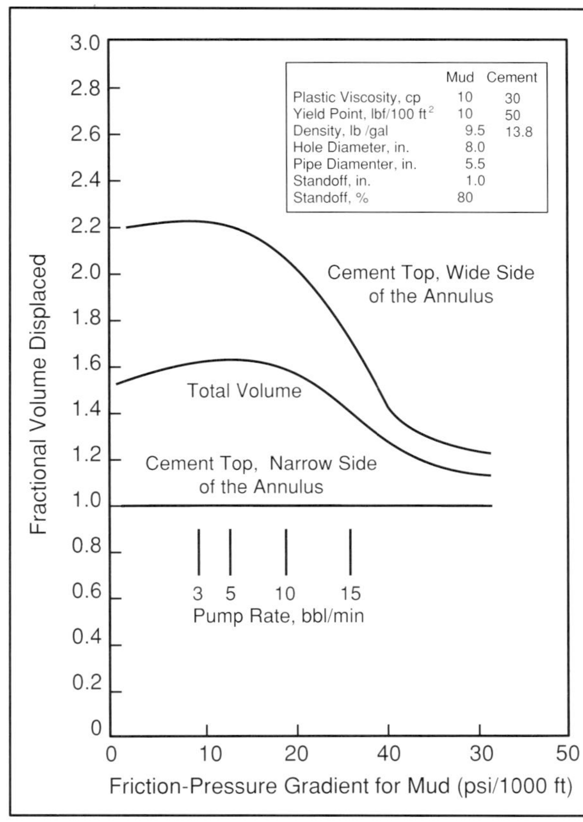

Figure 5–21—Effect of flow rate on displacement of a mud by a cement slurry from an eccentric annulus (from Graham, 1972).

Graham's theoretical developments led him to draw completely different conclusions from those of McLean et al.(1967). Fluids with low yield points and low plastic viscosities, displaced at the highest possible rate, were recommended. However, specific conditions were imposed on the mud rheology—$\tau_{y\,(m)} < 5$ and $\mu_{p\,(m)} < 12$. Graham claimed that this technique has been used successfully in the field, but no field data were reported.

Jamot (1974) extended Graham's model by introducing the effect of gravitational forces. The deformation of the fluid interface due to eccentricity was shown to be minimized at low displacement rates. The best results were obtained when the density of the mud was significantly lower than that of the displacing fluid (typically > 4.2 lb/gal [0.5 g/cm³]). Care was taken to minimize the gel strength of the mud, and to use viscous displacing fluids. On the other hand, turbulent flow was shown to be preferred when the density differences were small (typically < 1.7 lb/gal [0.2 g/cm³]). In between, both flow regimes showed equivalent efficiencies, and laminar flow gave the poorest results in all cases (Fig. 5–22).

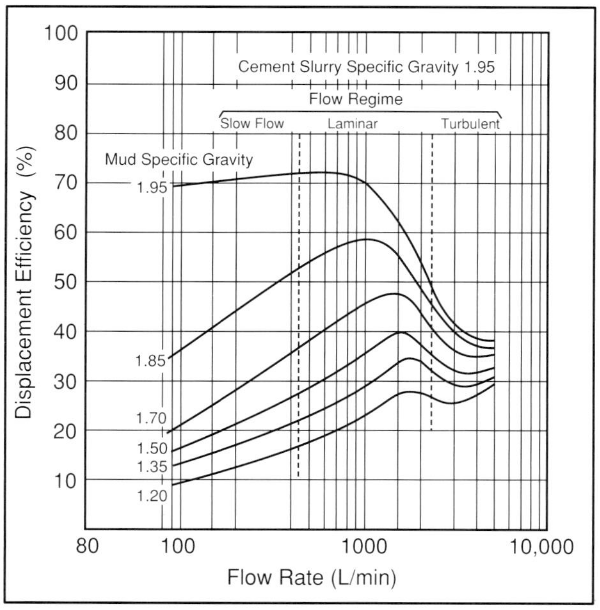

Figure 5–22—Effect of flow rate and flow regime on the displacement of mud of various densities by a cement in an eccentric annulus (STO = 80%) (after Martin et al., 1978).

Using a slightly different approach, Martin et al. (1978) considered the flow as two-dimensional, the values of all parameters being averaged along directions perpendicular to the cylindrical surfaces. The flow equations were solved by making an analogy with those governing the flow of two immiscible fluids in porous media. Jamot's (1974) recommendations regarding the optimum flow regime and density ratio were largely confirmed. The density and viscosity ratios were also claimed to have a similar effect, but this statement was purely qualitative.

One of the originalities of the Martin et al. (1978) study was to investigate the displacement of a given fluid by two others, (e.g., a cement slurry in the presence of a spacer). At low displacement rates, their model demonstrated that the spacer could have no effect or even an unfavorable effect on mud displacement. This was based on a purely hydrodynamic point of view—the fluids were assumed to be separated by sharp interfaces, with no mixing zones or possible chemical interactions. To fulfill this role, they found it essential for the spacer to have a density and rheology between those of the mud and cement. When this was not the case, the spacer would tend to flow preferentially. In extreme cases, the flow was confined to either the wide side or the narrow side of the annulus; consequently, the cement slurry would directly contact the mud.

5–4.1.4.2 Experimental Studies

Using a simulated borehole, Clark and Carter (1973) performed an interesting experimental study on the effect of high eccentricities on mud removal by cement slurries. They attempted to correlate the displacement efficiency with the friction pressure of the displacing fluid. Very poor results were usually obtained when the cement was in laminar flow—the efficiency increased slowly with frictional pressure. It is worthwhile to mention that they allowed the mud to gel and encounter filtration in their experimental device, which may have strongly affected the results (Section 5–3.2).

Much better results were obtained when the fluids were pumped under partial turbulent-flow conditions (Fig. 5–23) for the same frictional pressure. They also observed that, for a given pressure drop, the results improved as the viscosity of the displacing fluid decreased. This was due to the extension of the turbulent-flow regime in the annulus. These results do not contradict those of McLean et al. (1967), because the results were compared for the same friction pressure. McLean et al. (1967) performed their comparison at a constant flow rate. These points also agree with some of the authors' unpublished data showing that, when displacing a mud with water in an eccentric annulus, the lower the standoff the higher the displacement rate necessary to obtain an excellent efficiency within a reasonable amount of time (Fig. 5–24). Their data also showed the contact time to be a key parameter when displacing muds with thin cements at high rates (partial turbulent-flow conditions), while this parameter had little effect when using thick slurries in laminar flow. In addition, these results showed again that, when displacing a mud at high flow rates with a thin fluid, the annular pressure drop is not the only driving

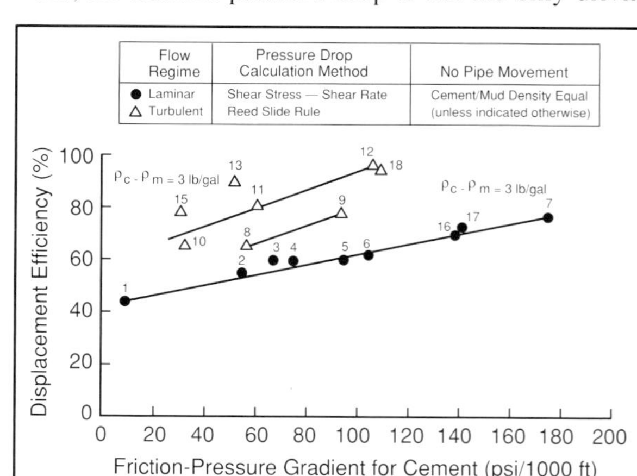

Figure 5–23—Effect of cement/mud density difference and cement rheology on displacement efficiency (after Clark and Carter, 1973).

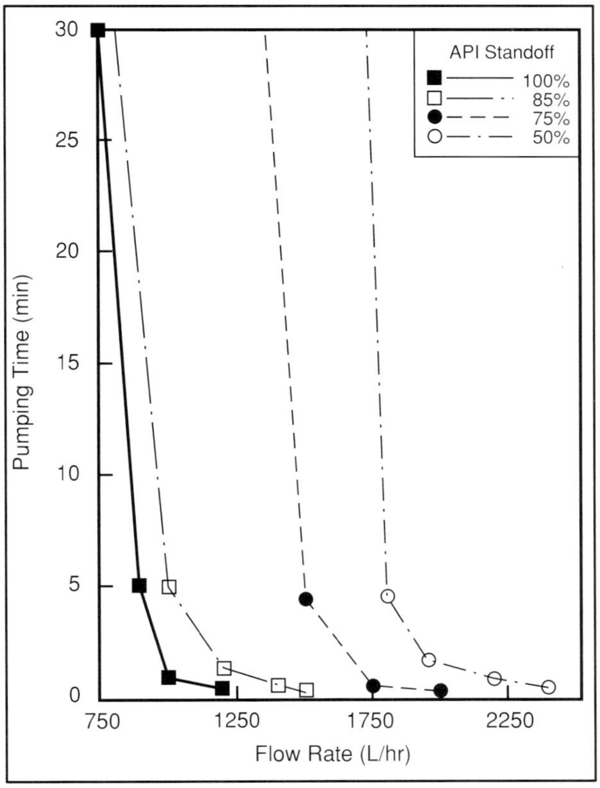

Figure 5–24—Effect of pipe standoff on the pumping time necessary to remove all the mud with water from an annulus (ρ_m = 1,045 kg/m^3, μ_p = 24 cp, τ_y = 11 Pa, D_o = 5cm, D_i = 4cm).

force. Other mechanisms are involved such as drag stresses, erosion at the fluid interfaces, and dilution of one fluid into another.

Clark and Carter (1973) also performed a few experiments with the density of cement 3 lb/gal (0.36 g/cm^3) higher than that of the mud. Although the corresponding gravitational forces were in all cases higher than the frictional forces, no noticeable effect on the efficiency was observed. Since the mud was allowed to gel and filter through a permeable medium, they concluded that gravitational forces were not effective for removing the immobile mud.

In a paper concerning the cementing of multiple casings, Childers (1968) claimed that turbulent flow was unnecessary and difficult to achieve, and that the relative rheological properties between the mud and the cement were the controlling factors in successful jobs. Using the same argument as McLean et al. (1967), Childers (1968) proposed to design the yield stress of the cement on the basis of the following equation.

$$\tau_{y(c)} = \frac{e_{max}}{e_{min}} \times \tau_{y(m)} \quad , \qquad (5\text{–}27)$$

where e_{max} and e_{min} are the maximum and minimum separation between the walls of the casing and the wall of the borehole, respectively.

Unfortunately, the required cement yield stress would be far too high, and other forces would be necessary to prevent channeling of the cement through the mud gravitational forces, high flow rates (which is in contradiction with the results of the McLean et al. [1967] model), and casing movement (a point which will be addressed later). The results of a survey of 13 successful jobs were reported, where the following relative properties were adopted. The cement slurry densities and yield points were, respectively, 2.3 lb/gal (0.28 g/cm^3) and 17 lb/100 ft^2 (8.14 Pa) higher than those of the muds. A plastic viscosity ratio higher than one was claimed to be favorable, but no supporting data were shown.

More recently, Lockyear et al. (1988) published some interesting experimental results which are backed up by theoretical arguments. For efficient mud displacement in an eccentric annulus, the friction pressure during the displacement should meet the following condition.

$$\frac{dP^*}{dz}\frac{Di}{4} \times \left[1 - \left(\frac{3D_i - D_o}{2D_i}\right)^2\right] > \tau_y \quad , \qquad (5\text{–}28)$$

an expression which can be further simplified to

$$\frac{dP^*}{dz}\frac{D_o - D_i}{4} > \tau_y \quad . \qquad (5\text{–}29)$$

Equation 5–29, which imposes that the mud is flowing on the narrow side of the annulus, was verified by their experimental results (Table 5–7 and Fig. 5–25). They also claimed that the velocity of the displacing fluid should be nonzero as well on the narrow side of the annulus far away from the interface, a condition McLean et al. (1967) found not to be necessary. As discussed earlier in Section 5–2, satisfying such conditions does not guarantee optimum circulation efficiency, because both fluids may flow completely around the annulus but with a large difference in the interfacial velocity between the narrow side and the wide side. Lockyear et al.(1989) gave only a partial answer to this problem. In the absence of density differences, and for standoff values of 50%, they observed a sharp transition between severe and minimal channeling for an average Reynolds number of 1,500 for the displacing fluid, with standoff values around 50%. This confirms that good displacement may be obtained,

Experiment No.	B	C	D	E	G
Mud Type	KCl/P	KCl/P	KCl/P	KCl/P	KCl/P
PV/YP(cp/lbf/100 ft^2)	37/10	44/12	39/13	41/13	38/8
Density (SG)	1.62	1.62	1.62	1.63	1.69
10-s /10-min Gel*	11/12	11/12	13	13	5
Gel During Experiment	12	12	12	14	9
Spacer Type	Type A	Type A	Type A	Type A	Type B
PV/YP(cp/lbf/100 ft^2)	33/9	39/5	43/11	56/18	28/2
Density (SG)	1.62	1.62	1.62	1.63	1.69
10-s Gel*	6	7	9	16	1
Volume Pumped (bbl)	6.0	5.0	5.3	6.4	6.0
Cement Type	Neat G	Neat G	Neat G	Neat G	Neat G
PV/YP(cp/lbf/100 ft^2)	21/42	33/40	21/20	33/37	47/38
Density (SG)	1.9	1.9	1.9	1.89	1.89
10-s Gel*	21	22	17	28	23
Volume Pumped (bbl)	7.0	5.6	2.8	3.2	7.0
Deviation (°)	0	0	0	0	0
Mean Standoff (%)	40	40	60	60	50
Narrow Side Gap at Shoe (mm)	8	8	17	17	14
Casing Size (in.)	7	7	7	7	7
Hole Size (in.)	8.625	8.625	8.625	8.625	8.625
Displacement Rate (BPM)	8.1	2.0	5.2	2.1	8.2
Maximum Annular Pressure Drop During Displacement (psi/100 ft)	26.0	4.4	10.5	16.3	59
Calculated Minimum Gap to Achieve Full Mud Displacement (mm)	1.9	12	5	4	1
Fluid in Narrow Side at Shoe	Cement	Mud	Cement	Cement	Cement

* Gel strength is given in lbf/100 ft^2.

Table 5–7—Experimental conditions for various tests shown in Fig. 5–25 (after Lockyear, 1988).

even in eccentric annuli, when the Reynolds number of the displacing fluid is in the upper laminar or turbulent-flow range.

5–4.1.4.3 Mobile Mud Displacement in Eccentric Annuli—Conclusions

The best conditions for optimizing mud removal are not very well defined for concentric annuli, and the problem is even worse for eccentric annuli. However, there are two major schools of thought.

- The yield point, density, and eventual plastic viscosity of the displacing fluid should be higher than the corresponding properties of the displaced fluid.
- Based on the effects of turbulent flow or partially turbulent flow, thin displacing fluids should be pumped at a rate such that at least partial turbulent flow is obtained.

With few exceptions (e.g., in the absence of density differences), the theoretical approaches tend to favor the first approach. This is not surprising, because most models did not take into account the mechanisms which are known to underlay the turbulent-flow technique—erosion and dilution. On the other hand, the experimental studies agree with one or the other, the great majority supporting the second approach.

When looking carefully at the various experimental conditions used by different authors, it appears that the experimental studies favoring the first approach were performed between two impermeable walls—in the absence of filtration. In addition, the effect of mud gelation was minimized by not allowing it to remain static in the apparatus. Thus, the studies were focused on the mobile mud. The studies favoring the second approach took no precautions to prevent mud gelation and filtration; con-

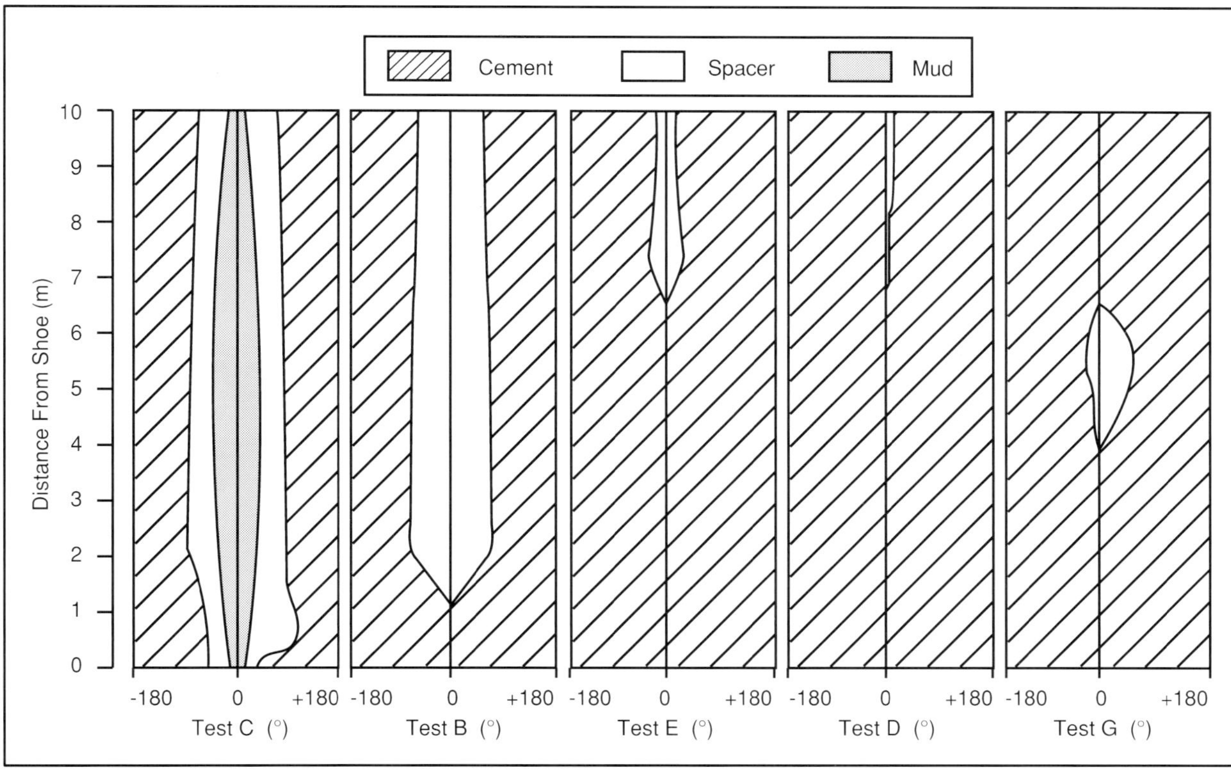

Figure 5–25—Distribution of cement, space, and mud in the annulus for a number of tests (0° represents the narrow side of the annulus) (after Lockyear et al., 1988).

sequently, there was some emphasis on the immobile mud. Therefore, the remaining question concerns the optimum conditions for removal of both types of mud. This subject is addressed in the next section.

5–4.2 Displacement of the Immobile Mud

As mentioned earlier, a mud which has been altered by gelation and filtration is difficult to mobilize during circulation. In this section, the displacement of such muds is considered. As with mud circulation, the process is not well understood, because gelled or dehydrated muds are so poorly characterized.

Martin et al. (1978) attempted to model this phenomenon. Using some simplifying assumptions to describe the buildup and breakdown of gel strength, they showed the displacement efficiency to be strongly affected by gelation. The effect is qualitatively illustrated in Fig. 5–26. When the drilling mud exhibits a low gel strength, the best results are obtained at low displacement rates, provided the density of the displacing fluid is higher than that of the mud. Under the same conditions, if the mud exhibits a high gel strength, turbulent flow is preferred. When the density of the displacing fluid is less than that of the mud, turbulent-flow displacement is best for low-gel-strength muds; high-gel-strength muds are difficult to remove regardless of the displacement rate.

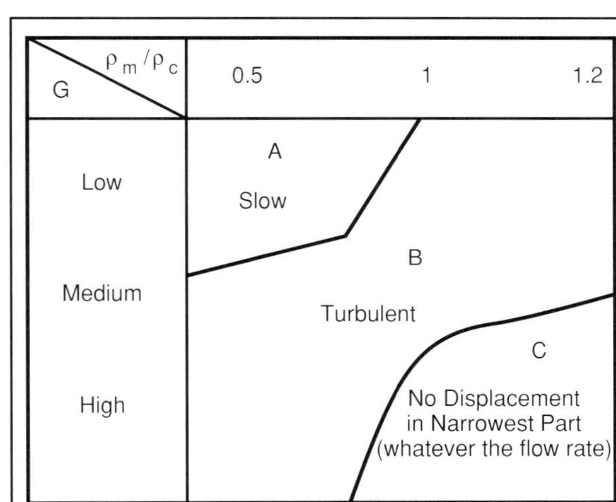

Figure 5–26—Effect of density differential and mud gel strength on mud displacement. G is the ratio of the 10-min gel strength to the initial gel strength of the mud (after Martin et al., 1978).

Unfortunately, the theoretical results of Martin et al. (1978) are not supported by experimental data. After performing displacement experiments, McLean et al. (1967) and Lockyear and Hibbert (1988) related the flow resistance of gelled mud on the narrow side of an eccentric annulus to the gel strength of the mud. McLean et al. (1967) found no correlation between the two, while

Lockyear and Hibbert (1988) found the opposite. In view of these contradictions, which in fact may be due to a poor characterization of the rheological properties of drilling muds, this area certainly requires more attention before even qualitative conclusions can be drawn.

The situation regarding the effect of the mud cake is even worse. Very little is known about the erosion of mud cakes by displacing fluids, although it is generally admitted that mud cakes are eroded by displacing fluids at high Reynolds numbers. Haut and Crook (1981) characterized the ability of a mud to be displaced by a single parameter, the *mud mobility factor*, which considers mud gelation and mud dehydration.

$$\textit{Mud Mobility Factor} = \frac{1}{V_{FL} \times \tau_{gel-10}}, \quad (5\text{–}30)$$

where

V_{FL} = fluid-loss rate (mL/30 min), and

$\tau_{gel\text{-}10}$ = 10-min gel strength (lbf/100 ft^2).

In their experiments, the muds were allowed to gel prior to the displacement. A good correlation was found between the efficiency of the displacement and the mud mobility factor multiplied by the square of the average velocity of the displacing fluid (Fig. 5–27). These results confirm the intuitive idea that removal of the immobile mud requires much more energy than that for removing mobile mud. The problem with this approach, sometimes called the "pump as fast as possible method," cannot always be adopted because of fracture pressure limitations; therefore, other solutions for removing the immobile mud have been developed. As described in Section 5–2.2.7, casing movement coupled with various types of casing hardware is effective. Spacers and washes are also useful, as will be explained in the next section.

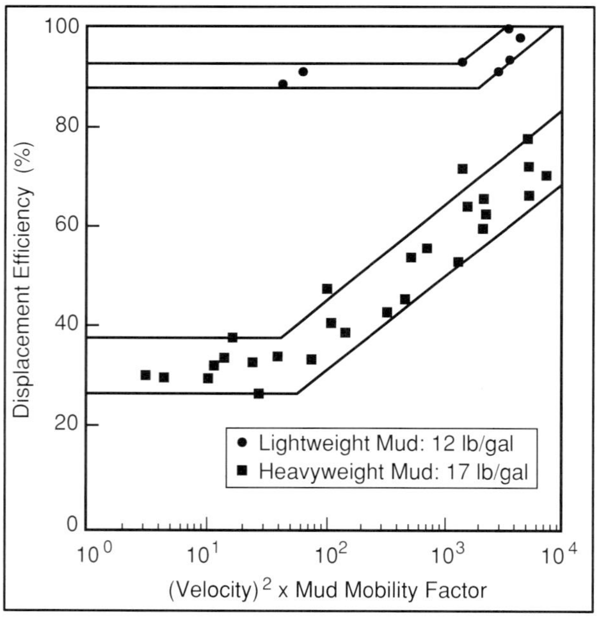

Figure 5–27—Effect of displacement velocity and mobility factor on the displacement efficiency of muds by water (from Haut and Crook, 1981).

5–4.3 Effect of Casing Movement and Casing Hardware

Pipe movement during cement placement helps to remove the mud which would otherwise be trapped on the narrow side of an eccentric annulus. The basic principle is the same as during mud circulation; however, the physics involved is more complicated, and published models including the effect of casing movement are currently limited to the circulation process only. On the experimental side, McLean et al. (1967) reported a few conclusions concerning the effect of casing movement on mud displacement between impermeable walls. They observed casing rotation to provide a better means of removing the mud than reciprocation. However, as mentioned earlier, they emphasized that lateral motion of the casing was not allowed in their experiments, which is likely to happen in the field.

Mechanical devices (such as scratchers, scrapers, or cable wipers) were also shown to improve the efficiency of the displacement process when used in combination with casing movement (Jones and Berdine, 1940; Teplitz and Hassebroek, 1946). These devices are attached to the pipe, and they contribute to the erosion of the gelled and/or dehydrated mud which would otherwise remain static in the annulus.

With the improvement of the necessary equipment, casing movement is now a more common practice. Casing reciprocation has been used successfully in a great number of critical operations (Kolthoff and Scales, 1982; Holhjem, 1982). Typical amplitudes used for casing reciprocation are of the order of 20 to 40 ft, a full cycle being completed in one to five minutes. The main drawbacks of this type of movement are threefold.

- Pipe reciprocation induces pressure surges and swabs which may adversely affect well control, especially with small annular clearances.
- There is a risk of the casing becoming stuck at the edge of the upstroke.
- The movement amplitude is reduced downhole because of pipe stretch or buckling. Excessive casing pull may be required, especially in highly deviated wells.

Pipe rotation was found to improve the quality of primary cement jobs, specifically liner jobs (Landrum et al., 1985; Buchan, 1986), without presenting the above drawbacks. Rotary power tongs or power swivels were used, and the rotation rate usually varied between 10 to 40 RPM. The key to the success of this technique was a good control of the torque; for this reason, power swivels were preferred over other systems.

Although not a common practice, it is worthwhile to mention that some operators use both movements (i.e., rotation and reciprocation) simultaneously with excellent results. This being said, one must remember that casing movement is not the panacea for all mud-displacement problems. Since the effects of casing movement have been characterized only qualitatively, other methods for improving primary cement jobs should not be omitted.

5-5 SPACERS AND WASHES

During a cementing job, the cement slurry must displace all of the drilling mud from the annulus. However, contact between the drilling mud and cement slurry often results in the formation of an unpumpable viscous mass at the cement/drilling mud interface. Under such circumstances, the drilling mud and the cement slurry are said to be *incompatible*.

When incompatibility exists between fluids being displaced in the annulus, the displacing fluid (i.e., the cement slurry) tends to channel through the viscous interfacial mass, leaving patches of contaminated mud sticking to the walls of the casing and formation. This may lead to insufficient zonal isolation, necessitating expensive remedial cementing prior to stimulation treatment of the formation. The very viscous cement/mud mixture can also cause unacceptably high friction pressures during the cement job, with the obvious danger of fracturing a fragile formation. In extreme cases, total plugging of the annulus can occur, preventing the completion of the cement job.

To avoid such problems one or more intermediate fluids (or *preflushes*), which are compatible with both the cement slurry and drilling mud, are often pumped as a buffer to prevent or at least minimize contact between them. Preflushes, pumped into the borehole in front of the cement slurry, are designed to clean the drilling mud from the annulus and leave the annular surfaces receptive to bonding with the cement. Thus, they must eliminate the mud from the casing and formation walls (Crinkelmeyer et al., 1976; Sauer, 1987). To accomplish all of these tasks, the rheological and chemical properties of preflushes must be carefully designed.

Washes are fluids with a density and a viscosity very close to that of water or oil. They act by thinning and dispersing the mud. Because of their very low viscosity, they are particularly useful for displacement in turbulent flow. As discussed earlier in this chapter, the turbulent-flow regime can lead to very efficient mud displacement. The walls of the casing and the formation are also swept clean by the turbulent fluid. The simplest form of a wash is fresh water (Warembourg et al., 1980; Haut and Crook, 1981; Smith and Crook, 1982; Sauer, 1987). However, for a more efficient thinning and dispersion of the mud, *chemical washes*, which contain a mixture of dispersants and surfactants, are more commonly used (Evanoff and Cook, 1988). The dispersants are often of the same types applied in cement slurries—polynaphthalene sulfonates (Wieland and Woods, 1975), lignosulfonates, tannates, etc.

If an oil-base mud is involved, surfactants must be present in the chemical wash. Not only do surfactants help disperse the mud, they also leave the casing water wet, and receptive to bonding with the cement system. Nonionic or anionic surfactants are usually preferred (Goode et al., 1983). Examples of nonionic surfactants include ethoxylated nonylphenols (Weigand and Totten, 1984), fatty acid esters, and ethoxylated fatty alcohols (Bannister, 1981). Examples of anionic surfactants used for this purpose are alkyl sulfonates and alkyl aryl sulfonates (Bannister, 1981), and sulfonated ethoxylated fatty alcohols (Wiegand and Totten, 1984). Optionally, chemical washes may contain a small concentration of friable and pliable hydrocarbon oil-soluble resin particles. The particles leave a thin filter cake on the formation wall, minimizing the loss of the chemical wash to the formation, and helping to reduce the fluid loss from the cement slurry (Sharpe and Free, 1977; Bannister, 1978; Bannister, 1987). Sodium chloride (NaCl) and potassium chloride (KCl) are sometimes added to chemical washes to protect freshwater-sensitive formations.

Spacers are preflushes with carefully designed densities and rheological properties (Warembourg et al., 1980). They have a much higher solid particle content than washes, and are generally more effective buffers for avoiding contact between the cement slurry and the drilling mud. Some may be pumped in turbulent flow, and thus share the same cleaning action as washes. The particles in spacers are also thought to have a scrubbing effect on the annular surfaces. Other spacers are designed to be pumped in laminar flow (Crinkelmeyer et al., 1976).

One of the simplest forms of what can be called a spacer is the *scavenger slurry* (Brice and Holmes, 1964)—a low-density cement slurry with a low fluid-loss rate, which can easily be pumped in turbulent flow. The principal drawback of scavenger slurries is that they are frequently incompatible with drilling muds.

It is generally accepted that the best mud removal is obtained if the density of the spacer is higher than the density of the drilling mud, but lower than that of the cement slurry (McLean et al., 1967; Zuiderwijk, 1974; Martin et al., 1978; Weigand and Totten, 1984; Sauer, 1987). The buoyancy effect assists in the removal of the mud. Weighting agents (generally insoluble minerals with a high density) are used to adjust the density of the spacer fluid. To achieve efficient suspension of the weighting agent, a viscosifier is also included.

The preferred flow regime for a spacer is turbulent flow (Zuiderwijk, 1974; Haut and Crook, 1981; Sauer, 1987; Evanoff and Cook, 1988), because it leads to better mud removal and annular cleaning; however, a compromise must be reached. The viscosity should be as low as possible, to allow turbulent flow at reasonable pumping rates. On the other hand, the viscosity must be sufficiently high to effectively suspend the weighting agent. Hydroxypropylcellulose polymers may be used to satisfy these conflicting requirements. Such materials impart sufficient viscosity to suspend the weighting agent(s) during mixing and pumping on the surface; however, when a critical temperature is reached during pumping downhole, the polymer is no longer soluble, and the viscosifying effect is lost, permitting turbulent-flow displacement in the annulus (Bannister, 1981). The turbulent action keeps the weighting material suspended.

In many cases, the pumping rates necessary for turbulent flow cannot be applied, because of limitations imposed by the available pumping equipment, or when the resulting friction pressures would present a danger of fracturing the formation. In such cases, a laminar-flow spacer is used. The best results are obtained if not only the density, but also the rheological properties of the spacer fall between those of the mud and cement slurry. The spacer will not channel through the mud, and the cement will not channel through the spacer.

Spacers are more complicated chemically than washes. Below is a summary of some common ingredients.

Viscosifiers are necessary to suspend the weighting agent(s) and control the rheological properties. They can be subdivided into two classes.

- Water-Soluble Polymers
  —Polyacrylamides (Belousov et al., 1987).
  —Guar and guar derivatives (Weigand and Totten, 1984; Wieland and Woods, 1975).
  —Cellulose derivatives (CMC, HEC, HMC, HPC) (Thomas, 1981; Wiegand and Totten, 1984; Bannister, 1987).
  —Xanthan gum and other biopolymers (Wiegand and Totten, 1984; Sehault and Grebe, 1987; Parcevaux and Jennings, 1985).
- Inorganic Clays
  —Bentonite, attapulgite, kaolinite, and sepiolite (Beirute, 1981; Thomas, 1981; Weigand and Totten, 1984; Evanoff and Cook, 1988).

Dispersants enhance the compatibility of the spacer with water-base muds and cement slurries, and disperse the weighting agent in the spacer. The most common dispersant is polynaphthalene sulfonate (Wiegand and Totten, 1984; Guillot et al., 1986).

Fluid-loss-control agents are usually water-soluble polymers—guar gum (Wieland and Woods, 1975), poly(ethyleneimine) (Wieland and Woods, 1975), cellulose derivatives (Weigand and Totten, 1984; Guillot et al., 1986), and polystyrene sulfonate (Guillot et al., 1986). Sometimes the same polymer functions as both a viscosifier and fluid-loss-control agent (Wieland and Woods, 1975). The inorganic clays discussed above also have a beneficial influence on fluid-loss control.

Weighting agent(s) are used to obtain the desired spacer density—silica flour, fly ash, calcium carbonate, barite, hematite, and ilmenite (Thomas, 1981).

Surfactants increase the compatibility of spacers with oil-base muds, and leave the casing water wet (Sauer, 1987). The same nonionic or anionic surfactants described above for washes are usually appropriate.

Optionally, NaCl or KCl may be used to protect or prevent the dissolution of massive salt formations or fresh-water-sensitive shales (Wieland and Woods, 1975; Smith and Crook, 1982).

A special problem is posed by oil-base muds. As has been explained earlier, a special mixture of surfactants (generally anionic and nonionic) can be added to a water-base spacer or wash to render it compatible with the mud, and to leave the casing and formation water wet. Oil-base spacers or washes also exist. The simplest form is a wash made from oil (the same oil as used for the mud), pumped between the mud and an ordinary water-base spacer (Motley et al., 1974; Bannister, 1987). The oil may contain mutual solvents (Goode et al., 1983) and a mixture of surfactants—nonionic to water wet the casing and the formation (e.g., substituted amides and amines, and oil-wetting surfactants such as quaternary fatty ammonium salts to clean the oil-base drilling mud from the walls [Motley et al., 1974]). Aluminum aliphatic orthophosphate esters and other aluminum soaps can be used as thickening agents for high-viscosity, oil-base spacers (Hill et al., 1973; Motley et al., 1974).

An original idea for the removal of oil-base muds has been described by Oliver and Singer (1986)—a water-free mixture of surfactants and an alcohol. Excellent compatibility with the mud and the cement slurry has been obtained.

Spacers and washes can also be used in combination. If pumped in the order mud-wash-spacer-cement, the wash can thin the mud and make it easier for the spacer to displace (Sauer, 1987).

5–6 CEMENT MIXING

Mixing is one of the most important practical cementing problems. The goal of the mixing process is to effect a correct proportioning of solids and carrier fluid, and to prepare a slurry with properties similar to those expected from prejob laboratory testing. This goal must be met; otherwise, the relevance of the careful job planning calculations to determine the optimum displacement rate, friction pressure, etc., is questionable. In addition, the thickening time and fluid-loss rate of the cement slurry may change dramatically. Such a situation can severely compromise the removal of the drilling mud.

Cement slurry properties obtained on location are not routinely compared to those predicted in the laboratory. When such measurements are performed, significant differences are often found. The same has been shown to be true for spacer fluids (Benge, 1989). Such differences may result from density errors. Slurry properties are also sensitive to the mixing conditions. Both concepts are discussed in this section.

5–6.1 Density Error

During the design phase of a cement job, an extensive program of laboratory tests is usually performed on candidate cement systems. The final result is a cement system with an optimum thickening time, compressive strength, fluid-loss rate, and rheology. In addition, the free-water development and sedimentation are minimized. In the laboratory, test slurries are always mixed at the exact densities proposed for the cement job. It is assumed the slurries will be mixed in the field at the anticipated density.

The density accuracy of field mixing equipment and the sensitivity of cement systems to density error are topics of increasing interest in the cementing industry. In 1989, Grant et al. reported the results of a density-error sensitivity study. The typical density variations they observed during continuous mixing and batch mixing are shown in Fig. 5–28. The design density was 15.8 lb/gal

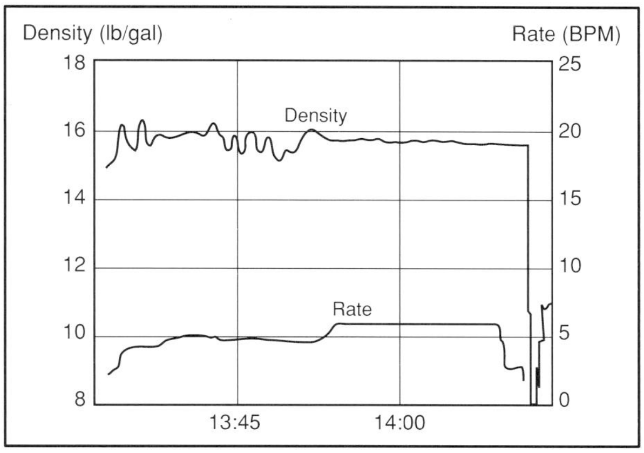

Figure 5–28—Density variations while cementing (from Grant et al., 1989).

(1.89 g/cm³). During continuous mixing, the density varied from 15.1 lb/gal (1.81 g/cm³) to 16.3 lb/gal (1.95 g/cm³). The density was much more consistent during the batch-mixing period.

The sensitivity study was performed on 18 slurries ranging from low-density lead systems to normal density tail systems. The effects of density error on thickening time, fluid-loss rate, free water, and compressive strength development (time to reach 500 psi [3.5 MPa]) were measured at three temperatures—140°F (60°C), 160°F (71.1°C), and 180°F (82.2°C). The study also included a comparison of equivalent systems containing liquid additives or dry-blended solid additives.

The magnitudes of density error varied from –0.8 lb/gal to +0.4 lb/gal. The effects of density error on thickening time, free-water development, and strength development are excerpted in Figs. 5–29, 5–30, and 5–31,

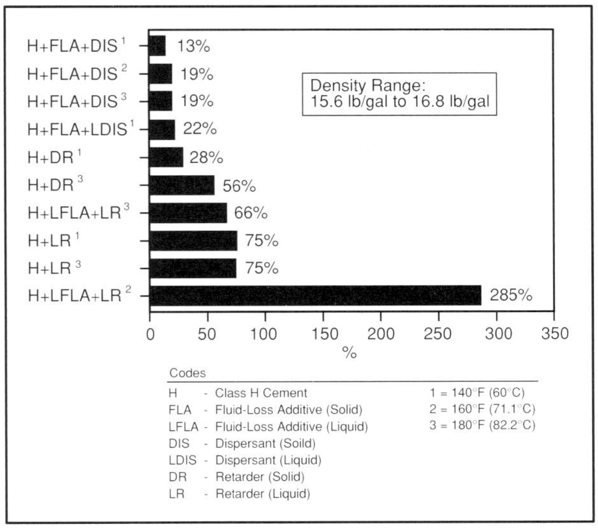

Figure 5–29—Comparison of tail slurries: change in thickening times (from Grant et al., 1989).

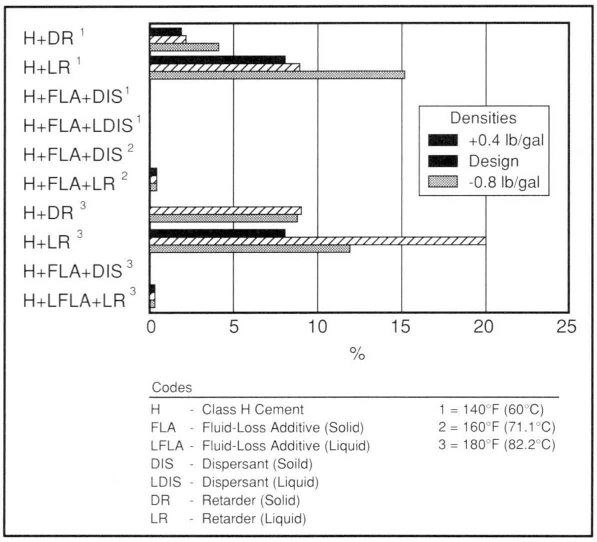

Figure 5–30—Comparison of tail slurries: free-water percents (from Grant et al., 1989).

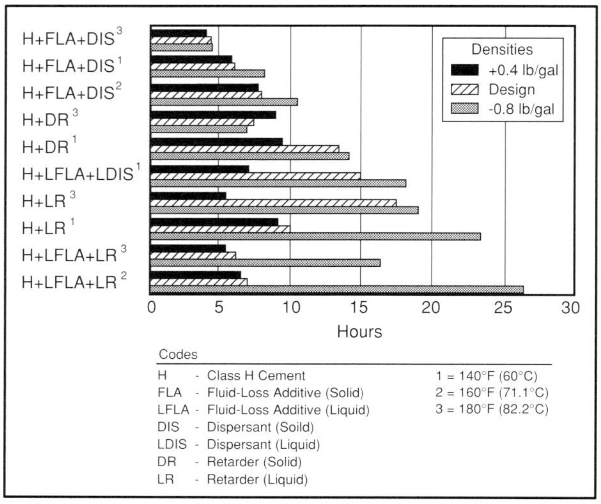

Figure 5–31—Comparison of tail slurries: time to reach 500-psi compressive strength (from Grant et al., 1989).

respectively. Scrutiny of the data reveals that the magnitude of the performance fluctuation was highly dependent upon the system. In addition, the systems containing liquid additives were generally more sensitive to density error than their solid additive counterparts. The concentrations of solid additives by weight of cement are independent of density error, while those for liquid additives are not.

This study effectively demonstrated the importance of density accuracy with regard to cement-system performance and, by implication, mud removal. To reliably deliver a cement system in the field which performs as designed in the laboratory, the system's sensitivity to density error must be minimized, or the mixing equipment must be improved to provide better density accuracy on a routine basis.

5-6.2 Mixing Energy

Cement slurry mixing deviates from classical solid/liquid mixing, because Portland cement is a reactive material. The rate of hydration is affected by mixing conditions. Thus, is it necessary to consider both the physical and the physico-chemical aspects of cement slurry mixing. The following points must be considered.

- How is the rate of cement hydration affected by different mixing conditions?
- How does mixing affect slurry performance characteristics such as yield value, thickening time, and fluid-loss rate?
- What is the most important parameter of cement mixing?

These topics are discussed below.

5–6.2.1 Physical Process

Cement is a powder; therefore, it is characterized physically by its particle-size distribution, specific surface area, etc. (Table 5–8) (Chapter 2). Cement powder consists of agglomerates and aggregates, and different interparticulate forces exist. The most basic are van der Waals forces, which are attractive. In addition, there are forces between particles which carry an adsorbed film of liquid. However, these are probably significant only at high relative humidity, perhaps after a long period of cement storage.

Cement Class	Median Diameter (μm)	Blaine Surface Area (m^2/g)	BET Surface Area (m^2/g)
H	20	0.28	0.8
G	24	0.3	0.7

Table 5–8—Physical characteristics of Class H and Class G cements.

The mixing process involves a number of distinctly different stages—

- wetting,
- deflocculation of aggregates and agglomerates, and
- stabilization of the resulting suspension or paste.

Wetting requires the replacement of the air on the surface of each particle by water; however, it is first necessary to effect a complete breakdown of cement agglomerates and aggregates. The difficulty in achieving deagglomeration and wetting of cement particles may be appreciated when one realizes that, for a 50 kg bag of cement, the surface area to be wetted is approximately 50,000 m². However, the critical step is the deflocculation (Vidick, 1989).

In the laboratory, well cement slurries are normally mixed by the standard API procedure (Appendix B). The mixer is a commercial blender which consists of a cup with a propeller at the bottom which can rotate at very high speeds (12,000 RPM). The mixing produced by this machine can be classified as turbulent mixing.

The mechanical work provided by the mixer during time t is

$$E = T\omega t, \tag{5–31}$$

where

E = mixing energy (kJ),
ω = rotational speed (radians/s),
t = mixing time (s), and
T = torque (N m).

The torque T can be calculated by

$$T = k\rho\omega, \tag{5–32}$$

where

ρ = slurry density (kg/m³), and
k = 6.1×10^{-8} m⁵/s (found experimentally).

Thus, the energy per mass of slurry is

$$\frac{E}{M} = \frac{k\omega^2 t}{V}, \tag{5–33}$$

where

M = mass of slurry (*kg*), and
V = slurry volume (m³).

These equations can be used to define the mixing energy applied in the laboratory; however, as will be explained later, the same concepts can also be used to describe the energy applied by field mixers. The energy of the API mixing procedure, which calls for mixing at 4,000 RPM for 15 seconds, followed by a 35-second period at 12,000 RPM, is 5.9 kJ/kg of slurry. Different levels of mixing can be obtained by changing the rotational speed and/or the mixing time.

The variation of the plastic viscosity of a neat 15.8-lb/gal (1.9-g/cm³) Class G cement slurry with mixing energy exerted by the commercial blender is shown in Fig. 5–32. Two different zones can be observed. First, at low mixing energy, the plastic viscosity decreases strongly with increasing mixing energy. In the second zone, further application of mixing energy no longer produces a large plastic viscosity variation. In the first zone, the interactions between the cement particles are stronger than the shear stress produced by the mixing. After a threshold energy is attained, complete breakdown of the cement agglomerates occurs; consequently, in the second zone, an additional increase in energy does not strongly affect the plastic viscosity of the slurry. This type of curve is useful, because it gives the minimum energy required to deflocculate and stabilize a cement slurry.

The principal feature of turbulent mixing is the presence of eddies which aid or are responsible for the mixing process. According to Kolmogoroff's theory (Harnby et al., 1985), the eddies vary in size, having a maximum scale L, which corresponds to the size of the mixing equipment, and a minimum scale, I, which can be calculated by

$$I = \left(\frac{\mu^3 V}{P\rho^2}\right)^{1/4}, \tag{5–34}$$

where

I = minimum eddy size (m),
μ = dynamic viscosity (Pa s),
ρ = slurry density (kg/m³),
P = mixing power (W), and
V = slurry volume (mL).

Using the viscosity data of Fig. 5–32, it is possible to calculate this minimum eddy size for the commercial blender at different rotational speeds (Table 5–9). The typical median particle diameter for a Class G cement is about 30 μm; accordingly, the values obtained at 6,000 and 12,000 RPM seem to represent the best dispersion

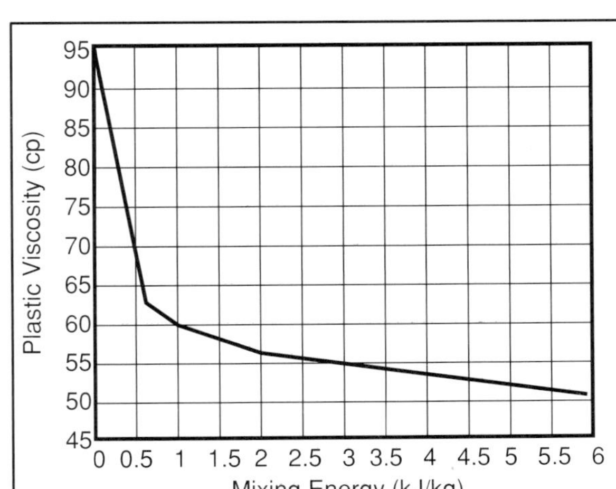

Figure 5–32—Evolution of plastic viscosity as a function of mixing energy for a 15.8-lb/gal (1.9-kg/m³) Class G cement slurry.

state attainable with the blender. It must be noted that this equation does not give the time period necessary to obtain a given eddy size.

Rotational Speed (RPM)	l (μm)
1,000	235
6,000	86
12,000	61

Table 5–9—Minimum eddy size for the Waring blender at different rotational speeds.

5–6.2.2 Chemical Process

As explained in Chapter 2, several processes occur during the preinduction period of cement hydration—

- dissolution of anhydrous cement phases (resulting in a supersaturation with respect to different hydrated phases),
- precipitation of hydrates from the solution, and
- growth of hydrates.

The dissolution of the anhydrous cement phases results in the production of Ca^{2+} ions, which are in turn consumed as C-S-H gel and/or ettringite are precipitated. The evolution of the Ca^{2+} concentration as a function of short mixing time for a neat 15.8-lb/gal (1.9-g/cm^3) Class G cement slurry is illustrated in Fig. 5–33. The Ca^{2+} ions are consumed faster than they are produced. Since no maximum can be measured, it seems reasonable to assume that the first hydration of the anhydrous cement phases is essentially instantaneous, and that the mixing conditions have no influence on this step.

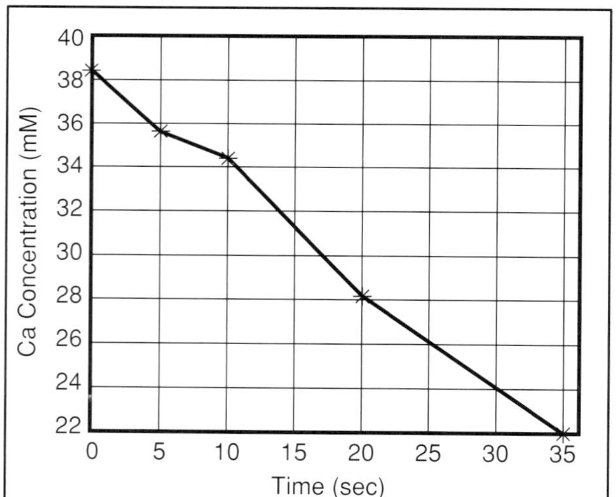

Figure 5–33—Evolution of Ca^{2+} concentration as a function of mixing time for a neat 15.8-lb/gal (1.9-kg/m^3) Class G cement slurry.

As discussed in Chapter 2, a supersaturated solution of C-S-H, gypsum, ettringite, and portlandite is produced after the initial hydration. The rate of nucleation of these hydrates is dependent on the mixing conditions—time and rotational speed. Very few data exist on this subject, but they are useful to demonstrate tendencies. Studies concerning nucleation and crystallization from solutions (McCabe and Smith, 1976; Gohar and Cournil, 1986) indicate that fast stirring speeds accelerate the precipitation. Crystals can originate from collisions between embyros formed by the collision of molecular clusters. For a given mixing time, the probability of such collisions is increased by a fast stirring speed. In fact, owing to the presence of cement particles, this phenomenon can be seen as a secondary nucleation called *contact nucleation* (Gohar and Cournil, 1986).

The transfer of the cement slurry from the commercial blender (high shear mixing) to a consistometer (low shear mixing) is analogous to the transition between field mixing to pumping. In theory, the initial high shear mixing process should affect further hydrate precipitation during the low shear period (McCabe and Smith; 1976). The action of contacting solids—in this case, it could be fast rotation of the paddle—deflects or dislodges particles ranging in size from embryos to small crystals larger than a critical size S_c. Particles at least as large as S_c survive and grow, while smaller ones dissolve. S_c is the size defined by the Kelvin equation relating the solubility of a substance to its size.

The shape of a typical precipitation curve as a function of time at high shear is shown in Fig. 5–34. Again, two zones are evident. In Zone 1, which represents less total mixing energy, very few hydrate nuclei have been formed, and thus the rate of precipitation does not change. In Zone 2, a threshold quantity of nuclei has been formed to induce a higher rate of precipitation. In a Portland cement slurry, this would correspond to a faster hydration rate and a larger hydrate surface area.

The above hypothesis has been validated with a Portland cement slurry containing a dispersant (sodium polynaphthalene sulfonate [PNS]). As explained in Chapter 3, this material adsorbs onto the cement particle surfaces, causing repulsions between particles, and lowering the yield value. The evolution of the yield value as a function of time in an atmospheric consistometer for two slurries prepared with 0.8% PNS by weight of cement (BWOC) is shown in Fig. 5–35. A dramatic increase of the yield value with time was observed for the slurry mixed according to the API procedure. In the first case, there was insufficient PNS to cover the growth of hy-

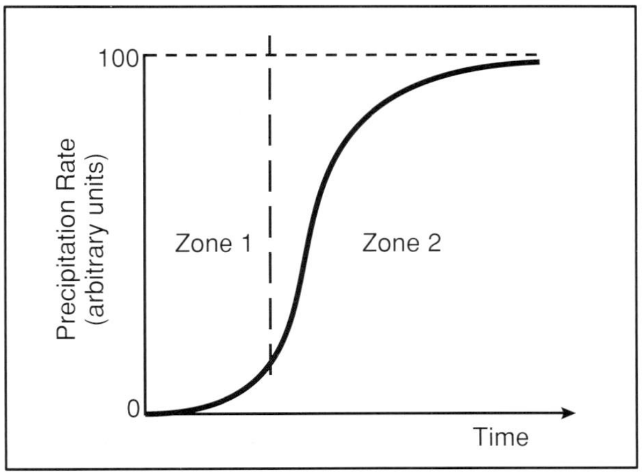

Figure 5–34—Typical curve of precipitation rate as a function of time for supersaturated solutions.

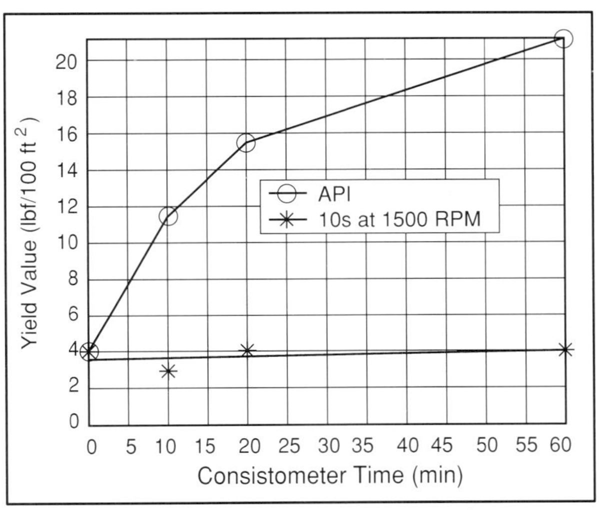

Figure 5–35—Yield value as a function of stirring time (15.5-lb/gal [1.85-kg/m^3] Class G slurry + 0.7% PNS at 65°C).

drates formed during the mixing procedure. In the second case, insufficient shear was applied to initiate rapid nuclei formation; consequently, sufficient PNS exists in solution to fully saturate the surfaces and maintain a low yield value.

5–6.2.3 Influence of Cement Mixing on Cement Slurry Properties

As can be surmised from the preceding discussion, mixing conditions can greatly affect many aspects of cement slurry performance, including—

- yield value,
- fluid-loss rate, and
- thickening time.

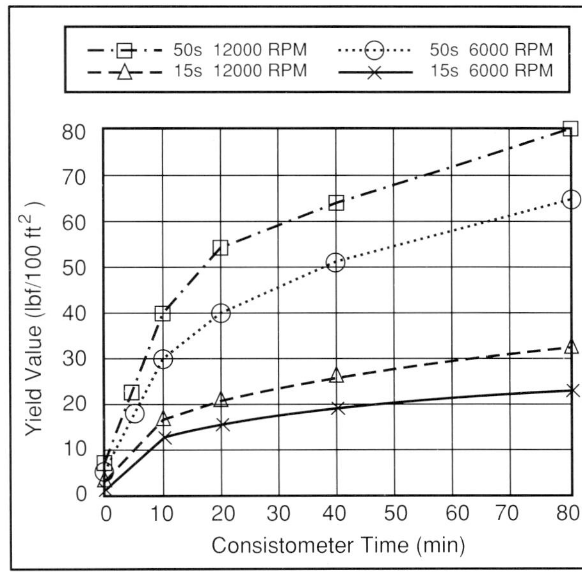

Figure 5–36—Influence of mixing time and speed on yield value (15.8 lb/gal Class G + PNS at 25°C) (from Vidick, 1989).

Yield Value

The influence of mixing parameters on the yield value has been evidenced using dispersed slurries. The evolution of the yield value as a function of time spent in an atmospheric consistometer is shown in Fig. 5–36. The curves show that the mixing time is more important than the rotational speed during the mixing procedure. At longer mixing times, a larger number of hydrates have been formed, which in turn adsorb a greater amount of dispersant. At shorter mixing times, much less dispersant is used. In the laboratory, one would choose a dispersant concentration sufficient to obtain a desired yield value after following the API mixing procedure. If the same system were to be mixed in the field according to a mixing procedure of shorter duration, a danger of obtaining an overdispersed slurry would exist.

Fluid-Loss Rate

The evolution of the fluid-loss rate with mixing energy for a Class G cement containing a cellulosic fluid-loss additive is shown in Fig. 5–37. It appears that this property is a function of both the rotational speed and mixing time. The variation of the plastic viscosity as a function of mixing energy for the same slurry is presented in Fig. 5–32. The plastic viscosity and the fluid-loss rate follow the same tendency, and the breakpoint for both curves occurs at the same mixing energy. Thus, no fluid-loss control is obtained without sufficient

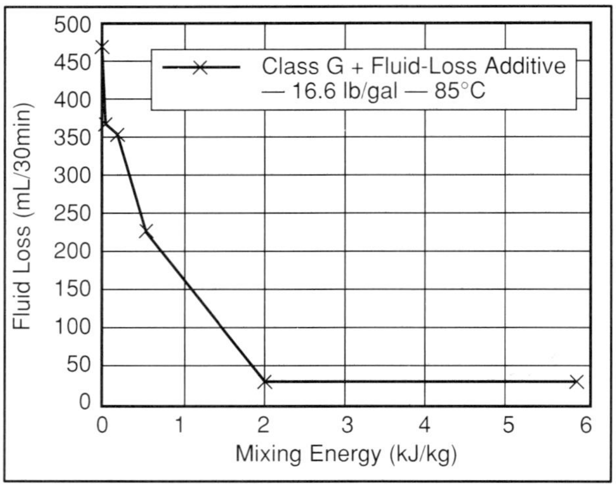

Figure 5–38—Fluid loss as a function of mixing energy (from Vidick, 1989).

deflocculation. At higher mixing energies, no improvement in fluid-loss control is observed. Notice that the API mixing energy far exceeds that necessary to obtain excellent fluid-loss control..

The same effect has been observed with another type of fluid-loss additive (Fig. 5–38). In this case, the minimum mixing energy was 2 kJ/kg. The effect of mixing time on fluid-loss control at a constant mixing energy of 2.2 kJ/kg is shown in Fig. 5–39. After 10 seconds, the fluid-loss rate was constant.

Thickening Time

The thickening time of a cement system has been found to be dependent on the mixing energy. The evolution of the thickening time with mixing energy for a retarded slurry at 150°F (65°C) is shown in Fig. 5–40. Once again, as soon as the slurry had been deflocculated, the optimum

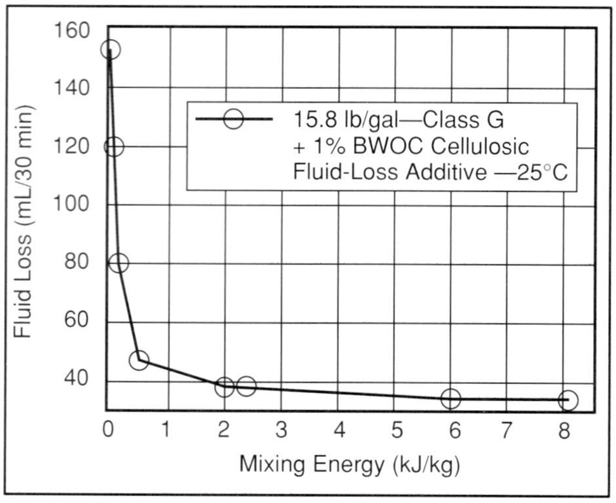

Figure 5–37—Evolution of fluid loss with mixing energy (from Vidick, 1989).

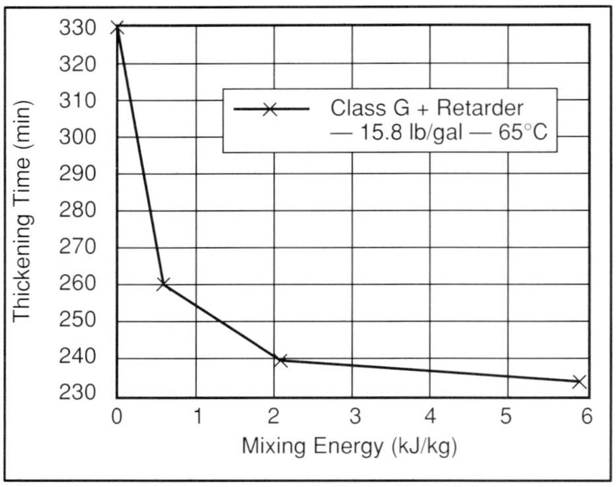

Figure 5–40—Thickening time as a function of mixing energy (from Vidick, 1989).

thickening time was obtained, and further increases in mixing energy had no effect

5–6.2.4 Field Mixing

Field mixing processes can be divided into two classes with respect to the mechanisms of flow—these are called continuous mixing and batch mixing. A discussion of the equipment appears in Chapter 10.

Continuous mixing is a process whereby materials are fed through the process zone at a given rate, and the resulting mixture is discharged at the same rate. Batch mixing involves the mixing of all material simultaneously in a container before discharge. Obviously, these classes are extreme ends of a spectrum of possible mixing techniques.

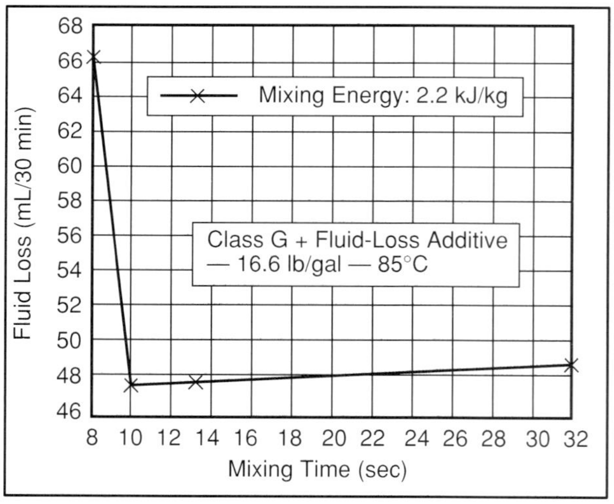

Figure 5–39—Influence of mixing time on fluid-loss control at constant mixing energy (from Vidick, 1989).

The mixing energy provided by a field mixer is the sum of the mechanical work provided by flow through orifices, rotating agitators, and centrifugal pumps. Orban et al. (1986) proposed that the mixing energy concept could be a method for comparing field and laboratory mixing. They defined the total mechanical energy to be

$$\frac{E}{M} = \sum \frac{Pt}{\rho V} = \sum \frac{P}{\rho Q} \; , \qquad (5\text{–}35)$$

where

P = power,

Q = flow rate,

t = time,

V = slurry volume, and

ρ = slurry density.

The mixing energy applied by field mixers was also expressed as a fraction or multiple of the standard API mixing energy, and called the Specific Mixing Energy (SME).

$$SME = \frac{Mixing\ Energy}{API\ Mixing\ Energy} \qquad (5\text{–}36)$$

Continuous Mixing

For this process, the most widely used device is the jet mixer (Chapter 10). The principal disadvantage of this type of mixing is that slurry homogenization decreases with an increasing rate. However, the simplicity of this type of mixing makes it reliable and easy to perform. The mixing energy obtained with this type of equipment is generally low, approximately one-fifth of that obtained by the API procedure. The effect of low mixing energy on the plastic viscosity and yield value of neat Class G cement slurries has been demonstrated by Orban et al. (1986), and is shown in Fig. 5–41.

Batch Mixing

Batch mixing is used to prepare a definite volume of slurry before pumping. The goal is to obtain a slurry having exactly the designed properties before it is pumped. The mixing energy in this situation can be increased, but great care should be taken because of the influence of mixing time on the slurry properties. For example, as discussed earlier, the yield value tends to increase as cement hydration progresses (Orban et al., 1986).

Chokes

As shown above, the properties of field-mixed slurries can be quite different from those obtained with the same ingredients in the laboratory. The use of high-pressure chokes can, in some cases, improve slurry homogeneity and allow more predictable results. The principle of this equipment is to supply sufficient mixing energy to completely deflocculate the slurry. Chokes are simple mechanical devices which can be used with standard equipment. Pumping through chokes generates a pressure drop given by

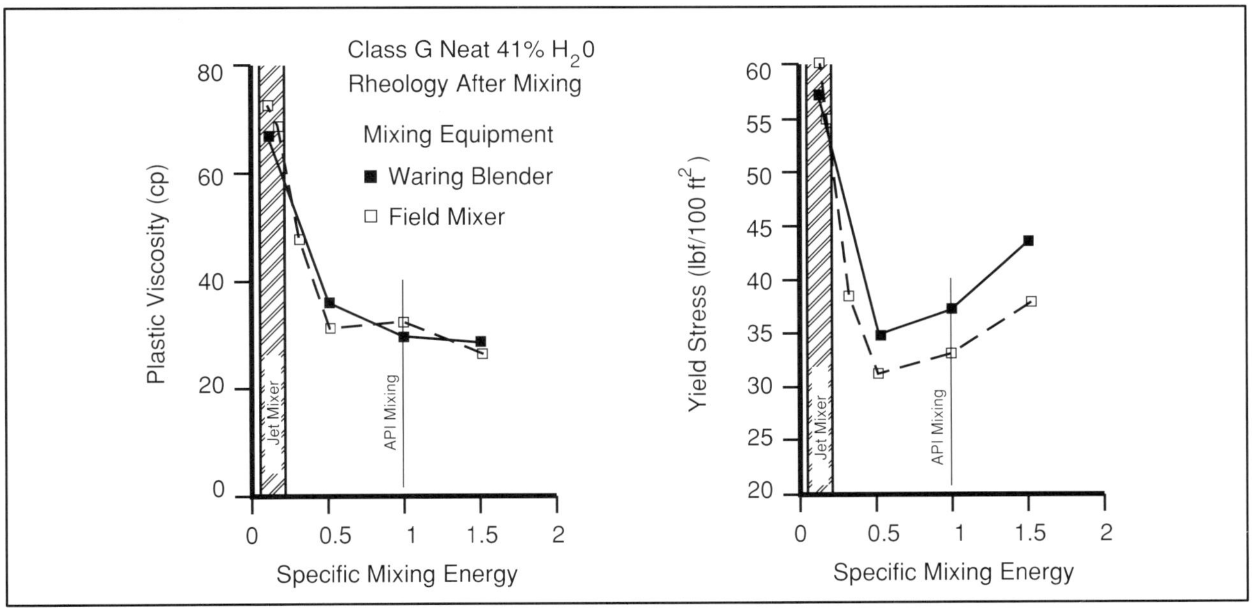

Figure 5–41—Similarity between field and laboratory mixing as a function of SME (SME = Mixing Energy/API Mixing Energy) (from Orban et al., 1986).

$$P = 0.24 \frac{\rho Q^2}{N^2 C_d^2 D^4}, \tag{5–37}$$

where

P = pressure drop (psi),

N = number of chokes,

Q = total flow rate through N chokes (BPM),

ρ = fluid density (lb/gal),

C_d = choke discharge coefficient (dimensionless), and

D = choke diameter (in.).

The pressure drop generated by the choke results in high velocities which create a powerful turbulent zone of high mixing energy. Generally, field mixing provides bad homogenization of the slurry at high pumping rates; therefore, the use of chokes is most suitable for such conditions.

5–7 CONCLUSIONS

This overview of the mud removal process demonstrates the complexity of the problem facing the industry. Although the main factors responsible for poor mud displacement during primary cementations were identified more than 40 years ago, a complete understanding of the process as a whole has not yet been attained. Consequently, there is no consensus today on the subject. In addition, the relationships between the properties of laboratory-prepared and field-prepared cement systems have not been adequately characterized. Nevertheless, the modeling and experimental work performed thus far have allowed the industry to define simple qualitative guidelines for improving primary cement jobs.

- Mud gel strength, yield point, and plastic viscosity should be reduced to a minimum value before removing the drillpipe. However, one must be careful not to impair its ability to suspend the weighting agent.
- The best possible centralization should be obtained through a proper centralization program.
- In cases where mud removal is expected to be difficult, such as—

  —presence of hole irregularities,

  —mud with high gel strength,

  —mud with poor fluid-loss control, and

  —poor centralization,

the pipe should be equipped with scratchers, scrapers, or cable wipers, and pipe movement should be planned.

- Prior to pumping the preflushes, sufficient time should be allowed to circulate at least two annular volumes of mud at the highest rate possible, without losing returns. A better procedure involves using tracers to monitor the volume of circulatable mud, and circulating until this volume represents at least 85% of the hole volume.
- The mud and spacer should be separated by a preflush, which must be compatible with both.
- If possible, a chemical wash should be used. The volume of wash should be such that a contact time of at least eight minutes across the zone of interest is allowed.
- If the necessary volume of chemical wash is such that formation pressures cannot be controlled, one should attempt to apply the same procedure with a turbulent-flow spacer or a combination of a chemical wash and turbulent-flow spacer. The density of the spacer should preferably be between the mud density and the lead-cement density.
- If turbulent-flow-displacement techniques cannot be applied, the density and rheological properties of the spacer should lie between those of the mud and lead slurry. The spacer volume should correspond to at least 500 ft of annular length.
- The properties of the field-mixed cement slurries must resemble those observed in the laboratory during prejob testing. To accomplish this, the field-mixed systems must be prepared at the prescribed densities, and sufficient mixing energy must be applied to obtain adequate slurry homogenization.

NOMENCLATURE

D	m	inner diameter of a pipe
D_o, D_i	m	outer and inner diameter of an annulus, respectively
e	m	thickness of a rectangular slot or local annular gap
fr	—	Fanning friction factor
g_z	m s^{-2}	component of the gravity acceleration in the main direction of the flow
k	Pa s^n	Consistency Index of a power law fluid, or constant in other rheological models
L	m	length of a pipe, annulus, or coaxial cylinder viscometer geometry
n	—	Power Law Index of a power law fluid or constant in other rheological models

P^*	Pa	total pressure
p	Pa	frictional pressure
Q	$m^3\ s^{-1}$	volumetric flow rate
r	m	distance from pipe axis or from the plane of symmetry of a rectangular slot
r_y	m	shortest distance from rotational axis of a coaxial cylinder viscometer where shear stress is zero
R	m	inner radius of a pipe
R_o, R_i	m	outer and inner radius of an annulus, respectively
Re	—	Reynolds number
Re_{BG}	—	Bingham plastic Reynolds number
Re_{MR}	—	Metzner and Reed Reynolds number for a pipe
Re_{AN}	—	generalized Reynolds number for a narrow annulus
Re_1, Re_2	—	critical Reynolds number for the upper limit of the laminar-flow regime and the lower limit of the turbulent-flow regime, respectively
Re_{1ecc}	—	critical Reynolds number for the upper limit of the laminar-flow regime on the wide side of an eccentric annulus
STO	—	API standoff (%)
t	s	time
t^*	—	number of annular volumes
v	$m\ s^{-1}$	velocity of a fluid particle
V	$m\ s^{-1}$	volumetric flow rate per unit of section area
Vol	m^3	volume of an annulus
w	m	width of a rectangular slot
z	m	axial coordinate in the main direction of flow
α	—	annulus diameter ratio D_i/D_o
ε	—	eccentricity of an annulus
$\dot{\gamma}$	s^{-1}	shear rate
$\dot{\gamma}_a$	s^{-1}	average shear rate in a coaxial cylinder viscometer
$\dot{\gamma}_w$	s^{-1}	average shear rate at the wall of a pipe or of a narrow concentric annulus
$\gamma\ N_W$	s^{-1}	Newtonian shear rate at the wall of a pipe or of a narrow concentric annulus
η	Pa s	shear-rate-dependent viscosity or viscosity of a Newtonian fluid
κ	—	diffusivity
μ_p	Pa s	plastic viscosity of a Bingham plastic fluid
ρ	$kg\ m^{-3}$	fluid density
τ	Pa	shear stress
τ_{gel}	Pa	fluid gel strength
τ_w	Pa	shear stress at the wall of a pipe or of a narrow concentric annulus
τ_y	Pa	fluid yield stress
ξ	—	dimensionless shear rate
ψ	—	dimensionless shear stress

REFERENCES

Bannister, C. E.: "Evaluation of Cement Fluid Loss Under Dynamic Conditions," paper SPE 7592, 1978.

Bannister, C. E.: "Aqueous Treatment Fluid and Method of User," Can. Patent No. 1,185,777 (1981).

Bannister, C. E.: "Aqueous Treatment Fluid and Method of Use," U.S. Patent No. 4,656,834 (1987).

Bannister, C. E.: "Aqueous Chemical Wash Composition," U.S. Patent No. 4,681,165 (1987).

Beirute, R. M.: "High-Temperature Cement Mud Spacer," U.S. Patent No. 4,276,182 (1981).

Beirute, R. M. and Flumerfelt, R. W.: "Mechanics of the Displacement Process of Drilling Muds by Cement Slurries Using an Accurate Rheological Model," paper SPE 6801, 1977.

Belousov, G. A., Muratov, V. K., Byvaltsev, A. N., and Skorikov, B. M.: "Spacer Fluid for Separating Drilling Fluid and Cement Slurry," *Neft. Khoz.* (1987) **8**, 25–29.

Benge, G.: "Field Study of Offshore Cement Spacer Mixing," paper SPE 19864, 1989.

Brice, J. W. and Holmes, R. C.: "Engineered Casing Cementing Programs Using Turbulent Flow Techniques," *JPT* (1964) 503–508.

Buchan, L.: "Innovative Technique Improves Liner Cementation in North Sea Wells: An Operator's Experience," paper SPE 15896, 1986.

Childers, M. A.: "Primary Cementing of Multiple Casing," *JPT* (July 1968), 775–783.

Clark, C. R. and L. G. Carter: "Mud Displacement With Cement Slurries," *JPT* (July 1973) 775–783.

Cowthral, J. L.: "Technology Used to Improve Drilling Performance and Primary Cementing Success in Katy Field," paper SPE 10956, 1982.

Crinkelmeyer, O. W., Puntney, A. W., and Sharpe, J. R.: "Use of Water-Base Spacer With Thixotropic Cement Systems Improves Cement Jobs," paper SPE 6367, 1976.

Evanoff, J. I. and Cook, C.: "Optimizing Cement Design for Improved Job Results," paper SPE 17441, 1988.

Flumerfelt, R. W.: "An Analytical Study of Laminar Non-Newtonian Displacement," paper SPE 4486, 1973.

Gohar, P., and Courmil, M.: "Agglomeration: Etude Experimentale et Simulation Numèrique Realise'e sur un Système Liquide-Solide Pulvè rulent," *J. Chim. Phys.* (1986) **83**, No. 4.

Goode, D. L., Phillips, A. M., Williams, D. L., and Stacy, A. L.: "Removal of Oil-Phase Muds From Wells in the Anadarko Basin," paper SPE 11568, 1983.

Graham, H.L.: "Rheology Balanced Cementing Improves Primary Success,' *Oil & Gas J.*(Dec. 18, 1972) 53–59.

Grant, W. H., Rutledge, J. R., and Christy, R. H.: "Field Limitations of Liquid-Additive Cementing Systems," paper SPE 18616, 1989.

Griffin, T. J. and Root, R. L.: "Cementing Spacers and Washes Improve Production," *Oil & Gas J.* (Sept. 1977) 115–123.

Guillot, D., Parcevaux, P., and Jennings, D. B.: "Aqueous Composition for Universal Spacer and Its Use in the Field of Drilling Wells, Notably Oil and Gas Wells," Eur. Patent No. 273,471 (1986).

Harnby, N., Edwards, M. F., and Nienow, A. W.: *Mixing in the Process Industries*, Butterworths, London (1985).

Haut, R. C. and Crook, R. J.: "Primary Cementing: The Mud Displacement Process," paper SPE 8253, 1979.

Haut, R. C. and Crook, R. J.: "Laboratory Investigation of Lightweight, Low-Viscosity Cementing Spacer Fluids," paper SPE 10305, 1981.

Hill, D. G., Smith C. F., and Kucera, C. H.: "Displacement of Drilling Fluids From Boreholes," U. S. Patent No. 3,749,173 (1973).

Holhjem, A.: "Reciprocation of Casing While Cementing From a Floating Drilling Unit," paper EUR 364, 1982.

Hooper, A. P. and Grimshaw, R.: "Non–linear Instability at the Interface Between Two Viscous Fluids," *Phys. Fluids* (1985) **28**, No. 1.

Howard, G. C. and Clark, J. B.: "Factors to be Considered in Obtaining Proper Cementing of Casing," *Drill. and Prod. Prac.*, API (1948) 257–272.

Iyoho, A. W. and Azar, J. J.: "An Accurate Slot Flow Model for Non-Newtonian Fluid Flow Through Eccentric Annuli," paper SPE 9447, 1981.

Jamot, A.: "Dèplacement de la boue par le laitier de ciment dans l'espace annulaire tubage-paroi d'un puits," *Rev. Assn. Fr. Tech. Pet.* (March–April 1974) No. 224, 27–37.

Jones, P. H. and Berdine, D.: "Oil Well Cementing: Factors Influencing Bond Between Cement and Formation," *Drill. and Prod. Prac.*, API, Dallas (Mar. 1940) 45–63.

Keller, S. R., Crook, R. J., Haut, R. C., and Kulakofski, D. S.: "Problems Associated With Deviated Wellbore Cementing," paper SPE 11979, 1983.

Kolthoff, K. W. and Scales, G. H.: "Improved Liner Cementing Techniques for Alaska's Prudhoe Bay Field," paper SPE 10756, 1982.

Landrum W. R., Porter, J. E., and Turner, R. D.: "Rotating Liners During Cementing in the Grand Isle and West Delta Areas, Louisiana," *JPT* (July 1985) 1263-1266.

Lockyear, C. F. and Hibbert, A. P.: "A Novel Approach to Primary Cementation Using a Field-Scale Flow Loop," paper SPE 18376, 1988.

Lockyear, C. F., Ryan, D. F., and Gunningham, M. M.: "Cement Channeling: How to Predict and Prevent," paper SPE 19865, 1989.

Martin, M, Latil, M., and Vetter, P.: "Mud Displacement by Slurry During Primary Cementing Jobs. Predicting Optimum Conditions," paper SPE 7590, 1978.

McCabe and Smith: *Unit Operations in Crystal Chemistry*, McGraw-Hill Book Co., Inc., New York, 1976.

McLean, R. H., Manry, C. W., and Whitaker, W. W.: "Displacement Mechanics in Primary Cementing," *JPT* (Feb. 1967) 251-260.

Mitchell, R. F.: "Dynamic Surge/Swab Pressure Predictions," SPEDE (Sept. 1988) 325–333.

Motley, H. R., Morris, E. F, and Pavlich, J. P.: "Use of a Spacer Composition in Well Cementing," U.S. Patent No. 3,820,602 (1974).

Nauman, E. B. and Buffham, B. A.: *Mixing in Continuous Flow Systems*, John Wiley & Sons, New York, 1983.

Oliver, J. E. and Singer, A. M.: "Improved Well Cementing Process," Eur. Patent No. 238,675 (1986).

Orban, J. A., Parcevaux, P. A., and Guillot, D. G.: "Specific Mixing Energy: A Key Factor for Cement Slurry Quality," paper SPE 15578, 1986.

Parcevaux, P. and Jennings, J.: "An Aqueous Spacer Composition Compatible With Drilling Muds and Cement Slurries Including Saline Slurries and Application Thereof to Drilling Oil and Gas Wells," Eur. Patent No. 207,536 (1985).

Parker, P. N., Ladd, B. J., Ross, W. M., and Wahl, W. W.: "An Evaluation of a Primary Cementing Technique Using Low Displacement Rates," paper SPE 1234, 1965.

Sauer, C. W.: "Mud Displacement During Cementing: A State of the Art," *JPT* (Sept. 1987) 1091-1101.

Schlichting, H.: *Boundary Layer Theory*, McGraw-Hill Book Co., Inc., New York (1979).

Sehault, J. M. and Grebe, E. L.: "Spacer Fluid," Eur. Patent No. 0243067 (1987).

Sharpe, J. R. and Free, D. L.: "Method for Treating a Well Using a Chemical Wash With Fluid-Loss Control," U.S. Patent No. 4,127,174 (1977).

Smith, R. C.: "Successful Primary Cementing Can Be a Reality," *JPT* (Nov. 1984) 1851-1858.

Smith, T. R.: "Cementing Displacement Practices: Application in the Field," paper SPE/IADC 18167, 1989.

Smith, T. R. and Crook, R. J.: "Investigation of Cement Preflushes for a KCl-Polymer Mud," paper CIM 82.33.71, 1982.

Speers, R. A. et al.: "Drilling Fluid Shear Stress Overshoot Behavior," *Rheol. Acta.* (1987) **26**, No. 5, 447-452.

Teplitz. A. J. and Haasebroek, W. E.: "An Investigation of Oil-Well Cementing," *Drill. and Prod. Prac.,* API, Dallas (1946).

Thomas, D. C.: "A Spacer System Useful in Brine Completion of Wellbores," U.K. Patent No. 2073284A (1981).

Vidick, B.: "Critical Mixing Parameters for Good Control of Cement Slurry Quality," paper SPE 18895, 1989.

Walton, I. C. and Bittleston, S. H.: "The Flow of a Bingham Plastic Fluid in a Narrow Eccentric Annulus," *J. Fluid Mech.* (1990).

Warembourg, P. A., Kirksey, J. M., and Bannister, C. E.: "Improving Cement Bond in the Rocky Mountain Area by the Use of Spacer, Wash and Thixotropic Cement," paper SPE 9031, 1980.

Weigand, W. A. and Totten, P. L.: "Fluid Spacer Composition for Use in Well Cementing," U.S. Patent No. 4,588,032 (1984).

Wieland, D. R. and Woods, B. L.: "Cement Preflush Method," U.S. Patent No. 3,878,895 (1975).

Zuiderwijk, J. J. M.: "Mud Displacement in Primary Cementation," paper SPE 4830, 1974.

6

Cement/Formation Interactions

Jean-François Baret, Gèrard Daccord, and John Yearwood

Dowell Schlumberger

6–1 FLUID LOSS—INTRODUCTION

Fluid-loss control agents have been added to well cement slurries for more than 20 years, and it is now recognized that the quality of cement jobs has improved significantly. Indeed, it is generally acknowledged that insufficient fluid-loss control is often responsible for primary cementing failures, because of excessive increases in slurry density or annulus bridging. In addition, formation invasion by cement filtrate may be very damaging and deleterious to production (Bannister and Lawson, 1985; Economides and Nolte, 1987). With respect to remedial cementing, the problem is to adjust the fluid-loss rate to the perforation size and the nature of the formation (Binkley et al., 1957; Cook and Cunningham, 1977). However, for both primary and remedial cementing, very little has been written to justify the level of fluid-loss control required to achieve a good cement job.

To properly address the quantitative evaluation of fluid-loss limits compatible with successful cementing operations, two different stages must be considered: (1) the placement or dynamic stage; and (2) the waiting-on-cement (WOC) or static stage (Hook and Ernst, 1969; Smith, 1984). During the first stage, the slurry is flowing and eroding the cement cake as it forms. Therefore, in the dynamic regime, the cement cake begins to form during a short transient period, and then stops growing (Hartog et al., 1983). In contrast, when the pumping is stopped the cake can grow freely.

From an operational point of view, the relevant parameter during placement is the decrease of slurry water content. During WOC, it is the continuous increase of cake thickness. Therefore, to define the acceptable amounts of fluid loss for these two periods, the criteria are quite different (Baret, 1988). Section 6–2 includes a discussion concerning how to determine, from an upper boundary of slurry density, the maximum amount of water which can be lost during the dynamic stage without impairing slurry properties. This boundary can be obtained by measuring the dependency of slurry rheology or thickening time upon density. During the static stage, the maximum acceptable cake thickness and volume of fluid loss are deduced from the most narrow annular gap assumed to exist (Bannister, 1978) (Section 6–3).

If the fluid-loss rate is to be controlled, chemicals must be added to the slurry. Different types of polymers or particulate materials are used as fluid-loss agents, and are described in Chapter 3.

6–2 DYNAMIC FLUID LOSS

The first critical parameter to consider is the density increase (or loss of water) which is tolerable for a proper cementing job. As can be seen in Figs. 6–1 and 6–2, the slurry properties are very sensitive to the water-to-cement (W/C) ratio, (i.e., density variations). While the thickening time increases almost linearly with the W/C ratio, the yield value of the slurry abruptly increases and

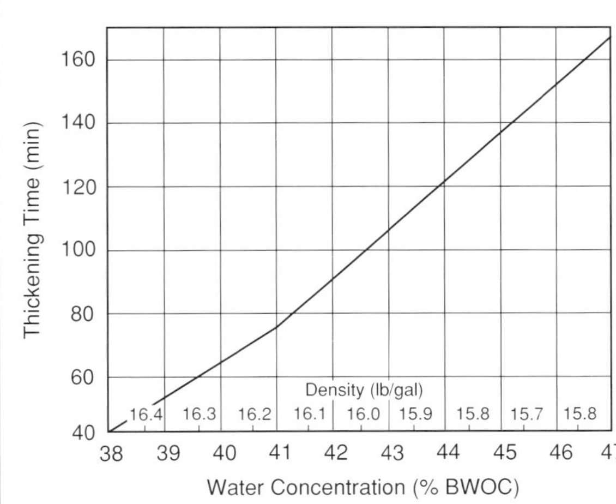

Figure 6–1—Thickening time of Class G cement slurries at 185°F (85°C) for different water concentrations.

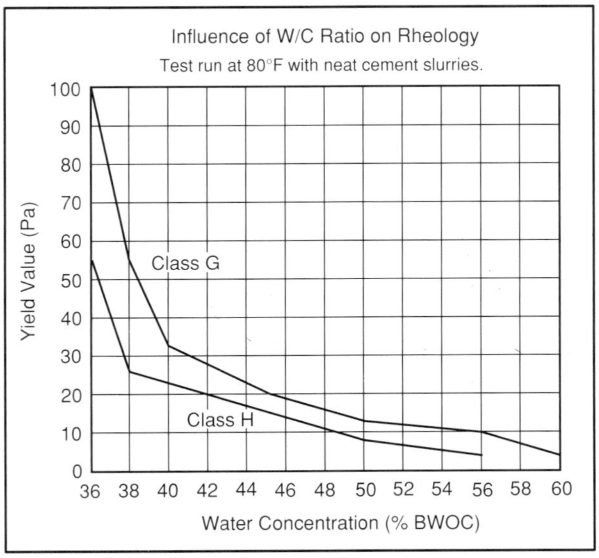

Figure 6–2—Yield value of two neat cement slurries vs water concentration (80°F [25°C]).

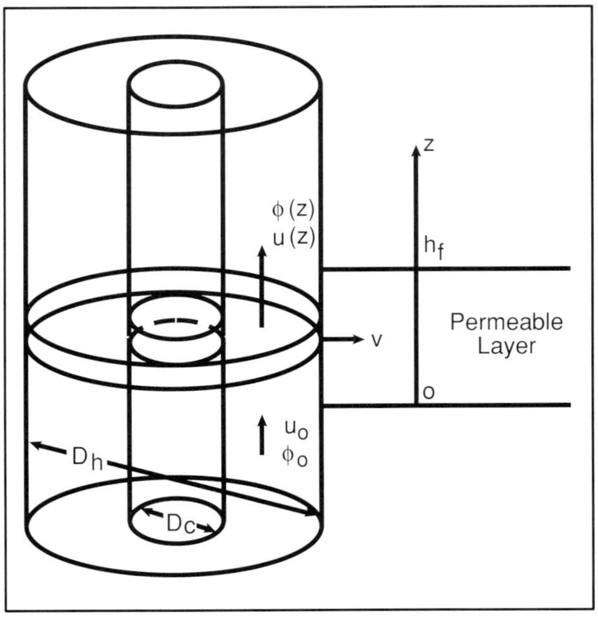

Figure 6–3—Schematic illustration of dynamic fluid loss.

reaches very high levels when the W/C ratio falls below 38% to 40%. Therefore, at high water contents, a 10% variation of slurry density may not have a significant influence on the yield value, but the effect upon thickening time is substantial. At lower W/C ratios, the yield value of the slurry can increase rapidly below a critical level.

The curves shown in Figs. 6–1 and 6–2 are examples corresponding to specific slurries. The thickening time and the yield value dependency upon slurry density will change significantly from cement to cement, and with the additives present in the slurry.

6–2.1 Density Change Due to Dynamic Fluid Loss

In this section, an equation is derived which calculates the change in slurry density due to fluid loss, for a slurry passing in front of a permeable layer. A schematic illustration is shown in Fig. 6–3. The slurry reaches the bottom of the layer with a water volume ϕ_{wo} and an upward velocity u_o. It is assumed that there is no settling, i.e., the solid (cement) phase has the same vertical velocity, $u(z)$, as the liquid phase (water), where z is the vertical coordinate. The conservation equations for the water, which can be lost to the formation, and for the cement solids, which cannot (except in the event of lost circulation) are shown below.

water:

$$\pi D_h v + \frac{\pi}{4}(D_h^2 - D_c^2)\left(u\frac{d\phi_w}{dz} + \phi_w\frac{du}{dz}\right) = 0 \qquad (6\text{–}1)$$

cement:

$$\frac{\pi}{4}(D_h^2 - D_c^2)\left(u\frac{d\phi_c}{dz} + \phi_c\frac{du}{dz}\right) = 0 \qquad (6\text{–}2)$$

with

$$\phi_w + \phi_c = 1 \qquad (6\text{–}3)$$

where

v = filtration velocity,
ϕ_w = water volume fraction, and
ϕ_c = cement volume fraction.

The dimensionless vertical coordinate Z is introduced as follows.

$$Z = \frac{4D_h}{(D_h^2 - D_c^2)}\frac{v}{u_o}z = \frac{\pi D_h v}{Q_o}z \qquad (6\text{–}4)$$

where

Q_o = annular flow rate at the entrance of the permeable layer.

The two conservation equations become

$$u_o + u\frac{d\phi_w}{dZ} + \phi_w\frac{du}{dZ} = 0, \qquad (6\text{–}5)$$

and

$$u\frac{d\phi_c}{dZ} + \phi_c\frac{du}{dZ} = 0, \qquad (6\text{–}6)$$

with the boundary conditions

$$u(Z=0)=0$$
$$\phi_w(Z=0)=\phi_{wo}, \text{ and}$$
$$\phi_c(Z=0)=\phi_{co}=1-\phi_{wo},$$

taking the origin of the vertical coordinate at the bottom of the permeable layer. For the sake of simplicity, it is assumed that the filtration velocity, v, is independent of the slurry composition and of the vertical position. Under these conditions, the solution for the above system of equations is shown below.

$$u = u_o(1 - Z), \ \phi_c = \frac{\phi_{co}}{1 - Z}, \qquad (6\text{–}7)$$
$$\phi_w = \frac{\phi_{wo}}{1 - Z},$$

and

$$Z = \frac{\phi_{wo}^{-z} - \phi_w}{1 - \phi_w} = 1 - \frac{\phi_{co}}{\phi_c} = 1 - \frac{u}{u_o}. \qquad (6\text{–}8)$$

If the height of the permeable formation* is h_f, then the water and cement volume fractions at the top of the layer are given by Eq. 6–7, with $Z=\pi D_h v h_f / Q_o$.

The slurry density, ρ_s, is related to the water volume fraction by the relation

$$\rho_s = \rho_c(1 - \phi_w) + \rho_w \phi_w \qquad (6\text{–}9)$$

with ρ_w and ρ_c being the water and cement densities, respectively. Another useful quantity is the water/cement ratio (by weight), $F_{w/c}$. It is related to the previous quantities by the relation

$$F_{w/c} = \frac{\rho_w}{\rho_c} \frac{\phi_w}{1 - \phi_w} \ or \ \phi_w \frac{F_{w/c}\,\rho_c}{\rho_w + F_{w/c}\,\rho_c}. \qquad (6\text{–}10)$$

Thus, if the minimum admissible W/C ratio is $F_{w/c}^{min}$ e.g., for keeping the rheology below maximum values, the corresponding maximum filtration velocity, v_{max}, is given by Eqs. 6–4, 6–8, and 6–10:

$$v_{max} = \frac{Q_o}{\pi D_h h_f} \frac{\phi_{w_o} - \phi_w^{min}}{1 - \phi_w^{min}},$$

with $$\phi_w^{min} = \frac{F_{w/c}^{min}\rho_c}{\rho_w + F_{w/c}^{min}\rho_c}.$$

* A formation is said to be permeable from a production point of view if its permeability is larger than 10 md; even with lower values of permeability, the water leakoff from the slurry into the formation may be significant considering the large areas involved.

6–2.2 Cake Permeability and Dynamic Fluid Loss

Throughout the following, Darcy's law in its linear formulation is assumed to be applicable. This means that the permeability of the cake is assumed to be constant with respect to flow rate, pressure, and thickness, and implies in particular that the cake is homogeneous and *incompressible*. Also, the size of the annulus is assumed to be small in comparison with the hole diameter, and the formation's resistance to flow in the permeable region is neglected. This last assumption is not likely to be very stringent for permeabilities above 10 md.

6–2.2.1 Without a Mud Cake

In this case, a borehole is considered in which the mud cake has been completely removed and replaced by a cement cake. Darcy's law applied to the cement cake states that the filtration velocity through the cake, $v = Q/A$, is proportional to the pressure gradient, $\Delta P/e_c$ (ΔP is the differential pressure across the cake and e_c is the cake thickness), and to the cake permeability, k_c, and inversely proportional to the filtrate viscosity, μ:

$$v = \frac{k_c}{\mu}\frac{\Delta P}{e_c} = \frac{k_c}{e_c}\frac{\Delta P}{\mu} \qquad (6\text{–}11)$$

The factor e_c/k_c is the cake resistance to flow. Therefore, once the maximum filtration velocity is determined, a measurement of μ and an estimation of ΔP will allow the calculation of the required cake resistance to flow.

6–2.2.2 With a Mud Cake

In this case, a mud cake and a cement cake are superposed. As shown in Fig. 6–4, Darcy's law becomes

$$\Delta P = \mu v \left(\frac{e_c}{k_c} + \frac{e_m}{k_m}\right) \qquad (6\text{–}12)$$

where the subscript m refers to the mud cake. It is now the sum of the two cake resistances which is deduced from the maximum admissible filtration velocity.

6–3 STATIC FLUID LOSS

Once pumping is stopped, there is no more bulk annular flow and the cement cake can grow. Ultimately, it may grow so large that it fills the annular gap completely, and bridging occurs. If the vertical flow through the cake is low enough to neglect friction pressures, the pressure difference between the top and bottom of the cake is the water hydrostatic pressure, $\rho_w g h_f$, instead of the slurry hydrostatic pressure, $\rho_s g h_f$. For example, the pressure resulting from a 33-ft (10-m) interval of cement cake originating from a 15.8-lb/gal (1.90-g/cm^3) slurry is approximately 13 psi (90 kPa), which is not large.

In contrast, if the filtration rate into the formation occurs at a rate such that the vertical pressure drop is high, the pressure will decrease sharply. This is simply due to Darcy's pressure drop through a long and low-permeability porous medium. Two consequences may result:

- Mechanical rupture of the cake. The cake has a horizontal cross-sectional of $\pi(D_h^2+D_c^2)/4$ and a lateral area of $\pi(D_h-D_c)h_f$. It will mechanically withstand a differential pressure ΔP if its shear strength is larger than $\Delta P(D_h-D_c)/4h_f$. Although there are no published data concerning cement-cake shear strength, it is generally agreed in the industry that a cement cake is sufficiently strong to withstand the differential pressure normally encountered.
- Loss of zonal isolation. If a high-pressure zone exists below the cake bridge, i.e., below a relatively low-pressure porous layer where leakoff occurs, formation fluids can flow into the annulus and prevent proper zonal isolation.

In addition to these direct potential problems, annulus bridging will increase the risk of microannulus formation resulting in noncompensated cement shrinkage. Therefore, annulus bridging should be avoided at all costs (Stout and Wahl, 1960; Beach et al., 1981).

6–3.1 Without a Mud Cake

For incompressible cement cakes, the fluid-loss volume is proportional to the cake volume, or the fluid-loss volume per unit area V_{fl}/A is proportional to the cake thickness, e_c: $e_c=RV_{fl}/A$ (Bannister and Lawson, 1985). The R values have been measured for different 15.8 lb/gal (1.90 g/cm^3) slurries, and found to vary between 1.5 and 2.5 (Christian et al., 1976; Desbrières, 1988). Expressing Darcy's law using R, a maximum cake permeability can be deduced from the maximum value of the cake thickness, $e_c^{max}=(D_h-D_c)/2$—

$$k_c^{max} \approx \frac{(D_h - D_c)^2 \mu}{2R\Delta P t_t}, \qquad (6\text{–}13)$$

where:

t_t = thickening time.

Equation 6–13 is independent of the size of the permeable zone. During placement, the global effect of the fluid loss must be taken into account. The dynamic fluid loss is directly proportional to the permeable area of the well and the slurry properties, whereas during the static stage the fluid loss has only local effects on the cement. Even if there is, for the whole well, only one narrow permeable section, (e.g., a few meters high) the cement cake may bridge in front of it and impair hydrostatic pressure transmission. On the other hand, this same narrow permeable zone would have very little contribution to dynamic fluid loss.

The cake permeability limit obtained from Christian et al., (1976) may be too large because the annular gap may (in some places) be smaller than D_h-D_c due to borehole irregularities. Standoff or eccentricity reduces the annular gap on one side, but increases it on the other; thus, bridging could first occur on the narrow side, whereas sufficient pressure transmission would still be achieved on the wider side. Since a continuous ring of cake occupying a whole horizontal portion of the annulus is to be avoided, eccentration is a favorable situation from this very particular point of view.

The hydrostatic pressure differential ΔP is not constant throughout the WOC period, decreasing sharply during the transition period. Therefore, the time t_t considered by Christian et al. (1976) is more precisely the length of the induction period rather than the thickening time.

6–3.2 With a Mud Cake

In the previous discussion, it was considered that only a cement filter cake is present at the slurry/formation interface. If there is also a mud cake, the pressure drop across the cement cake is reduced (Fig. 6–4). In addition, the fluid-loss volume is no longer proportional to the square

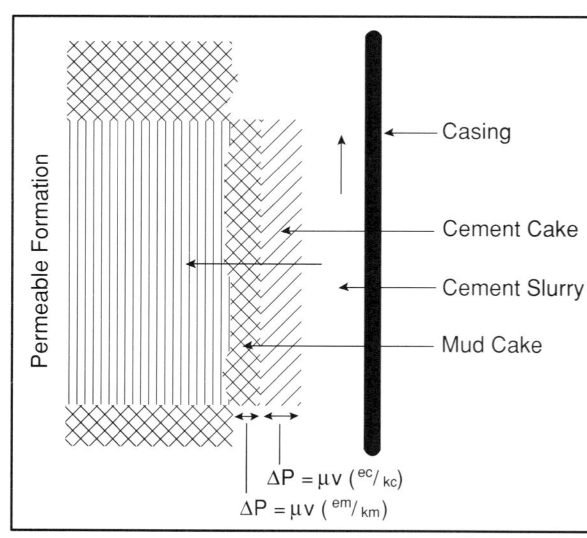

Figure 6–4—Cement filter-cake deposition on a mud cake in front of a permeable zone.

root of time, but instead is a hyperbola in the plane $(V_{fl}/A, \sqrt{t})$. Effectively,

$$\Delta P_{total} = \Delta P_{mud\ cake} + \Delta P_{cement\ cake} = \mu v \frac{e_m}{k_m} + \mu v \frac{e_c}{k_c} \quad (6\text{–}14)$$

with

$$e_c = R\frac{V_{fl}}{A} \text{ and } v = \frac{d}{dt}\left(\frac{V_{fl}}{A}\right) = \frac{1}{R}\frac{de_c}{dt}. \quad (6\text{–}15)$$

Hence,

$$\Delta P = \frac{\mu}{R}\left(\frac{e_m}{k_m} + \frac{e_c}{k_c}\right)\frac{de_c}{dt}, \quad (6\text{–}16)$$

with the solution

$$\frac{V_{fl}}{A} = \frac{e_c}{R} = \sqrt{\frac{2k_c\Delta P}{R\mu}t}\left[\sqrt{1 + \left(\frac{e_m k_c}{k_m R}\right)^2 \frac{R\mu}{2k_c\Delta P t}} - \frac{e_m k_c}{k_m R}\right]. \quad (6\text{–}17)$$

The equation of the hyperbola asymptote is the usual square root term for a cement filter cake alone, $V_{fl}/A = \sqrt{2k_c\Delta Pt/R\mu}$, plus a constant term, $-e_m k_c/k_m R$, which represents the mud-cake resistance relative to that of the cement.

Another way to write this expression is to express the time with respect to the cake thickness. For a given allowed thickness (annular gap, $e_c^{max} = (D_h - D_c)/2$), the time elapsed before bridging depends linearly on the mud-cake resistance—

$$t = \frac{\mu}{R\Delta P}\left[\frac{(e_c^{max})^2}{2k_c} + \frac{e_m}{k_m}e_c^{max}\right]. \quad (6\text{–}18)$$

6–4 COMPARISON BETWEEN STATIC AND DYNAMIC REQUIREMENTS ON FLUID-LOSS CONTROL

Figure 6–5 includes graphs showing the variation of the maximum allowable cement-cake permeability for proper cementing with respect to mud-cake resistance, in both the static and dynamic regimes (Fordham et al., 1988). The curves corresponding to the static stage are obtained from Eq. 6–19.

$$k_c = \frac{1}{2}\frac{(e_c^{\ max})^2}{\frac{R\Delta P t_t}{\mu} - e_c\frac{e_m}{k_m}} \quad (6\text{–}19)$$

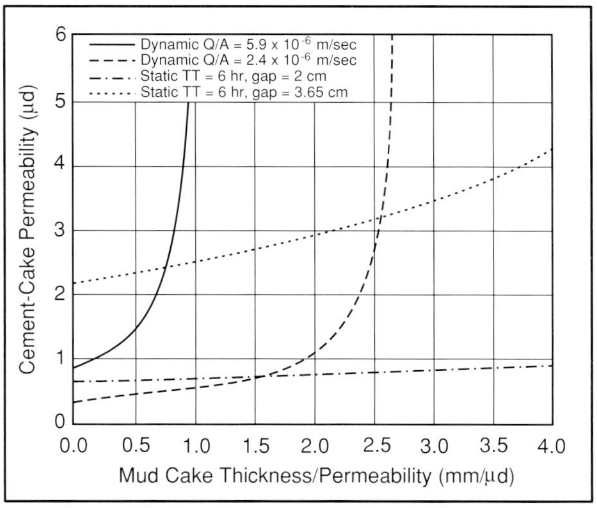

Figure 6–5—Cement cake permeability required to prevent excessive density increase or annulus bridging (assuming 1mm cement cake).

Equation 6–20 gives the curves corresponding to the dynamic stage

$$k_c = \frac{(e_c^{max})}{\frac{\Delta P}{\mu v} - \frac{e_m}{k_m}}. \quad (6\text{–}20)$$

The dynamic curves have been calculated for a water loss equal to 10% of the total slurry water. Assuming a 7-in. (18 cm) casing placed in a 9⅞-in. (25-cm) open hole of 2,000 ft (610 m), 94 bbl (15 m^3) of slurry are required to fill the annular space. For a neat slurry containing 44% water BWOC (15.8 lb/gal), 10% of the total water is 5 bbl (0.8 m^3). The differential pressure between the slurry and formation is assumed to be 1,000 psi, the filtrate viscosity 1 cp, and the cement-cake thickness 0.04 in. (1 mm). The pumping rate is assumed to be 4 bbl/min. The fluid loss occurs across 435 ft (133 m) of permeable formation for the first curve, and 1,067 ft (325 m) for the second curve. For the 435-ft case, good fluid-loss control of the cement slurry is required up to a mud-cake resistance of about 1 mm/µd, whereas in the 1,067-ft case it is required up to 3 mm/µd.

For the static regime, the curves have been drawn for two annular gaps, assuming a WOC duration of six hours. For a mud-cake resistance of 2 mm/µd, a cement cake permeability lower than 3 µd is required to avoid the bridging of a 1.5-in.(3.65-cm) gap, whereas less than 1 µd is required for an annular gap of 0.8 in. (2 cm).

6–5 FLUID LOSS DURING REMEDIAL CEMENTING

Poor primary cementations are usually the cause of expensive remedial jobs. Nondisplaced drilling mud may leave pockets or communication channels behind the casing. Such channels can often be cemented using squeeze techniques (Chapter 13). A properly designed slurry for this type of operation should allow the complete filling of perforation cavities, leaving a minimum node buildup into the casing (Fig. 6–6a). This is achieved by controlling the fluid-loss rate of the slurry (Binkley et al., 1957; Cook and Cunningham, 1977).

Because of its extremely high filtration rate, a neat slurry is usually not suitable for squeezing. The deposition of a thick filter cake inside the casing would prevent the removal of excess cement by reverse circulation, and the withdrawal of the tubing. On the other hand, if the fluid-loss rate is too low, the slurry would leave a thin and weak filter cake which could be easily removed during the reversing process, or later when subjected to negative differential pressures (Fig. 6–6b). Determining the optimum fluid-loss rate depends mainly on three parameters—the dimensions of the perforations, the permeability of the zone to be treated, and the time. A discussion of these parameters and recommendations concerning fluid-loss rates are presented in Chapter 13.

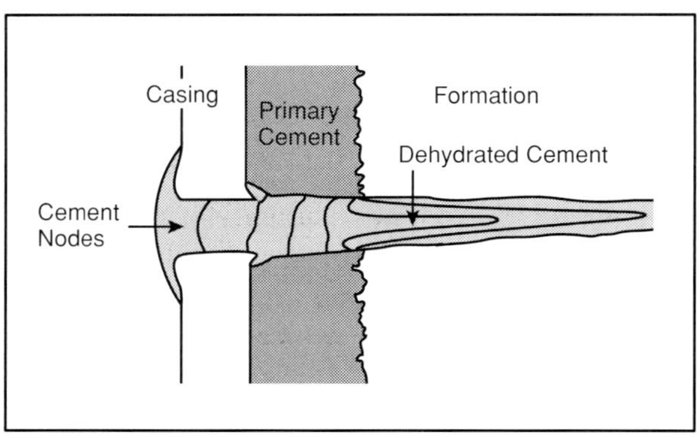

Figure 6–6a—Perforation channel properly filled by a cement filter cake.

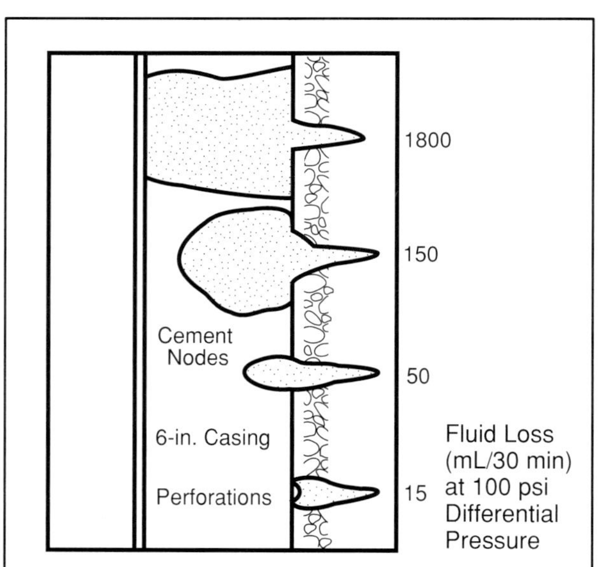

Figure 6–6b—Schematic of cement filter cake buildup vs fluid-loss rate.

6–6 FORMATION DAMAGE

The formation can be damaged by all fluids used in a well from drilling to stimulation. Cement slurry is no exception. Although the contact time with the formation is short compared to drilling fluids, damage to particularly sensitive zones can occur if proper fluid-loss control is not achieved. The cement particles do not endanger formation permeability, because even highly porous formations are able to retain enough particles to build a filter cake rapidly. However, cement filtrate has a high pH (12 to 12.5) and contains many ions, in particular about 20 mM of calcium, which can be responsible for significant formation permeability impairment. Several mechanisms are known. Calcium can destabilize clay minerals by an ion exchange effect (Cunningham and Smith, 1968). When mixed with connate brines that contain high concentrations of calcium, the high pH filtrate can provoke the precipitation of calcium carbonate, lime, or calcium silicate hydrates (Records and Ritter, 1978; Krueger, 1986). Similarly, the potassium dissolved in the cement filtrate can form potassium carbonate precipitates. Moreover, the water itself can have deleterious effects in the case of oil sands, because it can cause shaly impurities in the sand to swell, and thus reduce its permeability (Cutforth, 1949).

6–7 FLUID LOSS—CONCLUSIONS

Excessive slurry water loss endangers a cementing operation in two ways.

- *During placement,* because slurry density may increase beyond an acceptable limit. This increase may become very important when the area of the permeable formation is large and the contact time is long (low pump rate). The cement-cake permeability, required to limit this density increase, sharply varies with mud-cake resistance. With a thin and permeable mud cake, a low-permeability cement cake is required. With a sufficiently impervious mud cake, a permeability reduction due to cement cake is no longer required.

- *During WOC*, because bridging may occur. Bridging is a local process, and is more likely to occur in a narrow annulus. The cement-cake permeability required to limit cement-cake thickness varies slowly with mud-cake resistance. It does not make much difference whether the mud cake has low or high permeability. This is especially true for narrow annuli or long thickening times. Thus, impervious cement cakes are always needed.

In the above discussion, maximum values for the cement-cake permeability have been determined. This parameter is not routinely determined at present. A clear relationship between the cement-cake permeability and the required API fluid-loss rate has not been developed; consequently, field experience in a particular area is still the best guide. However, there are rules of thumb regarding certain critical situations.

- If there is a gas zone below a permeable formation, bridging is likely to favor gas migration. In this case, the cement slurry should yield very low API fluid-loss values, in the range of 20 to 40 mL/30 min.
- With high-density slurries, any decrease in water content may critically impair the placement operation, especially at low pump rates. Here again, API fluid loss has to be very low (less than 50 mL/30 min).

6–8 LOST CIRCULATION—INTRODUCTION

Lost circulation (or lost returns) is defined as the total or partial loss of drilling fluids or cement slurries into highly permeable zones, cavernous formations, and natural or induced fractures during drilling or cementing operations (Goins, 1952). Lost circulation must not be confused with fluid loss, which has been previously described. Figure 6–7 depicts how the fluid-loss process is more related to primary porosity, whereas lost circulation can occur in formations with both primary and secondary porosities. Lost circulation is a problem which is best attacked before the cementing process is initiated. Therefore, the treatment of lost circulation during drilling is included in the following discussion.

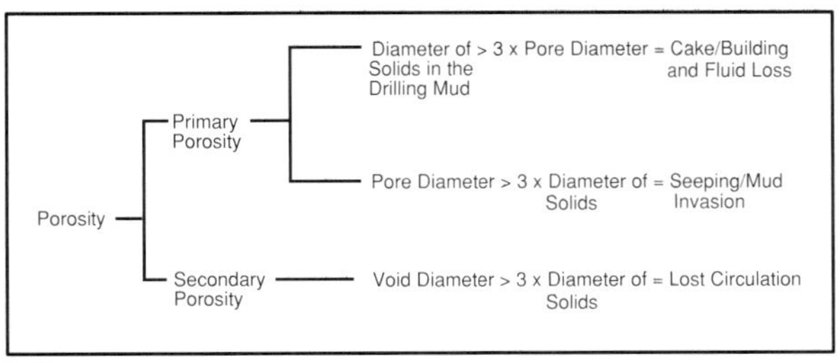

Figure 6–7—Fluid loss vs lost circulation.

6–9 CONSEQUENCES OF LOST CIRCULATION

Lost circulation can be an expensive and time-consuming problem. During drilling, this loss may vary from a gradual lowering of the mud level in the pits to a complete loss of returns.

The major consequences of lost circulation include the following.

- The possibility of a blowout because of a drop in the mud level.
- The possibility of sticking the drillpipe because of poor cuttings removal.
- No zonal isolation due to insufficient cement fill-up.
- Excessive cost because of loss of mud, increased rig time, and remedial cementing operations.
- Losses to the producing zone resulting in extensive formation damage.
- The loss of the well.

To effectively solve lost circulation with the correct technique, it is necessary to know the severity of the losses, the type of lost-circulation zone, and the drilling history of the well just before the losses occurred.

6–10 CLASSIFICATION OF LOST-CIRCULATION ZONES

A standard severity classification for lost circulation is shown in Table 6–1. In addition, it is common to classify lost-circulation zones into four categories.

- Unconsolidated or highly permeable formations.
- Natural fractures or fissures.
- Induced vertical or horizontal fractures.
- Cavernous and vugular formations.

Type of Losses	Severity
Seeping (minor)	< 10 bbl (1.5 m^3)/hr
Partial (medium)	10 to 500 bbl (1.5 to 75 m^3)/hr
Complete (severe)	Total, unable to keep the hole full.

Table 6–1—Severity classification for lost circulation.

Seeping losses can occur with any type of lost-circulation zone, when the solids in the mud are not sufficiently fine to seal the formation face. Partial losses frequently occur in highly permeable gravels, small natural fractures, or as a result of fracture initiation. Complete losses are usually confined to long gravel sections, large natural fractures, wide induced fractures, or cavernous formations. Table 6–2, from Howard and Scott (1951), is a summary of some characteristic features associated with each type of lost-circulation zone.

6–10.1 Highly Permeable Formations

To permit the penetration of whole mud or cement, the matrix of a porous formation must have a permeability greater than 10d; however, significant seepage losses can be experienced in consolidated sandstones of lower permeability. Such formations are typically found at shallow depths.

6–10.2 Natural Fractures or Fissures

Hard consolidated formations may contain natural fractures which take mud when penetrated. For a natural horizontal fracture to exist, the overburden must be self-supporting, but this is not the case for a vertical natural fracture. To widen a horizontal fracture, the overburden must be lifted; whereas for a vertical fracture, only the fracture propagation pressure need be exceeded. A sudden loss of returns in hard consolidated formations is indicative of natural fractures.

6–10.3 Induced Fractures

If the borehole pressure exceeds the formation parting pressure, open fractures will be created permitting the loss of mud or cement. There are three typical circumstances when this can occur.

- An immovable mud ring may develop in the annulus. The resulting circulating pressure increase may initiate a hydraulic fracture.
- When drilling through an undercompacted formation, typically found offshore.
- When drilling from a mountaintop, it is possible to drill through formations where the overburden pressure is low, and fracturing occurs easily.

Well irregularities, high mud weight, and rough handling of the drilling tools may also help induce fractures.

Simpson et al. (1988) suggested that lost circulation due to fracture initiation is more common when using oil-base instead of water-base mud. They believed this to be true because of the failure to consider the compressibility of the oil under downhole conditions. They also pointed out that induced fractures do not "heal" readily when oil-base mud is present. Upon partial loss of water-base mud, an accepted practice is to let the hole soak for a period of time.

Filtration from the mud allows the fractures to be filled with mud solids, often permitting full circulation to be restored with no reduction in mud weight. However, filtration from oil-base mud is often too slow to be helpful. Once fractures are initiated with an oil-base mud,

Porous Sands and Gravels	Natural Fractures	Induced Fractures	Cavernous Zones
1. Gradual lowering of mud level in pits. 2. Losses may become complete, if drilling is continued. 3. Since rock permeability must exceed about 10d before whole mud can penetrate, and oil and gas sand permeability seldom exceeds about 3.5d, it is improbable that loose sands are the cause of mud loss to an oil or gas sand unless the loss can be attributed to the ease with which this type of formation fractures.	1. May occur in any type rock. 2. Loss is evidenced by gradual lowering of the mud in the pits. If drilling is continued and more fractures are exposed, complete loss of returns may be experienced. 3. Fracture must have a finite supported width to take mud.	1. Occur where fractures are horizontal in any formation under mud rings. 2. Loss is usually sudden and accompanied by complete loss of returns. Conditions are conducive to forming induced fractures when mud weight exceeds 10.5 lb/gal. 3. Loss may follow any sudden surge of pressure or trip. 4. When loss of circulation occurs and adjacent wells have not experienced lost circulation, induced fractures should be expected.	1. Normally confined to limestone. 2. Loss of returns may be sudden and complete. 3. Bit may drop several inches to several feet just preceding loss. 4. Drilling may be rough before loss.

Table 6–2—Identifying features of lost-circulation zones (after Howard and Scott, 1951; Messenger, 1981.)

fracture extension can be expected until the borehole pressures can be reduced or the fracture openings can be sealed.

6–10.4 Cavernous Formations

Large voids or caverns are sometimes encountered when drilling through certain limestone and dolomite formations as well as the caprock of salt domes. Sudden and complete losses are typical of this type of zone.

6–11 LOST CIRCULATION WHILE DRILLING

According to Messenger (1981), it is possible to classify the available solutions into three main categories:

- bridging agents in the drilling fluid,
- surface mixed systems, and
- downhole mixed systems.

There is an optimum technique for solving each particular type and severity of a lost-circulation problem.

6–11.1 Bridging Agents in the Drilling Fluid

When the loss of mud is first detected, immediate consideration should be given to the possibility of reducing and maintaining the mud weight at the minimum necessary to control the formation pore pressure. Reduced mud pressure will help combat losses no matter what types of formations are exposed. A continuing partial loss of returns is indicative of seepage, and can usually be solved by decreasing the equivalent mud circulating density, or by adding Lost-Circulation Materials (LCMs) to the drilling mud. The equivalent mud circulating density can be reduced by decreasing the weight of the mud and/or its downhole rheological properties. According to their physical nature and their mechanism of action, LCMs can be classified into four different groups:

- granular,
- lamellar,
- fibrous, and
- encapsulated fluid-absorbing particles.

Howard and Scott (1951) performed a series of experiments comparing the fracture sealing capacity of these groups vs their concentration in drilling mud (Fig. 6–8). They found that granular LCMs were more effective than the laminar or fibrous materials for sealing larger fractures. Table 6–3 is a list of typical commercial materials, their particle-size distributions, and the normal concentrations used.

The granular LCMs form two types of bridges—one at the formation face, and one within the formation matrix. The latter type of sealing is preferred, because a more permanent bridge forms within the formation, and the granular particles are not easily dislodged by pipe movement in the wellbore. The effectiveness of granular LCMs depends primarily on a proper particle-size distribution, with larger particles first forming a bridge across or within the void, and the smaller particles bridging the openings between the larger particles (Gatlin and Nemir, 1961). This process continues until the void spaces become smaller than the drilling mud solids. The problem finally becomes one of filtration. A blend of large, medium, and small particles, or one of large and small particles, is most commonly used. Such systems are usually more successful in high solids ratio systems, such as cement slurries. In 1976, Abrams showed that the median particle size of the bridging additive should be equal to or slightly greater than one-third the median pore size of the void. In addition, the minimum concentration of the bridging solids was shown to be five percent by volume of solids in the final mud mix.

Fibrous materials are best used for controlling losses to porous and highly permeable formations, because they are able to form a mat-like bridge over the pore openings. The mat reduces the size of the openings to the formation, permitting the colloidal particles in the mud to rapidly deposit a filter cake. Flake LCMs are also designed to bridge and form a mat on the formation face, also providing the best results when treating losses to permeable and porous formations.

Blends of granular, flake, and fibrous LCMs are effective in solving actual field problems. This strategy provides a gradation of particle size as well as a variation of material types for sealing different classes of lost-circulation zones.

Nayberg and Petty (1986) performed a laboratory study comparing the effectiveness of fibers, flakes, granules, and thermoset rubber in controlling mud losses to simulated medium-size (0.13 in. or 3.3 mm) fractured formations. They claimed that a blend of medium- and fine-grained (10- to 200-mesh) particles of thermoset

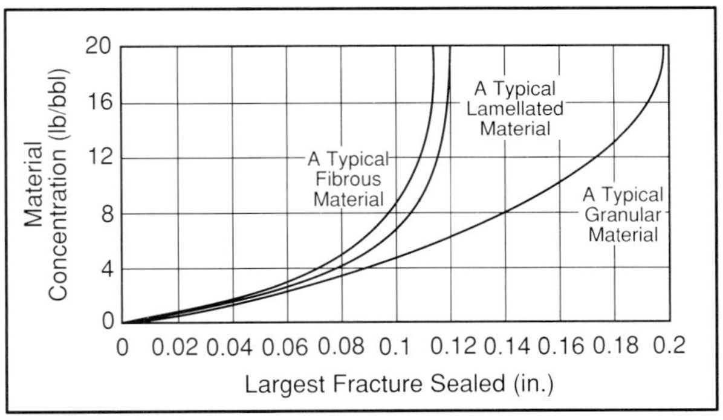

Figure 6–8—Effect of concentration of lost-circulation materials when sealing fractures (after Howard and Scott, 1951).

Material	Type	Description	Concentration (lb/bbl)
Nut Shell	Granular	50%—3/16 + 10 mesh	20
		50%—10 + 100 mesh	
Plastic	Granular	50%—10 + 100 mesh	20
Limestone	Granular	50%—10 + 100 mesh	40
Sulfur	Granular	50%—10 + 100 mesh	120
Nut Shell	Granular	50%—10 + 16 mesh	20
		50%—30 + 100 mesh	
Expanded Perlite	Granular	50%—3/16 + 10 mesh	60
		50%—10 + 100 mesh	
Cellophane	Lamellated	3/4-in. Flakes	8
Sawdust	Fibrous	1/4-in. Particles	10
Prairie Hay	Fibrous	1/2-in. Fibers	10
Bark	Fibrous	3/8-in. Fibers	10
Cottonseed Hulls	Granular	Fine	10
Prairie Hay	Fibrous	3/8-in. Particles	12
Cellophane	Lamellated	1/2-in. Flakes	8
Shredded Wood	Fibrous	1/4-in. Fibers	8
Sawdust	Fibrous	1/16-in. Particles	20

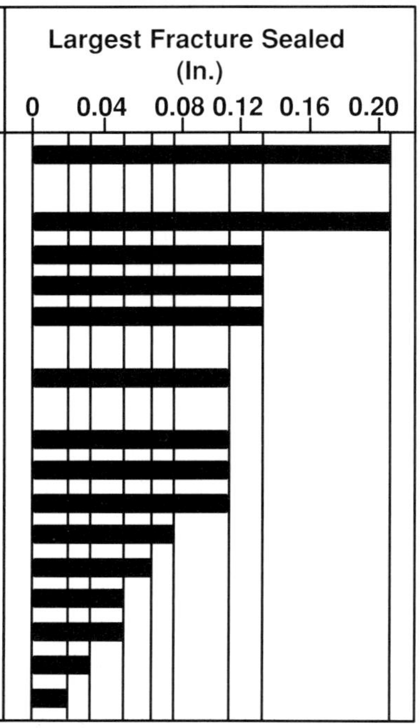

Table 6–3—Typical lost circulation materials (LCMs) (after Howard and Scott, 1951).

rubber performed better than the conventional lost-circulation materials. An interesting observation was that granular LCMs sometimes exhibited a "channeling" phenomenon. When a high pressure differential and an insufficient mud solids concentration existed, a bridge at the formation face, or within the formation matrix, could not develop.

The first patent concerning the use of encapsulated particles to control lost circulation was that of Armentrout (1958). The technique consists of encapsulating bentonitic particles within a low-permeability polymeric coating. When the encapsulated bentonite is pumped down the wellbore, water from the mud seeps into the capsules. The bentonite swells and ultimately ruptures the coating. The swollen bentonite then seals the voids in the lost-circulation zone. Walker (1987) followed this by describing a technique where the lost-circulation additive is a highly water-absorbent polymer encapsulated by a protective casing. The casing can be a material which dissolves after a period of time in contact with the wellbore fluid, or a waxy substance which melts at a temperature between the bottomhole static and circulating temperatures. The polymer then absorbs water, forming a semisolid, nonflowing mass which seals the zone. The water-absorbent polymers include alkali metal polyacrylates or saponified copolymers of a vinyl ester, which have the capacity to absorb more than 100 times their weight of water. Another patent by Delhommer and Walker (1987) described a very similar technique for oil-absorbing polymers, permitting the use of such systems in oil-base mud.

6–11.2 Surface-Mixed Systems

6–11.2.1 Cement Plugs

Neat cement slurries are effective for solving seeping or minor loss, with the advantage of providing high final compressive strengths. Slurries with a limited degree of fluid-loss control can be used to solve seeping, partial, or total losses, and contain a mixture of clays, diatomaceous earth, and LCMs. The size of the LCM is increased as the losses become more severe. Low-density cement systems can be used for any type of lost-circulation problem. They have the added advantage of reducing the hydrostatic pressure.

Thixotropy is a term used to describe the property exhibited by a system that is fluid under shear (i.e., pumping or agitation), but develops a gel structure when the shear is stopped (Chapter 4). In practical terms, thixotropic systems are fluid during mixing and displacement, but rapidly form a rigid, self-supporting gel structure when pumping ceases. When a thixotropic slurry enters a lost-circulation zone, the velocity of the leading edge decreases and a gel structure starts to form (Chapter 7). As the gel strength develops, resistance to flow increases until the entire zone is plugged (Childs et al., 1985). Such systems are very effective for solving severe lost circulation to naturally fractured formations.

6–11.2.2 Other Surface-Mixed Systems

Systems which do not contain Portland cement usually involve a gelling agent with an activator. After a given period of time, or due to an increase in temperature, the components react to form a nonflowing mass. The advantage of such systems is the ability to predict when the mixture will change from a liquid to a solid. In general, they are most applicable to partial lost-circulation problems in high-permeability sandstones, or for sealing microfissures.

Sharp (1966) first described the use of an aqueous solution of sodium silicate and urea which, at temperatures above 145°F (63°C), reacts to form a hydrosol of silicic acid. With time, the hydrosol converts to a silica gel within the formation, providing a firm structure which is essentially impervious to fluid. The gel time may be preselected by varying the relative concentrations of the reactants; however, the downhole temperature must exceed 145°F (63°C) for the reaction to proceed at a useful rate. This feature permits the preblending of the mixture several hours before the operation commences.

Elphingstone et al. (1981) described the use of halogenated hydrocarbons, more specifically sodium trichloroacetate, as an activator for aqueous silicate solutions. The addition of silica flour (325-mesh) was recommended to increase the viscosity of the final gel.

Smith (1986) claimed the use of reducing sugars, such as lactose and fructose, as thermally responsive activators for silicate solutions. For applications where the well temperature is below 120°F (49°C), the addition of small amounts of a reactive salt (such as calcium chloride) was suggested to provide short gelling times without having to increase the concentration of the reducing sugar.

Yearwood et al. (1988) and Vidick et al. (1988) described the use of an internally activated low-viscosity silicate solution which, depending on the fluid design and the temperature, gels rapidly after a given period of time. The final gel is strong and permanent, with very little free-water development at temperatures up to 355°F (180°C). To demonstrate the sealing capacity of this system, a series of laboratory experiments was performed using core plugs with different permeabilities. The results demonstrated that, once the gel has formed in the formation matrix, the system is able to withstand differential pressures greater than 1,500 psi/ft (Table 6–4).

Vidick et al. (1988) presented an equation to relate the gelling time of the silicate systems to the active matter content and the temperature. Equation 6–21 helps not only to predict the gelling time at a particular temperature for a given active matter content, but also to calculate the gel time variations resulting from slight bottomhole temperature fluctuations.

$$GT = K_T x^{n_T} \exp\left[-E_a/RT\right] \qquad (6\text{–}21)$$

where

GT = gelling time (min),

K_T = a constant (function of the temperature),

x = active matter content (% by volume),

n_T = coefficient related to the active matter content,

E_a = activation energy (Kcal/mole),

R = gas constant (1.99 Kcal/mole °K), and

T = temperature (°K).

The values for n_T and K_T at four temperatures are given in Table 6–5.

In cases where the lost-circulation zone is also a zone of interest, either for production or injection purposes, it may be necessary to design the plugging material for eventual removal during the completion of the well. Such systems are generally acid soluble, consisting of

Core Nature	Average Perm. to Water (darcies)	Saturating Fluid	Test Temp. (°F)	Extrusion Pressure Resistance for One Foot of Plugged Core (psi)
20/40 Frac Sand	6	Fresh Water	105	1200
Porous Sandstone	2	Fresh Water	105	>1500
Porous Sandstone	2	Diesel Oil	105	>1500
Porous Sandstone	2	Brine	105	>1500
Fissured Limestone	1	Fresh Water	140	>1500
Fissured Limestone	1	Brine	140	>1500
20/40 Frac Sand	6	Fresh Water	175	>1500
20/40 Frac Sand	6	Brine	175	>1500

Table 6–4—Performance of internally activated silicate system in core flow test (Yearwood, et al., 1988).

Temperature (°F)	n_τ	K_τ (min)
104	−16.94	e^{62}
140	−17.14	e^{61}
176	− 3.84	e^{18}
266	−11.34	e^{51}

Table 6–5—Values for n_τ and K_τ at different temperatures (after Vidick et al., 1988).

bridging agents slurried in a viscous fluid, or cementitious materials.

Typical bridging materials include ground calcium carbonate particles with diameters ranging from 0.0003 in. (8 μm) to 0.01 in. (254 μm). They are used at concentrations up to 10 lb (4.5 kg) per barrel of carrying fluid. Assuming a relatively homogeneous sandstone formation where the sand grains are of similar size, it is possible to predict the required particle size of calcium carbonate to form a bridge in the pore throats of the formation matrix, thereby reducing the loss of fluid. These values are given in Table 6–6.

An acid-soluble cementitious product is Sorel cement, a mixture of magnesium oxide, magnesium chloride, and water. Alsdorf and Dittmar (1987) pointed out that this type of cement is not applicable at elevated temperatures, because control of the setting time is difficult. They rec-

Sand Grain Size (in.)	Sand Grain Size (μm)	Diameter of Pore Throat Opening (μm)	Approximate Permeability (md)	Bridging Particle Size (μm)
0.00025	6.46	1.0	1	0.33
0.00125	31.67	4.9	24	1.63
0.0015	38.10	5.90	35	1.97
0.0017	43.18	6.68	45	2.23
0.0021	53.34	8.26	68	2.75
0.0024	60.96	9.44	89	3.15
0.0029	73.66	11.40	130	3.80
0.0035	88.90	13.76	189	4.59
0.0041	104.14	16.12	260	5.37
0.0049	124.46	19.27	370	6.42
0.0058	147.32	22.80	520	7.60
0.0069	175.26	27.13	740	9.04
0.0082	208.28	32.24	1040	10.80
0.0097	246.38	38.14	1460	12.70
0.0116	294.64	45.61	2080	15.20
0.0138	350.52	54.26	2940	18.10
0.015	381.00	58.97	3480	19.70
0.0164	416.56	64.48	4160	21.50
0.0195	495.30	76.67	5880	25.60
0.0232	589.28	91.22	8320	30.40
0.0276	701.04	108.5	11,800	36.20
0.0328	833.12	128.9	16,600	43.00
0.0390	990.60	153.3	23,500	51.10
0.046	1168.00	181.0	32,800	60.30
0.055	1396.00	216.0	46,700	72.00
0.065	1650.00	255.0	65,000	85.00
0.078	1980.00	307.0	94,200	102.00
0.093	2361.00	365.0	133,000	122.00
0.110	2793.00	432.0	187,000	144.00
0.131	3326.00	515.0	266,000	172.00
0.156	3960.00	613.0	376,000	204.00
0.185	4697.00	727.0	529,000	242.00
0.221	5610.00	868.0	753,000	289.00
0.263	6677.00	1034.0	1,070,000	345.00
0.312	7921.00	1226.0	1,500,000	409.00

Table 6–6—Optimizing the particle size of the bridging material according to the formation permeability.

ommended the use of a material containing about 60% by weight of ground milk-of-lime grit, plus calcium or magnesium chloride and water. By varying the percentage of the calcium or magnesium chloride, it is possible to vary the thickening time from one to four hours at temperatures up to 195°F (90°C). The maximum compressive strength is obtained after 5 to 24 hours. The final product is completely soluble in 5% hydrochloric acid.

6–11.3 Downhole-Mixed Systems

Downhole-mixed systems consist of two or more fluids which, upon making contact in the wellbore or the lost-circulation zone, form a viscous plug or a precipitate which seals the zone. It is common practice to prevent the mixing of the fluids until they are in front of the lost-circulation zone, by pumping a spacer or by pumping one fluid down the drillstring while the other fluid is simultaneously pumped down the annulus. These systems are not suitable for total lost-circulation situations, where the actual displacement rates are not known, because it is very difficult to control the mixing of the fluids.

For partial losses, Iljas (1983) found better success by using Mud-Diesel-Oil-Bentonite (M-DOB) plugs instead of LCMs. M-DOB plugs are a combination of diesel oil and bentonite, and are sometimes called "gunk plugs." When this mixture contacts water or water-base mud, a mass with high gel strength is formed. Soft, medium, and hard plugs may be formed by controlling the proportions of the ingredients. The DOB slurry is pumped down the drillpipe, and the mud down the annulus.

M-DOB plugs suffer from several drawbacks.

- They break down with time.
- They are difficult to apply in long openhole intervals.
- When losses are severe, it is impossible to achieve a reliable pumping rate down the annulus; therefore, the degree of mixing cannot be controlled.
- No compressive strength is developed.

Gaddis (1975) increased the gel strength of the M-DOB plug by blending a water-soluble polymer with the bentonite in diesel oil. On contact with water, the polymer hydrates, and the clay flocculates to form a stiff cement-like plug. For severe losses, Messenger (1981) and Iljas (1983) suggested a better version of the M-DOB plug—the Mud-Diesel-Oil-Bentonite-Cement plug (M-DOB2C). The advantage of this system is the development of compressive strength. The ratios of mud and DOB2C required to produce mixtures of various hardnesses are shown in Table 6–7. Many downhole-mixed systems use a combination of two or more surface-mixed systems to provide an effective plugging material. For example, an M-DOB plug can be followed by a cement plug, thereby improving its strength and permanence.

Mud	DOB2C	Hardness
1	1	Soft
1	1.5	Medium
1	2	Medium hard
1	2.5	Hard
1	3	Very Hard

Table 6–7—Hardness of different combinations of mud with DOB2C (after Iljas, 1983).

In 1972, Biles described a technique for sealing highly permeable channels. A sodium silicate solution is allowed to mix with a solution containing divalent cations, forming a precipitate. This technique successfully sealed permeable formations, but the precipitate was not sufficiently strong to seal naturally fractured formations. Russell (1983) refined this technique by employing a preflush of an extended Portland cement slurry, followed by a sodium silicate solution, and a neat or thixotropic cement slurry. Both laboratory and field results showed a dramatic strength improvement when the sodium silicate and cement slurry intermixed. This was apparently due to the high availability of calcium ions from the cement, and the instantaneous dehydration of the cement slurry due to the reaction.

Murphey (1983) proposed the use of potassium silicate instead of sodium silicate, because the latter may tend to gel prematurely when mixed with brine. He described a common practice for solving total lost circulation in fractured and cavernous formations—pumping alternating batches of silicate and divalent cation solutions, with small freshwater spacers as separating fluids. The entire sequence is then followed by a Portland cement slurry.

6–12 LOST CIRCULATION DURING CEMENTING

Before initiating a conventional primary cementing operation, the lost-circulation problem should be eliminated or significantly reduced by the techniques described above. If this is not possible, or if losses are anticipated during the primary cementing job, there are two possible options as described by Nayberg and Linafelter (1984). The first is to maintain the downhole pressures during the job below the maximum equivalent mud circulating density, either by reducing the density of the cement slurry, minimizing the height of the cement column, or limiting the casing and annular friction pressures during the placement of the cement slurry. The second option is to pump a plugging material as a spacer in front of the cement slurry, add lost-circulation materials to the cement slurry itself, or use special additives which impart thixotropic properties to the cement slurry. When trying to prevent cement losses to highly fractured or vugular formations, it is often necessary to use a combination of techniques.

6–12.1 Downhole Pressure Reduction

Computer simulators can calculate the estimated downhole pressures at any particular depth in the well, and at any time during the cementing operation (Chapter 11). This enables the operator to know (for a particular well completion) exactly which cement slurry parameters and job procedures are required to prevent lost circulation and maintain adequate hydrostatic pressure in front of permeable zones. The most relevant parameter is the cement slurry density, which may be reduced by adding one or more cement extenders. Chapter 3 provides a detailed discussion of extenders, and the optimum slurry density range for each.

The rheological properties of a cement slurry may also be adjusted to provide lower friction pressure losses during placement. This is especially critical in narrow annuli where viscous slurries can cause very high friction pressures. Another technique mentioned by Nayberg and Linafelter (1984) is to lighten the hydrostatic column above the top of the cement by injecting nitrogen into the mud.

The downhole pressures exerted on lost-circulation zones can also be decreased by using mechanical devices such as stage collars or external casing packers (ECPs). Stage collars permit the casing string to be cemented in two or three stages, lowering the dynamic and hydrostatic pressures (Chapters 10 and 12).

To reduce the risk of cement fallback if losses do occur, a special stage collar with a packoff adaptation can be used which, when expanded, provides a seal between the casing and the formation to prevent downward fluid movement. Cement baskets can also be placed just below the stage collar to provide the same effect. Turki and Mackay (1983) described the placement of ECPs immediately above the lost-circulation zone to reduce the hydrostatic pressure. A typical application would be a two-stage job with an ECP just above the lost-circulation zone, and a stage collar just above the ECP. After the first stage is performed, the ECP is expanded to seal the annulus, preventing the transmission of hydrostatic pressure to lower zones (Fig. 6–9). However, if the size of the hole is larger than anticipated, the ECP may fail to provide a perfect seal because of insufficient lateral expansion.

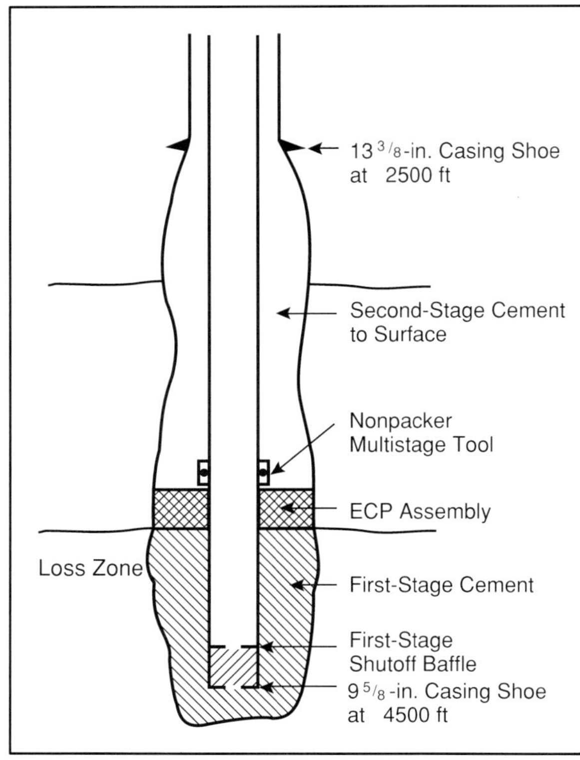

Figure 6–9—Cementation using an external casing packer.

Turki and Mackay (1983) also mentioned the "Hydrostatic Cementing Technique" for attempting to obtain zonal isolation across cavernous lost-circulation zones. A conventional first-stage job is performed, followed by pumping a predetermined quantity of cement slurry down the annulus. Most of the slurry is lost to the cavernous formation. However, after the hydrostatic pressure of the cement slury equilibrates with the formation pressure of the lost-circulation zone, a portion will remain in the annulus. When the cement sets, the cavern is bridged, and cement exists at some height above the cavern. The application of this technique was recommended only when lost circulation cannot be significantly reduced by conventional means, or when open holes are excessively washed out.

6–12.2 Preflushes

Murphey (1983) described the use of a potassium silicate solution as a preflush, to enable the formation to support a greater than normal hydrostatic pressure. The preflush penetrates the highly permeable formations, permitting contact with calcium ions in the formation, and resulting in the formation of a gel. If insufficient calcium ions are present in the formation, a second preflush of a calcium chloride solution can be pumped. The high concentration of calcium ions in the cement slurry ensures immediate sealing of the formation.

6–12.3 Lost-Circulation Materials for Cement Slurries

Nayberg and Petty (1986) and Turki and Mackay (1983) agreed that the effectiveness of LCMs in cement slurries is limited to minor or partial losses in highly permeable formations, and not for solving total lost circulation in naturally fractured or cavernous formations. They suggested that the only truly effective solution is foamed cement (Chapter 14).

When LCMs are used in the cement slurry, care must be taken to ensure that the materials are inert to the cement composition. Also, the size and concentration of the materials should be selected to avoid bridging or plugging of the downhole equipment. The morphologies of the materials are the same as those used in drilling fluids. The authors are not aware of any reported use of encapsulated additives in cement slurries for solving lost-circulation problems. Table 6–8, from Smith (1987), is a typical list of LCMs for cement slurries, their properties, and typical effective concentrations.

The most common LCMs for cement slurries are of the granular type, designed to bridge at the formation face or within the matrix. Gilsonite, a naturally occurring black asphaltite hydrocarbon with a particle size between 8 and 60 mesh, is widely used. Gilsonite is not suitable for high-temperature applications, because of its low melting point (220°F [104°C]). Crushed coal, with a standard mesh size of 14 to 200, and a melting point of approximately 1,000°F (538°C), is applied in the same manner as gilsonite, and can be used in high-temperature wells. Shells from walnuts, pecans, etc., are also available in fine, medium and coarse grades; however, care should be exercised at concentrations above 4 lb/sk to avoid the plugging of downhole equipment.

Cellophane flake with diameters of 3/8 to 3/4 in. (9.5 to 19 mm) is the most common flake material. At concentrations above 2 lb/sk, bulk loading and mixing of the cement slurry becomes extremely difficult.

Fibrous materials are seldom used in cement slurries, because they can plug cementing equipment. In addition, some organic chemicals may be present that may retard the thickening time of the cement slurry.

6–12.4 Thixotropic Cement Systems

The self-supporting property of thixotropic cements is useful across formations with low fracture gradients. When ordinary slurries pass over a weak zone, the increase in hydrostatic pressure can cause formation breakdown. As a result, the top of the cement falls to a point below the desired level of fill-up. Thixotropic slurries do not fall back, because some of the hydrostatic pressure is transmitted to the formation face and casing walls. Several thixotropic cement compositions exist, and their chemistries are described in Chapter 7.

6–13 LOST CIRCULATION—CONCLUSIONS

Lost-circulation problems, either during drilling or cementing, can be solved if the correct technique is applied for each individual case. Choosing the correct solution from the wide variety of available remedies described above can be a difficult task; however, certain general guidelines can be followed. Messenger (1981) summarized the most important factors to consider.

- The location of the loss zone must be determined accurately; otherwise, the remedy will be placed in the wrong zone. Many loss zones thought to be at the bit are actually further up the hole at the first point of loss.
- Lost-circulation materials and techniques must be systematically matched to the type and severity of the loss zone. For example, using LCMs in the drilling mud to stop total losses to a vugular limestone will normally never work. One has a much better chance for success with a combination of surface- and

Type	Material	Nature of Particles	Amount Used	Water Required
Granular	Gilsonite	Graded	5 to 50 lb/sk	2 gal/50 lb
	Perlite	Expanded	½ to 1 ft^3/sk	4 gal/ft^3
	Walnut Shells	Graded	1 to 5 lb/sk	0.85 gal/50 lb
	Coal	Graded	1 to 10 lb/sk	2 gal/50 lb
Lamellated	Cellophane	Flaked	⅛ to 2 lb/sk	None
Fibrous	Nylon	Short-Fibered	⅛ to ¼ lb/sk	None

Table 6–8—Materials commonly added to cement slurries to control lost circulation (after Smith, 1987).

downhole-mixed systems with low densities, thixotropic behavior, and good strength development.

- Consulting records of prior experience with lost circulation in a particular field often points the way to an effective solution.
- Above all, careful prejob planning can prevent the occurrence of lost circulation. It is important to obtain, if possible, sufficient well information to perform a computer simulation of the cement job (Chapter 11).

NOMENCLATURE

D_c = casing outside diameter, m
D_h = hole diameter, m
e_c = cement-cake thickness, m
e_m = mud-cake thickness, m
$F_{w/c}$ = water/cement ratio by weight, dimensionless
g = acceleration of gravity, m^2/sec
k_c = cement-cake permeability, m^2
k_m = mud-cake permeability, m^2
R = cake to filtrate volume ratio, dimensionless
t_t = thickening time, sec
u = slurry vertical velocity, m/sec
u_o = slurry vertical velocity below the permeable layer, m/sec
v = filtration velocity, m/sec
v_{max} = maximum filtration velocity, m/sec
z = vertical coordinate, m
Z = vertical coordinate, dimensionless
ϕ_w = water volume fraction of the slurry, dimensionless
ϕ_{wo} = initial water volume fraction of the slurry, dimensionless
ϕ_c = cement volume fraction of the slurry, dimensionless
ϕ_{co} = initial cement volume fraction of the slurry, dimensionless
μ = filtrate viscosity, Pa-sec
ρ_w = water density, kg/m^3
ρ_c = cement density, kg/m^3
ρ_s = slurry density, kg/m^3

REFERENCES

Abrams, A.: "Mud Design to Minimize Rock Impairment due to Particle Invasion," paper SPE 5713, 1976.

Alsdorf, H. and Dittmar, A.: "Material for Sealing Borehole Walls," U.S. Patent No. 4,670,056 (1987).

Armentrout, A. L.: "Material for Recovering Lost Circulation in Wells," U.S. Patent No. 2,836,555 (1958).

Bannister, C. E.: "Evaluation of Cement Fluid-Loss Behavior Under Dynamic Conditions," paper SPE 7592, 1978.

Bannister, C. E. and Lawson, V. M.: "Role of Cement Fluid Loss in Wellbore Completion," paper SPE 14433, 1985.

Baret, J. F.: "Why Cement Mud-Loss Additives are Necessary," paper SPE 17630, 1988.

Beach, H. J., O'Brien, T. B., and Goins, W. C. Jr.: "Here's How Gulf Improves its Formation Cement Squeezes by Using Low-Water-Loss Cements," *Proc.*, Spring Meeting API Div. Prod. South. Dist., Shreveport, LA (1981).

Biles, J. W.: "Selective Plugging Method," U.S. Patent No. 3,658,131 (1972).

Binkley, G. W., Dumbauld, G. K., and Collins, R. E.: "Factors Affecting the Rate of Deposition of Cement in Unfractured Perforations During Squeeze-Cementing Operations," paper SPE 891–G, 1957.

Bradford, B. and Reiners, B.: "Analysis Gives Successful Cement Squeeze," *Oil & Gas J.* (April 1, 1985) 71–74.

Childs, J., Sabins, F., and Taylor, M. J.: "Method of Using Thixotropic Cements for Combating Lost Circulation," U.S. Patent No. 4,515,216 (1985).

Christian, W. W., Chatterji, J., and Ostroot, G. W.: "Gas Leakage in Primary Cementing—A Field Study and Laboratory Investigation," *JPT* (Nov. 1976) 1361–1369.

Cook, C. and Cunningham, W. C.: "Filtrate Control—A Key in Successful Cementing Practices," *JPT* (Aug. 1977) 951–956.

Cunningham, W. C. and Smith, D. K.: "Effect of Salt Cement Filtrate on Subsurface Formation," *JPT* (March 1968) 259–264.

Cutforth, H. G.: "Low Water-Loss Cement Slurry and Method of Cementing a Well Therewith," U.S. Patent No. 2,598,675 (1949).

Delhommer, H. J. and Walker, C. O.: "Encapsulated Oil Absorbent Polymers as Lost Circulation Additives for Oil-Base Drilling Fluids," U.S. Patent No. 4,704,213 (1987).

Desbrières, J.: "Influence of Polymeric Additives on Cement Filter Cake Permeability," *Proc.*, Third Intl. Symp. Chem. Oil Indus., Manchester, UK, (1988) Royal Soc. Chem Spec. Publ. No. 67, 62–67.

Elphingstone, E. A., McLaughlin, H. C., and Smith, C. W.: "Temperature Gelation Activated Aqueous Silicate Mixtures and Process of Forming Impermeable Gels," U.S. Patent No. 4,293,440 (1981).

Fordham, E. J., Ladva, H., K. J., Hall, C., Baret, J. F., and Sherwood, J. D.: "Dynamic Filtration of Bentonite Muds Under Different Flow Conditions," paper SPE 18038, 1988.

Gaddis, P. G.: "Method of Making High-Viscosity Aqueous Mediums," U.S.Patent No. 3,909,421 (1975).

Gatlin, C. and Nemir, C. E.: "Some Effects of Size Distribution on Particle Bridging in Lost Circulation and Filtration Tests," *JPT* (June 1961) 575–578.

Goins, W. C. Jr.: "How to Combat Circulation Loss," *Oil & Gas J.* (June 9, 1952) 71–74.

Hartog, J. J., Davies, D. R., and Stewart, R. B.: "An Integrated Approach for Successful Primary Cementations," *JPT* (Sept. 1983) 1600–1610.

Hook, F. E. and Ernst, E. A: "The Effect of Low-Water-Loss Additives, Squeeze Pressure, and Formation Permeability on the Dehydration Rate of a Squeeze Cementing Slurry," paper SPE 2455, 1969.

Howard, G. C. and Scott, P. P. Jr: "An Analysis and the Control of Lost Circulation," *Trans.*, AIME (1951) **192**, 171–182.

Iljas, R.: "Lost Circulation and Control in Reefal Limestone Depositions," *Proc.*, Twelfth Annual Indonesian Pet. Assoc. Convention, Jakarta (1984) **2**, 1–10.

Krueger, R. F.: "An Overview of Formation Damage and Well Productivity in Oilfield Operations," *JPT* (Feb. 1986) 131–152.

Messenger, J.: Lost Circulation, *A Practical Approach to Preventing, Assessing, and Solving Lost Circulation Problems*, PennWell Publishing Co., Tulsa, OK (1981).

Murphey, J. R.: "Rapidly Dissolvable Silicates and Methods of Using the Same," U.S. Patent No. 4,391,643 (1983).

Nayberg, T. M. and Linafelter, R. L.: "Controlling Cement Circulation Loss to Both High-Permeability and Fractured Formations," paper SPE 12905, 1984.

Nayberg, T. M. and Petty, B. R.: "Laboratory Study of Lost Circulation Materials for Use in Both Oil-Base and Water-Base Drilling Muds," paper IADC/SPE 14723, 1986.

Records, L. R. and Ritter, J. R.: "Results of Field Use of Very Low-Water-Loss Oil Well Cements for Better Production Capacity of Oil and Gas Wells," paper SPE 7010, 1978.

Reservoir Stimulation, M. J. Economides and K. G. Nolte (eds.), Schlumberger Educational Services, Houston (1987) 12–5.

Russell, J.: "Remedial Cementing in a Low-Pressure Formation," *Drilling* (May 1983) **44**, 82–83.

Sharp, L. G.: "Sealing of Deep Permeable Earth Formations," U.S. Patent No. 3,244,230 (1966).

Simpson, J. P., Salisbury, D. P., and Jewell, R. A.: "How to Combat Oil-Base Mud Losses," *World Oil* (Jan. 1988) 30–32.

Smith, R. C.: "Successful Primary Cementing Can Be a Reality," *JPT* (Nov. 1984) 1851–1858.

Smith, D. K.: *Cementing*, Monograph Series, SPE, Richardson, TX (1987) **4.**

Smith, W. H.: "Gelling Aqueous Silicate Compositions," European Patent Application No. 0,230,725,A1 (1986).

Stout, C. M. and Wahl, W. W: "A New Organic Fluid-loss Control Additive for Oil Well Cements," paper SPE 1455–G, 1960.

Turki, W. H. and Mackay, A. S.: "Primary Cementing Across Massive Lost Circulation Zones," paper SPE 11490, 1983.

Vidick, B., Yearwood, J. A., and Perthuis, H.: "How to Solve Lost Circulation," paper SPE 17811, 1988.

Walker, C. O.: "Encapsulated Water-Absorbent Polymers as Lost-Circulation Additives for Aqueous Drilling Fluids," U.S.Patent No. 4,664,816 (1987).

Yearwood, J. A., Vidick, B., and Boissier, J. C.: "A New Technique for Solving Lost-Circulation Problems and Zone Plugging," paper CIM 88–39–105, 1988

7

Special Cement Systems

Erik B. Nelson and Philippe Drecq

Dowell Schlumberger

7–1 INTRODUCTION

As the technology of well cementing has advanced, certain problems have been encountered for which special cement systems have been developed. This chapter presents cement technologies specific to such problems as slurry fallback, lost circulation, microannuli, cementing across salt formations, and corrosive well environments. Special technologies also exist for problems such as high temperature and annular gas migration, and are presented in separate chapters (Chapters 9 and 8, respectively).

7–2 THIXOTROPIC CEMENTS

Thixotropy is a term used to describe the property exhibited by a system that is fluid under shear, but develops a gel structure and becomes self-supporting when at rest (Shaw, 1970). In practical terms, thixotropic cement slurries are thin and fluid during mixing and displacement, but rapidly form a rigid self-supporting gel structure when pumping ceases. Upon reagitation, the gel structure breaks and the slurry is again fluid and pumpable. Then, upon cessation of shear, the gel structure reappears and the slurry returns to a self-supporting state. This type of rheological behavior is continuously reversible with truly thixotropic cements.

As a rule, thixotropic slurries behave as Bingham plastic fluids under stress (Chapter 4); consequently, their behavior is defined by a yield value (τ_y) and a plastic viscosity (μ_p) (Clement, 1979). The τ_y is a theoretical value concerning the behavior of a fluid under conditions of shear. With thixotropic slurries, the τ_y would be the shear stress necessary to initiate movement, i.e., measured at zero shear rate.

For a nonthixotropic fluid, the yield value remains the same whether the shear rate is increasing or decreasing. There is no change in the physical structure of the fluid during the static period, and the pressure needed to put the fluid in movement does not change with time. In the case of a thixotropic fluid, the yield point is exhibited only upon the withdrawal of shear stress. If there is a lapse of time, a greater force than that indicated by the yield point will be required to put the fluid back into motion, as indicated in Figs. 7–1, 7–2, and 7–3. The difference between the "gel strength" and the yield point gives a measure of the "degree of thixotropy" of the fluid.

Thixotropic cement systems have several important applications. They are often used in wells where excessive fallback of the cement column is a common occurrence (Wieland et al., 1969). Such wells have weak zones which fracture under low hydrostatic pressure. Self-supporting cements reduce the hydrostatic pressure to the formation as gel strength increases, and fallback is prevented.

Another important application is the treatment of lost circulation during drilling (Chapter 6). When a thixotropic slurry enters the thief zone, the velocity of the leading edge decreases and a gel structure begins to develop. Eventually, the zone becomes plugged because of the increased flow resistance. Once the cement sets, the zone is effectively consolidated.

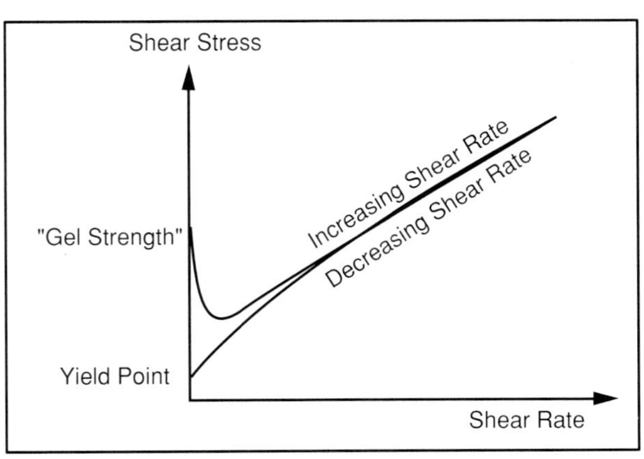

Figure 7–1—Generalized rheological behavior of thixotropic fluids.

Figure 7–2—Thixotropic behavior.

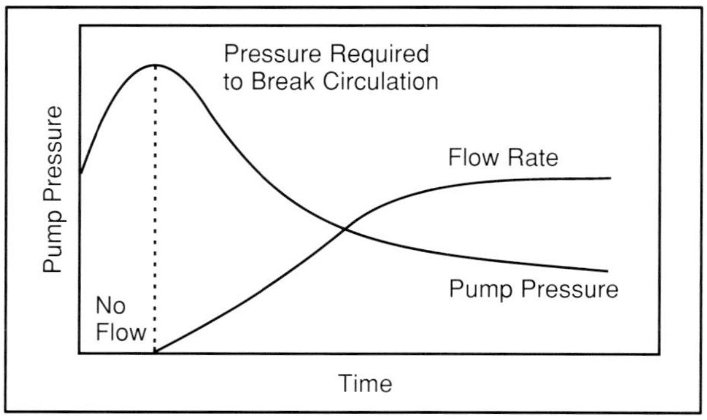

Figure 7–3—Pump pressure and flow rate for a thixotropic fluid.

Other uses for thixotropic cement systems include the following: to repair split or corroded casing; as lead slurries for remedial cementing in situations where it is difficult to obtain a squeeze pressure (Spangle and Calvert, 1972) (Chapter 13); as a grout, in circumstances where it is desirable for the slurry to become immobile quickly; and to prevent gas migration in certain situations (Chapter 8).

Thixotropic cement slurries have another notable characteristic. After each static-dynamic cycle, the gel strength and yield point tend to increase. During cementing operations this could pose a problem because, after repeated stops, excessive pump pressure may be required to restart movement. For this reason, most operators try to avoid a prolonged shutdown when pumping these systems.

Several thixotropic cement systems currently exist. The chemistry and special operational considerations of each are described below.

7–2.1 Clay-Base Systems

Portland cement systems containing water-swellable clays (such as bentonite) develop gel strength, and exhibit some degree of thixotropic behavior (Messenger, 1980). Such systems have also been shown to control gas migration in certain circumstances (Chapter 8). The concentration of bentonite and the slurry density can be varied from 0.05% to 2.0% BWOC and 11.5 and 21.0 lb/gal (1.4 to 2.5 g/cm^3), respectively.

7–2.2 Calcium Sulfate-Base Systems

The most widely used material to prepare thixotropic cement slurries is calcium sulfate hemihydrate ($CaSO_4 \cdot \frac{1}{2}H_2O$ or, in cement notation, $C\overline{S}H_{1/2}$) (also called plaster of Paris). When this material is added to Portland cement, it first hydrates to form gypsum ($CaSO_4 \cdot 2H_2O$ or $C\overline{S}H_2$), then reacts with tricalcium aluminate (C_3A) to form a calcium sulfoaluminate hydrate mineral called "ettringite." The chemical equation for the reaction is shown below (Kalousek, 1973).

$$3CaSO_4 \cdot 2H_2O + 3CaO \cdot Al_2O_3 \xrightarrow{H_2O} 3CaO \cdot Al_2O_3 \cdot 3CaSO_4 \cdot 32H_2O \quad \textit{(ettringite)} \qquad (7\text{–}1)$$

Ettringite occurs as needle-shaped, pseudo-hexagonal uniaxial crystals, and is deposited upon the surfaces of the cement grains. The presence of the ettringite crystals promotes greater physical association between the cement particles, resulting in the formation of a loose network or gel. Upon agitation, the network is easily disrupted, and the slurry returns to a fluid state.

Most Portland cements can be used to prepare thixotropic cements with calcium sulfate hemihydrate. Depending upon the cement, the optimum $C\overline{S}H_{1/2}$ concentration varies between 8% and 12% BWOC. Cements with a C_3A content less than 5% should not be used, because insufficient ettringite would crystallize to impart thixotropy. The water requirement for calcium sulfate hemihydrate-containing slurries is higher than that for conventional systems; consequently, the slurry densities are lower. Representative data for such systems are presented in Table 7–1.

System	Density (lb/gal)	% Water	Water (gal/sk)	% Calcium Sulfate Hemihydrate	% $CaCl_2$	Yield (ft^3/sk)
1	15.6	46	5.20	0	0	1.18
2	14.9	60	6.78	12	3	1.48
3	14.6	64	7.20	10	2	1.50
4	14.6	64	7.20	12	3	1.54
5	14.2	70	7.90	10	0	1.60

System	Well Conditions (°F) BHCT	BHST	Thickening Time (hr:min) at BHCT	Compressive Strength (psi) at BHST 7 hr	18 hr	24 hr	96 hr
1	113	170	4:00+	350	1750	2900	—
2	70	60	3:10	100	—	500	1150
3	80	95	2:08	485	1165	1295	—
4	80	95	1:50	500	1250	1350	—
5	125	200	3:15	950	1750	2200	—

Table 7–1—Slurry properties and performance of thixotropic slurries containing calcium sulfate hemihydrate.

Thixotropic cements containing calcium sulfate hemihydrate are not compatible with most fluid-loss additives. To provide adequate fluid-loss control, such slurries are usually preceded by a spacer with a low fluid-loss rate (Warembourg et al., 1980).

Calcium sulfate hemihydrate systems have additional attributes besides thixotropy. Such systems are highly sulfate resistant, because the C_3A is effectively neutralized (Chapter 2). Also, after setting, ettringite continues to form; as a result, a significant amount of bulk expansion occurs within the cement matrix. This phenomenon, and the benefits derived from it, are addressed in detail later in this chapter.

7–2.3 Aluminum Sulfate/Iron (II) Sulfate System

An additive composed of a blend of $Al_2(SO_4)_3$ and $FeSO_4$ also relies upon the formation of ettringite to impart thixotropy to cement slurries (Nelson, 1983). It was developed for use with Portland cements which contain less than 5% C_3A. The material is also effective with non-Portland cements, such as Class J cement. It can be supplied in liquid form, which is convenient for offshore operations.

The aluminum sulfate reacts with calcium hydroxide in the cement slurry to form ettringite.

$$2Al(OH)_4^- + 3SO_4^{2-} + 6Ca^{2+} + 12OH^- \xrightarrow{H_2O} 3CaO \cdot Al_2O_3 \cdot 3CaSO_4 \cdot 32H_2O \quad (7\text{–}2)$$

The kinetics of the above reaction are much faster than those observed with calcium sulfate hemihydrate. Aluminum sulfate is a powerful cement accelerator, and a strong irreversible gel structure would develop if it were added alone. Iron (II) sulfate, a weak cement retarder, is included in the system to inhibit the aluminum sulfate and preserve thixotropy throughout the pumping time. Because of the fast kinetics of this system, very little ettringite is formed after the cement sets. Thus, significant cement expansion is not observed except at curing temperatures below 100°F (38°C).

7–2.4 Crosslinked Cellulose Polymer Systems

Thixotropic cements can prepared by the addition of water-soluble crosslinkable polymers and a crosslinking agent (Childs et al., 1985). Hydroxyethylcellulose (HEC), carboxymethylhydroxyethylcellulose (CMHEC), polyvinyl alcohol, and various sulfonate polymers can be crosslinked with certain titanium or zirconium chelates. The optimum polymer/crosslinker combination, and the relative concentrations of each, vary depending upon the temperature of the well.

7–3 EXPANSIVE CEMENT SYSTEMS

Good bonding between cement and pipe and between cement and formation is essential for effective zonal isolation. Poor bonding limits the desired production, and reduces the effectiveness of stimulation treatments (Chapter 1). Communication between zones can be caused by inadequate mud removal, poor cement/formation bonding because of excessive mud filter-cake buildup, expansion and contraction of the casing as a result of internal pressure or thermal stress, and cement contamination by drilling or formation fluids (Parker and Wahl,

1966; Beirute and Tragresser, 1973). Under such circumstances, a small gap or "microannulus" is frequently present at the cement/casing or the cement/formation interface.

Cement systems which expand slightly after setting are recognized as a means of sealing microannuli and improving primary cementing results. The improved bonding is the result of mechanical resistance or tightening of the cement against the pipe and formation. Good bonding can be obtained even if mud is left on the casing or formation surfaces.

The reader may recall from Chapter 2 that Portland cement manufacturers limit the amount of certain alkaline impurities to avoid expansion of the set cement, a condition called "unsoundness." In an unrestrained environment such as a road or building, expansion of the set cement can result in cracking and failure. In a wellbore environment, the cement is restrained by the casing and, when competent, the formation; consequently, once the cement has expanded to eliminate void spaces, further expansion is translated into a reduction of internal cement porosity.

7–3.1 Ettringite Systems

Most expansive well cement systems rely upon the formation of ettringite, discussed in the preceding section, after the cement has set. Ettringite crystals have a greater bulk volume than the components from which they form; consequently, expansion occurs because of the internal pressure exerted upon crystallization. Currently, there are four commercial expanding cement systems in the ettringite category.

Type K cement is a blend of Portland cement, calcium sulfate, lime, and anhydrous calcium sulfoaluminate (Klein and Troxell, 1958). This cement is composed of two separately burned clinkers which are interground. Type K cement systems typically expand by 0.05% to 0.20%.

Type M cement is either a blend of Portland cement, refractory calcium aluminate cement (Chapter 9) and calcium sulfate, or an interground product made with Portland cement clinker, calcium aluminate cement clinker, and calcium sulfate (Root and Calvert, 1971).

Type S cement is a commercially prepared blend of high C_3A Portland cement with 10.5% to 15% gypsum. It has a limited shelf life.

The fourth method of preparing an ettringite-base expansive cement is the addition of calcium sulfate hemihydrate to a Portland cement containing at least 5% C_3A. This formulation is similar to that of Type S; however, because the blend is prepared as needed before a cement job, shelf life is not a concern. As discussed in the previous section, such systems are also thixotropic. If not desired, the thixotropy can be defeated by the addition of a cement dispersant. The expansion performance of Portland cement/calcium sulfate hemihydrate systems is illustrated in Fig. 7–4.

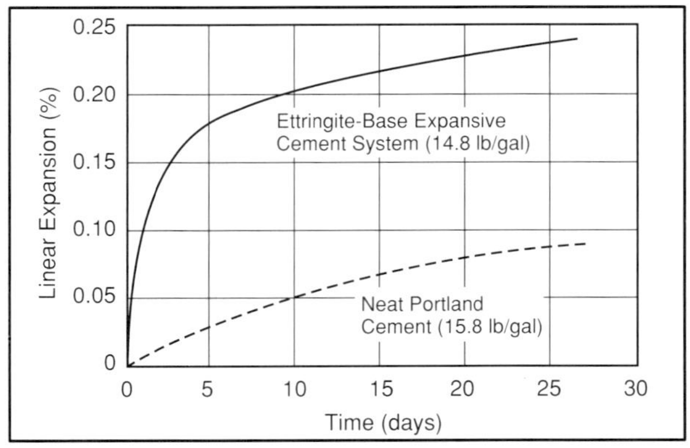

Figure 7–4—Comparison of expansion between neat Portland cement and an ettringite-base expansive cement system.

A major limitation of ettringite-base systems is their inability to provide significant expansion at curing temperatures above about 170°F (76°C) (Bour et al., 1988). Ettringite is not stable at higher temperatures, and converts to a more dense calcium sulfoaluminate hydrate and gypsum according to the following chemical equation (Lea, 1970).

$$3CaO \cdot Al_2O_3 \cdot 3CaSO_4 \cdot 32H_2O \rightarrow 3CaO \cdot Al_2O_3 \cdot CaSO_4 \cdot 12H_2O + 2CaSO_4 \cdot 2H_2O + 18H_2O \quad (7\text{–}3)$$

7–3.2 Salt Cements

The preparation of cement slurries containing high concentrations of NaCl and/or Na_2SO_4 was among the earliest methods for achieving expansion in well cements (Carter et al., 1965). After setting, cement expansion occurs because of internal pressure exerted by the crystallization of the salts within pores, and chlorosilicate reactions (Smith, 1987). Typical expansion performance of such systems at ambient conditions is shown in Fig. 7–5.

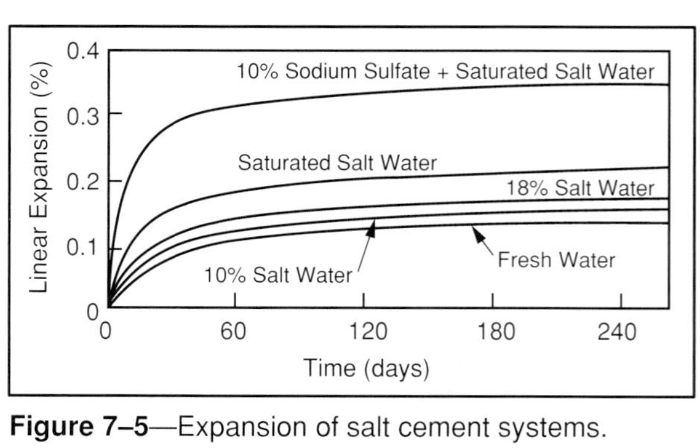

Figure 7–5—Expansion of salt cement systems.

These systems are equally effective at temperatures up to 400°F (204°C).

7–3.3 Aluminum Powder

Zinc, magnesium, iron, and aluminum powders can be used to prepare expansive cements (Carter et al., 1965). Finely powdered aluminum reacts with the alkalis in the cement slurry to produce tiny bubbles of hydrogen gas. This technique is effective in shallow well applications, because the expansive pressure of the bubbles is not exceeded by the formation pressure. The performance of such systems is illustrated in Table 7–2.

Volume Expansion % (80°F)		
Aluminum (%)	Curing Pressure 0 psi	3000 psi
0.00	—	—
0.05	11.84	0.712
0.10	17.90	0.917
0.25	24.00	1.64
0.50	56.51	2.64
1.00	57.19	5.17

Table 7–2—Expansive effect of powdered aluminum in cement (after Carter et al., 1965).

The reaction is strongly affected by the fineness and concentration of aluminum, temperature, and pressure. Thus, careful slurry design is necessary to obtain optimum results. More recently, the pressurization effect of aluminum powder systems has been applied to prevent gas migration (Chapter 8).

7–3.4 Calcined Magnesium Oxide

Magnesium oxide provides an expansive force within the cement matrix as a result of hydration to magnesium hydroxide. The hydrated material occupies more space than the original ingredients.

$$\text{MgO (periclase)} + H_2O \rightarrow \text{Mg}(OH)_2 \text{ (brucite)}$$
$$\text{S.G.} = 3.58 \qquad\qquad \text{S.G.} = 2.36 \qquad (7\text{–}4)$$

The MgO must be calcined at very high temperatures (dead-burnt), between 2,012° and 2,372°F (1,100° and 1,300°C); otherwise, the hydration occurs before the cement sets, and no significant cement expansion is observed (Spangle, 1988).

Cement systems containing MgO have been shown to provide excellent expansive performance at curing temperatures as high as 550°F (288°C). However, at temperatures below about 140°F (60°C), the hydration reaction proceeds too slowly to be of practical benefit. The concentration of MgO required to provide adequate expansion varies between 0.25% and 1.00% BWOC, depending upon temperature. Fig. 7–6 shows the expansion performance of a Class G cement system containing 1.0 % MgO (BWOC), and illustrates that the amount of expansion increases with increasing temperature.

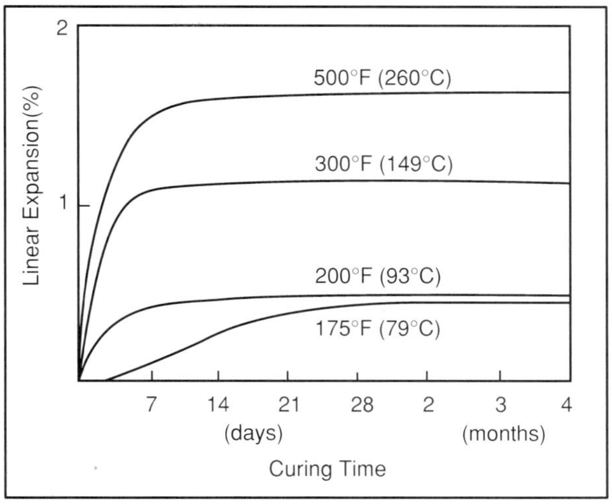

Figure 7–6—Expansion of cement containing 1% calcined MgO (BWOC).

7–4 FREEZE-PROTECTED CEMENTS

Permafrost zones in Alaska and northern Canada present some unique cementing difficulties. Permafrost is defined as any permanently frozen subsurface formation. The depths of such formations vary from a few feet to 2,000 ft (600 m). Below the permafrost, the geothermal gradients are normal. Permafrost sections vary from unconsolidated sands and gravels with ice lenses to ice-free, consolidated rock.

When permafrost exists, thawing of the formation must be avoided during drilling and completion. Melting can cause the thawed earth to subside, particularly in the upper 200 ft (60 m) of the well (Thorvaldson, 1962). The cement system should have a low heat of hydration, and be able to develop sufficient compressive strength (without freezing) at temperatures as low as 20°F (–3°C). Casing strings must be cemented to surface, or a non-freezing fluid placed in the annulus, to prevent casing damage because of the expansion of water upon freezing.

Conventional Portland cement systems are not satisfactory in permafrost conditions, because they freeze before developing sufficient compressive strength. It is possible to add salt, alcohol or other freeze-depressing materials to the mix water; however, this has been shown to have adverse effects upon the quality of the set cement (Morris, 1970). Two types of cement systems have been shown to perform successfully in this severe environment: (1) calcium aluminate cement, and (2) gypsum/ Portland cement blends (Benge et al., 1982).

As described in Chapter 9, calcium aluminate cement is a special-use material of limited production, and is used to cement in-situ combustion thermal wells. Such cements also set and gain strength rapidly at low and near-freezing temperatures (Maier et al., 1971). Fly ash is often added as a diluent to reduce the cement's heat of hydration, and for economy. The typical performance of 50:50 fly ash:calcium aluminate cement systems is shown in Table 7–3.

0.53 ft³ (3.96 gal) water per 74 lb of blend						
Sodium Chloride* (%)	Slurry Weight (lb/gal)	Slurry Volume (ft³)	Curing Temp. (°F)	Curing Time (hr) 8	16	24
0	14.8	0.95	20	—	—	355
5	14.9	0.97	20	—	—	310
10	15.0	0.96	20	—	—	145
0	14.8	0.95	25	20	45	90
5	14.9	0.97	25	NS**	210	495
10	15.0	0.96	25	NS	NS	NS
0	14.8	0.95	40	50	658	1560
0	14.8	0.95	50	115	2250	2475
0	14.8	0.95	60	145	1800	2900

*Based on weight of mixing water.

**Not set.

Table 7–3—Performance of 50:50 calcium aluminate/fly ash cement systems.

Gypsum/Portland cement blends, with sodium chloride as a mix-water freezing depressant, are used extensively for permafrost cementing. The gypsum sets and gains strength rapidly at freezing temperatures, and protects the slower setting Portland cement from freezing. Such cement systems also have a lower heat of hydration than that of calcium aluminate cement; therefore, they are particularly applicable to unconsolidated permafrost formations. The typical performance of a 50:50 blend of gypsum and Portland cement, with 12% NaCl BWOW, is shown in Table 7–4. The effect of freeze/thaw cycling upon compressive strength is illustrated in Table 7–5. No degradation of strength is observed.

Temp (°F)	Compressive Strength (psi) 16 hr	24 hr	3 days	7 days
20	300	350	1100	2000
40	340	450	1350	1950
60	700	800	1400	2450
80	750	1000	1850	2650

Table 7–4—Typical compressive strength data for a 50:50 gypsum/Portland cement blend.

Day	°F	psi	Day	°F	psi
1	40	860	14	100	2750
2	40	970	15	50	3100
3	30	1250	16	15	3480
4	15	1450	17	50	2850
5	50	1790	18	100	2820
6	100	1990	19	160	2740
7	160	2100	20	160	2680
8	160	2270	21	160	2690
9	160	2360	22	160	2670
10	160	1980	23	160	3380
11	160	2520	24	160	2750
12	160	2420	25	160	2710
13	160	2460	26	100	3000

Table 7–5—Compressive strengths of a 50:50 gypsum/Portland cement blend after freeze/thaw cycling.

7–5 SALT CEMENT SYSTEMS

Cement systems which contain significant quantities of sodium chloride (NaCl) or potassium chloride (KCl) are commonly called "salt cements." Salt has been used extensively in well cements for three principal reasons.

- In certain areas, salt is present in the available mix water, e.g., offshore.
- Salt is a common and inexpensive chemical which, when used as an additive, can modify the behavior of the cement system.
- Addition of large quantities of salt has proved to be necessary when placing cements across massive salt formations or water-sensitive zones.

Although NaCl is most frequently used in salt cements, the use of KCl has been reported for the protection of particularly sensitive clay formations (O'Brien and Chenevert, 1973). The effects of KCl and NaCl upon the performance of cement slurries are essentially the same; however, according to Smith (1987), KCl imparts excessive slurry viscosity at high concentrations.

7–5.1 Salty Water as Mixing Fluid

In the absence of fresh water, salt brackish water or seawater is frequently used for mixing cement slurries. Such waters are advantageous because of their availability and economy.

Brackish waters from ponds, etc., vary significantly, and should be thoroughly tested in the laboratory prior to use on location. The most important species to monitor are Cl^-, SO_4^{2-}, Ca^{2+}, Mg^{2+}, and various organic compounds resulting from the decomposition of plant material. Such impurities have significant effects upon the

Components (mg/L)	Gulf of Mexico	Cook Inlet, Alaska	Trinidad, W.I.	Persian Gulf (Kharg Is.)	Gulf of Suez	Sable Island	Sea-rite Lake Prod.	North Sea	Standard Seawater ASTM D-1141
Chloride	19,000	16,600	19,900	23,000	22,300	18,900	19,952	17,970	19,359
Sulfate	2,500	2,000	2,580	3,100	3,100	2,260	2,738	2,810	2,702
Bicarbonate	127	140	78	171	134	140	144	181	142
Carbonate	12	0	27	24	11	—	—	—	—
Sodium and Potassium	10,654	9,319	11,170	13,044	12,499	10,690	11,276	10,270	11,155
Magnesium	1,300	1,080	1,300	1,500	1,570	1,199	1,326	1,270	1,297
Calcium	400	360	408	520	464	370	419	390	408
Total Dissolved Solids	33,993	29,499	35,283	41,359	40,078	33,559	35,600	32,890	35,169
pH	8.2	8.0	8.3	8.2	8.2	7.3	—	—	8.2
SG	1.026	2.023	1.027	1.031	1.030	1.022	—	—	1.025

Table 7–6—Seawater analyses.

performance of Portland cement systems, including gelation and/or overretardation (Kieffer and Rae, 1987). All laboratory cement-slurry design experiments should be performed with a sample of the location water.

Seawater is the basic mixing fluid for offshore cementing operations. Lyman and Fleming (1940) and McIlhenny and Zeitoun (1969) characterized seawaters from various locations and, as shown in Table 7–6, found them to be reasonably uniform. Smith and Calvert (1974) confirmed seawater to be suitable for preparing well cements, and stated that the performance is "predictable to a safe degree."

Comparative laboratory testing has identified the following effects of sea water upon the performance of Portland cement systems.

- Reduced Thickening Time (Table 7–7)
- Higher Fluid-Loss Rate
- Higher Early Compressive Strength at Low Temperatures (Table 7–7)
- Slight Dispersing Effect
- Higher Shear-Bond Strength
- Increased Tendency for Slurry Foaming During Mixing

	Thickening Time (hr:min) at 6000 ft	Compressive Strength (psi) at 100°F after 24 hr
Class A mixed with fresh water	2:32	1780
Class A mixed with seawater	2:05	2150

Table 7–7—Thickening time/compressive strength of cement mixed with seawater/fresh water.

As discussed in Chapter 3, the presence of salt depresses the ability of bentonite to extend a cement slurry. Thus, either prehydration of the bentonite or the use of attapulgite is necessary (Smith and Calvert, 1974).

7–5.2 Salt as a Cement Additive

Salt is an extremely versatile cement additive. Depending upon its concentration in the slurry, salt can behave as an accelerator or a retarder (Chapter 3). Salt is also used to disperse cement slurries (Chapter 3), induce cement expansion (Section 7–3.2), and prepare freeze-protected cements (Section 7–4). Marginally, salt can be used as a weighting agent (Slagle and Smith, 1963), and to increase the electrical conductivity of cement. For further details, the reader is referred to the indicated sections of this book.

7–5.3 Cementing Across Shale and Bentonitic Clay Formations

Approximately 87% of petroleum reservoirs contain clay minerals and silica fines (Hill, 1982). Therefore, any change in the original medium of these clays may induce destabilization, clay swelling or fines migration, resulting in formation damage. For this reason, freshwater cement slurries are not appropriate for primary cementing across certain shale or bentonitic clay formations. This problem was first identified when remedial cementing across such formations was found to be more successful if saline formation waters were used to mix the slurry (Slagle and Smith, 1963). In addition, laboratory studies have shown significant formation permeability reductions as a result of exposure to low-salinity fluids (Hewitt, 1963; Jones, 1964; Mungan, 1965).

Slagle and Smith (1963) tested the visual integrity of clay formations after immersion into cement slurries of varying salinity. The results showed salt-saturated cements to be most compatible with formations containing montmorillonite, illite, and chlorite. However, NaCl concentrations as low as 10% BWOW were often sufficient to prevent significant damage. Cunningham and Smith (1967) showed saline cement filtrate to reduce the cleavage of nonswelling shales, and restrict the swelling and migration of water-sensitive clays. Lewis et al. (1987) demonstrated improved bonding between salt cements and sensitive formations.

It is important to mention a paper by Beach (1982), showing that the cement slurry salinity must be chosen with care. Significant long-term deterioration was observed when the ionic concentration of the cement was not comparable to that of the formation. Disequilibrium causes ionic diffusion, and the Portland cement binder is apparently disrupted. In the same vein, Economides and Nolte (1987) recommended that cement slurries for sensitive formations should contain a minimum of salt (in equilibrium with the formation salinity), exhibit sufficient fluid-loss control (to minimize cement filtrate invasion), and not be overdispersed (to minimize invasion by a large amount of free water).

7–5.4 Cementing Across Massive Salt Formations

The presence of salt domes and massive evaporite sequences has long been problematic in terms of drilling, completion, and long-term production. The high water solubility and plasticity of such zones increase the difficulty of obtaining a successful primary cementation. The cement slurry can dissolve large quantities of formation material, resulting in a modification of performance (Ludwig, 1951). Plastic salt zones can also encroach upon the casing before the cement sets. Non-uniform formation movement exerts point-loading on the casing string, sometimes resulting in casing failure and collapse (Cheatham and McEver, 1964). Salt cements are used routinely to reduce the severity of these problems; however, some controversy exists regarding their efficacy.

The first recorded use of salt in well cements was during the 1940s, when wells were completed across salt domes in the U.S.A. Gulf Coast. Later, this became standard practice in the Williston basin (North Dakota and Montana), certain areas in the North Sea, etc. The concentration of NaCl usually varied from 18% to 37% BWOW. While such practices prevented the dissolution of the formation, the high salt concentrations were antagonistic to the performance of other cement additives, especially dispersants and fluid-loss additives (which were originally developed for fresh water systems). In addition, the high salt concentrations tended to over-retard the cement system; thus, formation encroachment and casing damage could occur before the cement set. Two approaches have been followed to solve these difficulties: eliminating salt from the cement system, and developing additives which are compatible with salt cements.

Salt-free cement (Goodwin and Phipps, 1982), or cements containing very low salt concentrations (3% BWOW) (Bryant, 1984), have been successfully applied in the Williston basin. No casing collapse was reported with such systems, compared to a 20% failure rate with salt-saturated cements. To prevent excessive dissolution of the formation, low displacement rates were recommended.

An intermediate approach was proposed by Ford et al., (1982). Semi-saturated cement systems (18% NaCl BWOW), in combination with holding the casing in tension, improved the success rate of primary cement jobs in the Williston basin.

The above approaches may improve initial results; however, considering the previously discussed long-term effects of ionic disequilibrium, cement failure may ultimately occur. The rate of ionic diffusion would be determined by the difference in salt concentration between the cement and formation, and the permeability of the cement (Kumar et al., 1985).

Experiments performed by Drecq (1987) illustrated that low displacement rates would not necessarily prevent significant formation dissolution. Three NaCl blocks of equal dimensions were submerged for 60 min in cement slurries with various salt concentrations. The temperature was 140°F (60°C), and slight agitation was provided. As shown in Fig. 7–7, significant salt erosion was observed, except when the cement was salt-saturated.

In addition, Rae and Brown (1988) revealed that contamination of a fresh water cement system by as little as 10% salt can alter the thickening time by 30%, increase the slurry viscosity by 100%, and increase the fluid-loss

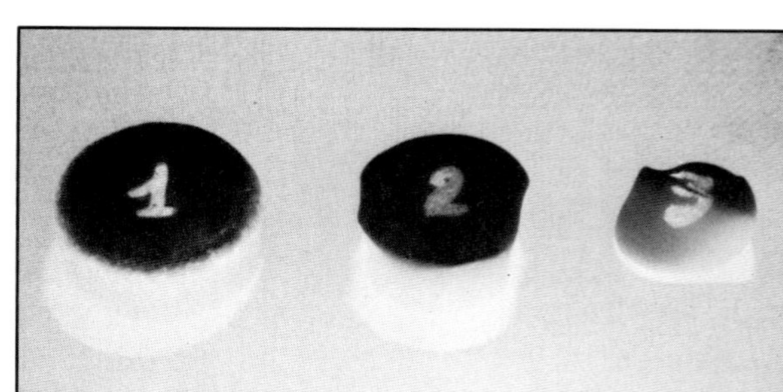

Figure 7–7—Salt block appearance after 60 minutes at 140°F in cement slurries of different salinities (after Drecq, 1987).

rate by nearly 500%. Yearwood, et al. (1988) confirmed these findings.

Since the late 1970s, research has been performed to develop salt-saturated cement systems (37.2% NaCl BWOW) without the disadvantages discussed earlier. Such systems could be relied upon to maintain formation integrity, and develop strength with sufficient speed to prevent casing collapse. In 1978, Messenger patented the use of certain hydroxycarboxylic acids as dispersants for salt cement slurries. Fluid-loss additives for salt cement systems were invented by Chatterji and Brake (1981) and Nelson (1986) (Chapter 3). Such additives improved the placement characteristics of saturated salt slurries, but the problem of overretardation and delayed compressive strength development remained to be solved.

In 1988, cement systems containing up to 30% NaCl (BWOW), with excellent placement characteristics but, more importantly, appropriate thickening times and compressive strength development, were reported by Rae and Brown (1988), Yearwood et al. (1988), and Whisonant, et al. (1988) Typical performance data are presented in Tables 7–8 and 7–9. Successful field results have been reported in various locations around the world. As of this writing, the system compositions are proprietary.

NaCl (% BWOW)	Cellulose/ Organic Acid (%BWOC)	BHST (°F)	Density (lb/gal)	Compressive Strength at 3000 psi 8 hr	24 hr
5	—	200	15.8	1850	2600
15	—	200	15.8	1700	2300
30	—	200	16.6	2020	2650
30	0.8/0.1	200	16.6	0	2500
30	—	200	16.6	1350	2100
30	0.8/0.1	200	16.2	0	2040
30	—	230	16.2	1450	2150

Table 7–8—Typical compressive strength performance of proprietary salt cement systems.

NaCl (% BWOW)	Cellulose Organic Acid (% BWOC)	BHCT (°F)	Fluid Loss (mL/30 min)	Rheological Properties at BCHT T_y (lbf/100ft²)	μ_p (cp)
30	0.8/0.1	120	178	22	244
30	—	120	64	0.2	33.1
30	0.8/0.1	160	285	31.3	105
30	—	160	52	0.1	17.5

Table 7–9—Rheology and fluid-loss performance of proprietary salt cement systems.

7–6 LATEX-MODIFIED CEMENT SYSTEMS

Latex is a general term describing an emulsion polymer. The material is usually supplied as a milky suspension of very small spherical polymer particles (200 to 500 nm in diameter), often stabilized by surfactants to improve freeze/thaw resistance and prevent coagulation when added to Portland cement. Most latex dispersions contain about 50% solids. A wide variety of monomers, including vinyl acetate, vinyl chloride, acrylics, acrylonitrile, ethylene, styrene, butadiene, etc., is emulsion polymerized to prepare commercial latices.

The first use of latices in Portland cements occurred in the 1920s, when natural rubber latex was added to mortars and concretes. Since then, latex-modified concretes have become commonplace because of the following improvements in performance (Ohama, 1987).

- Improved Workability
- Decreased Permeability
- Increased Tensile Strength
- Reduced Shrinkage
- Increased Elasticity
- Improved Bonding Between Cement/Steel and Cement/Cement Interfaces

As discussed in Chapter 2, an absolute volume shrinkage is observed as a result of Portland cement hydration. Upon setting, stresses are created within the cement matrix resulting in the formation of microcracks (Fig. 7–8).

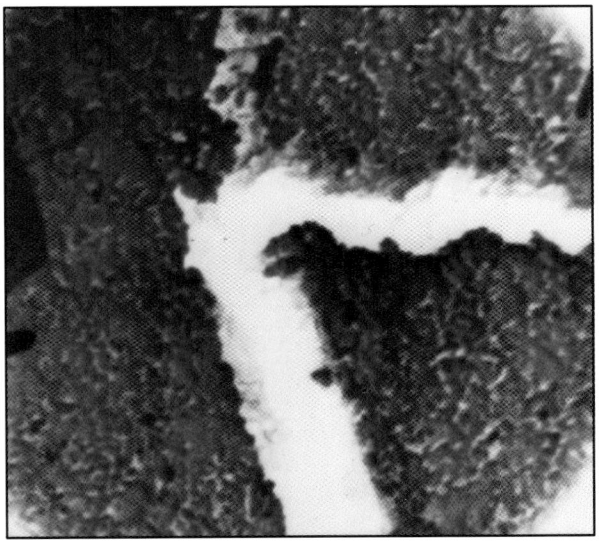

Figure 7–8—Photograph of microcracks in set Portland cement (after Kuhlmann, 1985).

The propagation of the cracks lowers the tensile capacity of the set cement and increases its permeability. In latex-modified systems (Fig. 7–9), the latex particles coalesce to form a plastic film which surrounds and coats the C-S-H gel. Because of its elasticity and high bonding strength, the latex bridges the microcracks, and restrains their propagation; as a result, the tensile strength of the set cement increases and the permeability decreases.

Figure 7–9—Photograph of latex-modified Portland cement, 1200X (after Kuhlmann, 1985).

7–6.1 Behavior of Latices in Well Cement Slurries

The use of latices in well cements occurred much later. In 1957, Rollins and Davidson reported improved performance when latex was added to the mix water. In addition to the attributes mentioned above, the following additional benefits were cited:

- better bonding to oil-wet and water-wet surfaces,
- less shattering when perforated,
- increased resistance to contamination by well fluids,
- lowered fluid-loss rate, and
- improved durability.

When latex is added as part of the liquid phase of a Portland cement system, a slurry of normal color and consistency is obtained; however, because of the solids content of the latex, such slurries contain 20% to 35% less water. After curing, the set product consists of hydrated cement connected by a "film" of latex particles (Kuhlmann, 1985). It is this film of latex particles which imparts the physical and chemical properties described above (Parcevaux and Sault, 1984; Drecq and Parcevaux, 1988). While the slurry is still liquid, the latex particles impart excellent rheological properties because of a lubricating action. In addition, the latex particles provide excellent fluid-loss control by physically plugging small pores in the cement filter cake (Drecq and Parcevaux, 1988) (Chapter 3).

7–6.2 Early Latex-Modified Well Cement Systems

In 1958, Eberhard and Park patented the use of vinylidene chloride latex in well cements. Improved performance was claimed for systems containing up to 35% latex solids BWOC. Later, polyvinyl acetate latex was identified as a suitable material (Woodard and Merkle, 1962). The preferred concentration of latex solids varied from 2.5% to 25% BWOC. The polyvinyl acetate system has been used successfully for many years; however, it is limited to applications at temperatures less than 122°F (50°C).

7–6.3 Styrene-Butadiene Latex Systems

An improvement in latex cement technology occurred when Parcevaux et al. (1985) identified styrene-butadiene latex as an effective additive for the prevention of annular gas migration (Chapter 8). Additional refinements have been made by Sault et al. (1986).

Styrene-butadiene latices impart the same beneficial effects described above; however, they are effective at temperatures as high as 350°F (176°C). Fig. 7–10 is a plot of fluid-loss rate versus latex concentration for various well cement slurries. The results illustrate that nor-

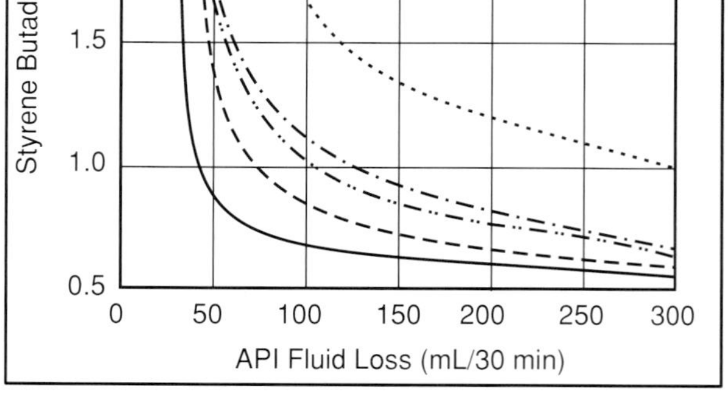

Figure 7–10—API fluid loss of latex-modified slurries (185°F, 85°C).

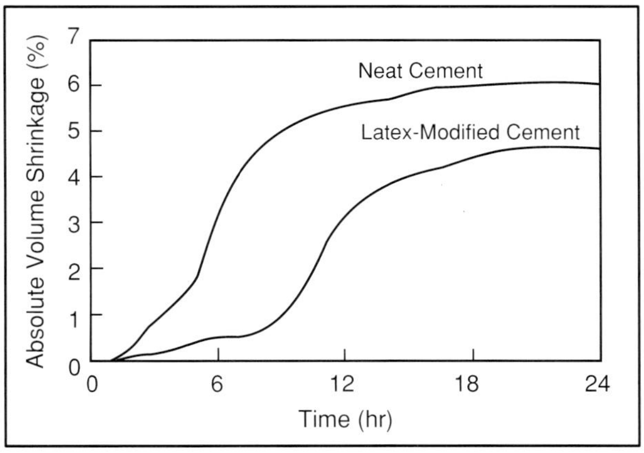

Figure 7–11—Absolute volume shrinkage of normal density Portland cements (from Parcevaux and Sault, 1984).

mal-density neat slurries require less latex to achieve a given fluid-loss rate. More latex is required for slurries containing extenders or weighting agents, especially those with a lower solids content (extended with sodium silicate). Figure 7–11 illustrates the decreased volumetric shrinkage observed with a latex-modified Portland cement system cured at 158°F (70°C).

7–7 CEMENTS FOR CORROSIVE ENVIRONMENTS

Set Portland cement is a remarkably durable and forgiving material; however, there are limits beyond which it will rebel. In a wellbore environment, Portland cement is subject to chemical attack by certain formations and by substances injected from the surface. As discussed in Chapter 9, saline geothermal brines containing CO_2 are particularly deleterious to the integrity of the set cement. In addition to geothermal well cementing, one must also pay close attention to cement durability in wells for chemical waste disposal and for enhanced oil recovery by CO_2-flooding.

7–7.1 Cements for Chemical Waste Disposal Wells

Zonal isolation is of paramount importance in a chemical waste disposal well. If not properly confined, injected waste fluids could contaminate fresh water strata and corrode the exterior of the casing. To ensure the maintenance of zonal isolation throughout the life of such wells, the cement and the tubular hardware in the well must be chemically resistant to the waste fluids (Runyan, 1974).

The chemically resistant casings used in waste disposal wells include modified polyester and epoxy fibercast, or metal alloys such as Carpenter 20, Incoly 825, and Hastalloy G. The cement systems are chosen depending upon the nature of the injected waste material.

Modified Portland cements are generally appropriate for disposal wells involving weak organic acids, sewage waters or solutions having a pH of 6 or above (Ostroot and Ramos, 1971). The durability of the set cement is improved by adding pozzolans, increasing the density by addition of dispersants, or adding liquid latices to the slurry. These methods substantially reduce the permeability of the set cement.

Portland cement systems are not compatible with strong inorganic acids such as sulfuric, hydrochloric, and nitric. In such environments, organic polymer cements, usually epoxy-base, must be used to provide sufficient chemical resistance (Cole, 1979). Such systems are also known as "synthetic cements."

Epoxy cements are prepared by mixing an epoxy resin such as bisphenol A (Fig. 7–12) with a hardening agent. Depending upon the desired end properties, the hardening agent can be an anhydride, aliphatic amine or polyamide (Sherman et al., 1980). A solid filler such as silica flour is often used to build density, and to act as a heat sink for the exotherm which occurs during the cure. Depending upon the circulating and static well temperatures, various catalysts and accelerators can also be added to control the placement and setting times.

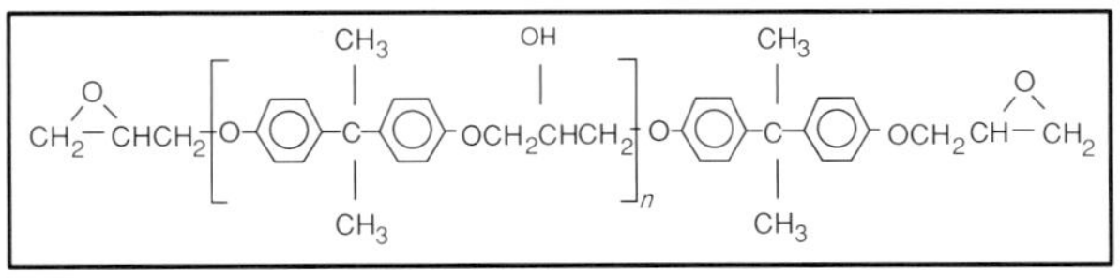

Figure 7–12—Chemical structure of bisphenol-A.

Epoxy resin cement systems are characterized by their corrosion resistance, and high compressive and shear bond strength. They are compatible with strong acids and bases (up to 37% HCl, 60% H_2SO_4, and 50% NaOH) at temperatures up to 200°F (93°C) during extended exposure periods. Epoxies are also resistant to hydrocarbons and alcohols, but not to chlorinated organics or acetone. Typically, the compressive strengths range between 8,000 to 10,000 psi (56 to 70 MPa), and shear bond strengths can be as much as nine times higher than those of Portland cement (Bruckdorfer, 1985).

Non-aqueous spacers are required on all epoxy cement jobs. Gelled oil, diesel or alcohol systems remove mud and water from the pipe and formation, as well as oil-wet all bonding surfaces.

7–7.2 Cements for Enhanced Oil Recovery by CO_2-Flooding

Carbon dioxide EOR has seen a surge of activity in the last several years. Most of these projects are located in

Texas and the Gulf Coast region. Corrosion owing to CO_2 in production operations is well documented (Newton and Hausler, 1984), and studies of Portland-base well cement corrosion by CO_2 have been conducted by Onan (1984) and Bruckdorfer (1986). It is well known that carbon dioxide-laden waters can destroy the structural integrity of set Portland cements (Biczok, 1967). The basic chemistry describing this process is as follows.

$$CO_2 + H_2O \rightleftharpoons H_2CO_3 \rightleftharpoons H^+ + HCO_3^- \quad (7\text{–}5)$$

$$Ca(OH)_2^+ + H^+ + HCO_3^- \rightarrow CaCO_3 + 2H_2O \quad (7\text{–}6)$$

$$\text{C-S-H gel} + H^+ + HCO_3^- \rightarrow CaCO_3 + \text{amorphous silica gel} \quad (7\text{–}7)$$

In Eq. 7–5, approximately 1% of the dissolved carbon dioxide reacts with water to form carbonic acid. As the carbon dioxide-laden water diffuses into the cement matrix, the dissociated acid is free to react with the free calcium hydroxide (Eq. 7–6) and the C-S-H gel (Eq. 7–7). As carbon-dioxide-laden water continues to invade the matrix, other equilibria are established.

$$CO_2 + H_2O + CaCO_3 \rightleftharpoons Ca(HCO_3)_2 \quad (7\text{–}8)$$

$$Ca(HCO_3)_2 + Ca(OH)_2 \rightleftharpoons 2CaCO_3 + H_2O \quad (7\text{–}9)$$

In the presence of excess carbon dioxide (Eq. 7–8), calcium carbonate is converted to water-soluble calcium bicarbonate, which can migrate out of the cement matrix. In Eq. 7–9, the dissolved calcium bicarbonate can react with calcium hydroxide, forming calcium carbonate and "fresh water." The liberated water can then dissolve more calcium bicarbonate. The net result is a leaching of cementitious material from the cement matrix, an increase of porosity and permeability, and a decrease of compressive strength. Downhole, this translates to a loss of casing protection and zonal isolation.

Carbon dioxide corrosion of Portland cements is thermodynamically favored, and cannot be prevented. An easy solution to this problem would be synthetic cement; unfortunately, such systems are not economically feasible for most CO_2-flooding projects. Instead, measures are taken to lower the degradation rate of Portland cement systems.

The cement matrix permeability can be reduced by lowering the water-to-cement ratio and/or adding pozzolanic materials. As discussed in Chapter 3, pumpable Portland cement slurries with densities up to 18.0 lb/gal (2.16 g/cm^3) can be prepared with the addition of a dispersant. After setting, the water permeability of such systems is usually less than 0.001 md; consequently, invasion of carbon-dioxide-laden water is inhibited, and the rate of corrosion is slowed. The addition of pozzolans (such as fly ashes) also results in a permeability reduction (Chapter 3), and effectively eliminates Eq. 7–6 above. When such measures are taken, the rate of corrosion can be reduced by as much as 50%.

The long-term efficacy of the modified Portland cement systems in CO_2-flood wells remains to be seen. At best, such systems only postpone the inevitable. More research is needed to develop truly stable, yet economically realistic, cements for this difficult environment.

7–8 CEMENTITIOUS DRILLING FLUIDS

Many well completion problems such as lost circulation, excessive fluid loss, and annular fluid migration could be prevented, if the drilling fluid were cementitious. Indeed, good zonal isolation could be easily achieved, because mud removal by an incompatible cement slurry would no longer be a concern. A few techniques have been developed; however, the practice is not yet widespread.

In 1971, Harrison and Goodwin developed a bentonite-extended Portland cement system which, when retarded by D-gluco-D-glucoheptolactone, could be used indefinitely as a drilling fluid. Upon completion of drilling, a polyvalent metal salt such as $CaCl_2$ was added to the fluid, and the setting process was activated. Other techniques have involved radiation-activated polymer mud systems (Novak, 1985), and heat-activated, cement-base muds (Tsao and Binder, 1985).

REFERENCES

Beach, H. J.: "Consequences of Salting Well Cements," paper SPE 10032, 1982.

Beirute, R. and Tragresser, A.: "Expansive and Shrinkage Characteristics of Cements Under Actual Well Conditions," *JPT* (Aug. 1973) 905–909.

Benge, O. G., Jones, R. R., Dresher, T. D., and Dolan, R. T.: "A New Low-Cost Permafrost Cementing System," paper SPE 10757, 1982.

Biczok, I.: *Concrete Corrosion—Concrete Protection*, Chemical Publishing Co., Inc., New York (1967) 287–298.

Bour, D. L. Daugherty, D., and Sutton, D. L.: "New Expansive Cement System for High Temperature," *Proc.* Southwestern Petroleum Short Course, Lubbock, TX (1988).

Bruckdorfer, R. A.: "Carbon Dioxide Corrosion in Oilwell Cements," paper SPE 15176, 1986.

Bruckdorfer, R. A.: Unpublished Data, 1985.

Bryant, G. A.: "Successful Alternatives to Conventional Cement Designs in the Williston Basin," paper SPE 12904, 1984.

Carbon Dioxide Corrosion in Oil and Gas Production, Selected Papers, Abstracts, and References, L. E. Newton, Jr. and R. H. Hausler. (eds.), National Association of Corrosion Engineers, Texas (1984).

Carter, L. G., Waggoner, H. F., and George, C. R.: "Expanding Cements for Primary Cementing," *JPT* (May, 1966) 551-58.

Chatterji, J. and Brake, B. G.: "Water-Loss Reducing Additives for Salt Water Cement Slurries," British Patent No. GB 2,080,812 (1982).

Cheatham, J. B. and McEver, J. W.: "Behavior of Casing Subjected to Salt Loading," paper SPE 828, 1964.

Childs, J., Sabins, F., and Taylor, M. J.: "Method of Using Thixotropic Cements for Combating Lost Circulation Problems," U. S. Patent No. 4,515,216 (1985).

Clement, C.C.: "A Scientific Approach to the Use of Thixotropic Cement," *JPT* (March 1979) 344–346.

Cole, R. C.: "Epoxy Sealant for Combating Well Corrosion," paper SPE 7874, 1979.

Cunningham, W. C. and Smith, D. K.: "Effect of Salt Cement Filtrate on Subsurface Formations," paper SPE 1920, 1967.

Drecq, P.: Unpublished Data, 1987.

Drecq, P. and Parcevaux, P. A.: "A Single Technique Solves Gas Migration Problems Across a Wide Range of Conditions," paper SPE 17629, 1988.

Eberhard, J. F. and Park, A.: "Portland Cement—Vinylidene Chloride Polymer Composition, Method of Making, and Method of Using," U. S. Patent No. 2,819,239 (1958).

Ford, R. E., Turcich, T. A., Pierson, R. A., Divan, D. J., and Ramsey, L. K.: "Obtaining Quality Primary Cement Jobs in the Williston Basin," paper SPE 10874, 1982.

Goodwin, K. J. and Phipps, K.: "Salt-Free Cement—An Alternative to Collapsed Casing in Plastic Salts," paper SPE 10885, 1982.

Harrison, H. T. and Goodwin, K. J.: "Method of Drilling and Cementing a Well Using an Aqueous Hydraulic Cement Slurry," U. S. Patent No. 3,605,898 (1971).

Hewitt, C. H.: "Analytical Techniques for Recognizing Water-Sensitive Reservoir Rocks," *JPT* (Aug. 1963) 813–818.

Hill, D. G.: "Clay Stabilization—Criteria for Best Performance," paper SPE 10656, 1982.

Jones, F. O.: "Influence of Chemical Composition of Water on Clay Blocking of Permeability," *JPT* (April 1964) 441–446.

Kalousek, G. L.: *Development of Expansive Cements,* Klein Symposium on Expansive Cement Concretes, American Concrete Institute Publication SP–38, (1973).

Kieffer, J. and Rae, P.: "How Gelation Affects Oil Well Cements," *Pet. Eng. Intl.* (May 1987) 59, 46–48.

Klein, A. and Troxell, G.E.: "Studies of Calcium Sulfoaluminate Admixtures for Expansive Cements," *Proc.*, ASTM (1958) **58**, 986–1008.

Kuhlmann, L. A.: "Latex-Modified Concrete for the Repair and Rehabilitation of Bridges," *Intl. J. of Cement Composites and Lightweight Concrete* (1985) **7**, No. 4, 241–247.

Kumar, A., Komarneni, S., and Roy, D. M.: "Diffusion of Ca^{2+} and Cl^- Through Sealing Materials," *Cement & Concrete Res.* (1985) **5**, 110–114.

Lea, F. M.: *The Chemistry of Cement and Concrete*, Chemical Publishing Co. Inc., New York, 1971.

Lewis, W. J., and Rang, C. L.: "Salt Cements for Improved Hydraulic Isolation and Reduced Gas Channeling," paper SPE 16386, 1987.

Ludwig, N. C.: "Effects of Sodium Chloride on Setting Properties of Oil Well Cements," *Drill. & Prod. Prac.*, API (1951) 20–27.

Lyman, J. and Fleming, R. H.: "Composition of Sea Water," *J. Marine Res.* (1940) No. 3, 134–136.

Maier, L. F., Carter, M. A., Cunningham, W. C., and Bosley, T. G.: "Cementing Materials for Cold Environments," *JPT* (Oct. 1971) 1215–1220.

McIlhenny, W. F. and Zeitoun, M. A.: "A Chemical Engineer's Guide to Seawater," *Chem. Eng.* (1969) No. 24, 81–86; No. 25, 251–256.

Messenger, J. U.: "Cementing Against Evaporites," U. S. Patent No. 4, 089, 376 (1978).

Messenger, J. U.: "Treating Wells to Mitigate Flow-After-Cementing," U. S. Patent No. 4,235,291 (1980).

Morris, E. F.: "Evaluation of Cement Systems for Permafrost," paper SPE 2824, 1970.

Mungan, N.: "Permeability Reduction Through Change in pH and Salinity," *JPT* (Dec. 1965) 1449–1453.

Nelson, E. B.: "Pumpable Thixotropic Cement Slurries For Use in Cementing Pipes in a Well," U. S. Patent No. 4,415,367 (1983).

Nelson, E. B.: "Sulfonated Poly (Vinyl Aromatics) As Fluid-Loss Additives for Salt Cement Slurries," U. S. Patent No. 4,601,758 (1986).

Novak, L. H.: "Drilling Mud Composition Which May Be Converted to Cement Upon Irradiation," U. S. Patent No. 4, 547, 298 (1985).

O'Brien, D. E. and Chenevert, M. E.: "Stabilizing Sensitive Shales with Inhibited Potassium-Based Drilling Fluids," *JPT* (Sept. 1973) 1089–1100.

Ohama, Y.: "Principle of Latex Modification and Some Typical Properties of Latex-Modified Mortars and Concretes," *ACI Materials J.* (Nov.–Dec. 1987) 511–518.

Onan, D. D.: "Effects of Supercritical Carbon Dioxide on Well Cements," paper SPE 12593, 1984.

Ostroot, G. W. and Ramos, J.: "Deep-Well Acid Disposal—Planning and Completion," Underground Waste Management Symposium (Dec. 1971).

Parcevaux, P. A. and Sault, P. H.: "Cement Shrinkage and Elasticity: A New Approach for a Good Zonal Isolation," paper SPE 13176, 1984.

Parcevaux, P. A., Piot, B. M., and Vercaemer, C. J.: "Cement Compositions for Cementing Wells, Allowing Pressure Gas-Channeling in the Cemented Annulus to Be Controlled," U. S. Patent No. 4,537,918 (1985).

Parker, P. N. and Wahl, W. W.: "Expanding Cement—A New Development in Well Cementing," *JPT* (May 1966) 359–364.

Rae, P. and Brown, E.: "New Material Improves the Cementation of Salt Formations," *Proc.,* Southwest Petroleum Short Course, Lubbock, TX (1988) 38–48.

Reservoir Stimulation, M. J. Economides and K. G. Nolte (eds.), Schlumberger Educational Services, Houston, 1987.

Rollins, J. T. and Davidson, R. D.: "New Latex Cement Solves Special Well Problems," *Pet. Eng.* (Feb. 1957) **29**, No. 2, B48–51.

Root, R. L. and Calvert, D. G.: "The Real Story of Cement Expansion," paper SPE 3346, 1971.

Runyan, E. E.: "Cementing of Well Casings for Pollution Control," paper SPE 4812, 1974.

Sault, P. H., Parcevaux, P. A., and Piot, B. M.: "Cement Composition for Cementing Wells Enabling Gas Channeling in the Cemented Annulus to be Inhibited by Right-Angle Setting," European Patent No. 0,189,950 (1986).

Shaw, D. J.: *Introduction to Colloid and Surface Chemistry,* Butterworth & Co. Ltd., London (1970).

Sherman, S., Gannon, J., Buchi, G., and Howell, W. R.: "Epoxy Resins," *Kirk-Othmer Encyclopedia of Chemical Technology,* John Wiley and Sons, New York, (1980) **9**, 267–290.

Slagle, K. A. and Smith, D. K.: "Salt Cement for Shale and Bentonitic Sands," *JPT* (Feb. 1963) 187–194.

Smith, D. K.: *Cementing,* SPE, Dallas (1987) **4**.

Smith, R. C. and Calvert, D. G.: "The Use of Sea Water in Well Cementing," paper SPE 5030, 1974.

Spangle, L. B. and Calvert, D. G.: "Improved Primary and Remedial Cementing With Thixotropic Cement Systems," paper SPE 3833, 1972.

Spangle, L. B.: "Expandable Cement Composition," European Patent No. 254,342, (1988).

Thorvaldson, W. M.: "Low Temperature Cementing," paper presented at the 1972 CIM Annual Meeting, Calgary.

Tsao, Y. H., and Binder, G. G. Jr.: "Method of Drilling and Cementing a Well Using a Drilling Fluid Convertible in Place into a Settable Cement Slurry," U. S. Patent No. 4,519,452 (1985).

Warembourg, P. A., Kirksey, J. M., and Bannister, C. E.: "Improving Cement Bond in the Rocky Mountain Area by the Use of Spacer, Wash and Thixotropic Cement," paper SPE 9031, 1980.

Whisonant, B. J., Rae, P., and Ramsey, L. K.: "New Materials Improve the Cementation of Salt Formations in the Williston Basin," paper SPE 17512, 1988.

Wieland, D. R., Calvert, D. G., and Spangle, L. B.: "Design of Special Cement Systems For Areas With Low Fracture Gradients," paper SPE 2556, 1969.

Woodard, G. W. and Merkle, G. H.: "Composition of Hydraulic Cement and Polyvinyl Acetate and Use Thereof," U. S. Patent No. 3,0158,520 (1962).

Yearwood, J., Drecq, P., and Rae, P.: "Cementing Across Massive Salt Formations," paper Petroleum Society of CIM 88–39–104, 1988.

8

Prevention of Annular Gas Migration

Philippe Parcevaux, Phil Rae, and Philippe Drecq

Dowell Schlumberger

8–1 DEFINITION AND TERMINOLOGY

Annular fluid migration may occur during drilling or well completion procedures, and has long been recognized as one of the most troublesome problems of the petroleum industry. It consists of the invasion of formation fluids into the annulus, because of a pressure imbalance at the formation face. The fluids may migrate to a lower pressure zone, or possibly to the surface (Fig. 8–1). Within this category of problems, gas migration is the most frequent, and no doubt the most critical and dangerous (Bearden et al., 1964; Carter and Slagle, 1970; Sutton and Faul, 1984).

Gas migration—also called gas communication or gas leakage (Carter and Slagle, 1970), annular gas flow (Garcia and Clark, 1976), gas channeling (Parcevaux et al., 1983), flow after cementing (Webster and Eikerts, 1979), or gas invasion (Bannister et al., 1983)—is a potential problem on almost any gas-bearing or gas storage well. However, the severity of the problem ranges from the most hazardous, e.g., the blowout situation when well control is lost because of a severe pressure imbalance during drilling or cementing, to the most marginal, e.g., a residual gas pressure of a few psi at the wellhead. In addition, less easily detected downhole interzonal communication can occur.

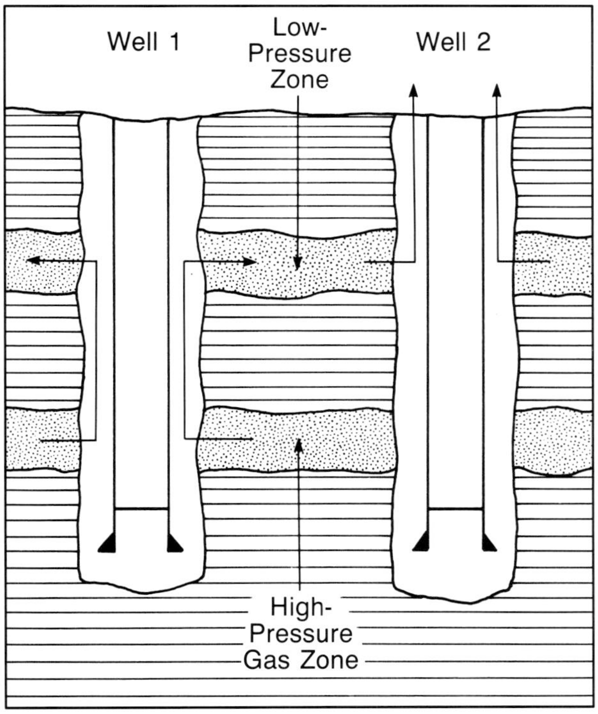

Figure 8–1—Two scenarios of annular gas migration.

The investigation of well control during drilling, which is well described in the drilling literature (Moore, 1974), is beyond the scope of this chapter, which concentrates on the problem of gas migration after primary cementing. However, the specificity of gas migration during cementing vs that which can occur during drilling is outlined.

8–2 PRACTICAL CONSEQUENCES OF GAS MIGRATION

The potential consequences of gas migration following primary cementing are numerous, but not always immediately detectable. At the extreme, those that manifest themselves at the surface, e.g., gas pressure or gas flow at the wellhead, may lead to well abandonment. More frequently, remedial cementing is performed until gas flow is shut down, and gas pressure is reduced to a level compatible with the operator's safety policy and local regulations. However, the efficiency of squeeze cementing in such circumstances is very poor for three essential reasons: (1) gas channels are difficult to locate, especially if they are submillimetric; (2) gas channels may be too small to be fillable by cement; and (3) the pressure exerted during the squeeze job is sometimes sufficient to break downhole cement bonds, or even to initiate formation fracturing, worsening downhole communication problems. A thorough discussion of remedial cementing appears in Chapter 13. Furthermore, cement repair operations are expensive, especially in high-cost operation areas (Cooke et al., 1982). Therefore, preventing the gas migration problem is definitely preferable to repairing it.

Interzonal gas migration, with no surface manifestations, is very difficult to detect (Fig. 8–1). In such cases, the subsequent production of gas may be impaired, undesired refilling of an upper depleted zone may occur (possibly followed by gas migration to the surface on another well), or the efficiency of stimulation treatments may be reduced (Cooke et al., 1982). Such downhole channeling can be evaluated by special methods such as noise logs (Garcia and Clark, 1976) or acoustic logs (Catala et al., 1984; Rang, 1987). Hydraulic communication testing is not recommended. If such potentially destructive testing is not properly designed and controlled, it may induce communication across properly cemented zones, or aggravate minor defects of the cement job. Interpretation of the cement job in gas wells is discussed in greater detail in Chapter 16.

8–3 PHYSICAL PROCESS OF GAS MIGRATION

Gas migration is a complex problem involving fluid density control, mud removal, cement slurry properties, cement hydration, and cement/casing/formation bonding. Since the problem was recognized in the early 1960s, when a major gas communication problem occurred in gas storage wells in the U.S.A. (Stone and Christian, 1974), considerable effort has been exerted to find solutions.

Extensive research has been performed to understand the fundamental components of the physical process. As a result, a vast quantity of literature has appeared, which includes the analysis of field case studies or field experiments for making practical recommendations (Vidovskii et al., 1971; Stone and Christian, 1974; Garcia and Clark, 1976; Cooke et al., 1982; Lukkien, 1982), laboratory physical investigations for understanding the fundamentals of the problem (Guyvoronsky and Farukshin, 1963; Bulatov et al., 1970; Carter and Slagle, 1970; Carter et al., 1973; Webster and Eikerts, 1979; Sabins et al., 1982; Bannister et al., 1983; Parcevaux, 1984), the development of technical "solutions" to the problem (Levine et al., 1979; Tinsley et al., 1979; Cheung and Beirute, 1982; Parcevaux et al., 1983; Stewart and Schouten, 1986; Sykes and Logan, 1987), the application in the field of new products and techniques (Kucyn et al., 1977; Watters and Sabins, 1980; Cheung and Myrick, 1983; Seidel and Greene, 1985; Sepos and Cart, 1985; Matthews and Copeland, 1986), and the establishment of empirical qualitative prediction techniques (Sutton et al., 1984; Rae et al., 1989). Surprisingly, successful numerical simulations of the process, or scaled laboratory experiments that could allow a generalized and quantitative prediction of gas migration, have not been reported.

The difficulty in understanding and modeling the gas migration phenomenon arises from the fact that the material through which the gas can channel, i.e., an annular column full of cement slurry (with possibly some spacer and drilling fluid left in the hole), evolves with time. The physical state of the slurry progresses from liquid immediately after placement, to gel after some time left static, to permeable weak solid when setting, and finally to impermeable solid after hardening. It is thus convenient, when reviewing the physical process of gas migration from a phenomenological viewpoint, to detail each of the above stages with respect to gas intrusion in the cemented annulus.

8–3.1 Mud Removal

When the gas migration problem was first recognized, it was perceived to be principally a matter of poor mud removal and/or poor bonding at the casing/cement/formation interfaces (Carter and Evans, 1964; Carter and Slagle, 1970). Although other important causes have since been discovered, proper mud removal still remains a prerequisite for controlling annular fluid migration. Regardless of the quality of the cement formulation itself, continuous mud channels in the annulus between two permeable zones will favor annular flow.

For detailed information on mud displacement mechanics and guidelines, the reader is referred to Chapter 5. Proper mud removal techniques to minimize gas leakage were outlined as early as 1973 by Carter et al. They are related to the following:

- Mud conditioning,
- Casing centralization,
- Casing movement, namely rotation or reciprocation, during mud circulation and possibly during cement placement,
- Choice of proper preflushes and spacers, in terms of compatibility with mud and cement, density, rheology, fluid-loss control, and solids control,
- Choice of proper fluid volumes (contact times), and
- Determination, by a computer simulation, of adequate flow rates according to downhole conditions, with preference to high rates and turbulent flow.

8–3.2 Density Control

Gas control during and immediately after cement placement is very similar to well control during drilling. For this reason, one of the first approaches to the problem was simply to increase fluid densities. However, such an approach is limited by the dangers of losing circulation or fracturing an interval if fluid densities are too high. In 1970, Carter and Slagle recommended circulation of the well prior to cementing to help remove any trapped gas bubbles which, if not removed prior to cement placement, would lower the hydrostatic head of the fluid column.

The principal difference between well control during drilling and that of cementing is the free-fall or U-tubing phenomenon that occurs during the cement job. Because of the density differences between the mud, preflushes, spacer, and cement slurry/slurries, the hydrostatic pressure exerted at the formation face is not constant during the job (Beirute, 1984; Smith et al., 1985). If the hydrostatic pressure falls below the formation gas pressure at any time, a "gas kick" could be induced which, by further relieving the hydrostatic pressure, may lead to an irreversible gas entry process. Consequently, the cement job design should be performed with a computerized free-fall simulator, to assure that the pressure at critical zones is maintained between the pore and the fracturing pressure at all times during, and immediately after, the cement job. An example is shown in Fig. 8–2 (Drecq and Parcevaux, 1988).

If a free-fall simulator is not available, and the density of the drilling fluid is high (above 15 lb/gal), small

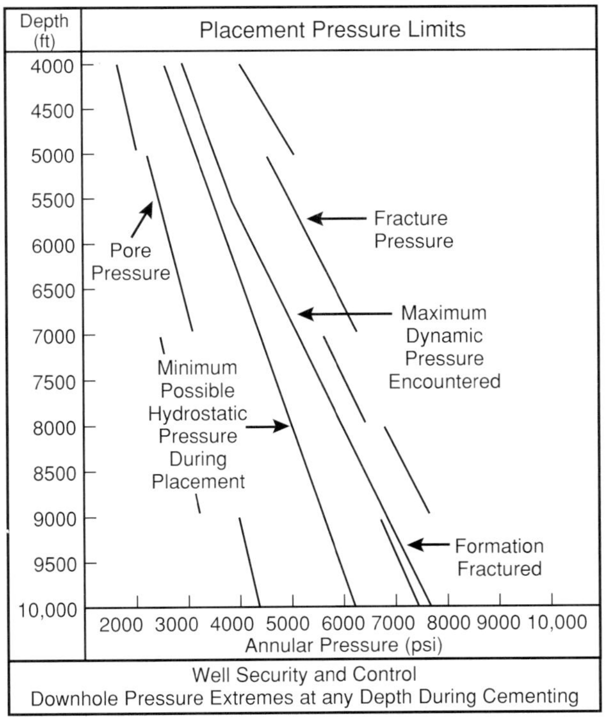

Figure 8–2—Computer-aided program output (from Drecq and Parcevaux, 1988).

density differentials from mud to spacer to cement should minimize the free-fall phenomenon. For this reason, the use of low-density preflushes may be proscribed in high-pressure wells. Provided the hydrostatic head of the fluid column in the annulus is greater than the formation gas pressure, no gas migration should occur, apart from that which occurs through a negligible dissolution and diffusion process at the molecular level.

One final point should be made concerning density control during the cementing operation. Many large cement jobs are performed on a continuous-mix basis (i.e., "on the fly"). Density fluctuations may occur during the course of the job, resulting in the placement of a nonuniform column of cement in the annulus (Granberry et al., 1989). Such a condition may cause solids settling, free-water development, or perhaps premature bridging in some parts of the annulus. Therefore, if possible, batch mixing is recommended when the potential for annular gas migration exists.

8–3.3 Fluid-Loss Control

The negative influence of fluid loss from the cement slurry into the formation was recognized by Carter and Slagle (1970) as the second most important factor contributing to gas migration in a wellbore. At that time, the respective influences of fluid-loss control and cement slurry gelation were not fully understood. However, it was pointed out that bridging or gelation because of fluid loss could restrict the transmission of hydrostatic pressure.

Before the cement slurry sets, the interstitial water is mobile; therefore, some degree of fluid loss always occurs when the annular hydrostatic pressure exceeds that of the formation (Parcevaux, 1987). The process slows when a low-permeability filter cake forms against the formation wall, or can stop altogether when the annular and formation pressures equilibrate. Once equilibrium is obtained, any volume change within the cement will provoke a sharp pore-pressure decline; consequently, because of the low compressibility of the cement, severe gas migration may be induced. Poor fluid-loss control in front of a gas-bearing zone accelerates the decrease of cement pore pressure.

In 1975, Christian et al. derived a method for calculating the fluid-loss control needed to prevent bridging of the cement across permeable formations, during and after cement placement. They concluded that reducing the API fluid-loss rate to less than 50 mL/30 min would result in less gas invasion and lower cement permeability. In 1977, Cooke and Cunningham also described a procedure for analyzing gas leakage potential based on a similar fluid-loss rate computation. However, Webster and

Eikerts (1979) judiciously pointed out that since this work was not based upon flow equations, the relative importance of fluid loss may have been overemphasized by neglecting the positive influence of the drilling mud filter cake, and mud-particle invasion into the formation. Nevertheless, Baret (1988) recently confirmed the critical importance of fluid loss by more precise direct computations based upon Darcy's flow (Chapter 6). He determined that even in the presence of drilling mud filter cake, API fluid-loss rates as low as 10 mL/30 min could sometimes be required to prevent bridging.

It is important to mention that poor fluid-loss control across permeable formations further up the hole will impair full transmission of the hydrostatic pressure to the gas zone. In 1976, Garcia and Clark reported that gas migration was observed if fluid loss occurred high in the hole such that hydrostatic head was no longer transmitted from the column above the bridging point to the bottom of the hole. Bannister et al. (1983) concluded that cement filter-cake deposition at the point of gas invasion could hinder gas flow because of its low permeability.

8–3.4 Free-Water Development

The effect of cement free-water separation was studied and discussed by Tinsley et al. (1979), and by Webster and Eikerts (1979). The former concluded through pilot-scale experiments that, although undesirable, free water is not an influential factor with respect to annular gas flow. The latter group studied the problem by constructing a nine-foot-long acrylic model, inclined up to 70°, and connected to a gas entry source and several pressure sensors (Fig. 8–3). They observed that, in deviated holes, the free water can coalesce to form a continuous channel on the upper side of the hole; as a result, a privileged path is created by which the gas may migrate. Thus, cement slurries which develop essentially no free water were recommended.

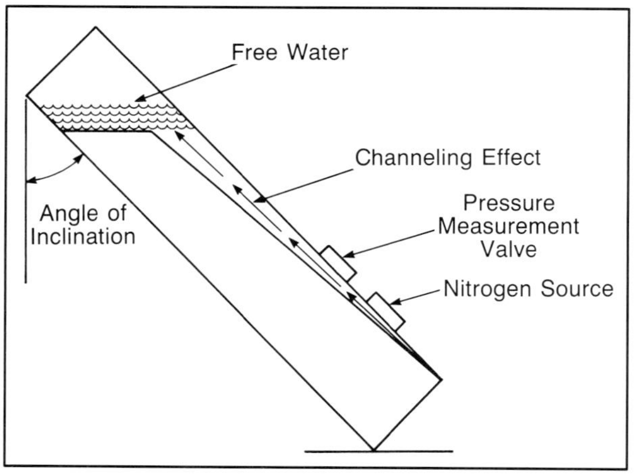

Figure 8–3—Schematic diagram of model showing fully developed water channeling (from Webster and Eikerts, 1979).

Despite their observations in the laboratory-scale model, Webster and Eikerts experienced difficulty establishing a clear relationship between the importance of the water channel and the angle of deviation. The large difference between the free water measured at room temperature using the API method (Appendix B), and that which can develop at downhole conditions, was also emphasized. This discrepancy led to the development of an "Operating Free-Water Test" by API Committee 10, where the cement slurry is heated in a pressurized consistometer prior to the measurement of free water. Angular deviation is not covered by the present API standards; however, most service and operating companies are developing in-house procedures for measuring the free water under such circumstances. Webster and Eikerts (1979) and Bergeron and Grant (1989) recommended that testing be performed at a 45° angle, the most severe test condition.

8–3.5 Cement Hydrostatic and Pore-Pressure Decrease

Despite the work described above to identify the principal causes of annular gas migration, the problem often persists even when the annular fluid densities are such that the initial hydrostatic head is much higher than the gas pressure, no free water is present, and fluid-loss control is extremely well controlled. Continued research concerning gas migration has identified the overwhelming importance of Portland cement physicochemistry.

8–3.5.1 Pressure Decrease due to Gelation

As early as 1970, Carter and Slagle noticed that the thixotropy or gelation of wellbore fluids was relevant with regard to the lowering of hydrostatic head, but no explanation was provided. Experiments to quantify the effect of gelation on hydrostatic pressure transmission gave inconclusive results (Carter et al., 1973). Some pressure restriction was observed at low curing pressure, but experiments at higher pressures (500 to 1,000 psi or 3.5 to 7 MPa) indicated no pressure change. This was most probably related to deficiencies of the experimental design (Section 8–4).

It is interesting to note that hydrostatic pressure reduction during cement hydration had been demonstrated in the laboratory, and confirmed by field measurements much earlier by Guyvoronsky and Farukshin (1963), and by Vidovskii et al. (1971) in the USSR. Similar field measurements were performed by Cooke et al. (1982), where the use of external casing sensors permitted the observation of downhole temperature and pressure fluctuations, as well as the transmissibility of applied surface pressure (Fig. 8–4). From this information, it was possible to derive the extent of vertical fluid movement into the wellbore, to locate the top of the cement column, and to measure the cement setting time at different depths.

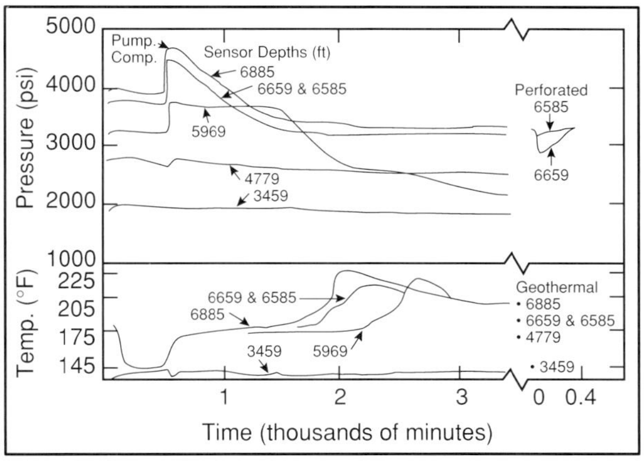

Figure 8–4—Annular pressure and temperature measurements from external casing sensors (from Cooke et al., 1982).

8–3.5.2 Hydrostatic Pressure Restriction due to Cement Hydration

In 1979, a significant contribution was made by Levine et al., who measured the hydrostatic pressure transmission of cement slurries in a 47-foot-long cell with no external pressure source (Fig. 8–5). They demonstrated that the hydrostatic pressure gradient gradually decreases to that of the mix water. Later, when the cement slurry begins to set, the hydrostatic pressure quickly approaches zero (Fig. 8–6). The hydrostatic pressure reduction is the result of shrinkage within the cement matrix due to hydration and fluid loss. At this point, the pore pressure cannot be reestablished by the fluid column above.

8–3.5.3 Hydrostatic Pressure and Slurry Gel Strength

In 1982, Sabins et al. related the kinetics of hydrostatic pressure reduction to the cement slurry gel-strength development, fluid-loss volume, volume reduction because of hydration, and the slurry compressibility factor. This work resulted in an empirical method for the prediction

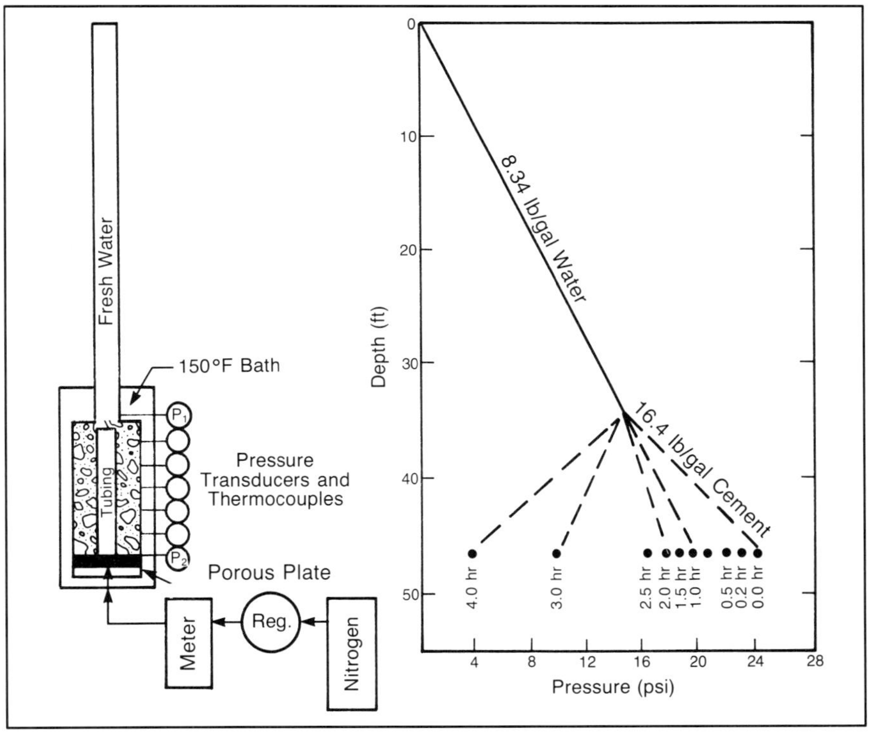

Figure 8–5—Schematic diagram of apparatus to measure hydrostatic pressure transmission of cement slurries (from Levine et al., 1979).

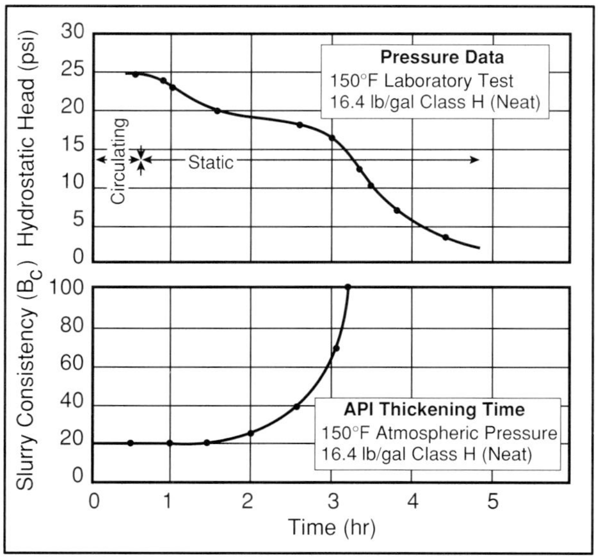

Figure 8–6—Annular gas flow test results (from Levine et al., 1979).

of gas migration, and the following equation was derived (Section 8–6).

$$P_h - P_l = \left(\frac{4 \cdot SGS \cdot L}{D_h - D_p}\right) \; or \; \left(\frac{FLVR + HVR}{CF}\right), \quad (8\text{–}1)$$

where

$P_h - P_l$ = hydrostatic pressure change across column length,

D_h and D_p = hole and pipe diameters, respectively,

SGS = static gel strength,

L = cement column length,

$FLVR$ = fluid-loss volume reduction,

HVR = hydration volume reduction, and

CF = slurry compressibility factor.

In 1979, Tinsley et al. had introduced the concept of "transition state," an intermediate period during which the cement behaves neither as a fluid nor as a solid, and the slurry loses its ability to transmit hydrostatic pressure. The concept of transition state was quantified by a transition time starting with the first measurable gel strength (about 21 lb/100 ft^2 or10 Pa), and ending when gas could no longer percolate within the gelled cement. They showed that a gel strength range from 250 to 500 lb/100 ft^2 (120 to 240 Pa) was sufficient to prohibit "gas percolation." Gas percolation can be considered as a particular type of gas migration, where gas in the form of macroscopic bubbles invades the slurry, and rises due to buoyancy effects in accordance with Stokes' Law. Cement slurries behave as non-Newtonian fluids; therefore, this process involves the breaking of the slurry's gel strength. However, gas may also flow at the microscopic level within the pores of the gelled cement structure (Section 8–3.5.4), or directly along the cement/pipe and cement/formation interfaces (Section 8–3.6). Any or all of these processes may contribute to the overall phenomenon of gas migration, and this limits the applicability of Eq. 8–1.

8–3.5.4 Gas Migration Through the Cement Pore Structure

The concept of gas migration through the pore structure of a very permeable gelled or set cement, as well as the potential gas percolation within the gelling slurry that can occur beforehand, was first introduced by Guyvoronsky and Farukshin (1963). During the period of hydrostatic pressure reduction, the cement matrix permeability was measured to be as high as 300 md. In 1982, Cheung and Beirute proposed a gas migration mechanism, based on laboratory experiments, by which the gas first invades cement pore spaces, and eventually permeates the entire cement matrix; consequently, the hydration process is prevented from closing the pore spaces. This mechanism was further refined by Parcevaux (1984), who studied the pore-size distribution of cement slurries during thickening and setting. He demonstrated the existence of free porosity composed of well-connected pores which begin to appear upon the initiation of the setting period. The same author went on to confirm (Parcevaux et al., 1983; Parcevaux, 1984) that gas migration is driven by an unsteady permeability effect through the cement pores. After an initial enlargement of the cement pores, a pseudosteady state is achieved when communication has been established throughout the cement column, and gas channels have reached a stable size.

In 1986, Stewart and Schouten confirmed and expanded upon the earlier results of Levine et al. (1979). They concluded that when a stable cement slurry (i.e., featuring negligible particle settling) enters the transition state, it begins to gel, and the hydrostatic pressure decreases ultimately to that of its water phase. When initial setting commences, this pressure, now a pore pressure, decreases further. In the same paper, Stewart and Schouten questioned the validity of static gel strength for describing the potential pressure restriction in Eq. 8–1, arguing that this equation assumes the slurry acts as a coherent "one phase body." Such an assumption is valid for pumping applications, but not for cases where the slurry is depressurized internally by fluid loss or hydration.

8–3.5.5 Pore-Pressure Decrease Described by Soil Mechanics Theory

Most recently, Parcevaux (1987) and Drecq and Parcevaux (1988) further formalized the pressure reduction process, by taking advantage of the similarities between a gelling cement column and a layer of soil undertaking some consolidation. Once again it is to be noted that Soviet scientists had previously reached similar conclusions (Grachyov and Leonov, 1969) after an experimental study.

Using the theory of soil mechanics, and assuming that the cement slurry behaves as a virgin sedimentary soil before significant hydration occurs, the state of stress in the slurry can be described by Terzaghi's law (Vyalov, 1986).

$$T = T' + u, \qquad (8\text{–}2)$$

where

T = total stress exerted at a given linear depth z,

T' = intergranular or effective stress related to the gel strength development, and

u = interstitial (pore) or hydrostatic pressure.

T is constant and equal to the full overburden pressure because of the fluid column.

$$T = g_0\int^H \rho_s(z)\ cos\ \theta(z)\,dz, \qquad (8\text{–}3)$$

where

H = total linear depth,

θ = angular deviation, and

ρ_s = specific gravity of the slurry at depth z.

The effective stress T' is related to the static gel strength determined in the laboratory, e.g., using the method described by Sabins et al. (1982) or by Hannant and Keating (1985), through the classic shear stress equation

$$T' = \frac{4 \cdot L \cdot SGS}{D_h - D_p}, \qquad (8\text{–}4)$$

where

T' = shear stress (Pa),

L = length (m),

SGS = static gel strength (Pa), and

$(D_h - D_p)$ = width of the annular gap (m).

Equations 8–3 and 8–4 can thus be combined to obtain

$$u = T - T' = \rho_f gHcos\theta - \frac{4 \cdot L \cdot SGS}{D_h - D_p}. \qquad (8\text{–}5)$$

The hydrostatic pressure u exerted by the slurry in front of the formation varies as a function of the static gel strength T'. However, the exact value of u at time t may be different from that given by Eq. 8–5 because of kinetic effects.

When gelation occurs during the induction or dormant period, there is no significant hydration of the cement grains, but essentially a buildup of intergranular forces mainly because of interparticle electrostatic forces and the precipitation of chemical species (Chapter 2). In a first approximation, the total stress T remains the same, but a transfer from u to T' occurs. Eventually, T' increases to a point where the cement becomes self-supporting. At this time, the interstitial pressure drops to the water gradient, as shown by Eqs. 8–6 and 8–7.

$$u = \rho_w gHcos\,\theta, \text{ and} \qquad (8\text{–}6)$$

$$T' = (\rho_s - \rho_w)gHcos\,\theta \qquad (8\text{–}7)$$

where ρ_w = the specific gravity of the interstitial water.

8–3.5.6 Pore-Pressure Reduction Below the Water Gradient due to Shrinkage

Later, when the cement system enters the setting period and hydration accelerates, intergranular stresses increase above the value given in Eq. 8–7, because of the intergrowth of calcium silicate hydrates. Were no volume change to occur at this stage, the pore pressure u would remain at the level given by Eq. 8–6, and the cement would behave as a porous formation. However, this is not the case. Cement hydration is responsible for an absolute volume reduction of the cement matrix, also called cement chemical contraction, which was first identified by Le Chatelier in 1887. For normal Portland cement, he showed a volumetric shrinkage of 4.6%. The shrinkage is well documented in the civil engineering literature (Setter and Roy, 1978), and occurs because the volume of the hydrated phases is less than that of the initial reactants.

The shrinkage of pure cement phases was studied as early as 1935 by Powers, who found it to increase along the series C_2S-C_3S-C_4AF-C_3A from 1% for C_2S up to 16% for C_3A. He found the absolute shrinkage, SH, of Portland cement pastes to vary between 2.3% and 5.1%, according to

$$SH = a[C_3S] + b[C_2S] + [C_3A] + d[C_4AF] \qquad (8\text{–}8)$$

Powers assumed that for each type of cement, the shrinkage is a linear function of the percentages of the four major clinker phases. The values *a*, *b*, *c*, and *d* are coefficients with values varying with the age (degree of hydration) of the specimen. In 1982, Geiker and Knudsen found the rate and magnitude of the chemical shrinkage to increase slightly with the water-to-cement ratio, but the ultimate degree of shrinkage to decrease with increasing curing temperature. This total chemical contraction is split between a bulk or external volumetric shrinkage, less than 1%, and a matrix internal contraction representing from 4% to 6 % by volume of cement slurry, depending upon the cement composition (Parcevaux and Sault, 1984). Thus, when considering cement shrinkage, a distinction should always be made between the two types. In most cases, data reported in the literature refer to total chemical contraction.

Shrinkage values less than 4% were reported by Chenevert and Shreshta (1987); however, their experimental design suggests that the phenomenon measured was not the total chemical contraction, but a combination of bulk shrinkage and reabsorption of cement free water. Chemical contraction is a time-dependent parameter (Fig. 8–7), which begins during the initial setting, and levels off after the final set (Stewart and Schouten, 1986).

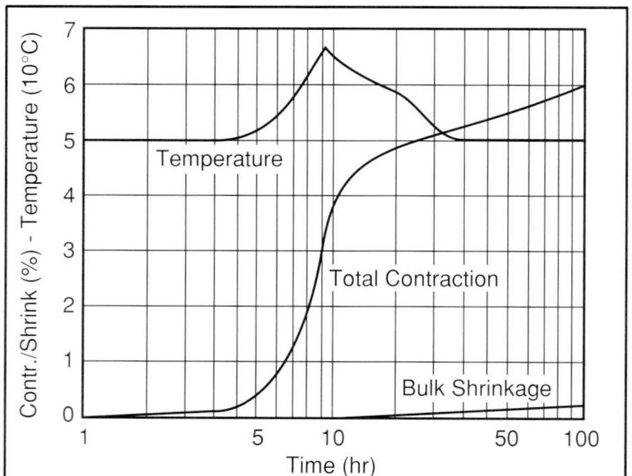

Figure 8–7—Typical contraction and shrinkage (after Parcevaux, 1987).

Chemical contraction is also responsible for a secondary porosity, mainly composed of free and conductive pores (Parcevaux, 1984). At the same time, interstitial water is trapped within the pores through physicochemical and capillary forces, and can no longer move when only submitted to its own hydrostatic pressure gradient. The combination of chemical shrinkage and secondary porosity is responsible for the sharp decrease in cement pore pressure from the water gradient to the formation pressure, or less than the atmospheric pressure if the system is isolated, as observed by Levine et al. (1979), or described by Stewart and Schouten (1986).

8–3.6 Gas Migration After Cement Setting

After setting, during the hardening phase, a normal density cement becomes a solid of very low permeability, at the microdarcy level. As a result, gas can no longer migrate at any detectable rate within the partially water-saturated pores of the cement matrix. It should be noted that low-density cement systems with high water-to-cement ratios can exhibit fairly high permeabilities (0.5 to 5.0 md). Therefore, it is possible for gas to flow, albeit at low rates, within the matrix of such cements, and to eventually reach the surface. Such events may take weeks or months to manifest themselves as measurable phenomena at the surface, where they usually appear as slow pressure buildups in the shut-in annulus.

8–3.6.1 Shear and Hydraulic Bond Strengths

Regardless of the cement system, gas can still migrate at the cement/formation or cement/casing interface if a microannulus has developed, or along paths of weakness where the bond strength is reduced. Cement-to-formation and pipe bonds have long been a subject of discussion. Indeed, good bonding is the principal goal of primary cementing. Surprisingly, however, few papers have been published on this fundamental subject (Chapter 1).

In an attempt to determine the minimum waiting-on-cement (WOC) time in the laboratory, Bearden and Lane (1961) set up a simple laboratory procedure for determining the cement-to-pipe mechanical shear bond strength (Fig. 8–8). They concluded that this shear bond strength, within experimental error, is almost independent of the specimen dimensions. They also pointed out that the shear bond is proportionally related to a number of factors. First, a positive relationship exists between shear bond and cement tensile strength. This relationship is dependent upon the cement system composition, the curing temperature and pressure, and time. Second, cement/casing shear bond strength is reduced significantly if the casing is mud-wet. Finally, the bonding strength is related to the physical nature of the pipe surface.

In 1962, Evans and Carter presented laboratory equipment which directly measured the hydraulic bond strength against the pipe or formation (Figs. 8–9a and 8–9b). Although they did not find a correlation between the shear and hydraulic bond strengths, both properties were found to vary as a function of the same external parameters. Both decrease with decreasing surface roughness, with lack of mud removal, and with oil-wet surfaces. A change in internal casing pressure or temperature, a consequence of stimulation stresses or cement hydration, causes a corresponding change in bond strength. Finally, they concluded that hydraulic bond failure is primarily a function of pipe expansion or contraction, and of the viscosity of the pressurizing fluid. In this last parameter, the hydraulic bond strength with respect to gas was found to be 5% of that obtained with water, with failure propagation rates in excess of 20 ft/min.

A separate study concerning shear bond and tensile strength, conducted by Becker and Peterson (1963), reached similar conclusions. They showed that the bonding of cement to the casing and formation is related to adhesive forces at the interfaces; therefore, the shear bond strength is related to wettability of the surfaces and to the degree of hydration of the cement.

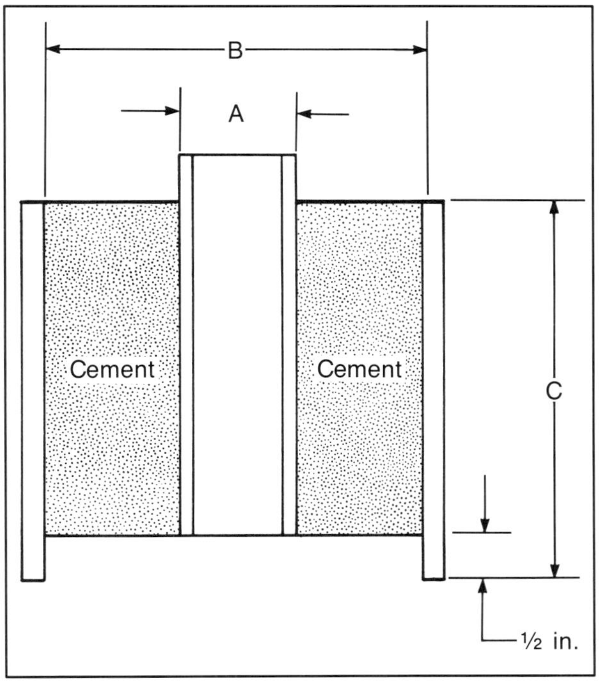

Figure 8–8—Apparatus used to determine cement support coefficient (after Bearden and Lane, 1961).

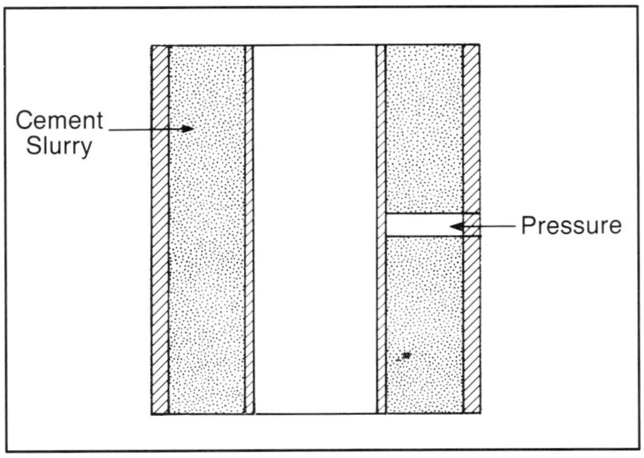

Figure 8–9a—Hydraulic bond test to pipe (after Evans and Carter, 1962).

Much later, Parcevaux and Sault (1984) performed a combined investigation of the shear and hydraulic bond strengths to pipe, total chemical cement contraction, and cement stress/strain relationships. They characterized the nature of the bond by measuring the shear bond stress and the interfacial permeability, and showed that lower chemical contraction and higher cement deformability promote better bonding. In addition, the bond was not influenced by cement compressive strength. No evidence of full microannulus development was found, implying that cement shrinkage by itself does not lead to the development of a microannulus, but instead to the development of some unbonded surface area. Thus, the development of a true microannulus could only be due to a stress imbalance between one of the two considered interfaces, as mentioned by Carter and Evans (1964).

It is fair to say that the absolute values of hydraulic and shear bond strengths found by these various authors are of little interest in themselves, as they can vary by several orders of magnitude as a function of experimental conditions (Evans and Carter, 1962). Thus, such numbers cannot be used for making any computation related to the stability of the casing/cement and casing/formation interfaces, but only for making relative comparisons between various cement formulations.

8–3.6.2 Gas Migration as a Function of the Cement-to-Pipe and Cement-to-Formation Bond

The investigations discussed in the previous section lead to the conclusion that the principal potential causes for a

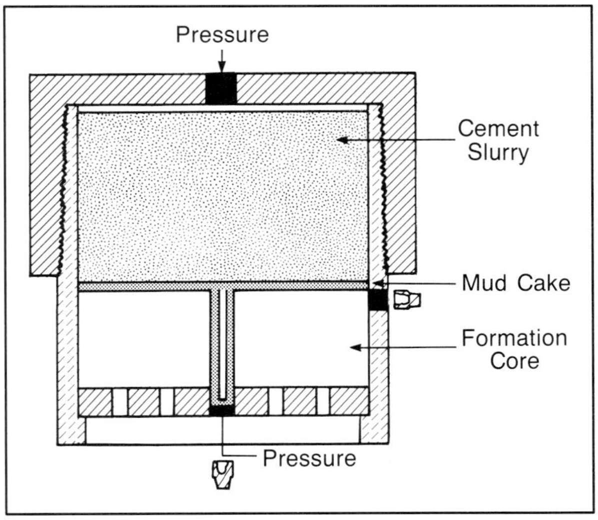

Figure 8–9b—Hydraulic bond test to formation (after Evans and Carter, 1962).

bonding defect at the cement-to-casing or cement-to-formation interface are the following:

- Lack of casing and formation roughness,
- Cement bulk volumetric shrinkage,
- Mud film or mud channel at the interface,
- Free-water channel or layer in deviated wells,
- Excessive downhole thermal stresses,
- Excessive downhole hydraulic stresses, and
- Excessive downhole mechanical stresses.

Mud removal and free water have been dealt with in detail in Sections 8–3.1 and 8–3.4, and their influence remains the same at any stage of cement hydration, so no more need be said at this point. Furthermore, it should be noted that very little can be said concerning the mechanics of gas migration in hard cement, considering the small amount of research performed as of this writing. This undoubtedly constitutes a topic requiring additional investigation.

8–3.6.2.1 Bulk Shrinkage and Surface Roughness

Thermal effects arising during cement hydration have been studied by Gotsis et al. (1984), who showed that tensile stresses at the interface may arise at the early stage of hydration when cement undergoes a bulk volumetric shrinkage (up to 0.5% in their experiments). However, they believed that this effect is minimal on long cement columns where consolidation in the plastic state, and early stage creep, may compensate for the shrinkage effect. As discussed earlier, the total chemical shrinkage of cement slurries represents several percent by volume. The bulk shrinkage portion (external volumetric reduction) occurring after initial set, which could be responsible for a bonding defect, is generally only a few tenths of one percent (Wu et al., 1983; Gotsis et al., 1984).

In a wellbore, for cement placed across 7-in. to $8^1/_2$-in. casing, a homogeneous volumetric bulk shrinkage of 0.5% would result in a retraction of about 20 µm. This is in the same order of magnitude as an average cement particle, probably too small to induce a significant continuous microannulus (Drecq and Parcevaux, 1988). However, local bonding defects could result. Such defects can be reduced by increasing the roughness of the casing. Although not negligible, local bonding defects are not a fundamental factor governing gas migration at the casing or formation interfaces.

8–3.6.2.2 Thermal and Hydraulic Downhole Stresses

Downhole deformations can occur as a result of thermal stresses (cement hydration, wellbore cooldown treatments, steam injection, cold fluid injection, etc.) or hydraulic stresses (replacement of casing fluid density, communication tests, squeeze pressure, stimulation treatment pressure, etc.).

The effect of pressure changes on casing dimensions and stability is well documented in the literature (Carter and Evans, 1964; Cain et al., 1965; Durham, 1987). The well-known relationship concerning the expansion of pipe diameter vs internal pressure is shown in Fig. 8–10.

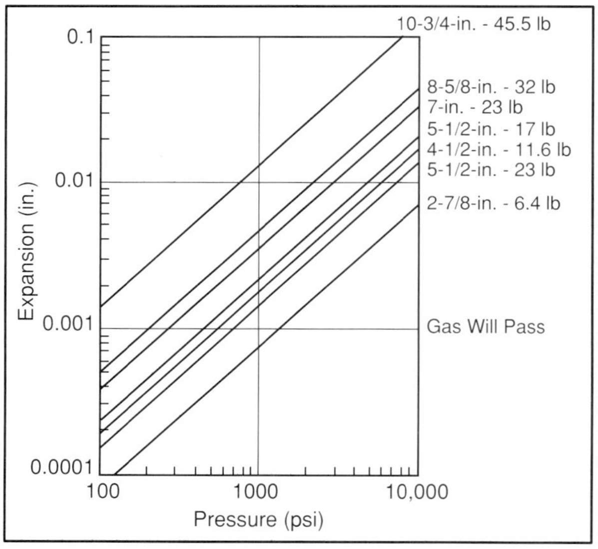

Figure 8–10—Expansion of pipe diameter vs internal pressure (after Carter and Evans, 1964).

Cain et al. (1965) presented a study of the effects of pressure and temperature on casing and cement, in an attempt to improve the cementing of steam injection wells, where casing problems, pipe growth, cement bond breakdown, and cement failure had been reported. The coefficients of linear thermal expansion for cement and steel were found to be comparable, approximately 7×10^{-6}/°F. In addition, thick-shell stress equations for the casing and cement were found useful for calculating the stress conditions in the cement because of temperature differentials, and the limits of pressure or temperature the cement could withstand.

The magnitude of the hydraulic effect was illustrated by Matthews and Copeland (1986). In a liner, 14.5 lb/gal (1.74 g/cm^3) drilling mud was replaced by KCl water, resulting in an internal pressure reduction of 3,900 psi

(27 MPa); as a result, a pipe diameter reduction of 0.008 in. (203 μm) occurred, and gas migration was observed.

The results of these studies show that downhole deformations resulting from thermal and hydraulic stresses constitute a major drive for gas migration at the hard cement casing and formation interfaces. These factors, which are generally not taken into account today, should be considered carefully. It is apparent that extensive efforts to ensure an excellent primary cement job, including the incorporation of special gas migration prevention agents (see Section 8–5), can be rendered useless by ignoring such factors.

8–3.6.2.3 Downhole Mechanical Stresses

The influence of mechanical stresses on gas migration appears not to be referenced in the literature, and this section is derived from discussions with field personnel. Occasionally, gas migration on an intermediate string occurs several days after cementing, and after drilling has resumed. In such a situation, the influence of mechanical stresses generated by drilling cannot be overlooked, especially in cases where weak formations are present behind the cemented string. Field reports indicate that improved results are likely to be achieved by measures such as using quick-setting cement or high-strength cement.

8–4 GAS MIGRATION TESTING

Gas migration laboratory testing has not been standardized by the API; thus, no laboratory procedure is currently recognized worldwide for characterizing the ability of a cement system to prevent or reduce gas migration. In addition, apparently no major oil or service companies have released proprietary testing equipment from R&D laboratories to the field laboratory level. The principal reason for this lack of standardization lies in the complexity of the problem, and also in the fact that the various mechanisms have only recently been accepted by the industry.

A large variety of different experimental prototypes is described in the literature which attempt to simulate the gas migration process. Two main types of experimental simulators exist: large-scale pilot devices, which reproduce the process as it occurs in the wellbore, and small-scale, bench-type models, which can be used to derive the fundamental laws of a particular physical process under investigation. To date, none of the simulators described in the literature permits the derivation of a physical model which quantitatively describes gas migration over a wide range of conditions.

8–4.1 Large-Scale Simulators

The earliest large-scale simulator was first described by Carter and Slagle (1970), and later upgraded by Carter et al. (1973). In 1976, Garcia and Clark constructed a device specifically to study the influence of uneven cement setting. Levine et al. (1979) described a simulator for studying hydrostatic pressure profiles within a cement column at rest (Fig. 8–5). The apparatus built by Tinsley et al. (1979) investigated the influence of fluid loss and compared different cement systems (Fig. 8–11). Finally, the equipment described by Bannister et al. (1983) evaluated the influence of filter-cake formation from cement fluid loss, and the conductivity to gas of a setting cement (Fig. 8–12).

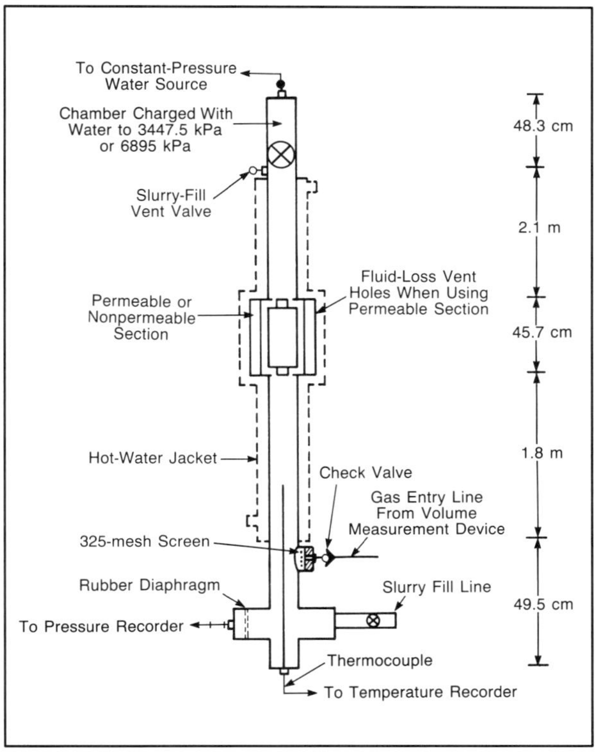

Figure 8–11—Schematic diagram of test fixture used to study gas leakage (after Tinsley et al., 1979).

8–4.2 Bench-Scale Simulators

Three bench-scale devices are described in the literature. The first, described by Cheung and Beirute (1982), used a modified API fluid-loss cell to investigate the hydrostatic pressure decrease and subsequent gas migration in a setting cement column (Fig. 8–13). This device could be adapted for routine use; however, at this scale, three factors can unduly affect the gas migration process. Fluid loss could result in the formation of an impenetrable filter cake at the gas inlet or outlet. Free water development could artificially delay the pore pressure decrease by reabsorption during hydration. Finally, considering the length of the cement column versus the external applied pressure, such an experiment can only consider gas mi-

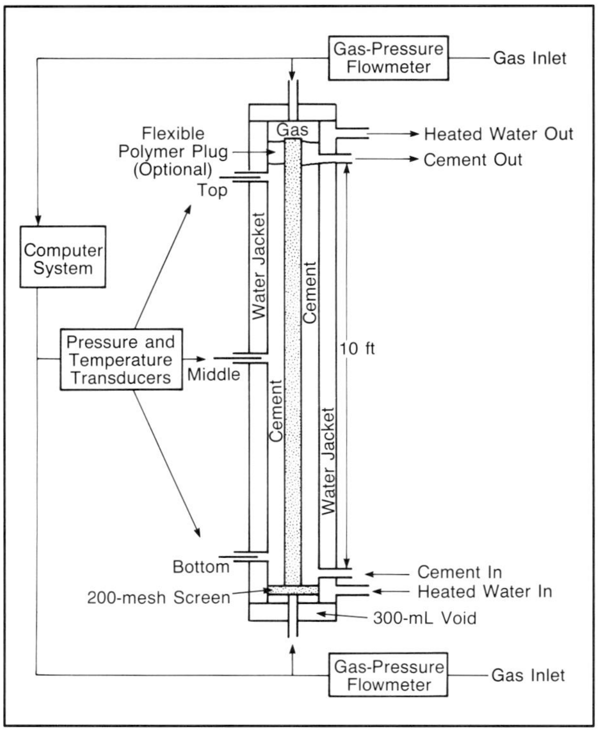

Figure 8–12—Annular gas flow laboratory testing apparatus (after Bannister et al., 1982).

gration across a short interval. Another notable apparatus reported by Stewart and Schouten (1986) investigated gas migration in set and hard cement, using a U-tube apparatus shown in Fig. 8–14 (Richardson, 1982).

Parcevaux (1984) and Drecq and Parcevaux (1988) described a small-scale simulator which eliminated some of the limitations of earlier devices. As illustrated in Fig. 8–15, the artificial effects of fluid loss and free water were eliminated, and the external curing pressure was computer-controlled to maintain a differential pressure of close to zero between the top and bottom of the cell. This model was an attempt to investigate the process of gas migration during cement setting without side artifacts.

8–5 GAS MIGRATION SOLUTIONS

Over the years, a number of methods to control gas migration have been proposed. Historically, these methods have reflected the level of knowledge at the time of development. In addition to the basic "good cementing practices" which facilitate mud removal, a prerequisite for controlling gas migration, at least a dozen different techniques have been applied.

8–5.1 Physical Techniques

It has long been known that a number of physical techniques can, under certain circumstances, help control gas

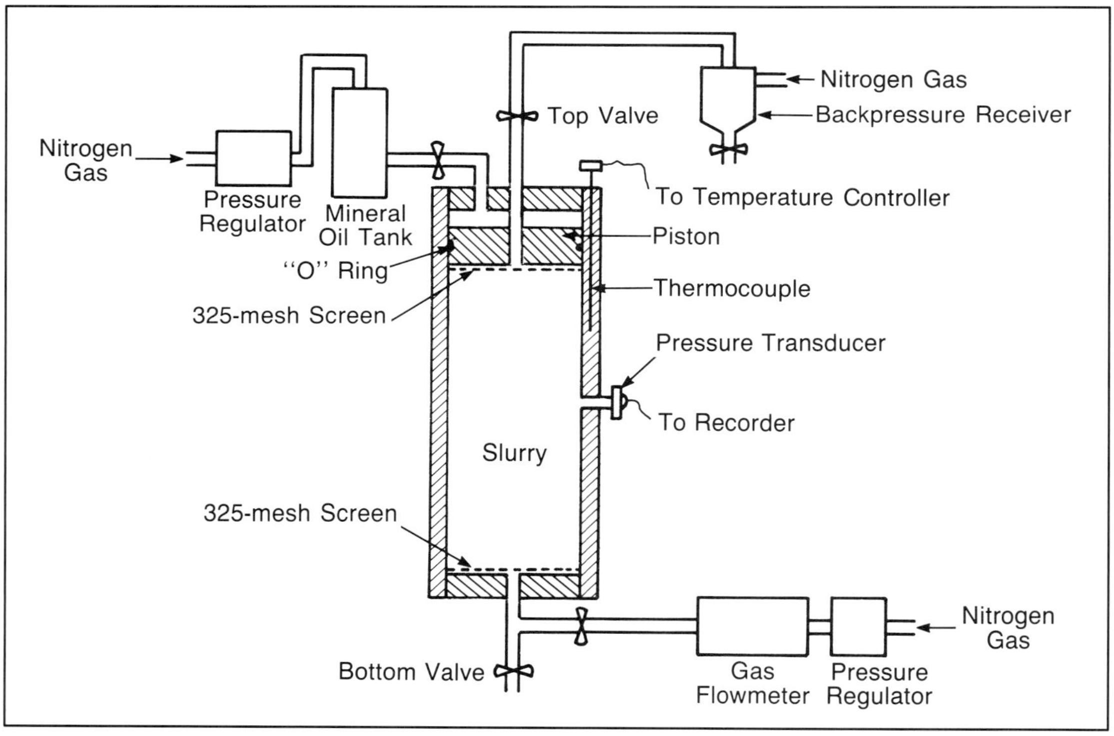

Figure 8–13—Gas flow simulator (after Cheung and Beirute, 1982).

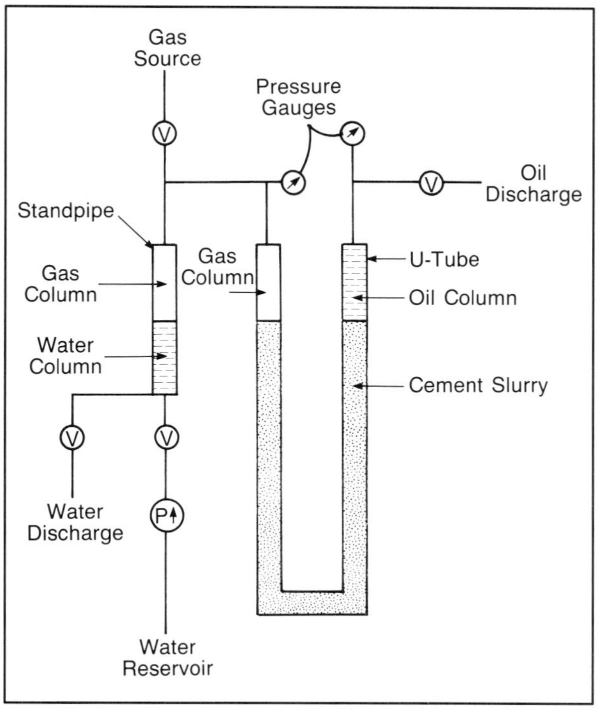

Figure 8–14—U-Tube gas migration tester (after Richardson, 1982).

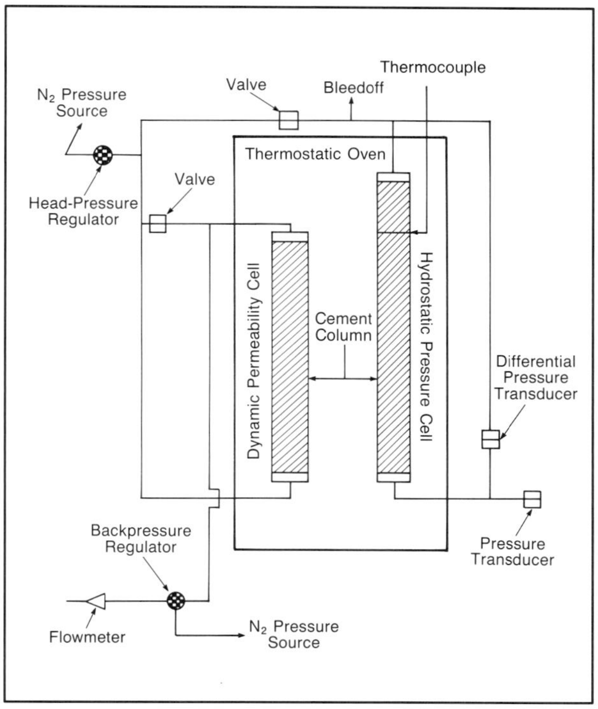

Figure 8–15—Dynamic permeability apparatus (after Parcevaux, 1984).

migration. These include the application of annular backpressure, the use of external casing packers (ECPs), and the reduction of cement column height (including multistage cementing). Each attempts to delay the occurrence of downhole pressure restriction at the gas-bearing formation face until the cement is sufficiently hard and impermeable.

Such techniques are certainly valid under a variety of conditions, but well conditions often limit their application. For example, the presence of weak zones may restrict the use of annular backpressure, because of the risk of inducing lost circulation (Levine et al., 1979). ECPs (Fig. 8–16), which can be inflated by mud or cement slurry, control gas migration by forming a positive barrier in the annulus (Suman, 1984; Baker, 1986). However, ECPs require a competent formation against which to seal, and they complicate the execution of the job. Because of the small clearance between the uninflated element and the borehole, such tools have been known to suffer mechanical damage while running casing, or circulating at high rates. Also, it is not uncommon for the packers to set prematurely because of unexpected pressure fluctuations during the course of the job. Parcevaux (1984) pointed out that ECPs can exacerbate some problems, since they effectively isolate the lower portion of the annulus shortly after cement placement. Slurry volume reduction below the packer, from fluid loss or chemical contraction, can result in gas invasion of the cement in this interval at an even earlier time. This could permit undesirable crossflow between zones located below the packer.

The technique of reduced cement column height stems originally from the work of Levine et al. (1979), described in Section 8–3.5.2. Viewing the mix-water gradient as a natural step in the pressure reduction, and through a very simple graphical method (Fig. 8–17), they proposed the minimization of the cement column height above the gas zone. The job would be designed such that the pressure sum of an equivalent height of water plus the hydrostatic above the cement would always exceed the formation pressure. There is little doubt that this approach can help the design process in a gross sense; e.g., severe risks of underbalance may be avoided. It has indeed been applied with success across some depleted sands, but it is clearly not stringent enough. As noted in the same paper, as cement changes from liquid to solid, the hydrostatic pressure falls to values far below the water gradient because of fluid loss and chemical contraction.

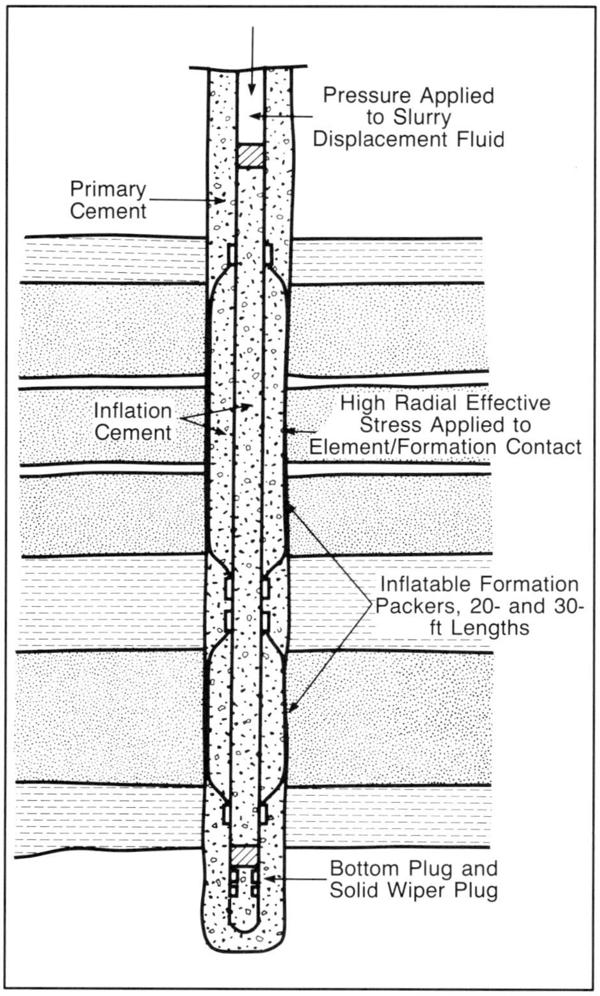

Figure 8–16—Use of external casing packers (after Suman, 1984).

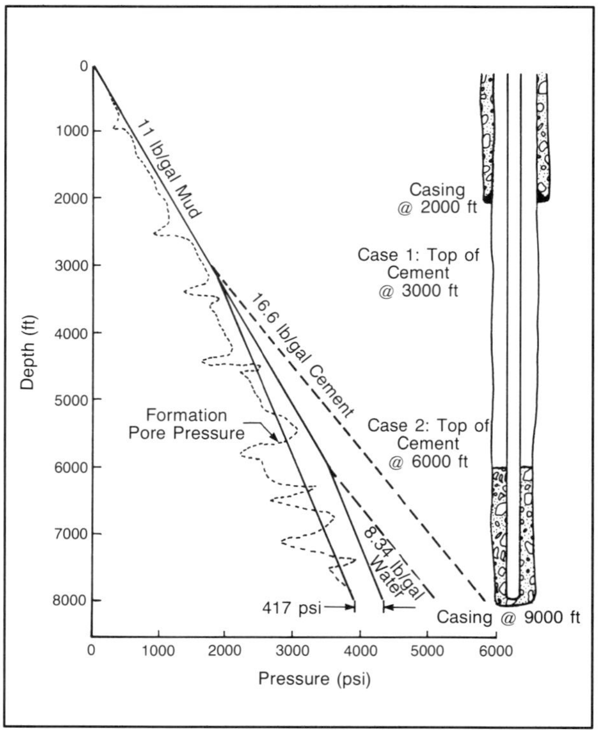

Figure 8–17—Comparison of cement column height adjustments (from Levine et al., 1979).

An elastomeric seal ring, which Bol et al. (1986) described, presents an additional line of defense for interfacial migration. The success rate may be improved in wells where downhole stresses, such as density changes or thermal cycling, induce casing deformation. However, it is important to note that this device cannot solve the problem of gas flow through the cement matrix; thus, it should be used in concert with other techniques.

8–5.2 Fluid-Loss and Free-Water Control

Fluid loss and free water (in deviated wells) have been identified as promoting the occurrence of gas migration (Sections 8–3.3 and 8–3.4). To minimize the impact of these parameters on gas flow, both must be reduced to fairly low levels, approximately 50 mL/30 min and 0.25%, respectively (Webster and Eikerts, 1979; Baret, 1988).

Latices, anionic synthetic polymers, and some cellulosic derivatives (at low temperature) are able to provide such low fluid-loss rates, without inducing free-water separation. However, most of them affect other cement slurry properties, including gel-strength development and thickening time, in a deleterious fashion. Defossé (1983) described a series of metallic salts which depress free-water development, yet are not antagonistic to other aspects of slurry performance. This subject is covered in greater detail in Chapter 3.

8–5.3 Compressible Cements

Compressible cement slurries have been developed in an attempt to maintain cement pore pressure above the formation gas pressure. In theory, this should prevent any movement of gas from the formation into the cemented annulus. Compressible cements fall into two main categories—foamed cements and in-situ gas generators—and it is important to draw a clear distinction between them.

Foamed cements become nearly incompressible at high pressures, because of the relative incompressibility of gases under such conditions (Fig. 8–18). Therefore, their ability to compensate for volume reduction during the transition state is probably restricted to situations close to the surface, where gas expansion is significant.

The in-situ gas generators are designed to maintain cement pore pressure by virtue of chemical reactions which produce gas downhole. The produced gases may

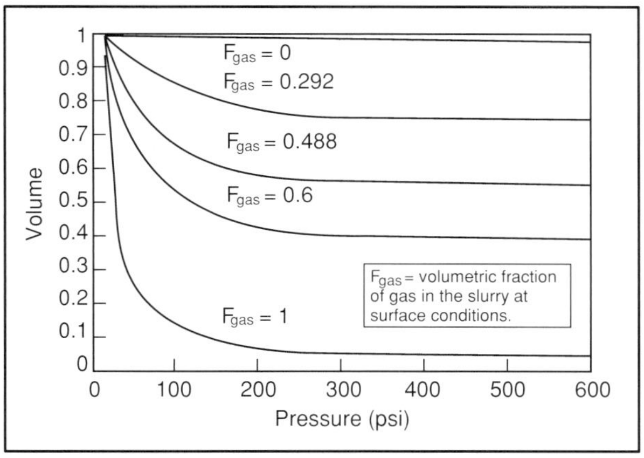

Figure 8–18—Compression of foamed cement slurries.

be hydrogen (Bulatov, 1970; Sutton, 1982) or nitrogen (Richardson, 1982; Burkhalter et al., 1984). To the authors' knowledge, the field application of nitrogen to control gas migration has not been reported. Hydrogen-generating agents such as aluminum powder have been used in the USSR (Kucyn et al., 1977) and elsewhere (Tinsley et al., 1979; Watters and Sabins, 1980). It is important to note that gas migration cannot be prevented by the gas-generating agents alone. Fluid-loss control agents and dispersants are necessary to minimize interstitial water leakoff.

The principal drawback of these systems, other than the safety hazard from those which generate hydrogen, is the inability of a gas at typical downhole pressures to achieve the 4% to 6% volumetric expansion necessary to maintain pore pressure. Strictly applying Boyle's law, the volume of gas required to offset just the chemical contraction would be excessive at high pressure. Gas-generating systems must also be carefully stabilized; otherwise, gas bubbles may coalesce and create channels for formation gas to follow. These criticisms notwithstanding, it is clear that this technology has been used with success.

8–5.4 Expansive Cements

Expansive cements have been advocated in places where a microannulus has been identified as the gas migration pathway, and successful field results have been reported (Seidel and Greene, 1985). As discussed in Chapter 7, there are two principal techniques for inducing expansion in Portland cement: crystal growth and gas generation. The latter operates on the same principle as the compressible cements mentioned above with the exception that the concentration of gas-generating material (typically aluminum) is reduced (Sutton and Prather, 1986). The former, on the other hand, relies upon the nucleation and growth of certain mineral species within the set cement matrix. The bulk volumetric expansion is usually controlled to be less than one percent (Griffin et al., 1979).

There is little doubt that the controlled expansion of a cement can help to seal small gaps between the cement sheath and the casing or formation, but it is unlikely to be effective in sealing large channels created by gas migration. Attempts to increase the expansive properties of Portland cement can result in unsoundness, an uncontrolled expansion which disrupts and fractures the set cement. One must also be aware that, although these cements undergo a bulk dimensional expansion, they still exhibit a net chemical contraction, and experience the same hydrostatic and pore-pressure decreases as nonexpansive cements.

8–5.5 Thixotropic and High-Gel-Strength Cements

Carter and Slagle (1970) identified slurry gelation as a major potential cause of gas migration. However, the work of Sabins et al. (1982) and Childs and Sabins (1985) indicated that high gel strength development by the cement may help resist gas percolation; for this reason, they proposed thixotropic and high-gel-strength cements.

As discussed in Chapter 7, thixotropic cements may be prepared by a number of methods, including the addition of bentonite, certain sulfate salts, or crosslinkable polymers to a Portland cement slurry. In all cases, the transmitted hydrostatic pressure of a thixotropic system should revert to the gradient of its interstitial water, and remain as such until the setting period begins. Therefore, thixotropic systems are unlikely to be effective in situations where the gas-zone pressure exceeds the water gradient, unless additional backpressure is held on the annulus.

It is true that the very high gel strength of thixotropic cements can offer considerable resistance to physical deformation and percolation by a large gas bubble. However, as discussed earlier, the bubbles may often be smaller than the pore spaces in the setting cement. Under such circumstances, gas migration may occur without slurry deformation, and gel strength is no longer a relevant factor.

Thixotropic cement slurries tend to have high fluid-loss rates; therefore, the risk of dehydration and bridging must be considered. Sykes and Logan (1987) found the influence of fluid loss to be greater than that of gel strength immediately after placement, and they recommended designing the slurry to be well dispersed until after the bulk of fluid-loss volume reduction has occurred. Some degree of fluid-loss control for thixotropic slurries can also be obtained by the use of low fluid-loss spacer fluids (Bannister, 1978).

Successful field results have been obtained in shallow, low-temperature applications (Sepos and Cart, 1985). Stehle et al. (1985) reported good results at higher temperatures (250° to 280°F or 120° to 140°C) when cementing liners and long strings.

8–5.6 "Right-Angle-Set" Cements

"Right-angle-set (RAS)" cement slurries can be defined as well-dispersed systems which show no progressive gelation tendency, yet set very rapidly because of rapid hydration kinetics. Such systems maintain a full hydrostatic load on the gas zone up to the commencement of set, and develop a very low-permeability matrix with sufficient speed to prevent significant gas intrusion. It should be pointed out that the "transition time" involved here is not the same as that described by Sabins et al. (1982), nor is the mechanism similar to that of high-gel-strength systems (Kieffer and Rae, 1987). A true set occurs, involving the deposition and recrystallization of mineral hydrates.

RAS slurries are sometimes characterized as such by standard high-temperature, high-pressure thickening time tests, as shown by Drecq and Parcevaux (1988). An RAS slurry maintains a low consistency until setting, when the slurry viscosity increases to more than 100 Bc within a few minutes. The increase in consistency is accompanied by a temperature increase resulting from the exothermic cement hydration reactions taking place (Fig. 8–19).

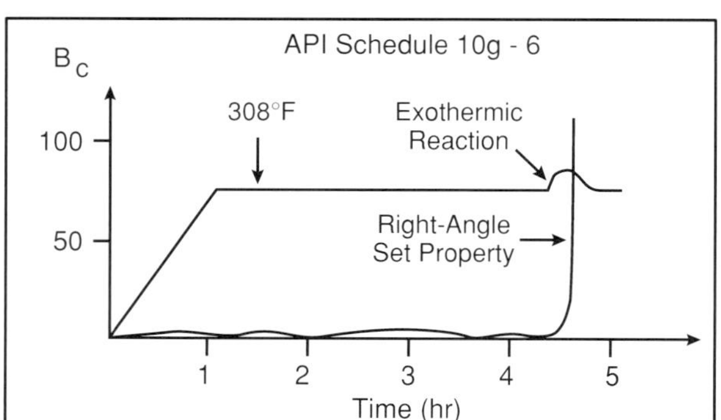

Figure 8–19—Pressurized consistometer output from Right-Angle-Set (RAS) cement system (after Drecq and Parcevaux, 1988).

Mainly because of cement hydration kinetics, it is difficult to design RAS cement systems for circulating temperatures below 250°F (120°C). Regardless of temperature, it is probable that the shear imparted during the API thickening time test varies significantly from that which occurs during a cementing operation. The presence of additives such as fluid-loss control agents and dispersants exacerbates the problem, because such materials often have set-retarding tendencies.

8–5.7 Impermeable Cements

Gas migration can be prevented by reducing the matrix permeability of the cement system during the critical liquid-to-solid transition time described earlier. Several methods have been developed.

The first approach involved the use of water-soluble polymers to viscosify the interstitial water of the cement slurry. Since at least a part of gas migration involves the displacement of cement pore water, viscosification of the water tends to limit gas mobility. This approach is also appropriate for fluid-loss control (Chapter 3); unfortunately, viscosification of the cement slurry is a major side effect of this technique, with resultant mixing difficulties, higher displacement pressures, and increased risk to weak formations. This method is currently limited to low-temperature applications, because the efficiency of the viscosifiers decreases with temperature.

Cheung and Beirute (1982) described the use of an impermeable cement which operates by immobilizing the fluids within the pore spaces of the cement. Since the cement mix water cannot be displaced, gas cannot move within the pore spaces of the cement slurry. According to Williams et al. (1983), the system is composed of a combination of bridging agents and polymers. Such systems have been applied throughout the 140° to 350°F (60° to 180°C) BHST range (Cheung and Myrick, 1983).

Latex additives for prevention of gas migration were first described in a 1982 patent application by Parcevaux et al. (issued 1985). Subsequent refinements of this technology (Bannister et al., 1983, Parcevaux and Sault, 1984) have extended its applicability to a wide range of well conditions, and its field application is well-established (Evans, 1984; Peralta, 1984; Matthews and Copeland, 1986; Rae, 1987; and Drecq and Parcevaux, 1988).

As described in Chapters 3 and 7, latices are aqueous dispersions of solid polymer particles, including surfactants and protective colloids, which impart stability to the dispersion. Most latices have film-forming capabilities; thus, when contacted by a gas, or when the particle concentration exceeds a given threshold value, latex particles coalesce to form an impermeable polymer barrier. In a wellbore situation, the gas first invades the portion of the cemented annulus across the gas zone, and contacts the dispersed latex particles in the slurry. As shown in Fig. 8–20, the latex coalesces within the pore spaces, blocking further progress up the annulus.

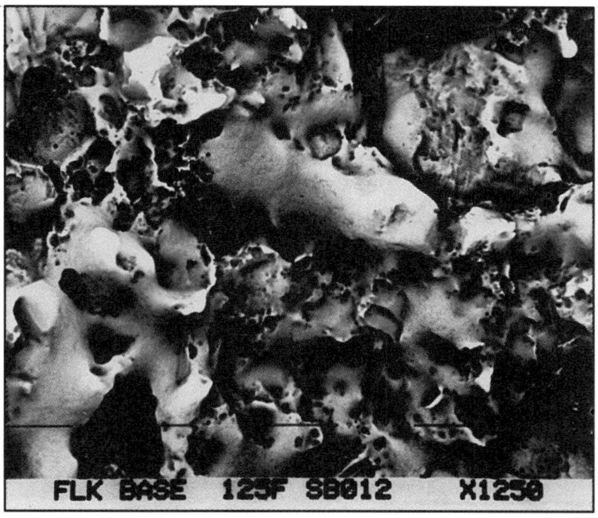

Figure 8–20—Latex film in cement after coalescence.

Latices have a number of other beneficial properties when used in cement slurries (Parcevaux, 1987). The small, spherical latex particles act as lubricants, imparting excellent rheological properties. Fluid-loss control is provided by a mechanical plugging mechanism. The shrinkage-compensating and bonding actions of latices have long been recognized by civil engineers, and such attributes translate to improved shear-bond strength and elastic deformability in well cements (Parcevaux and Sault, 1984).

More recently, Blomberg et al. (1986) described yet another technique which uses fine mineral particulates to prepare low-density, low-permeability cements. The preferred particulate in this application is silica fume (also called microsilica), a byproduct in the production of silicon and ferrosilicon. The average particle size of this material is 1 μm; consequently, it is able to fill pore spaces and plug pore throats. Field success has been reported (Grinrod et al., 1988) for shallow, low-pressure gas.

8–5.8 Surfactants

Marrast et al. (1975) described the use of surfactants in cement slurries and preflushes. These surfactants may, under the right circumstances, entrain invading gas downhole and create a stable foam. This foam then presents significant resistance to flow, thereby limiting upward migration. Stewart and Schouten (1986) reported the technique to be effective, particularly when combined with the use of the elastomeric seal rings, described earlier.

8–6 GAS MIGRATION PREDICTION

As detailed in the preceding section, a varied assortment of techniques exists for the prevention of gas migration. Few are applicable universally, but most have been proved effective under certain circumstances. As a general rule, universality and cost are directly related; consequently, systematic well analysis techniques have been developed to qualitatively determine the relative risk of gas migration, and to identify the most cost-effective remedy.

The best-known predictive technique is that described by Sutton et al. (1984), which calculates a "Gas Flow Potential (GFP)." This is defined as the ratio of another variable, the Maximum Pressure Restriction (MPR), to the well's hydrostatic overbalance pressure (OBP).

$$GFP = \frac{MPR}{OBP} \tag{8–9}$$

The MPR, in turn, is defined as

$$MPR = 1.67 \frac{L}{(D_h - D_p)}, \tag{8–10}$$

where

L = cement column length (ft),

D_h = diameter of the open hole (in.), and

D_p = outside diameter of the pipe (in.).

The GFP factor can vary between 0 and infinity, and the severity of the potential gas migration problem is rated, based on unpublished rules, as follows.

GFP	Severity Rating
<4	Minor
4 to 8	Moderate
>8	Severe

The GFP concept is based on the premise that gas flow in the cemented annulus occurs via percolation through the cement slurry, and that gel-strength development can arrest the invasion. The above equation is in fact a modified version of the standard shear stress equation used to calculate the pressure required to break circulation. The technique assumes that a static gel strength of 500 lb/100 ft^2 (240 Pa) indicates sufficient resistance to the macroscopic shear forces developed by migrating gas bubbles. Stewart and Schouten (1986) showed that gel strengths considerably below this value could inhibit gas percolation, but most gas migration was conclusively shown to occur after the cement's initial set. At initial set, cements can exhibit gel strengths far in excess of 240 Pa, indicating that the primary path for gas flow is within the evolving cement matrix porosity.

More recently, Rae et al. (1989) described an alternative technique for predicting postplacement gas flow. Their method, driven by a phenomenological approach, is based on the derivation of four factors whose components are considered fundamental to the occurrence of the gas migration. These four factors independently examine the contributions made by the formation and annular configuration, fluid hydrostatics, mud removal, and slurry performance. Well parameters such as the reservoir productive capacity, annular geometry, pore pressures, hydrostatic head, mud removal efficiency, cement hydration kinetics, and fluid loss are entered into the calculations.

The first of the factors, the "Formation Factor," is a dimensionless term which represents the ratio of the formation productive capacity, kh, with a critical volume, V_c. The latter is equated to the porosity (created in the setting cement by chemical contraction during the early stages of the transition period) from the top of the gas zone to the point of pressure balance in the annulus. The porosity is estimated at two percent at this stage of transition and the gas is assumed to permeate the annulus in a uniform fashion. Mathematically, the Formation Factor can be expressed by

$$FF = \frac{kh}{V_c} = \frac{467.7\, kh\rho}{OBP(D_h^2 - D_p^2)}, \qquad (8\text{–}11)$$

where

k = zone permeability (md),

h = zone height (ft),

ρ = cement slurry density (lb/gal),

OBP = overbalance pressure (psi),

D_h = hole diameter (in), and

D_p = pipe diameter (in).

Increasing values of kh/V_c indicate increasing risk of postplacement gas flow, assuming other factors remain constant.

The concept of the "Hydrostatic Factor" is based on the work of Levine et al. (1979). They observed that the hydrostatic pressure exerted by cement slurries tends to approach that of the interstitial water as gel strength increases. Only after the initial set does the pressure decay to a value below the interstitial water gradient. This, of course, corresponds to the stage of structural consolidation and permeability decline of the cement matrix and the consumption of pore water by the hydrating cement grains. When cementing to the surface, gas zones with pressures greater than the hydrostatic of water can theoretically flow as soon as the cement gels. Where a mud column remains above the cement, this must be taken into account as an additional pressure head which is summed with that of the cement interstitial water. Thus, the Hydrostatic Factor is represented by the ratio of the gas-zone pore pressure with that of the annular pressure at the commencement of true transition, i.e. at the initial set. Mathematically, this can be represented by

$$HF = 19.281 \times \frac{P_g}{((R_m \cdot h_m) + (R_s \cdot h_s) + (8.32 \cdot h_c))} \qquad (8\text{–}12)$$

where

P_g = gas-zone pore pressure (psi),

R_m = mud density (lb/gal),

R_s = spacer density (lb/gal),

h_m = mud height in the annulus (ft),

h_s = spacer height in the annulus (ft), and

h_c = cement height in the annulus (ft).

Again, higher values of the Hydrostatic Factor indicate a higher risk of gas migration in a given well situation.

The third factor relates to mud removal and, although subjective in nature at present, recent developments in the understanding of the displacement process promise to offer better quantification of this parameter in the future (Chapter 5). Today, the Mud Removal Factor is assessed according to a set of standard industry guidelines (Table 8–1), and then rated on a 10-point scale, 1 being excellent, 10 being very poor.

The fourth factor is the "Slurry Performance Number." It was developed to rank cement systems according to their hydration kinetics and fluid loss, factors which are fundamental to the process of gas migration. The SPN attempts to provide, with conventional test equipment and procedures, a relative value for the cement interstitial water loss during the critical time when the cement begins to change from a liquid to a solid. It is based on the fact that, as a first approximation, the fluid loss varies linearly with the square root of time and, therefore, the theoretical volume of fluid loss during the setting process is given by

$$SPN = \frac{API \cdot [(t_{100Bc})^{1/2} - (t_{30Bc})^{1/2}]}{(30)^{1/2}}, \qquad (8\text{–}13)$$

where

API = API fluid-loss value of slurry (mL/30 min),

t_{100Bc} = time to 100 Bc consistency (min), and

Excellent	Moderate
• Hole in excellent condition before cementing —Circulate one hole volume —No gas —Condition mud	• Hole in good condition before cementing —Circulate one hole volume —Condition mud
• Greater than 67% standoff	• Greater than 50% standoff
• Rotation/reciprocation of casing	
• Minimal U-tubing —Compatible fluids —Use of spacers/washes	• Reduce U-tubing —Compatible fluids —Use of spacers/washes
• Engineered displacement regime	• Engineered displacement regime
• 10-min spacer contact time at selected flow regime	• 10-min spacer contact time at selected flow regime
• Two bottom plugs when possible	

Table 8–1—Mud removal guidelines.

t_{30Bc} = time to 30 Bc consistency (min).

It must be emphasized that this equation is not claimed to represent the actual performance of the slurries under static downhole conditions where the mud cake influences the leakoff. What is claimed is that the SPN provides a method of comparing slurry performance on a relative basis, and provides a useful tool in both the design and evaluation of cement programs for gas wells. Slurries with high SPNs are very poor candidates for gas migration control. Those with low API fluid-loss rates and short critical hydration periods offer a much greater probability of success.

This method is based neither on a single experimental investigation nor on numerical simulations, but on a pragmatic compilation of the state of the art. A statistical analysis of data from a wide variety of gas wells in the United States, Canada, Latin America, Europe, Africa, Middle East and Far East has allowed the calculation of semi-empirical relationships between the four factors. Rae et al. (1989) claimed that the wide range of field conditions through which this method has been established justifies its use in most real cases.

The following actual field case serves to illustrate the utility of this approach, and highlights the danger of using an oversimplistic method to predict postplacement gas flow. Figure 8–21 shows the well configuration, which is basically a 7-in production liner hung from a $9^5/_8$-in.-longstring. The two pay intervals lie at depths of 5,400 to 5,590 ft (1,646 to 1,703 m) and 6,100 to 6,420 ft (1,859 to 1,957 m). The upper reservoir contains insignificant gas, while the lower possesses a sizable gas cap extending from 6,100 to 6,260 ft (1,859 to 1,908 m). The reservoir pressures are 1,850 psi (265 MPa) and 2,530 psi (365 MPa) for the upper and lower zones, respectively; both zones have permeabilities in excess of 200 md. These wells are completed with the intention of producing only from the upper zone, because the field is in a remote location and lacks gas-gathering facilities at present.

Using the Gas Flow Potential equation of Sutton et al. (1984), the lower zone appears to pose little risk of gas flow (GFP = 1.66). The technique proposed by Rae et al., (1989) suggests that this well presents a high risk of postplacement gas flow mainly because of the high productive capacity of the intervals in question. In fact, this prediction is borne out by actual results. Wells in this field suffer from severe crossflow, and oil produced from the upper zone has shown gas/oil ratios (GOR) of 20,000 scf/bbl. This crossflow has been further confirmed by noise and temperature logs. Conventional cement slurries used on earlier wells were ineffective in

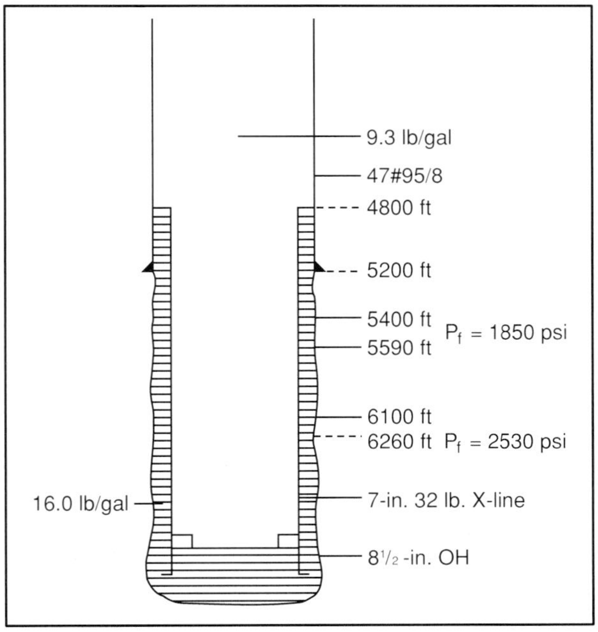

Figure 8–21—Example of well configuration from Rae et al. (1989) gas migration prediction method.

controlling the gas flow and the field was finally shut in, because of government regulations, to allow extensive remedial work to be performed.

The above example illustrates the value of selecting a cement system appropriate for the specific well conditions. Before using any predictive technique, however, it is important to appreciate its limitations. It should not be forgotten that the prediction offered by any approach is based on a number of assumptions, whether they be physical or statistical. Thus, the approach of Sutton et al. (1984) presupposed a percolation model in which gel strength is the only parameter considered, while that of Rae et al. (1989) considered gas flow through the evolving cement matrix, first as a gel and next as a very permeable porous structure.

Neither model can predict the appearance of gas flow some weeks or months after the cement job, and this should be considered because of other unrelated factors, described in Section 8–3.7. The fact that gas migration is a complex physical phenomenon comprised of several facets renders its physical modeling a formidable problem. Furthermore, it is a nonsteady-state phenomenon involving changing pressure fields and fluid saturations, and an evolving matrix structure. Heterogeneities within the cement paste, or boundary effects at the casing or formation, can induce singular events (such as nonuniform gas breakthrough) which are, by definition, unpredictable. Therefore, no one can claim to be capable of predicting the occurrence of gas migration, nor its definitive solution, on an absolute basis.

8–7 CONCLUSIONS

At present the mechanisms of gas migration are well understood, and an extensive amount of literature is available covering virtually all aspects of the subject. It should be clear from the above discussion that gas migration is an extremely complex problem requiring a considerable effort to prevent. Some solutions have been applied successfully in certain areas, but have failed when extended to other locations with different conditions. For this reason, when faced with a gas-migration problem, one should consider the well conditions carefully, and select a technique that has proved successful in similar conditions.

When dealing with such a complex problem, there is always the potential for overdesign. However, the risks associated with failure are of sufficient magnitude that an additional safety factor is justified. There is no doubt that preventing gas migration is much less costly than attempting to cure it.

REFERENCES

Baker Production Technology: "External Casing Packers—Applications, Calculations and Considerations," *Noroil* (Aug. 1986) 34–137.

Bannister, C. E.: "Evaluation of Cement Fluid-Loss Behavior Under Dynamic Conditions," paper SPE 7592, 1978.

Bannister, C. E. *et al.*: "Critical Design Parameters to Prevent Gas Invasion During Cementing Operations," paper SPE 11982, 1983.

Baret, J.-F.: "Why are Cement Fluid-Loss Additives Necessary?" paper SPE 17630, 1988.

Bearden, W. G. and Lane, R. D.: "Engineered Cementing Operations to Eliminate WOC Time," *Drill. & Prod. Prac.*, API (1961) 17–26.

Bearden, W. G. *et al.*: "Control and Prevention of Inter-Zonal Flow," paper SPE 903, 1964.

Becker, H. and Peterson, G.: "Bond of Cement Compositions for Cementing Wells," *Proc.*, Sixth World Petroleum Cong., Frankfurt, Germany (1963).

Beirute, R. M.: "The Phenomenon of Free Fall During Primary Cementing," paper SPE 13045, 1984.

Bergeron, H. A. and Grant, W. H.: "Cement Quality Control Program Shows Substantial Savings," paper SPE/IADC 18621, 1989.

Blomberg, N., Dingsovr, E. O., Svenkerud, P., and Vassoy, B.: "Boue de ciment hydraulique pour la cimentation des puits de pétrole," French Patent No. 2,587,988 (1986).

Bol, G. M., Meijs, F. H., Schouten, F. C., Stewart, R. B., and DeRoo, P. C.: "Preventing Fluid Migration Around a Well Casing," European Patent Application No. 197,609 (1986).

Bulatov, A. I., Obosin, O. N., and Kuksov, A. K.: "Occurrence of Channels in the Annular Spaces of Wells After Cementing," *Gazov. Prom.* (1970) 15, No. 2, 3–6 (translated from Russian).

Burkhalter, J. F., Childs, J. D., and Sutton, D. L.: "Well Cementing Process and Gasified Cements Useful Therein," U.S. Patent No. 4,450,010 (1984).

Cain, J. E., Shryock, S. H., and Carter, L. G.: "Cementing Steam Injection Wells in California," paper SPE 1320, 1965.

Carter, L. G. and Evans, G. W.: "A Study of Cement–Pipe Bonding," paper SPE 764, 1964.

Carter, L. G. and Slagle, K. A.: "A Study of Completion Practices to Minimize Gas Communication," paper SPE 3164, 1970.

Carter, L. G., Cooke, C., and Snelson, L.: "Cementing Research in Directional Gas Well Completions," paper SPE 4313, 1973.

Catala, G., Stowe, I., and Henry, D.: "A Combination of Acoustic Measurements to Evaluate Cementations," paper SPE 13139, 1984.

Chenevert, M. E. and Shreshta, B.: "Shrinkage Properties of Cement," paper SPE 16654, 1987.

Cheung, P. R. and Beirute, R. M.: "Gas Flow in Cements," paper SPE 11207, 1982.

Cheung, P. R. and Myrick, B. D.: "Field Evaluation of an Impermeable Cement System for Controlling Gas Migration," paper SPE 11983, 1983.

Childs, J. and Sabins, F.: "Methods of Using Thixotropic Cements for Combatting Gas Migration Problems," U.S. Patent No. 4,524,828 (1985).

Christian, W. W., Chatterji, J., and Ostroot, G. W.: "Gas Leakage in Primary Cementing—A Field Study and Laboratory Investigation," paper SPE 5517, 1975.

Cooke, C. and Cunningham, W. C.: "Filtrate Control—A Key in Successful Cementing Practices," *JPT* (1977) 951–956.

Cooke, C. E., Jr., Kluck, M. P., and Medrano, R: "Field Measurements of Annular Pressure and Temperature During Primary Cementing," paper SPE 11206, 1982.

Defossé, C. A.: "Compositions de laitiers de ciment pour cimentation de puits petroliers, permettant de controler l'eau libre, et le procede de cimentation correspondant," French Patent No. 2,540,097 (1983).

Drecq, P. and Parcevaux, P. A.: "A Single Technique Solves Gas Migration Problems Across a Wide Range of Conditions," paper SPE 17629, 1988.

Durham, K. S.: "How to Prevent Deep-Well Liner Failure," *World Oil* (Nov. 1987) 47–49.

Evans, G. W. and Carter, L. G.: "Bonding Studies of Cementing Compositions to Pipe and Formations," API Division of Production (1962).

Evans, H. P.: "An Effective Technique for Primary Cementation of Gas Wells (La tecnica efectiva para cementacion primaria en pozos de gas)," *Proc.*, Fourth Venezuela Min. Energia Minas et al Latin Amer. Drilling Cong., Caracas (1984) **1**.

Garcia, J. A. and Clark, C. R.: "An Investigation of Annular Gas Flow Following Cementing Operations," paper SPE 5701, 1976.

Geiker, M. and Knudsen, T.: "Chemical Shrinkage of Portland Cement Pastes," *Cement & Concrete Res.* (1982) **12**, No. 5, 603–610.

Gotsis, C., Roy, D. M., Licastro, P. H., and Kaushal, S.: "Thermal and Thermomechanical Analysis of a Cylindrical Cementitious Plug Hydrating in a Borehole," American Concrete Inst. Publication SP 95–4 (1984).

Grachyov, V. V. and Leonov, E. G.: "Study of Pore and Skeletal Pressure of Cement Slurry Column During the Period of Setting," *Burenie* (1969) No. 3, 17–21 (translated from Russian).

Granberry, V. L., Grant, W. H., and Clarke, J. W.: "Monitoring Blended Cement Quality and Design With a Mobile Cement Testing Laboratory," paper IADC/SPE 17179, 1989.

Griffin, T. J., Spangle, L. B., and Nelson, E. B.: "New Expanding Cement Promotes Better Bonding," *Oil & Gas J.* (June 25, 1979) 143–144.

Grinrod, M., Vassoy, B., and Dingsoyr, E. O.: "Development and Use of a Gas-Tight Cement," paper IADC/SPE 17258, 1988.

Guyvoronsky, A. A. and Farukshin, L. K.: "Hydrostatic Pressure of Cement Slurry," *Neftyanik* (1963) No. 10, 30–32 (translated from Russian).

Hannant, D. J. and Keating, J.: "Equipment for Assessing the Development of Structure in Fresh Cement Pastes by the Measurement of Shear Modulus," *Cement & Concrete Res.* (1985) **15**, 605–612.

Kieffer, J. and Rae, P.: "How Gelation Affects Oil Well Cements," *Pet. Eng. Intl.* (May 1987) 46–48.

Kucyn, P. V. et al.: "Prevention des manifestations du gaz entre le tubage et les parois du puits et de l'eruption incontrolee au cours du forage," *Gazov. Prom.* (1977) **2**, 48 (translated from Russian).

LeChatelier, H.: *Recherches Experimentales sur la Constitution des Mortiers Hydrauliques*, second edition, Dunod, Paris (1887).

Levine, D. C., Thomas, E. W., Bezner, H. P., and Talle, G. C.: "Annular Gas Flow After Cementing: A Look at Practical Solutions," paper SPE 8255, 1979.

Lukkien, H. B.: "Subsea Shallow Gas Presents Unique Problems and Solutions," *Oil & Gas J.* (Aug. 2, 1982) 120–122.

Marrast, J., Blondin, E., and Hinssieux, L.: "Well Cementing Process," U. S. Patent No. 3,926,257 (1975).

Matthews, S. M. and Copeland, J. C.: "Control of Annular Gas Flow in the Deep Anadarko Basin," paper SPE 14980, 1986.

Moore, P.: *Drilling Practices Manual*, PennWell Publishing Co., Tulsa, OK (1974).

Parcevaux, P: "Pore Size Distribution of Portland Cement Slurries at Very Early Stage of Hydration," *Cement & Concrete Res.* (1984) **14**, No. 3, 419–430.

Parcevaux, P.: "Mechanisms of Gas Channeling During Primary Cementation—Methods for Prevention and Repair," Chemische Produkte in der Erdolgewinnung,Clausthal-Zellerfeld, (Sept. 6, 1984).

Parcevaux, P. A.: "Gas Migration and GASBLOK™ Technology," *Drilling & Pumping J.* (Aug. 1987) 11–22.

Parcevaux, P. et al.: "Annular Gas Flow, a Hazard Free Solution," *Pet. Inform.* (July 15, 1983) 34–36.

Parcevaux, P. A. and Sault, P. H.: "Cement Shrinkage and Elasticity: A New Approach for a Good Zonal Isolation," paper SPE 13176, 1984.

Peralta, M.: "Control de la invasion de gas durante las operaciones de cementacion," *Petroleo Internacional* (July–Aug. 1984) 30–37.

Powers, T. C., *Ind.& Eng. Chem.* (1935) 790.

Rae, P.: "Preventing Gas Migration," *Noroil* (March 1987).

Rae, P., Wilkins, D., and Free, D.: "A New Approach for Predicting Gas Flow After Cementing," paper SPE/IADC 18622, 1989.

Rang, C. L.: "Evaluation of Gas Flows in Cement," paper SPE 16385, 1987.

Richardson, E. A.: "Nitrogen Gas Stabilized Cement and a Process for Making and Using It," U.S. Patent No. 4,333,764 (1982).

Sabins, F.L., Tinsley, J. M., and Sutton, D. L.: "Transition Time of Cement Slurries Between the Fluid and Set States," *SPEJ* (Dec. 1982) 875–882.

Seidel, F. A. and Greene, T. G.: "Use of Expanding Cement Improves Bonding and Aids in Eliminating Annular Gas Migration in Hobbs Grayburg San Andres Wells," paper SPE 14434, 1985.

Sepos, D.J. and Cart, B.W.: "New Quick Setting Cement Solves Shallow Gas Migration Problems and Reduces WOC Time," paper SPE 14500, 1985.

Setter, N. and Roy, D.M.: "Mechanical Features of Chemical Shrinkage of Cement Paste," *Cement & Concrete Res.* (1978) **8**, No. 5, 623–634.

Smith, R. C., Beirute, R. M., and Holman, G. B.: "Postanalysis of Abnormal Cementing Jobs Using a Cementing Simulator," paper SPE 14201, 1985.

Stehle, D., Sabins, F., Gibson, J., and Theis, K.: "Conoco Stops Annular Gas Flow With Special Cement," *Pet. Eng. Intl.* (April 1985) 21–24.

Stewart, R. B. and Schouten, F. C.: "Gas Invasion and Migration in Cemented Annuli: Causes and Cures," paper IADC/SPE 14779, 1986.

Stone, W. H. and Christian, W. W.: "The Inability of Unset Cement to Control Formation Pressure," SPE 4783, 1974.

Suman, G. O.: "Well Completion Method," U. S. Patent No. 4,440,226 (1984).

Sutton, D. L. and Faul, R.: "Annular Gas Flow Theory and Prevention Methods Described," *Oil & Gas J.* (Dec. 10, 1984) 84–112.

Sutton, D. L. and Prather, D. A.: "New Expansion Additive Gives Good Results With Low C_3A Cements," *Proc.*, Southwest Petroleum Short Course, Lubbock, TX (1986) 39–48.

Sutton, D. L., Sabins, F. L., and Paul, R.: "New Evaluation for Annular Gas-Flow Potential," *Oil & Gas J.* (Dec. 17, 1984) 109–112.

Sutton, D. L.: "Well Cementing Process and Gasified Cements Useful Therein," U.S. Patent No. 4,340,427 (1982).

Sykes, R. L. and Logan, J. L.: "New Technology in Gas Migration Control," paper SPE 16653, 1987.

Tinsley, J. M., Miller, E. C., Sabins, F. L., and Sutton, D. L.: "Study of Factors Causing Annular Gas Flow Following Primary Cementing," paper SPE 8257, 1979.

Vidovskii, A. I., Bulatov, A. I., Akhmetov, R. A., and Perevertov, Y.P.: "Change in Pressure of a Column of Cement Slurry Behind the Casing in a Well During Time of Setting and Hardening," *Burenie* (1971) No. 9, 27–29.

Vyalov, S. S.: *Rheological Fundamentals of Soil Mechanics*, Elsevier Science Publishing Co., New York (1986) 267–283.

Watters, L. T. and Sabins, F. L.: "Field Evaluation of Method to Control Gas Flow Following Cementing," paper SPE 9287, 1980.

Webster, W. W. and Eikerts, J. V.: "Flow After Cementing —Field and Laboratory Study," paper SPE 8259, 1979.

Williams, D., Cheung, R., Norman, M., and Woodroof, R., Jr.: "Annular Gas Migration Can be Controlled," *Oil & Gas J.* (Jan. 31, 1983) 146–151.

Wu, X., Roy, D. M., and Langton, C. A.: "Early Stage Hydration of Slag Cement," *Cement & Concrete Res.* (1983) **13**, 277–286.

9

Thermal Cements

Erik B. Nelson

Dowell Schlumberger

9–1 INTRODUCTION

High-temperature wells present special cement system design challenges. The physical and chemical behavior of well cements changes significantly at elevated temperatures and pressures. One must also pay close attention to the chemical and physical properties of the formations with which the cement will come into contact. Corrosive water zones and very weak formations are not uncommon in thermal wells. Without careful modification of slurry design, the set cement may lose strength and gain permeability, potentially resulting in the loss of zonal isolation.

Thermal cementing encompasses three principal types of wells: deep oil and gas wells, geothermal wells, and thermal recovery wells. In this chapter, each scenario is discussed separately, because the cement system design parameters can differ significantly.

Before discussing the cement system design for the various types of thermal wells, it is necessary to understand the hydrothermal chemistry of the cements used to complete thermal wells: Portland cement, Class J cement, silica-lime systems, and high-alumina cement. In this chapter, the special chemical notation for cement compounds is used. The reader is referred to Chapter 2 for an explanation of the customary abbreviations.

9–2 HIGH-TEMPERATURE CHEMISTRY OF PORTLAND CEMENT

As discussed in Chapter 2, Portland cement is essentially a calcium silicate material, the most abundant components being tricalcium silicate (C_3S) and dicalcium silicate (C_2S). Upon addition of water, both hydrate to form a gelatinous calcium silicate hydrate called "C-S-H gel," which is responsible for the strength and dimensional stability of the set cement at ordinary temperatures. In addition to C-S-H gel, a substantial amount of calcium hydroxide (CH) is liberated.

C-S-H gel is the early hydration product even at elevated temperature and pressure, and is an excellent binding material at well temperatures less than about 230°F (110°C). At higher temperatures, C-S-H gel is subject to metamorphosis, which usually results in decreased compressive strength and increased permeability of the set cement. This phenomenon, known as "strength retrogression," was first reported in the petroleum literature by Swayze (1954) as a result of the growing trend toward deep well completions.

C-S-H gel often converts to a phase called "alpha dicalcium silicate hydrate (α-C_2SH)." α-C_2SH is highly crystalline and much more dense than C-S-H gel. As a result, a shrinkage occurs which is deleterious to the integrity of the set cement. This effect is illustrated in Fig. 9–1,

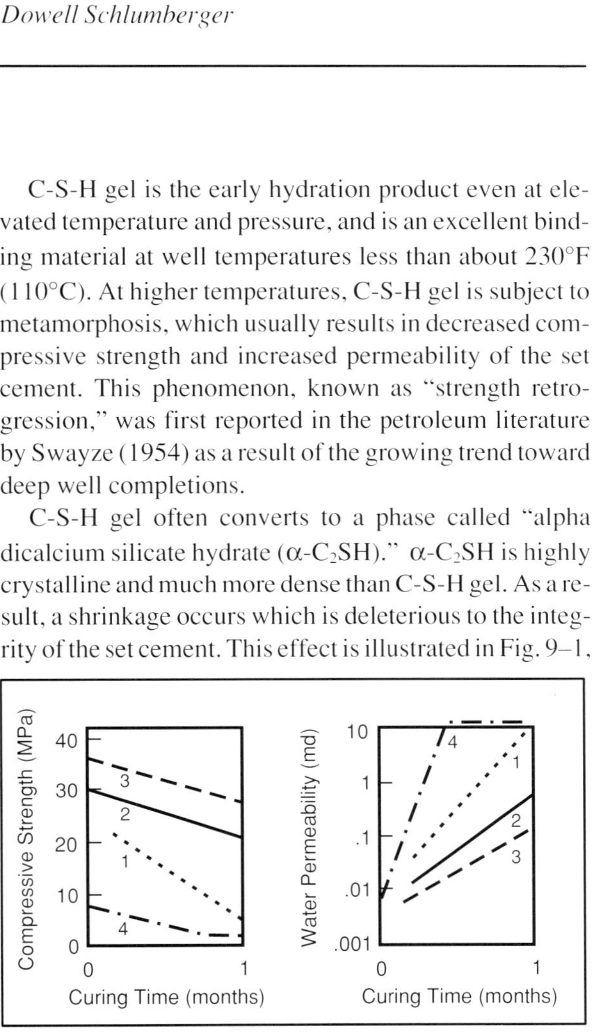

Figure 9–1—Compressive strength and permeability behavior of neat Portland cement systems at 230°C (from Nelson and Eilers, 1985).

which depicts the compressive strength and water permeability behavior of conventional Portland cement systems cured at 446°F (230°C). Significant loss of compressive strength occurred within one month; however, the levels to which strength falls are sufficient to support casing in a well (Suman and Ellis, 1977). The real problem lies in the severe permeability increases. To prevent interzonal communication, the water permeability of well cements should be no more than 0.1 md. Within one month, the water permeabilities of the normal density Class G systems (1, 2) were 10 to 100 times higher than the recommended limit. The permeability of the high-density Class H system (3) was barely acceptable. The deterioration of the lower density extended cement (4) was much more severe.

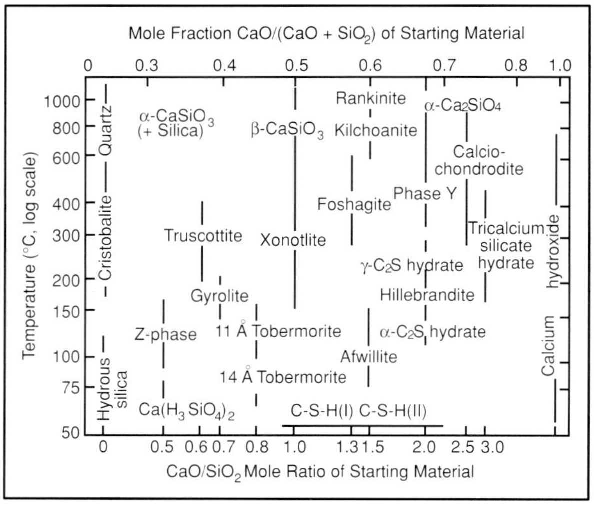

Figure 9–2—Formation conditions of various calcium silicates (from Taylor, 1964).

The strength retrogression problem can be prevented by reducing the bulk lime-to-silica ratio (C/S ratio) in the cement (Menzel, 1935; Kalousek, 1952; Carter and Smith, 1958). To accomplish this, the Portland cement is partially replaced by ground quartz, usually as fine silica sand or silica flour. In some areas, special cements are available where quartz has been interground with Portland cement clinker (Italcementi, 1977; Berra et al., 1988). Figure 9–2 is a diagram depicting the conditions for the formation of various calcium silicate compounds, many of which occur geologically (Taylor, 1964). The C/S ratio is plotted vs the curing temperature. C-S-H gel has a variable C/S ratio, averaging about 1.5. The conversion to α-C_2SH at 230°F (110°C) can be prevented by the addition of 35% to 40% silica (BWOC), reducing the C/S ratio to about 1.0. At this level, a mineral known as tobermorite ($C_5S_6H_5$) is formed; fortunately, high strength and low permeability are preserved. As the curing temperature increases to about 300°F (150°C), tobermorite normally converts to xonotlite (C_6S_6H) and a smaller amount of gyrolite ($C_6S_3H_2$) with minimal deterioration. Tobermorite sometimes persists to 482°F (250°C) in Portland cement systems because of aluminum substitution in the lattice structure (Kalousek and Chow, 1976).

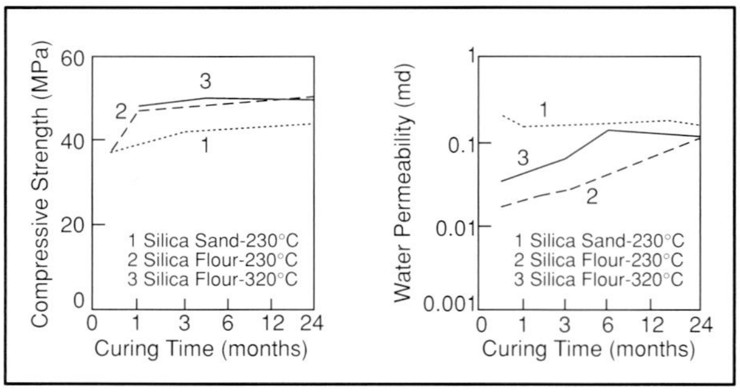

Figure 9–3—Compressive strength and permeability behavior of 16.0-lb/gal Class G systems stabilized with 35% silica (from Nelson and Eilers, 1985).

The improved performance of "silica-stabilized" Portland cements at elevated temperatures is illustrated in Fig. 9–3. Normal density Class G cements, stabilized with silica sand or silica flour, were cured at 446° and 608°F (230°and 320°C).

At 480°F (250°C) the phase truscottite ($C_7S_{12}H_3$) begins to appear (Luke and Taylor, 1984). As the curing temperature approaches 750°F (400°C), both xonotlite and truscottite are near their maximum stable temperatures, and dehydration of the residual CH to C occurs. At higher temperatures, the xonotlite and truscottite dehydrate, resulting in the disintegration of the set cement.

In addition to the compounds cited above, other phases such as pectolite (NC_4S_6H), scawtite ($C_7S_6\bar{C}H_2$), reyerite ($KC_{14}S_{24}H_5$), kilchoanite (C_3S_2H approximately), and calcio-chondrodite (C_5S_2H approximately) may appear in Portland cement systems cured at elevated temperatures. These phases can affect the performance of the set cement, even when present in small quantities.

Cements containing significant amounts of truscottite are usually characterized by low permeability (Gallus et al., 1978). The formation of pectolite, a sodium calcium silicate hydrate, is accompanied by cement expansion (Nelson and Eilers, 1982); in addition, pectolite appears to render cements more resistant to corrosion by highly saline brines (Nelson and Kalousek, 1977; Nelson et al., 1981). Scawtite has been shown to enhance cement compressive strength when present in minor amounts (Eilers et al., 1983). In general, set cements which consist predominantly of calcium silicate hydrates with C/S ratios less than or equal to 1.0 tend to have higher compressive strengths and lower water permeabilities.

9–3 CLASS J CEMENT

Class J cement (a provisional API designation) was developed in the early 1970s for cementing wells with static temperatures in excess of 260°F (126°C) (Maravilla,

1974). This cement is advantageous from a logistical point of view, because the addition of silica is not required.

Like Portland cement, Class J cement is a calcium silicate material; however, no aluminate phases or C_3S are present. The composition is essentially β-C_2S, α-quartz, and CH. As discussed in Chapter 2, the hydration rate of β-C_2S is relatively slow; consequently, retarders are rarely necessary with Class J cement at circulating temperatures less than 300°F (149°C). The C/S ratio of Class J cement is adjusted such that tobermorite and xonotlite (scawtite also occurs frequently) are obtained upon curing (Kalousek and Nelson, 1978; Sasaki et al., 1984). In addition, the sulfate resistance of Class J cement is very high because of the absence of C_3A. Despite these attributes, the availability of Class J cement is very limited today.

9–4 SILICA-LIME SYSTEMS

The silica-lime system consists of a simple mixture of ground α-quartz and hydrated lime. At temperatures above 200°F (94°C), lime reacts with the silica to form calcium silicate hydrates such as tobermorite (Hook et al., 1971), provided the two materials are blended in the correct stoichiometric ratio.

Silica-lime blends are reported to behave more predictably than Portland cement-base systems, because of the absence of many impurities. The blends respond to common cement retarders, and extenders or weighting agents can be added to vary the slurry density from 12.5 to 20 lb/gal (1.50 to 2.40 g/cm³).

9–5 HIGH-ALUMINA CEMENT

High-alumina cement is a special material manufactured primarily for applications where a refractory binder is required (Robson, 1962). In wells, it is used where the in-situ combustion process is employed (Section 9–8), and is also useful for cementing across permafrost zones (Chapter 7). The primary cementitious constituent is monocalcium aluminate (CA). As illustrated in Fig. 9–4, there are three initial metastable hydrates which occur when water is added to CA: CAH_{10}, C_2AH_8 and C_4AH_{13}.

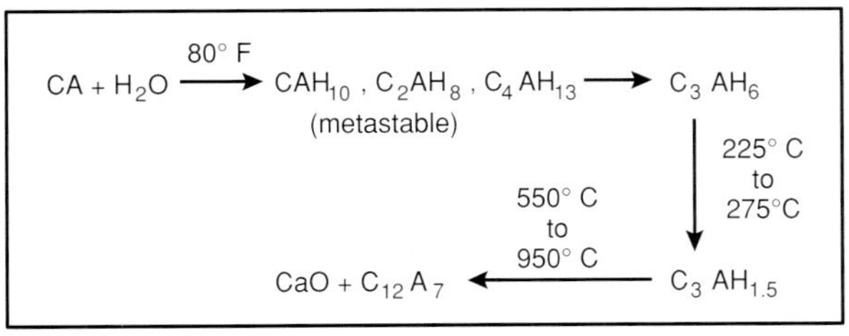

Figure 9–4—Sequence of reactions of high-alumina cement at various curing temperatures.

They ultimately convert to C_3AH_6 (Quon and Malhotra, 1979). Unlike Portland cement, set calcium aluminate cement does not contain calcium hydroxide.

C_3AH_6 is probably the only stable hydrated calcium aluminate at temperatures below 437°F (225°C). At higher temperatures, the water content begins to drop, and at 527°F (275°C) $C_3AH_{1.5}$ is found. As the temperature continues to increase, decomposition of $C_3AH_{1.5}$ occurs with the liberation of C. Between 1,022°F (550°C) and 1,742°F (950°C), a recrystallization occurs ultimately resulting in C and $C_{12}A_7$.

It should be noted that high-alumina cement is not used in ultrahigh-temperature wells for greater retention of compressive strength. At temperatures up to 930°F (500°C), the proportional strength loss is often greater than that experienced by unstabilized Portland cements. High-alumina cement is used because of its stability to wide-ranging temperature fluctuations, owing mainly to the absence of calcium hydroxide. Figure 9–5 illustrates

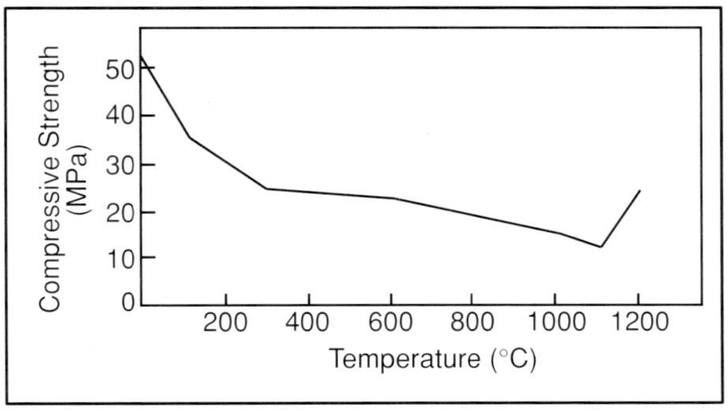

Figure 9–5—Compressive strength of high-alumina cement/crushed firebrick concrete after four months' exposure from 20° to 1,200°C (from Heindl and Post, 1954).

the effect of curing temperature upon a high-alumina cement extended with 70% crushed firebrick (Heindl and Post, 1954). The initial strength loss between room temperature and 212°F (100°C) is primarily due to the conversion of the initial hexagonal calcium aluminate hydrates to the cubic C_3AH_6. With further heating, the strength continues to drop because of dehydration and the formation of C and $C_{12}A_7$. Strength improves above 1,830°F (1,000°C) as the $C_{12}A_7$ crystals intergrow and form a tightly bonded "ceramic" network. In thermal wells, such a high temperature is not generally exceeded; thus, it is important to ensure that the minimum compressive strength obtained is sufficient for maintenance of well integrity.

The strength and durability of high-alumina cements between 440° and 1,830°F (225° and 1,000°C) are primarily controlled by the initial water-to-cement ratio. Depending upon the application, the amount of added water should be the minimum to obtain a pumpable slurry. The use of dispersants is particularly helpful. A higher proportion of cement with respect to an aggregate extender is also necessary. For most applications, at least 50% of the solids should be cement.

A great variety of materials may be used as extenders in calcium aluminate cement slurries, provided they have suitable stability at high temperatures, and do not decompose or show anomalous thermal expansions or inversions. Silica sand should not be employed if temperatures exceeding 572°F (300°C) are anticipated. Because of crystalline modification, thermal expansion of quartz is relatively high at these temperatures, and thermal cycling could eventually disrupt the cement. The most commonly used extender for these temperatures is crushed aluminosilicate firebrick. Other materials which have been found suitable include calcined bauxite, certain fly ashes, diatomaceous earth, and perlite.

9–6 DEEP OIL AND GAS WELLS

Wells with depths exceeding 15,000 ft (4,570 m), with bottomhole temperatures above 230°F (110°C), are common throughout the world. In recent years, several wells with depths exceeding 25,000 ft (7,600 m) have been completed (Arnold, 1980; Wooley et al., 1984). Such wells represent a large investment of time and money; therefore, obtaining a successful well completion is of paramount importance.

The procedures for cementing deep wells are basically the same as those for shallower wells; however, such wells are generally considered critical, because of the more severe well conditions and higher complexity of the casing programs (Smith, 1987). Higher temperatures, narrower annuli, overpressured zones, and corrosive fluids are commonly encountered. Consequently, the cement system design can be complex, involving an elaborate array of retarders, fluid-loss additives, dispersants, silica, and weighting materials. One must be certain that the cement system can be properly placed, and will maintain zonal isolation throughout the life of the well. Portland cement is used in virtually all deep oil and gas well completions.

Typical casing programs and cementing procedures for deep wells are given in Chapter 12. Detailed information regarding the various types of cement additives is found in Chapter 3. In this section, information is presented concerning the design of appropriate cement systems for deep high-temperature wells.

9–6.1 Thickening Time and Initial Compressive Strength Development

In deep wells, at least three to four hours of pumping time are usually required to allow adequate placement time. However, there are several complicating factors which need to be mentioned.

As the length of the casing string or liner increases, the problem of achieving a cement seal becomes more severe (Suman and Ellis, 1977). Static temperature differentials in excess of 38°C (100°F) have been noted in many cases between the top and bottom of the cement column. Sufficient retarder must be added to the cement slurry to allow adequate placement time at the maximum circulating temperature; consequently, such a slurry may be overretarded at the top of the cement column, resulting in a very long waiting-on-cement (WOC) time. If high-pressure gas exists behind the string or liner, the risk of gas invasion into the cement is high (Chapter 8).

When designing cement slurries for deep, hot wells, it is very important that accurate static and circulating temperature information be used. These data may be obtained from drillstem tests, logs, special temperature recording subs, or circulating temperature probes run during hole conditioning (Jones, 1986). Computer programs have also been developed to better predict well temperatures (Wedelich et al., 1987). The circulation of fluids in the well for several hours prior to cementing can significantly decrease well temperatures; thus, there is a danger of overestimating the circulating temperature, and overretarding the slurry.

The cement slurry is exposed to high pressures in deep wells and, as shown in Fig. 9–6, a significant accelerating effect is observed (Bearden, 1959). Earlier compressive strength development and higher ultimate compressive strength are also observed as curing pressure increases (Handin, 1965; Metcalf and Dresher, 1978). Therefore, when designing a proper cement slurry composition in the laboratory, performing the tests at the anticipated pressure is recommended (Appendix B).

In general, the higher the circulating temperature, the higher the sensitivity of Portland cement systems to subtle chemical and physical differences between the slurry ingredients. Therefore, all laboratory tests should be performed with samples of the water, cement, and additives which will be used during the job.

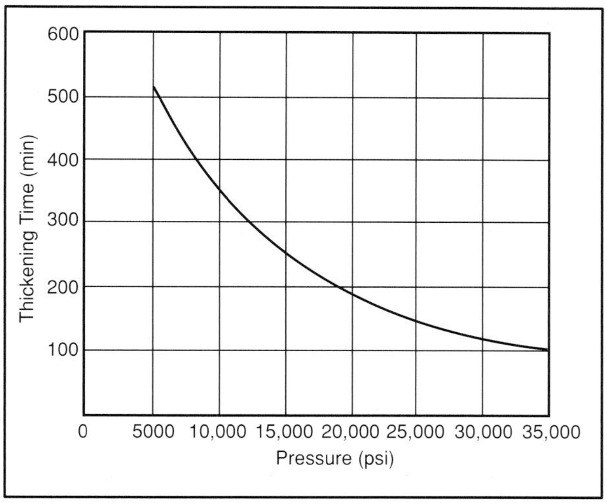

Figure 9–6—Effect of pressure on pumpability of cement. (Cement: API Class H with 0.3% retarder; bottomhole circulating temperature: 200°F) (after Smith, 1976).

9–6.2 Cement Slurry Rheology

The narrow annuli associated with deep well completions increase the difficulty of achieving a good bond between the cement and the pipe and formation. The risk of cement contamination by drilling fluids is increased by the small clearance between the casing and open hole. Proper centralization is difficult to achieve. As discussed in Chapter 5, the rheology of the completion fluids is a crucial aspect. In many cases, the cement slurry is designed to be pumped in turbulent flow; therefore, the use of dispersants is common. When designing highly dispersed slurries, one must be careful to avoid sedimentation and free water development. This is especially important when the borehole is highly deviated (Chapter 15).

9–6.3 Cement Slurry Density

Deep wells often involve cementing across high-pressure formations. To maintain control of the well, the hydrostatic pressure of the wellbore fluids must meet or exceed the formation pressure at all times. Consequently, cement slurries with densities as high as 22 lb/gal (2.64 g/cm^3) are often placed. When large quantities of weighting materials are present in the slurry, sedimentation is again a major concern.

9–6.4 Fluid-Loss Control

As discussed in Chapter 3, fluid-loss control is necessary to preserve the chemical and physical characteristics of the cement slurry, and to prevent the development of a cement filter cake which could cause bridging in the annulus. For most primary cementing operations, an API fluid-loss rate between 50 to 100 mL/30 min is generally considered to be adequate.

9–6.5 Long-Term Performance of Cements for Deep Wells

Once the cement system is successfully placed in the annulus, it is important to ensure that adequate casing support and zonal isolation will be provided throughout the life of the well. As discussed earlier in this chapter, the most important method for stabilizing Portland cements to a thermal environment is the addition of sufficient silica to produce C-S-H phases conducive to high strength and low permeability.

A typical slurry composition for a deep, hot well would consist of Class H or Class G cement, 35% to 40% silica (BWOC), a dispersant, a fluid-loss additive, a retarder, and a weighting agent. The long-term performance of such cement systems would be very similar to that shown in Fig. 9–3.

When high-density slurries are unnecessary, or if lower density slurries are required to prevent lost circulation or formation breakdown, extenders such as fly ash, diatomaceous earth, bentonite, perlite, etc., are commonly used. The long-term performance of typical systems in laboratory tests is illustrated in Figs. 9–7 and 9–8. All systems contained 35% silica flour (BWOC). In Fig. 9–7, the systems have been cured at 450°F (232°C) under saturated steam pressure for up to two years, and compressive strength and permeability measurements have been performed at periods ranging from one day to 24 months. Figure 9–8 presents data for systems cured at 600°F (315°C). It is important to note the nonlinear time scale and the logarithmic permeability scale.

System 1 contained Type F fly ash as an extender and was the heaviest of the four. Despite the density advantage and the highest initial compressive strength, the performance of System 1 over a two-year period was no better than the lower density systems at 450°F (232°C), and was the poorest of the four at 600°F (315°C). This delayed degradation of fly-ash-containing systems was probably the result of alkali contaminants in the fly ash. Such contaminants can slowly react and form substituted calcium silicate hydrates, notably reyerite, with deleterious effects (Eilers and Root, 1976). It is important to mention that cement degradation associated with fly ash has not been observed at curing temperatures below 450°F (232°C).

Systems 2 and 3 were extended with perlite and bentonite. System 2 performed well at both 450° and 600°F (232° and 315°C) with regard to compressive strength. The permeability of System 2 varied back and forth across the 0.1-md line. System 3 was the least dense of

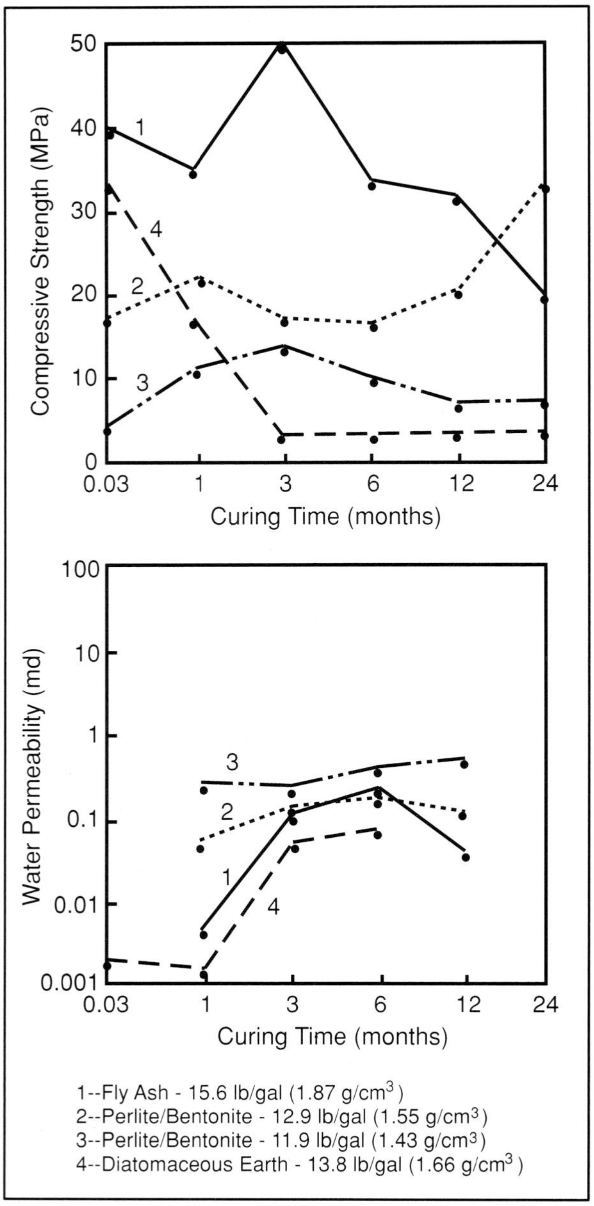

Figure 9–7—Compressive strength and permeability performance of conventionally extended Portland cement slurries—232°C (after Nelson and Eilers, 1985).

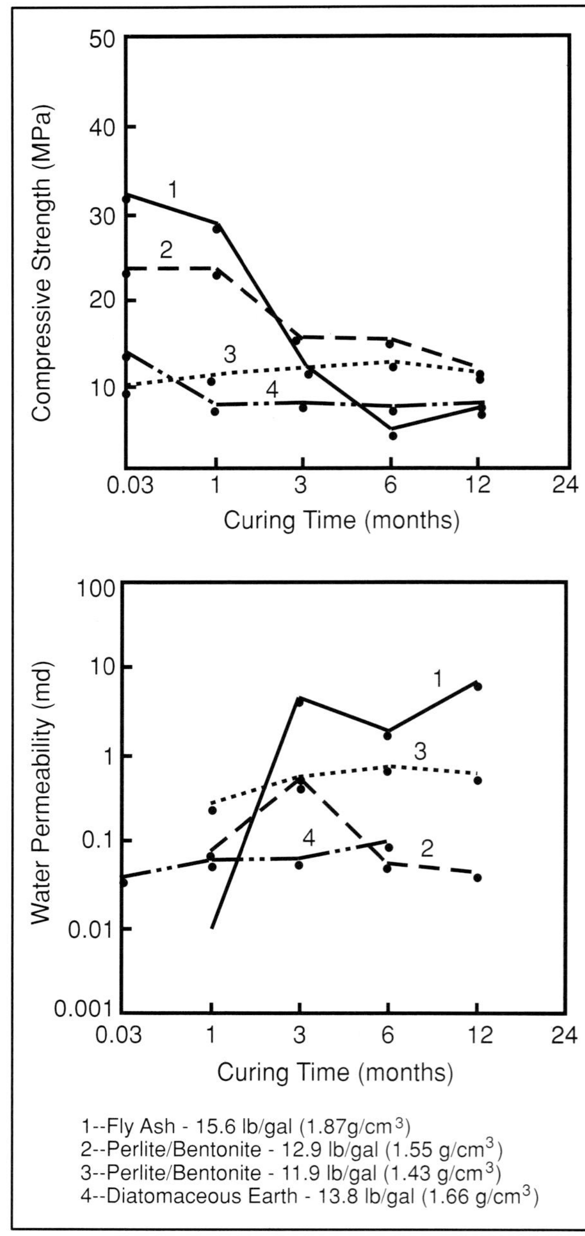

Figure 9–8—Compressive strength and permeability performanace of conventionally extended Portland cement slurries—315° C (after Nelson and Eilers, 1985).

the four. The compressive strength performance was adequate at both curing temperatures, but the permeabilites were too high. System 4, containing diatomaceous earth, was a rather poor performer in the strength category, yet had low permeability.

Figure 9–9 shows the typical performance of a normal density neat Class J system. Its behavior is similar to that observed with normal density silica-stabilized Portland cement systems.

The behavior of these systems illustrates that high compressive strength and low water permeability are not necessarily linked. Although water permeability is not as convenient to determine as compressive strength (Appendix B), one should do so before the application of a cement in severe downhole conditions. In addition, the data suggest that conventionally extended Portland cement systems with densities below about 12.5 lb/gal (1.5 g/cm^3) may not be able to perform suitably in high-temperature wells, except perhaps as "filler" systems which are not placed across producing zones.

If competent cement systems with densities less than 12.5 lb/gal (1.5 g/cm^3) are necessary, microsphere-extended (Chapter 3) or foamed cements (Chapter 14) may

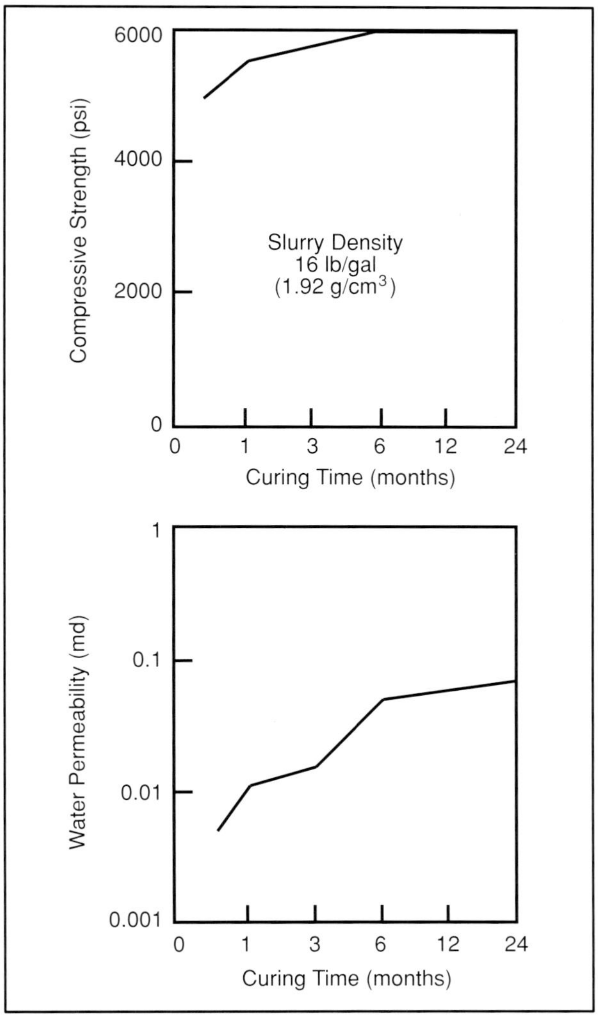

Figure 9–9—Compressive strength and permeablity behavior of Class J cement at 230°C.

be appropriate. However, when contemplating the use of ceramic or glass microspheres, one must be certain that they can withstand the hydrostatic pressure. Ceramic microspheres and most grades of glass microspheres can withstand no more than 3,000 psi (20.7 MPa), which eliminates them from consideration in most deep well completions. However, glass microspheres with hydrostatic crush strengths as high as 10,000 psi (69.0 MPa) are available. Foamed cement, occasionally used in deep high-temperature wells, is more common in geothermal and steamflood wells.

9–7 GEOTHERMAL WELL CEMENTING

Projects to extract geothermal energy exist throughout the world. Virtually any location with thermal anomalies is a potential site for geothermal well drilling. Some of the more notable geothermal projects are located in California, Utah, New Mexico, Mexico, the Philippines, Indonesia, New Zealand, Iceland, and Italy.

At present, geothermal wells are usually completed in much the same manner as conventional oil and gas wells; however, the environment with which the cements must contend is frequently much more severe. The bottom hole temperature in a geothermal well can be as high as 700°F (370°C), and the formation brines are often extremely saline and corrosive. The failure of wells in several geothermal fields has been directly attributed to cement failure (Kennerly, 1961; Radenti and Ghiringelli, 1972; Shen, 1989); as a result, extensive research has been conducted to identify cement formulations which perform suitably under such conditions.

9–7.1 Well Conditions Associated With Geothermal Wells

With the exception of hot, dry rock completions with circulating temperatures as high as 500°F (260°C) (Carden et al., 1983), the majority of geothermal wells is not cemented under "geothermal" conditions, because the fluids circulated during drilling cool the formation. The maximum circulating temperatures during the cement job seldom exceed 240°F (116°C); therefore, the design of cement systems with adequate thickening times is usually not a problem. Most geothermal wells are less than 10,000 ft (3,050 m) in depth. Downhole pressures are seldom above the water gradient.

The drilling programs for geothermal wells usually call for setting surface and production casing above the reservoir. In some cases, a slotted liner is hung through the producing zone, but cementing the liner is not considered critical. It is very important to cement the casings to the surface; otherwise, creep or elongation will occur because of thermal expansion when the well is brought into production (Shryock, 1984).

The nature of an economical geothermal reservoir is such that large quantities of hot water or steam must be produced from each well. Therefore, the reservoirs are usually naturally fractured and have effective permeabilities that are probably greater than one darcy. The integrity of the formations ranges from poorly consolidated to highly fractured, and the fracture gradients tend to be low; thus, lost circulation is a common problem. For this reason, low-density cement systems are required by most geothermal operators (Nelson, 1981).

The chemistry of the reservoir fluids varies from fresh water to saline brines with greater than 200,000 mg/L total dissolved solids. The fluids extracted from dry steam fields contain relatively few salts and low concentrations of noncondensible gases, the most noticeable being H_2S.

The saline brines often contain significant quantities of carbonate and sulfate.

9–7.2 Performance Requirements and Design Considerations

Geothermal wells arguably present the most severe conditions to which well cements are exposed. As a result, the performance requirements are among the most stringent. At present, geothermal well cements are usually designed to provide at least 1,000-psi (7.0-MPa) compressive strength, and no more than 0.1-md water permeability (API Task Group on Cements for Geothermal Wells, 1985). In addition, the set cement often must be resistant to degradation by saline brines.

When the cement is to contact highly saline and corrosive geothermal brines, the particle size of the added silica is an important consideration. As explained in Chapter 3, there are two forms of silica commonly used in well cementing: silica sand, with a particle size of approximately 175 to 200 µm, and silica flour, with an average particle size of about 15 µm. Silica sand is usually preferred by field personnel, because its lower surface area facilitates easier slurry mixing. However, in certain geothermal environments, silica sand cannot be relied upon to provide adequate stabilization.

Eilers and Nelson (1979) investigated the effect of silica particle size on the performance of Class G cement formulations cured at various temperatures in a geothermal brine. The salinity of the brine was 25,000 mg/L TDS. Figure 9–10 shows the relationships between the silica particle size and several parameters—compressive strength, water permeability and cement phase composition. The slurry density was 15.8 lb/gal (1.90 g/cm^3). A decrease in compressive strength and an increase in water permeability occurred when the average particle size of the added silica exceeded about 15 µm. Xonotlite was also replaced by kilchoanite as the predominant cement phase. Figure 9–11 shows that the silica particle-size effect is significantly more pronounced with lower density cement compositions.

High concentrations of sodium chloride depress the rate at which silica enters solution (Fournier, 1979); as a result, when the silica particle size is large, the rate of dissolution of silica is insufficient to allow the formation of the desired calcium silicate hydrates (C/S ratio <1). The kinetics of dissolution can be affected by the particle size of the solute. Reducing the particle size of the silica increases its surface area; consequently, a sufficient supply of silica is available.

More recently, Grabowski and Gillott (1989) studied the effects of silica fume, with an average particle size of approximately 0.1 µm (Chapter 3), upon Portland cement systems at elevated temperatures and pressures. Maintaining a constant SiO_2 concentration (40% BWOC) and water-to-solids ratio (0.5), samples were prepared containing silica fume, combinations of silica fume and silica flour, and silica flour. Curing was performed at 450°F (230°C) and 400 psi (2.75 MPa) for 7 days, using samples aged under ambient conditions for periods up to 270 days. The systems containing silica fume developed less compressive strength, but lower permeability, than equivalent systems containing only silica flour (Fig. 9–12). The major phase found in all of the samples was xonotlite (scawtite was detected in the samples containing only silica flour); however, the microstructures were different. The samples containing silica flour exhibited short parallel needles of xonotlite. As the quantity of silica fume increased, the texture of the

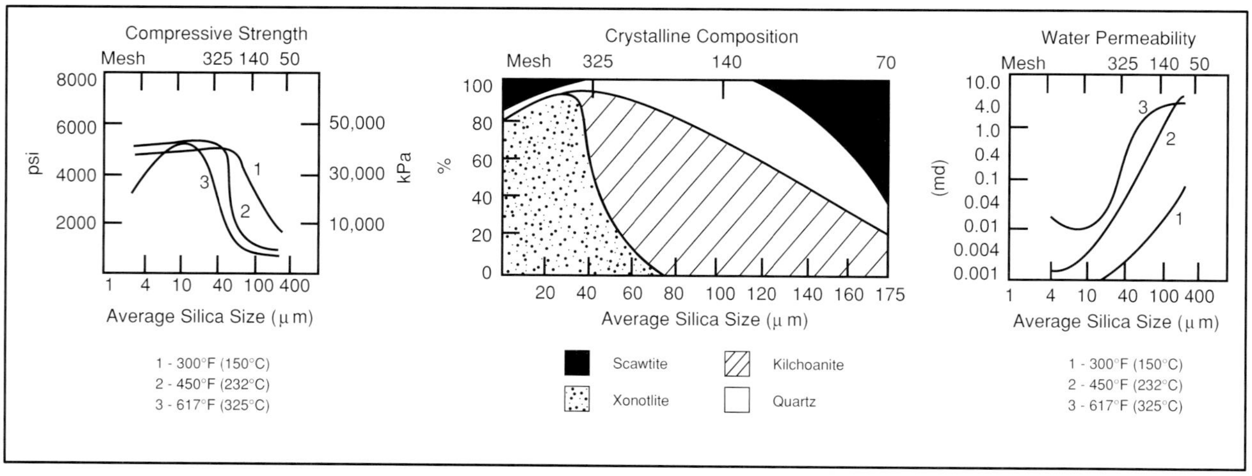

Figure 9–10—Effect of silica particle size on the performance of Class G cement cured in geothermal brine (from Eilers and Nelson, 1979).

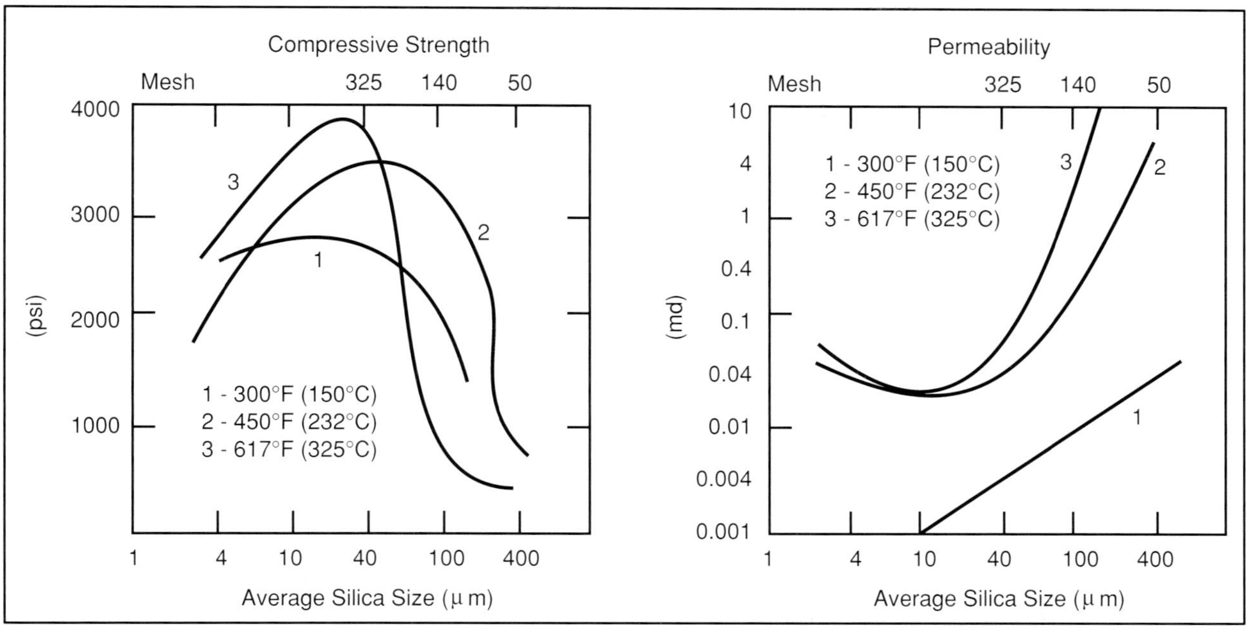

Figure 9–11—Effect of silica particle size on performance of 13.5-lb/gal Class G perlite/bentonite system cured in geothermal brine (from Eilers and Nelson, 1979).

xonotlite was granular. In general, the samples with needle-shaped xonotlite crystals exhibited the higher permeabilities.

The presence of carbonate in certain geothermal brines presents a serious difficulty for Portland cement systems (Chapter 7). Calcium silicate hydrates are not stable in such a chemical environment, even at ordinary temperatures (Taylor, 1964). Upon exposure to carbonate solutions, calcium silicate hydrates are eventually converted to a mixture of calcium carbonate and amorphous silica. This phenomenon has been observed in well cements by numerous researchers (Onan, 1984; Bruckdorfer, 1986; Shen, 1989). High alumina cements are also known to suffer from degradation in the presence of carbonate. At present there appear to be no published data regarding the behavior of high alumina cements in a carbonaceous environment at elevated temperatures; however, a study is currently in progress (Kukacka, 1989).

The principal defense against such degradation has traditionally been the placement of low-C/S ratio cement systems with very low permeability, and successful results have generally been obtained. However, such systems have recently been shown to be inadequate for geothermal wells with formations containing very high concentrations of CO_2 (Hedenquest and Stewart, 1985). A recent study by Milestone et al. (1986, 1987) demonstrated that tobermorite and xonotlite are among the least resistant cement phases to carbonation, and the deterioration is accelerated when bentonite is present in the cement. They discovered that reducing the silica flour concentration from 35% to 20% (BWOC) improves the cement's resistance to CO_2. When less silica is present, weaker and more permeable calcium silicate hydrates are obtained; however, a substantial quantity of calcium hydroxide also remains in the system. Upon substantial carbonation, the calcium hydroxide reacts to form a protective layer of calcite, the permeability decreases, and further attack is inhibited.

Another method for preventing cement degradation by corrosive geothermal brines would be the placement of cements which are chemically inert to such an environment. Such systems, commonly referred to as "synthetic cements," are used routinely to complete wells for CO_2-flooding projects or chemical waste disposal (Chapter 7). Epoxy-base polymer systems are most commonly used for such applications; unfortunately, they would suffer thermal degradation at the temperatures encountered in geothermal wells.

Research has been performed with polymers which are stable to high temperatures. Zeldin and Kukacka (1980) developed an organosiloxane polymer cement which was proven suitable as a geothermal cement in an API study. A coal-filled furfuryl alcohol-base cement system for geothermal wells was invented by Eilers (1985). No commercial use of these technologies has been reported.

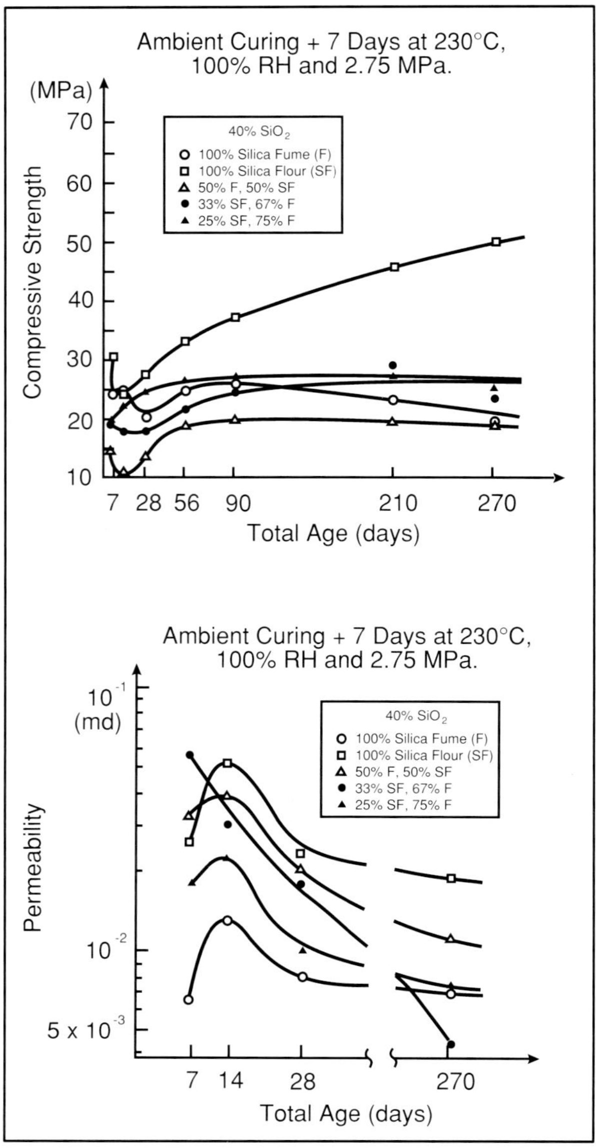

Figure 9–12—Compressive strength and permeability behavior of silica-stabilized Portland cements containing various amounts of silica fume (after Grabowski and Gillot, 1989).

9–7.3 Geothermal Well Cement Compositions

Most Portland cements and Class J cement have been shown to perform suitably in geothermal wells. Normal density cement systems are best at providing sufficient compressive strength and, more importantly, low water permeability. The cement system designs for geothermal wells differ from those for conventional high-temperature oil and gas wells in two principal ways: the exclusive use of silica flour instead of silica sand for stabilization, and the avoidance of fly ash as an extender.

Because of the presence of weak formations and low fracture gradients, lower density cements are often required. Therefore, much research has been performed to develop low-density systems that will perform adequately. The typical extenders used to prepare low-density geothermal cements are bentonite, perlite, and diatomaceous earth. Additional silica flour, up to 100% by weight of cement, is sometimes included in lower density systems to ensure proper stabilization (Gallus et al., 1979).

Table 9–1 lists the compositions of both normal and low-density systems which are often used as geothermal cements. The compressive strength and water permeability upon long-term exposure to actual geothermal conditions are shown in Figs. 9–13 and 9–14, respectively.

More recently, ultralow-density foamed cements (Rickard, 1985; Sugama et al., 1986) and microsphere-extended systems have been used to cement geothermal wells. Such systems have been used successfully in thermal recovery wells (Section 9–7.1); however, very limited data have been published regarding the long-term stability of these systems to corrosive brines, and research is continuing (Kukacka, 1989). In the meantime, it would be prudent to restrict the use of ultralow-density systems to applications where formation fluids are relatively clean.

9–8 THERMAL RECOVERY WELLS

The application of heat to stimulate oil production has been practiced for over 50 years. Methods such as in-situ combustion (fireflood), downhole heaters, hot fluid injection, and steam stimulation have been used. In-situ combustion and steam injection are the most popular methods practiced today. These techniques have been the salvation of many oil fields with high-viscosity crudes, and essentially involve the trading of heat for viscosity reduction (Kastrop, 1965).

Like geothermal wells, the formations associated with steam recovery and fireflood wells are frequently problematic. Weak and unconsolidated zones with low fracture pressures and high permeability are often present; as a result, severe lost circulation and fluid-loss problems are often encountered.

Thermal recovery wells are usually less than 3,000 ft (915 m) in depth, and are frequently deviated (30° to 50°). The circulating temperatures during primary cementing operations are often less than 104°F (40°C), and accelerators such as calcium chloride or sodium chloride are often added to promote early cement strength development.

Thermal recovery wells are always cemented to surface and, when heat is initially supplied, the temperature rise should be controlled to prevent undue thermal shock to the casing and cement. Nevertheless, because of

Sample Code	Parts by Weight	Components	Slurry Weight
1	100	API Class G cement	
		(64.2C, 21.5S, 3.9A, 3.8F)	1.81 g/cm³
	35	Silica flour	(15.1 lb/gal)
	1	Lignin-sugar	
	54	Water	
2	100	API Class J cement	
		(37.3C, 54.2S, 1.1A, 1.0F)	1.85 g/cm³
	0.4	Lignin-sugar	(15.4 lb/gal)
	44	Water	
3	100	API Class F cement	
		(63.9C, 21.1S, 3.1A, 54.F)	1.81 g/cm³
	40	Silica flour	(15.1 lb/gal)
	0.7	Lignin-sugar	
	63	Water	
4	30	API Class J cement	
	40	Pozzolan	1.65 g/cm³
	30	Blast furnace slag	(13.7 lb/gal)
	0.5	Carboxymethylcellulose	
	60	Water	
5	100	API Class G cement	
		(64.2C, 21.5S, 2.9A, 3.8F)	1.62 g/cm³
	35	Silica flour	(13.5 lb/gal)
	8.5	Perlite	
	2	Bentonite	
	1	Lignin-sugar	
	116	Water	
6	100	API Class G cement	
		(64.2C, 21.5S, 3.9A, 3.8F)	1.68 g/cm³
	35	Silica flour	(14.0 lb/gal)
	10	Diatomaceous earth	
	1	Lignin-sugar	
	91	Water	
7	100	API Class G cement	
	40	Silica flour	1.86 g/cm³
	0.8	Dispersant	(15.5 lb/gal)
	0.8	Fluid-loss agent	
	0.4	Retarder	
	60.3	Water	
8	100	API Class G cement	
	100	Silica flour	1.63 g/cm³
	0.3	Retarder	(13.6 lb/gal)
	85.1	Water	
9	100	API Class G cement	
	80	Silica flour	1.85 g/cm³
	0.5	Fluid-loss agent	(15.4 lb/gal)
	0.3	Retarder	
	76.8	Water	
10	100	API Class G cement	
	40	Silica flour	1.89 g/cm³
	1	Retarder	(15.7 lb/gal)
	59.2	Water	

A = Al_2O_3, C = CaO, F = Fe_2O_3, M = MgO, S = SiO_2

Table 9–1—Compositions of typical geothermal cement systems (from API Task Group on Geothermal Well Cements, 1985).

thermal expansion, high levels of stress are built up in the pipe and the cement sheath (Pollock et al., 1966); therefore, the strongest possible pipe/cement and cement/formation bonds are necessary. Failure of the bonds could allow interzonal communication and pipe expansion. The ultimate result would be casing failure by buckling or telescoping (Humphrey, 1960). A substantial amount of work has been performed to devise cementing techniques which minimize the effects of thermal expansion. Such methods include the placement of thermal packers (Smith, 1966), and the inclusion of a sliding sleeve in the casing string which can move freely in response to thermal stress (Greer and Shryock, 1967). A third procedure involves holding the casing in tension during the cement job to minimize the expansion when thermal stress is eventually applied (Farouq Ali and Meldau, 1979).

The cement must also be able to withstand the elevated temperature exposure and thermal cycling associated with steamflood and fireflood wells. To maximize the delivery of heat to the pay zones, an insulating cement is desirable in thermal recovery wells; however, the presence of such cements places additional thermal stress on the casing (Leutwyler, 1966). Thermal conductivity is more dependent upon the cement density than cement composition (Nelson, 1986). At equivalent density, the thermal conductivity of a foamed cement is only marginally different from that of a conventionally extended cement. Typical laboratory data are shown in Fig. 9–15.

9–8.1 Steam Recovery Wells

Steam recovery may be either steamflooding or cyclic steam stimulation (Gates and Holmes, 1967). Steamflooding consists of injecting steam into an injection well and on through the formation to a production well. Cyclic steam stimulation of production wells involves the injection of steam into the production well for a short period of time, and returning the well to production (Earlougher, 1968). Steam recovery techniques are practiced extensively throughout the world (Chu, 1983). The most important steamflood fields are located in central and southern California, Alberta, Saskatchewan, Venezuela, Holland, West Germany and Indonesia. Reservoir temperatures seldom exceed 600°F (315°C); therefore, Portland cement is used in virtually all well completions.

The characteristics of steamflood wells and the associated performance requirements of cementing materials are often at cross purposes. A strong cement with low permeability is required, and normal to high-density slurries are best at providing these qualities. Unfortunately, because of the lost circulation and thermal conductivity considerations, such slurries are generally

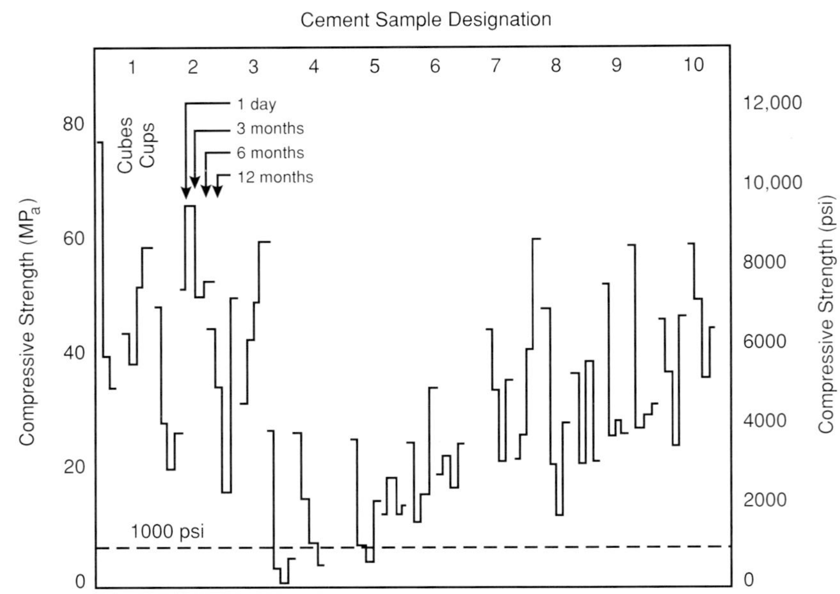

Compressive strength of cement cube and sandstone cup samples after aging periods of 1 day and 3, 6, and 12 months. Cup samples were cured and aged downhole. Cubes were laboratory cured under water at 392°F (200°C) for 1 day, then exposed downhole for 3, 6, and 12 months in the Cerro Prieto geothermal field, Mexico. Downhole temperature was 417°F (214°C).

Figure 9–13—Compressive strength performance of typical performance of typical geothermal well cements under actual conditions (from API Task Group on Geothermal Well Cements, 1985).

unsuitable. Therefore, much research has been performed to devise low-density slurries with the desired properties described above.

Conventionally extended Portland cement systems, containing perlite, bentonite, diatomaceous earth, etc., generally perform adequately in steamflood wells, provided the slurry density is above 12.5 lb/gal (1.5 g/cm^3). Their long-term performance is very similar to that exhibited by such systems in deep wells (Fig. 9–8).

The formations in steamflood wells are often so incompetent that cement systems with densities less than 12.5 lb/gal (1.5 g/cm^3) are required to avoid lost circulation or formation damage. Thus, silica-stabilized foamed cements (Smith, 1983) and microsphere-extended systems (Ripley et al., 1980) are very common in steamflood well completions today. Previously, multistage cementing was necessary to successfully complete these wells.

Typical slurries using glass or ceramic microspheres are prepared with a silica-stabilized Portland cement base slurry. The long-term performance of glass microsphere systems cured at 450° and 600°F (232° and 315°C) is shown in Fig. 9–16. The slurry densities vary from 10.0 to 12.0 lb/gal (1.20 to 1.45 g/cm^3).

The performance of silica-stabilized ceramic microsphere systems at 450° and 600°F (232° and 315°C) is shown in Fig. 9–17. Initially, these systems were generally stronger and less permeable than their glass microsphere counterparts. However, between one and two years of curing, significant deterioration was noted at both temperatures (Nelson, 1987). X-ray diffraction analysis of the systems revealed the coincident appearance of reyerite and certain aluminosilicate hydrate phases. Ceramic microspheres are derived from fly ashes, and the delayed (reyerite-related) deterioration of normal density fly ash cement systems has been discussed earlier in this chapter. Based upon these recent data, the efficacy of ceramic microspheres in thermal well completions has been called into question.

Typical foamed cement systems for thermal wells are prepared from a normal density base slurry of Portland cement, at least 35% silica flour, a surfactant, and a foam stabilizer. The long-term performance at 450° and 600°F (232° and 315°C) of three foamed cement systems with densities ranging from 9.0 to 12.0 lb/gal (1.08 to 1.44 g/cm^3) is shown in Fig. 9–18. Comparison of the foamed cement data with those of equal density microsphere systems reveals the foams to have significantly higher compressive strength. The water permeabilities of the foamed cements are also higher (>0.1 md), and more variable with curing time.

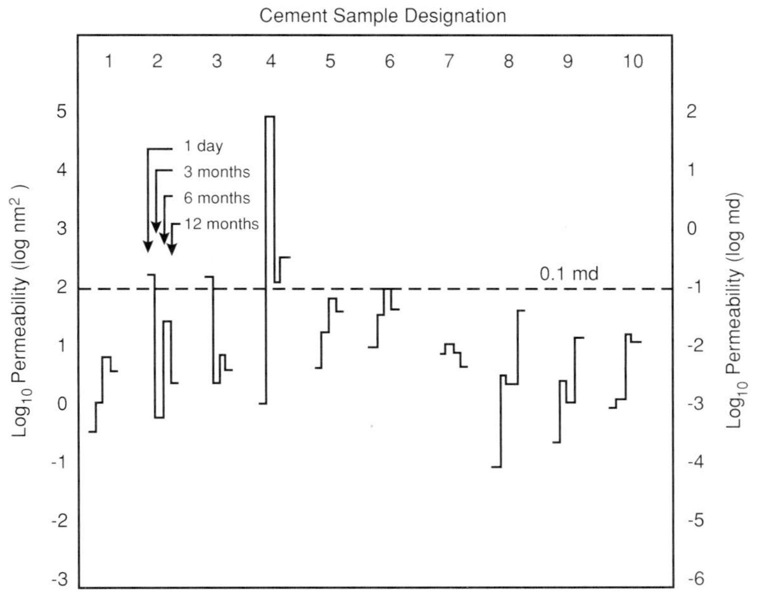

Water permeabilities of cement samples taken from slurry-filled sandstone cup holders after curing 1 day and 3, 6, and 12 months downhole in the Cerro Prieto geothermal field, Mexico. Downhole temperature was 417°F (214°C).

Figure 9–14—Permeability performance of typical geothermal cements cured under actual conditions (from API Task Group on Geothermal Well Cements, 1985).

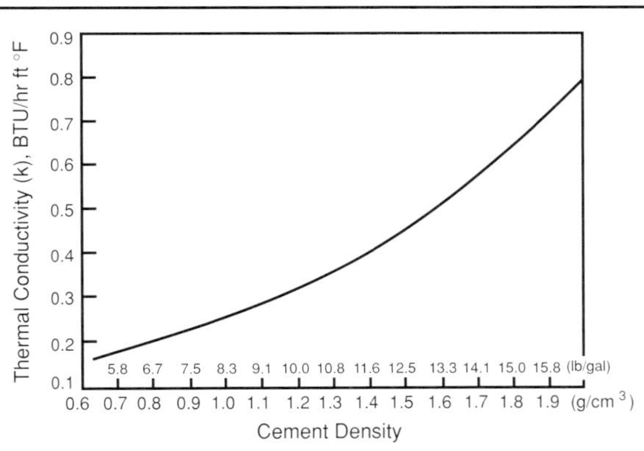

Figure 9–15—Typical cement density/thermal conductivity relationship (from Nelson, 1986).

Foamed cements have also been shown to be resistant to repetitive thermal cycling, which occurs when the cyclic steam stimulation technique is applied (Harms and Febus, 1984). Compressive strength and permeability data for systems cycled between 550° and 100°F (288°and 94°C) are shown in Table 9–2.

9–8.2 In-Situ Combustion Wells

In-situ combustion recovery, or fireflood, consists of initiating combustion in an injection well, and then propagating the combustion front by the injection of air through the reservoir to the production wells (Chu, 1981). In such wells, the cement is exposed to maximum temperatures between 700° and 1,700°F (371°and 926°C) near the burning zone. Such temperatures exceed the stable range of Portland cement; therefore, high-alumina cement is necessary.

Fireflood wells are physically similar to and are usually found in the same locations as steam injection wells. Thus, the formation conditions and cement performance requirements are basically the same. Usually, most of the casing can be cemented with Portland cement systems, with calcium aluminate cement placed opposite and about 100 ft (31 m) above the pay zone as a tail slurry.

The performance of two normal density, calcium aluminate cement systems is depicted in Fig. 9–19. Data are given for systems cured at 100° and 220°F (38° and 93°C) and heated in a refractory furnace at 600°, 1,000° and 1,500°F (315°, 538° and 815°C). The compressive strengths of the aluminate systems at the lower temperatures are adequate, yet considerably lower than similar density Portland cement systems. This is primarily because of the previously described conversion of the initial aluminate hydrates to C_3AH_6. The water-permeability values are extremely low as well.

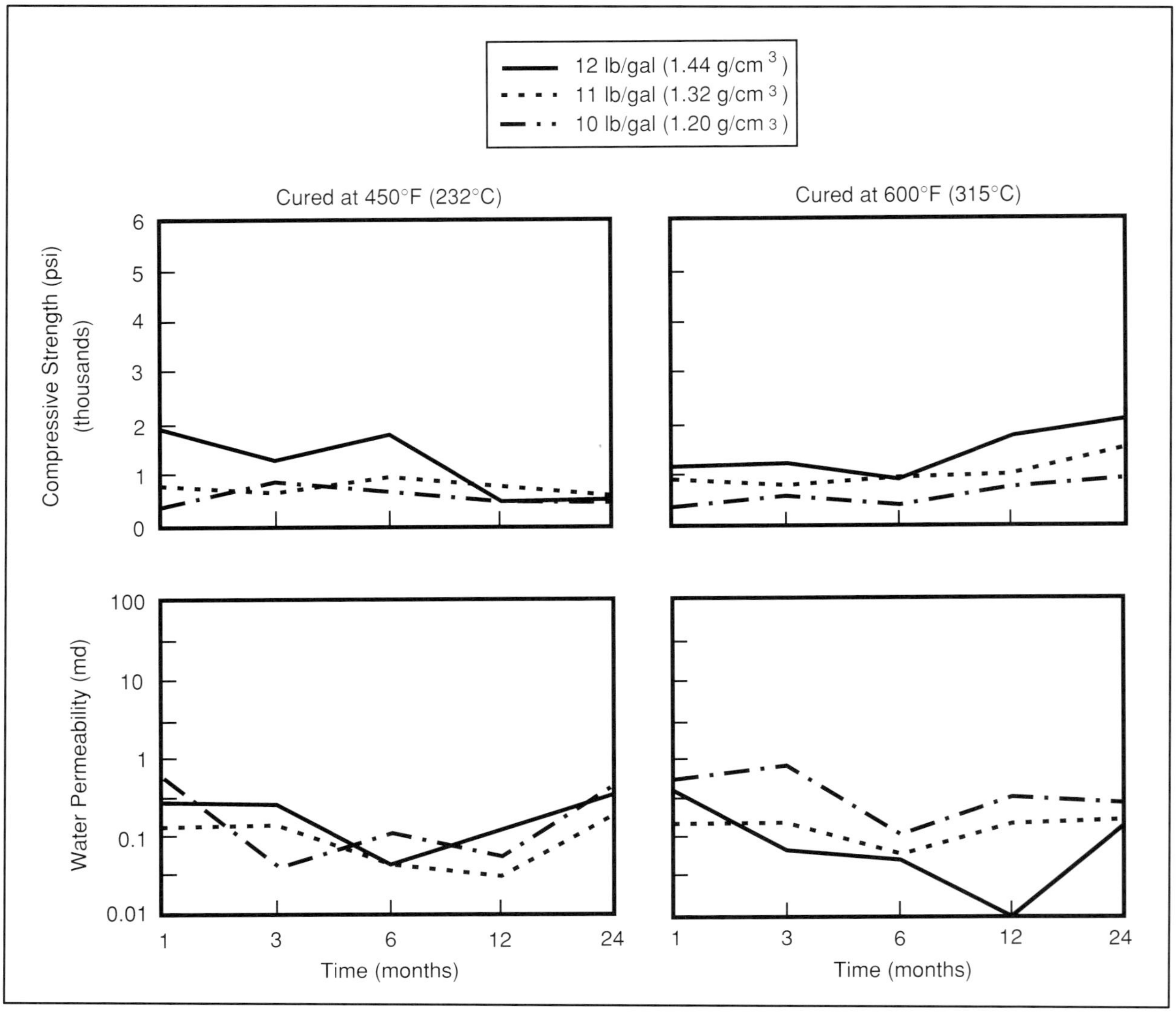

Figure 9–16—Long-term performance of glass microsphere systems.

The performance of foamed calcium aluminate cements has also been investigated (Nelson and Eilers, 1985). Figure 9–20 shows the compressive strength and water permeability of three systems cured for 7 and 28 days at 1,250°F (677°C) in a refractory furnace. Two foams, with densities of 11.0 and 9.0 lb/gal (1.32 and 1.08 g/cm^3) were prepared from a neat calcium aluminate cement-base slurry. Another foam, with a density of 11.0 lb/gal (1.32 g/cm^3), contained fly ash. The compressive strength was adequate; however, the water permeabilities were excessive.

9–9 CONCLUSIONS

The preceding discussion has demonstrated that thermal cements encompass a wide variety of wellbore conditions and complex chemical processes. Many factors must be considered to determine the optimum cement composition for a particular situation. Nevertheless, there are several basic points which the engineer must remember when contemplating this problem.

- When static temperatures exceed 230°F (110°C), 35% to 40% silica BWOC must be added to Portland cements; otherwise, strength retrogression will occur.
- If saline geothermal brines are present, fine silica flour (<15 µm particle size) should be added to Portland cement as a stabilizer. Silica sand does not reliably provide adequate protection.
- If high concentrations of CO_2 are present, Portland cement degradation can be inhibited by reducing the silica concentration to 20% BWOC.
- Most common cement extenders are compatible with thermal cements; however, if the static temperature

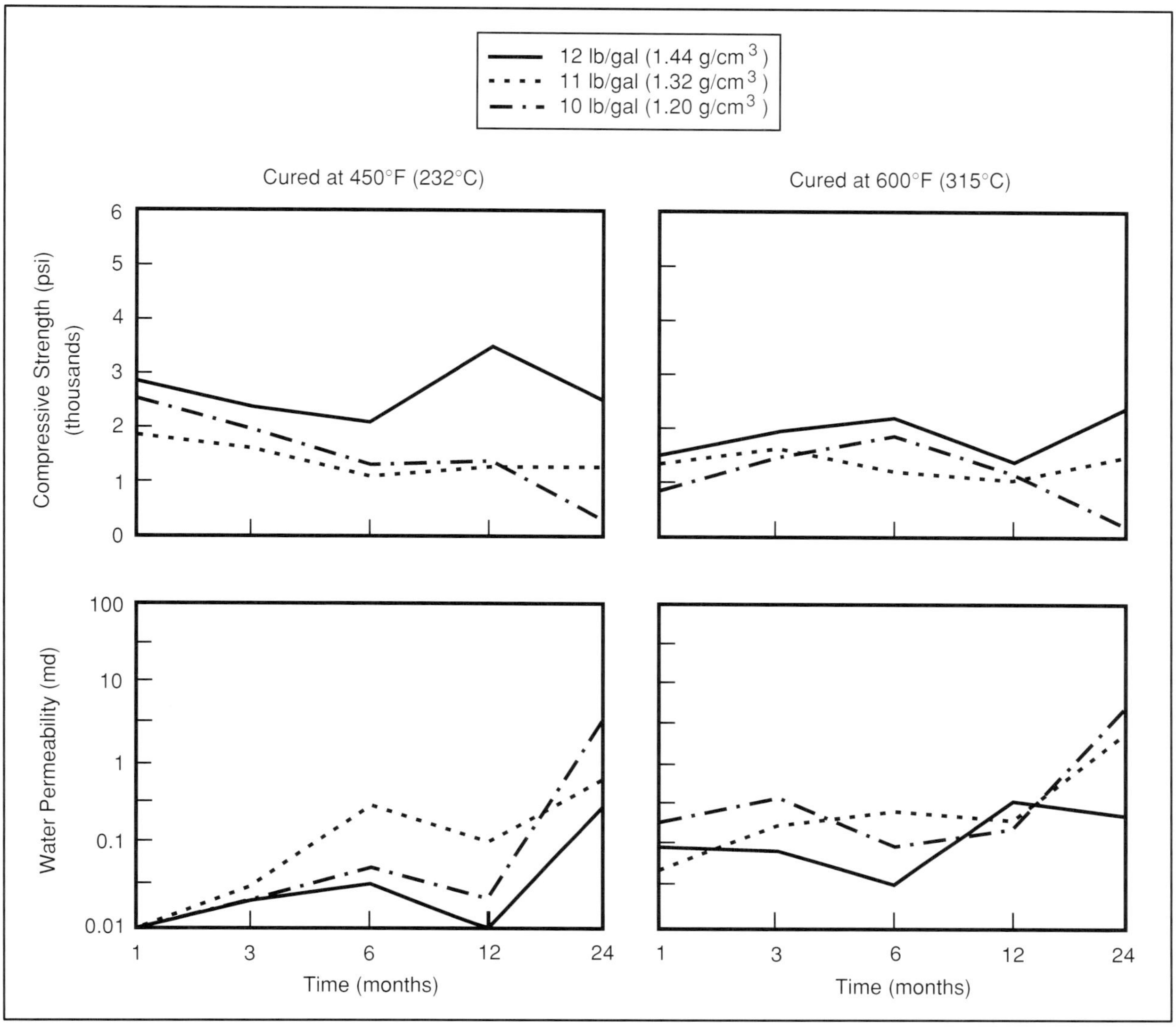

Figure 9–17—Long-term performance of ceramic microsphere systems.

exceeds 450°F (232°C), fly ash should not be used in Portland or Class J cement systems. Bentonite, perlite and diatomaceous earth are suitable.

- Microsphere cement systems can be used in thermal wells, provided the base slurry is stabilized to high temperatures, and the collapse pressure (usually 3,000 psi or 20.7 MPa) is not exceeded. Ceramic aluminosilicate microspheres may not be suitable at temperatures above 450°F (232°C).
- Foamed cement, made from a stabilized base slurry, can be used with confidence in most thermal wells. In geothermal wells, where corrosive fluids are produced, the long-term stability of foamed cements has not been proven.
- If the cement will be exposed to temperatures exceeding 750°F (400°C), Portland cement should not be used. High-alumina cement is suitable.
- Silica is deleterious to the stability of high-alumina cements at temperatures exceeding 572°F (300°C). Crushed aluminosilicate firebrick or fly ash is suitable.
- During laboratory testing, accurate static and circulating temperatures must be used to obtain an optimum thickening time and compressive strength at the wellsite.
- High pressure strongly affects the behavior of thermal cement systems; therefore, laboratory testing must be performed at the anticipated bottomhole pressure.

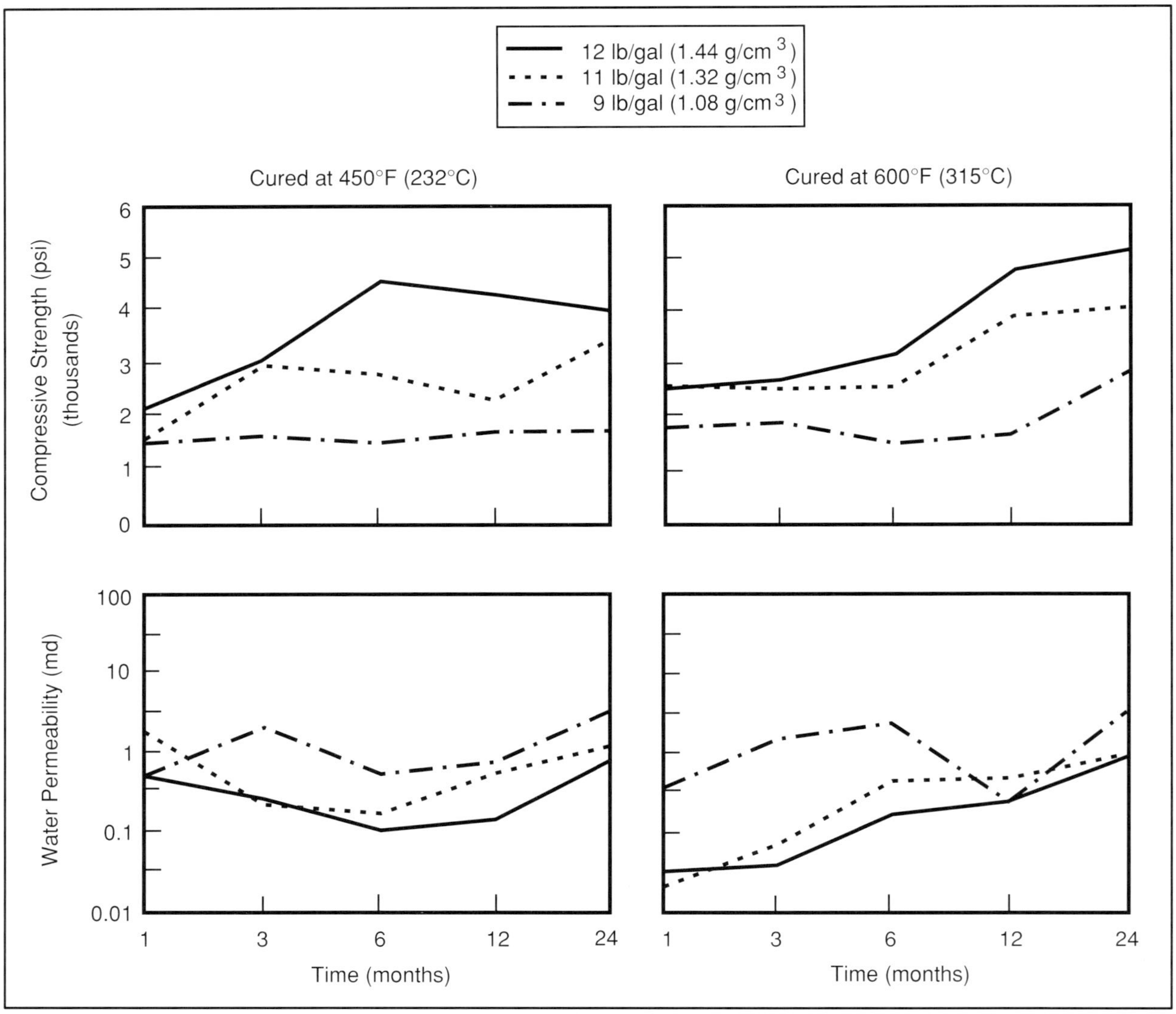

Figure 9–18—Long-term performance of foamed cement systems.

- Thermal cements are sensitive to subtle chemical changes; therefore, laboratory testing should always be performed with samples of the cement, additives, and location water which will be used during the job.
- The common assumption that high compressive strength is automatically linked with low permeability is false. Permeability should be measured in the laboratory before a cement system is placed in a thermal well.

REFERENCES

API Task Group on Cements for Geothermal Wells: "API Work Group Reports Field Tests of Geothermal Cements," *Oil & Gas J.* (Feb. 11, 1985) 93–97.

Arnold, D. R.: "Planning and Progress on Drilling a Record-Depth Well in the Rocky Mountains," *JPT* (April 1980) 694–702.

Bearden, W. G.: "Effect of Temperature and Pressure on the Physical Properties of Cement," *Oil-Well Cementing Practices in the United States*, API, New York (1959) 56.

Berra, M., Fabbri, F., Faceotti, M., Pezzuoli, M., Ricciardulli, R., Romano, G. and Tarquini, B.: "Behaviour of a Cementing Hydraulic Binder Under Severe Geothermal Conditions," *Geothermics* (1988) **17**, 785–813.

Bruckdorfer, R. A.: "Carbon Dioxide Corrosion in Oilwell Cements," paper SPE 15176, 1986.

Carden, R. S. et al: "Unique Aspects of Drilling and Completing Hot Dry Rock Geothermal Wells," paper IADC/SPE 11373, 1983.

Properties of Foamed Cement	Foamed Cement Density 10 lb/gal	11.5 lb/gal	13 lb/gal
Compressive strength after 20 days at 550°F	1210 psi	1680 psi	2260 psi
Compressive Strength after 100 days at 550°F[1]	1630 psi	1550 psi	2440 psi
Compressive strength after 160 days at 550°F[2]	1240 psi	2020 psi	2430 psi
Air permeability after 100 days	2.4 md	1.0 md	0.9 md

[1]Cycled to 100°F twice.
[2]Cycled to 100°F three times.

Table 9–2—Effect of thermal cycling upon performance of foamed cements for steamflood conditions. Surface slurry: 15.4 lb/gal Class G, 40% silica flour, 3% lime) (from Harms and Febus, 1984).

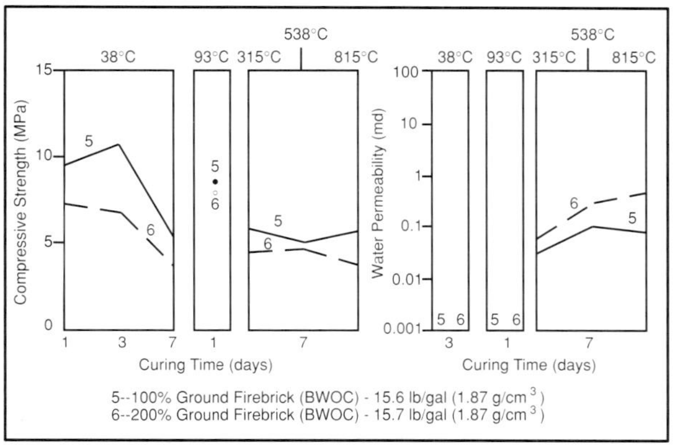

Figure 9–19—Compressive strength and permeability performance of calcium aluminate cement systems at various temperatures (from Nelson and Eilers, 1985).

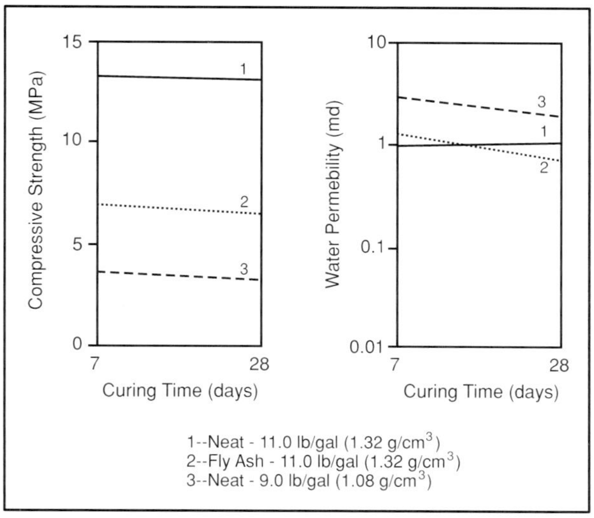

Figure 9–20—Performance of foamed calcium aluminate cement systems at 677°C (after Nelson and Eilers, 1985).

Carter, G. and Smith, D. K.: "Properties of Cementing Compositions at Elevated Temperatures and Pressures," *Trans.*, AIME (1958) **213**, 20–27.

Chu, C.: "State-of-the-Art Review of Fireflood Field Projects," paper SPE/DOE 9772, 1981.

Chu, C.: "State-of-the-Art Review of Steamflood Field Projects," paper SPE 11733, 1983.

Earlougher, R. C.: "Some Practical Considerations in the Design of Steam Injection Wells," paper SPE 2202, 1968.

Eilers, L. H.: "Process for Cementing Geothermal Wells," U.S. Patent No. 4,556,109 (1985).

Eilers, L. H. and Nelson, E. B.: "Effect of Silica Particle Size on Degradation of Silica-Stabilized Portland Cement," paper SPE 7875, 1979.

Eilers, L. H. and Root, R. L.: "Long-Term Effects of High Temperature on Strength Retrogression of Cements," paper SPE 5028, 1976.

Eilers, L. H., Nelson, E. B., and Moran, L. K.: "High-Temperature Cement Compositions—Pectolite, Scawtite, Truscottite, or Xonotlite: Which Do You Want?" *JPT* (July 1983) 1373–1377.

Farouq Ali, S. M. and Meldau, R. F.: "Current Steamflood Technology," *JPT* (Oct. 1979) 1332–1342.

Fournier, R.: Private Communication, U.S. Geological Survey, Menlo Park, CA, 1979.

Fundamentals of Thermal Oil Recovery, J. E. Kastrop (ed.): Petroleum Engineer Publishing Co., Dallas (1965).

Gallus, J. P., Pyle, D. E., and Watters, L. T.: "Performance of Oil-Well Cementing Compositions in Geothermal Wells," paper SPE 7591, 1978.

Gallus, J. P., Pyle, D. E., and Moran, L. K.: "Physical and Chemical Properties of Cement Exposed to Geothermal Dry Steam," paper SPE 7876, 1979.

Gates, C. F. and Holmes, B. G.: "Thermal Well Completions and Operation," *Proc.*, Seventh World Pet. Conf. (1967) paper PD–11, 419–429.

Grabowski, E. and Gillott, J. E.: "Effect of Replacement of Silica Flour with Silica Fume on Engineering Properties of Oilwell Cements at Normal and Elevated Temperatures and Pressures," *Cement & Concrete Res.* (1989) **19**, 333–344.

Greer, F. C. and Shryock, S. H.: "New Technique Improves Steam Stimulation Completions," paper SPE 1944, 1967.

Handin, J.: "Strength of Oil Well Cements at Downhole Pressure-Temperature Conditions," *JPT* (Dec. 1965) 341–347.

Harms, W. M. and Febus, J. S.: "Cementing of Fragile-Formation Wells With Foamed Cement Slurries," paper SPE 12755, 1984.

Hedenquist, J. W. and Stewart, M. K.: "Natural CO_2-Rich Steam-Heated Waters in the Broadlands-Ohaaki Geothermal

System, New Zealand: Their Chemistry Distribution and Corrosive Nature," Geothermal Resources Council, International Symposium on Geothermal Energy (1985).

Heindl, R. A. and Post, Z. A.: "Refractory Castables—II: Some Properties and Effects of Heat-Treatments," *J. Amer. Ceramic Soc.* (1954). **37**, No. 5, 206–216.

Hook, F. E., Morris, E. F., and Rosene, R. B.: "Silica-Lime Systems for High-Temperature Cementing Applications," paper SPE 3447, 1971.

Humphrey, H. C.: "Casing Failures Caused by Thermal Expansion," *World Oil* (Nov. 1960) 105–108.

Italcementi S.p.a.: "GEOTERM," Customer Brochure (1977).

Jones, R. R.: "A Novel Economical Approach for Accurate Real-Time Measurement of Wellbore Temperatures," paper SPE 15577, 1986.

Kalousek, G. L. and Chow, S. Y.: "Research on Cements for Geothermal and Deep Oil Wells," paper SPE 5940, 1976.

Kalousek, G. L. and Nelson, E. B.: "Hydrothermal Reactions of Dicalcium Silicate and Silica," *Cement & Concrete Res.* (1978) **8**, 283–290.

Kalousek, G. L.: "The Reactions of Cement Hydration at Elevated Temperatures," *Proc.*, Third Intl. Cong. Chem. Cement, London (1952) 334–354.

Kennerly, R. A.: *New Zealand J. Sci.* (1961) **4**, 453–468.

Kukacka, L. E.: Personal Communication, Brookhaven National Laboratory, 1989.

Leutwyler, K.: "Casing Temperature Studies in Steam Injection Wells," *JPT* (Sept. 1966) 1157–1162.

Luke, K. and Taylor, H. F. W.: "Equilibria and Non–equilibria in the Formation of Xonotlite and Truscottite," *Cement & Concrete Res.* (1984) **14,** 657–662.

Maravilla, S.: "A Hydrothermal Setting Cement for Cementing Ultradeep, Hot Wells," *JPT* (Oct. 1974) 1087–1094.

Menzel, C. A.: "Studies of High Pressure Steam Curing of Tamped Hollow Concrete Block," *J. Amer. Concrete Inst.* (1935) **7**, 64

Metcalf, A. S. and Dresher, T. D.: "How Pressure Affects the Set Properties of Various Cement Systems," paper SPE 7186, 1978.

Milestone, N. B., Sugama, T., Kukacka, L. E., and Carciello, N: "Carbonation of Geothermal Grouts—Pt. 1: CO_2 Attack at 150°C," *Cement & Concrete Res.* (1986) **16**, 941–950.

Milestone, N. B., Sugama, T., Kukacka, L. E., and Carciello, N: "Carbonation of Geothermal Grouts—Pt. 2: CO_2 Attack at 250°C," *Cement & Concrete Res.* (1987) **17**, 37–46.

Milestone, N. B., Sugama, T., Kukacka, L. E., and Carciello, N: "Carbonation of Geothermal Grouts—Pt . 3: CO_2 Attack on Grouts Containing Bentonite," *Cement & Concrete Res.* (1987) **17**, 295–306.

Milestone, N. B., Kukacka, L. E., and Carciello, N: "CO_2 Corrosion of Geothermal Cement Grouts," *Proc.*, Eighth Intl. Cong. Chem. Cement, Rio de Janeiro (1986).

Nelson, E. B.: "Improved Cement Slurry Designed for Thermal EOR Wells," *Oil & Gas J.* (Dec. 1, 1986) 39–44.

Nelson, E. B.: Unpublished Data, 1987.

Nelson, E. B. and Eilers, L. H.: "High-Temperature Expanding Cement Composition and Use," U.S. Patent No. 4,328,036 (1982).

Nelson, E. B. and Eilers, L. H.: "Cementing Steamflood and Fireflood Wells—Slurry Design," *J. Cdn. Petr. Tech.* (Sept.–Oct. 1985) 58–63.

Nelson, E. B., Eilers, L. H., Moran, L. K., Spangle, L. B., and Simpson, B. E.: "Evaluation and Development of Cement Systems for Geothermal Wells," paper SPE 10217, 1981.

Nelson, E. B., Eilers, L. H., and Kalousek, G. L.: "Formation and Behavior of Calcium Silicate Hydrates in a Geothermal Environment," *Cement & Concrete Res.* (1981) **11**, 371–381.

Nelson, E. B. and Kalousek, G. L.: "Effects of Na_2O on Calcium Silicate Hydrates at Elevated Temperatures," *Cement & Concrete Res.* (1977) **7**, 687–694.

Onan, D. D.: "Effects of Supercritical Carbon Dioxide on Well Cements," paper SPE 12593, 1984.

Pollock, R. W., Beecroft, W. H., and Carter, L. G.: "Cementing Practices for Thermal Wells," *J. Cdn. Pet. Tech.* (July–Sept. 1966) 130–134.

Quon, D. H. H. and Malhotra, V. M.: "Performance of High Alumina Cement Concrete at Elevated Temperature," *J. Cdn. Ceramic Soc.* (1979) **48**, 7–16.

Radenti, G. and Ghiringelli, G.: *Geothermics* (1972) **1**, No. 3, 119–123.

Rickard, W. M.: "Foamed Cement for Geothermal Wells," *Trans.*, Geothermal Resources Council (1985) **9**, pt. 1, 147–152.

Ripley, H. E., Harms, W. M., Sutton, D. L., and Watters, L. T.: "Ultra-Low Density Cementing Compositions," paper Petroleum Society of CIM 80–31–19, 1980.

Robson, T. D.: *High-Alumina Cements and Concretes*, John Wiley and Sons, Inc., New York (1962) 184–185.

Sasaki, S., Kobayashi, W., and Okabayashi, S.: "Strength Development of $2CaO{\cdot}SiO_2$—Silica Cement Under High-Temperature and High-Pressure Condition," paper SPE 13406, 1984.

Shen, J. C.: "Effects of CO_2 Attack on Cement in High Temperature Applications," paper SPE 18618, 1989.

Shryock, S. H.: "Geothermal Well Cementing Technology," paper SPE 12454, 1984.

Smith, D. K.: *Cementing*, Henry L. Doherty Series, SPE Richardson, TX (1987).

Smith, F. M.: "Thermal Expansion of Cemented Casing," *Proc.*, 25th Pennsylvania. State U. Pet. Prod. Conf. (1966) 249–273.

Smith, T. R. et al.: "Foamed Cement Application in Canada," paper Petroleum Society of CIM 83–24–31, 1983.

Sugama, T., Kukacka, L. E., and Galen, B. G.: "Advanced High-Temperature Lightweight Foamed Cements for Geothermal Well Completions," Brookhaven National Laboratory, Report No. BNL–38087–DE87005263 (1986).

Suman, G. O. and Ellis, R. C.: *Cementing Handbook*, Gulf Publishing Co., Houston (1977).

Swayze, M. A.: "Effects of High Temperatures and Pressures on Strengths of Oil Well Cements," *Drill.& Prod. Prac.*, API, Dallas (1954) 72–81.

The Chemistry of Cements, H. F. W. Taylor (ed.), Academic Press, Ltd., London, (1964) 106–122.

Wedelich, H. et al.: "Key Factors That Affect Cementing Temperatures," paper SPE/IADC 16133, 1987.

Wooley, G. R. et al.: "Cementing Temperatures for Deep-Well Production Liners," paper SPE 13046, 1984.

Zeldin, A. N. and Kukacka, L. E.: "Polymer Cement Geothermal Well Completion Materials," Brookhaven National Laboratory, Report No. BNL 51287, (1980).

10

Cementing Equipment and Casing Hardware

Paul Buisine and H. Steve Bissonnette

Dowell Schlumberger

10–1 CEMENTING MATERIALS

Before describing the design and function of cementing equipment, one must be familiar with the physical and chemical properties of the various cementing materials. A thorough discussion is presented in Chapters 2 and 3; however, a rapid review of the principal points is useful for this discussion.

10–1.1 Cement

As discussed in Chapter 2, Portland cement is used during almost all well cementing jobs. It is a finely divided and highly reactive powder. The vigorous hydration of Portland cement during initial mixing, as well as the changeable slurry properties during placement, complicates the design of cement mixing and pumping equipment.

Portland cement is usually stored in silos at a central storage location. Alternatively, it is packaged in U.S. (94-lb) or metric (50-kg) sacks, and in so-called "big bags" (1 to 1.5 metric tons generally), or in larger quantities (truck, railway car, or ship).

10–1.2 Water

Fresh water is normally used for cementing onshore wells, and seawater for offshore locations. However, one must be aware that fresh waters are often not very "fresh." Inorganic salts and organic plant residues are frequently present in significant quantities. Such materials are known to affect the performance of Portland cement systems.

10–1.3 Dry Cement Additives

The relevant properties of dry additives with regard to cementing equipment and logistics are summarized in Table 10–1.

10–1.4 Liquid Additives

Offshore, liquid additives are usually preferred. Such materials are more compatible logistically, because their mixing requires less space. Since liquid additives are preblended with the mix water, the resultant slurry tends to be more homogeneous than one mixed from a dry-blended powder mixture. Table 10–2 is a summary of the relevant points.

10–2 BASIC EQUIPMENT

Figure 10–1 is a schematic flow diagram of cement slurry preparation, and indicates the steps performed at the central storage location and at the wellsite. Each function in the flow diagram also represents a major piece of equipment. Some items may be combined into a multipurpose "cementing unit." In the following discussion, the design and operation of each class of equipment are presented in detail.

10–2.1 Cement and Dry Additive Blending

Figure 10–2 is an overview of the delivery and blending processes which occur at the central storage location. Cement is delivered to the central storage location. Upon delivery in bags, the cement is usually bulked. The transfer may be accomplished by several techniques—pneumatic loading bottles (Fig. 10–3), mechanical screw elevators, or combined systems (for unloading from ships). These transfer systems may also be used to load dry additives.

The bulk cement is stored in pneumatic or atmospheric silos. Transfer systems are available to move the cement from one silo to another, or to a blender, road transport unit, or supply boat. When the transfer system is pneumatic, several silos are connected permanently to save time and labor. In humid climates, an air dryer may be installed in the system.

The cement and dry additives are usually combined in a pneumatic blending tank (10 to 20 ton capacity) at the

Type of Material	Chemically Inert	Chemically Active
Form	Insoluble powder or finely cut material	Soluble powder or finely cut material
Examples (function)	Powdered coal (extender) Hematite (weighting agent) Barite (weighting agent)	Lignosulfonate (retarder) Sodium silicate (extender) Calcium chloride (accelerator)
Maximum concentration (order of magnitude)	100% (BWOC)	10% (BWOC)
Influence of accuracy in concentration on slurry quality	Concentration acts directly on system density; no unexpected effect.	Materials may have secondary effects.
Handling	Blended with the dry cement in a special blender, at the central storate location, up to several days before the pumping job. The blended material is then transported to the wellsite. (If the amount required is small and the material easily scattered in the water, it is treated as a soluble material.)	Normally dry-blended with the cement. Sometimes added to the mix water on location in an open horizontal tank, just prior to the job.
Localization	Mainly on land	Only on land

Table 10–1—Important properties of dry additives with respect to cementing logistics.

Type of Material	All liquid additives are chemically active. Frequently, they are water solutions of the corresponding dry additive.
Examples (function)	Lignosulfonate solution (retarder) Naphthalene sulfonate solution (dispersant) Latex (gas migration prevention)
Concentration (order of magnitude)	Up to 25 liters/100 kg cement (3.0 gal/U.S. sk)
Influence of Accuracy in Concentration on Slurry Quality	Materials may have detrimental secondary effects.
Handling	Blended with the mix water on location in a horizontal open tank, shortly before the job. Also can be blended on the fly during the job. The second method is preferred.
Localization	Mainly offshore

Table 10–2—Important properties of liquid additives with respect to cementing logistics.

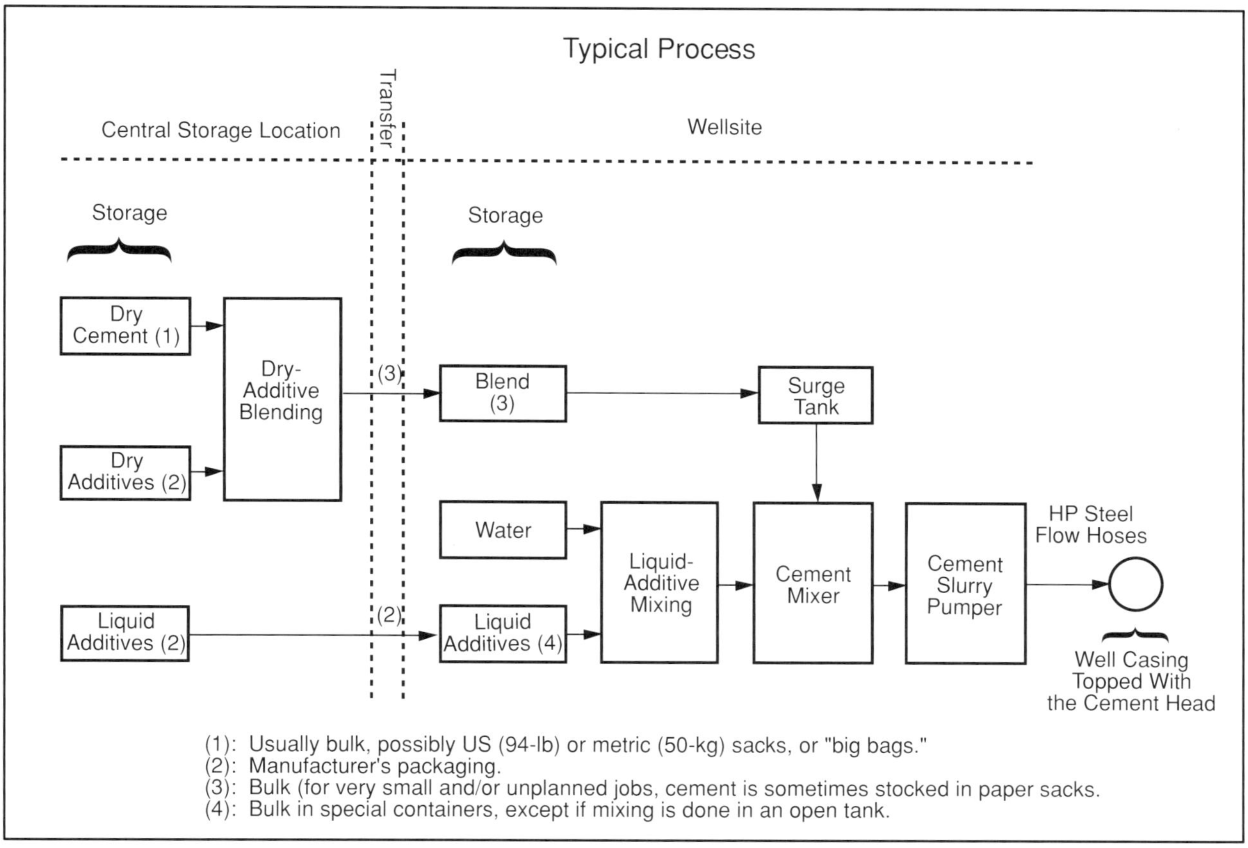

Figure 10–1—Typical process.

central bulk material plant (Fig. 10–4). Bulk materials are usually air-blown, and sacked materials are poured into the tank through a hopper located on top. The sacked additives may also be poured into pneumatic additive bottles (1- or 2-ton capacity), and then blown into the blending tank. Pressurized air is supplied by one or more air compressor units.

The bulk materials are loaded into the blending tank first for weighing purposes. A weigh cell is permanently integrated into the tank frame. The sacked additives are loaded last. Air, pressurized to about 35 psi (2.5 bars), is injected through nozzles into the mass of the materials until thorough blending is accomplished. Then the blend is transferred pneumatically to a bulk material transport, for delivery to the wellsite. To obtain the required amount of cementing materials, more than one batch must frequently be prepared.

Although very popular, this method has two drawbacks—

- Some segregation (due to shaking) can take place during the transportation to the rig site, especially if the distance is long and the road surface is rough (e.g., corrugated desert track). This problem is magnified if the blended materials have significantly different particle sizes or densities. If small quantities of soluble or easily dispersible additives are needed, mixing the dry additives with water at the rig site is often preferred.
- If the cement job is canceled or postponed for a long period after the blend has been prepared, a storage or reuse problem may arise. Liquid additives prevent such situations.

10–2.2 Transportation of Bulk Materials or Blends to the Wellsite

The equipment used to deliver the cementing materials to the wellsite varies according to the location. The various types of transports are discussed below.

Land Rigs: Trucks or semitrailer transports are generally used for land operations. As shown in Fig. 10–5, truck-mounted vertical tanks or semitrailer-mounted vertical or horizontal tanks are the most common.

Limited Access Locations: Generally, cement in sacks or big bags is transported by helicopter to the rig location. Occasionally, aluminum containers are used.

Offshore Rigs: Supply boats or cementing vessels are used for such locations. They normally have built-in

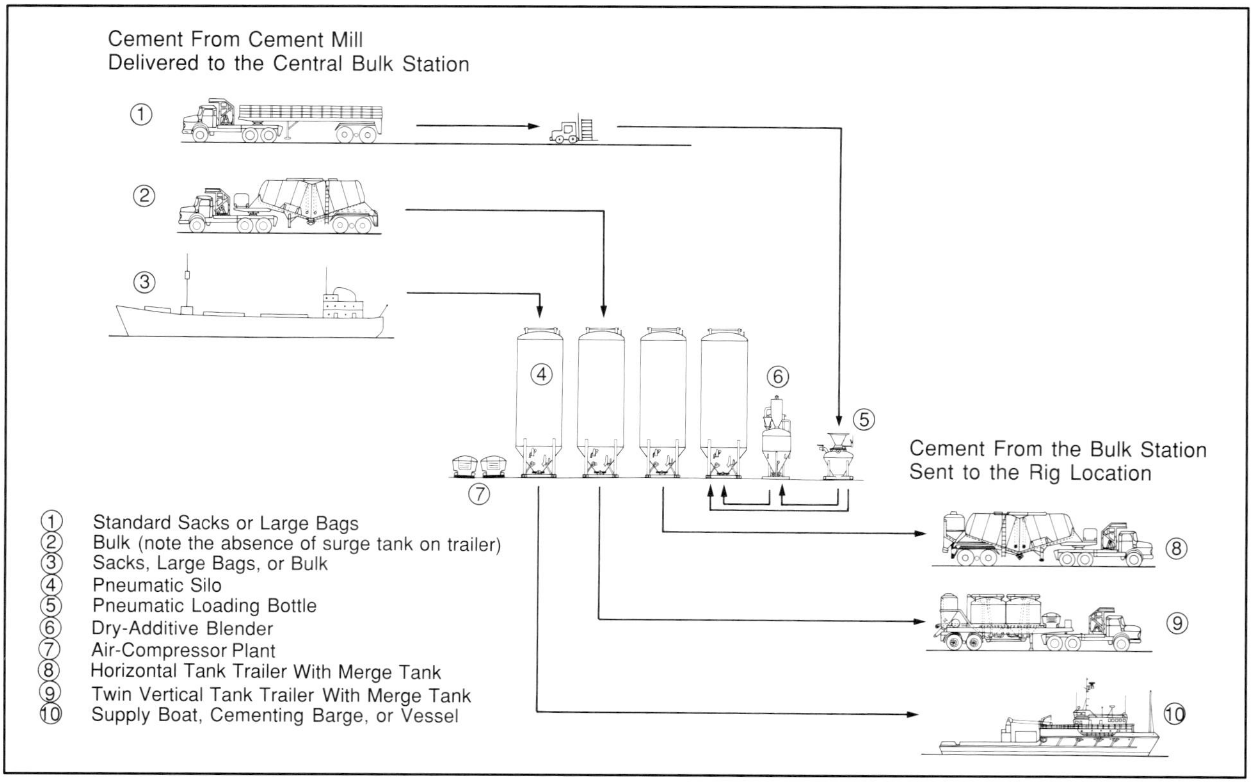

Figure 10–2—Delivery, storage, and distribution of cement and dry additives.

Offshore Rigs: Supply boats or cementing vessels are used for such locations. They normally have built-in tanks, pneumatic unloading equipment, and a supply of hoses. Sometimes, mobile skid-mounted tanks (with a low center of gravity) and mobile unloading equipment are used. It is important to note that the pneumatic equipment must be sufficiently powerful to blow heavy materials such as barite (specific gravity: 4.33) up to the drilling rig tanks, a vertical distance of 130 to 200 ft (40 to 60 m). The most commonly used air compressors deliver 250 to 350 ft^3/min with a pressure rating of 2 to 3 bars.

10–2.3 Wellsite Storage of Cement or Blends

As discussed above, pneumatic bulk trucks or trailers transport neat or preblended cement to the wellsite from

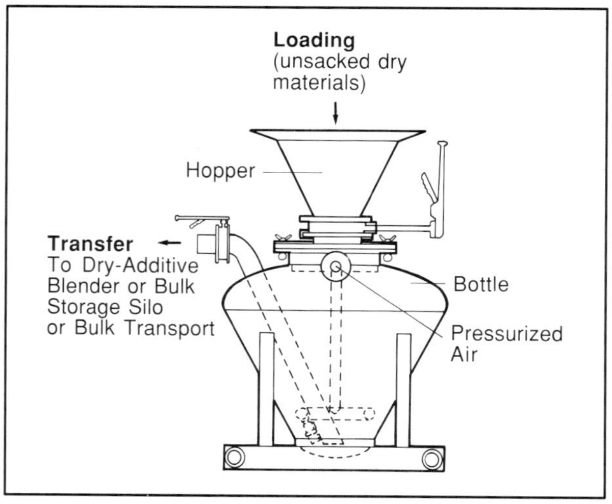

Figure 10–3—Pneumatic dry-material loading bottle.

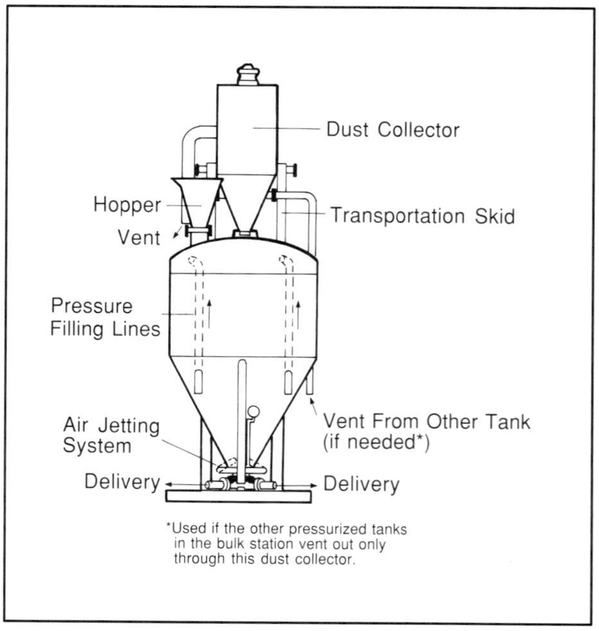

Figure 10–4—Dry-additive blender.

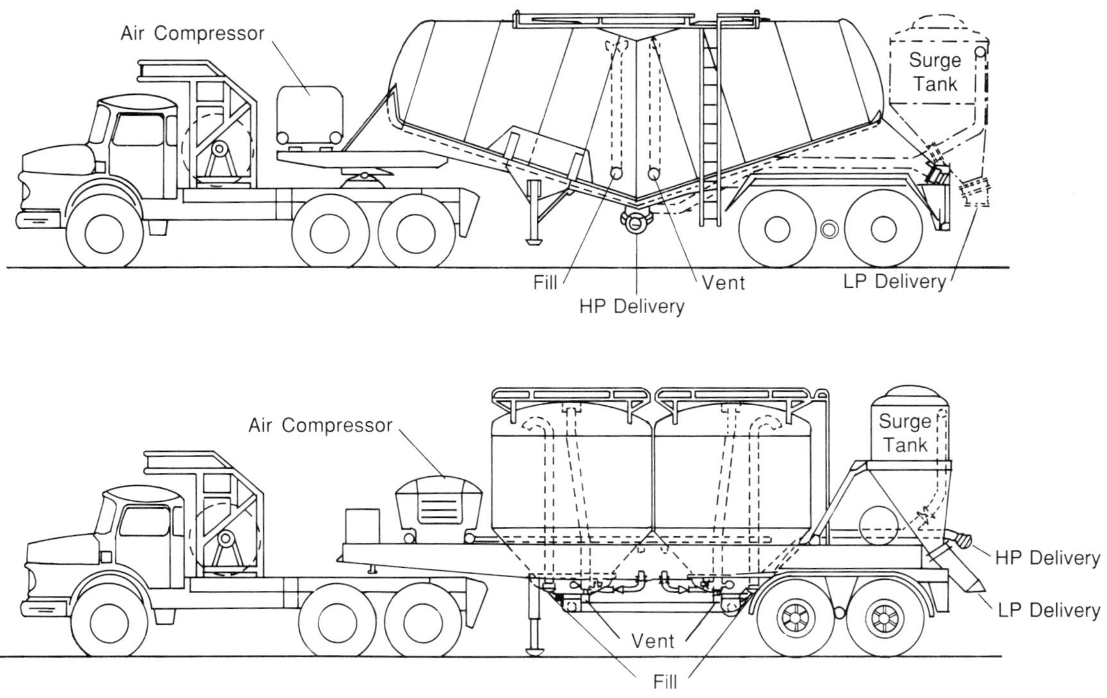

HP Delivery: Material transferred to storage tank or to mixing unit surge tank.
LP Delivery: Material air-blown to surge tank on rear of unit, and dumped into cement mixer hopper.

Figure 10–5—Bulk material transports.

which are brought to the rig site for the cement job, or are a permanent part of the drilling rig equipment. Such tanks are similar to those used at central storage locations, but their dimensions allow transport on standard or specially designed (with a built-in hydraulic laying/raising system) trailers. When empty, the tanks do not exceed the weight limits specified by various countries. A large variety of storage tanks exists within two principal categories—atmospheric and pressurized. Both are equipped with a set of skids for proper installation on imperfectly leveled ground, and easy winching onto trailers.

The atmospheric tank is always operated in a vertical position. Air at low pressure (about 3 psi or 0.2 bars) is blown into a gutter fixed to the slant bottom of the tank, and the roof of the gutter is made of a porous material. The air passes through the porous partition, and fluidizes the cement blend. The cement blend glides along the slant bottom to a chute gate, and to the hopper of a slurry mixing system. As illustrated in Fig. 10–6, atmospheric tanks are made in the shape of a parallelepiped.

Pressurized tanks use air at about 44 psi (3 bars) pressure, and can operate horizontally or vertically. Figure 10–7 is a schematic diagram of a typical unit. As shown

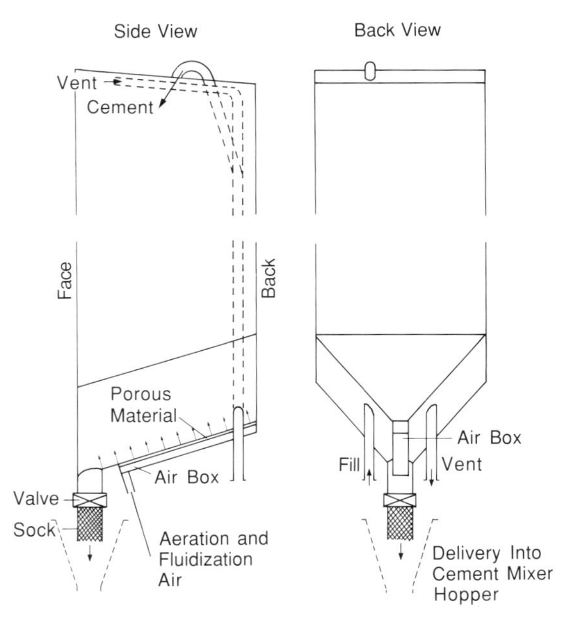

Figure 10–6—Atmospheric transportable bulk tank (typical piping arrangement).

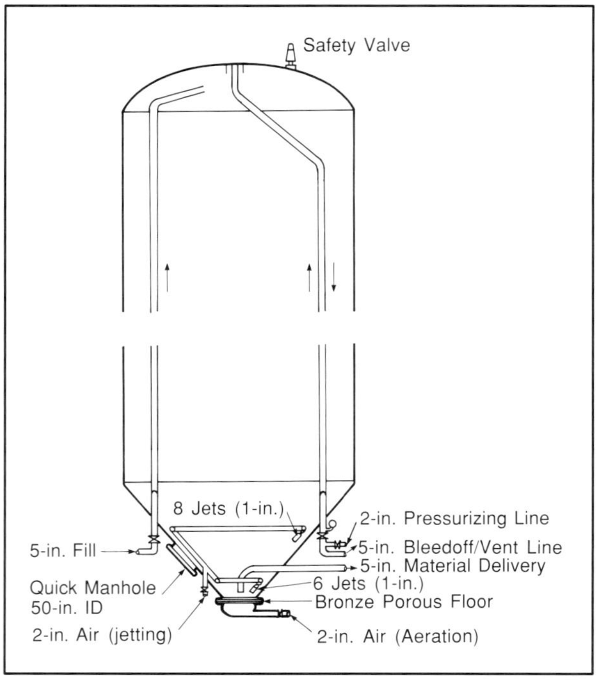

Figure 10–7—Pressurized bulk tank (typical piping arrangement).

in Fig. 10–8, the vertical tanks are generally cylindroconical in shape, while horizontal models are more complex. In the first stage, pressure-reduced air is blown from the bottom through the mass of cement for aeration and fluidization. Then air at 44 psi (3 bars) is injected into the tank, and the cement flows out through a discharge line to a surge tank, which feeds the cement mixer. For versatility, some vertical pressurized tanks are also equipped to release the cement directly to a hopper at atmospheric pressure (Fig. 10–9).

The bulk trailers are sometimes used as additional storage. Indeed, they can serve all storage needs on the rig site, provided they are equipped with their own surge tank, described later.

10–2.4 Metering of Water

Contrary to what one might think, the simple method of employing a flowmeter is not used. A set of twin 10-bbl (or sometimes 20-bbl) tanks is preferred. A "displacement tank" (Fig.10–10), which is divided equally by a partition, is also used.

Both sides of the partitioned tank are filled with mix water from the rig storage. If offshore, the freshwater or seawater distribution system is used. Each batch of mix water is then used successively to feed the cement mixer.

The additives may be preblended with the water in the storage tanks, or while the water is passing through the displacement tank. In the second case, a liquid-additive metering system (described later) is required.

For precise placement of the slurry in the wellbore, the volume of the displacement fluid must be accurately measured. After the cement slurry has passed through the mixing system, the displacement fluid usually passes through the displacement tanks for volume measurement, and is pumped by the cementing unit instead of the rig's mud pumps.

10–2.5 Liquid-Additive Storage and Mixing

The simplest method of mixing liquid additives (and dry additives at less than 3% BWOC) with water consists of pouring the required amount of each additive into a tank of water. One should measure the additives and water accurately to obtain the correct concentration; the preparation of a slight excess of solution is also advisable. The mixing can be achieved with a paddle mixer, circulation pump, jetting system, etc., or a combination of them.

The premix method has several disadvantages. An extra tank, which must be clean and sufficiently large, is required. Such tanks are not always available, and sufficient space to accommodate them may not exist on the rig site, especially offshore. If the job is canceled or postponed, the costly solution may have to be thrown away. Also, if a larger-than-expected volume of slurry becomes necessary during the job, the volume of the premixed additive solution may be inadequate. Thus, methods which allow continuous ("on-the-fly") mixing are often preferred. On-the-fly methods employ a semimanual or automatic metering system which delivers the correct amount of additives to each side of the displacement tank.

10–2.5.1 Liquid-Additive Metering System With Metering Tanks

All liquid-additive metering systems consist of two principal parts—a storage/transfer unit and a metering unit.

The storage/transfer unit generally includes four storage tanks of various capacities (usually between 6.2 to 25 bbl or 1,000 to 4,000 L). Well cement slurries typically contain two or three additives. The storage and transfer unit allows the independent metering of additives according to the requirements of a particular job.

Each storage tank is equipped with its own air-operated diaphragm pump and agitation system (recirculation, as illustrated, or air-operated stirrer) to avoid segregation of the additive components. Therefore, the operation of the unit requires a source of clean and dry air at 120 to 145 psi (8 to 10 bars). The configuration of the unit varies depending on whether it is designed for use on land (skid or trailer-mounted) or offshore (containerized).

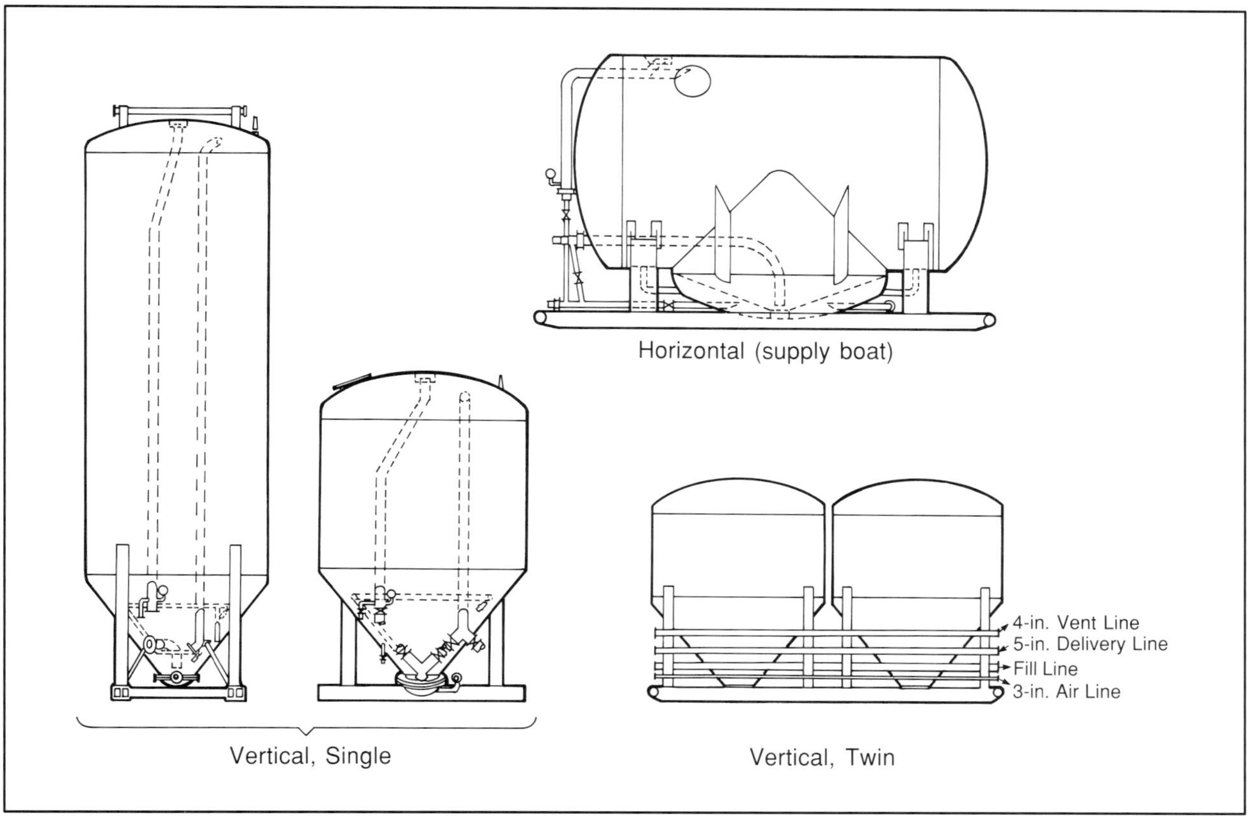

Figure 10–8—Pressurized bulk tanks (various configurations).

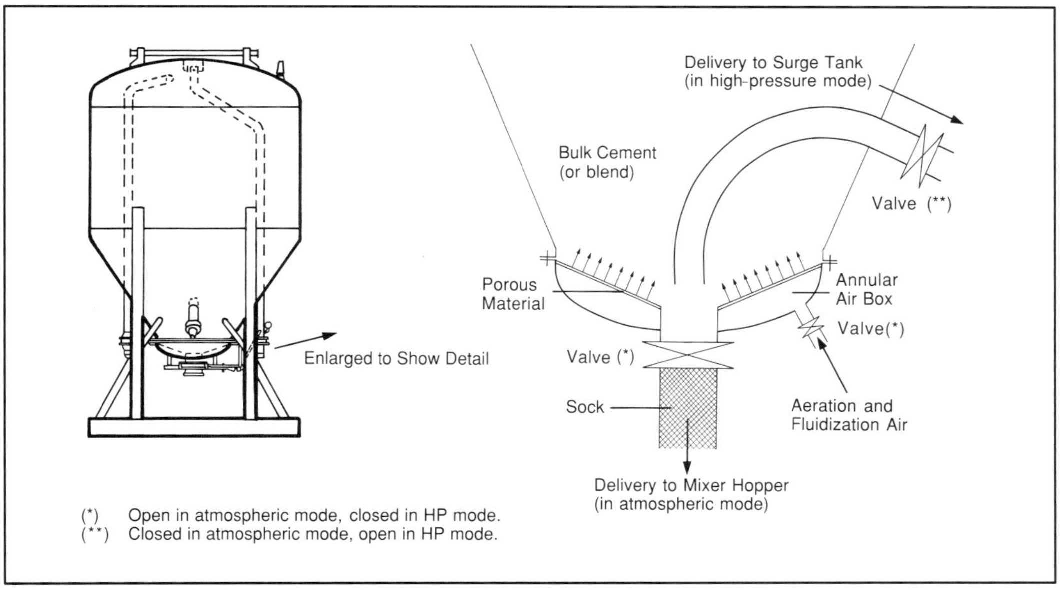

Figure 10–9—Pressurized bulk tank (typical piping arrangement).

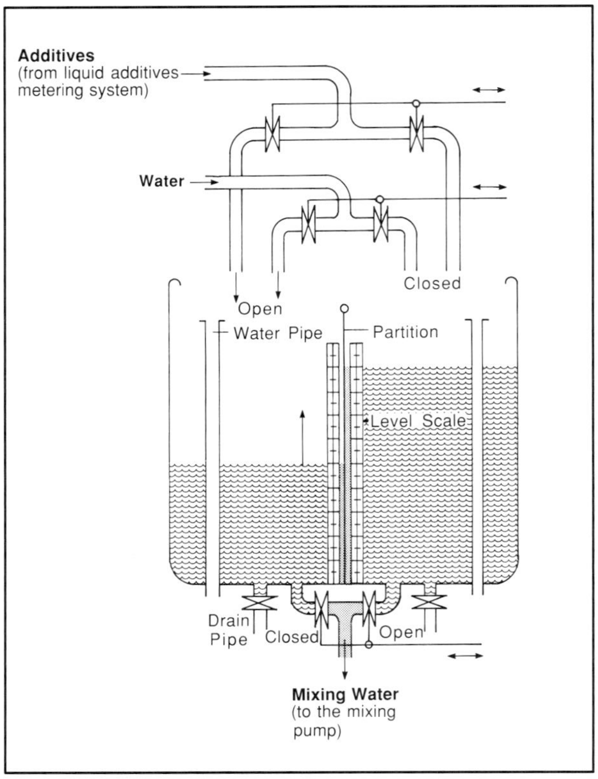

Figure 10–10—Displacement tank system.

The metering unit generally consists of a set of three (Fig. 10–11) or four 25-gal or 10-L tanks, with visible level scales. To prepare a batch (10 bbl or 20 bbl according to the displacement tank), the proper amounts of the selected additives are introduced into the metering tanks. The additives are then released into one displacement tank half, which is being filled with water. Finally, the mixture is agitated to obtain a homogeneous solution. The same operation is repeated for the following batch in the other displacement tank half and so on. The repetitions of the operation may be automatically or semi-automatically controlled.

10–2.5.2 Liquid-Additive Metering System Without Metering Tanks

The liquid-additive-system metering rack (Fig. 10–12) is used to provide accurate (±2%) delivery of up to four additives into displacement tanks. The operator depresses a push button to initiate the delivery. The metering rack behaves as four "smart valves," installed between the additive pumps and the displacement tanks. The valves are controlled by a microcomputer using data from electromagnetic flowmeters.

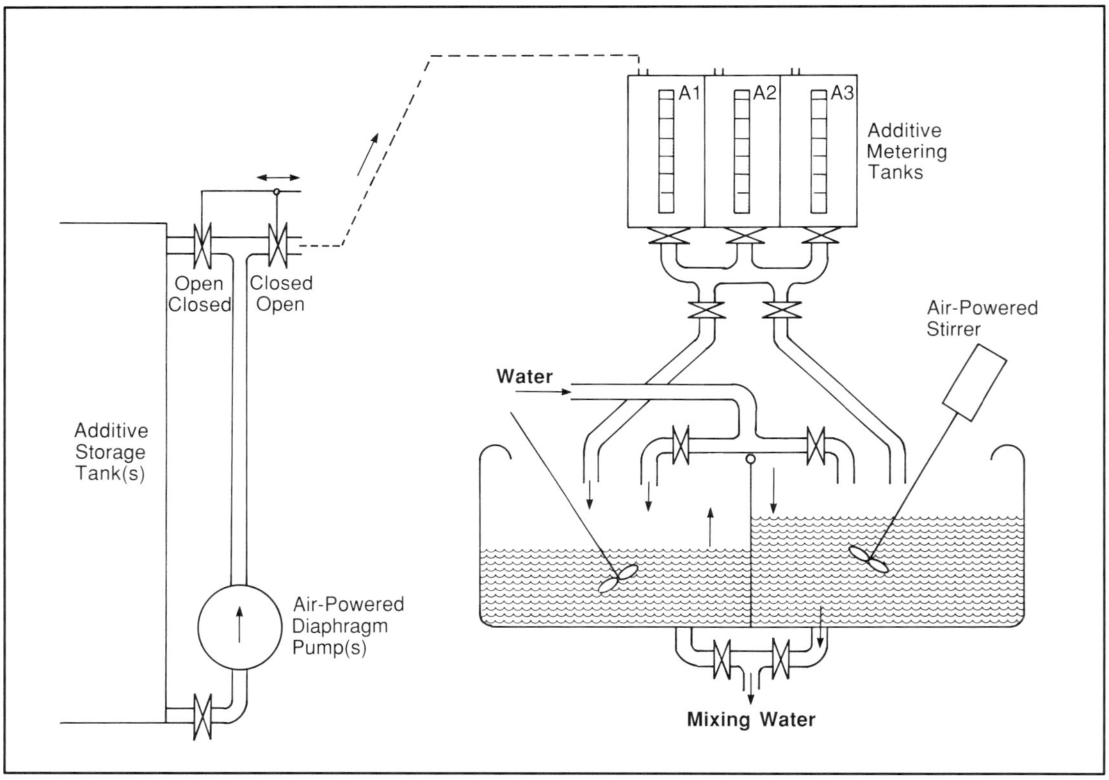

Figure 10–11—Liquid-additive metering system (with metering tanks).

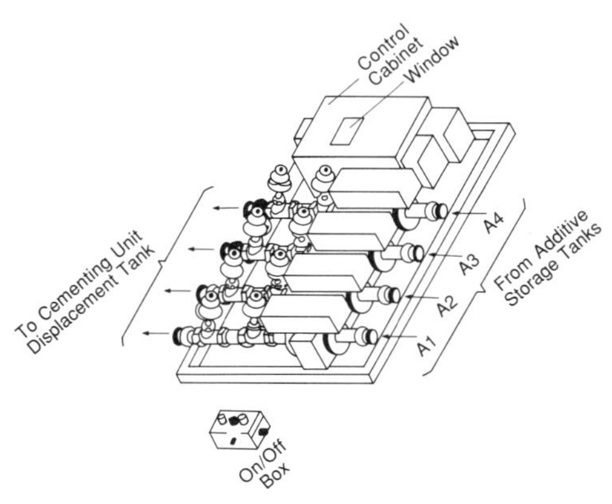

Figure 10–12—Metering rack hardware.

10–2.6 Surge Tanks

For smooth cement mixer operation, the supply of cement (or blend) should be steady, and the pressure at the mixer bowl should remain constant. The bulk cement is moved from the storage tank toward the cement mixer, driven by the differential pressure created between the tank and the end of the line. If the line is longer than approximately 23 ft (7 m), the cement tends to separate from the conveying air into slugs, giving pulsar flow. To smooth the flow and allow for operational requirements, such as changing from one storage tank to another, a surge tank is used.

As shown in Fig. 10–13, the surge tank has a cylindroconical shape. It has a capacity of about 70 ft^3 (2,000 L), and connects the end of the transfer line to the top of the mixer bowl. This device maintains the pressure above the mixer bowl at atmospheric plus the hydrostatic head of the material. A dust separator is also installed where the conveying air is vented to the atmosphere.

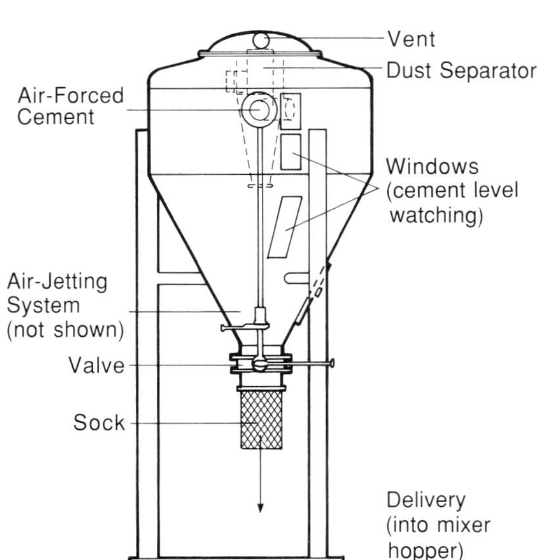

Figure 10–13—Surge tank.

In some cases, the surge tank is maintained at higher than atmospheric pressure. The advantage is a higher rate of delivery available from the surge tank, and the drawback is a lower differential pressure to feed the surge tank from the bulk system. A pressurized surge tank is also used with certain types of recirculating cement mixers to maintain a steady delivery rate. The tank is pressurized by controlling the flow or air in the vent line. This type of surge tank is used almost exclusively offshore, where the bulk system is required to transport the cement long distances.

10–2.7 Cement Mixing

The cement mixer is a device in which a flow of pressurized water (possibly containing additives) meets a flow of cement (possibly containing additives), and a cement slurry is formed at a prescribed rate. Several types of mixing systems exist, and are described individually below.

10–2.7.1 Conventional Jet Mixer

The conventional jet mixer consists of a hopper, a mixing bowl, a discharge gooseneck, and a slurry tub. The maximum slurry-generating capacity of the conventional jet mixer, evaluated in rate of dry material, is slightly higher than 2,200 lb (one metric ton)/min. Figure 10–14 shows a configuration for sacked cement, and a system for pneumatically delivered cement is illustrated in Fig. 10–15.

The cement is delivered to the hopper. The water is injected into the bowl through jets for mixing with the cement, and into the gooseneck for adjusting the slurry density. The jets are chosen according to the operating pressure, slurry fabrication rate, and the type of dry materials. The movement of cement down through the hopper is assisted by the high-pressure flow of water through the jets. The resulting pressure drop pulls the dry cement into the stream of water. To reinforce this effect, the gooseneck can be given a venturi tube profile. Further along at the gooseneck, turbulent flow mixes the cement particles with the water, and the result is a cement slurry.

The slurry density is adjusted by using the bypass system to change the water-to-cement ratio . As the bypass is opened, the suction effect decreases, and reduces the amount of cement drawn out of the hopper. At the same time, the water bypassing the jets enters the slurry. The combined effect is a decrease in slurry density. Conversely, if the bypass is closed, the density increases.

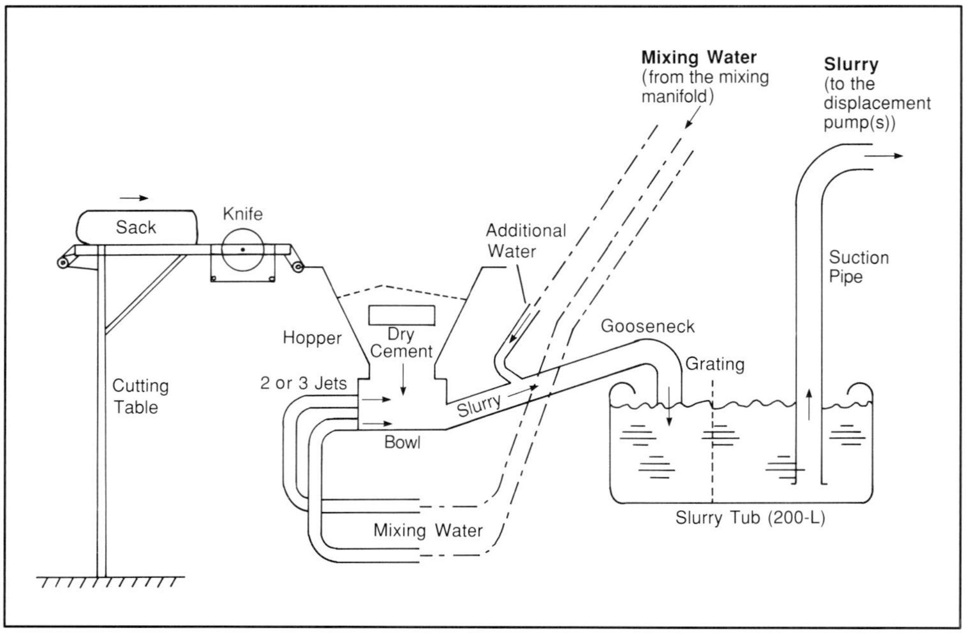

Figure 10–14—Conventional jet mixer (sacked cement).

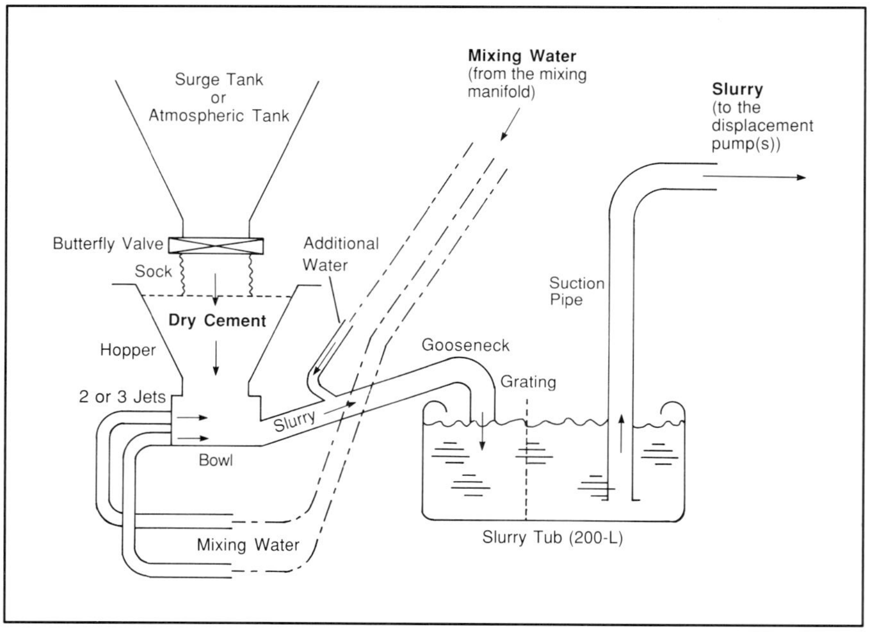

Figure 10–15—Conventional jet mixer (bulk cement).

The conventional jet mixer can be operated at low (175 to 200 psi or 12 to 14 bars) or high (880 to 1,180 psi or 60 to 80 bars) water pressure. In the first case, the mix-water pump is a centrifugal pump. In the latter case, it is a reciprocating pump, usually identical (except perhaps in plunger size) to the displacement pump. The "double HP pump cementing units," which are the most widely used throughout the world, are equipped to mix at either low or high pressure. The low-pressure method is preferred for two main reasons—less horsepower is required and, since both HP pumps are available to displace the slurry, a higher fabrication and displacement rate can be achieved. The benefit of the high-pressure method is that the jets or the bowl/gooseneck assembly are less apt to become plugged with dirty mixing water or poor quality cement. This method is mainly limited to emergency use today.

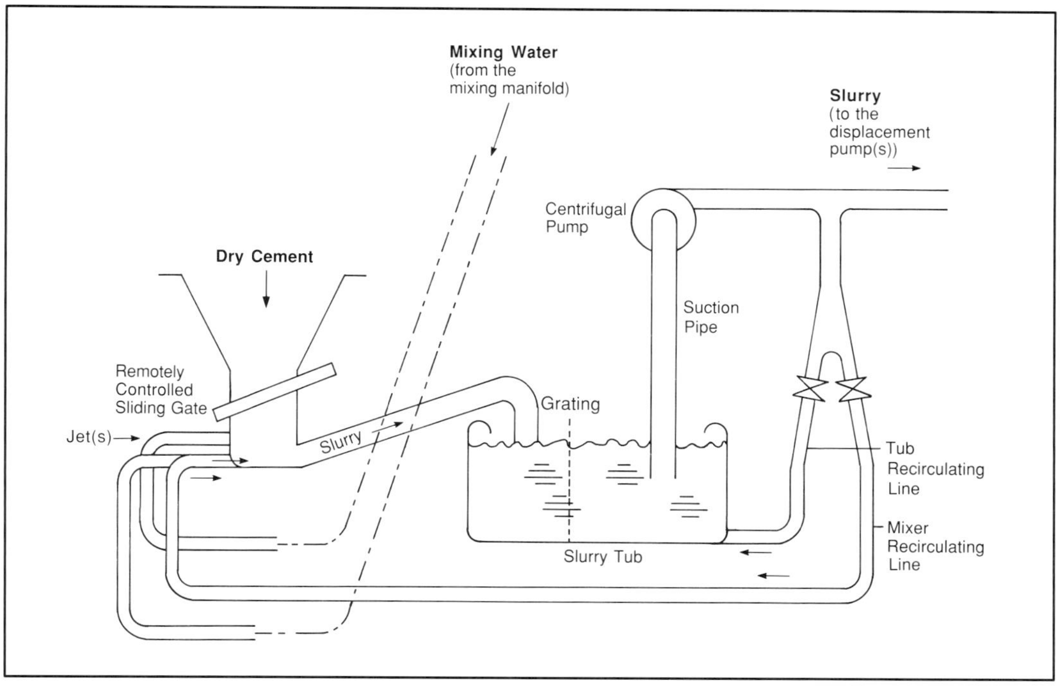

Figure 10–16—Recirculation jet mixer (bulk or sacked cement).

10–2.7.2 Recirculation Jet Mixer

The maximum capacity of the recirculation jet mixer (Fig. 10–16) is slightly over 4,400 lb (two metric tons)/min. The recirculation jet mixer differs from the conventional type in several ways.

- A remotely controlled sliding gate is present between the hopper and the bowl.
- The slurry density is adjusted by operating the sliding gate.
- The slurry is removed from the slurry tub by a recirculation jet, fed by a centrifugal pump. The centrifugal pump force feeds the displacement pumps, and recirculates some slurry through the mixing system.
- Water is permanently injected ahead of the recirculation jet.

Recirculation through the mixer heart and the tub improves the homogeneity and rheology of the slurry. Adjustment of the slurry density is also easier.

10–2.7.3 Recirculation Mixer Without Conventional Jets

This equipment includes a variety of mixers without conventional jets (Fig. 10–17). The maximum capacity of most mixers, evaluated by rate of dry material, is close to 4,400 lb (two metric tons)/min. They all consist of the following.

- A sophisticated metering system to mix cement with water, and a device to mix the resulting slurry with previously mixed slurry from the mixing tub. It is important that water and cement meet before contacting the previously mixed slurry.
- A centrifugal pump or similar device (located at the bottom of the tub) to improve the initial mixing by shearing, ensure recirculation through the mixer, and feed pressurized slurry to the downhole pump.
- A mixing tub which can be split into two sections. A film-like flow is created over the common partition which assists the release of entrapped air. Both sections can be equipped with a stirrer to further improve the mixing.

The density of the slurry is remotely controlled by metering the cement and/or water, depending upon the model. Usually the water rate is kept constant, and the slurry density is controlled by altering the rate at which cement is delivered to the mixer. Normally, the cement is transferred directly from a pressurized tank without passing through a surge tank.

10–2.7.4 Cement Mixing Units

As discussed above, the batch-blending system and the liquid-additive metering system have been designed to solve the proportioning problems encountered with cementing materials. However, the slurry properties are affected not only by the proportions between cement,

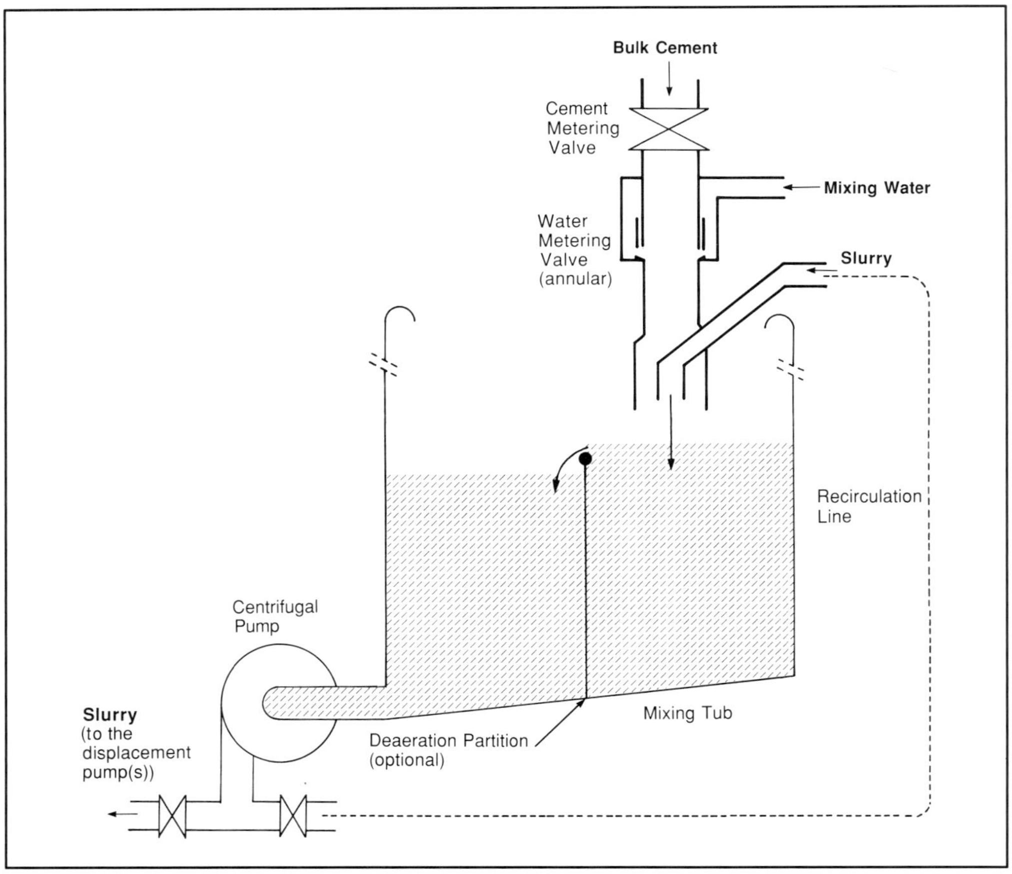

Figure 10–17—Recirculation mixer (without jets).

water, and additives, but also by the shearing which occurs during mixing.

Proper operation of a mixing unit should solve the problems of proportioning between the cement blend and the mix water. The correct proportion will give the slurry the expected slurry density and other properties. Continual verification of slurry density is essential; however, some density fluctuations during the slurry mixing are unavoidable (Chapter 5). The longer the mixing time and the larger the slurry volume, the better the homogeneity of the resulting slurry.

Finally, the slurry should be given the proper amount of shearing, which is a function of mixing energy and mixing time (Chapter 5). Since a centrifugal pump is an ideal shearing device, it is advisable to increase the volume of slurry being recirculated.

Recirculating mixers are available in a variety of configurations (skid, truck-, or trailer-mounted, diesel- or electric-powered, sometimes soundproofed) and sizes. They all have certain common features (Figs. 10–18 and 10–19).

- A surge tank with a capacity ranging from 1.5 to 4 m^3.
- A conventional or recirculation jet mixer.
- One or two holding tanks with a capacity ranging from 6.3 bbl (1 × 1,000L) to 50 bbl (2 × 8 m^3). The size of the largest units is limited by transportability.
- Two recirculating centrifugal pumps (only one on the smallest units), with a maximum displacement rate of up to 25 bbl (4 m^3)/min. Both can either circulate the slurry in the holding tank(s) to improve shearing and homogeneity, or feed the slurry to the downhole pump.
- A pair of paddle stirrers, hydraulically or electrically driven, to maintain homogeneity.
- A manifold, sufficiently versatile to be used in a variety of combinations.

In some particular cases such as very small jobs, or when the proportions of additives and the slurry density are very critical, the total volume of slurry needed to complete the job (including the usual excess) is prepared before pumping downhole. The liquid additives are not metered as described earlier; instead, they are released directly into the tank, or added through a jet mixer. No special practice exists regarding the dry additives. The

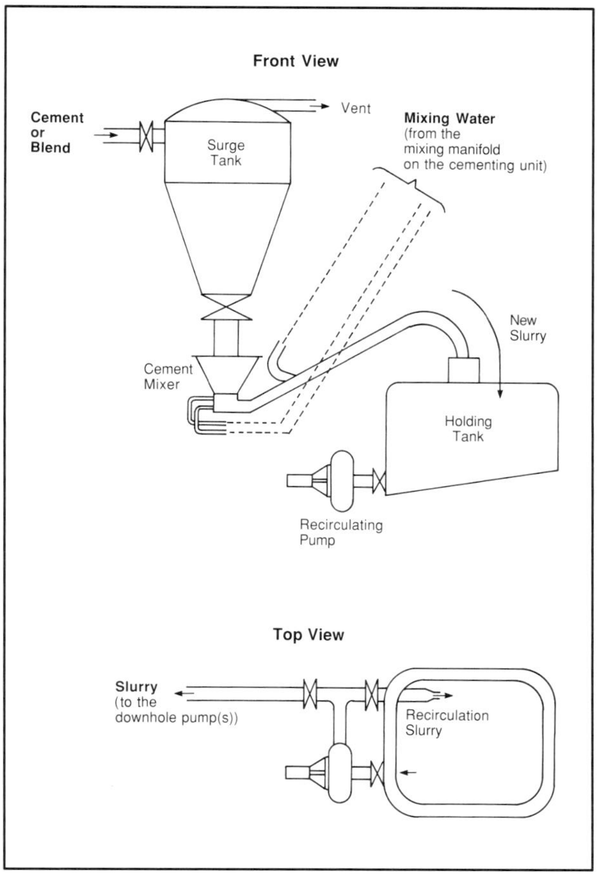

Figure 10–18—Single six-barrel tank mixing unit (with conventional jet mixer).

volumes or weights of additives are usually measured in the conventional manner.

10–2.8 High-Pressure Pumps

All current cementing pumps are of the reciprocating type, with three plungers (triplex), and spring-loaded suction and discharge valves (Fig. 10–20). The transformation of the rotating motion of the input shaft into the reciprocating motion of the plungers is generally accomplished by a crankshaft/connecting rods system, or sometimes by a swash plate/connecting rods system. These pumps include an internal fixed-ratio speed reducer. Depending on the make and the model, the plunger stroke varies from 5 to 10 in. (12.5 to 25 cm).

The global efficiency is not better than 85% to 90%. If adequately pressurized, the volumetric efficiency can reach 98% with water at 80% of the maximum speed. The construction is particularly rugged, allowing the pumps to handle the heaviest and most abrasive slurries.

10–2.8.1 Convertibility

Depending upon the manufacturer, the "size" of a pump can be altered by changing the fluid-end assembly, or the plungers and packing system using adapters to convert the fluid-end body. Size alteration changes the pressure and flow ratings without modifying the maximum available horsepower. The plungers used in cementing usually have a diameter between 3 and 6 in. (7.6 and 15.2 cm).

10–2.8.2 Hydraulic Horsepower

Depending, again, on the make and the model, the maximum horsepower varies between 200 and 500 hhp.

10–2.8.3 Versatility

These heavy-duty pumps, which can handle gravel-laden fluids, can also perform fracturing jobs. In the 200- to 500-hhp range, the same pumps are used for either cementing or stimulation.

10–2.8.4 Maximum Flow Rate and Pressure

By combining the makes, models, and sizes, a large variety of specifications can be found. One should bear in mind that during most cement jobs, only one pump is used to pump downhole. Usually the maximum rate is approximately 8 bbl/min. This limitation is based upon the maximum allowable rate for the 2-in. treating line most commonly used for cementing.

The pump pressure usually does not exceed 1,030 psi or 70 bars (cement squeezes excepted). As a matter of fact, if the density of the cement slurry is equal to that of the drilling mud, the pumping pressure is simply a consequence of the friction losses in the surface equipment (steel flow hoses and cement head) and below the surface.

10–2.8.5 Drive

The pumps which equip mobile units are driven by a diesel engine, associated with an automatic or manual transmission. Those which are permanently installed on an offshore rig are frequently driven electrically (usually a directly coupled DC motor).

10–2.9 Controls and Instruments

At or before the start of a job, some control devices on the mixer are selected (e.g., chokes on a jet mixer) or set in position (e.g., a mix-water valve on most recirculating mixers) according to the composition and density of the slurry, and the desired injection rate. During the job, the final adjustment is made with either the cement (blend) or the mix-water metering valve, depending upon the type of equipment. Adjustment of the downhole pump

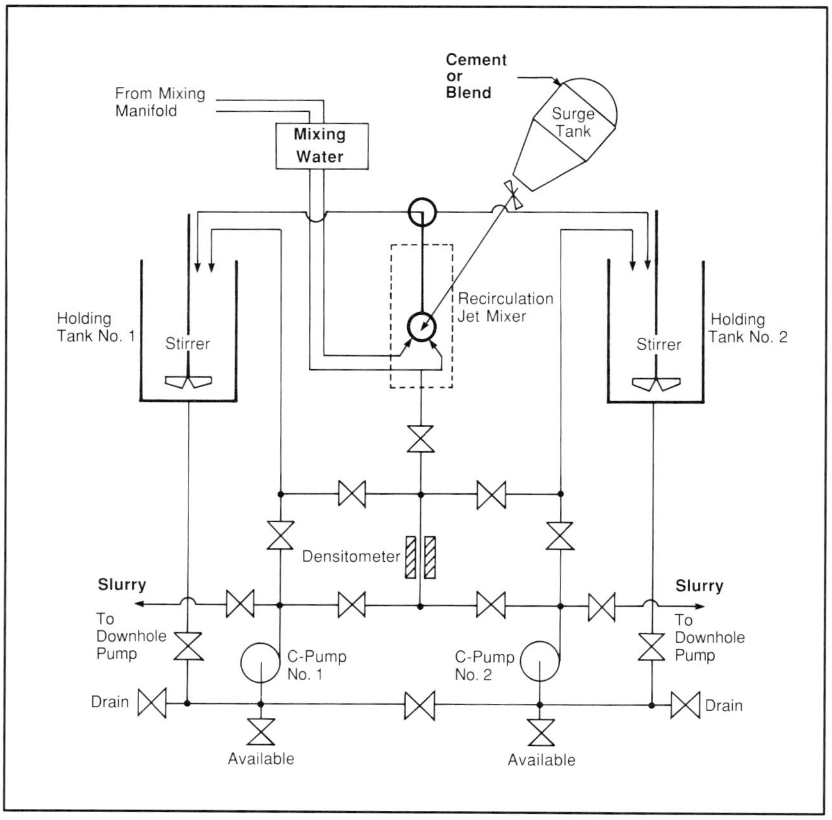

Figure 10–19—Twin-tank mixing unit (with recirculation jet mixer).

rate may also be necessary to maintain a constant level in the slurry tank, or maintain the pumping pressure within fixed limits (e.g., squeeze jobs).

Cement jobs require the measurement of many parameters. A discussion of the various types of equipment is given below.

Mix Water: The volumes of water are measured by the means of the displacement tanks.

Cement (blend) and Slurry: The volumes of mixed slurry and dry cement are determined by combining the mix-water volume and the slurry density.

Flow Rate: The slurry rate is observed at the downhole pump-stroke counter. A flowmeter is used if a continuous recording of job parameters is being made.

Pressure: The pumping pressure is read at a gauge or mechanical recorder. An electronic pressure transducer is used if the various parameters are recorded by a central unit.

Slurry Density: The slurry density is traditionally measured manually by a pressurized mud balance (Fig. 10–21). More sophisticated systems are becoming more common (e.g., a continuous U-tube weighing balance (Fig. 10–22) and radioactive densitometer connected to a central recording unit (Fig. 10–23)).

Slurry Rheology: No such measurements are currently performed routinely at the wellsite.

Compressive Strength: This measurement is rarely made at the wellsite. Slurry samples are normally taken to a central laboratory for postjob tests.

As illustrated in Fig. 10–24, central recording units are available which continuously record vital pumping parameters. The recorders significantly improve onsite job monitoring, while simultaneously storing data on digital tape for postjob evaluation.

The model shown in Fig. 10–24 displays data on four readouts, and records them in the form of a log. The stored data can be replayed through office computer systems to produce tabular job data sheets and enhanced graphics for in-depth analysis. Slurry density, flow rate, and pressure can be monitored for various configurations of pumping equipment. The microprocessor calculates and displays cumulative volumes, and the total volume is printed on the log every five minutes, or at will.

One can imagine that the continuous monitoring of key pumping parameters could be used to automatically control slurry mixing. As of this writing, equipment

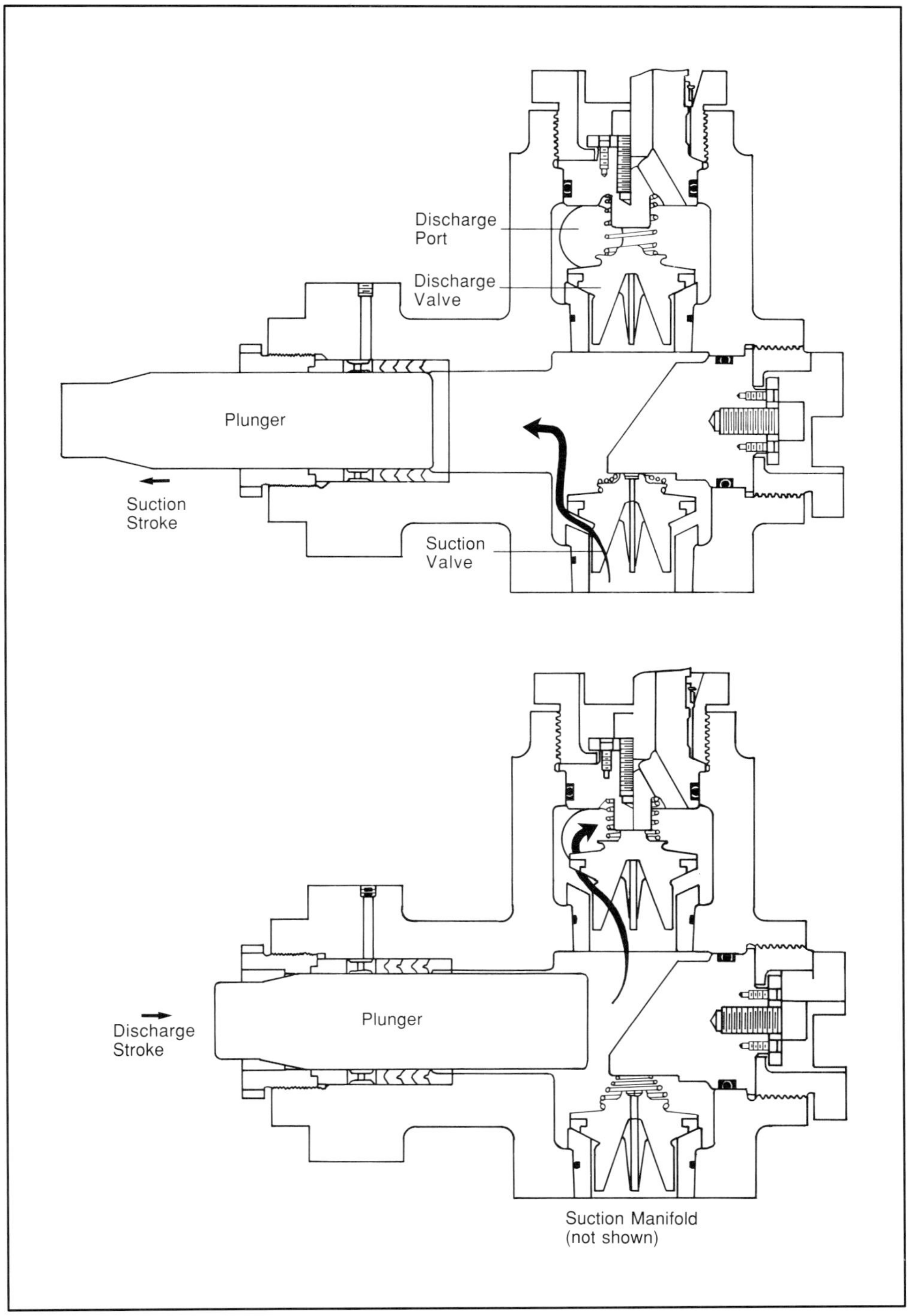

Figure 10–20—Reciprocating plunger pump fluid end (section).

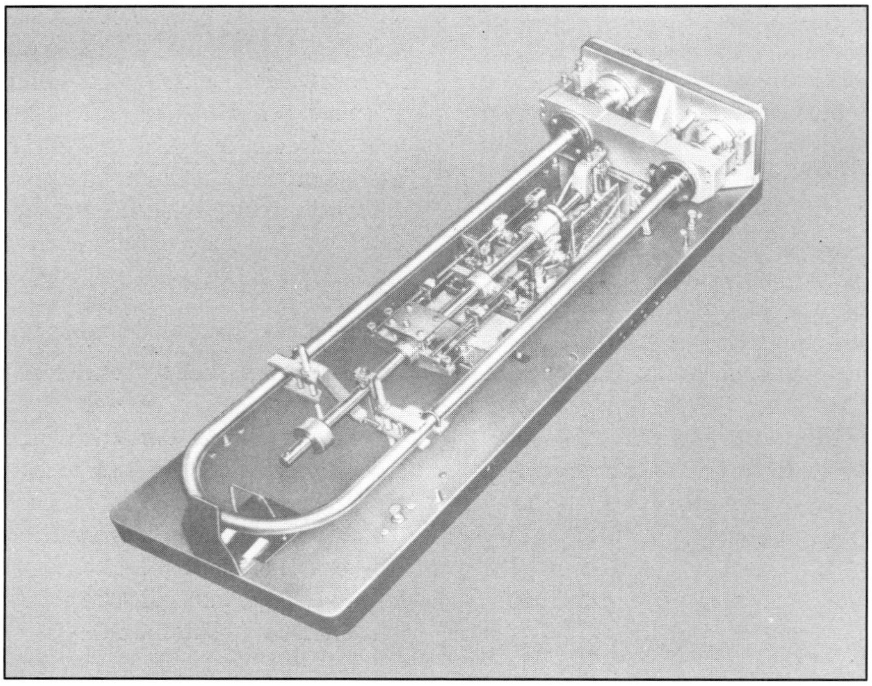

Figure 10–22—Continuous U-tube weighing balance.

Figure 10–21—Pressurized mud balance (available from Halliburton Services).

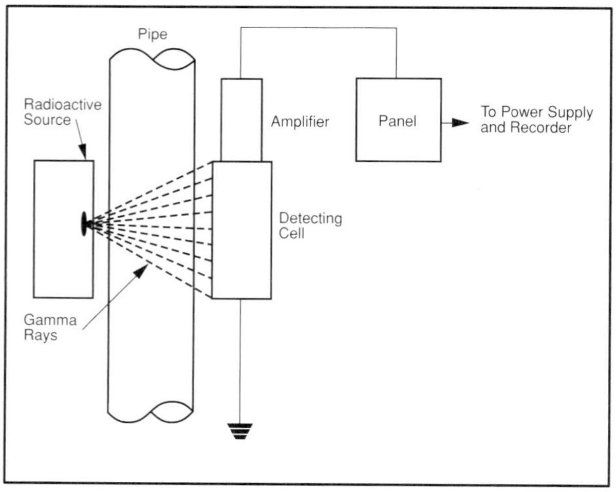

Figure 10–23—Schematic diagram of radioactive densitometer.

applying such a principle is beginning to appear in the field. In the near future, automatic cement mixing will undoubtedly become a routine procedure.

10–2.10 Steel Flow Hoses and Cement Head

A "cement head" (Section 10–5.14) is screwed into the top casing collar or DP tool joint, depending on the type of cement job. The discharge side of the downhole pump and the cement head are connected by a series of articulated or straight sections of high-pressure steel pipe (Fig. 10–25).

10–3 CEMENTING UNITS

The various components of cementing units, which fabricate and inject the cement slurry, have been described individually in Section 10–2. Figure 10–26 illustrates the combination of the components to assemble a basic cementing unit. A variety of configurations and compositions exists. They are designed according to the type of rig to be serviced, and the redundancy, versatility, and mobility required.

The various configurations are described below, according to the type of rig to be serviced.

Skid-Mounted Units: Illustrated in Fig. 10–27, skid-mounted units are most applicable to isolated land rigs, offshore rigs, cementing barges (lakes and rivers), and open-sea cementing vessels.

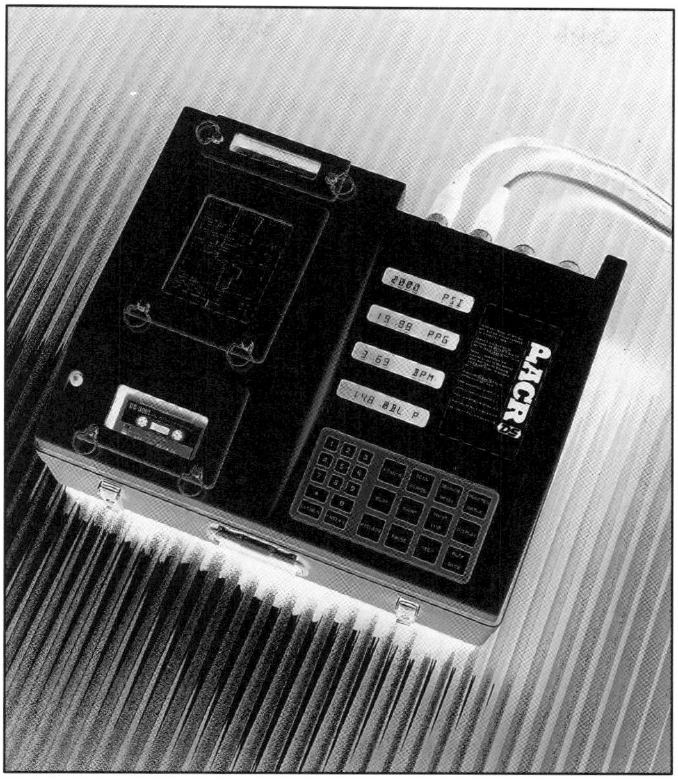

Figure 10–24—Central recording unit.

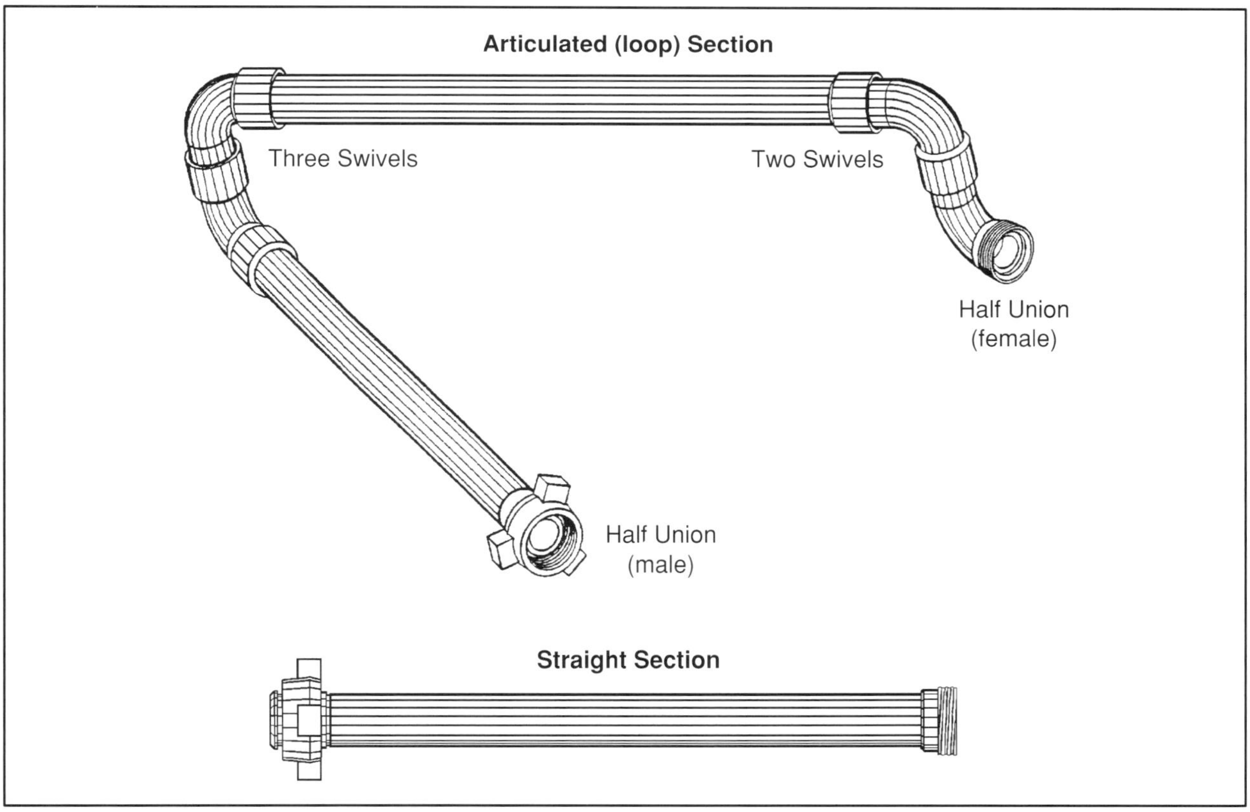

Figure 10–25—Steel flow hoses.

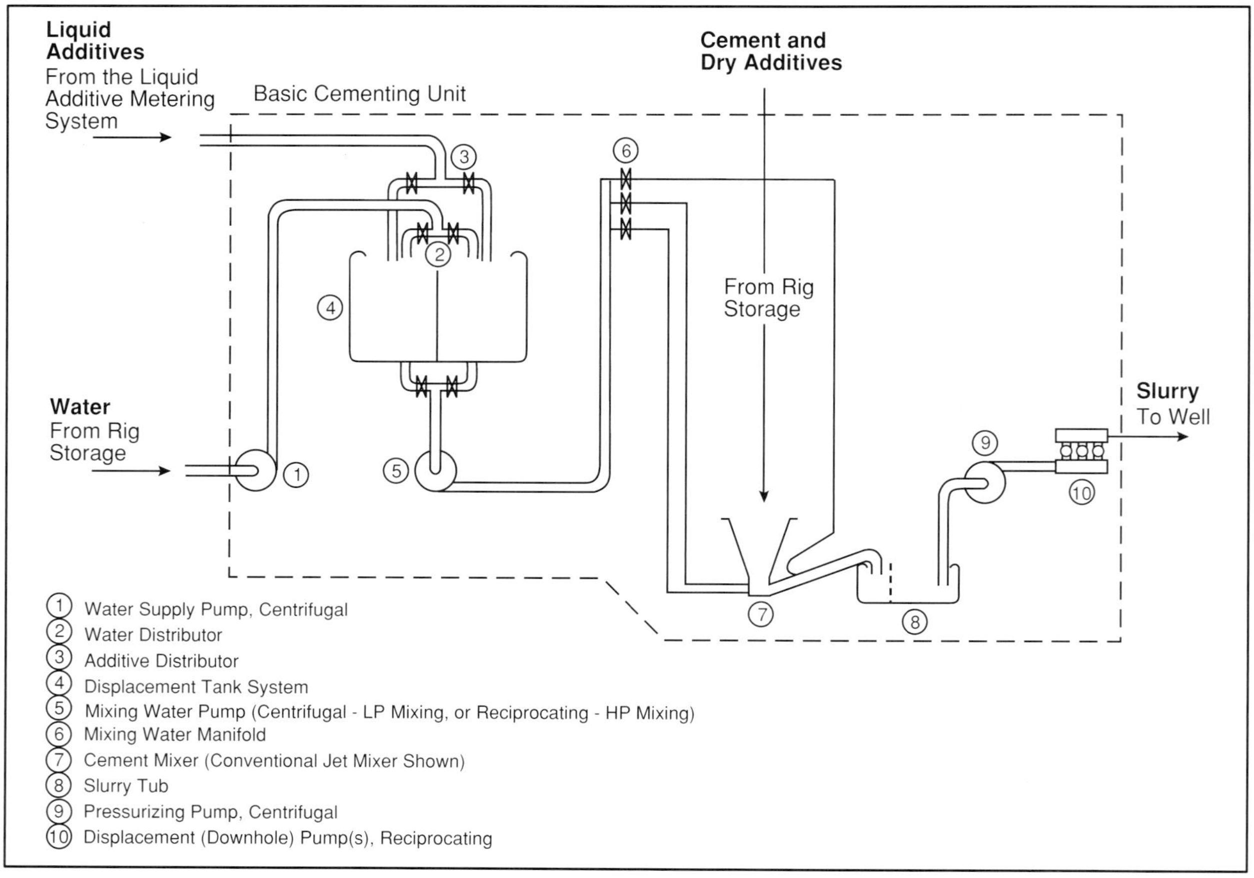

Figure 10–26—Mixing and pumping equipment on rig site (typical setup).

Truck-Mounted Units: Shown in Fig. 10–28, such units are suitable for almost any land rig. However, the chassis must be adapted to the type of surface upon which the unit will travel. The "standard" unit is designed to travel in a country where roads are available, and must conform to local road regulations. The "off-road" unit is built for more difficult terrain. The "desert" unit can be driven over soft surfaces, even sand dunes.

Semitrailer-Mounted Units: Like the truck-mounted units, semi trailer-mounted units are appropriate for almost any land rig; however, they are superior in a number of ways. They can be drawn by many types of tractors, which provides a logistical advantage. A heavy tractor-drawn unit with five axles has a better spanning ability than the corresponding truck with only three. The maximum authorized payload is superior to that of the truck, which allows the loading of more equipment on the same chassis.

Helicopter Units: Such units are intended for rigs totally inaccessible by land or river. The units, the mixing equipment, and the cement silos are custom designed to be transported by helicopter. They can be dismantled into smaller components, incorporating lifting frames, and are often made of lighter materials to reduce weight.

Traditionally, all vital items on a cementing unit are duplicated. This redundancy is necessary, because a well can be severely damaged or lost if it becomes impossible to complete a job after it has commenced. The extra equipment serves as an "insurance policy" to protect the operator's investment. Single-pump units exist, but are very rare.

For economic reasons, a cementing unit is designed to meet the requirements (including road regulations) of as many locations as possible. In Europe, for example, the required specifications vary from one country to another. In addition, soundproofing is more frequently demanded because of the proximity of wells to residential areas (Fig. 10–29).

Special attention should be paid to safety and environmental requirements. Environmental release of wastes (liquid or dry chemicals, cement slurries, or pump or engine oil spillages) is prevented by better equipment design, the use of recovery receptacles, and the attention of personnel.

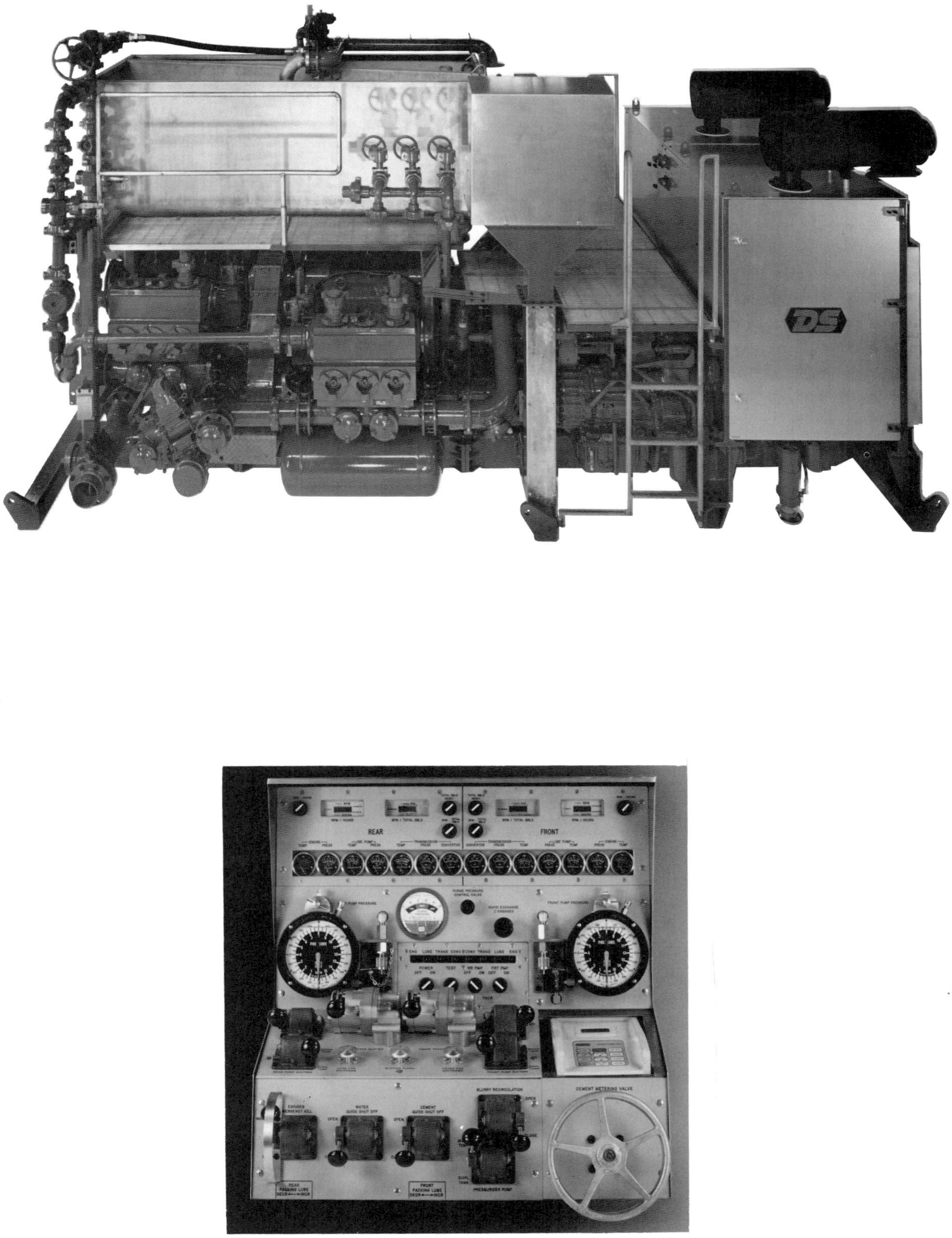

Figure 10–27—Typical cementing skid and control console.

Figure 10–28—Typical cementing truck.

The safety requirements to which the equipment should comply depend upon the location, and are especially dependent upon possible sources of inflammable or explosive gases. Whenever the unit can be placed more than 98 ft (30 m) away (as on most land rig sites) there are no special requirements. Standard equipment can often be used without modification. This distance condition is often difficult to satisfy on the offshore rigs, where every compartment or deck location is classified according to the potential risk of explosion or fire. The classification is made by official bodies according to standards that may vary slightly from one country to another; however, operators usually adhere to the most stringent regulations.

For example, Table 10–3 is a summary of the Det Norske Veritas (DNV) requirements for diesel engines to be located in a hazardous area, classified as "Zone 2," in which an explosive gas mixture may exist for a short time only under abnormal conditions. Diesel engines are banished from Zones 0 and 1, which are more sensitive areas. The DNV is the Norwegian certification body, and their standards serve as a reference in the North Sea. Diesel engines must often be adapted further to meet fire-protection standards. The equipment required to adapt an engine is sophisticated, entirely made of high-quality stainless steel, and very bulky. Needless to say, it is extremely expensive. Electric motors for Zone 2 areas are typically confined in a closed shelter, which is pressurized with air taken from a safe area. An overpressure is maintained so that no gas from the hazardous area can enter the shelter. When a cementing unit is to be operated offshore in a nonhazardous area, the drilling and service companies often opt for "protected" diesel engines, which provide increased security at a more reasonable cost. Table 10–4 is a list of the devices which should be installed on a standard diesel engine to ensure this protection.

10–4 INTRODUCTION TO CASING HARDWARE

This section focuses on equipment used on or within the casing string to enhance casing placement and cementing operations. Within the oil industry, this equipment is commonly referred to as casing hardware and cementing tools. Because of the vast array of these products, only the most common or basic types will be discussed at length, with the emphasis placed on application, principles of operation, and basic design characteristics.

Casing hardware consists of a wide variety of mechanical devices which are used to enhance primary cementing operations. They are permanently placed on or within the casing string, and may be purchased with or without service. Some common types of casing hardware include guide shoes, floating and auto-fill shoes and collars, stage collars, and external attachments such as scratchers and centralizers.

Cementing tools are generally retrievable devices, and may require some form of operation from the surface. They are typically associated with remedial cementing operations, and are supplied with service. Packers, bridge plugs, and retainers are examples of cementing tools which are commonly used for squeeze or plugback cementing.

10–5 CASING HARDWARE

A typical application of casing hardware for a primary cement job of moderate depth is shown in Fig. 10–30. The lower end of the casing is protected by a *guide shoe*. A *float* or *auto-fill-collar* is placed one or two joints above the shoe to provide, among other functions, a seat

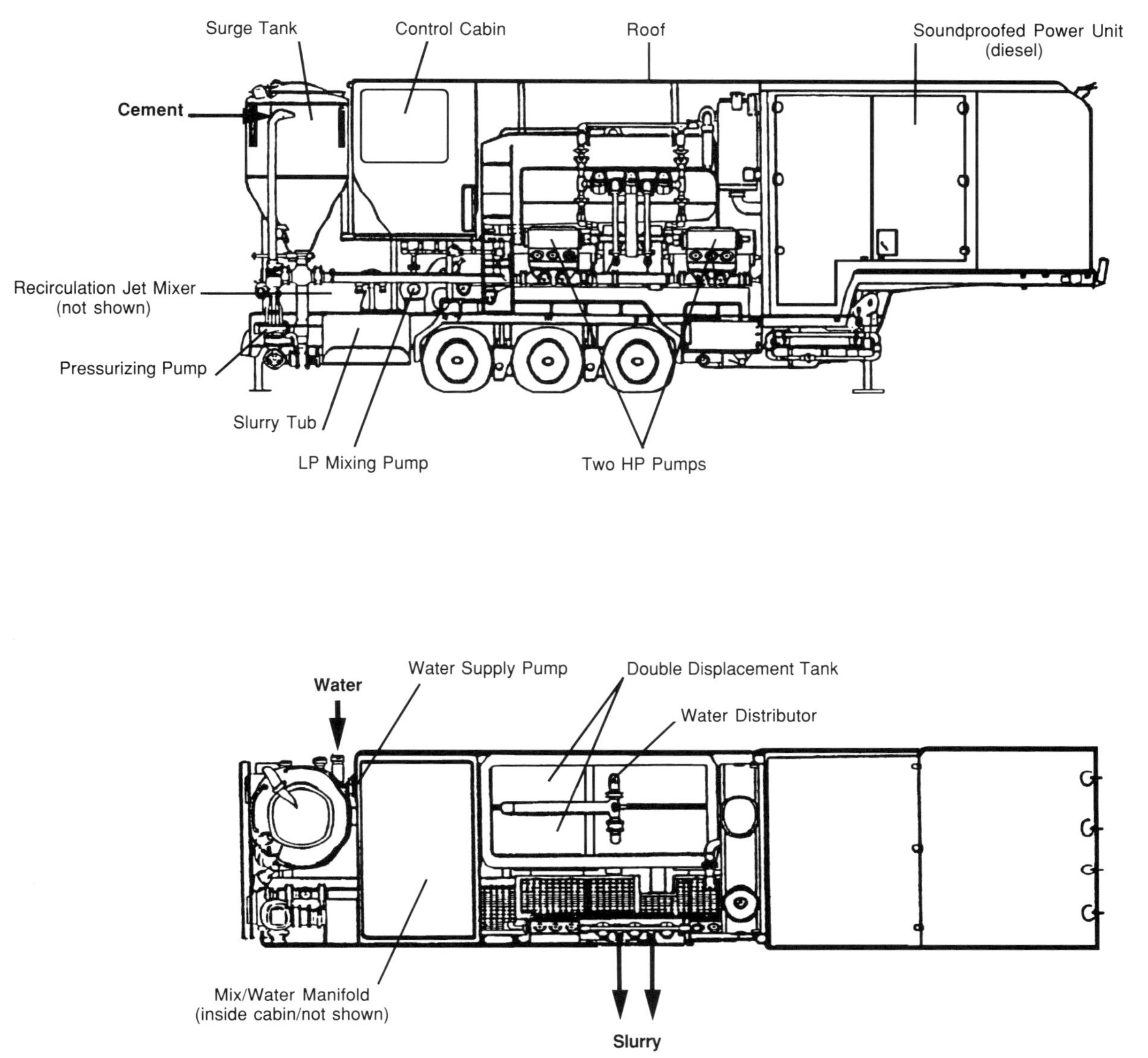

Figure 10–29—Semitrailer-mounted cementing unit (Europe).

to land *cement plugs* and to halt cement placement. The short section of casing bound by the shoe and float collar is called the *shoe joint*, and is provided as a buffer within the casing to retain contamination which may build up in the tail end of the cement slurry. The length of the shoe joint may exceed two joints of casing or 80 ft (24 m), to ensure the placement of good quality cement around the shoe. Plugs act as barriers to separate the cement slurry from the drilling mud and displacement fluids. *Centralizers* are placed in critical sections to improve casing centralization and cement placement.

1. Special water-cooled manifold rated to cool exhaust gas to 200°C (392°F) maximum, and with a surface temperature not exceeding 200°C at any point.
2. Oversized radiator.
3. Inlet air combustion, slam-shut valve
4. Inlet air flame trap.
5. Exhaust gas spark arrestor, DNV type approved.
6. Overspeed valve which closes the engine blower flapper valve when speed exceeds the normal maximum by 10%.
7. High-water-coolant temperature valve which shuts down the engine when water temperature exceeds 95°C (204°F). Fuel rack actuated.
8. Low-water-coolant level valve which shuts down the engine.
9. High-exhaust-gas temperature valve which shuts down the engine when the gas temperature exceeds 200°C.
10. Special control panel.

Table 10–3—Equipment required on diesel engine for operation in Zone 2.

1. Overspeed valve which closes the engine blower flapper valve when speed exceeds the normal maximum by 10%.
2. High-water-coolant temperature valve which shuts down the engine when water temperature exceeds 95°C. Fuel rack actuated.
3. Low-oil-pressure valve which shuts down the engine when engine oil pressure is below a value to be settled with the manufacturer. Fuel rack actuated.
4. High-oil-temperature valve which shuts down the engine when oil temperature exceeds 130°C (266°F). Fuel rack actuated.
5. Special control panel.

Table 10–4—Diesel engine protection kit.

10–5.1 Guide Shoes

Guide shoes are the most basic form of casing shoes, containing no check valves or flow-control devices. They are used to protect the lower edge of the casing. Most types offer a rounded nose to guide the casing through doglegs or restrictions in the hole. However, the regular pattern guide shoe does not have a rounded nose, and is not recommended for deviated holes. It simply reinforces the lower edges of the casing through its heavy wall con-

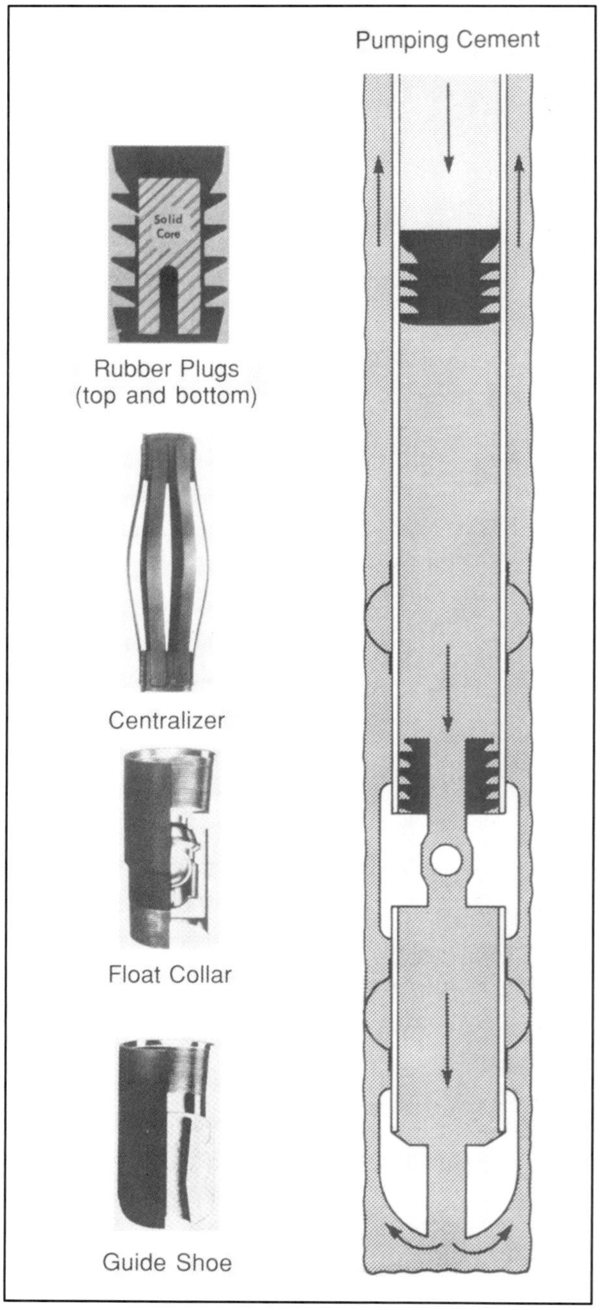

Figure 10–30—Typical casing hardware for primary cement job.

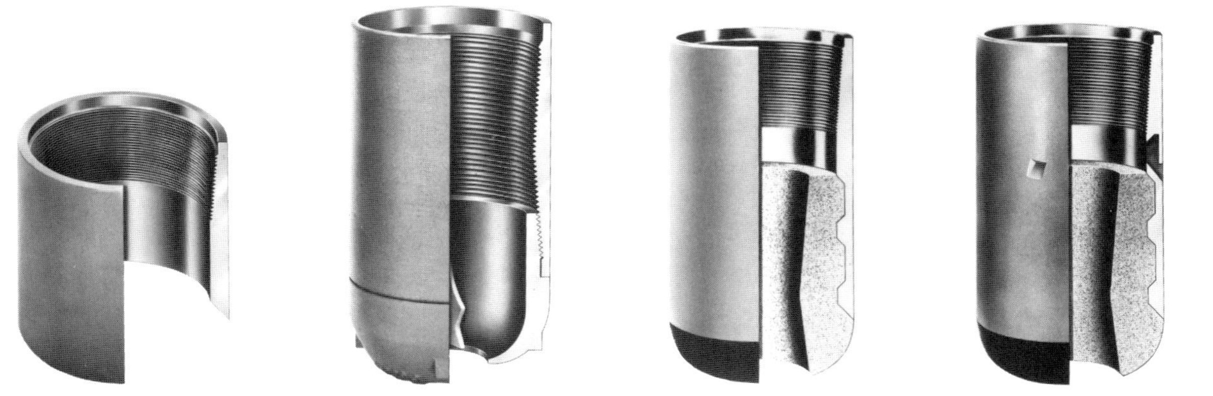

Figure 10–31—Guide shoes.

struction, and provides an inward bevel to guide subsequent drilling tools back into the casing.

Figure 10–31 shows several guide shoes which include contoured ends, baffles, and side ports. The nose and internal members are constructed of drillable materials such as cement or aluminum. The case is generally made of a steel used for casing collars, typically K55 or N80. The nose of the aluminum shoe includes helical fins which induce a swirling action to clean and lift debris from around the shoe, and improve cement slurry placement. Shoes with side ports provide a secondary flow path, which allows the casing to be set on the bottom while cementing. Side ports may also improve mud removal and washdown operations when circulation is needed to prevent sticking.

Guide shoes are generally used in shallow to moderate depth holes with a float, or auto-fill collar. They are often used below auto-fill collars, because their large bore allows the discarded auto-fill components to pass through (Section 10–5.3).

10–5.2 Float Equipment

As the demand for larger and heavier casings increased, so did the concern for derrick stress and fatigue. Float equipment reduces derrick stress by inducing flotation or increased casing buoyancy. Float equipment (Fig. 10–32) consists of specialized casing shoes and collars which contain check valves to prevent wellbore fluids from entering. As the casing is lowered, the hook load or hanging weight is reduced by the weight of fluid displaced. The casing is filled from the surface, and the hook load or amount of buoyancy is controlled by monitoring the weight indicator. The frequency of filling is generally once every 5 to 10 joints; however, some large-diameter or thin-wall casings may require more frequent filling to prevent casing collapse. In addition to proper filling, the casing should be lowered at a slow, steady rate to prevent pressure-surge damage.

Once the casing is landed, it is filled and circulation is established to begin conditioning the hole. The circulation of at least one hole volume of mud is typically required; however, to optimize hole and mud conditions for cementing, some drilling programs call for more than 20 hours of circulating (Chapter 5). Such large volumes, in addition to the pumping and displacing of cement, tend to cause excessive wear and increase the frequency of float-valve failure.

After the cement is displaced, the float valve must prevent backflow into the casing. Should the float valve fail, surface pressure and containment are necessary. Applying surface pressure is undesirable, because it expands the casing while the cement hardens. When the pressure is released, the casing relaxes causing a microannulus between the casing and cement. Although small, the microannulus compromises zonal isolation.

Other reasons for selecting a float valve also exist.

- Float valves are simpler, requiring no tripping operation to initiate the check valve function.
- Since all displaced fluids must flow up the annulus, the mud may be more continuously agitated and conditioned.
- Well kicking may be more clearly indicated and controlled.
- The casing may be filled with a clean, well-conditioned mud for cementing.

Pressure surges are generated each time the casing is raised and lowered, and are the product of inertia and

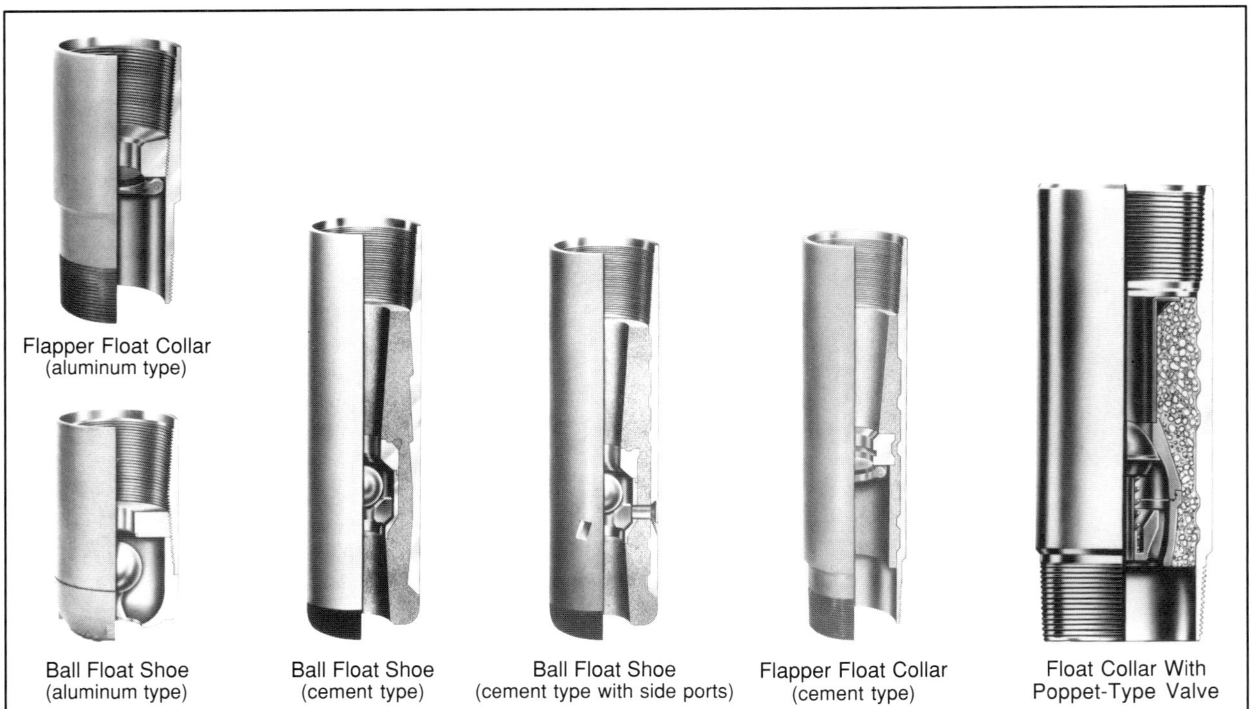

Figure 10–32—Float equipment.

flow resistance of the displaced fluid. Pressure surges combined with hydrostatic differentials may exceed the casing collapse resistance or the formation fracture pressure, causing loss of mud or permanent formation damage. External attachments such as centralizers and reciprocating scratchers may increase the flow resistance, and should be considered when determining a safe lowering speed.

Lowering speeds which create acceptable annular flow velocities during drilling are generally considered safe. Equation 10–1, derived from the Bingham Plastic Model, may be used to estimate a safe maximum lowering speed at a particular depth. The effects of hole abnormalities and external attachments are neglected. Turbulent flow is assumed, and a worst-case friction factor of 0.016 is used.

$$V_p = [25.6 \cdot P_s \cdot (D_h - D_p)/f \cdot L \cdot \rho]^{0.5} \times [(D_h^2/D_p^2) - 1], \qquad (10\text{–}1)$$

where

V_p = maximum velocity of casing to prevent damage to casing or formation;

f = 0.016 (mud friction factor);

L = depth (ft);

ρ = density (lb/gal);

D_h = hole diameter (in.);

D_p = pipe diameter (in.);

P_s = the lesser of Ps_f or Ps_c (psi) —

Ps_f = 0.5 L (G_f – 0.052 ρ) Formation Protection;

Ps_c = 0.5 (P_{cm} – 0.052 ρ) Casing Protection;

G_f = fracture gradient; and

P_{cm} = minimum casing collapse resistance (psi).

As a general rule, a safe practical lowering rate is two feet (0.6 m) per second or less.

10–5.3 Automatic Fill-Up Equipment

Automatic fill-up shoes and collars contain check valves similar to those used in float equipment. However, the check valves are modified to a normally open position to allow filling and reverse circulating (Figs. 10–33 and 10–34). The casing fills continuously, which saves time and reduces the pressure surges associated with float equipment. The valves are usually designed to reduce casing overflow by regulating the fill-up rate to a desired casing run-in speed. At an average run-in rate of one joint per minute, the fluid level inside the casing should remain one or two joints below the annulus level. Overflow may still occur should the annular flow resistance exceed the valve and internal flow resistance. This condition is most likely to occur in slim hole conditions, or when hole caving bridges and restricts flow in the annulus. To remove bridges or to relieve sticking, circulation in either direction is permitted.

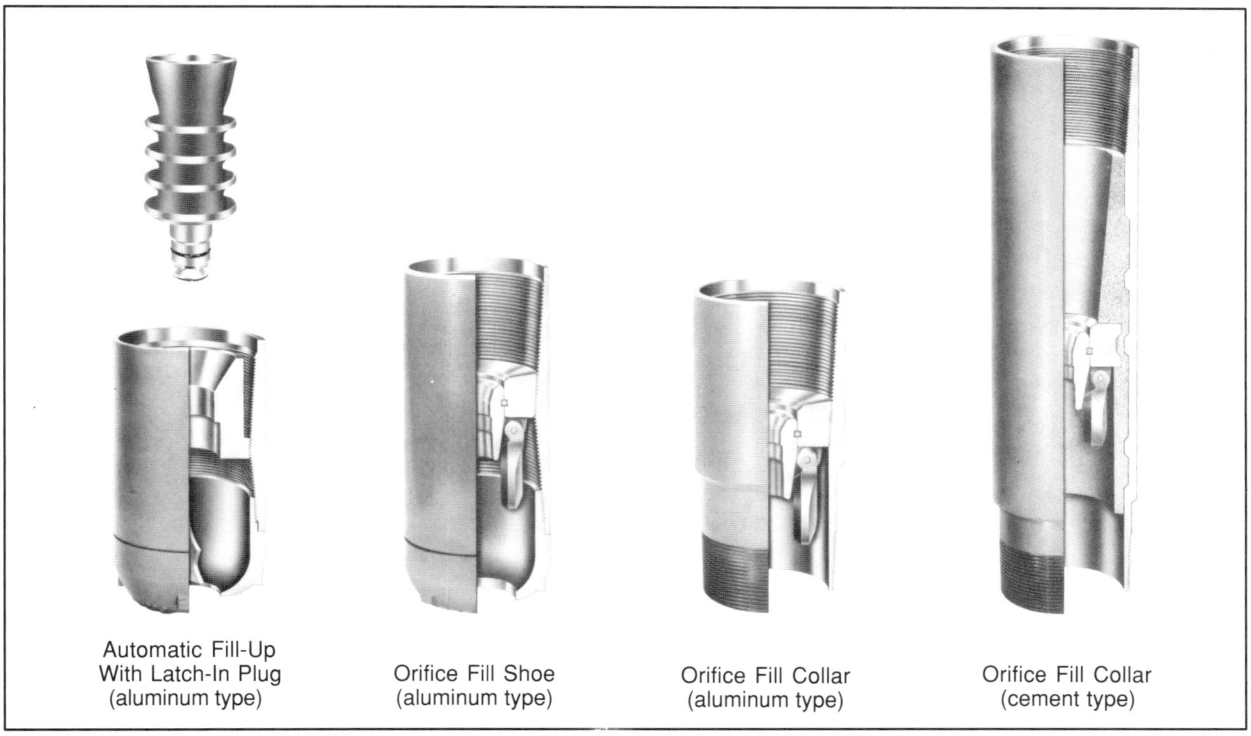

Figure 10–33—Automatic fill-up shoes.

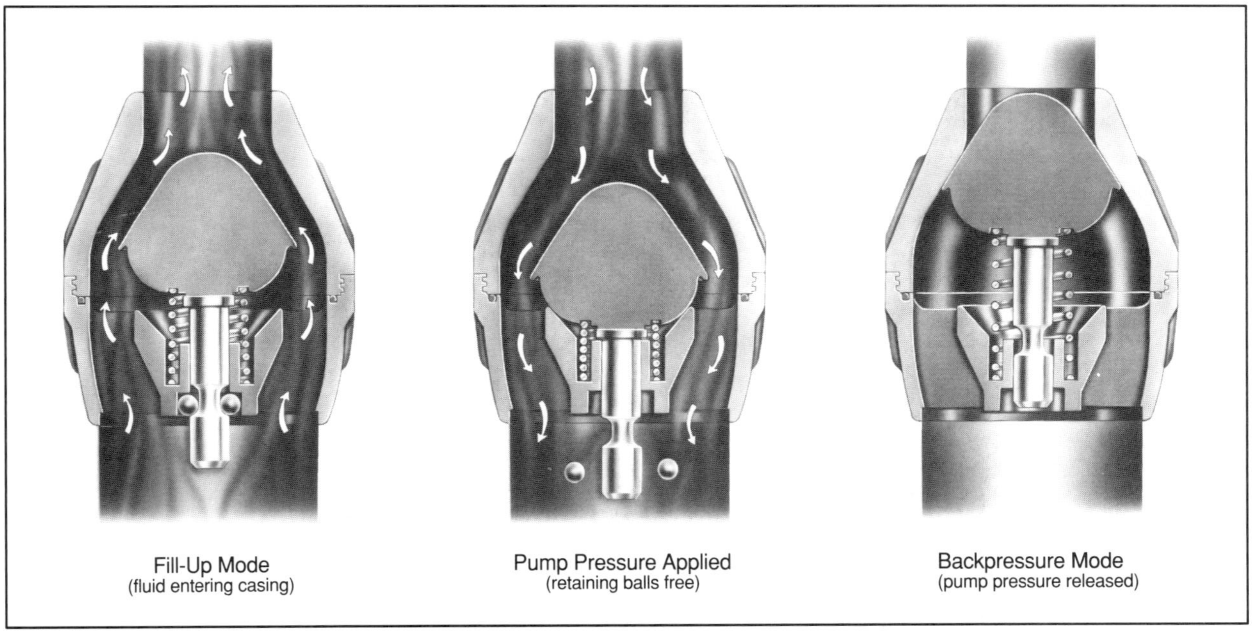

Figure 10–34—Poppet-type valve (drawing courtesy of Davis Lynch).

Auto-fill equipment must be tripped or converted to begin functioning as a one-directional check or float valve. Conversion is generally performed after the casing is in place, but also may be done while running to prevent overflow, or to control the hook load. To prevent premature tripping, the maximum flow rate down the casing may be limited by the tripping rate of certain valves.

Orifice fill (flapper) valves are converted by ejecting the orifice tube allowing the spring-loaded flapper to close (Fig. 10–33). This operation usually requires the use of a small metal tripping ball. To save time, the ball is generally dropped into the casing and allowed to free fall while running the last 5 to 10 joints. The free-fall rate is estimated at 200 ft (61 m)/min. The ball may be pumped down; however, should it set while pumping, the conversion may occur without indication. With the ball properly seated, the orifice tube may be discharged by applying 300- to 800-psi pressure, depending on the manufacturer. Some manufacturers list an optional flow rate for converting without a ball. This option is most useful when the hole deviation is greater than 30°, as difficulties in seating the ball may be encountered.

The *poppet or plunger* auto-fill valve (Fig. 10–34) contains a spring-loaded plunger which is held open to allow filling. The plunger is released to check reverse flow by establishing a minimum required flow rate through the valve. The minimum flow rate is generally between 4 and 8 bbl/min. Poppet collars are usually designed to retain the tripping mechanism, so two poppet units (shoe and collar) may be used to provide added seal insurance.

Auto-fill equipment is recommended when the hook load is not a major concern, or when hole conditions may be deteriorating, requiring reverse circulating and the ability to run casing as quickly as possible. Orifice and poppet valves are not recommended for use with drilling fluids containing large or heavy concentrations of lost-circulation materials. The use of many reciprocating scratchers and other external attachments may increase the annular flow resistance and cause overflowing. While lowering the casing, sudden stops should be avoided to prevent the premature conversion of the valve.

10–5.4 Differential Fill Equipment

Differential fill shoes and collars combine the benefits of floating and auto-fill equipment (Fig. 10–35). They are designed to automatically fill and regulate the fluid level within the casing. Most differential fill units (shoe or collar) will keep the casing approximately 90% full with respect to the annular fluid level. When both a shoe and collar are used, the casing should remain about 81% full.

Differential fill equipment is often used on longstrings to reduce surge pressures and the possibility of formation damage, which is normally associated with float equipment. They also save time, which lowers the probability of sticking. The fluid-level regulating feature reduces the hook load and prevents overflowing, provided the annulus is not restricted. Circulation may be established in either direction without harming the valve. The valve will resume operation when the casing and annulus fluid level reach the designed differential.

The typical differential valve (Fig. 10–36) regulates fill-up through the action of a floating differential piston. The piston slides up to open and down to close, and is designed such that the upper pressurized area is approximately 10% larger than the lower. The forces acting to shift the piston are produced by hydrostatic pressures acting on the upper and lower surfaces. Since the upper area is larger, less pressure above is required to balance the forces across the piston. When the pressure above (casing hydrostatic) exceeds 90% of the pressure below (annular hydrostatic), the piston will slide down to halt the filling. Likewise, when the pressure below exceeds 90% of the pressure above, filling will resume. This cycle is continuously repeated as the casing is lowered. However, cycling may not begin until the hydrostatic pressure is sufficient to overcome frictional forces. When two valves are used, the upper valve senses pressures regulated by the lower valve, and the combined effect should result in an 81% fill.

The inoperative flapper valves may be converted to begin functioning as a float valve at any time. Converting most valves requires a tripping ball operation like that described for orifice fill equipment. Circulating prior to dropping the ball may help clear the seat of debris. To verify proper tripping pressure, the ball should be allowed to fall to the seat before pumping. The pressure required to trip most valves is generally between 500 and 800 psi. Since only the ball is discharged, a shoe and collar may be used, and both may be tripped with a single ball. An orifice shoe also may be used below a differential collar, provided the tripping ball is compatible with both units or the orifice may be opened with flow.

The following are some additional tips and precautions—

- To reduce float valve wear during long circulating and conditioning periods, the tripping operation may be delayed until just prior to pumping cement.
- Because of restrictions in the fill-up path, the casing should be lowered at a moderate rate (generally two feet per second) to reduce pressure surging.
- Lost-circulation materials may tend to slow or prevent filling, which may increase surging or lead to collapse. Periodically, circulating and monitoring the weight indicator may be necessary.
- Hole deviation and casing size may prohibit the use of a weighted tripping ball. Some manufacturers offer ball guides for deviations over 20°. Others trap or preload the ball, which should allow use at any deviation; however, circulation is also prevented before

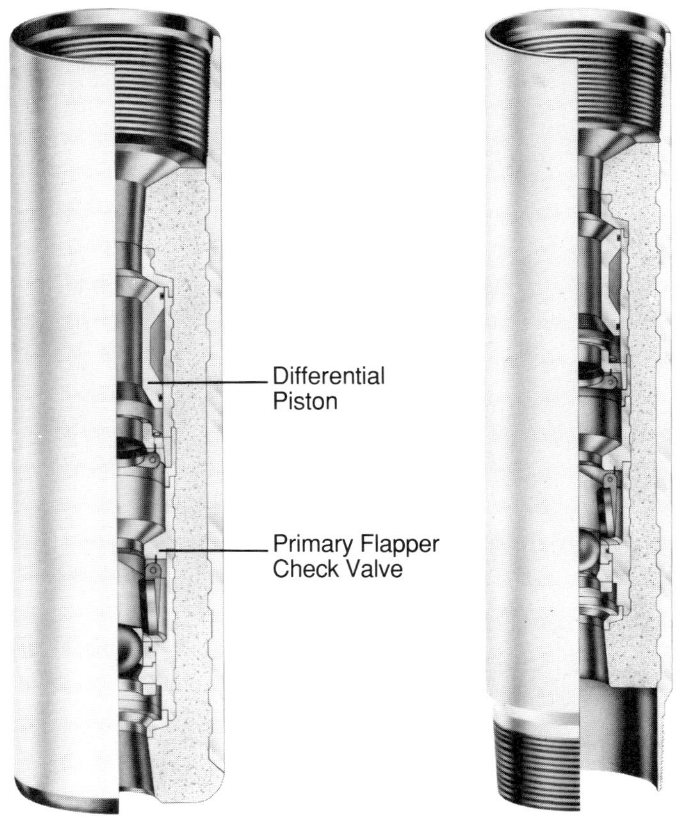

Figure 10–35—Differential fill shoe and collar.

tripping. The maximum allowable deviation should be provided by the supplier.

10–5.5 Casing Insert Equipment

Insert equipment offers an economical means of providing floating and auto-fill valves for low to moderate depth and pressure applications (Fig. 10–37). They are generally available for 10¾-in. (273-mm) and smaller casings, supplied with API 8 round long or short couplings. They are made of aluminum or cast iron, and contain ball, flapper, or latch-in valve mechanisms.

Insert valves are installed within a coupling, and are trapped between the above and below joints of casing. Their strength is limited to material trapped with the coupling. The insert valve must be fully made up to prevent damage and interference with proper casing engagement. When landing a plug, the rate must be reduced to prevent pressure spikes which may exceed the strength of the insert.

10–5.6 Inner-String Cementing Equipment

Inner-string cementing, described in Chapter 12, is a technique typically used with large-diameter casings, where drillpipe is placed inside the casing as the conduit for pumping fluids from the surface to the casing annulus. Inner-string cementing equipment (Fig. 10–38) provides a means to receive and seal the drillpipe downhole. This equipment is also commonly referred to as "stab-in equipment," and is generally available with latch or nonlatch receptacles. The shoes and collars are basically larger versions of the types previously discussed, with the addition of a seal receptacle and beveled surface. They are commonly available for 10¾-in. (273-mm) and larger casings.

Casing running operations correspond to the type of valve used. The lowering speed should be slow enough to prevent surging. Float equipment may require more frequent filling to prevent collapse.

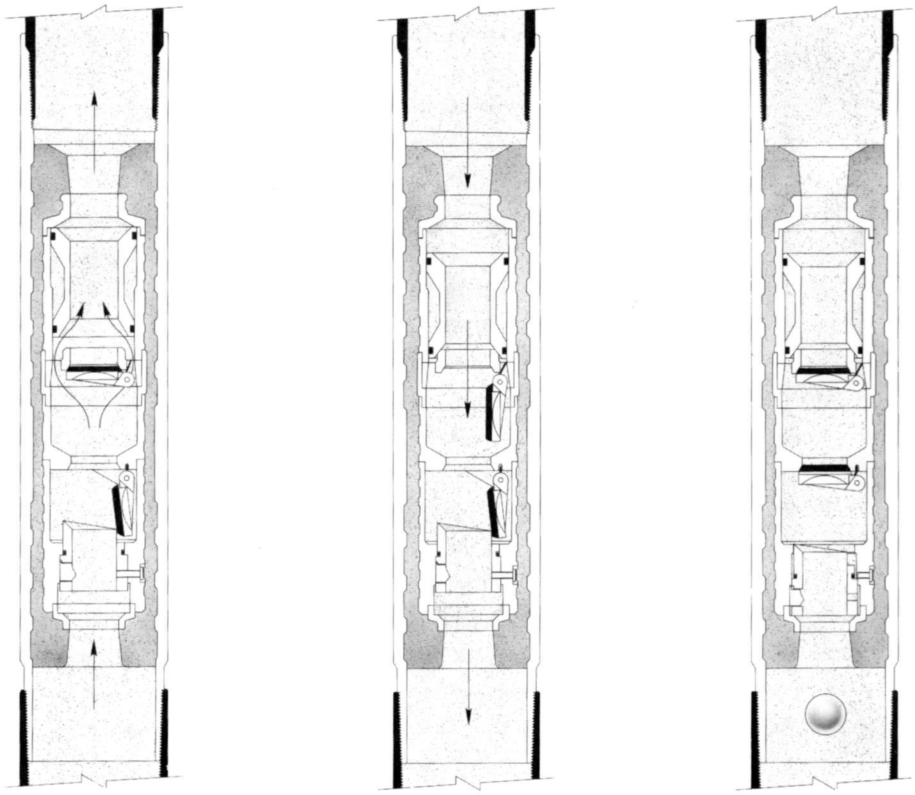

Figure 10–36—Operation of differential valve.

Once the casing has reached the desired depth, the stab-in seal unit and centralizer are connected to the drillpipe, and run into the casing (Fig. 10–39). The drillpipe is lowered until the seal unit engages the receptacle. Additional weight must be applied to nonlatch equipment to counteract the lifting force created while cementing. The maximum lifting force may be estimated by multiplying the maximum expected pumping pressure by the end area of the seal unit. A simple rule of thumb, which will generally provide adequate pipe weight, is to apply the larger of the following—15,000 lb, or 2,000 lb per 100 ft of depth. Achieving this weight may require the use of drill collars or heavyweight drillpipe. Latch-in seal units include an additional locking mechanism to counteract the lifting force. No additional weight is necessary; however, rotation is generally required to initiate the release.

The inner string isolates the casing interior from the pumping and hydrostatic pressures created while cementing. Care should be taken to prevent creating pressure differentials which may exceed the collapse resistance of the casing. To help prevent collapsing, pressure may be applied to the inside of the casing with the use of a packoff head.

Some reasons and benefits for using stab-in equipment are listed below—

- Greatly reduces displacement volumes and time;
- Reduces wasted overdisplaced cement when cementing to the surface;
- Reduces the need for extended working time cements; and
- Less contamination occurs because of reduced area and turbulent velocities in the drillpipe.

10–5.7 Valve Types and Characteristics

The three most common types of check valves used in casing hardware are the ball, flapper, and poppet valves. They have been adapted for floating and auto-fill applications, and are commercially available from several suppliers. There are many specialized variations—one such is the Rentry shoe (Fig. 10–40), which is used with a downhole camera for guideless completions. For special applications and a closer look at specific valve designs, product literature should be requested from the manufacturer.

Although the API recently established standards covering the performance characteristics of casing hardware valves (10F), typical valve performance data, such as minimum restriction, C_v or flow resistance, pressure limitations, and erosion or wear resistance, are not commonly published. Some of these characteristics may be obtained from the manufacturer; however, a comparison may not be valid because of differences in the test conditions. For this reason, the following comments will be more qualitative than quantitative.

Materials used in the valve mechanisms may be more or less drillable and compatible with certain types of bits. The major valve components are generally made of plastics, aluminum, and cast iron. Rubber is used for seals and coatings, and steel parts such as springs and pins are generally considered too small to pose a problem. The plastics are typically thermosetting materials such as phenolics which are stable to temperatures near 450°F (232°C). They are easily drilled, and are compatible with most bits. The us se of some thermoplastic materials may be limited to applications below 270°F (132°C). Aluminum is easily drilled, but has a tendency to foul or smear on some diamond or fine cutter bits. Neither aluminum nor cast iron is recommended for Polycrystalline Diamond Compacted (PDC) bits. Although cast iron is more difficult to drill, the quantity used in most cement-filled equipment is considered too small to pose a problem.

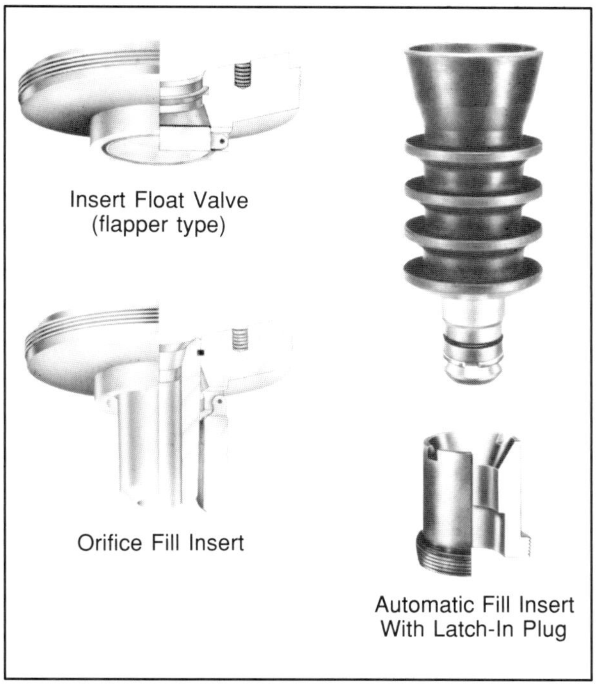

Figure 10–37—Casing insert equipment.

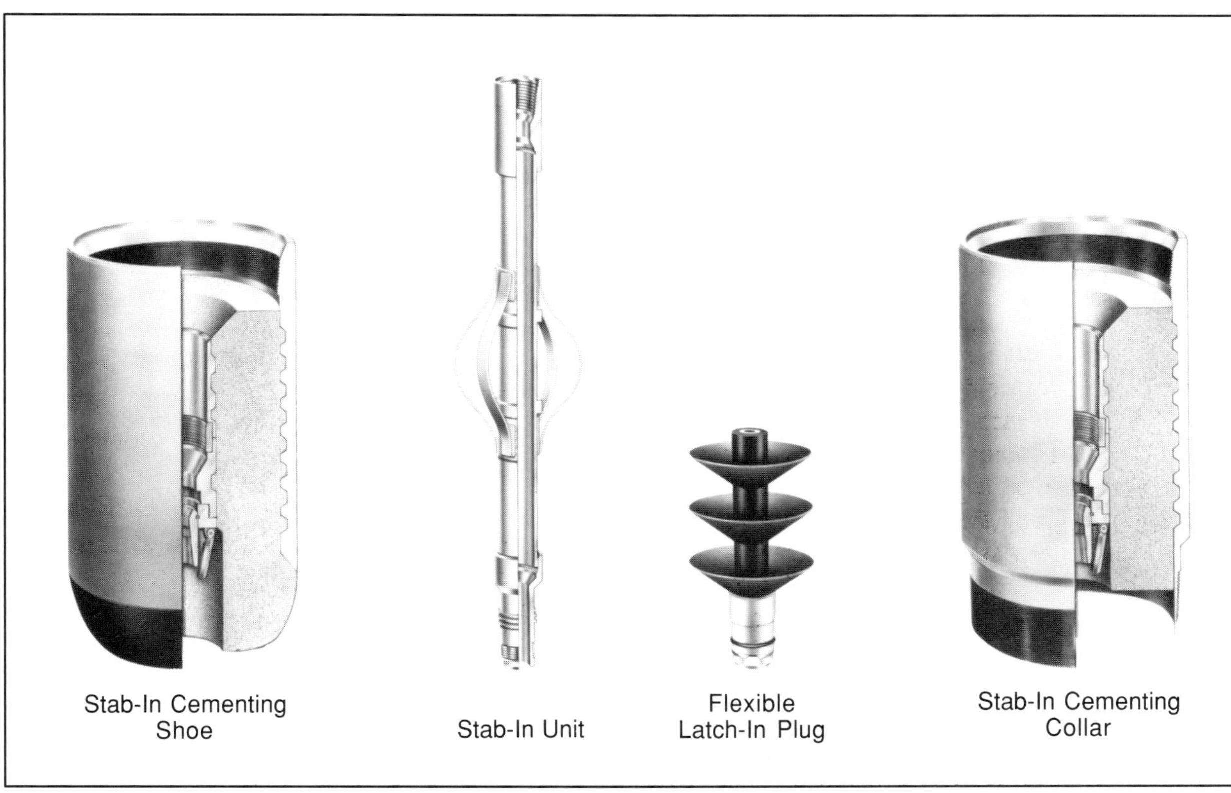

Figure 10–38—Inner-string cementing equipment.

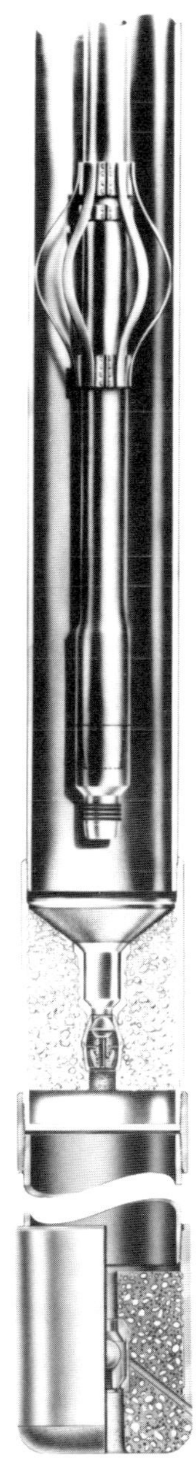

Figure 10–39—Operation of stab-in seal unit and centralizer (drawing courtesy of Davis Lynch).

Aluminum and cast iron are unaffected by temperatures normally encountered in drilling operations.

Ball valves consist of the typical ball and cage arrangement (Fig. 10–41). The ball is generally made of plastic and may be rubber coated. The cage may be made of plastic or aluminum. Although the ball is buoyant in most drilling fluids, hole deviation may require reverse flow to return the ball to the seat. Ball check valves may not be effective in preventing slight reverse flow or gas migration in highly deviated holes. The minimum clearance between the ball and the cage may be one-half inch or smaller. Solids such as mud scale or cuttings should be avoided to prevent the possibility of wedging the ball, or plugging the valve. Likewise, ball valves are not recommended for use in muds containing heavy concentrations of lost-circulation materials (LCMs). The pressure resistance is more than adequate for most applications. The temperature and wear resistance depend greatly upon the materials used and the construction of the valve. Wear resistance is generally fair, but should be qualified before used in extended circulating applications (eight hours or more).

Flapper valves are composed of a spring-loaded flapper hinged to a plate with an integral seat (Fig. 10–42). The flapper and plate are generally made of aluminum or cast iron. The spring-loaded seating action is virtually unaffected by hole deviation. The flapper opens to provide a large unobstructed flow path which will pass tripping balls and extraneous materials without harm. The temperature resistance is generally good because of the use of metal components. Pressure resistance is generally more than adequate for most conditions. The wear resistance varies with size and manufacture, and should be specified for extended circulating applications.

Poppet valves (Fig. 10–43) are composed of a spring-loaded poppet (plunger) housed in a cage much like that of a ball valve. Plastic and aluminum are generally used, and the poppet is often rubber coated. Like the ball valve, the poppet valve has a restricted flow path. Solid debris which may plug or bind the poppet should be avoided. The spring-loaded seating action is virtually unaffected by hole deviation. The temperature resistance depends on the type of materials used, and should be specified by the supplier. The pressure resistance is typically good, and more than adequate for most applications. The wear resistance is also good, but should be specified before use in extended circulating time applications.

10–5.8 Shoe and Collar Case Design

The case is the outer steel portion of a cementing shoe and collar. It becomes an integral part of the casing, and must be capable of meeting chemical, dimensional, and strength requirements. Aside from a threading standard (API Standard 5B), there are no specific industry standards covering shoe or collar cases. However, the following design criteria and guidelines are commonly used for general-purpose equipment.

Shoe and collar cases are usually made of low-alloy or carbon steels such as API K55 or N80, and are suitable for H_2S applications. To prevent interference and allow use over a broad range of casing weights, shoe and collar

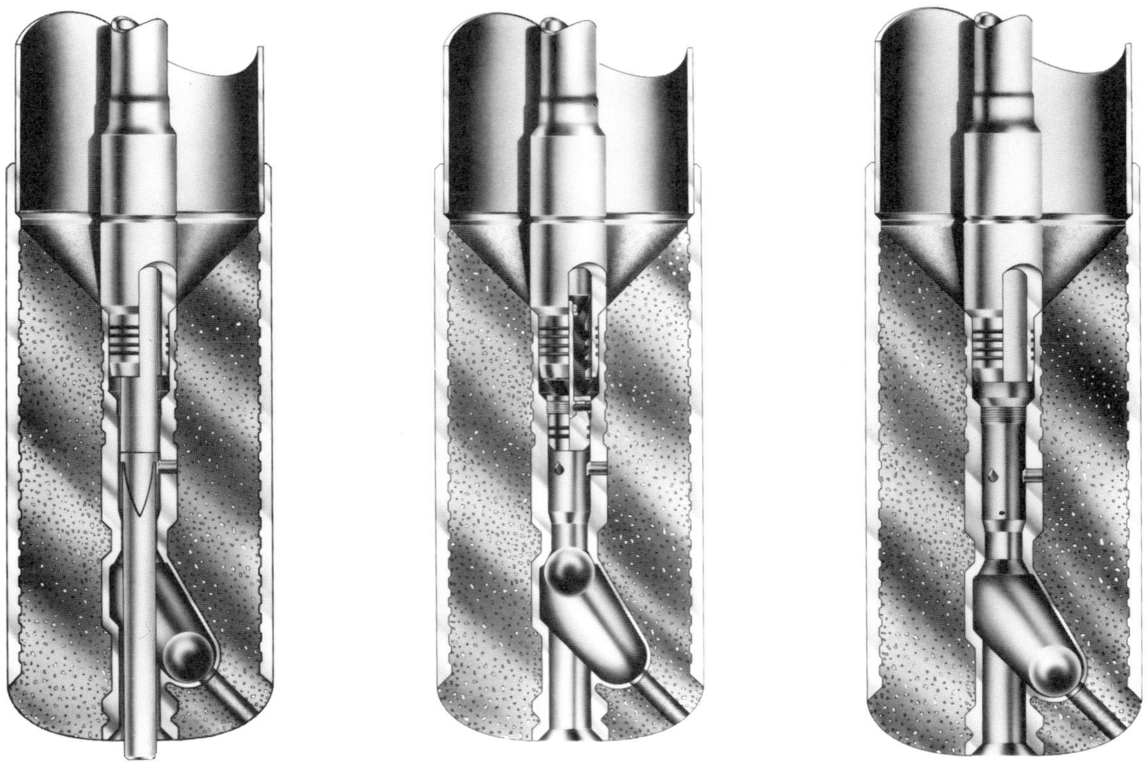

Figure 10–40—Rentry shoe operation (drawing courtesy of Davis Lynch).

internal and external diameters generally conform to the lightest weight API casing drift and coupling external diameters, respectively. This also provides a heavier than usual wall which allows use of low-strength steels, such as K55, to meet N80 burst and collapse pressures. Because of the lack of pressure differential across the shoe, general-purpose shoes are considered suitable for use with high-strength casings. Some manufacturers offer a heavyweight range to more closely match internal diameters, and to improve external thread strengths.

General-purpose equipment is most readily available with API 8-round and buttress connections. To allow makeup to either API long or short connections, 8-round equipment is usually provided with API long internal threads and short external threads where applicable. External thread joint strengths may be as low as lightweight K55 casing, but are generally adequate for bending and tensile stresses at the lower end of the casing. Some manufacturers offer double-box (internal thread) collars which eliminate the weak link, the external thread. Premium threads such as Vam, extreme line, and Hydril are not commonly stocked and may require an extended delivery time.

10–5.9 Stage Equipment

Stage equipment, consisting of stage collars and port collars, is placed within the casing string to provide a selectable intermediate passage to the annulus (Fig. 10–44). The collars are generally made of S95 and P110 equivalent materials, and may be available in special weight ranges to optimize strength and internal dimensions. Stage equipment is generally used to protect weak formations from excessive hydrostatic pressure, to cement two widely separated zones, and to reduce mud contamination. Stage-collar placement and cementing techniques are presented in Chapter 12.

Stage collars are typically hydraulically opened and closed using free-fall darts and pumpdown plugs to select and shift the appropriate internal sleeve (Fig. 10–45). The lower sleeve covers the ports initially. Once the first stage is complete, the lower sleeve is pumped down to uncover the ports by seating the free-fall (or pumpdown)

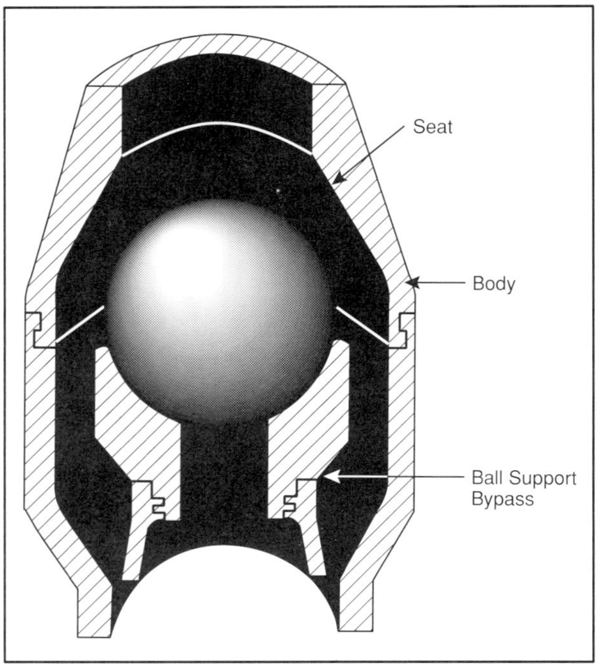

Figure 10–41—Ball-type valve (drawing courtesy of Davis Lynch).

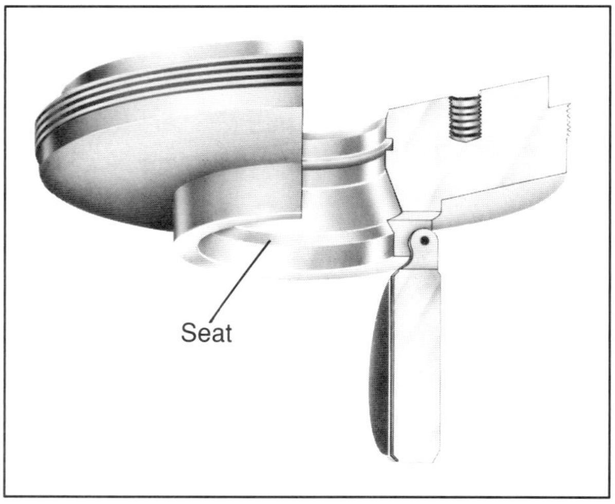

Figure 10–42—Flapper-type valve.

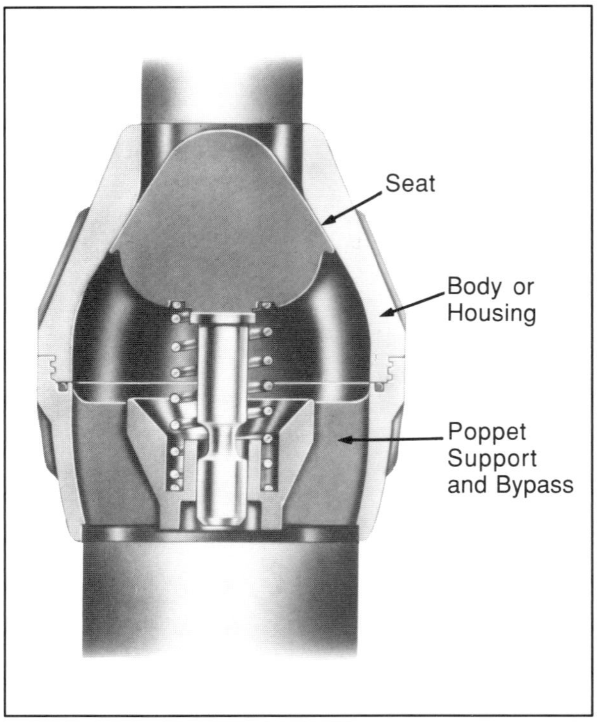

Figure 10–43—Poppet-type valve (drawing courtesy of Davis Lynch).

opening plug and applying pressure. The second stage is pumped and the ports are closed again by seating and applying pressure to the larger closing plug. Once closed, the stage collar cannot be reopened. The pressure required to open and close varies with manufacturers, but is generally between 800 and 1,400 psi. When two stage collars are used, a special upper stage collar is required, and care should be taken to release the correct plugs in the proper sequence. The internal diameter of the upper stage collar seats must be larger than the lower collar seats. For highly deviated holes, the free-fall dart should be replaced with a pumpdown plug as described in Chapter

12. Drilling is required to remove the plugs and aluminum seats.

Port collars are mechanically operated from the surface via a service tool connected to an inner string of drillpipe. They are available with sliding or rotational valve mechanisms, and may be opened or closed as often as necessary (Fig. 10–46). The sliding valves are generally opened with an upward motion and closed with a downward motion, and require a minimum of 10,000 lb to stroke. Port collars may be placed as often as necessary in the casing string and selected in any sequence. There are no plugs to use, nor drillout required. Some shifting tools may be fitted with cup type seals to form a conduit from the inner string to the ports.

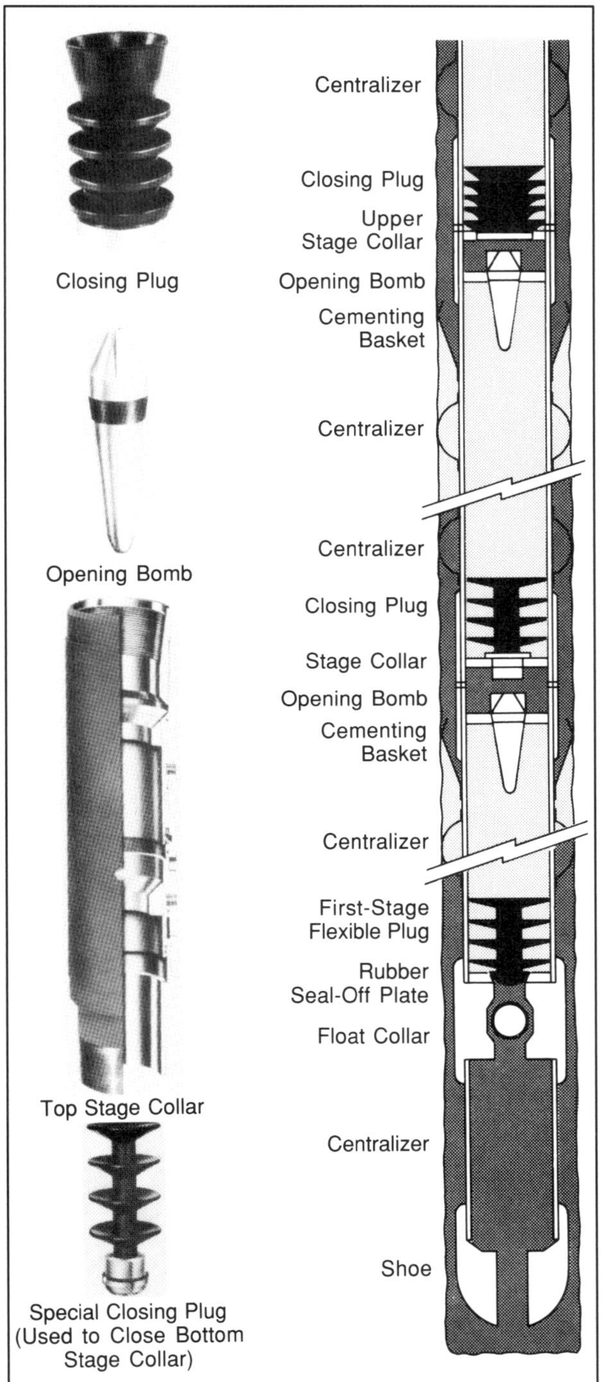

Figure 10–44—Stage equipment.

Stab-in stage collars, which combine the benefits of inner-string (stab-in) cementing with stage cementing, are available for 13⅜-in. and larger applications. Stab-in stage collars may be mechanically or hydraulically shifted via the inner string. Unlike port collars, stab-in stage collars are intended for multiple openings.

Centralizers and baskets, or external casing packers (ECPs), are often used with stage equipment. Baskets or external casing packers are placed below the stage collar to help support the hydrostatic pressure of the next stage, and to prevent cement from falling through lower density fluids below. ECPs are popular because of the improved sealoff and centralization benefits provided.

Stage collars and port collars must be handled with care because of the close tolerance between the case internal sleeves. Tongs or makeup equipment should not be placed in the body of a stage collar.

10–5.10 Annular Packoff Equipment

Annular packoff equipment (packer shoes, collars, and external casing packers) is used to protect areas of the formation from excessive hydrostatic pressure and/or contaminating fluids. The equipment has expanding rubber elements which pack off against the formation to create an impermeable annular barrier. The rubber elements also centralize the casing when expanded.

Packer shoes and collars are most often used in openhole completions to protect the formation below the casing from cement contamination and hydrostatic pressures. They contain check valves for floating or auto-fill applications, and prevent the backflow of cement.

Packer shoes and collars are hydraulically set with the use of a tripping ball (Fig. 10–47). With the casing in place, the tripping ball is dropped and allowed to fall to a seat on the piston. As pressure is applied, a load is transferred to an external sleeve which compresses and expands the element. At a pressure of approximately 800 psi the piston shears free, uncovering the ports. The external sleeve contains a ratchet mechanism which permanently holds the set position. The cement is pumped and, unlike stage equipment, the ports are not closed.

Packer collars provide the same basic functions as packer shoes, but also include a means for hanging slotted or perforated casing.

Basket shoes (Fig. 10–48) are a form of annulus packoff which uses a basket instead of an expanding rubber

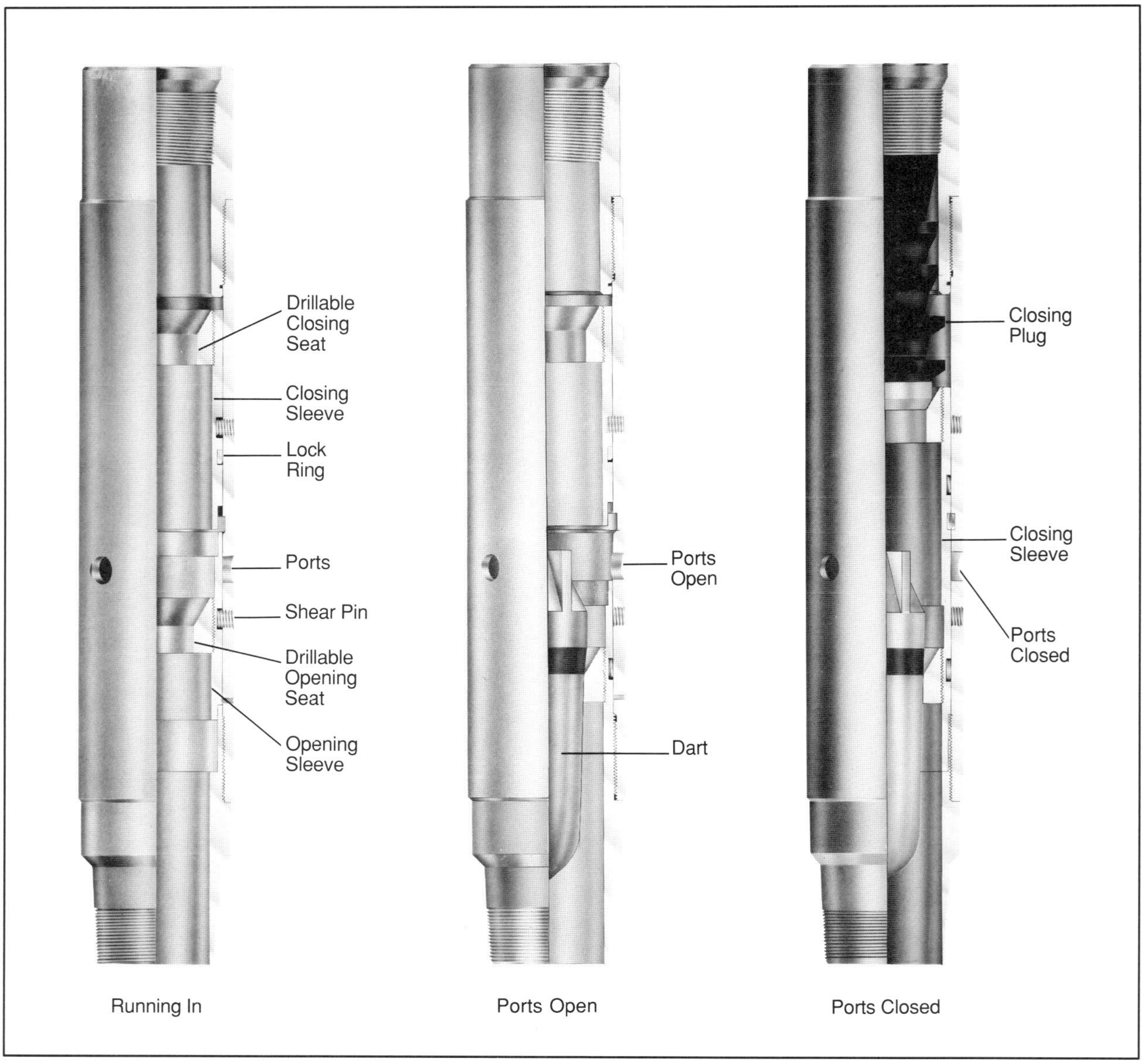

Figure 10–45—Stage-collar operation.

element. They provide the same function as a packoff shoe, but are limited to low differential pressure applications.

External casing packers (ECPs) are strictly packoff devices (Fig. 10–49). There are no ports to the annulus or internal float valves. They are generally used below stage collars or port collars to protect the formation below from excessive hydrostatic pressure or contamination. They are also used in an attempt to block gas and fluid migration, and to provide positive centralization.

ECPs may be packed off by either inflating or compressing the rubber element. The inflatable type is generally larger, and better capable of packing off oversized or irregular holes. The inflation process is generally set to begin at a predetermined setting pressure. The setting pressure should be sufficiently high to prevent premature packoff while conditioning or cementing.

An optional breakoff rod may be used to prevent premature setting by blocking the inflation port until broken free by a wiper plug. Once the element is inflated, an internal valve mechanism will hold the inflated position, and the surface pressure may be released. When possible, the element should be inflated with cement.

10–5.11 External Attachments

Casing centralizers are one of the simplest yet most beneficial mechanical aids used in primary cementing. They are designed to position the casing more centrally in the hole, and used to provide the following benefits.

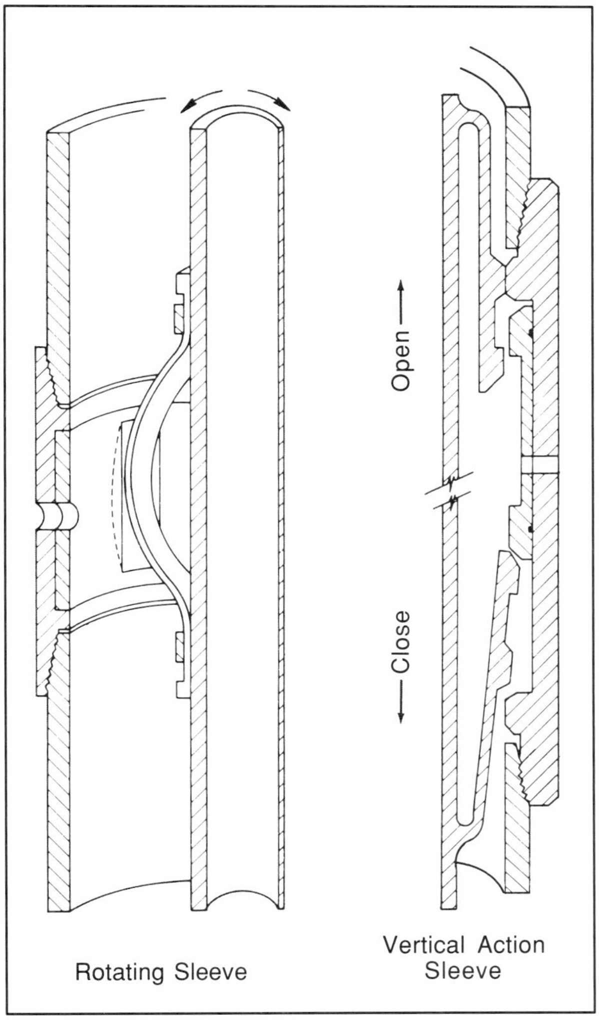

Figure 10–46—Operating principle of port collars.

- Reduce drag and differential sticking while running.
- Improve mud removal.
- Improve cement placement, creating a more uniform wall thickness.
- Improve performance of other external devices, such as scratchers and baskets.

To be effective, centralizers must be properly placed with respect to spacing and location within the string. Spacing is a function of several parameters such as casing size, hole size, deviation, and is discussed in more detail in Chapter 11. Most service companies offer a computer-aided service to determine optimum centralizer spacing and placement.

The type, fit, and strength of a centralizer are also important factors to consider. There are two basic types of centralizers—rigid and spring-bow (Fig. 10–50).

Rigid centralizers are built with a fixed bow height, and are sized to fit a specific casing or hole size. They provide a positive standoff when placed within casing or a well-gauged section of the hole.

Spring-bow centralizers are more commonly used. They are constructed with oversize spring-like bows which are flexible yet stiff enough to provide adequate standoff in various hole shapes and diameters. They are available with low or tall bow heights for slim or oversize holes, respectively. The bows may be placed in a helical pattern to aid in mud removal while running, and thin blades are sometimes attached to induce turbulence and aid mud removal. Such devices are known as *turbolizers* (Fig. 10–51).

Although most spring-bow centralizers are similar in appearance, subtle differences in bow thickness, or the lack of proper heat treatment, may considerably reduce their strength. To establish minimum strength requirements and a basis to determine the centralizer spacing, the American Petroleum Institute established the Casing Centralizer Specification 10D. This specification defines strength requirements and methods of testing for standard and close tolerance centralizers. Close tolerance centralizers are designed for casing-hole clearances less than 1½ in. (4 cm).

Most centralizers are available with solid or hinged collars. The hinged style is the simplest to install, and is most often used. Solid-collar centralizers are recommended for close tolerance conditions (typical of liner applications), because they are generally stronger and cause less flow resistance. Whenever possible, centralizers should be installed so that they are pulled by the casing rather than pushed. This reduces the chance of centralizer damage due to hanging, and is accomplished by placing a stop ring within the centralizer collars. In large or oversize hole conditions, centralizers may be placed over couplings, provided the couplings do not interfere with bow deflection. Close tolerance centralizers are generally designed with low profiles, and should not be placed over stop rings or couplings which may contact or interfere with the bows.

Scratchers are external devices designed to remove immobile mud from the wellbore. There are two general types—reciprocating and rotating scratchers. Reciprocating scratchers (Fig. 10–52), consisting of a collar with radial wires or cables, are designed to remove mud by pipe reciprocation. Rotating scratchers (Fig. 10–53) are straight bars containing similar wires or cables, and are attached lengthwise to the casing to remove mud while rotating.

Scratchers are most effective when the casing is well centralized and manipulated before and during cementing. To prevent buildup, scratchers should be spaced to ensure overlapping of areas worked by adjacent

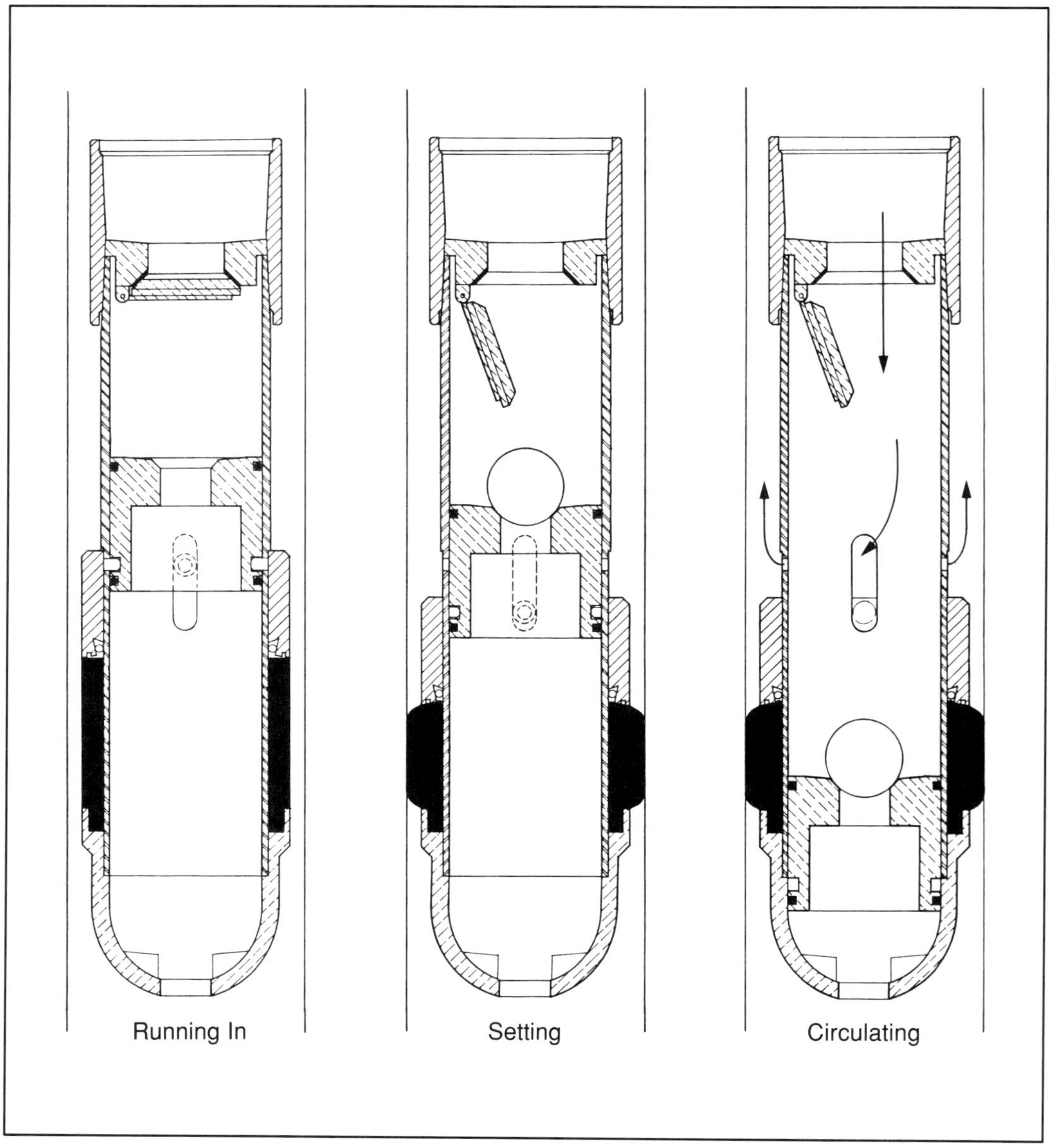

Figure 10–47—Operation of tripping ball (drawing courtesy of Arrow Oil Tools).

scratchers; circulation should be established prior to pipe movement. Reciprocating scratchers may be allowed to float two or three feet (0.6 to 0.9 m) to prevent disturbing the filter cake while running casing. Rotating scratchers are attached with special clamps or by tack welding.

Cementing baskets are a simple and economical type of annular packoff. They are used in low-differential-pressure applications to separate fluids and to help support the hydrostatic head of the cement. Baskets are typically constructed of many plastic or thin steel petals arranged in an overlapping pattern and reinforced by spring steel ribs (Fig. 10–54). The packoff or barrier is formed as the petals envelop to fit the hole, creating a sort of upside-down umbrella within the annulus.

Baskets are often used below stage collars to prevent cement from falling through less dense fluids, and to support hydrostatic pressure. They may also be used above a stage collar and at several positions throughout a weak zone, because their design freely permits flow toward the surface. Baskets are most effective when centralized and, when possible, placed in a well-gauged section of the hole. They are not hinged, and must be placed on a joint of casing before makeup. Baskets are generally allowed to float between couplings. Their travel may be limited with stop rings; however, vertical casing movement must be similarly limited to prevent damage to the basket. Although vertical motion should be avoided, casing rotation is allowed.

Stop rings are used to hold or limit the travel of external attachments, such as centralizers, baskets, or scratchers (Fig. 10–55). They should be used instead of welding, which may be harmful to the casing. They are available in solid or hinged designs and may be fastened with bolts, set screws, or hammer-up mechanisms. Stop rings for high-strength casing applications should contain hardened inserts to better grip the casing. Care should be taken to prevent dropping loose articles into the hole.

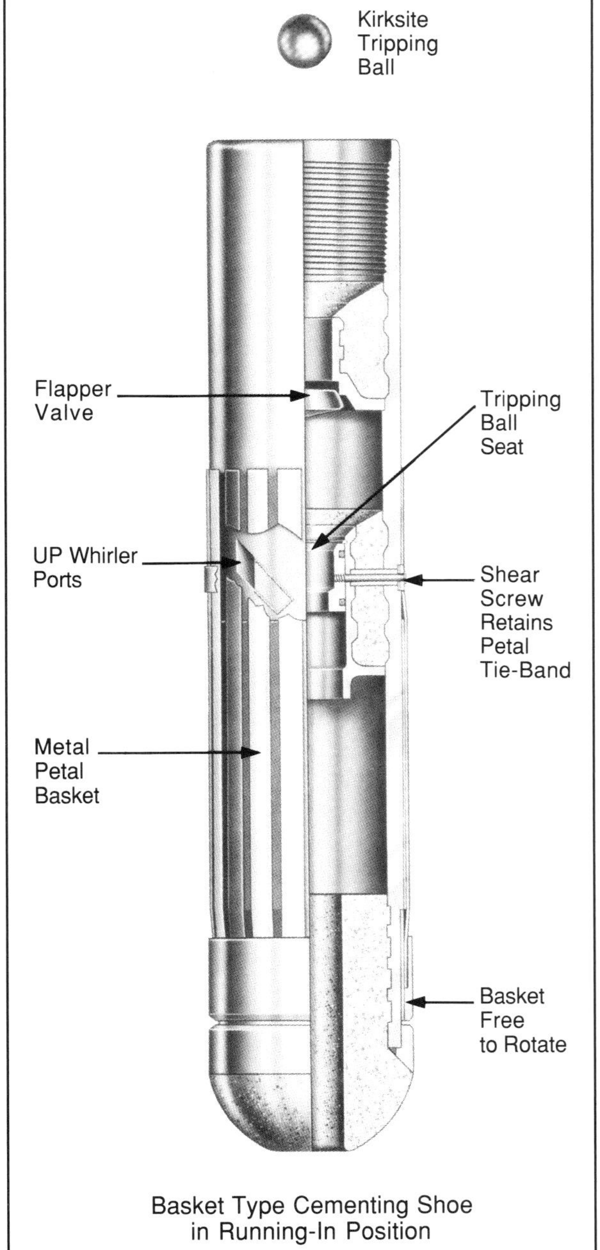

Figure 10–48—Basket shoe (drawing courtesy of Baker Packers).

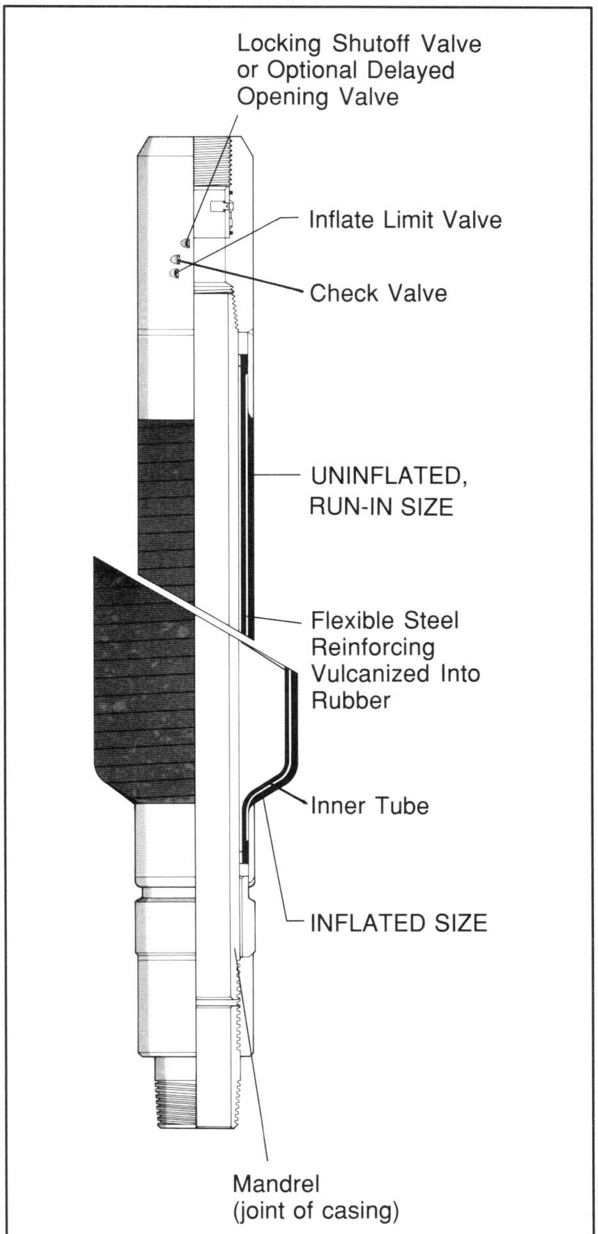

Figure 10–49—External casing packer (drawing courtesy of Lynes).

10–5.12 Cementing Plugs

Cementing plugs are semirigid barriers used to separate cement from drilling fluids, to wipe the casing, and to indicate when cement placement is complete. Plugs were once made of gunny sacks, wood, and leather. Present conventional designs include top and bottom plugs which are constructed of elastomers molded over drillable aluminum or plastic cores.

Although similar in external appearance, top and bottom plugs differ considerably in internal design and operation (Fig. 10–56). Bottom plugs were developed to

Figure 10–50—Types of centralizers (drawings courtesy of TRICO Industries).

Figure 10–51—Turbulence-inducing centralizers (drawings courtesy of Weatherford).

precede the cement, requiring an internal bypass or flow-through feature. This feature uses a hollow core and a thin membrane, which is designed to rupture and permit flow once the plug has seated. Bottom plugs also provide a seat for landing top plugs and sealing off displacement. To ensure compatibility, top and bottom plugs should be

Figure 10–52—Reciprocating scratchers (drawings courtesy of TRICO Industries).

from a common manufacturer. The use of bottom plugs with heavy concentrations of LCMs is cautioned, because the LCM may tend to bridge the float valve.

Top plugs are often used alone, and are designed to withstand the pressures and forces generated when landed abruptly. When used with bottom plugs, care must be taken to prevent inverse placement. Because of the exterior similarity, top and bottom plugs are generally color coded. Other less common plugs include the following—

- flex-plugs, which are used in tapered strings or with restrictions such as stage collars (Fig. 10–45);
- subsea plugs used with subsea completions (Fig. 10–67); and
- latch-in plugs used with latch-in equipment (Fig. 10–33).

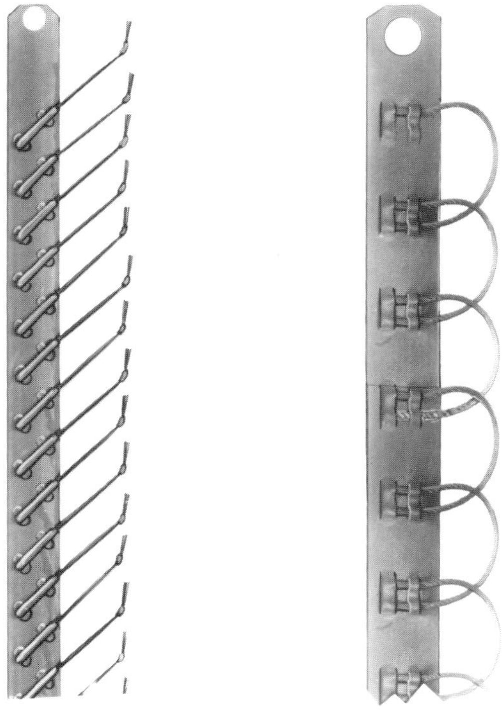

Figure 10–53—Rotating scratchers (drawings courtesy of TRICO Industries).

10–5.13 Liner Equipment

Liners are sections of casing (or tubing) which do not extend to the surface. They are generally used to reduce casing cost when only segments of casing are needed to provide well control (during drilling) or to repair damaged casing (during production).

Liners are lowered into position with drillpipe. They are set or released via setting tools, using mechanical and/or hydraulic operations. Liners are generally suspended using liner hangers to prevent helical buckling during cementing. Short liners less than 200 ft (61 m) in length may be cemented and set on the bottom using a simple and economical setdown system. Liner packers, which allow the circulation of fluids during cementing, are used to provide a seal between the liner top and adjacent casing. The type of liner equipment best suited for an application depends primarily on the clearance between the liner and host casing, the hanging weight of the liner, and whether or not the liner will be reciprocated or rotated during cementing.

10–5.13.1 Liner Hangers and Setting Tools

At present, there is a range of liner hangers and setting tools with different features. Generally, a liner hanger is composed of the following —

Figure 10–54—Cementing baskets (drawings courtesy of TRICO Industries and Baker Service Tools).

Figure 10–56—Cementing plugs.

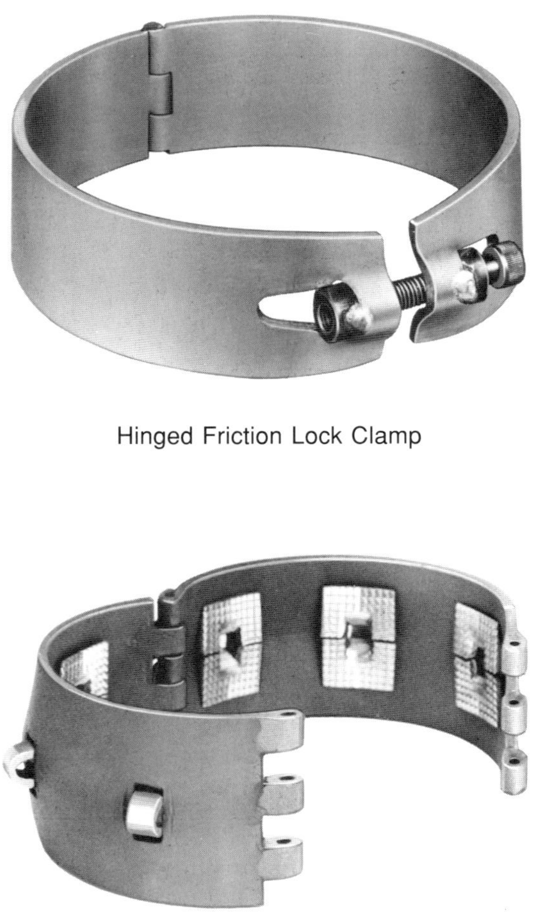

Figure 10–55—Stop rings (drawings courtesy of TRICO Industries).

- *Setting assembly.* This has either a mechanical or hydraulic mechanism. Mechanical set hangers can be operated either by rotation or reciprocation. In both cases, pipe movement will release a slip bowl from the retracted position by means of a J-slot or a dog-spring mechanism. The slips move into the hanging position, supported by an upper cone, thereby transferring the liner weight to the upper casing. The hydraulic set hangers are operated by hydraulic pressure. Some slip-releasing systems are actuated by dropping a ball (once the liner is in place) which will land in a catcher sub. Further pressure moves a piston assembly or shears pins, releasing the slips (Fig. 10–57).
- *Setting sleeve.* This is run above the setting assembly, and has a thread matching the liner setting tool. It is used to run the liner to depth (Fig. 10–58).
- *Tieback sleeve.* Screwed on top of the liner, it has an internal polished surface to provide a seat and seal for future tieback liners or casings (Fig. 10–59).
- *Liner packers.* Liner packers are separate tools that screw on top of the liner hanger. They are usually weight-operated, with special setting tools which operate the hanger at the same time. Liner packers can be set at any time after the liner hanger has been set. A tieback sleeve can be placed on top of the packer for future extensions of the liner (Fig. 10–60).
- *Liner hanger/packers* combine the functions of a hanger and packer. Packing elements can either be compressed in a separate operation, or at the time the hanger is set. The latter system is not used for cementing.
- *Setting tools* are used to run and cement liners. To support the liner as it is run, they usually have a left-hand threaded nut which engages coarse left-hand threads

in the liner hanger. Cup seals prevent upward flow through the tool during the cementing operation. The liner wiper plug is attached to the setting tool by pins which shear after the "pumpdown" plug has landed and pressure has been applied. This allows the displacement of the wiper plug down into the liner (Figs. 10–61 and 10–62).

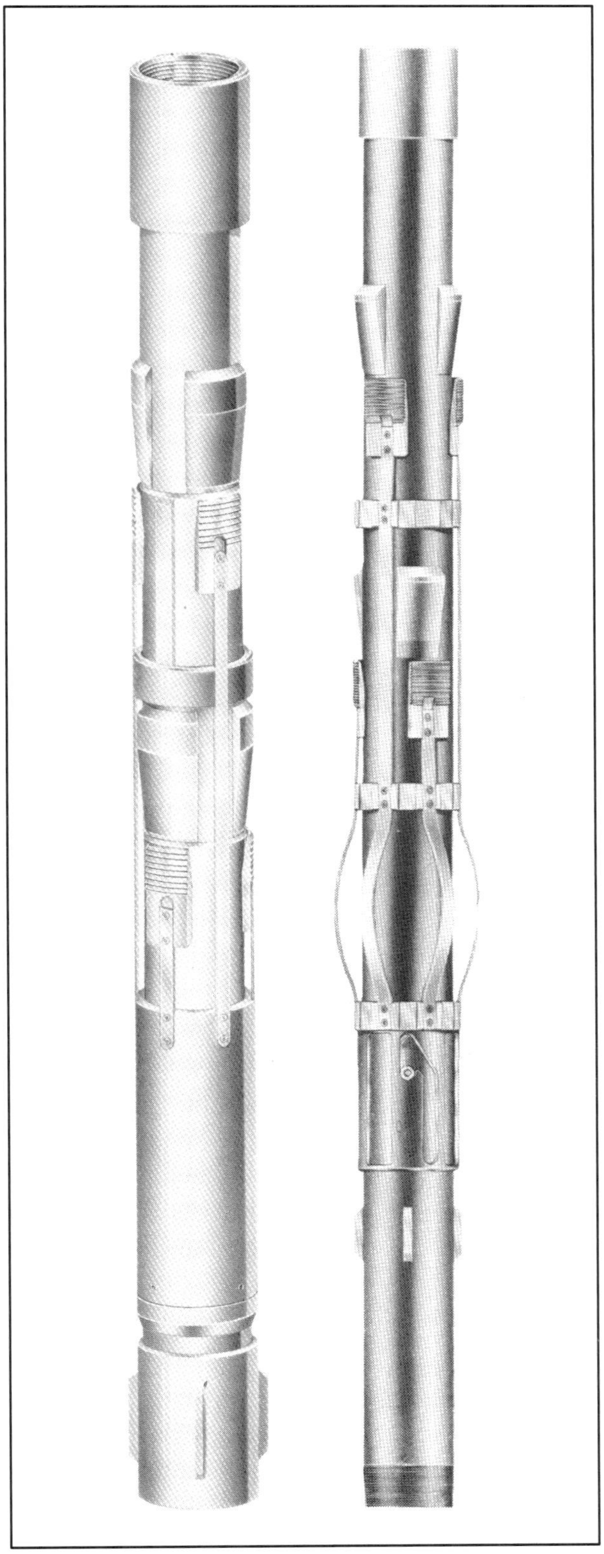

Figure 10–57—Hydraulically set liner hanger (left) and mechanically set J-slot hanger (right) (photo courtesy Baker Service Tools).

10–5.13.2 Liner Setdown Equipment

This equipment (Fig. 10–63) consists of a liner shoe, expansion joint, liner top, and latch-in plug. The nose cone of the shoe is generally made of drillable cast aluminum, the body is of drillable cast iron, and the outer case is of seamless steel tubing. The design includes two backpressure devices—a ball check valve and a latch-in wiper plug. The top portion of the body contains the square left-hand thread to accommodate the left-hand threaded sub of the expansion joint.

The liner shoe is placed on the bottom joint of the liner casing. This casing is run into the well with the last joint held in the slips or clamps. The liner top is placed on the top joint, and provides better entry of tools in the liner during subsequent operations.

The expansion joint is placed on the drillpipe tubing, and run inside the liner casing until contact is established with the liner shoe. Rotation of the tubing to the left connects it to the liner shoe.

The drillpipe, with liner casing attached, can be run to the total depth. Once circulation is established, the cement is mixed and pumped down the tubing, followed by the latch-in wiper plug. The wiper plug is pumped to shutoff, and approximately 1,000 lb of pressure above the amount required to displace the cement must be applied to seat the latch-in plug.

The tubing pressure is then released, and the tubing is rotated to the right to release the running tools from the liner shoe. Next, the tubing is raised 3 to 5 ft (0.9 to 1.5 m), and all excess cement is removed from the liner by

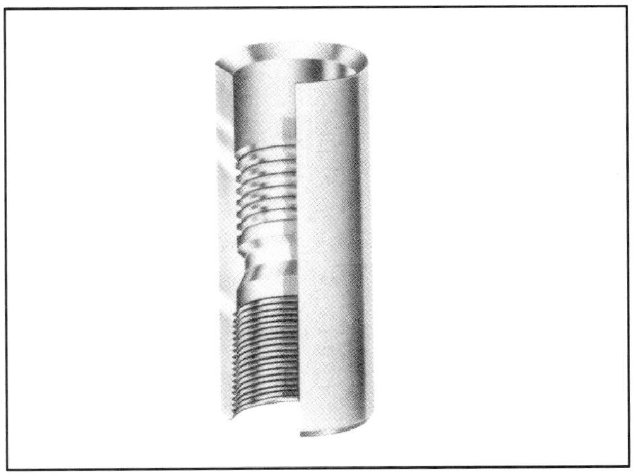

Figure 10–58—Liner setting sleeve (photo courtesy Baker Service Tools).

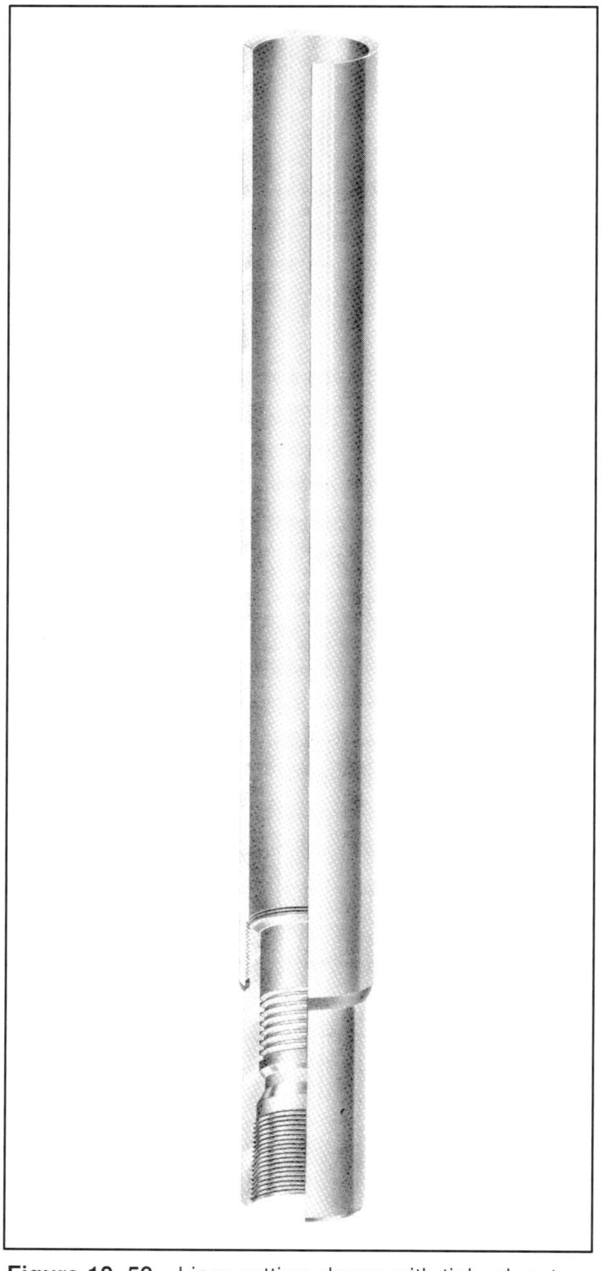

Figure 10–59—Liner setting sleeve with tieback extension (photo courtesy Baker Service Tools).

conventional or reverse circulation. Finally, the tubing is removed from the well and the operation is complete.

10–5.14 Cement Heads

10–5.14.1 Single-Plug Cement Head

Also called a single-plug container, the single-plug cement head is used to hold a bottom (or a top) plug until it is released and pumped down the casing during cementing. The normal procedure is for the bottom plug to be loaded and released in front of the spacer or the cement slurry. The cement head is then opened; the top plug is

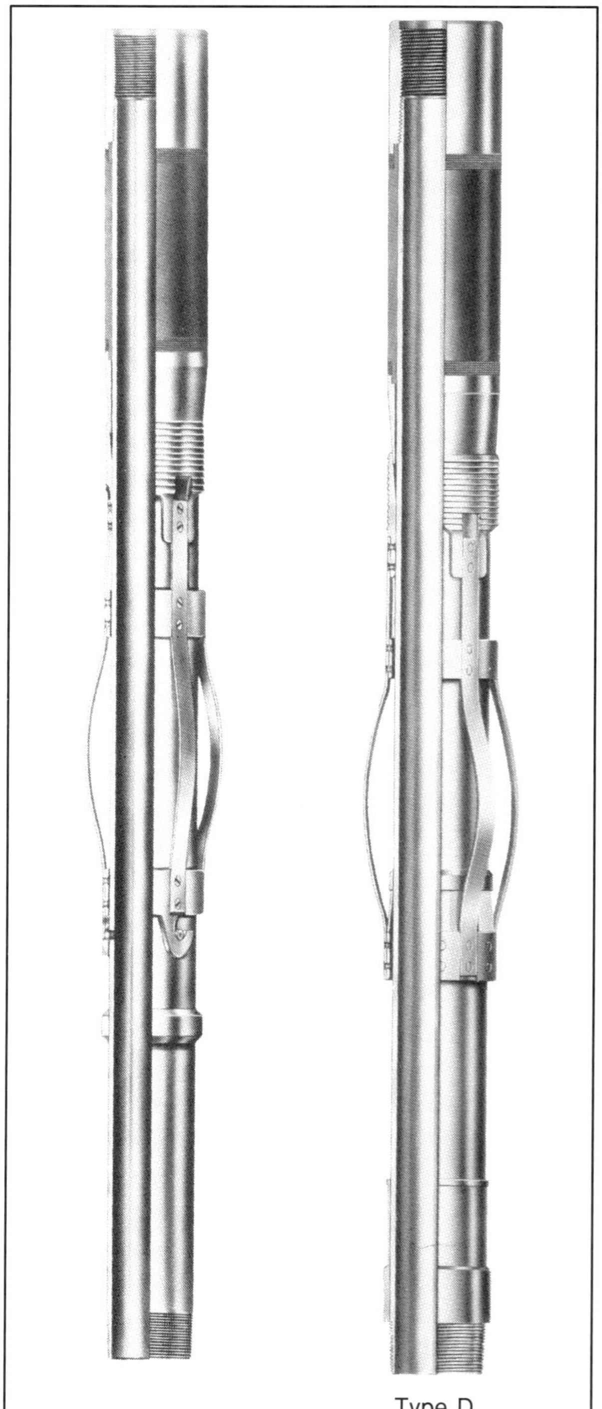

Figure 10–60—Hanger/packers (photo courtesy Baker Service Tools).

loaded between the two inlets, for release between the slurry and displacement fluid (Fig. 10–64).

There are different holding/releasing systems, depending on the manufacturer. The two most common are

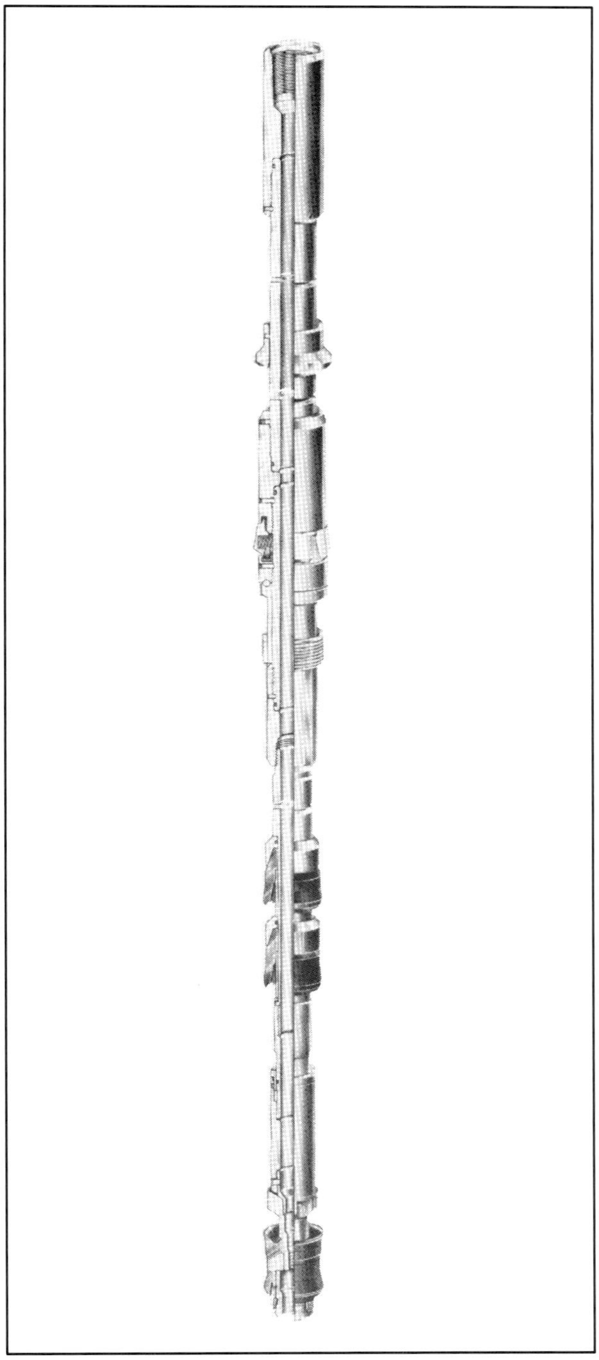

Figure 10–61—Type C-2 setting tool (photo courtesy Baker Service Tools).

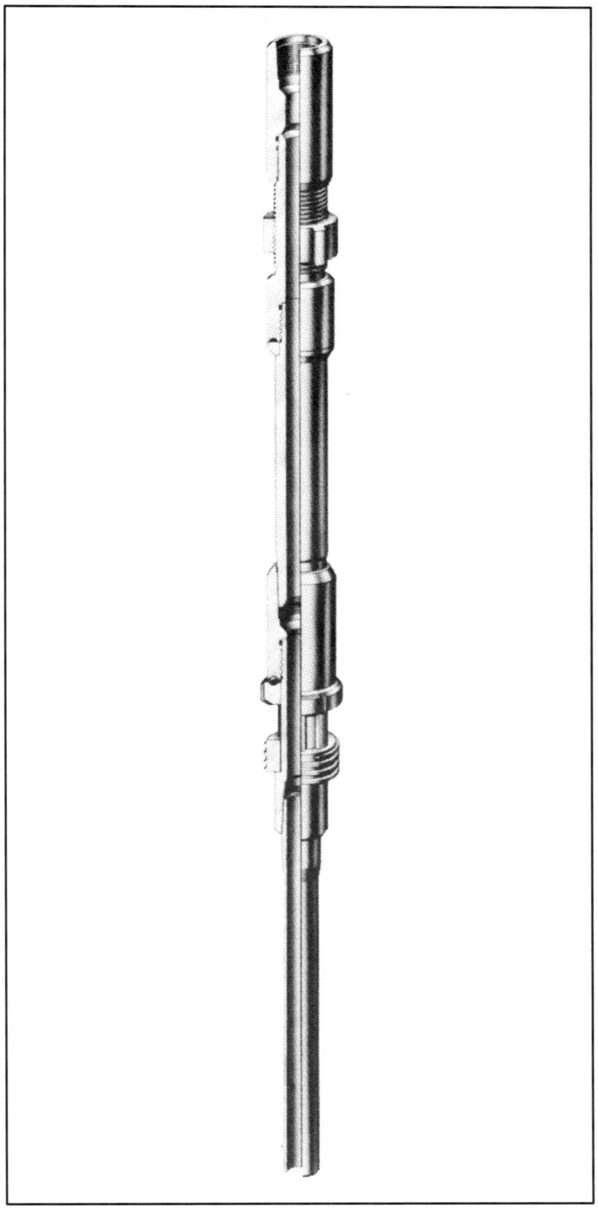

Figure 10–62—Type CS setting tool (photo courtesy Baker Service Tools).

the bar or pin system, and the bail system (Figs. 10–65 and 10–66, respectively).

Some types of cement heads are provided with a mechanical indicator to confirm the release of the plugs. High-pressure manifolds provide a means by which to pump fluid selectively above or below the cement plug.

10–5.14.2 Double-Plug Cement Heads

The main drawback of single-plug cement heads is the need to stop after mud circulation to load the plugs. This allows the buildup of mud gel strength which affects removal efficiency.

On the other hand, double-plug cement heads can hold both plugs, allowing a nonstop operation. Stopping is only necessary for changing over to the cementing lines (Fig. 10–67). The holding/releasing system of double-plug cement heads is the same as that used in single-plug heads. One difficulty with these heads is that the bails on

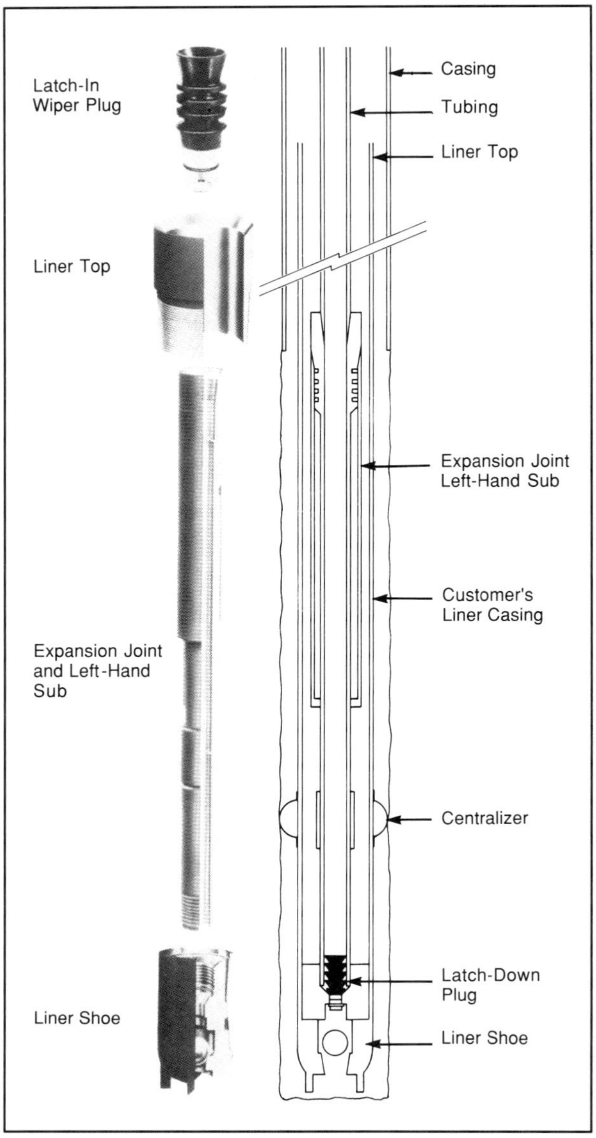

Figure 10–63—Setdown liner equipment.

small rigs are often not long enough to handle the double-plug cement heads.

10–5.14.3 Liner Cement Heads

Liner cement heads have the same basic design as casing cement heads, and are used to hold the "pumpdown" plug during liner operations. Bar or pin holding systems and mechanical indicators are commonly found with these heads (Fig 12-11 in Chapter 12).

10–5.14.4 Cementing Plug System for Floating Vessels

This system consists of two major components (Figs. 10–68 and 10–69)—

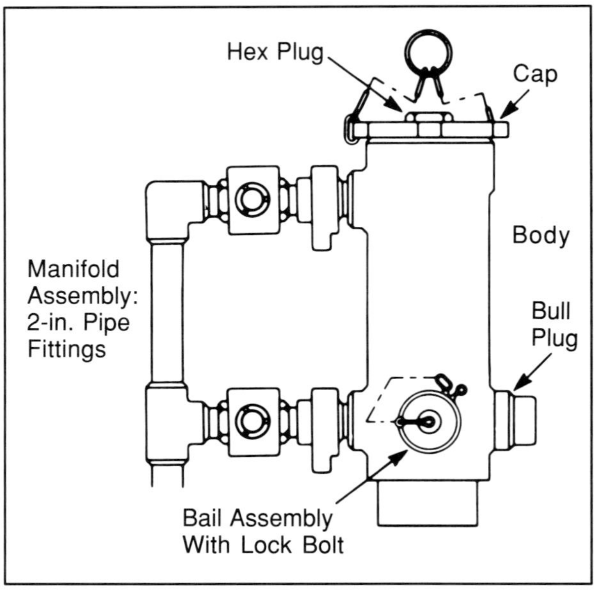

Figure 10–64—OCT cast, single-plug cement head.

- a subsea assembly (a) located in the casing below the casing hanger, and
- a cement head on the drilling vessel which adapts to the drillpipe and controls the cementing plug release (b).

The plug distribution consists of a launching ball and dart in the cement head, and top and bottom cementing plugs in the subsea assembly. In chronological order of use, (b) is the bottom plug launching ball which, when released before pumping the cement slurry, seats in the bottom plug (e). A 100- to 275-psi pressure increase allows the shearing of the connector pins (d), and permits the bottom plug (e) to travel down the casing until it bumps on the float collar or casing shoe.

Extra pump or hydrostatic pressure extrudes the ball (b) through its orifice seat, and cement displacement con-

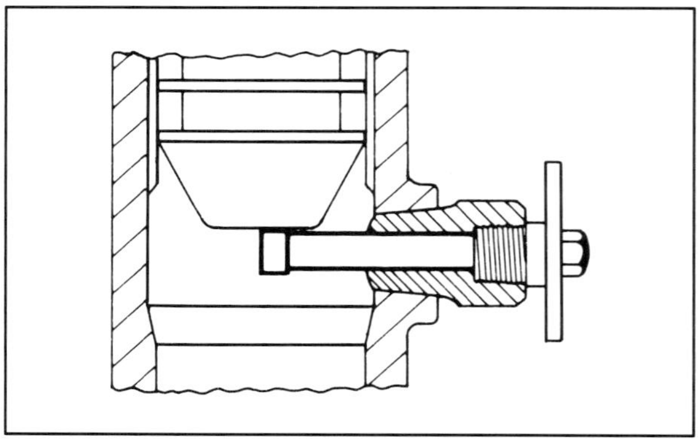

Figure 10–65—Pull-pin assembly of single-plug cement head.

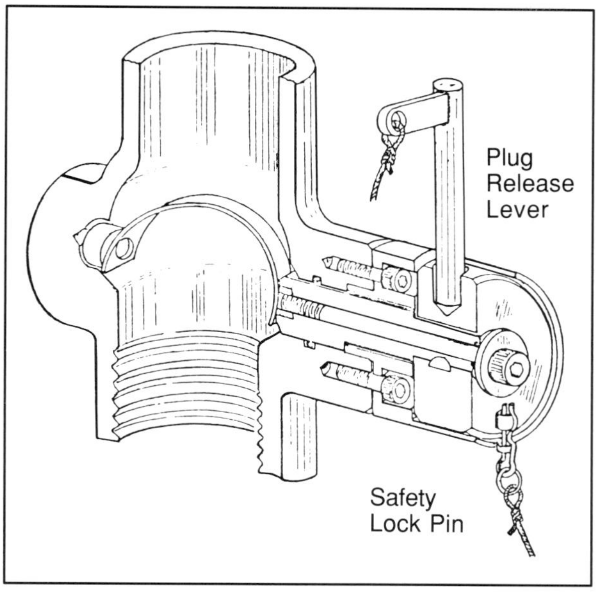

Figure 10–66—Bail-type assembly for cementing head. Operation: (1) Pull safety lock pin, (2) Rotate plug release lever to drop plug.

tinues. A ball catcher attached to the lower end of the bottomhole plug retains the ball.

Once the cement slurry has been pumped, the top plug launching dart (a) is released. It will seat into the body of the top cement plug (c). Increased circulation pressure of 350 to 500 psi will then shear the retaining pins and release the top plug (c) from the launching mandrel. Thus, the cementing operation continues. At the end of slurry

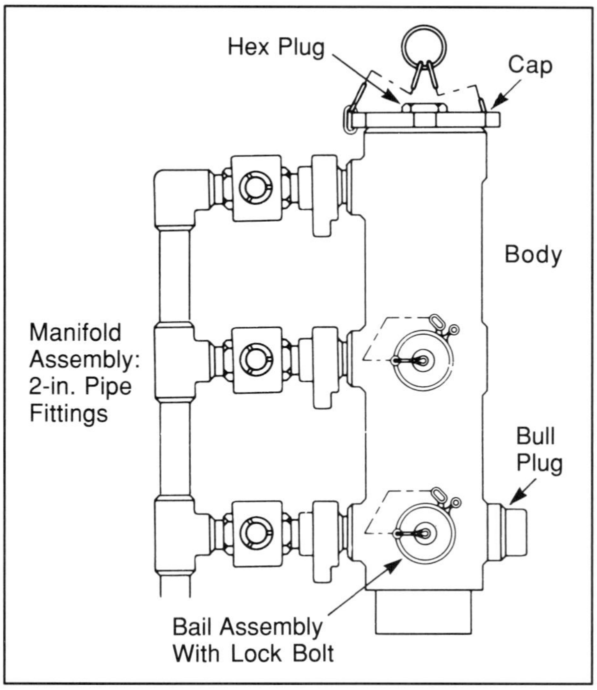

Figure 10–67—OCT cast, double-plug cement head.

displacement, the top plug (c) seats on the float collar or casing shoe.

In the subsea assembly before cementing, the top plug is pinned at the lower end of a running mandrel which has a swivel (g). This avoids any rotation of the cementing plugs inside the casing, which could damage the shear pins.

The upper part of the mandrel is screwed onto the lower part of the installation tool (h). This installation tool is a crossover that adapts to the casing hanger and serves to attach it to the drillpipe.

10–5.14.5 Casing Swivels

Casing swivels are designed to allow the rotation of casing during cementing operations. They do not have the capacity to support the weight of the casing, so they must be run above the casing elevators. Torque is applied to the rotary table while the casing is hanging in the slips. This does not allow simultaneous rotation and reciprocation, but can greatly improve the removal of mud from the narrow side of the annulus (Fig. 10–70).

10–5.14.6 Swages

Swages are normally used for cementing large-diameter casings at shallow depths when no stab-in method is applied. They are simply casing crossovers designed to allow the connection of casing to small-diameter pipes (Fig. 10–71). They can withstand more internal pressure than the casing with which they are intended to be used. However, they are not designed for dropping plugs, and should only be used on conductor or shallow surface casing.

10–6 REMEDIAL CEMENTING TOOLS

Remedial cementing tools are mechanical or hydraulic devices which are used downhole to assist in the placement of cement during plugback or squeeze cementing operations. They are generally used to isolate areas of the casing from treating pressures and cement. Some are available in retrievable or drillable designs, each being suited for a particular set of well conditions. Remedial cementing tools are generally provided with service. Details of a specific tool operation, or limitations, should be obtained from the service company or manufacturer.

10–6.1 Squeeze Packers

Squeeze packers are primarily used to isolate the upper portion of the casing and wellhead from cement and squeeze pressures, and to improve control and placement of fluids during squeeze cementing techniques. They are available in either drillable or retrievable types which differ considerably in appearance and operation (Figs. 10–72 and 10–73).

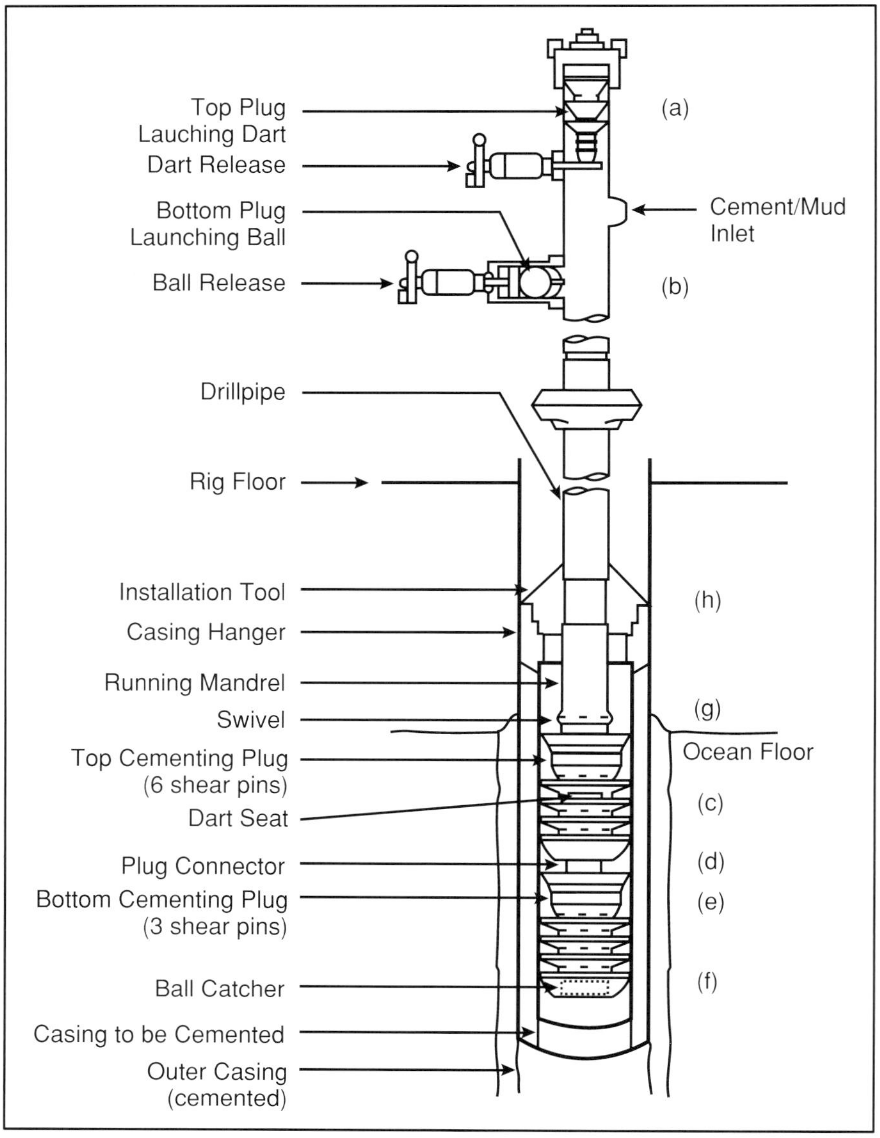

Figure 10–68—Single-stage subsea cementing system.

Retrievable squeeze packers may be set and released repeatedly, and are most useful in multiple setting operations such as selective testing and cementing of multiple zones. They are run on tubing, and are available in compression or tension set designs. Compression set packers are typically more versatile, and are recommended whenever sufficient tubing height is available. Generally 10,000 to 15,000 lb is the minimum weight or tension required to pack off the elements of either compression or tension packers.

Retrievable squeeze packers usually include many design features to improve performance and versatility. Most compression packers use hydraulic holddown slips to resist upward forces generated by treating pressures below the packer. Some packers provide a bypass valve which may be opened or closed while in the set position. This feature provides finer control of slurry placement by allowing circulation without unsetting the packer. Also, the provision of a bypass area beneath the elements reduces swabbing and piston effects while running, and allows reverse circulating of excess cement without causing excessive pressures. Packers with full-opening mandrels permit the use of through-tubing perforating guns, pressure recorders, or other wireline accessories.

Most compression packers are set by slightly lifting the tool, rotating to the right (one-quarter turn at the tool), and setting down to apply at least the minimum required force. This operation frees the lower slip assembly to slide over the lower cone and engage the casing wall. As the tubing is lowered, the elements are compressed until packed off against the casing. The mechanism used to select and hold the running (safety) or set position is called a J-slot, and consists of a lug confined to a slot shaped like the letter "J." The lug is typically mounted on the mandrel, whereas the slot is milled into the lower slip assembly. Figure 10–74 illustrates the positions of a manual and automatic "J"slots. The automatic "J" places the packer in the running or safety position while the manual "J" requires left-hand rotation to safety. Compression packers are released by simply raising the tubing.

Drillable squeeze packers, commonly referred to as *cement retainers,* may be set on wireline or tubing. They are generally made of cast iron, and are compact in size to minimize drilling time. A sliding sleeve or poppet valve

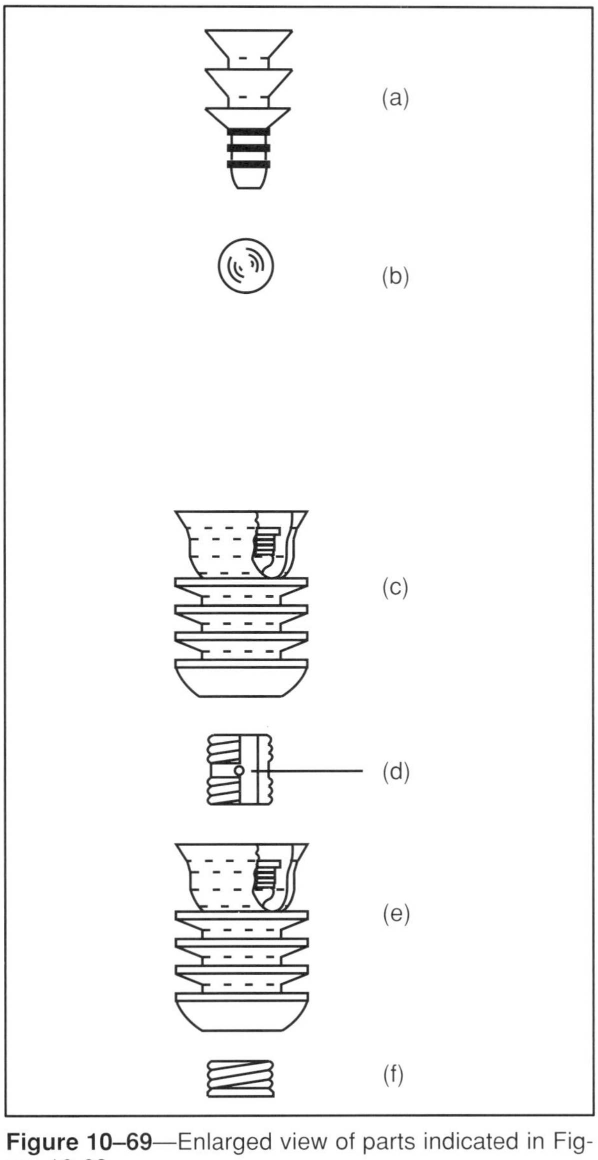

Figure 10–69—Enlarged view of parts indicated in Figure 10-68.

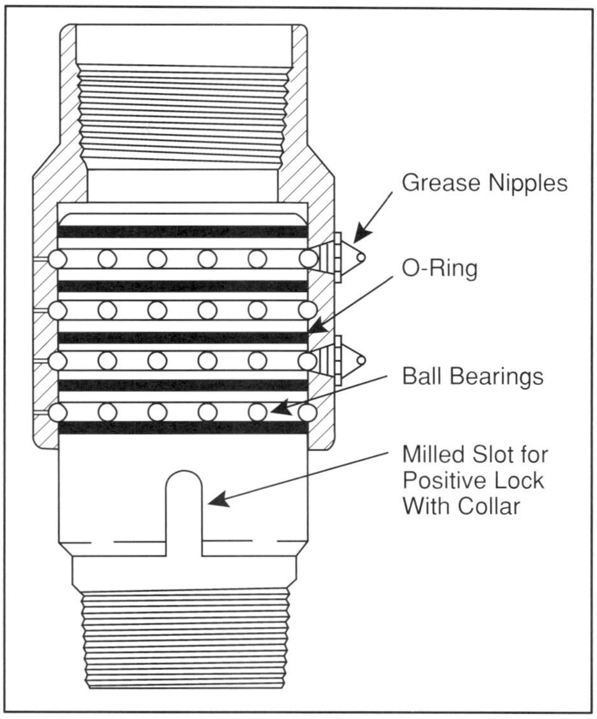

Figure 10–70—Casing swivel (drawing courtesy Industrial Rubber Inc.).

is provided to control slurry placement and preserve final squeeze conditions. Sliding sleeve valves are operated by raising and lowering the tubing, and prevent flow in either direction.

Cement retainers are often used instead of retrievable packers to prevent backflow of cement when dehydration is not expected, and to isolate the treated area from

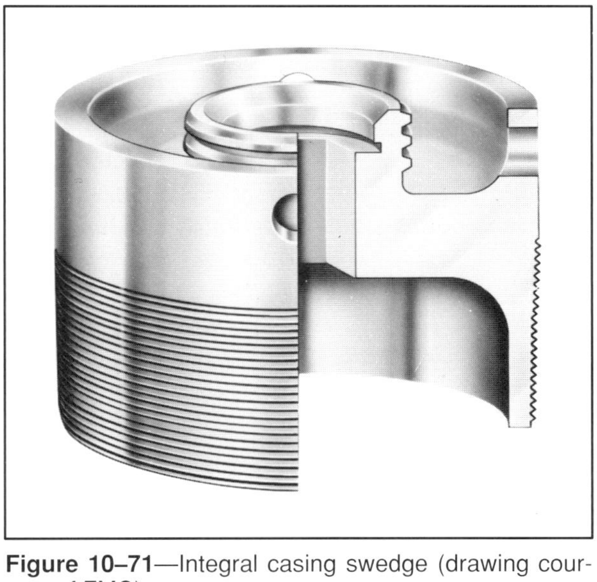

Figure 10–71—Integral casing swedge (drawing courtesy of FMC).

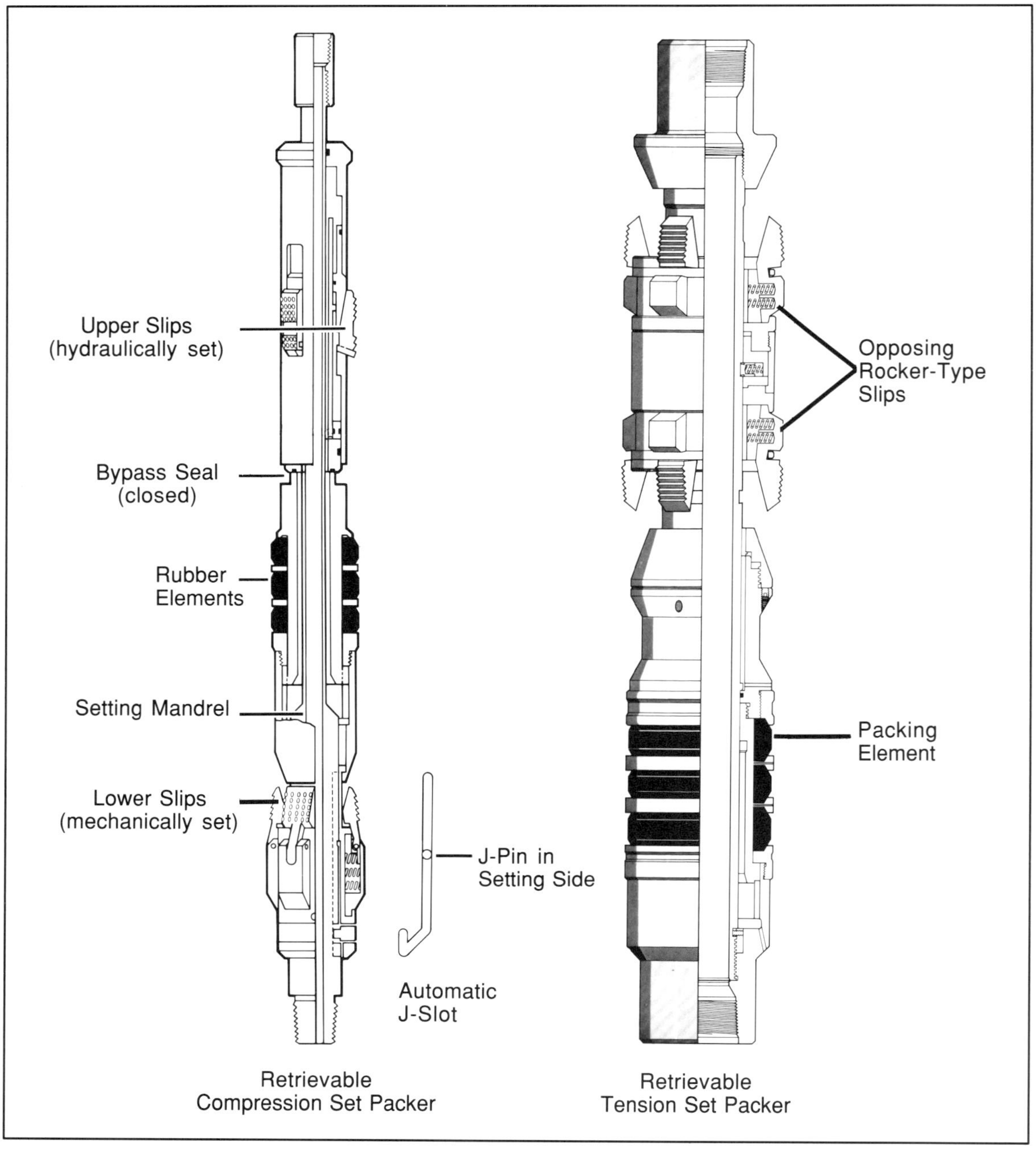

Figure 10–72—Retrievable squeeze packer.

pressures due to the reversing of excess cement from the tubing. Cement retainers are also better suited to situations where potential communication with upper perforations or casing problems may lead to cementing a retrievable packer.

To set a cement retainer on a wireline, an adapter is used to connect the cement retainer to the wireline setting tool. The cement retainer is lowered to the proper position and set by electrically firing a slow burning charge in the setting tool. When the cement retainer is completely packed off, the setting tool shears free and is retrieved with the wireline. The stinger is connected and run in the hole with tubing to perform the squeeze.

When set on tubing, the cement retainer is connected to a tubing setting tool. The valve is open to allow the tubing to fill as the cement retainer is lowered. Rotating the tubing to the right releases the upper slips and initiates packoff in some models. The tubing is then pulled to complete the packoff. When the proper setting tension is achieved, the setting tool shears free. The setting tension may range from 18,000 lb for 4½-in. sizes to 48,000 lb

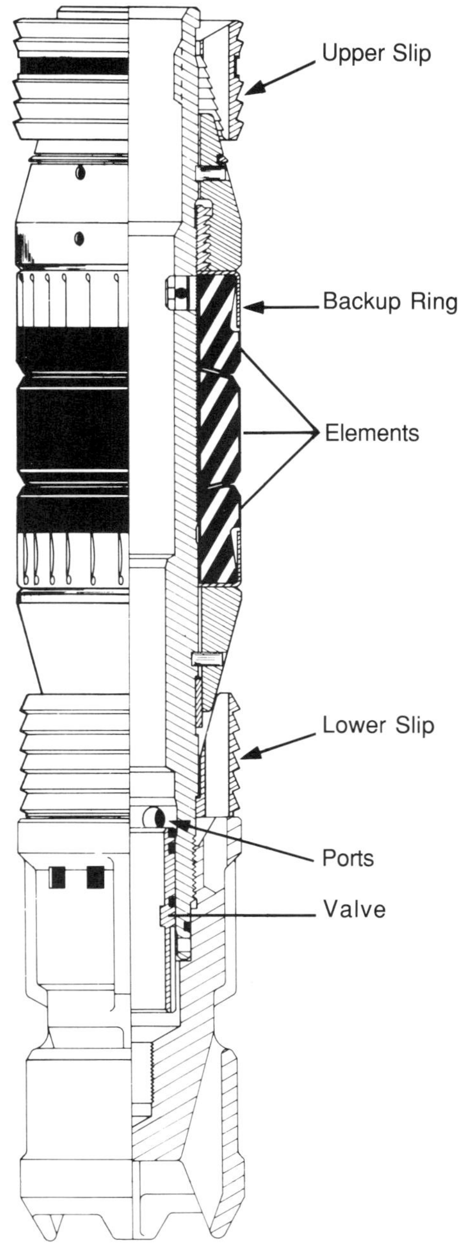

Figure 10–73—Drillable squeeze packer (cement retainer).

for 9⅝-in. sizes. The valve is pushed open by lowering the tubing, and closed by raising the tubing.

10–6.2 Bridge Plugs

Bridge plugs are normally used to isolate the casing below the zone to be treated. When set, bridge plugs act as solid barriers to prevent flow and resist pressure from above or below. They are available in drillable and retrievable designs for wireline or tubing set operations.

Retrievable bridge plugs are often used in multiple zone applications, because they can be set and released as often as necessary. They may be run in tandem with retrievable packers for single-trip straddle operations. However, tension packers are not recommended for tandem running operations with certain bridge plugs, because their combined operations may prevent freeing the tension packer, and lock the entire assembly downhole.

Retrievable bridge plugs are available in cup or packer configurations. Cup bridge plugs are generally used in shallow, moderate pressure applications (Fig. 10–75). They are simpler and more economical than packer models. However, the cups are in constant contact with the casing while it is being run, which causes wear and increases swabbing and piston effects.

Cup bridge plugs may be run on tubing or a sand line, and are released and retrieved via a retrieving sleeve. However, a special retrieving sleeve and procedure are required for sand-line operations. When released, they are automatically set by the application of pressure from above or below. Cup bridge plugs may be used in tandem with a tension packer.

Packer bridge plugs use a packer element design which is more durable and well suited for deep, high-pressure applications (Fig. 10–76). They have smaller external diameters, which permit faster running and reduce swabbing and piston effects.

Packer bridge plugs are coupled to the tubing via a retrieving sleeve, and are typically set by rotating to the right while lowering the tubing to apply weight (the minimum recommended weight is generally 10,000 lb). Some models also offer a left-hand rotational set which requires no additional weight, and may be used in shallow applications or to permit removal of surface equipment. They are released from the tubing by pulling a slight upstrain (1,000 lb) while rotating to the left (one-quarter turn at the tool). Packer bridge plugs are retrieved by lowering the retrieving sleeve while circulating to remove sand and debris from above the bridge plug. When solid contact is made, the tubing is raised to apply a slight

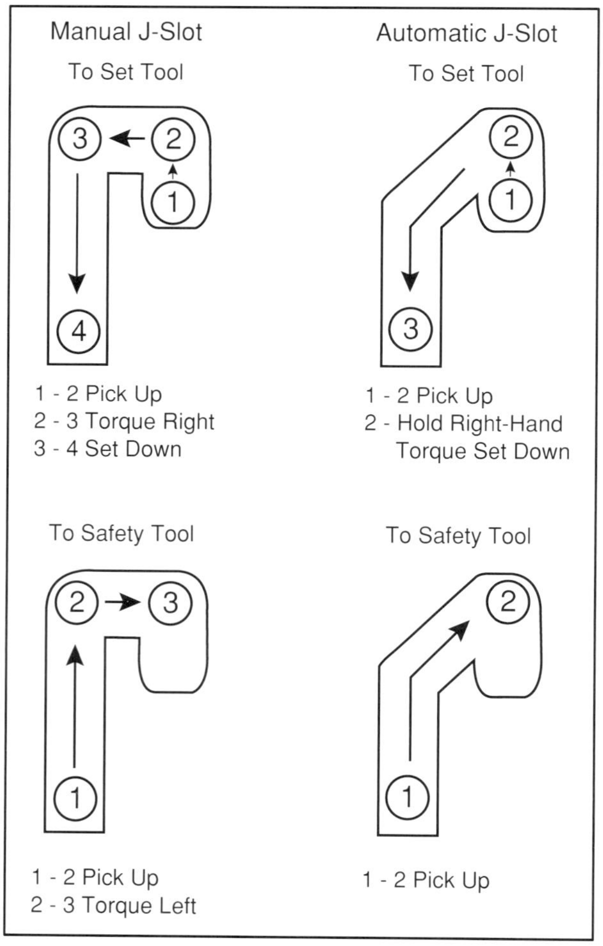

Figure 10–74—J-slot positions.

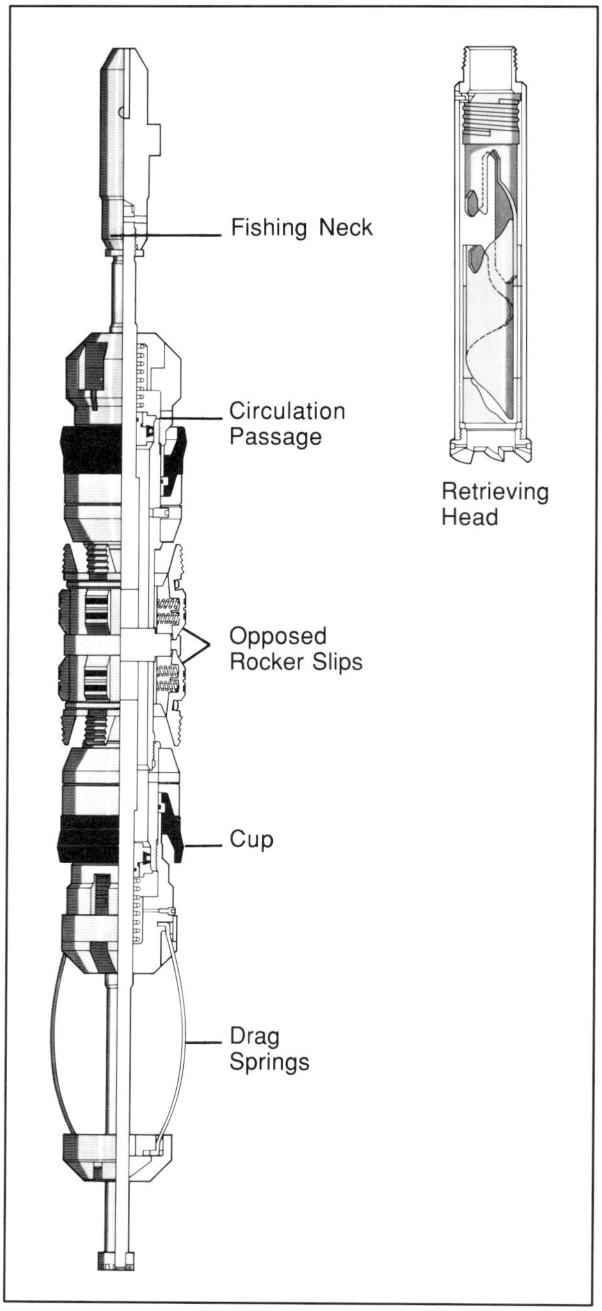

Figure 10–75—Bridge plug (cup type).

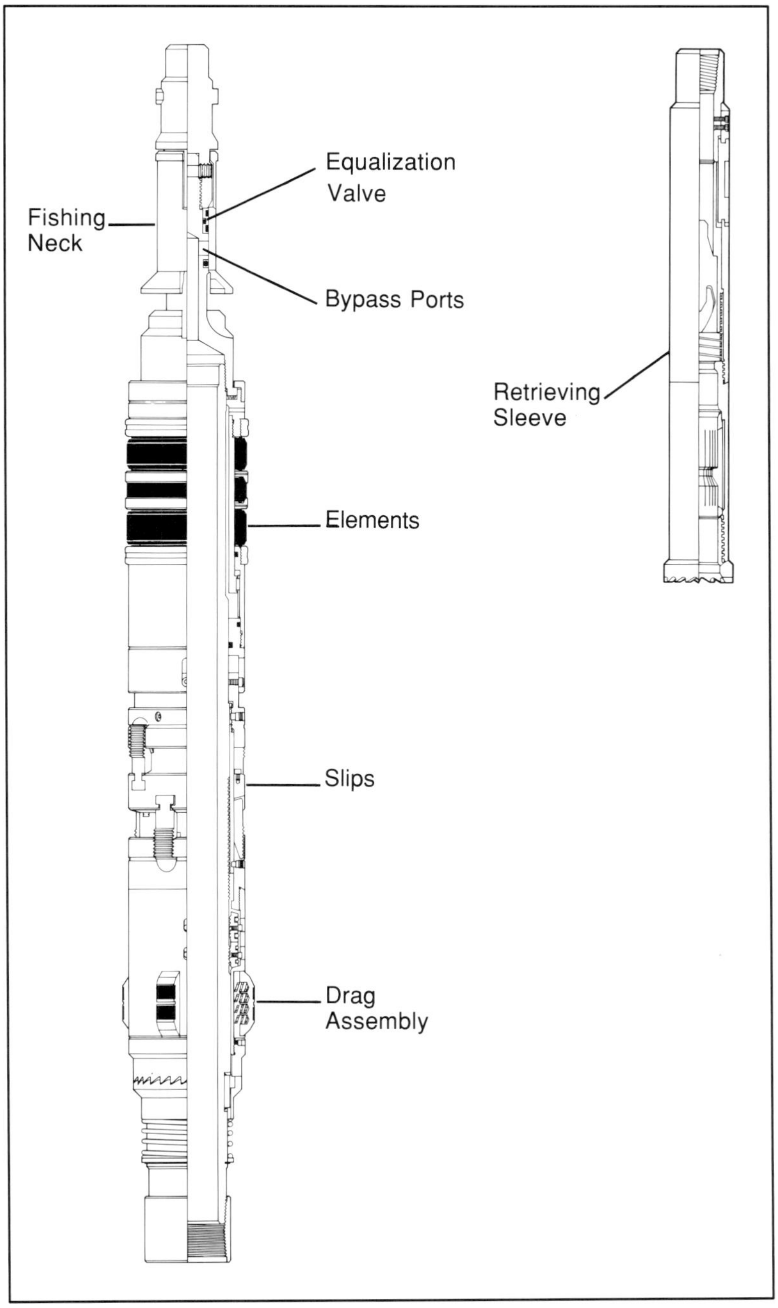

Figure 10–76—Bridge plug (packer type).

upstrain (2,000 to 3,000 lb) while rotating to the right. Once free, the bridge plug may be retrieved or moved to another position. Compression or rotational set bridge plugs should not be run in tandem with tension packers.

To prevent cement and perforation debris from interfering with the retrieval, sand should be placed on top of retrievable bridge plugs. Sand may also be used above drillable bridge plugs, to reduce damaging shock waves caused by closely placed perforating guns.

Drillable bridge plugs are used to create a temporary or permanent plug for squeezing or plugback applications (Fig. 10–77). They are often used to seal off nonproductive zones or wells to be abandoned. They are made of cast iron and are constructed similarly to cement retainers. The basic difference is that the mandrels are plugged, and do not contain check valves.

Drillable bridge plugs may be set on wireline or tubing. The setting tools and procedures are often the same used for cement retainers. Some drillable bridge plugs are designed to allow pressures above and below the plug to equalize before drilling through the top slips. This feature is most important when gas or high pressures may be expected to occur below the tool.

10–6.3 Tubing Testers and Unloaders

Tubing testers are basically downhole valves used to check the tubing for leaks (Fig. 10–78). They are typically used during squeeze cementing operations, because of the potential problems that exist while pumping cement under moderate-to-high differential pressures. A leaking connection may permit local cement dehydration, creating false squeeze indications or completely plugging the tubing.

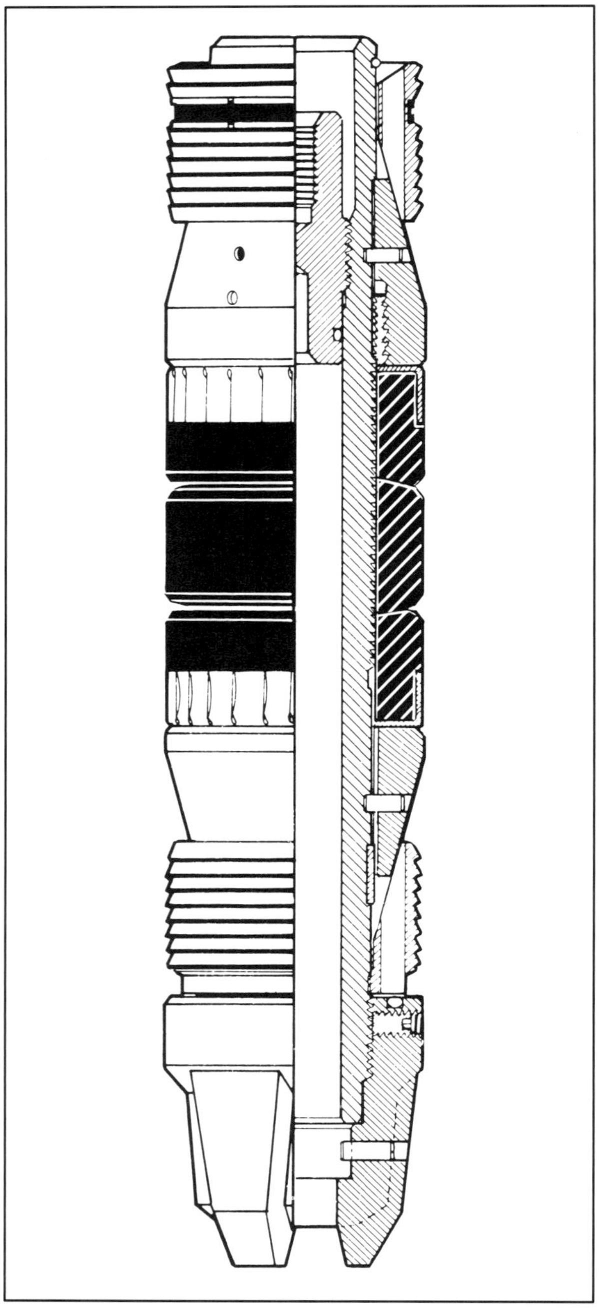

Figure 10–77—Drillable bridge plug.

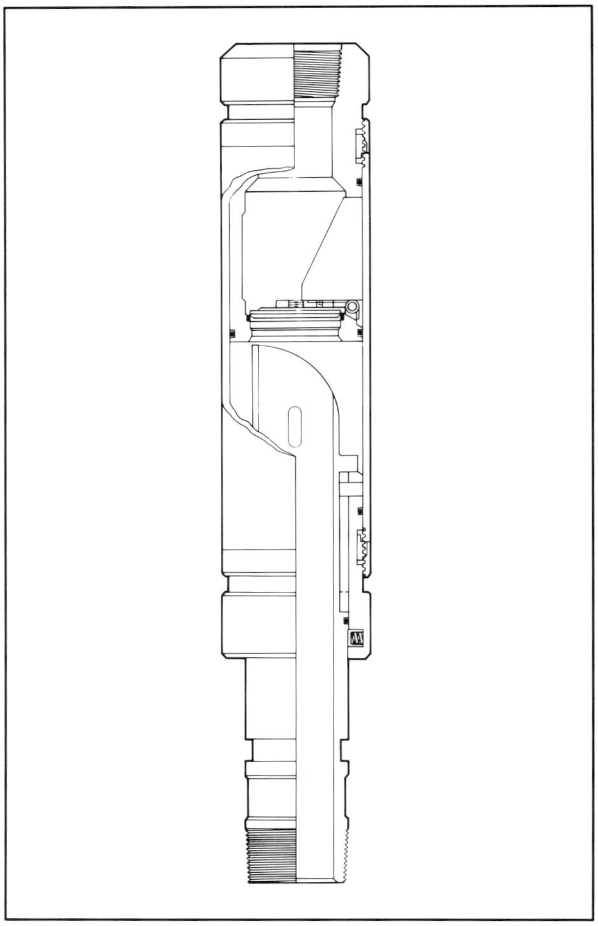

Figure 10–78—Tubing tester.

Tubing testers are typically equipped with a full opening flapper mechanism. The full opening bore permits the use of through-tubing perforating guns or other wireline tools. They are placed above a packer, and are run in the open position to allow filling. They are typically closed by rotating the tubing (one-quarter turn at the tool) to the right and lifting. They are reopened by simply lowering the tubing. A simple ball-seat sub may also be used to test the tubing. However, the ball seat somewhat restricts the inside diameter of the tubing and the ball must be reverse circulated to the surface prior to pumping.

Tubing unloaders, or *tubing bypass valves*, are placed in the tubing string to provide an alternate passage for circulating or spotting fluids (Fig. 10–79). They are often used with packers which are not equipped with built-in

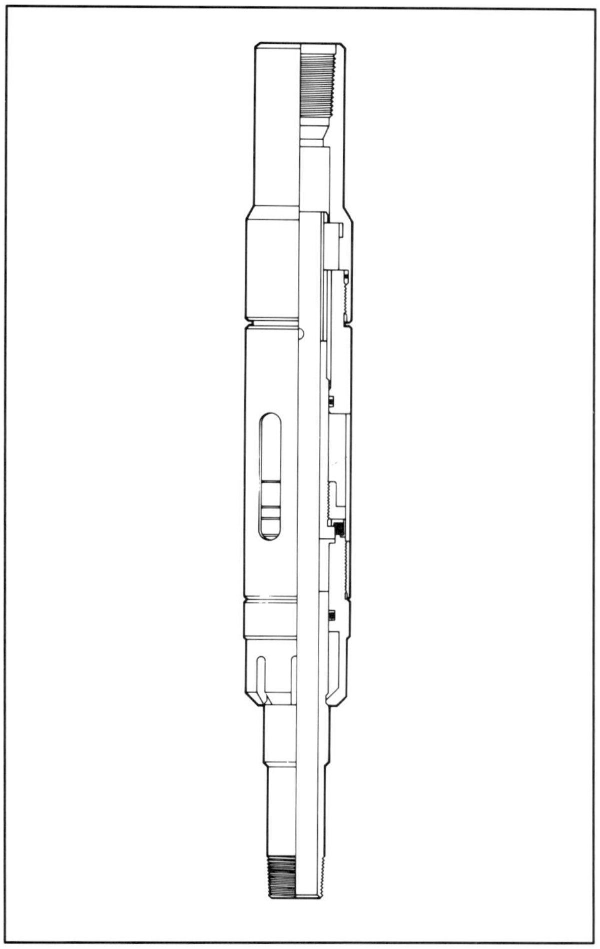

Figure 10–79—Tubing unloader or bypass.

bypass valves, and must be unset to permit circulation. They are operated by raising or lowering the tubing, and are available for tension or compression packer operations.

SUGGESTED READING

Baldridge, M.: "Quality Control Test of Float Equipment Simulates Downhole Conditions," *Oil & Gas J.*, (June 2, 1984) 80–83.

Casing Centralizers, API Specification, 10D, second edition, American Petroleum Institute, Dallas (1973).

Casing Hardware Catalog, Dowell Schlumberger, Tulsa, (1985).

Cementing Technology, Dowell Schlumberger: Nova Communications Ltd., London (1984).

Clark, E. H., Jr.: "Bottomhole Pressure Surges While Running Pipe," paper **ASME** No. 54-Pet-22, 1954.

Craft, Holden, and Graves, *Well Design: Drilling and Production*, Prentice-Hall, Inc., Englewood Cliffs, NJ (1962).

Floating Equipment and Cementing Aids, Stage and Stab-in Equipment, Bakerline, San Antonio (1984).

Fontenot, J. E. and Clark, R. K.: "An Improved Method for Calculating Swab/Surge and Circulating Pressures in a Drilling Well," paper SPE 4521, 1973.

Kimble, L. D., Bissonnette, H. S., and Bradford, B. B.: "Testing Improves Design and Performance of Casing Hardware," *Oil and Gas J.* (Feb. 10, 1986) 84–87.

Products and Services Catalog, Weatherford, Weatherford, TX (1985).

Sales and Service Catalog, Halliburton Services, Duncan, OK (1985).

Smith, D. K.: *Cementing*, Henry L. Doherty Series, SPE, Richardson, TX (1987).

Stringfellow, B.: "Test Find Hammering, Fluid Cutting, Erosion Cause Float Shoe Failures," *Oil & Gas J.* (Jan. 21, 1985) 66–70.

Suman, G. O. Jr., and Ellis, R. C.: *Cementing Handbook*, Gulf Publishing Co., Houston (1977).

Technical Sales Catalog, Davis Lynch, Houston (1985).

11

Cement Job Design

Phil Rae

Dowell Schlumberger

11–1 INTRODUCTION

The previous chapters have shown that there are many facets to a cementing operation. The engineer must consider data from a great many sources to arrive at the optimum cement job design for any given set of well conditions. This chapter shows how such data can be analyzed systematically.

11–2 PROBLEM ANALYSIS

In any job design, the factors which need to be first examined fall into three basic categories.

- Depth/Configurational Data
- Wellbore Environment
- Temperature Data

These data direct the selection of the preferred basic cement properties and displacement regime for a given well. The annular configuration suggests which flow regime is practical, and the required rheological properties. Wellbore conditions indicate whether special materials, due to the presence of gas, salt, etc., need be incorporated. The mud density indicates the minimum acceptable cement slurry density. These factors, together with the temperature data, guide the selection of additives for the control of slurry flow properties and thickening time. Each of the data categories is discussed in greater detail below.

11–2.1 Depth/Configurational Data

These include information concerning the vertical depth, measured depth, casing size (and weight), openhole size, and string type (i.e., full string, liner, tieback, multistage casing job, etc.). Depth data are particularly important because they strongly influence the temperature, fluid volume, hydrostatic pressure, and friction pressure. High angles of deviation can have a tremendous impact on many well parameters (Chapter 15), and may require the design of special systems for mud displacement and cement slurries exhibiting no free water (Keller et al., 1983).

In principle, openhole size is dictated by drill-bit size which, along with casing size and type, should be selected on the basis of the expected well conditions and the final expected completion configuration. In an actual well, the open hole is rarely "gauge." Some formations (e.g., those containing certain types of shale) are more liable to become eroded, or "washed out," than others. Wireline tools can be used to provide estimates of the openhole size (and, therefore, annular volume) with varying degrees of accuracy, depending upon the type of tool used (Table 11–1; Figs. 11–1 and 11–2). It is particu-

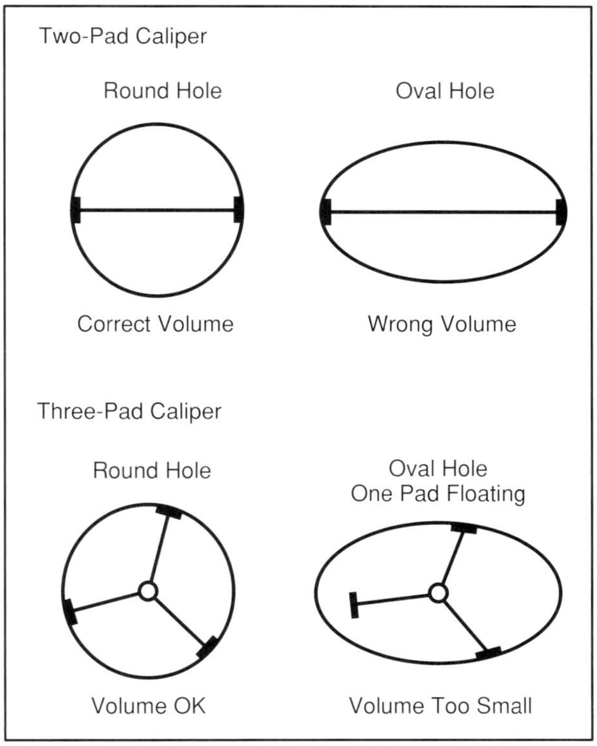

Figure 11–1—Two- and three-pad calipers.

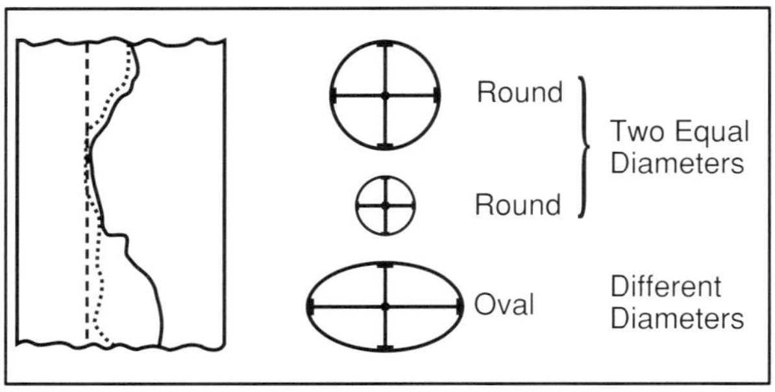

Figure 11–2—Four-arm caliper.

larly important to have an accurate estimate of the annular volume for the purpose of calculating material requirements, ensuring well security, etc. In many areas, wireline "caliper" logs are not available for the larger size open holes, and it is common practice to specify that a given percentage of "excess" cement slurry be pumped to ensure annular fill-up. This is an accepted field practice, but one that is not without its own dangers (Section 11–4).

Before selecting a particular casing nominal weight, it is necessary to consider the mechanical stresses to which the pipe will be exposed. Thus, one should consider pressure differentials across the pipe wall that may cause it to collapse or burst, as well as the longitudinal forces of stretch (because of deadweight) and compression (because of buoyancy). However, in many situations, additional factors play a major part. Thus, the use of 47-lbm/ft buttress thread 9⅝-in. (25-cm) casing as the longstring in a given location may be due more to the availability of that particular pipe than to its suitability for a specific set of well conditions.

11–2.2 Wellbore Environment

The specific problems posed by the nature of the openhole interval traversed by the casing string require careful evaluation. One must consider the presence of pay zones, of overpressured formations, or those with low fracture gradients, gas, massive salt zones, etc.

Pore pressures are important from a well-security standpoint, and information on this may be obtained by mud logging. If mud-logging facilities are not available on location, the mud weight provides a fair indication of the maximum pore pressure in any given interval. Obviously, if kicks have been taken in the course of drilling, this provides additional confirmation of the estimate. At the other extreme, the risk of formation fracture needs careful assessment, and a mean fracture pressure gradient ("frac gradient") is normally provided for each openhole interval. These values are normally based on "leakoff testing," which is performed upon drilling out the shoe of the previous casing string. Other sources for this type of information include data from stimulation or squeeze cementing treatments in offset wells.

Pay zones merit special attention for obvious reasons. It is important, for example, that they do not suffer unnecessary damage as a result of excessive leakoff from the cement slurry. It is also important that they be effectively isolated, both from each other and from nonproducing intervals, thus ensuring maximum long-term productivity from the well (Chapter 1). Finally, if the formations are known to contain gas, special cement slurries (along with other precautions) may be needed to ensure that gas does not migrate through the column of setting cement (Chapter 8). In such situations, the engineer must consider not only the target pay zones, but also the risk from other, often commercially unproductive, hydrocarbon-bearing zones in the same openhole interval.

The physical and chemical properties of the mud also need to be considered when designing a cement job (Sauer, 1985). Chemical washes, spacers, or other flush fluids must be compatible with the mud as well as the cement, and may need to contain special additives. Oil-base muds invariably require the use of surfactants in the fluids to improve compatibility, to remove the oil film from the formation surfaces, and to leave the surfaces water-wet (Carter and Evans, 1964). In some cases, where 100% mud removal cannot be assured, the cement slurry may be modified to ensure that it will not be adversely affected if contacted by the mud (Rae and Brown, 1988). Data on compatibility are obtained by laboratory testing in accordance with procedures defined by the API (Appendix B), operators, and service companies.

11–2.3 Temperature Data

Both bottomhole circulating temperature (BHCT) and bottomhole static temperature (BHST) need to be considered as well as the temperature differential (DT) between the bottom and top of the cement column. The first of these, BHCT, is the temperature to which the cement will, theoretically, be exposed as it is placed in the well. As such, it is the temperature which will be used for high-temperature, high-pressure thickening time testing of the proposed cement formulation(s). It is this figure which, by and large, directs the selection of specific retarders, etc., depending upon their efficacy under those given conditions. The BHCT is normally calculated in accordance with sets of temperature schedules published in API Spec 10 (1988). However, some operators prefer to work with temperatures actually measured in the well during circulation. One way of obtaining such temperatures is by the use of small thermosensitive temperature probes which are circulated in the mud and retrieved on exiting the well (Jones, 1986). Recently, computer simulators (which model the physics of heat transfer under dynamic conditions, etc.) have been introduced, but their use is not yet widespread (Wooley et al., 1984).

Bottomhole static temperature (BHST) is important principally for either the assessment of the long-term stability, or the rate of compressive strength development of a given cement system. It is normally calculated from the mean geothermal gradient in the area of interest, or may be estimated from measurements made during logging (adjusted accordingly for the time since the well was circulated).

The temperature differential between the top and bottom of the cement can be extremely important when embarking upon a cementing design. Cement which has been retarded for an adequate placement time at bottomhole circulating conditions may remain liquid or have poor strength development when circulated back to a shallower depth in the well. A good rule of thumb is to ensure that the static temperature at the top of the cement (TOC) exceeds the BHCT. Sabins et al. (1981) devised similar guidelines based on an experimental study of a number of cement formulations. Where it is not possible to meet these criteria, compressive strength tests need to be run, simulating conditions at the TOC. If these are not satisfactory, it may be necessary to execute the job in more than one stage. These rules of thumb provide a simple means of calculating a suitable depth for the location of the stage collar.

11–3 SLURRY SELECTION

A number of considerations come into play in the selection of a final slurry design for a specific well application. In many cases, the selection of slurry densities is dictated by factors other than simply pore and fracture pressures. Cements are often mixed at high density to achieve given values of compressive strength within a short time interval. In contrast, economics may necessitate the use of low-density extended or "filler" cements, which provide high slurry yield per sack at the expense of some of the mechanical properties of the set cement.

Well temperature is a key consideration in the slurry selection process. As discussed in Chapter 9, if static temperatures above 230°F (110°C) are anticipated, silica flour must be incorporated in conventional Portland cements to minimize strength retrogression. At the other extreme, cements used in Arctic or other low-temperature cementing applications are specially formulated so that they generate a low heat of hydration, thereby minimizing the melting of permafrost (Smith, 1976) (Chapter 7).

Fluid-loss additives are generally incorporated in slurries which traverse pay intervals, or in those where the annular gap is small. These materials reduce the rate of loss of the aqueous phase from the cement slurry. High water leakoff from a cement slurry can seriously affect its performance, particularly its viscosity, and can also cause damage to producing intervals (Suman and Ellis, 1977; Bannister, 1985). The incorporation of fluid-loss additives is notable, from the slurry design and performance standpoint, for several reasons. First, many fluid-loss additives are viscosifiers and, as a result, dispersants must be added to preserve mixability. Second, these additives often have secondary retarding effects which must be taken into account. Finally, their performance can be adversely affected by certain other additives, notably some of the sugar retarders. These and other cement additives are discussed in greater detail in Chapter 3.

Even in the absence of fluid-loss additives, dispersants are commonly incorporated in cement slurries. These materials act to reduce viscosity, thereby lowering turbulent flow pumping rates and minimizing displacement pressures. Thus, they help improve displacement efficiency, and are of particular value in situations where annular clearances are small and high friction pressures may pose some risk to weaker formations. When using dispersants, it is important to pay close attention to other aspects of slurry performance, because these materials can act in synergy with cement retarders, and produce unexpected increases in thickening time. Also, excessive dosing with dispersants can result in slurry instability which, in turn, can lead to high levels of free water and sedimentation.

Of the various groups of additives used in cement slurries, retarders are by far the most numerous. The selection of exactly which retarder to employ is based upon the circulating temperature (BHCT), the type (or even brand) of cement, and the exact slurry composition.

By and large, the compressive strength attained by a given cement system is of secondary importance when compared with the properties of the liquid slurry. This is probably because most well cement systems develop strengths which exceed those actually required under most circumstances. Certain industry and government regulatory bodies have issued guidelines and specifications for acceptable compressive strengths of cements used for certain applications (Tables 11–2, 11–3, and 11–4). Many of these deal specifically with shallower depths, where concerns center on the satisfactory isolation of fresh water supplies, etc. However, guidelines do exist for preferred strengths prior to drilling out (500 psi or 3.5 MPa) and perforating (2,000 psi or 14 MPa), and it is important to select a design which can meet these criteria. Strength can also become a critical consideration when cementing across some intervals, such as plastic salt, or pay zones which will require subsequent stimulation (Goodwin and Phipps, 1984; Rae and Brown, 1988).

Slurry design, inevitably, is an iterative process. The "first-guess" formulation, which is expected to meet the required performance criteria, is normally based upon experience and, more recently, on data bases and so-called "expert systems." However, the variability of cements demands that actual laboratory testing be performed to verify the predicted results and "fine tune" the design. Experienced engineers and laboratory personnel can dramatically reduce the number of testing hours needed to arrive at the formulation of choice for a given set of well conditions.

11–4 PLACEMENT MECHANICS

Good mud removal is the single, most important requirement for a successful primary cement job. This subject has been reviewed extensively by Haut and Crook (1979), Smith (1982), and Sauer (1985), and is discussed fully in Chapter 5.

Unfortunately, for the most part, cement is incompatible with drilling muds, causing gelation at the mud/cement interface and reducing displacement efficiency. For this reason, spacer fluids are usually pumped between the mud and cement, and careful selection of these fluids is mandatory. In many situations, it is possible to simply use water, or a water-like fluid, as a preflush ahead of the cement. In other situations, well security dictates that a weighted spacer fluid be run to maintain hydrostatic overbalance across active formations throughout the job.

Spacers are normally run at densities intermediate between those of the mud and cement, because buoyancy forces have been shown to favorably influence the mud removal process. Brice and Holmes (1963) were the first to identify the importance of contact time as a key parameter in the removal of mud, and their original recommendation of a minimum 10-minute contact time for fluids in turbulent flow has remained an industry standard.

The exact type of spacer depends upon the type of mud, the flow characteristics required (plug, laminar, or turbulent), the formations to be traversed, and the nature of the cement slurry which will follow it. Thus, freshwater-base spacers are used to remove freshwater-base muds, while salt-tolerant spacers may be required for salt-saturated muds. Oil-base muds are typically removed with spacers containing surfactants and/or organic solvents. Special care may be needed when dealing with low-toxicity, paraffinic, oil-base muds which require special surfactants. In all cases, compatibility testing between the various fluids helps to ensure that no unforeseen interactions occur which may undermine the spacer's performance downhole (Appendix B).

11–5 WELL SECURITY AND CONTROL

Each well offers an "envelope" of acceptable pressures within which the engineer must remain, if he is to design and eventually execute a successful cement job. The limiting pressure boundaries are normally the openhole pore pressure and the fracture pressure profiles; however, it is also important to consider the burst and collapse pressures of the well tubulars. Unless considerable computing power is available, it is impractical for the engineer to examine the pressures at each point in the well throughout the entire treatment. For this reason, a good approach is the so-called "worst-case scenario analysis," which allows the engineer to quickly evaluate the applicability of any particular design. This involves the identification of the "key" problem areas in a given well, which are typically the zone of highest pore-pressure gradient and the zone of lowest fracture pressure gradient. However, sections of the well where the annular configuration changes also need careful examination, because the contributions of friction and hydrostatic pressure undergo their greatest variability in such areas.

Normally, it is fair to say that the weak zones in a well will see their highest pressure just before the completion of the job (i.e., seconds before the top plug "bumps"). At this point, the longest column of high-density fluid will be in the annulus and friction will be at its highest level (ignoring any rate reduction in anticipation of bumping the plug). This, then, can be considered the worst case for zone breakdown.

Conversely, from a well-control standpoint, the worst situation occurs when the fluid of lowest density (typically a water-base wash or diesel oil) passes in front of an active zone. Depending upon the annular configuration (or openhole diameter) above, the zone in question need not be that of the highest pore pressure. Thus, a large washed out section considerably reduces the impact of a low-density fluid on the net hydrostatic pressure below it, while a tight interval can have the opposite effect.

In the event that the hole is gauge and no single zone exhibits an abnormally high pore pressure, a good rule of thumb is to select the shallowest active zone as that which poses the greatest risk to well security. Worst-case calculations should then focus on this zone. In these calculations, it is a good idea to ignore any frictional component which may be present at the time the low-density fluid passes the zone, i.e., only hydrostatic pressure should be considered.. This ensures that even in the event of a shutdown, the well remains secure in the absence of friction pressure (although it should be recognized that the fluids would generally continue to flow in such a situation due to U-tubing).

A fair degree of common sense is the engineer's greatest asset in job design. Considerable time can be saved by the ability to quickly identify those attributes of the well which are important and those which can be ignored. Thus, in small annuli, the effects of fluid density differential and friction pressure are significant and need to be calculated carefully. On the other hand, in large annuli, friction is usually negligible and need not feature in calculations. Instead, a safe estimated value (say 50 to 100 psi for the whole well) can be ascribed to friction pressure and used throughout.

There are a number of dangerous traps in job design into which the unwary may fall. One is the effect of excess cement. As mentioned above, "excess" is typically used where exact values for openhole size are unavailable or where field experience indicates the need for a given volume of excess cement to ensure adequate annular fill. On occasion, it may happen that the hole is closer to gauge than expected, and excess cement will then be circulated to a point in the well higher than originally intended. The resulting elevated hydrostatic pressure may induce losses, compromising the results of the job or endangering the well. Therefore, the selection of suitable excess volumes should bear this eventuality in mind. A similar, and no less dangerous situation, can arise if large volumes of low-density washes or flush fluids are used. Abnormally good hole geometry can raise a column of such a fluid to an unexpected height in the annulus, resulting in a loss of hydrostatic pressure and an increased risk to well security.

11-6 COMPUTER SIMULATORS

It has long been recognized in the industry that heavy fluids like cement slurry, pumped into a casing string, result in a phenomenon known as free-fall or U-tubing (Arnold, 1982; Beirute, 1984). This phenomenon, which arises from the natural tendency for the fluids inside and outside the casing to seek a pressure equilibrium, causes some interesting effects in the course of a cementing operation. The initial, internal pressure imbalance causes the cement pumped into the casing to "free-fall" from the cement head and draw a vacuum in the upper part of the casing. In most cementing operations, the rate of delivery, Q_{in}, of slurry (or displacement fluid) to the well is insufficient to keep the casing full during the early part of the job. This, then, results in a net efflux of fluid from the well and the rate of this efflux, Q_{out}, may be much greater than Q_{in}. After some time, as pressure equilibrium is approached, Q_{out} slows, falling below Q_{in} as the casing is gradually refilled. At some point, Q_{out} may actually fall to zero, i.e., the fluid column in the annulus may come to rest. Such events are easily misinterpreted as partial or complete loss of circulation. Finally, when the casing is again full of fluid, the rate of efflux and rate of delivery will match, i.e., $Q_{in} = Q_{out}$. However, they may not remain so for the remainder of the job. If a low-density wash is used, it will cause a reduction in annular pressure as it rounds the casing shoe. This will in turn cause a second period of free-fall, accompanied by another surge of high returns, etc. Both the onset and the end of the U-tubing phase can easily be detected by the measurement of surface pressure during the cement job.

Considering the importance of annular fluid velocities and pressures to the safe and successful execution of a cement job, it is clear that U-tubing, which affects both of these parameters, should be taken into account in any job design. Algorithms have been developed which permit fairly accurate simulations of the phenomena (Beirute, 1984; Wahlmeier and Lam, 1985). However, the numerical manipulation needed to accurately simulate the physics of well displacement is considerable. Fortunately, high-powered computers capable of handling the algorithms at practical speeds are now readily available to improve the efficiency and reliability of job design.

11-7 EXAMPLE OF JOB DESIGN PROCEDURE

The following example illustrates how the basic job design concepts discussed above can be combined with the power of computer simulators, to provide a realistic and technically competent well program.

The plan is to cement a 47-lbm/ft, 9⁵/₈-in. (25-cm) casing at a depth of 9,300 ft (2,835 m). The well is vertical, and the previous casing (68-lbm/ft, 13³/₈-in.) is set at 5,350 ft (1,631 m). The hole is reasonably gauge, with an openhole diameter of 12.5 in. (32 cm) for much of its length. Two shale sections show some washout (to a maximum of 15.5 in.[39.3 cm]), while two other intervals are tight (12.25 in. [31 cm]). Because of the hydrostatic limitations and the temperature differentials, a stage collar will be set just inside the shoe of the 13³/₈-in. (34-cm) casing.

There are several features of the openhole interval which require special attention. The most obvious is the presence of a major pay zone extending from 8,450 to 8,850 ft (2,576 to 2,697 m). This has the highest pore pressure and, therefore, probably poses the greatest risk to well control. However, a slightly shallower water-bearing formation may require careful examination, because its pore pressure is only slightly less. Fracture pressures for the entire open hole are fairly low, but a large depleted interval extending from 6,500 to 7,000 ft (1,981 to 2,134 m) exhibits the lowest fracture gradient in the well. This further restricts the choice of fluids. The mud in the hole is a water-base polymer system with a density of 11.4 lb/gal (1.37 g/cm^3), a system which maintains adequate coverage of the pay zone's 10.8-lb/gal (1.30-g/cm^3) equivalent mud-weight pore pressure. The rheological properties of the mud are reportedly good, with a Ty below 10 lbf/100 ft^2.

The reported BHST for the well is 238°F (114°C), which corresponds to a geothermal gradient of 1.7°F/100 ft. The calculated circulating temperature (BHCT) in the well (from API tables, Appendix B) is 168°F (76°C). The calculated static temperature at the TOC is 167°F (75°C), which is close to the BHCT. For this reason, the cement's compressive strength development should not be impaired. A summary of the well data can be found in Table 11–5.

From this information, we can draw several conclusions.

- Two cement slurries, a low-density lead and a normal density tail, will probably be required due to hydrostatic limitations. The tail slurry should be used to cover the pay zones and a reasonable length of annulus above them.
- Both slurries will require the incorporation of fluid-loss additives to avoid damage to pay zones, possible bridging, etc.
- Dispersants will probably be required due to the use of fluid-loss additives, and to the fact that the friction pressures generated by viscous slurries could pose a risk to weak zones.
- A cement retarder will be required to achieve adequate placement time.
- The tail slurry should contain silica flour to prevent strength retrogression (BHST > 230°F [110°C]). The lead slurry probably will not need it.
- It is unlikely that the cement slurries will be pumped in turbulent flow, because of the size of the annular gap and the presence of weak zones. This, and the risk of mud contamination, suggests the use of spacers or a combination of spacers and washes to achieve good mud removal. The spacer density should be intermediate between that of the mud and the lead cement. The required pump rate for turbulent displacement of the spacer is likely to be in the range of 6 to 10 BPM.
- With mixing and pumping time taken into account, the duration of the job is likely to be 2$^1/_2$ to 3 hours. With one hour for safety, we would normally look for a minimum thickening time of 3$^1/_2$ to 4 hours for the lead slurry, and somewhat less for the tail. However, considering that this is a two-stage job, the possibility exists that the cement may be lifted above the stage collar because of inaccuracies in the hole caliper. If the cement were to set, we would be unable to perform the second-stage cement job. Therefore, we must allow additional time for the stage tool to be opened, and for the volume of the annulus above the stage collar to be circulated to the surface. Allowing 15 minutes for the stage collar opening "bomb" to drop, and 30 minutes to circulate the annulus, this gives a required thickening time of 4 to 5 hours for the lead cement.

Based on these observations, the "first-guess" preferred job design would be as follows.

First-Stage Lead Slurry

API Class G Cement
+ Extender
+ Fluid-Loss Additive
+ Retarder
mixed at 12.5 lb/gal (1.50 g/cm^3) using rig water
Thickening time: 4 to 5 hours
API Fluid-Loss Rate: 150 to 300 mL/30 min

First-Stage Tail Slurry

API Class G Cement
+ 35% BWOC Silica Flour
+ Fluid-Loss Additive
+ Dispersant
+ Retarder
mixed at 15.8 lb/gal (1.90 g/cm^3) using rig water
Thickening time: 3 to 4 hours
API Fluid-Loss Rate: 50 to 150 mL/30 min

Mud Removal

Chemical Wash: 20 bbl
Turbulent Flow Spacer (12 lb/gal [1.44 g/cm^3]):
80 bbl
Total volume
100 bbl (sufficient for 10-minute contact time at a displacement rate of 10 BPM)

The casing should be well centralized and rotated/reciprocated throughout the job.

Laboratory testing optimizes the slurry formulations to meet the required performance specifications, and also provides data concerning the rheological properties of the slurries, spacers, and mud at both surface and downhole conditions. These data (Tables 11–6, 11–7, and 11–8) are then used in the final job design.

The annular cement fill (along with information on the position of other fluids in the annulus at the end of the job) is shown in Table 11–6. This table also indicates the safety margins above pore, and below fracture, pressures at this time.

It must be stressed here that these calculations are based purely on hydrostatic pressures, and are used to determine well security after placement. A graphical representation of these data is shown in Fig. 11–3.

A simulation of the actual operation, including shutdowns, rate changes, U-tubing, etc., is shown in Figs. 11–4 and 11–5. A job schedule table, representative of the expected rig procedures, upon which the simulation is based, is illustrated in Table 11–9. Figure 11–4 illustrates the fact that flow rates in and out of the well are not equivalent for a large part of the job. Sudden increases or decreases in rate, as a consequence of fluids of varying density moving from the casing into the annulus, can be predicted ahead of time. Knowledge of the magnitude of these fluctuations, and the times at which they are expected, can help allay fears that well security is threatened or that serious losses have occurred. Figure 11–6 provides data similar to that given by Fig. 11–3. In this case, frictional components are considered, as are all "worst-case scenarios" throughout the duration of the entire pumping operation. This one graph tells us all we need to know about well security and control.

11–8 PREPARING FOR THE JOB

A satisfactory job design is one where not only the chemical and physical requirements of slurry performance and displacement mechanics are met, but one which is capable of practical execution in the field. Therefore, it is an essential part of the design process to review equipment requirements and availability, bulk storage capacity, rig facilities, space availability, etc. The practicalities of using different dry cement blends or mixing various additive formulations for jobs featuring multiple slurries, spacers, and chemical washes must be considered. In many situations, the use of liquid additive systems and a single, basic, unblended cement may prove beneficial, particularly in locations where storage capacity and space are limited (e.g., offshore).

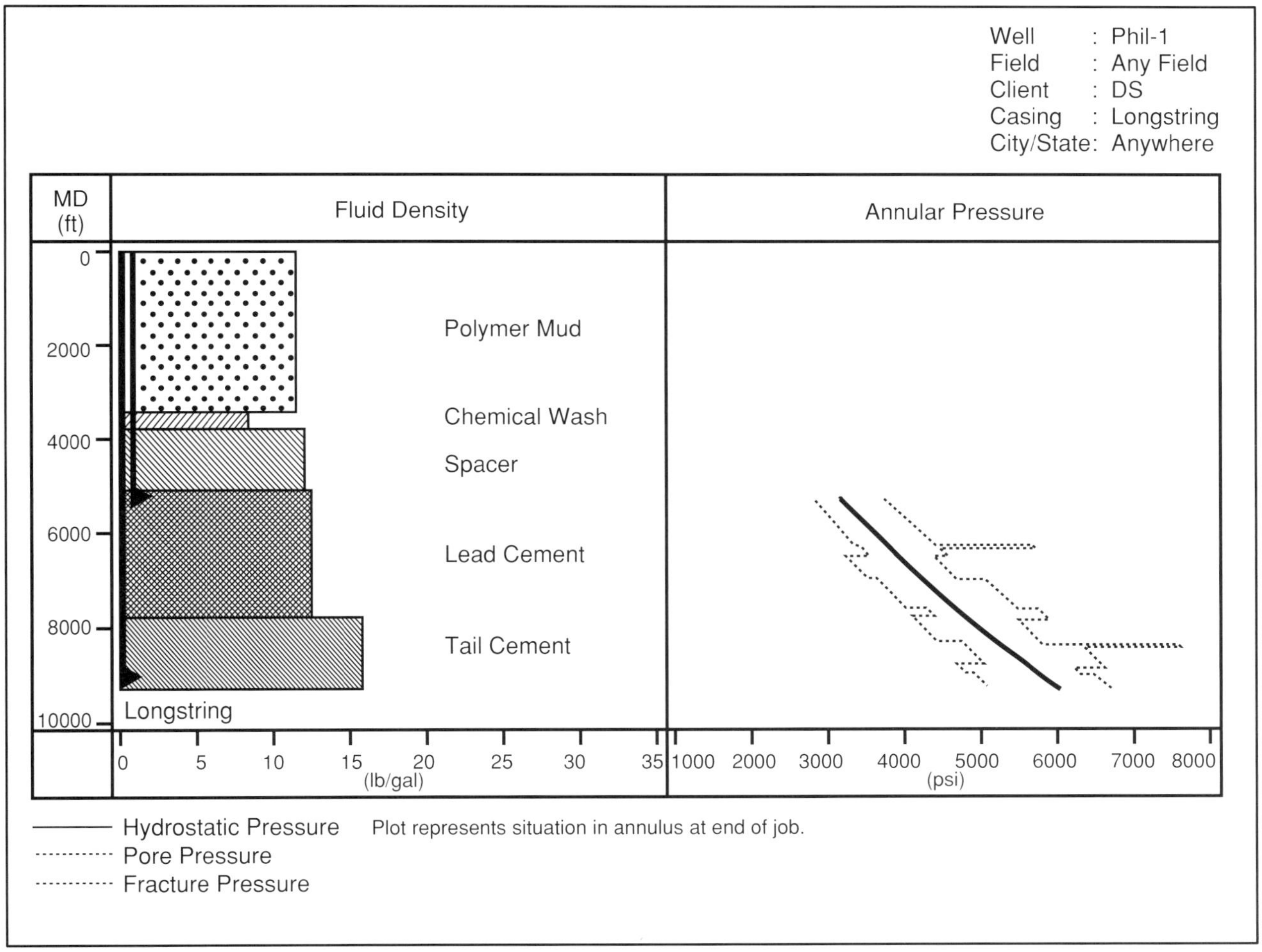

Figure 11–3—Downhole pressure-density plot.

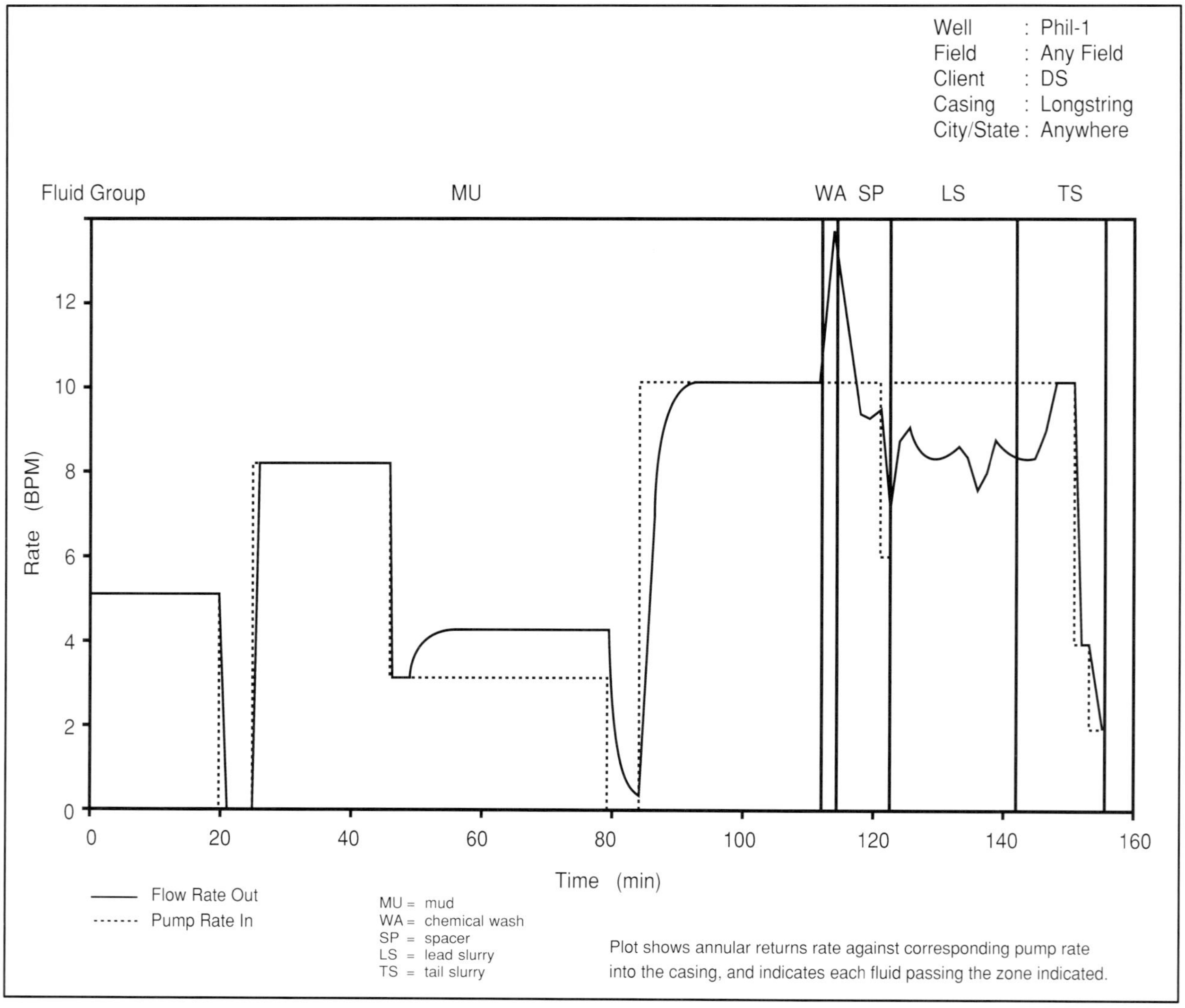

Figure 11-4—Flow-rate comparison at depth of 9300 ft.

As far as it is possible, confirmatory cement tests should be performed with the cement and mix water which will be used for the actual job. Many location waters contain dissolved salts which can detrimentally affect the performance of some slurries, and may even cause gelation or premature set (Kieffer and Rae, 1987). Cement may also become contaminated in the course of shipment, and should be sampled at the actual wellsite. Typical contaminants found in cement include bentonite (gel), barite, and sand, and all of these may affect its performance, particularly its mixability. Strict quality control (QC) procedures should be implemented by the service company and the operating company, to ensure that the materials to be used on the job perform properly. Additive drums, sacked materials, or silos bearing cement blends, should be clearly marked.

Special instructions or procedures which are important to the success of the job must be communicated to the wellsite. It is imperative that both the operating company and service personnel understand and agree upon the exact details of the job. Recommendations to reciprocate or rotate pipe, run a given number of centralizers per joint, or pump at certain rates for specific periods of time are made with good reason, and every effort should be made on location to follow these instructions. The use of computer-based data acquisition systems at the wellsite has allowed far better monitoring of the operation than was previously possible. Today, job recordings of rate, surface pressure, and fluid density can be plotted and overlaid with the original design simulations. This helps ensure that jobs are, indeed, executed as designed, and also helps identify anomalous well conditions in the course of

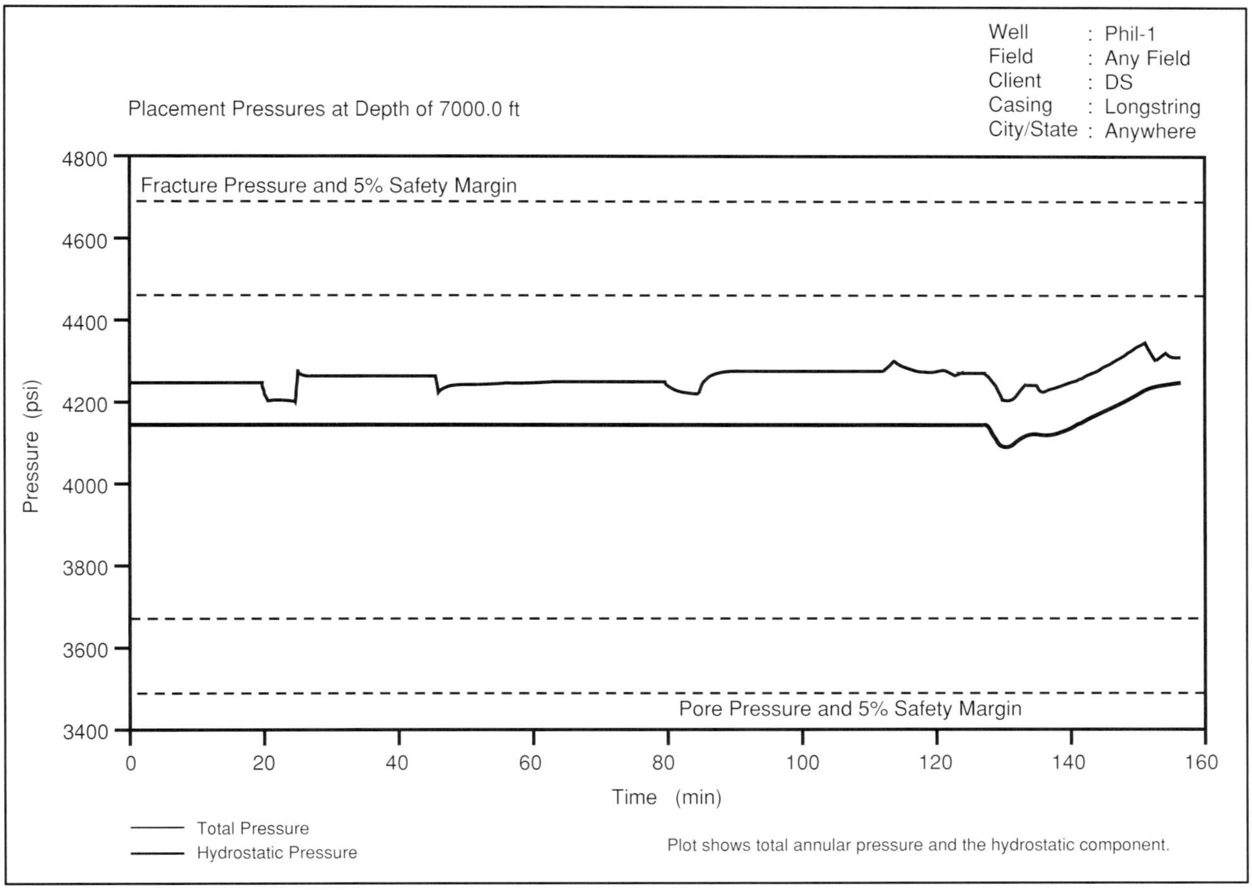

Figure 11–5—Placement pressures at depth of 7000 ft.

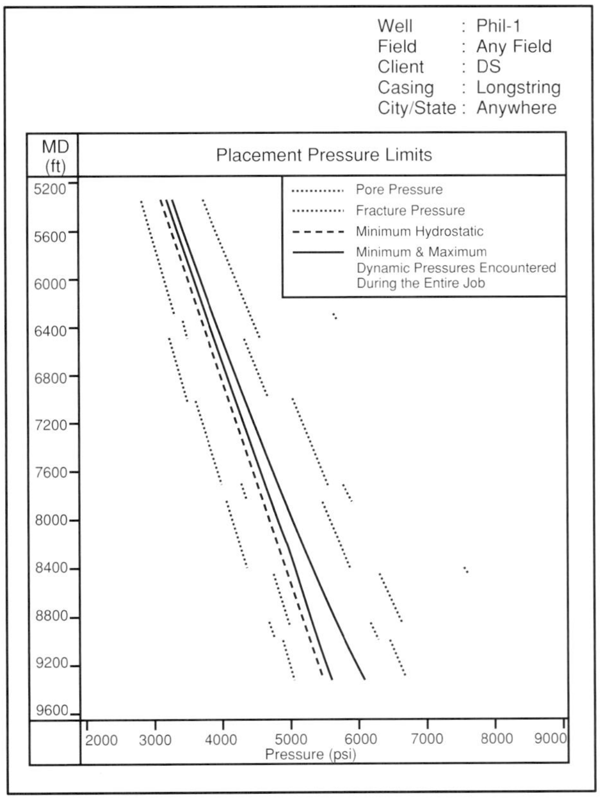

Figure 11–6—Well security and control.

the operation. Ultimately, such information can then be used to help "close the loop" between design and execution, permitting the engineer to modify future designs to achieve optimum results.

11–8 REFERENCES

American Petroleum Institute, *Specifications for Materials and Testing for Well Cements*, API Spec. 10, fourth edition, API, Dallas (1988).

Arnold, E. S.: "Cementing: Bridging the Gap from Laboratory Results to Field Operations," *JPT* (Dec. 1982) 1843–1852.

Bannister, C. E.: "The Role of Cement Fluid-Loss in Wellbore Completions," paper SPE 14433, 1985.

Beirute, R. M.: "The Phenomenon of Free Fall During Primary Cementing," paper SPE 13045, 1984.

Brice, J. W. and Holmes, B. C.: "Engineered Casing Cementing Programs Utilizing Turbulent Flow Techniques," paper SPE 742, 1963.

Carter, L. G. and Evans, G. W.: "A Study of Cement-Pipe Bonding," *JPT* (Feb. 1964) 157–160.

Goodwin, K. and Phipps, K.: "Salt-Free Cement: An Alternative to Collapsed Casing in Plastic Salt," *JPT* (Feb. 1984) 320–324.

Haut, R. C. and Crook, R. J.: "Primary Cementing: The Mud Displacement Process," paper SPE 8253, 1979.

Jones, R. R.: "A Novel Economical Approach for Accurate Real-Time Measurement of Wellbore Temperatures," paper SPE 15577, 1986.

Keller, S. R., Crook, R. J., Haut, R. C., and Kulakovsky, D. S.: "Problems Associated with Deviated Wellbore Cementing," paper SPE 11979, 1983.

Kieffer, J. and Rae, P.: "How Gelation Affects Oilwell Cements," *Pet. Eng. Intl.* (May 1987) **59**, 46–48.

Rae, P. and Brown, E.: "New Materials Improve the Cementation of Salt Formations," *Proc.*, Southwestern Petroleum Short Course, Lubbock, TX (1988).

Sabins, F. L., Sutton, D. L., and Cook, C. Jr.: "The Effect of Excessive Retardation on the Physical Properties of Cement Slurries," paper SPE 10221, 1981.

Sauer, C. W.: "Mud Displacement During the Cementing Operation: A State of the Art," paper SPE 14197, 1985.

Smith, D. K.: *Cementing*, Henry L Doherty Series, SPE, Richardson, TX (1976).

Smith, R. C.: "Checklist Aids Successful Primary Cementing," *Oil & Gas J.* (Nov. 1, 1982) 72–75.

Suman, G. O. Jr. and Ellis, R. C.: *Cementing Handbook*, Gulf Publishing Co., Houston (1977).

Wahlmeier, M. and Lam, S.: "Mathematical Algorithm Aids Analysis of 'U-Tubing' During Slurry Placement," *Oil & Gas J.* (Jan. 7, 1985) 80–86.

Wooley, G. R., Giussani, A. P., Galate, J. W., and Wedelich, H. F.: "Cementing Temperatures for Deep Well Production Liners," paper SPE 13046, 1984.

	No. of Arms	Phasing of the Arms	Max. Diameter	Remarks
Sonic	3	120°	16 in.	3 Arms Coupled 1 Reading
Microlaterolog Proximity	2	180°	20 in.	2 Arms Coupled 1 Reading
MicroSFL	4	90°	22 in.	4 Arms Coupled 2 × 2 1 Reading
Density	2	180°	Short Arm 16 in. Long Arm 21 in.	2 Arms Coupled 1 Reading
Dipmeter	4	90°	D Type 18 in. E Type 21 in.	4 Arms Coupled 2 × 2 2 Independent Readings
Borehole Geometry	4	90°	Standard 30 in. Special 40 in.	4 Arms Coupled 2 × 2 2 Independent Readings

Table 11–1—Characteristics of different calipers.

State	City and Zip Code	Regulatory Body	Date	Casing	Plugging
Alabama (inland wells)	University 35486	Oil and Gas Board of Alabama	1976	400-1-3-0.03 400-3X-0.02	400-1-3-0.04 to 0.07
Alaska	Anchorage 99501	State Oil and Gas Conservation Comm.	1981	Art 1. Sec. 30 20ACC 25.026	Art. 2 Sec. 105 20AAC 25.105
Arizona	Phoenix 85007	Oil and Gas Conservation Comm.	1982	Chap. 7, Art. 1 R-12-7-110 and 111	Chap. 7, Art. 1 R-12-7-126 and 127
Arkansas	El Dorado 71730	State Oil and Gas Comm.	1983	Rule B-15 and B-29	Rule B-8 and B-9
California	Sacramento 95814	Dept. of Conservation Div. of Oil and Gas	1983	Publication PRC-01 3220-3223	Publication PRO-01 3228-3232
Colorado	Denver 80203	Dept. of Natural Resources Oil and Gas Conservation Comm.	1983	317-327-404	332
Connecticut*	Hartford 06115	Consult State Geological Surveys	—		—
Delaware	Dover 19903	Dept. of Natural Resources and Environmental Control Water Resources Sec.	1971	Oil and Gas Regulations Sec. 2	Oil and Gas Regulations Sec. 6.04
Florida	Tallahassee 32301	Dept. of Natural Resources Oil and Gas Div.	1983	Oil and Gas Statute 16C-27.05 16C-29.07	Oil and Gas Statute 16C-29.09
Georgia	Atalnta 30334	Dept. of Natural Resources	1975	Rules 391 through 393 and 13.10	Rules 391 through 393 and 13.12
Idaho	Boise 83720	Oil and Gas Conservation Comm.	1963	Rules 8.3 through 8.12	Rules 32.1 through 32.5
Illinois	Springfield 62706	Dept. of Mines and Minerals Div. of Oil and Gas	1984	Rule VIII-6-B	Rule II-2-B RuleXI-5-A, B
Indiana	Indianapolis 46204	Dept. of Natural Resources Div. of Oil and Gas	1972	22-J	33
Iowa	Des Moines 50319	Dept. of Soil Conservation Mines and Minerals Div.	1983	Code of Iowa Chap. 84	Code of Iowa Chap. 84
Kansas	Topeka 67202	Corporation Comm. Oil and Gas Conservation Div.	1983	Conservation Rules 82-3-103 through 106	Conservation Rules 82-3-112 through 115
Kentucky	Lexington 40586	Dept. of Mines and Minerals Div. of Oil and Gas	1978	805 KRS 1:020	805 KRS 1:060, 1:070
Louisiana	Baton Rouge 70801	Dept. of Conservation	1982	29-B, Sec. V	29-B, Sec. XIX
Maine*	Augusta 04333	Consult State Geological Surveys	—		—
Maryland*	Annapolis 21401	Dept. of Natural Resources	—		—
Massachusetts*	Boston 02108	Consult State Geological Surveys	—		—
Michigan	Lansing 48909	Dept. of Natural Resources Oil and Gas Regulations	1983	Oct. 1961 R-299.1306	Oct. 1961 R-299.1801–09
Minnesota* (water wells)	St. Paul 55155	Dept. of Natural Resources	1980	Chap. 156A, 01–10	—
Mississippi	Jackson 39201	State Oil and Gas Board	1972	Rules 10-11-12 (Order 201–51)	Rule 28
Missouri	Rolla 65401	Missouri Oil and Gas Council	1982	Chap. 2 10CSR-50-2.040	Chap. 2 10CSR-50-2.060
Montana	Helena 59601	Dept. of Natural Resources and Conservation Oil and Gas Conservation Div.	1983	Rule 36.22.1001 through 36.22.1013	Rule 36.22.1301 through 36.22.1309
Nebraska	Sidney 69162	Oil and Gas Conservation Comm.	1983	Statute 57-905 Code 3-012	Statute 57-905, 906 Code 3-028
Nevada	Carson City 89710	Dept. of Conservation and Natural Resources	1979	Rules Part 2 Sec. 200–214	Rules Part 3 Sec. 300–308
New Hampshire**	Durham 03824	Consult State Geological Surveys	—		—

Table 11–2—Regulatory bodies and rules controlling the cementing of wells in the U.S. (from Smith, 1987).

State	City and Zip Code	Regulatory Body	Date	Casing	Plugging
New Jersey**	Trenton 08625	Consult State Geological Surveys	—		—
New Mexico	Santa Fe 87501	Energy and Minerals Dept. Oil Conservation Div.	1982	Rule 107 and 108	Rule 201, 202, and 1103
New York	Albany 12233	State Dept. of Environmental Cons. Bur. of Oil and Gas Regulations	1972	NYCRR Secs. 552 through 554	NYCRR Sec. 555
N. Carolina	Raleigh 27611-7687	Natural and Economic Resources Mining, Mineral Resources Oil and Gas Conservation	1976	G.S. 113-391-0007	G.S. 113-391-0009
N. Dakota	Bismarck 58505	North Dakota Industrial Comm. Oil and Gas Div.	1983	NDCC 38-08-04 43-02-03-21	NDCC 38-08-04 43-02-03-34
Ohio	Columbus 43224	Ohio Dept. of Natural Resources Div. of Oil and Gas	1982	Ohio Statutes—Rules 1501 : 9-11-03 through 09 and 1509.17	Ohio Statutes—Rules 1501 : 9-11-03 through 09 and 1509.17
Oklahoma	Okla. City 73105	Corporation Comm. Oil and Gas Conservation	1983	General Rules 3-206	General Rules 3-400 through 405 and 409
Oregon	Portland 97201	Dept. of Geology and Mineral Industries	1982	Administrative Rules 632-10-014	Administrative Rules 632-10-198
Pennsylvania	Harrisburg 17120	Dept. of Environmental Resources Oil and Gas Conservation	1983	General Provisions 79.12	General Provisions 79.17
Rhode Island**	Providence 02903	Consult State Geological Surveys	—		
S. Carolina**	Columbia 29211	Consult State Geological Surveys	—		
S. Dakota	Rapid City 57701	Board of Natural Resource Development Oil and Gas Conservation	1974	Chap. 52 : 02 : 03	Chap. 52 : 02 : 04
Tennessee	Nashville 37203	Dept. of Conservation State Oil and Gas Board	1972	State Order 2 1040-2-7	State Order 2 1040-2-9
Texas	Austin 78771	Railroad Comm. of Texas Oil and Gas Div.	1983	Rule 13	Rule 14
Utah	Salt Lake City 84101	Board of Oil, Gas and Mining Conservation of Oil and Gas	1982	Rule C-8	Rule D-1, D-2, and D-4
Vermont**	Montpelier 05602	Consult State Geological Surveys	—		
Virginia	Richmond 23241	Dept. of Labor and Industry Div. of Mines and Quarries Oil and Gas Conservation Comm.	1983	Code of Virginia 45.1-334 through 340	Code of Virginia 45.1-341 through 348
Washington	Olympia 98504	Dept. of Natural Resources Div. of Geology and Earth Resources	1982	WAC-344-12-087	WAC-344-12-131 and 133
W. Virginia	Charleston 25316	Dept. of Mines Office of Oil and Gas	1983	22-4-5 through 8	22-4-9, 10
Wisconsin	Madison 53701	Dept. of Natural Resources Water Well Regulations	1975	NR-112.085	NR-112.21
Wyoming	Cheyenne 82002	Oil and Gas Conservation Comm.	1982	Sec. III 320 through 323	Sec. III 312 through 315
Federal					
Alaska Federal	Reston, VA 22090	U.S. Dept. of the Interior	1980		
Atlantic	Reston	U.S. Dept. of the Interior Mineral Management Service Conservation Div., Atlantic Outer Shelf	1980	OCS, Order 2 3.1 through 3.6	OCS, Order 3 1.1 through 2.9
Gulf of Mexico	Reston	U.S. Dept. of the Interior Mineral Management Service Conservation Div., Atlantic Outer Shelf	1980	OCS, Order 2 3.1 through 3.6	OCS, Order 3 1.1 through 2.9
Pacific	Reston	U.S. Dept. of the Interior Mineral Management Service Conservation Div., Atlantic Outer Shelf	1980	OCS, Order 2 3.1 through 3.6	OCS, Order 3 1.1 through 2.9

* No commercial oil or gas—rules apply to water wells.

** No known rules in these states.

Table 11–2, continued—Regulatory bodies and rules controlling the cementing of wells in the U.S. (from Smith, 1987).

Country	Agency
Abu Dhabi	Ministry of Petroleum
Australia	Department of Mines
Austria	Oberste Bergbehörde
Canada	Ontario—Dept. of Mines and Northern Affairs
	Alberta—Oil and Gas Conservation Board
	Saskatchewan—Dept. of Mineral Resources
Colombia	Minister of Mines and Petroleum
France	Direction Génèrale des Mines
Germany	Bureau of Mines
Ireland	Offshore Operating Committee—London
Italy	National Mining Bureau for Hydrocarbons
Japan	Bureau of Mines
	Petroleum Mine Safety Regulations
Libya	Petroleum Ministry
Malaysia	Petroleum Ministry
Mozambique	Geology and Mines Dept.
The Netherlands	The Ministry of Mines
Norway	Petroleum Directorate
Turkey	Petroleum Admin.
United Kingdom	Dept. of Energy
Venezuela	Dept. of Hydrocarbons

Table 11–3—Countries other than U.S. known to have drilling and cementing regulations (from Smith, 1987).

	Type of Job*									
	Surface Pipe				Intermediate String				Production String	
State	1	2	3	4	1	2	3	4	1	2
Alabama (inland wells)	—	—	Up to 1,000 psi based on depth	12 hr	—	—	—	12 hr	1,500 psi or 0.2 psi/ft for 30 min	24 hr
Colorado	500 psi and 8 hr	500 psi and 8 hr	—	8 hr	—	—	—	500 psi and 8 hr	—	500 psi and 8 hr
Kansas	300 psi and 8 hr	300 psi and 8 hr	—	8 hr	—	—	—	—	—	—
Louisiana	12 hr	0	Up to 100 psi based on depth	12 hr	—	12 hr	800 to 1,500	12 hr	800 to 1,500 after 24 hr	—
Mississippi	12 hr	0	1 psi/ft up to 1,000 psi	12 hr	—	—	—	24 hr	0.2 psi/ft up to 1,500 psi	24 hr
Montana	8 to 18 hr**	8 to 18 hr**	Properly	8 hr and 300 psi	—	—	Properly	8 hr	8 hr	—
New Mexico	8 hr	0	600 to 1,500	18 hr**	8 hr	0	600 to 1,500	18 hr**	600 to 1,500	18 hr**
North Dakota	12 hr	0	None	12 hr	—	—	None	12 hr	None	12 hr
Oklahoma	8 hr	Undefined	Based on depth	8 hr	—	—	—	8 hr	1,500 psi max. based on depth	—
Texas	—	0	—	500 and 1,200 psi 72 hr	—	—	—	—	1,000 psi for 30 min	24 hr
Wyoming	300 psi and 8 hr	300 psi and 8 hr	Properly	300 psi and 8 hr	—	—	—	—	—	—
Federal Gulf of Mexico	12 hr	0	1,000	12 hr	1,500 psi or 0.2 psi/ft	—	—	12 hr	1,500 psi or 0.22 psi/ft for 30 min	12 hr

*Key to type of job

Surface pipe and intermediate string:

1. WOC with surface pressure, without float.
2. WOC with surface pressure, with float.
3. Pressure test.
4. WOC to drill out.

Production string:

1. Pressure test.
2. WOC to perforate.

** Operator can chose between 8 to 18 hours and a time based on the strength of the cement.

Table 11–4—Statewide WOC requirements for various states (from Smith, 1987).

Well Data	Well : Phil-1 Field : Any Field Client : DS Casing : Longstring City/State: Anywhere

String Interval	O.D (in.)	I.D. (in.)	Weight (lb/ft)	Depth (ft)		Pressure (psi)	
				MD	TVD	Collapse	Burst
1 casing	9⅝	8.681	47.00	9300.0	9300.0	4750.0	6870.0

Well Interval	Diameter (in.)	Depth ft		Pressure (psi)	
		MD	TVD	Pore	Frac
1 13⅜ Intermediate	12.415	5350.0	5350.0		
2	12.500	6300.0	6300.0	3339.3	4410.0
3 Shale	15.500	6350.0	6350.0		5715.0
4	12.500	6500.0	6500.0	3512.9	4550.0
5 Depleted Interval	12.250	7000.0	7000.0	3492.1	4690.0
6	12.500	7700.0	7700.0	4001.4	5544.0
7 Water Zone	12.250	7850.0	7850.0	4364.9	5887.5
8	12.500	8400.0	8400.0	4365.1	5880.0
9 Shale Cap	14.000	8450.0	8450.0		7605.0
10 Major Pay Interval	12.500	8850.0	8850.0	4966.9	6637.5
11	13.000	9000.0	9000.0	4770.5	6300.0
12	12.500	9300.0	9300.0	5026.1	6696.0

Table 11–5—Example of depth and configurational data for a longstring.

Slurry Fill								Well : Phil-1 Field : Any Field Client : DS Casing : Longstring City/State: Anywhere
	Fluid	Volume Required (bbl)	Top (ft)	Depth Ind	Bottom (ft)	Ind	Fill (ft)	Volume To Surface (bbl)
14	Polymer Mud	204.6	0.0	A	3425.9	A	3425.9	0.0
13	Chemical Wash	20.0	3425.9	A	3760.7	A	334.8	204.6
15	Spacer	80.0	3760.7	A	5100.0	A	1339.3	224.6
16	Lead Cement	166.8	5100.0	A	7800.0	A	2700.0	304.6
17	Tail Cement	102.0	7800.0	A	9220.0	C	1500.0	471.4
14	Polymer Mud	675.0	0.0	C	9220.0	C	9220.0	0.0

Indicator — C = Inside Casing A = Inside Annulus

Depth (ft)	Pore Pressure (psi)	Hydrostatic Pressure Internal (psi)	Hydrostatic Pressure External (psi)	Burst Pressure (psi)	Collapse Pressure (psi)	Fracture Pressure (psi)	Comment
3425.9		2029.6	2029.6	6870.0	4750.0		Hydrostatic OK
3760.7		2227.9	2174.3	6870.0	4750.0		Hydrostatic OK
5100.0		3021.3	3009.5	6870.0	4750.0		Hydrostatic OK
5350.0	2835.8	3169.4	3171.9	6870.0	4750.0	3745.0	Hydrostatic OK
6300.0	3339.3	3732.2	3788.9	6870.0	4750.0	4410.0	Hydrostatic OK
6350.0	3431.8	3761.8	3821.4	6870.0	4750.0	5715.0	Hydrostatic OK
6500.0	3512.9	3850.7	3918.9	6870.0	4750.0	4550.0	Hydrostatic OK
7000.0	3492.1	4146.9	4243.6	6870.0	4750.0	4690.0	Hydrostatic OK
7700.0	4001.4	4561.6	4698.3	6870.0	4750.0	5544.0	Hydrostatic OK
7800.0	4337.1	4620.8	4763.3	6870.0	4750.0	5850.0	Hydrostatic OK
7850.0	4364.9	4650.4	4804.4	6870.0	4750.0	5887.5	Hydrostatic OK
8400.0	4365.1	4976.2	5255.9	6870.0	4750.0	5880.0	Hydrostatic OK
8450.0	4742.4	5005.9	5297.0	6870.0	4750.0	7605.0	Hydrostatic OK
8850.0	4966.9	5242.8	5625.4	6870.0	4750.0	6637.5	Hydrostatic OK
9000.0	4770.5	5331.7	5748.6	6870.0	4750.0	6300.0	Hydrostatic OK
9220.0	4982.9	5462.0	5929.2	6870.0	4750.0	6638.4	Hydrostatic OK
9300.0	5026.1	5527.7	5994.9	6870.0	4750.0	6696.0	Hydrostatic OK

Table 11–6—Slurry fill data.

Performance Data	Well : Phil-1 Field : Any Field Client : DS Casing : Longstring City/State: Anywhere

16 Lead Slurry — Lead Cement	Fluid-Loss-Controlled Lead Slurry
Density — 12.500 lb/gal Base Fluid — 8.320 lb/gal Yield — 2.11 ft^3/sk Fann Readings N/A °F Rotor — Spring — Bob — Speed — Angle — 10 min Gel — 9 # min Gel — # min Consist — # **Rheological Model : Bingham Plastic** Index — τ_y — 6.0 # PV — 14.0 cp Fluid Loss at 168°F 210 cm^3 30 min **Thickening Time Using API Schedule** 30 Bc 4:37 hr 100 Bc 5:12 hr	Cement: Class G **Composition** 12.151 gal/sk Mix Fluid 0.420 gal/sk Extender 0.50 % BWOC Fluid-Loss Additive 0.050 gal/sk Retarder 0.020 gal/sk Antifoam Agent Mix-Water Type : Brackish Sack Weight : 94.00 lb **Liquid Additive Scheduling** Chemical Function — Quantity per 10-bbl Disp. Tank Extender 14.5 gal Retarder 1.7 gal Antifoam Agent 0.7 gal **Compressive Strength Using API Schedule** hr psi hr psi
17 Tail Slurry — Tail Cement	**Fluid-Loss-Controlled Tail Cement**
Density — 15.800 lb/gal Base Fluid — 8.320 lb/gal Yield — 1.52 ft^3/sk Fann Readings N/A °F Rotor — Spring — Bob — Speed — Angle — 10 min Gel — 2 # min Gel — # min Consist — # **Rheological Model : Bingham Plastic** Index — τ_y — 1.5 # PV — 45.0 cp Fluid Loss at 168°F 65 cm^3 30 min **Thickening Time Using API Schedule** 30 Bc 3:39 hr 100 Bc 3:54 hr	Cement: Class G **Composition** 6.354 gal/sk Mix Fluid 35.00 % BWOC Silica Flour 1.200 gal/sk Latex 0.100 gal/sk Fluid-Loss Additive 0.030 gal/sk Retarder 0.020 gal/sk Antifoam Agent Mix-Water Type : Brackish Sack Weight : 94.00 lb **Liquid Additive Scheduling** Chemical Function — Quantity per 10-bbl Disp. Tank Latex 79.3 gal Fluid-Loss Additive 6.6 gal Retarder 2.0 gal Antifoam Agent 1.3 gal **Compressive Strength Using API Schedule** 12 hr 2050 psi hr psi
# = $lbf/100ft^2$	

Table 11–7—Performance data for lead-cement system and tail-cement system.

Summary										**Well : Phil-1** **Field : Any Field** **Client : DS** **Casing :** **City/State: Anywhere**
Chemical Wash 8.320 lb/gal	WA Chemical Wash Power Law	K′	0.0000	@	n′	1.000				
Polymer Mud 11.400 lb/gal	MU XC Polymer Mud Bingham Plastic	τ_y	8.0	#	PV	15.1	cp			
Spacer 12.000 lb/gal	SP Spacer Bingham Plastic	τ_y	0.5	#	PV	12.0	cp			
Lead Cement 12.500 lb/gal	LS Fluid-Loss Controlled Lead Slurry Bingham Plastic	τ_y	6.0	#	PV	14.0	cp	Yield	2.11 ft^3/sk	
Tail Cement 15.800 lb/gal	TS Fluid Loss Controlled Tail Cement Bingham Plastic	τ_y	1.5	#	PV	45.0	cp	Yield	1.52 ft^3/sk	
# = $lbf/100ft^2$	@ = $lbf{\bullet}sec^{n'}/ft^2$									

Table 11–8—Summary of properties of all wellbore fluids.

Pumping Schedule					**Client : DS** **Casing : Longstring** **Well : Phil-1** **Field : Any Field** **County : Any County** **State : Any State** **Rig Name: Any Rig**
Fluid Pumped	**Pump Rate bbl/mn**	**Fluid Volume bbl**	**Stage Time min:sec**	**Elapsed Time min:sec**	**Comments**
					—START JOB— **Pre-job safety meeting — check data recorder — pressure test lines**
Chemical Wash	5.00	20.00	4:00	4:00	Pump Chemical Wash
Spacer	5.00	80.00	16:00	20:00	Pump Spacer
	0.00	0.00	5:00	25:00	Shutdown — drop bottom plug
Lead Cement	8.00	166.80	20:51	45:51	Start mixing lead cement
Tail Cement	3.00	102.00	34:00	79:51	Finish lead — mix tail cement
	0.00	0.00	5:00	84:51	Shutdown — drop top plug
Polymer Mud	10.00	360.00	36:00	120:51	Start displacement at 10 BPM
Polymer Mud	6.00	15.00	2:30	123:21	Slow rate — plug at stage tool
Polymer Mud	10.00	285.00	28:30	151:51	Pick-up rate to 10 BPM
Polymer Mud	4.00	8.00	2:00	153:51	Slow rate towards end displacement
Polymer Mud	2.00	6.96	3:28	157:19	Slow rate further — bump plug
					Pressure test casing—bleed-off pressure—check floats—job complete **—END JOB—**

Table 11–9—Job schedule table.

12 Primary Cementing Techniques

Leo Burdylo and George Birch

Dowell Schlumberger

12–1 INTRODUCTION

Primary cementing is a technique for placing cement slurries in the annular space between the casing and the boreholes. The cement then hardens to form a hydraulic seal in the wellbore, preventing the migration of formation fluids in the annulus. Primary cementing is therefore one of the most critical stages during the drilling and completion of a well. This procedure must be carefully planned and executed, because there is only one chance to complete the job successfully.

In addition to providing zonal isolation, the set cement sheath should anchor and support the casing string (preventing formation sloughing or caving into the wellbore) and protect the casing string against corrosion by formation fluids. Uncemented steel casing can rapidly corrode when exposed to hot formation brines, hydrogen sulfide, and carbon dioxide. It can also be subjected to erosion by the high velocity of produced fluids, particularly when solid particles such as formation sand are being transported. Lateral loads on poorly cemented casing strings can result in ovaling, buckling, or even complete collapse because of overloading at certain points. On the other hand, properly cemented casing is subjected to a nearly uniform loading approximately equal to the overburden pressure.

In principle, primary cementing techniques are the same regardless of casing string purpose and size. The cement slurry is pumped down inside the string to be cemented, exits the bottom, and displaces drilling mud while moving up the annulus. Details can vary from casing to casing; the differences in placement technique are discussed in this chapter. It is assumed that the reader is familiar with the related supporting material presented previously—Chapters 5 (Mud Removal), 8 (Prevention of Gas Migration), 10 (Cementing Equipment and Tools), and 11 (Cement Job Design) in particular. In addition, the reader is referred to Chapter 15 for a discussion of the special considerations related to deviated wellbore cementing, and Appendix C for primary cementing calculations.

12–2 CLASSIFICATION OF CASING STRINGS

A series of casing strings is necessary to complete a well, and produce the desired fluids successfully. The design of the casing program is contingent upon several factors—(1) depth, (2) the sizes of the holes in which the casing strings are to be set, (4) the mud-column and formation pressures, (5) the condition of the formation, and (6) the drilling objectives. The casing string must also be designed to withstand the mechanical and chemical stresses in the well (Lubinski, 1951; Bowers, 1955; API, 1959; Smith, 1987). In this section, the functions of the casing strings, the depths to which they are normally set, and special considerations for each are discussed.

12–2.1 Conductor Pipe

The conductor is usually the first and shortest casing string. Its purpose is to protect shallow sands from being contaminated by drilling fluids, and help prevent washouts which can easily occur near the surface because of loose, unconsolidated topsoils, gravel beds, etc. The conductor pipe also serves as a channel to raise the circulating fluid high enough to return to the mud system. It can be used for the attachment of a blowout preventer (BOP), should gas sands, for example, be encountered at shallow depths. The conductor pipe serves to protect the subsequent casing strings from corrosion, and may be used to support some of the wellhead load when the ground support may be inadequate.

At offshore locations, or during swamp barge operations, driving the conductor into the ground is a common practice. The drilling rig is equipped with a pile driver, and sections of conductor casing are welded together as they are driven into the ground. The setting depth is usually less than 300 ft (91 m), and is often determined by the limitations of the pile driver as the conductor begins to

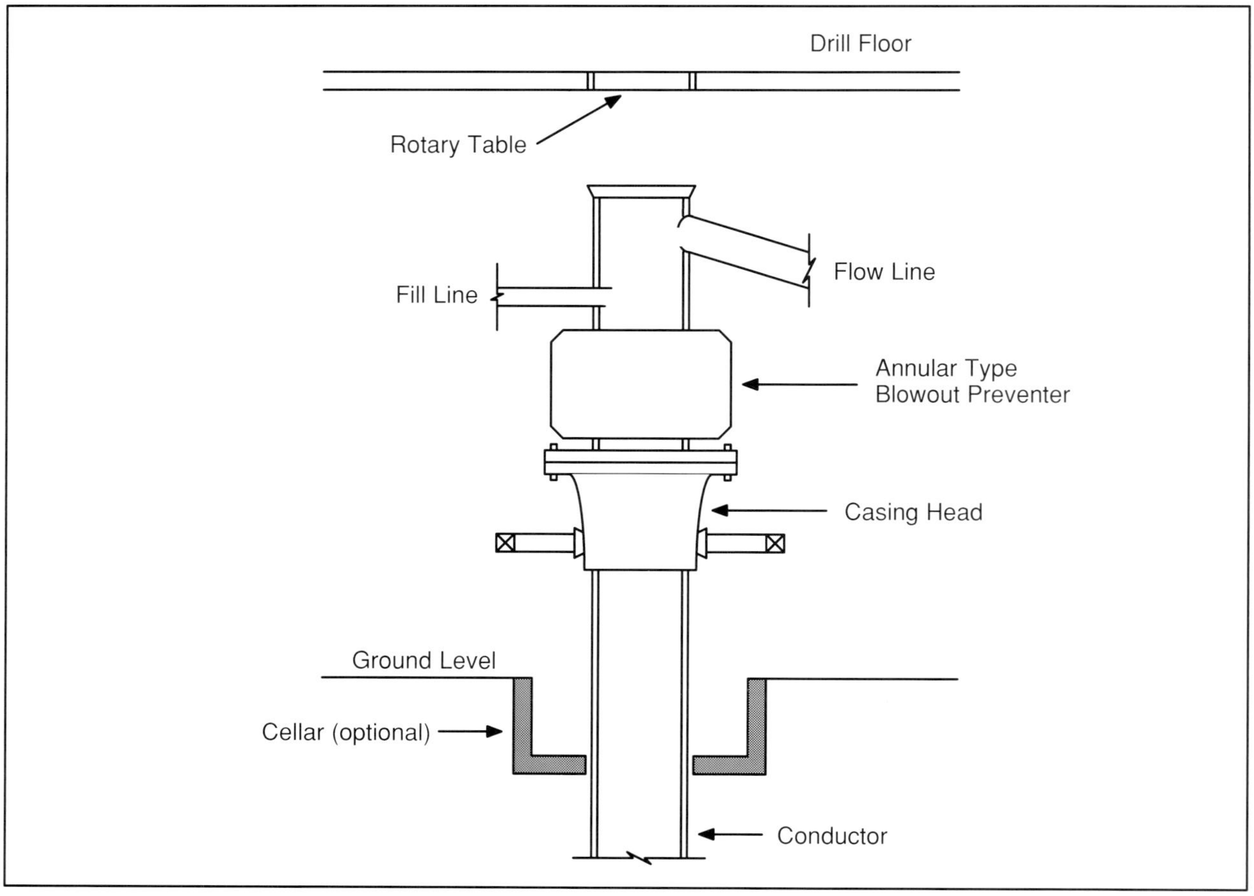

Figure 12–1—Conductor driven into the ground with the casing head welded on, and the annular blowout preventer connected and ready for drillout.

encounter firmer ground. Once driven to the maximum depth, the conductor is then cut to the appropriate height below the drill floor (and above the water line in offshore applications), and a casing head is welded into place (Fig. 12–1).

The hole for the conductor is sometimes drilled, and the pipe is made up and lowered in a manner similar to conventional casing. Most often, only a guide shoe may be welded to help lower the conductor into the well. Cementing of the conductor is performed through a swedge which is screwed to the top of the conductor. The cement slurry is pumped through the swedge and into the pipe. Since the length of the conductor is short, the annular and pipe volumes are relatively small, and cement slurry is pumped until returns are observed at the surface. Cement slurry is then displaced from the casing without the use of plugs.

In shallow casing jobs, washouts and lost circulation often prevent the cement from reaching the surface. Under these conditions, using normal procedures, the amount of cement to be used is estimated before the job, then mixed and pumped downhole. If the washouts prevent the cement from reaching the desired height, a "top-up" job must be performed (Section 12–3.2). If lost circulation occurs after the mixing is completed, the casing volume must be displaced, pumping large quantities of cement into the loss zone.

Large-diameter casings are also subject to large upward forces because of the pressure acting on the area of the cement head. If large enough, the upward forces may exceed the buoyed weight of the casing, and pump the casing out of the hole. To prevent such problems, a through-drillpipe (or "stab-in") cementing technique (Section 12–3.1) is often applied.

12–2.2 Surface Casing

The second string of casing, which serves to case off unconsolidated formations and aquifers found at relatively shallow depths, is known as surface casing (Fig. 12–2). In addition to maintaining hole integrity, the surface casing prevents the contamination of fresh groundwater by drilling fluids, subterranean brines, oil, or gas. Depending on the country, there are usually government

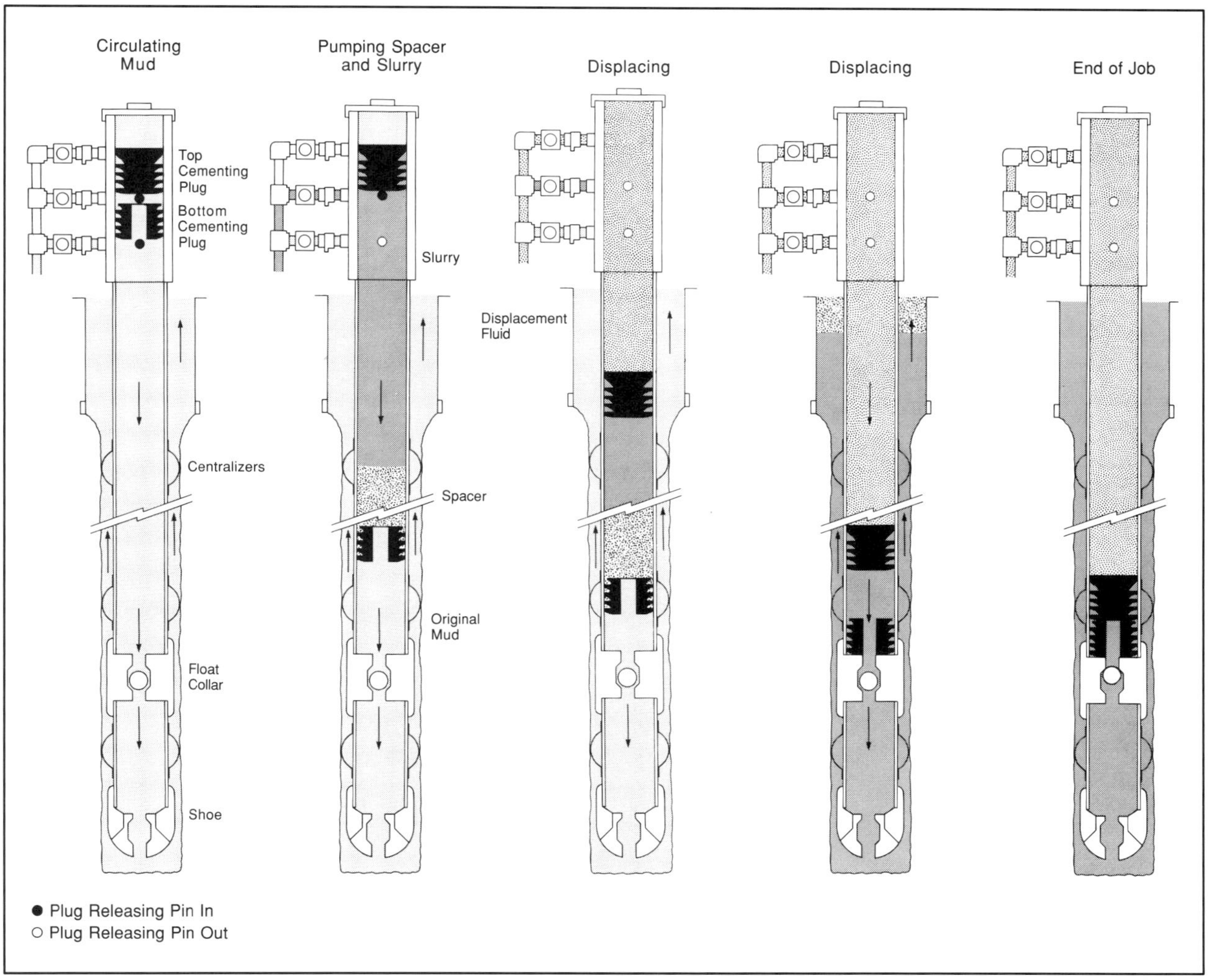

Figure 12–2—Typical one-stage primary cement job on a surface casing string.

regulations stipulating minimum requirements of the casing, and set cement properties (Chapter 11).

Quite often, the surface casing is the first string to which BOPs are connected (Fig. 12–3). Therefore, the selected casing must be strong enough to support a BOP and to withstand the gas or fluid pressures which may be encountered. Surface casing should have the strength to support further casing strings and production tubulars, and provide a solid anchor for the casing head when the well is put on production. Ordinarily, the burst pressure should be equal to one psi per foot of depth to which it is set. The sizes of the surface casing and the setting depths vary considerably; generally speaking, diameters range from 7 to 20 in. (18 to 50 cm), and depths can reach 5,000 ft (1,520 m).

A major problem associated with cementing surface casing is placing the required annular height of cement slurry (often to surface) when the hydrostatic pressures of the slurries often can exceed the formation fracture pressure. The use of low-density slurries and even foamed cement slurries is becoming more common in such circumstances (Chapters 3 and 14). Washouts are another frequent problem. The larger openhole sizes, particularly when enlarged due to washouts, often exceed the capability of caliper tools; as a result, accurate hole volumes may not be determinable.

The through-drillpipe stab-in cementing system can be used in some surface casing cementing operations, but often this is not possible when using smaller size surface casing, or when larger sizes are run beyond 3,000 ft (915 m). Drilling rig design constraints become the limiting factor in these applications.

Frequently, the primary cement job may have to be staged to successfully cement across severe lost-circulation zones or other troublesome intervals. Surface casing strings often must also deal with sloughing shales and shallow gas pockets (Chapter 8). Next to deep liner cementing, it is probably the most difficult casing string to successfully cement. Low formation temperatures prolong the thickening times of extended cement slurries,

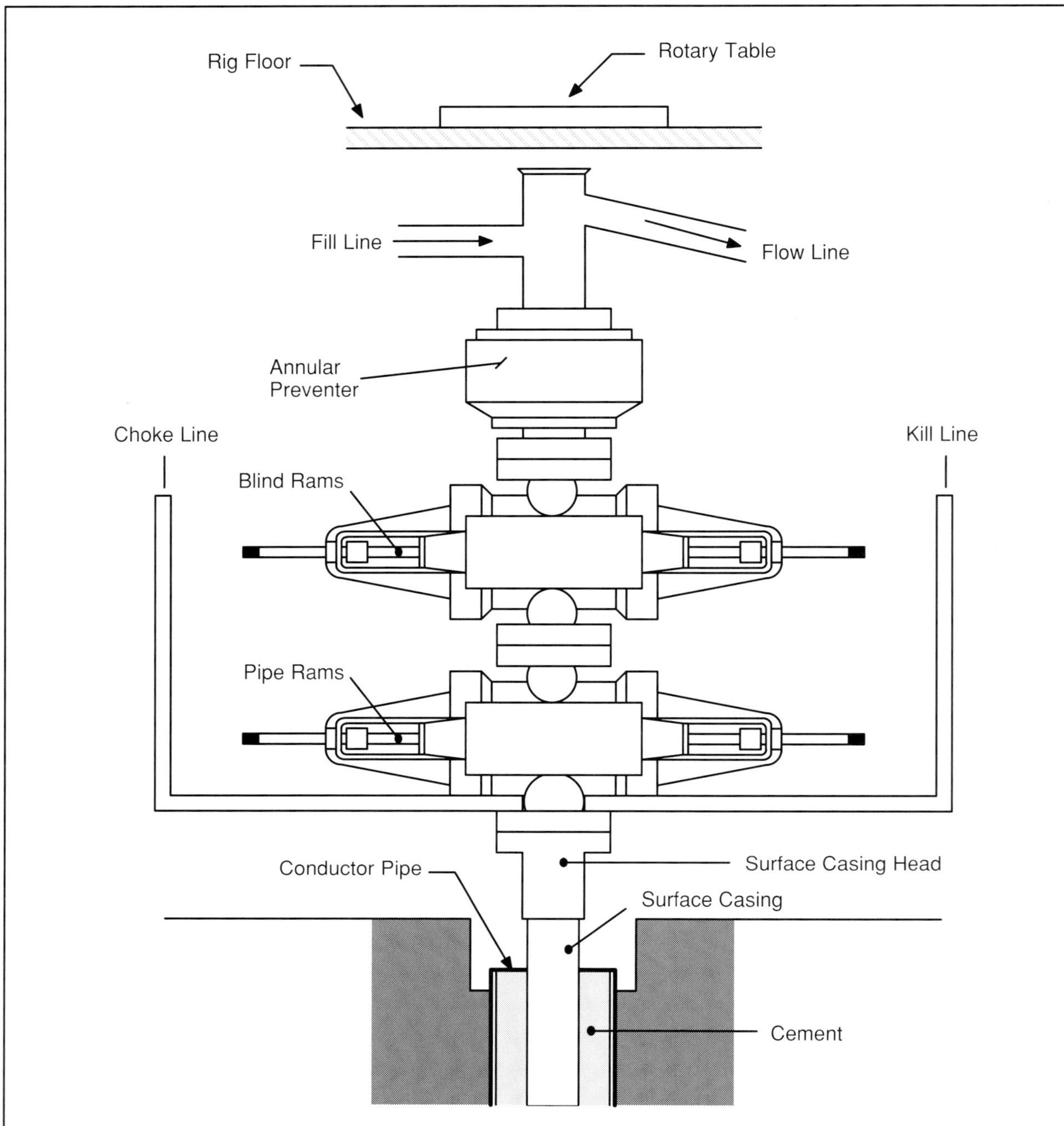

Figure 12–3—Surface casing support of blowout prevention equipment.

and the large annular cross-sectional area (even when considering gauge hole) often makes it difficult to achieve turbulent flow to assure efficient mud removal.

Plug flow should be used as an alternative displacement regime. Plug flow (plug-to-pipe ratio of 0.8) can be accomplished in large annular areas at relatively high pump rates, providing the cement slurries and spacers are designed with high yield points and low viscosities (Chapter 5).

12–2.3 Intermediate Casing

An intermediate casing is often necessary to maintain the borehole integrity as greater drilling depths are encountered. Typical casing sizes range from 6⅝ in. (17 cm) to 13⅜ in. (34 cm), and the depth can vary from 1,000 to 15,000 ft (305 to 4,570 m). Often, the intermediate casing string is the longest section of casing in the well. It is usually run to surface, and once again provides for the anchoring and connecting of BOPs for subsequent drilling.

Intermediate casing is generally employed to seal off weak zones that might fracture with the high-density mud usually needed when deepening a well, and prevent lost circulation. Occasionally, salt or anhydrite formations might cause drilling fluid contamination, or perhaps leach out to such an extent as to cause pipe sticking. Sometimes an intermediate string is used to seal off older producing zones to drill for deeper production. It can be used to protect the hole in deviated sections, and may also be necessary to hydraulically seal high-pressure (noncommercial) fluid zones which may be encountered well above the targeted pay zone.

An intermediate casing also affords better protection against well pressure than the surface string, owing to its smaller diameter. The setting depth of an intermediate string should be sufficient to reach formations which can hold the anticipated mud weight.

This casing string can be cemented in a single-stage primary cement job, but a multistage job is often performed because such a tall annular column of cement slurry would exert a hydrostatic pressure greater than the formation fracture pressure. Figure 12–2 is an illustration of a single-stage primary job. Figure 12–4 depicts a typical two-stage cement job on an intermediate casing string.

12–2.4 Production Casing

Setting this string of casing is one of the principal objectives when drilling a well. In many ways, the production string is the oil well. This string of casing serves to isolate the reservoir from undesirable fluids in the producing formation, and from other zones penetrated by the wellbore. It is the protective housing for the tubing and other equipment used in a well. Tubing may be pulled out of the hole for change or inspection, but the production string is cemented in place. In fact, the manner of cementing this string is usually subject to special attention to assure a pressure-tight bond between the formation and the casing. Common sizes range from $4^1/_2$ to $9^5/_8$ in. (11.5 to 24.5 cm). Depths can vary from 1,500 to over 25,000 ft (460 to over 7,620 m).

The production casing is normally run and cemented through a zone to be produced, and then perforated to allow communication with the formation. Sometimes it is set just above the zone, and an openhole completion is performed. The production casing is normally the last casing set in the well. It may be subjected to maximum well pressures and temperatures, and must be designed to withstand such conditions. The casing should be the best quality (usually the heaviest) pipe appropriate for the conditions involved. A small leak can develop into a blowout, so the threaded connections should be appropriate for the anticipated pressures. The casing joints should be carefully made up as the casing is run into the well to guard against future leaks.

The cementing of production casings is critical. The cement systems must be designed to safely keep the zone under control by providing adequate hydrostatic pressure. Preflushes and spacers run ahead of the slurry must also be checked to assure hydrostatic overbalance and to maintain well control at all times (Chapter 11).

Zonal isolation is imperative to protect the pay reservoir from fluid migration, and to isolate it hydraulically for any future stimulation treatments. In addition, the cement slurry must have adequate fluid-loss control to minimize the amount of filtrate lost to the zone. Good fluid-loss control avoids possible damage of the critical wellbore matrix, and also prevents premature dehydration of the slurry in the annulus, which could result in an annular bridging and a failed cement job. Fluid-loss rates should be less than 100 mL/30 min, and 50 mL/30 min should be strived for (Chapter 6).

An important property of the set cement is compressive strength, particularly across the pay zone. The set cement must also have low permeability to prevent fluid invasion. The rule of thumb for adequate zonal isolation is 1,000-psi (7.0-MPa) compressive strength, and less than 0.1-md water permeability. Strength retrogression must be prevented when the bottomhole static temperature (BHST) exceeds 230°F (110°C) (Chapter 9).

12–2.5 Tapered Casing Strings

It is common in casing-string design to vary the casing weights (internal diameters) within a nominal size range because of load considerations, cost savings, etc. These factors must be known when designing the cement job as burst and collapse ratings are affected, internal diameters vary, and thread connections may change within a particular string.

Another technique is to actually vary the nominal casing size. Combined strings such as $10^3/_4$ in. and 7 in. (27 cm and 18 cm) and $7^5/_8$ in. and 7 in. (19 cm and 18 cm) have been successfully used in completions on occasions. There could be various reasons for completing in this fashion, as the end result is similar to a liner type completion. The larger inside-diameter (ID) casing may be desirable in dual completions or in gas wells where additional tubular completion equipment is required, such as side-pocket injection mandrels, etc.

Figure 12–5 depicts a typical tapered string completion. This particular example is completed as illustrated

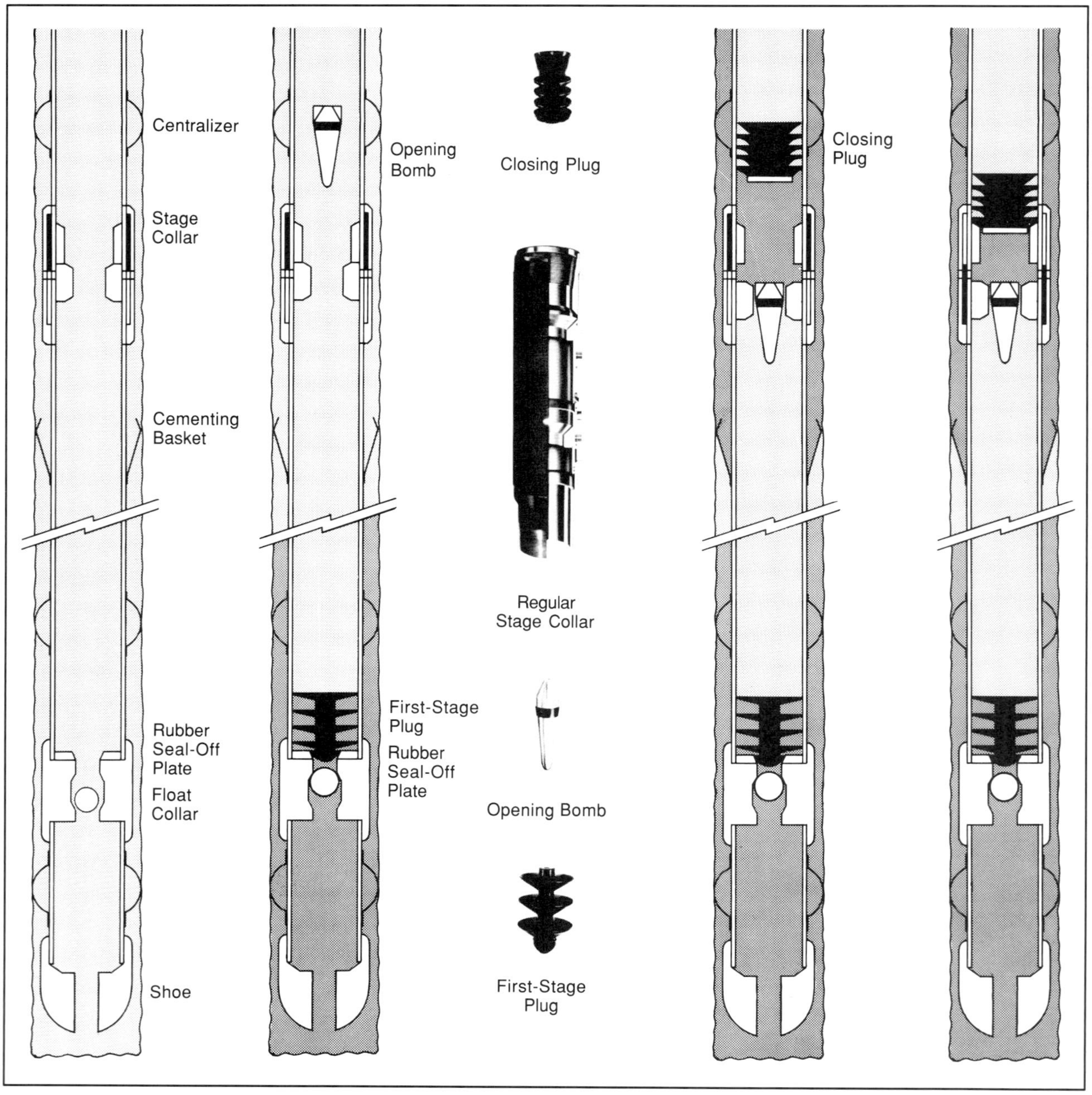

Figure 12–4—Two–stage cementing of an intermediate casing string.

to accommodate an additional tubing string which is run to the bottom of the 7⅝-in. (19-cm) casing, and allows for the circulation of hot oil in the upper portion of the string to prevent the waxing up of the crude produced from the zone penetrated by the 7-in. (18-cm) casing.

The only special considerations are for the displacement plug. The most common practice consists of modifying the displacement plug of the larger diameter casing. This is done by machining down the core of the plug to less than the ID of the smaller casing size. Thus, the wiper fins would aid displacement in the larger diameter casing. Because of their flexibility, they will fold and pass through the smaller casing size. First-stage wiper plugs of the type used in stage cementing could also be considered.

12–3 CEMENT PLACEMENT PROCEDURES

The vast majority of primary cement jobs is performed by pumping the cement slurry down through the casing and up the annulus. Other techniques also exist for solving various well-completion problems. For large-diameter casings, the traditional cementing technique is

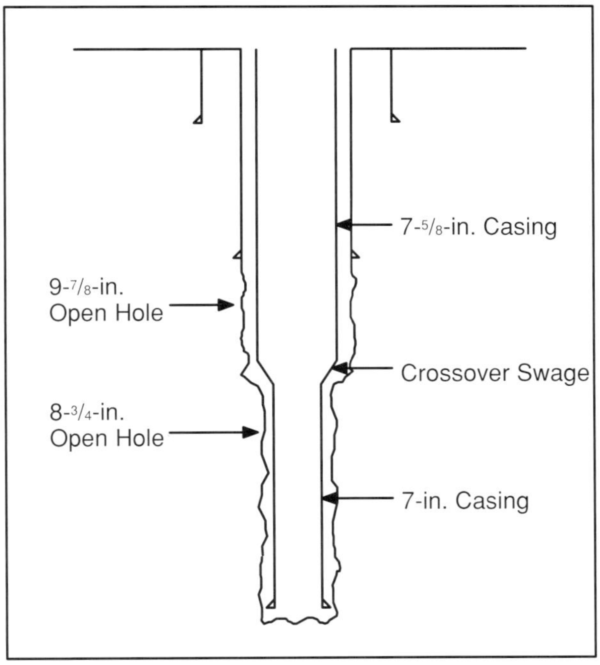

Figure 12–5—Typical tapered string completion.

frequently inadequate; consequently, cementing through the drillpipe or a grouting technique, where the cement is circulated into place by pumping the slurry down one or more small diameter pipes place in the annular gap, is performed. When cementing intermediate or production casings, well conditions and the length of interval to be cemented will decide the placement technique to be used. Usually, the maximum permissible downhole pressures determine whether a job should be performed in a single stage or in multistages. In this section, the most common procedures are described.

12–3.1 Cementing Through Drillpipe ("Stab-In" Cementing)

As discussed earlier, many problems related to the cementing of large casings can be prevented by performing the job through drillpipe. With this technique, the casing is run in place with a stab-in float shoe. The casing is set in the casing slips, thus suspending the string off bottom. Drillpipe made up with a stab-in stinger (Fig. 12–6) is then run in the casing until it is approximately 3 ft (1 m) above the float shoe. Circulation with the drilling fluid is then established, and returns are seen coming from the annulus between the drillpipe and the casing. Circulation is stopped, and the drillpipe is lowered, thus enabling the stinger to stab into and seal in the float shoe. Circulation again is broken, and returns should be observed flowing between the conductor pipe and the casing. Cement is mixed and pumped through the drillpipe and up the annulus until it reaches the surface. As soon as mud contamination is no longer evident in the cement returns, mixing can be stopped and the drillpipe volume displaced. If lost circulation is noticed before the cement reaches the surface, mixing should be stopped and the cement displaced, avoiding the pumping of large quantities of cement into the fractured zone. Care must be taken to avoid collapsing the casing because of excessive differential pressure between the outer annulus and the drillpipe/casing annular space.

Through-drillpipe cementing has several advantages. Accurate hole volumes (most often unknown in conductor or surface holes) are not required, as the cement slurry is mixed and pumped until returns are observed on the surface. This procedure optimizes the total volume of cement mixed and pumped. The subsequent volume displaced from the drillpipe is negligible. This method also eliminates the need for large-diameter swedges and/or cement heads, and also displacement plugs.

Various options are possible with the through-drillpipe stab-in technique. A backup check valve (float collar and float shoe) can be run as depicted in Fig. 12–6. Alternatively, a stab-in float shoe alone could be used. The types of available stab-in tools offer the possibility to latch into the float collar or shoe, thus preventing pump out of the stinger while cementing. Upon completion of the cementing operation, the drillpipe is rotated to the right several turns, and the coarse threads release the stab-in tool. Simpler stab-in tools are also commonly used which omit the latch-in design, and simply rely on the drillpipe weight to hold the stinger in place while cementing.

A further adaptation of through-drillpipe stab-in cementing is possible using a cementing mandrel as shown in Fig. 12–7, with drillpipe (or tubing) hanging freely to within 15 to 30 ft (4.6 to 9.2 m) of the shoe or collar. This type of arrangement offers all the previous advantages with the additional possibility of casing reciprocation. It also eliminates the possibility of casing collapse, because the pressures in the annulus and within the casing are equal.

12–3.2 Grouting ("Top-Up" Cementing)

When lost circulation occurs during large casing slurry displacement, the immediate solution is to recement down the annulus. A small-diameter tubing string is run down the annulus between the casing and the open hole (1⅞-in. [5-cm] tubing is a common size). Several joints can be screwed together, and pushed down the annulus as far as possible. The tubing string is then connected to the cementing unit through a high-pressure treating line, and circulation with drilling mud or water is established.

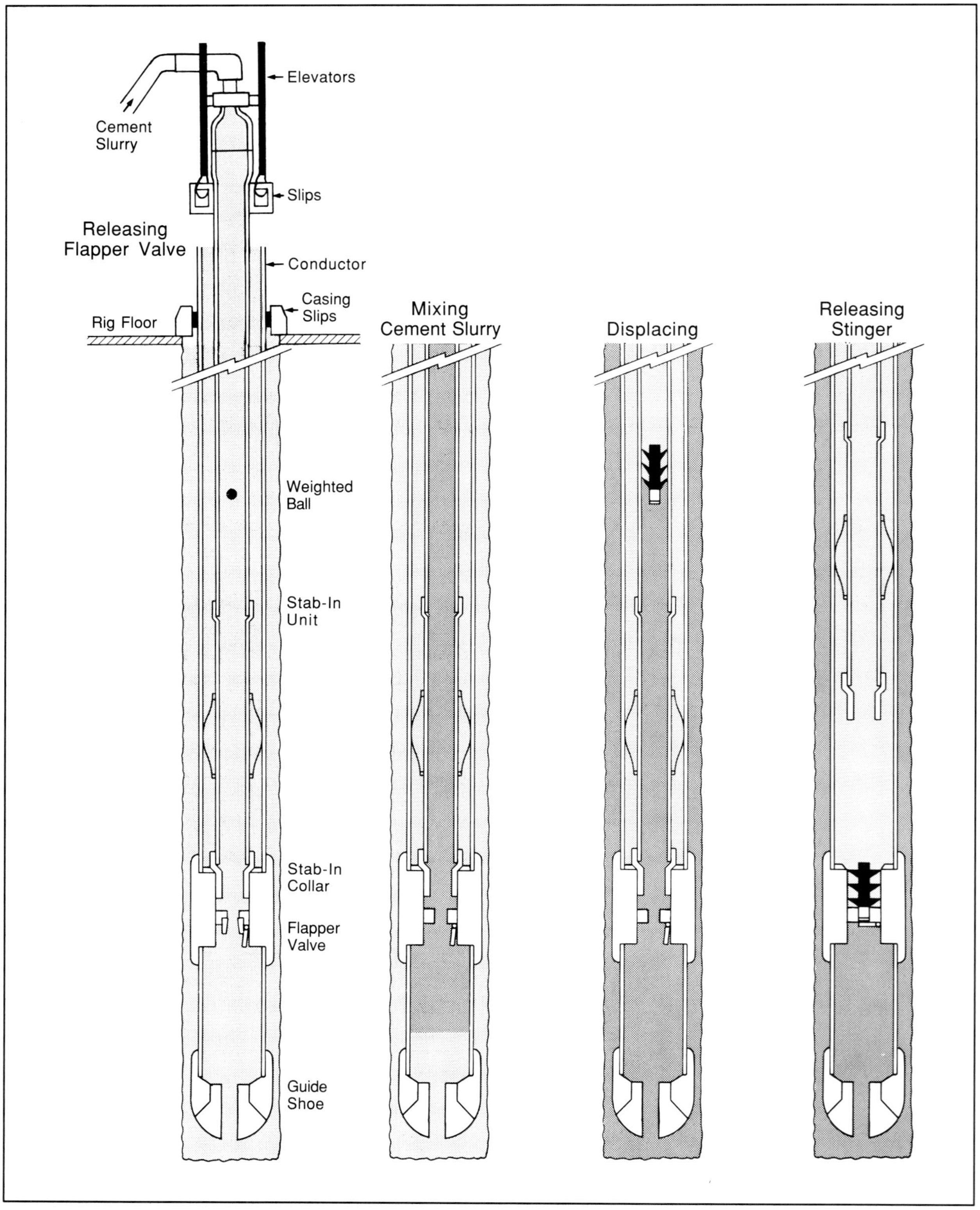

Figure 12–6—Through-drillpipe stab-in cementing.

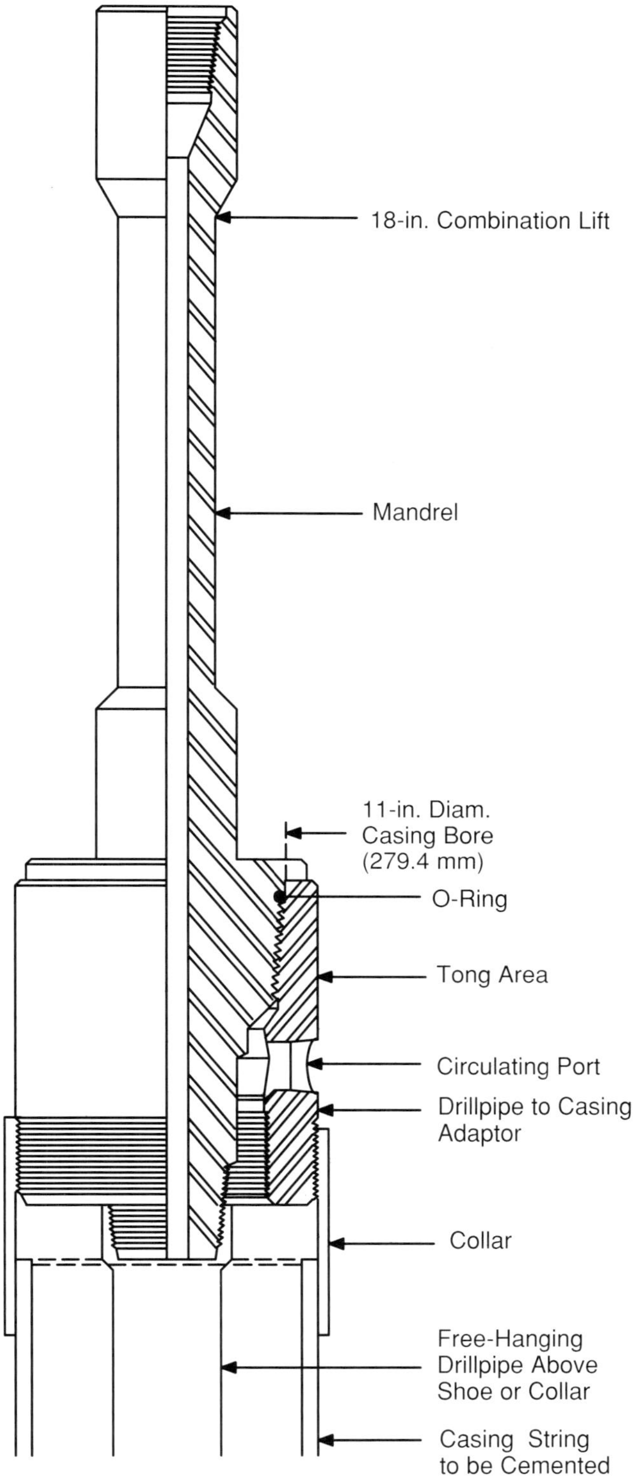

Figure 12–7—Cementing mandrel.

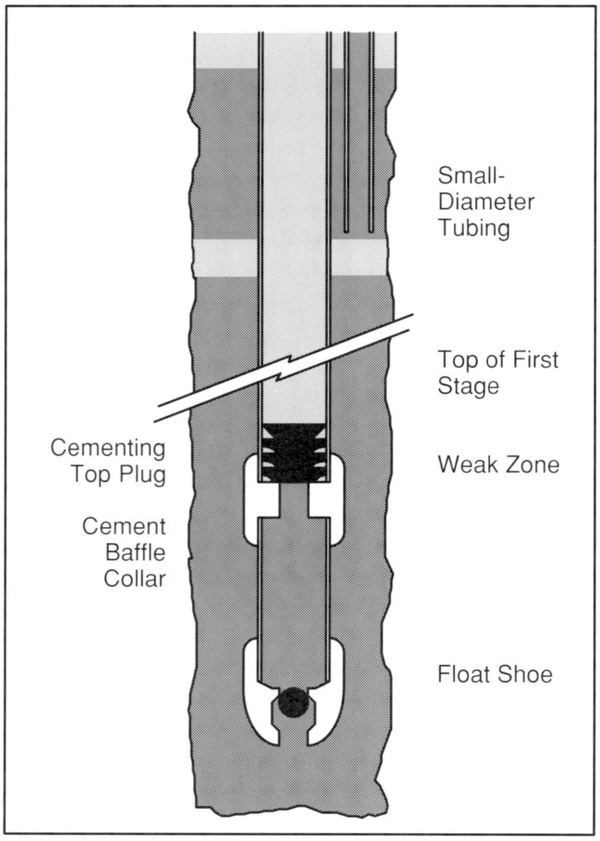

Figure 12–8—Top-up cementing.

Caution must be exercised as friction pressures will be high due to the small tubing ID. Cement slurry is then mixed and pumped in the conventional manner until cement slurry is circulated to the surface. The lines and tubing are flushed with water, and the tubing (if still hanging freely) is withdrawn from the annulus (Fig. 12–8).

The cement slurry can also be mixed and pumped directly into the annulus with the tubing string in place. In extreme cases, such cementations may have to be repeated several times until the cement slurry returns to the surface, and sufficient gel strength is built to support the slurry until it sets.

12–3.3 Single-Stage Cementing

With the development of new ultralow-density cement systems (Chapters 3 and 14), the need for multistage cementing has been reduced. A long column of microsphere-extended or foamed cement can often be placed in the annulus in one stage without the risk of breaking down weak formations.

12–3.3.1 Mud Conditioning

After the casing is in place, the mud is circulated as long as necessary to remove high-gel-strength mud pockets formed during the semistatic period of removing the drillpipe, logging, or running the casing (Chapter 5). Mud circulation is usually performed through the cement head to avoid stopping for an excessive period of time after the mud has been conditioned. Under static conditions, the development of mud gel strength can be fast, and may greatly reduce the mud removal efficiency.

If a single-plug cement head is used, circulation must be stopped prior to cementing to load both cement plugs. The bottom plug must be placed below the lower 2-in. (5-cm) inlet to allow room for the upper plug between the two inlets. If a double-plug cement head is used, both cement plugs can be loaded before starting mud circulation.

12–3.3.2 Bottom Cement Plug

The bottom cement plug serves two functions—(1) it prevents the intermixing of fluids, and (2) it sweeps clean the inner wall of the casing. The most obvious function of the bottom plug is to prevent slurry-mud contamination. However, if properly located, it can also contribute to the conservation of spacer properties. As discussed in Chapter 5, the ability of a spacer fluid to remove mud is crucially affected by its rheology. A small percentage of mud contamination could change its characteristics, reducing its effectiveness.

Another consideration with regard to the location of the bottom plug is the falling of heavy fluids (spacer or cement slurry) through lighter ones (chemical washes). This occurs during the displacement of such fluids in the casing, and its extent depends on the casing size and displacement rates. A bottom cement plug placed between the slurry and the wash will prevent unnecessary contamination due to the density differential effect.

When a spacer is used, sinking of the slurry through the spacer will not occur, because the difference in density is usually not very large. If a plug is run between the spacer and slurry, but not between the spacer and mud, the spacer will become contaminated with mud during the trip down the casing. In addition, the plug will sweep clean the casing wall, pushing ahead accumulated mud film which would contaminate the last part of the spacer. Once the bottom-plug diaphragm breaks, a mud-contaminated spacer will be in contact with the cement slurry—a situation the spacer was supposed to prevent. The ideal situation would be to use two bottom plugs to avoid intermixing of all fluids during the trip down the casing; however, with present plug containers, more than one shutdown would be necessary to load the plugs.

The following sequences are recommended when one bottom plug is used.

- Bottom Plug—Spacer—Cement Slurry
- Wash—Bottom Plug—Spacer—Cement Slurry
- Wash—Bottom Plug—Cement Slurry

12–3.3.3 Displacement Procedures

Dropping the top wiper plug is an easy operation, and should not take longer than the time needed to open and close the valves at the cement head. Cement heads are very reliable tools under normal working conditions, and are designed to minimize time delays. Stopping circulation for long periods of time (5 to 10 min) allows downhole fluids to develop high gel strength which could affect the final result in several ways.

- Additional applied pressure may be required to restart the movement of thixotropic fluids. In extreme cases, this pressure may overcome the fracture pressure, and lost circulation could be induced.
- Poor removal of regelled mud may occur, leading to poor bonding.

The spacer and slurries are then displaced through the casing, isolated between the two wiper plugs. In reality, because of U-tubing (Chapter 5), the top of the cement column may be at a considerable depth below the surface at the time the top plug is released. Depending upon the cement volume and density, and the mud density, the first part of the cement column might have already rounded the shoe. Such phenomena can be predicted by computer programs for cement job design (Chapter 11). The rate at which the cement is displaced into the annulus is not the same as the pumping rate; instead, it varies depending upon the different fluid densities and volumes. This phenomenon continues until the fluid level inside the casing reaches the surface. Continuous flow can then take place.

In general, there is a tendency to disregard the importance of displacement rates during the period of U-tubing. If turbulent flow has been programmed, the maximum pump rate possible is recommended during this period, as downhole fluid velocities are probably one-third to one-half of the surface pumping rate. If plug- or laminar-flow techniques are programmed, the surface rate must be controlled to maintain the desired flow regime. Again, computer programs can calculate the optimum pump rates according to the well geometry and fluid properties.

Once continuous flow takes place, the annular flow rate is equal to the pump rate, and the surface pressure begins to increase as the rest of the cement is placed behind the casing. The displacement then continues at the programmed rate until the top wiper plug bumps in the float collar. However, the pump rate is usually reduced at the end of the displacement to avoid too a sharp an increase in pressure when the plug reaches the collar.

Surface pressure is then released, and the wellhead is opened to test the functioning of the float equipment. If no returns are observed, the line is left open while waiting-on-cement (WOC) for the recommended curing time. If the float-collar valve fails, the fluid returned during the test must be pumped back into the well, leaving the casing pressurized until the cement gels and loses mobility. However, it is very important to release the casing pressure before the cement begins to develop compressive strength, to avoid the formation of a microannulus due to expansion and contraction of the casing.

12–3.4 Multiple-Stage Cementing

Multiple-stage cementing may be necessary for a variety of reasons.

- Downhole formations unable to support hydrostatic pressures exerted by a long column of cement,
- Upper zone to be cemented with (higher density, higher compressive strength) uncontaminated cement, and
- Cement not required between widely separated intervals.

Most of the reasons for multiple-stage cementing fall in the first category. At present, it is not uncommon to cement a longstring to the surface to protect the casing from corrosion. Alternatively, poorly plugged lost circulation zones below the last casing shoe often prevent cement slurries from reaching the surface. Two-stage cementing with the top of the first stage covering the weak zone will permit safe, complete filling of the total annular space.

Three standard multistage techniques are commonly employed.

- Regular two-stage cementing where the cementing of each stage is a distinct and separate operation,
- Continuous two-stage cementing with both stages cemented in one continuous operation, and
- Three-stage cementing where each stage is cemented as a separate operation.

The execution time of stage cementing increases the rig time. Consideration should also be given to the fact that most cement heads cannot accommodate the preloading of the plugs and bombs required in the sequence of operation. As a result, the cement head must be opened to release the opening bomb, assuming the first stage plug was preloaded. The shutoff plug could be loaded after the bomb is released, but caution should be exercised with the types of plugs and compatibility with the cement head. The fit of the plugs should always be carefully checked before the cement job to assure correct fitting of the plugs in the head.

12–3.4.1 Regular Two-Stage Cementing

In addition to conventional casing equipment (guide shoe, float collar, etc.), a stage cementing collar (Chapter 10) is run to the desired depth. There are several types of two-stage collars. It is important to be completely familiar with the operation of the selected type. It is also very important to follow the manufacturer's recommendations for operating the collars. Regardless of the type used, caution must be exercised in the initial handling of the stage collars, as the equipment is manufactured to close tolerances. Smooth sliding and sealing of the concentric sleeves is necessary for proper operation. Rough handling prior to or during installation can "egg" or misalign the moving parts, causing a failure during job execution. One must also be absolutely sure that the float collar and the stage collar are compatible. The first-stage wiper plug (if used) and the first-stage displacement plug must fit and seal against the float collar.

To explain the sequence of stage cementing operations, a brief explanation of the equipment is necessary (Fig. 12–4). Conventional stage equipment consists of the following.

- Stage cementing collar: basically a casing joint with ports which are opened and closed or sealed off by pressure-operated sleeves.
- Rubber sealoff plate: installed in the top float collar to assure a positive shutoff.
- First-stage plug: used to separate the slurry from the displacement fluid, it gives a positive indication of the end of displacement.
- Opening bomb: dropped after the first stage, it is allowed to gravitate to the opening seat in the stage collar; subsequent application of pressure will move the sleeve downward, opening the collar's ports.
- Closing plug: pumped to a shutoff on the closing seat.

Cementing the First Stage—The mixing and pumping of spacers and slurries during the first stage are similar to a single-stage job. After the slurry mixing, the first-stage plug is dropped and displaced until a positive indication of its landing in the float collar occurs. Some operators, when cementing production strings, displace the first stage using two fluids, leaving the casing below the stage collar filled with completion fluid and the upper casing filled with drilling mud. This mud is subsequently used to circulate the hole through the stage-collar ports.

Accurate hole volumes are necessary to determine the correct slurry height in the annulus. If the first-stage slurry covers the stage collar, it can be circulated out when the ports are opened. A caliper log should be mandatory on all multistage cementing jobs. Some types of stage collars allow the use of first-stage wiper plugs. First-stage displacement plugs are mandatory, and must be compatible with the original stage collar and the float collar. Plugs and stage collars from different manufacturers should never be mixed.

Cementing the Second Stage—After the first stage is completed, the opening bomb is dropped and allowed to fall by gravity to the lower seat in the stage collar. Once the bomb is seated, pressure is applied until the retaining pins are sheared, forcing the lower sleeve to move downward and uncover the ports. Usually 1,200 to 1,500 psi (8.4 to 10.5 MPa) will shear the retaining pins. A sudden drop in surface pressure indicates the opening of the ports. This operation could be performed at any time after the completion of the first stage, depending upon the design of the job. If a complete fill is scheduled, the cement from the first stage will be above the stage collar, and must be circulated out of the hole before it develops excessive gel strength.

Once the ports of the stage collar have been opened, the well must be circulated until the mud is conditioned for the second stage. For cementing the second stage, spacers and slurries are mixed as in any single-stage job. The closing plug is dropped after the slurry mixing, and is displaced to its seat in the stage collar. After the plug has seated, a minimum of 1,500 psi (10.5 MPa) above the second-stage displacing pressure is required to close the stage-collar ports. Pressure is usually released from the casing after the ports are closed.

Most second stages of two-stage jobs are performed using low-density filler slurries to allow circulation to the surface. Tail slurries are rarely used even if an openhole section is to be cemented. Protection of the weakest point in the casing string, the stage collar, can be improved by simply increasing the density of the last portion of the cement slurry.

In case of high incompatibility between the cement and mud, it may be desirable to run a wiper plug ahead of the slurry in the first stage. To do so, the following additional equipment must be used for a regular two-stage job.

- Flexible Plug: this special wiper plug is pumped ahead of the first-stage slurry.
- Bypass Insert: located above the float collar or float shoe, it provides a seat for the flexible plug, but allows continued circulation of slurry through its ports.
- Special Insert Collar: located one casing joint above the bypass insert, it provides a seat for the special first-stage plug which follows the cement.
- Special First-Stage Plug: provided with a special head to seal off in the insert collar, it replaces the first-stage plug in the regular stage equipment.

The sequence of operations is similar to the regular two-stage cementing procedure, except that the additional wiper plug is ahead of the first-stage slurry or spacer.

12–3.4.2 Continuous Two-Stage Cementing

Sometimes the situation demands that the cement be mixed and displaced without stopping to wait for an opening bomb to gravitate to the seat in the stage collar. This is known as the continuous-stage cementing method (Fig. 12–9).

The first stage of cement is mixed and pumped into the well. A wiper plug follows the cement to separate it from the displacement fluid. Following the plug, a measured volume of water or mud is pumped. This volume is calculated to displace the cement out of the casing below the stage collar. Allowance must be made for compression, pipe stretch, etc., so that the cement is not overdisplaced around the casing shoe. Overdisplacement is possible because a bypass insert is installed above the float collar on which the cement wiper plug lands. This insert prevents a shutoff when the plug lands, and permits some tolerance in the displacement fluid volume. After this displacement fluid has been pumped, the stage-collar opening plug is released.

The second stage of cement may be pumped immediately behind the opening plug. This slurry is followed by the closing plug. Displacement of this slurry will cause the opening plug to seat on the opening sleeve which will move downward when pressure is applied, opening the collar ports. Further pumping will displace the slurry through the ports and eventually land the closing plug on the closing seat. Application of 1,500 psi (10.5 MPa) above the circulation pressure will close the tool.

12–3.4.3 Three-Stage Cementing

Weak zones combined with gas channeling or potential casing corrosion problems in deep wells could require a three-stage cement job. The basic procedure does not differ from regular two-stage cementing; however, there is one additional stage (Fig. 12–10). The first stage is performed through the shoe, the second through a regular two-stage collar, and the final one through a top stage collar. The first stage is performed through the shoe in the conventional manner, using a first-stage plug to shut off at the float collar. The second stage could be performed at any time after the first, depending upon the cement program.

A regular opening bomb is used to open the ports of the lower stage collar. The well is circulated, and the spacers and slurries are pumped through the ports. The ports are closed using a special closing plug which replaces the regular closing plug. This flexible type of plug passes through the top stage collar, and seats on the lower collar, allowing the application of pressure to close the ports.

The final stage can also be carried out at any time after the second one. An opening bomb (larger than the one used for the second stage) is dropped and allowed to gravitate to the lower seat of the top stage collar. Ports are opened, and the final stage is performed as usual. A special closing plug is then used to close the collar ports.

12–4 LINERS

A liner is a string of standard casing which does not extend all the way to the surface, but is hung from inside the previous casing string. The overlap depends on the purpose of the liner, and could vary from 50 ft (15 m) for drilling liners to as long as 500 ft (152 m) for production liners. Liners can be classified as follows (Fig. 12–11).

- Production Liners: Run from the last casing to total depth, they replace production casing. Cementing is usually critical as zonal isolation is essential during production and any subsequent stimulation treatments that may be necessary.
- Drilling or Intermediate Liners: These are set primarily to case off and isolate zones of lost circulation, highly overpressured zones, sloughing shales, or plastic formations, so that drilling may be continued. Cementing these liners is often difficult due to the circumstances mentioned.

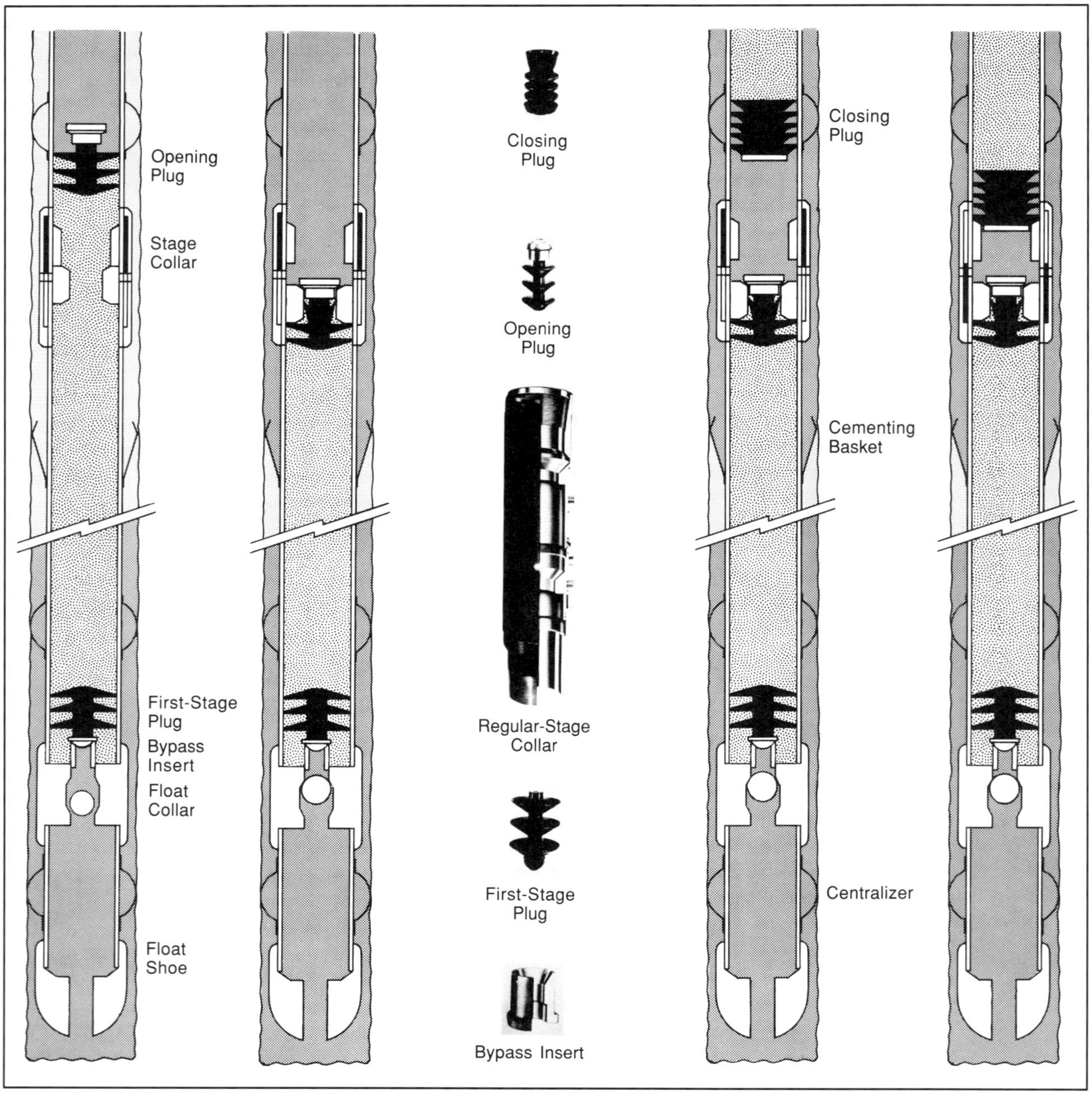

Figure 12–9—Continuous two-stage cementing.

- Tieback Stub Liners: These extend from the top of an existing liner to a point uphole inside another casing. They are generally used to repair damaged, worn, corroded, or deliberately perforated casing above the existing liner, and to provide additional protection against corrosion or pressure.

12–4.1 Running a Liner

The liner is assembled joint by joint at the rotary table, and lowered into the well, just as for a standard casing string. Float equipment is included, and sometimes a landing collar, one joint above the float collar, is used to provide a seat for the liner wiper plug. The dart and plug system must be compatible with the float collar.

Centralizers are critical in liner cementing. Because annular clearances are so small, the liner must be kept clear of the borehole wall. The mud displacement efficiency improves significantly with better centralization. Centralizers also help prevent the liner from differentially sticking while running in the hole, and also allow liner reciprocation and/or rotation during cement displacement.

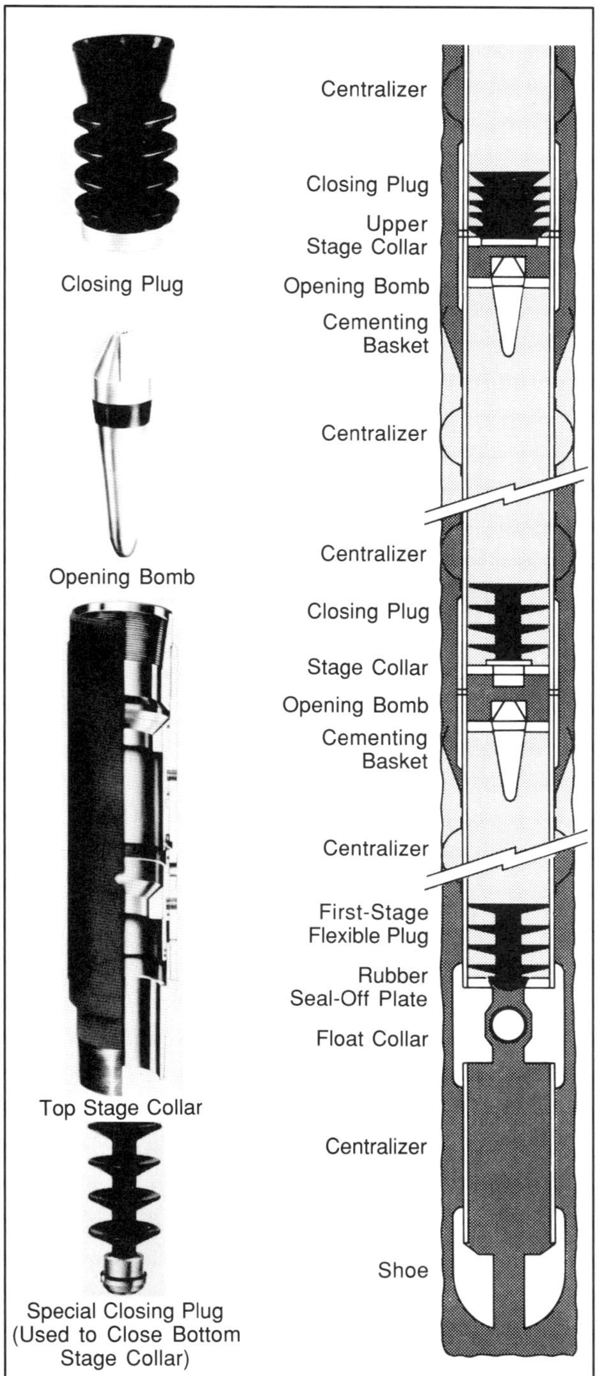

Figure 12–10—Three-stage cementing.

A liner hanger is installed at the top of the liner (Fig. 12–12). When set at the desired depth, this supports the weight of the liner string. Therefore, the liner is kept under tension which prevents it from buckling under its own weight. The liner hangers have slips which, when set, bite into the upper casing and provide the anchor to support the liner. The liner hanger remains permanently in place once the liner is cemented.

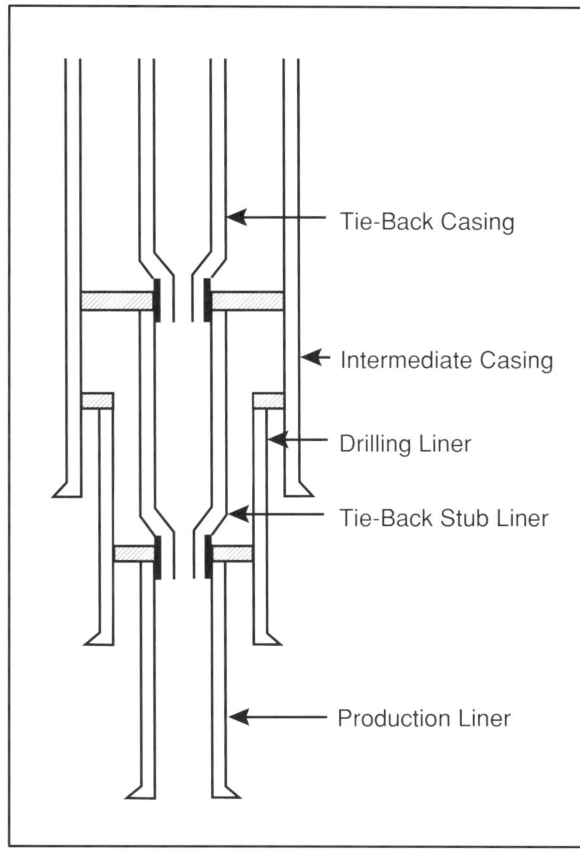

Figure 12–11—Types of liners.

Liners are usually run into the well using drillpipe and a special setting tool. This tool is retrievable, i.e., it is pulled out of the well with the drillpipe after the liner is run and cemented. It performs the following functions.

- It provides a pressure-tight seal between the drillpipe and the liner. Thus, fluids which are pumped into the drillpipe have to circulate down inside the liner and out of the shoe before returning up the annulus.
- It holds the weight of the liner as it is run into the well.
- It provides attachments for the liner wiper plug. The liner wiper plug, attached by shear pins, has a hole through its center to allow the passage of fluids and cement slurry until the "pumpdown drillpipe" plug closes it. Applied pressure will then shear the pins, and the wiper plug can be pumped down the liner behind the cement slurry.

With the liner at the desired depth, but before the hanger is set, connections are made and the liner and hole are completely circulated with the rig pumps. This

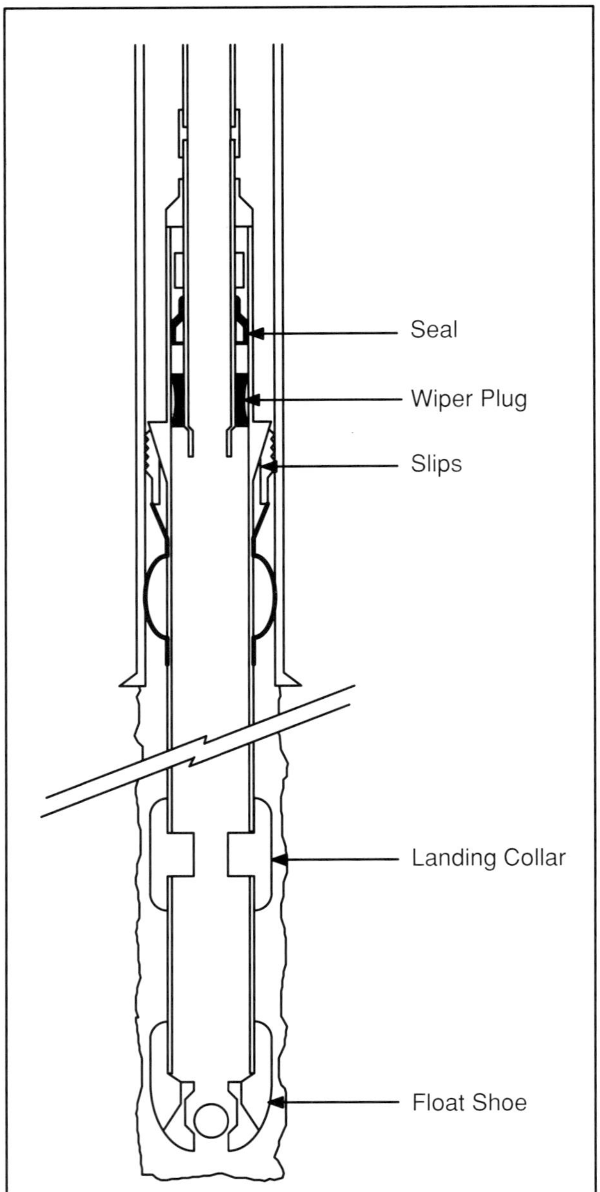

Figure 12–12—Liner setting tool and hanger assembly.

conditions the mud and ensures that circulation is possible before the liner is hung. In some deep liner setting assemblies, a circulation valve is included, which allows circulation to be established above the liner before closing the valve.

Once the mud is conditioned, the liner hanger is set, and the drillpipe and setting tool are then raised slightly to verify that the setting tool is released from the liner. The seal assembly holding the liner wiper plug is usually 10 to 15 ft (3.0 to 4.6 m) long to enable this operation to be performed without breaking the seal between the liner and the drillpipe. This operation must be performed to ensure that the drillpipe and the setting tool can be freed from the liner after the cement is in place.

12–4.2 Liner Cementing Procedures

12–4.2.1 Mud Removal

The success of any cement job depends upon the efficiency of the mud removal. Liner cementing can be one of the most difficult cases. Usually, the annular space is small, and the pipe may not be well centralized. This subject is covered thoroughly in Chapter 5; nevertheless, there are certain points which deserve reinforcement.

A 5-in. (13-cm) OD liner hung from a 7-in. (18-cm) casing inside a $6^1/_8$-in. (16-cm) drilled open hole will have a maximum clearance of $^9/_{16}$ in. (1.4 cm), if the liner is perfectly centered. In some parts of the hole, the annular clearance will be less because of a thin, nonremovable mudcake on the wall of permeable formations. Crooked hole and small clearances between the casing and formation often inhibit the use of centralizers, resulting in eccentration of liners. Under severe conditions, actual borehole contact occurs. Under these circumstances, it becomes much more difficult for cement slurries to remove mud.

It is for these reasons that pipe movement during displacement becomes critical. Bowman and Sherer (1988) reported that less than 20% of all liner jobs include plans to move the liner during cementing. There are many industry misconceptions about liner reciprocation and/or rotation—

- the fear of not becoming unlatched from the liner after cementing,
- a large/stronger drillstring may be required for fear of drill string parting during pipe movement,
- excessive drag caused by centralizers,
- swabbing or surging the pay zone,
- hole deterioration caused by moving pipe, which could lead to annulus bridging, and
- fear that the liner may become stuck and have to be cemented without the designed tension.

In fact, the advantages of liner movement during cementing far outweigh the drawbacks listed above. With the hole in good condition, and correctly selected centralizers on the liner, fewer problems would probably be experienced, and certainly better cementing results would be achieved. Bowman and Sherer (1988) stated that, in their study of over 300 liner jobs, the inability to release the liner setting tool had only occurred twice. One was caused by premature setting of the cement, and the other involved a very early tool design, which has been successfully modified.

In many instances, rotation has advantages over reciprocation. If the liner is in contact with the hole at any point, up-and-down motion would not remove the drilling fluid effectively. However, pipe rotation would allow slurry to be dragged behind the pipe, thus ensuring a sheath of cement around the liner. As stated by Bowman and Sherer (1988), the inability to rotate liners is often due to insufficient starting torque. Once this has been overcome, the torque required for rotation will probably be much less (assuming good centralizer design).

Because of the above problems, it is apparent that the use of adequate volumes of washes and spacers is even more critical in liner cementing than in casing cementing. Maximizing contact time generally increases the chances of a good cement bond, and washes and spacers should be used. In conventional casing strings, contact time can simply be improved by increasing the volume of the scavenger slurry. However, in a liner situation, slurry volumes can be critical, because of the formation hydrostatic limits.

Turbulent-flow-displacement techniques are definitely more efficient for mud removal than plug flow, but care must be taken not to exceed allowable downhole pressures. Fortunately, small annular clearances make it easier to accomplish turbulent flow at low pump rates (Chapter 5). If a job must be performed using plug- or laminar-flow regimes, spacer volumes may be designed to allow for the lower mud displacement efficiency of these techniques.

12–4.2.2 Regular Liner Cementing

The liner cement head and manifold are installed on the drillpipe with the "pumpdown" slurry displacement plug placed between the two inlets. The plug releasing stem holds the plug in the cement head (Fig. 12–13). After the cementing lines are rigged up and pressure tested, the chemical wash or spacer is pumped down the drillpipe. No bottom wiper plug is used ahead of the spacer or slurry. If possible, the cement slurry should be batch mixed to obtain a homogeneous slurry of desired density.

As shown in Fig. 12–14, once the slurry is mixed and pumped into the drillpipe, the pumpdown plug is dropped and displaced to the liner hanger. At this point, the pumpdown plug passes through the liner setting tool, and then latches into and seals the hole in the liner wiper plug. The surface pressure will rise as an indication of the plug landing. Further applied pressure of approximately 1,200 psi (8.4 MPa) will shear the pins holding the liner wiper plug in place. Once released, the two plugs move as one inside the liner as displacement is continued. When the internal volume of the liner has been completely displaced, the plugs seat on the float or landing collar, and a further pressure rise will occur, indicating completion of the job. Functioning of the float collar is tested after pressure is released by monitoring the returns.

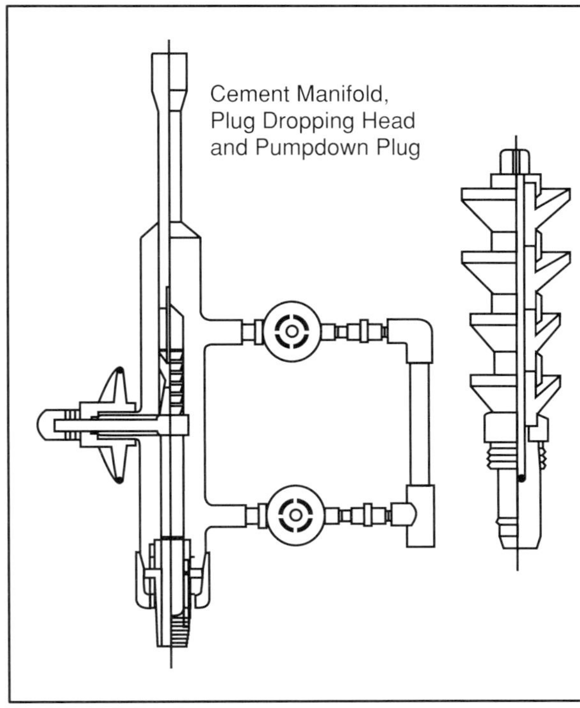

Figure 12–13—Liner cementing head.

If a packer-type liner hanger has been used, the packer between the liner and the upper casing is set at this time, the setting tool is pulled free from the liner hanger, and any excess cement is reversed out. If no packer is incorporated into the hanger, the reversing out depends on the quantity of excess cement expected, and whether lost circulation is anticipated. This is an important decision in liner cementing design, as proper isolation of the liner/casing annular space is critical.

The amount of cement excess must be carefully calculated by taking into account the well conditions and operator requirements. The following factors must be balanced.

- Sufficient excess cement must be planned for, if non-contaminated cement at the liner hanger is needed. A four-arm caliper should be run prior to the liner operation, and the slurry volume determined from the caliper logs. A recent study by Graves (1985) pointed out that hole volumes can change by as much as 31%.
- Displacement efficiency also becomes a key variable in determining cement slurry volumes; although

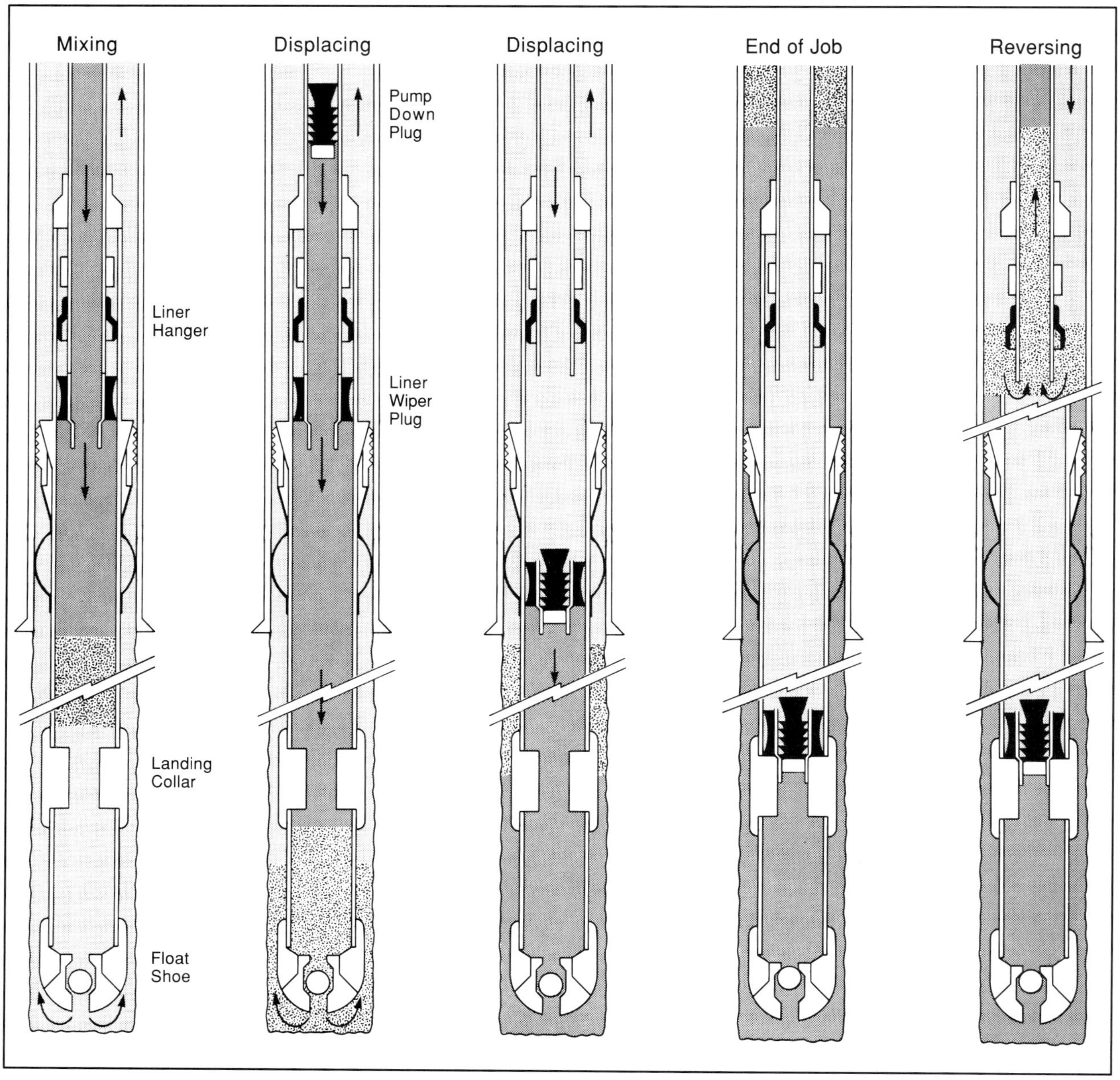

Figure 12–14—Liner cementing.

100% efficiency is the ideal, it is not uncommon to have 60% to 80% displacement efficiency in liner cementing (Smith, 1989). The volumes become more adversely affected the longer the pipe to be cemented.

- If excess slurry is to be reversed out, weak formations could pose a problem. The thickening time of such slurries must be extended to allow for the reversing operation.
- If reversing out is not scheduled, operators usually do not want to drill out long columns of cement; therefore, the excess slurry may have to be limited. This could definitely affect the quality of cement around the overlapped interval.

Once reversing out (or nonreversing) is completed, the setting tool and drillpipe are pulled leaving the cement to cure throughout the recommended WOC time. A checklist for running liners is published in API Bulletin D17, and is reproduced in Table 12–1.

The following procedure is taken from API *Bulletin D17*. The reader will note that it should be modified if the intent is to reciprocate or rotate the liner. The procedure is as follows:

1. Run drillpipe and circulate to condition hole for running liner. Temperature subs should be run on this trip if bottomhole circulating temperatures are not known. Drop hollow rabbit (drift) to check drillpipe ID for proper pumpdown plug clearance. On trip out of hole, accurately measure and isolate drillpipe to be used to run the liner. Tie off remaining drillpipe on other side of the racking board

2. Run_____ ft of ____ liner with float shoe and float collar spaced ____ joints above float collar. Volume between float shoe and plug landing collar is ____bbl. Sandblast joints comprising the lower 1,000 ft and upper 1,000 ft of the liner. Run thread-locking compound on float equipment and bottom eight joints of liner. Pump through the bottom eight joints to be certain that float equipment is working.

3. Fill each 1,000 ft of the liner while running, if automatic fill-up type equipment is not used.

4. Install liner hanger and setting tool assembly. Fill dead space (if packoff bushing is used in lieu of liner setting cups) between liner setting tool and liner hanger assembly with inert gel to prevent solids from settling around the setting tool.

5. Run liner on ____ (size, type connection, weight, and grade) drillpipe with ____ pounds minimum over pull rating. Run in hole at 1 to 2 minutes per stand in casing and 2 to 3 minutes per stand in open hole. Circulate last joint to bottom with cement manifold installed. Shut down pump. Hang liner five feet off bottom. Release liner setting tool and leave 10,000 pounds of drillpipe weight on setting tool and liner top.

6. Circulate bottoms-up with ____ barrels per minute to achieve ____ feet per minute annular velocity (approximately equal to previous annular velocities during drilling operations).

7. Cement liner as follows:________________.

8. If unable to continue circulation while cementing, due to plugging or bridging in liner and hole wall annular area, pump on annulus between drillpipe and liner to maximum ___ psi and attempt to remove bridge. Do not overpressure and fracture the formation. If unable to regain circulation, pull out of liner and reverse out any cement remaining in drillpipe.

9. Slow down pump rate just before pumpdown plug reaches liner wiper plug. Drillpipe capacity is ___ bbl. Watch for plug shear indication, recalculate or correct cement displacement, and continue plug displacement plus ____ bbl maximum over displacement.

10. If no indication of plug shear is apparent, plug calculated displacement volume plus ____ bbl (100% + 1% to 3%).

11. Pull out 8 to 10 drillpipe stands or above top of cement, whichever is greatest. Hold pressure on top of cement to prevent gas migration until cement sets.

12. Trip out of hole.

13. Wait-on-cement ____ hours.

14. Run ____-in. OD bit and fill cement to top of liner. Test liner overlap with differential test, if possible. Trip out of hole.

15. Run ____-in. OD bit or mill and drill out cement inside liner as necessary. Displace hole for further drilling. Spot perforating fluid (if in production liner) or other conditioning procedures as desired.

Table 12–1—Liner running procedures checklist (from Bowman and Sherer, 1988).

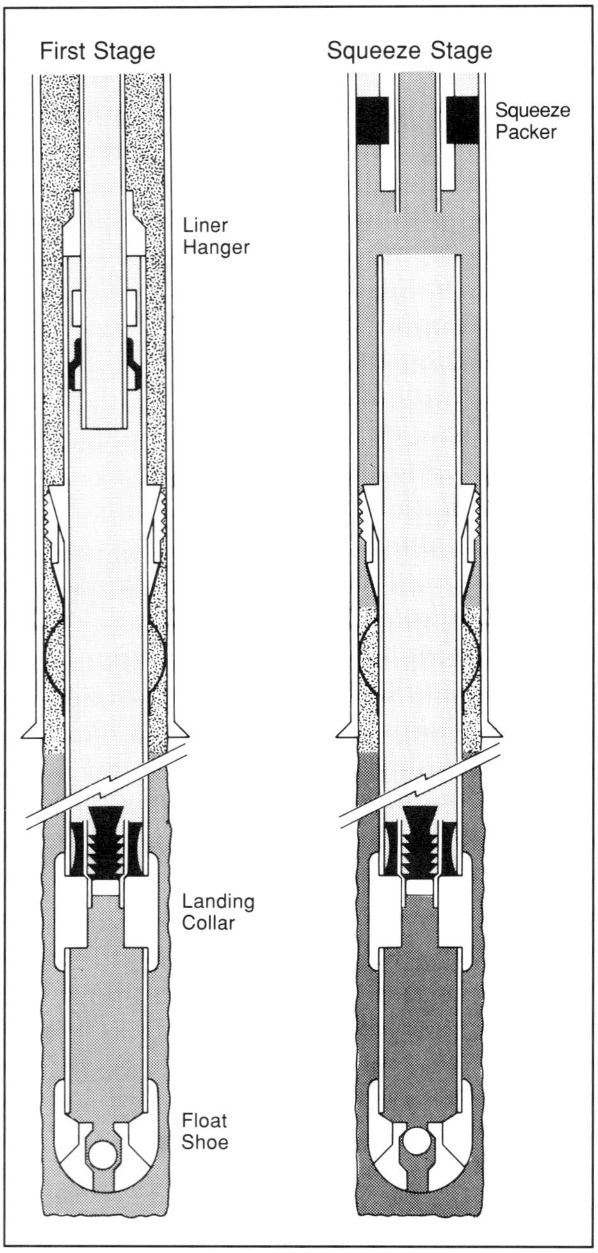

Figure 12–15—Liner cementing—planned squeeze.

12–4.2.3 Planned Squeeze

Long liners can be cemented in two stages when they traverse weak formations which would not withstand the hydrostatic pressure of a long column of cement. As illustrated in Fig. 12–15, the first stage is performed through the shoe using a limited, calculated amount of cement to cover the weak zone placing the cement top as close to the last casing shoe as possible. After the first stage is completed, the setting tool and drillpipe are pulled out of the hole, and the cement is allowed to cure. Drillpipe with a standard squeeze packer is then run into the hole, and the packer is set two or three joints above the liner hanger. The second stage is then performed by squeezing a premixed amount of cement around the liner hanger. To be able to squeeze cement into the annular space, the formation fracture pressure in the openhole section must be overcome.

The advantages of the method are—

- avoids damaging weak productive formations,
- uncontaminated cement is placed at the liner hanger, and
- no excess cement is necessary.

Disadvantages include—

- the complete annular space may not be cemented, and
- the technique is more expensive.

With the advent of ultralow-density microsphere and foamed cement systems, this procedure is obsolete.

12–4.2.4 Waiting-on-cement (WOC) for Liners

When long liners are to be set, there may be a considerable temperature differential between the bottom and top of the liner. A slurry designed to have sufficient thickening time at the total depth may take a very long time to set at the liner top. Drilling of cement must be done after the cement develops the minimum compressive strength to withstand the shock caused by drilling tools.

12–4.2.5 Tieback Liners

There are situations when it may be necessary to extend an existing liner further uphole, with a tieback "stub" liner, or to surface with a tieback casing string. Some of the reasons for running tieback stub liners or tieback casing are—

- to cover up damaged casing above the top of an existing liner,
- the need for a bigger casing on top of the existing liner to allow for multiple production strings,
- selective testing of multiple zones to design future production assemblies and production casing size, and
- cementing of troublesome intervals (high pressure, sloughing shales, etc.) before running the casing string to surface.

To accomplish this, special tools to connect the two liner strings must be used.

Tieback Sleeve—Installed on top of the liner hanger, the tieback sleeve provides a receptacle for the sealing nipple. Its internal surface is usually polished and beveled on the top to guide the entry of the different tools used during the operation.

Tieback Sealing Nipple—Run at the bottom of the tieback stub liner or casing, the tieback sealing nipple has multiple packing elements which provide a seal against the polished surface of the tieback sleeve.

Tieback casings are usually cemented by conventionally circulating the slurries. The job is performed before landing the seal nipple into the tieback sleeve. However, the cementing may also be conducted with the tieback casing in place, using a stage collar located above the sealing nipple.

Tieback liners must be cemented after their liner hangers have been set, and with the seal nipple landed into the sleeve (Fig. 12–16). A stage collar can be run on top of the seal nipple, in the open position. The liner wiper plug must be able to land on the upper seal and close the collar ports.

Apart from the special procedures given above, the considerations applicable to all cement jobs equally apply to tieback liner cementing. In most cases, hydrostatic pressures are not significant because cementing is done between casings and usually with extended slurries.

The use of washes ahead of cement slurries will prevent mud/cement contamination and help to remove the mud from the annular space. This is especially important in tieback liner cementing, where no bottom plug is used to separate the mud from cement inside the liner. If a completion fluid is in the hole, compatibility with the cement must be checked or large volumes of fresh water be pumped ahead of the slurry. Salts used in completion brines may drastically affect a cement slurry's thickening time, causing a premature set or, conversely, resulting in excessively long times for the development of early compressive strength.

12–5 SPECIAL OFFSHORE TECHNIQUES

As discussed in Chapter 10, the logistics of offshore cementing operations can be much different from those for land-based operations, but the cementing procedure employed on offshore drilling rigs or platforms fixed to the seabed is very similar to primary cementing operations on land. However, considerable differences exist in the plug release technique from floating drilling vessels, and where there are subsea completions.

Figure 12–17 illustrates the system, which consists of a special subsea assembly located in the casing below the casing hanger, and the cementing head on the floating drilling vessel which is screwed onto the drillpipe and controls the cementing plug release. The head contains a launching ball and dart, while the subsea assembly con-

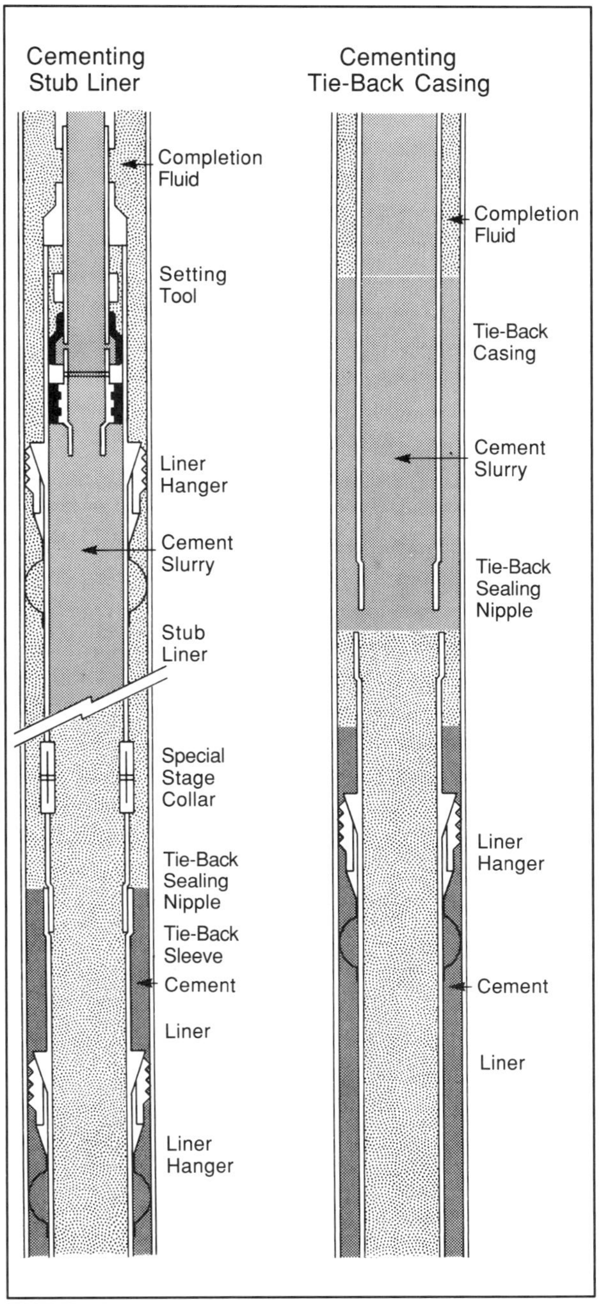

Figure 12–16—Tieback liner cementing.

tains the top and bottom casing plugs. Referring to Fig. 12–17, and by chronological order of usage, (b) is the bottom plug launching ball which, when released before pumping the cement slurry, seats in the bottom plug (e). A 100- to 275-psi (0.7- to 1.9-MPa) pressure increase allows the connector pins to be sheared (d), and permits the bottom plug (e) to travel down the casing until it bumps on the float collar and casing shoe.

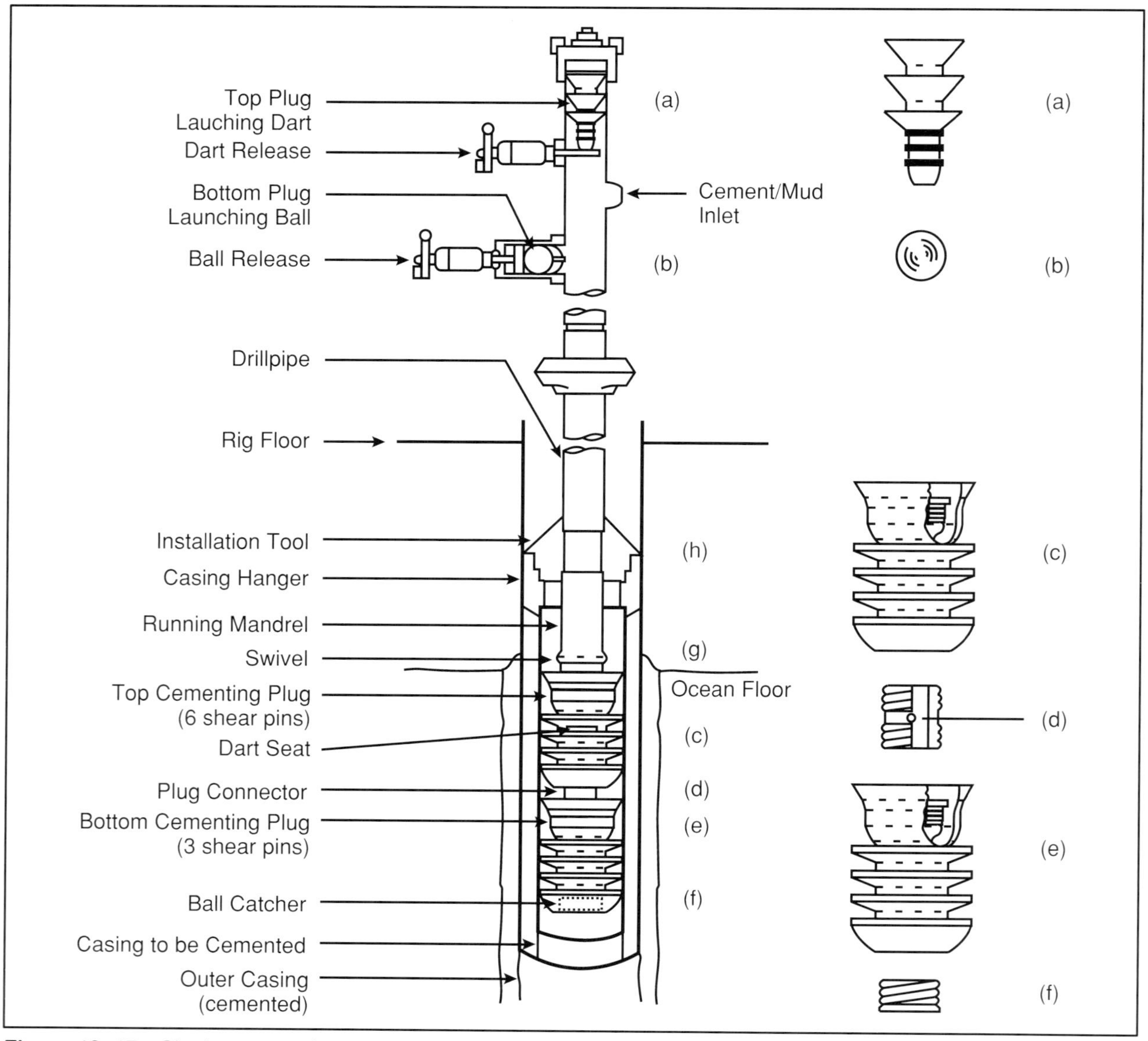

Figure 12–17—Single-stage subsea cementing system.

Extra pump or hydrostatic pressure extrudes the ball (b) through its orifice seat, and cement displacement continues. A ball catcher attached to the lower end of the bottomhole plug retains the ball.

Once the cement slurry has been pumped, the top plug launching dart (a) is released. It will seat into the body of the top cement plug (c). Increased circulation pressure will then shear the retaining pins and release the top plug (c) from the launching mandrel. The cementing operation thus continues. At the end of the slurry displacement, the top plug (c) bumps on the float collar or casing shoe.

In the subsea assembly before cementing, the top plug is pinned at the lower end of a running mandrel which has a swivel (g). This avoids any rotation of the cementing plugs inside the casing, which could damage the shear pins. The upper part of the mandrel is screwed onto the lower part of the installation tool (h). This installation tool is a crossover that adapts to the casing hanger, and serves to attach it to the drillpipe. One of the major limitations, and often a source of cementing failure, is the reduced flow area through the plug-retaining mandrel. This restricted flow area is susceptible to fluid erosion and failure due to high pump rates (often required for turbulent flow) and large fluid volumes of either mud or cement.

Other significant points to consider in subsea cementing are hydrostatic pressures and temperature. The additional column of fluid equal to the water depth can be a significant factor, as is the temperature at the sea bottom and the first several hundred meters of hole.

In addition to these, a complete understanding of the subsea wellbore geometry is necessary to assure full well control during the entire cementing operation. Figure 12–18 illustrates the general arrangement of the subsea BOP system.

12–6 OPERATIONAL CONSIDERATIONS

Planning is basic to successful primary cementing. It begins with accurate knowledge of the well conditions. The cement job is designed for these conditions, and job parameters must be monitored and recorded during job execution, so that the actual job can be compared to the design.

12–6.1 Calculations

Because of the difficulty of calipering large open holes, surface casing hole volumes are rarely known. The volume of cement slurry then has to be based on common field practice in the area. If this is not known, excess slurry volumes of 50% to 100% should be used. Excess slurry volumes of up to 200% are common in some areas.

Even when a caliper is run, and the theoretical volume is calculated, an excess volume is often required to assure proper fill-up. Local experience may dictate the excess cement required. Up to 50% in excess of the calipered hole volume may be used. In many countries, the volumes of slurries are governed by regulations which can be very comprehensive (Chapter 11).

12–6.2 Hole Condition

In addition to the physical parameters of the hole (i.e., depth, diameter, direction, etc.), the drilling log should be reviewed to identify potential problems which would affect the cement job. Hole washouts, lost circulation, tight spots, etc., all should be noted and compensated for in the design. A caliper log should be mandatory on most jobs. Drilling mud type and properties have a significant effect on hole condition, conditioning the hole prior to the cement job, and the cement slurry. Drilling muds must be designed for good primary cement jobs. If this is not always possible, then intermediate muds should be mixed specifically to satisfy cementing requirements.

12–6.3 Temperature

Knowledge of the bottomhole circulating temperature (BHCT) is vital. The cement slurry pumping time is a direct function of the hole temperature. Extended slurry pumping times can be as disastrous to primary cementing as can too short a time. Temperature also affects the cement and mud rheology; consequently, the flow regimes, U-tube effect, and friction pressures are all directly affected (Chapters 4 and 5). The BHCT must be determined if unknown. This can be done through logging, circulating temperature probes (Jones, 1986), or mathematical simulation of circulating temperature (Beirute, 1988; Mitchell and Wedelich III, 1989).

12–6.4 Pressure

Accurate knowledge of downhole pressure is necessary for well control and successful primary cementing. Slurry density is required for well control and set cement strength. Too high a density will lead to fractured formations and lost circulation. A typical intermediate casing string cement job, and the minimum and maximum hydrostatic pressures, are shown in Fig. 12–19. This type of plot should be generated for all primary cement jobs (Chapter 11).

12–6.5 Quality Control

A definite quality control program should be employed to test all materials prior to cementing. Laboratory conditions should closely simulate the job as is possible from known well conditions. Actual field batch samples of cement, additives, and mix water should be used for testing.

As API specifications for cements are necessarily broad in scope, additional testing should be performed whenever consistent quality of the cement is suspect. API rheology tests may help to identify potential problems. Liquid additives should also be checked and thoroughly blended with the mix water prior to cementing. Certain dry additives are prone to separation (particularly weighting agents), and care should be taken to verify that proper blending with the dry cement exists prior to the job (Gerke et al., 1985).

12–6.6 Casing Movement

As stated in the Amoco cementing guidelines (Smith, 1982), "The best aid to moving casing is the will to do it." Casing movement, reciprocation, rotation, or all three, positively improves the quality of primary cement jobs (Fig. 12–20). Casing movement breaks up areas of stagnant mud which can cause cement channeling. Scratchers and wipers are of little benefit, unless they are put to work by casing movement.

Casing, once landed, which cannot be moved prior to cementing, is a positive indication that something is wrong. Often, not much can be done at this point other than to cement the casing in place; however, the chances of a successful cement job are diminished before even mixing the slurry.

12–6.7 Cement Job Monitoring

The recording of critical parameters during cementing is paramount. Accurate knowledge of pressure, slurry rate,

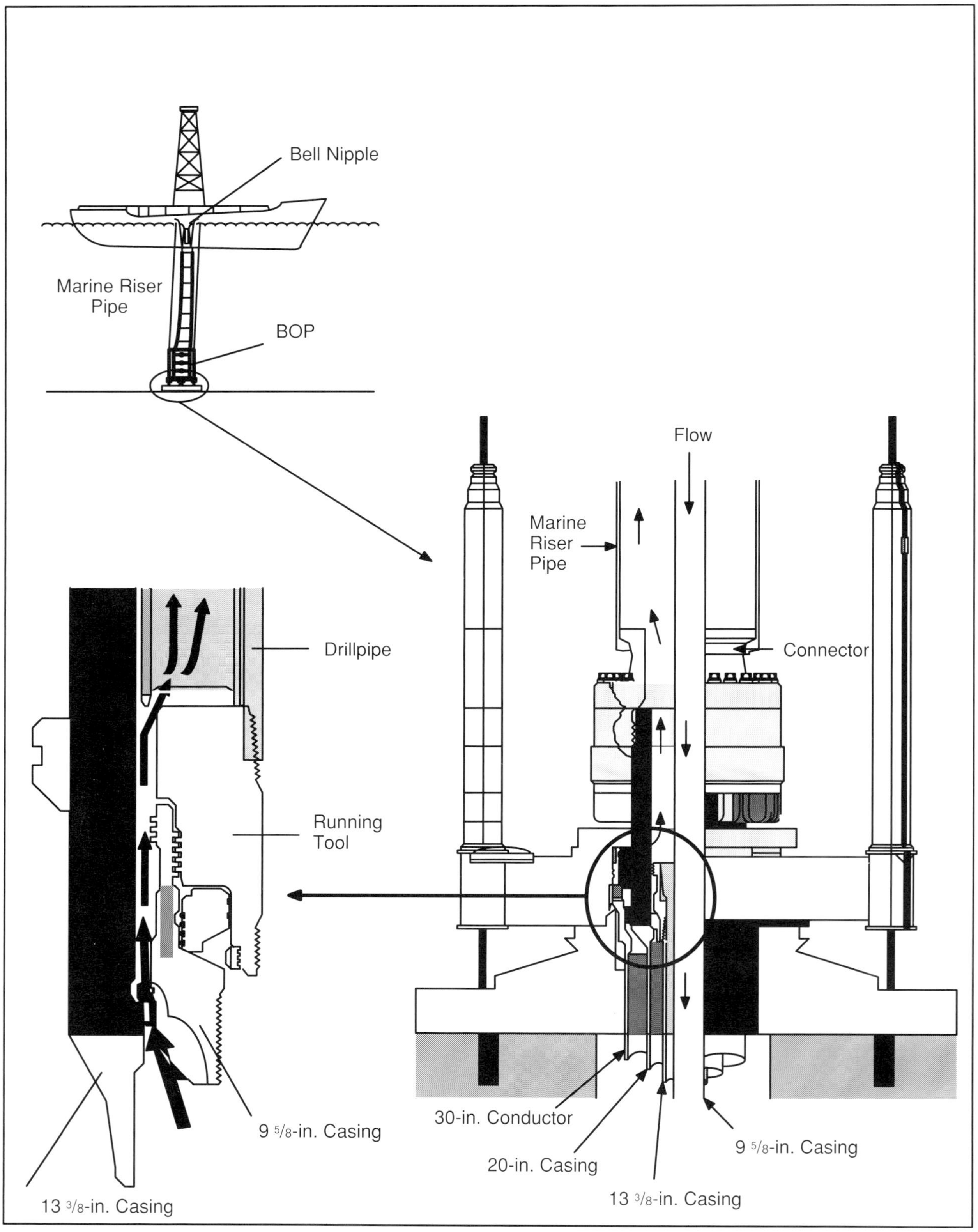

Figure 12–18—Wellbore geometry with a floating drilling vessel.

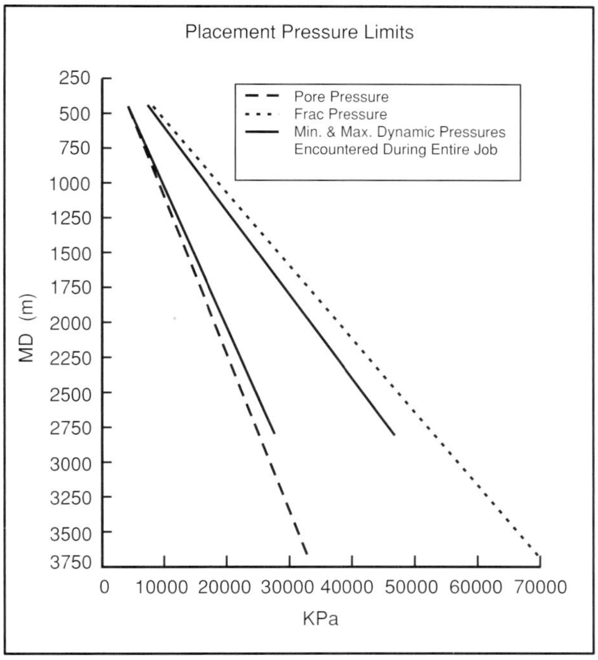

Figure 12–19—Pressure plot for intermediate casing string cement job.

and density, along with the integrated volume, are factors that must be known to the cementer and operating company representative in real time. These data should also be recorded so that future playback and analysis are possible, to evaluate and optimize the design of future jobs.

A typical recording device output is shown in Fig. 12–21. The recording of the vital parameters has greatly improved the success of primary cementing. All operations on the job are recorded, and subsequent review can be performed to evaluate the primary job, and compare the actual job to that designed. Recording devices also verify that the correct volumes and densities of preflushes, spacers, and cement slurry were pumped into the well.

The sensor package is equally important. Pressure is more routinely monitored with more accurate and faster responding electronic transducers replacing the traditional hydraulic gauges. Flowmeters working on an electromagnetic principle have been recently introduced, and are capable of measuring nonconductive oil-base fluids as well as conductive aqueous systems. Currently, most flow measurements are obtained from the drive shafts of

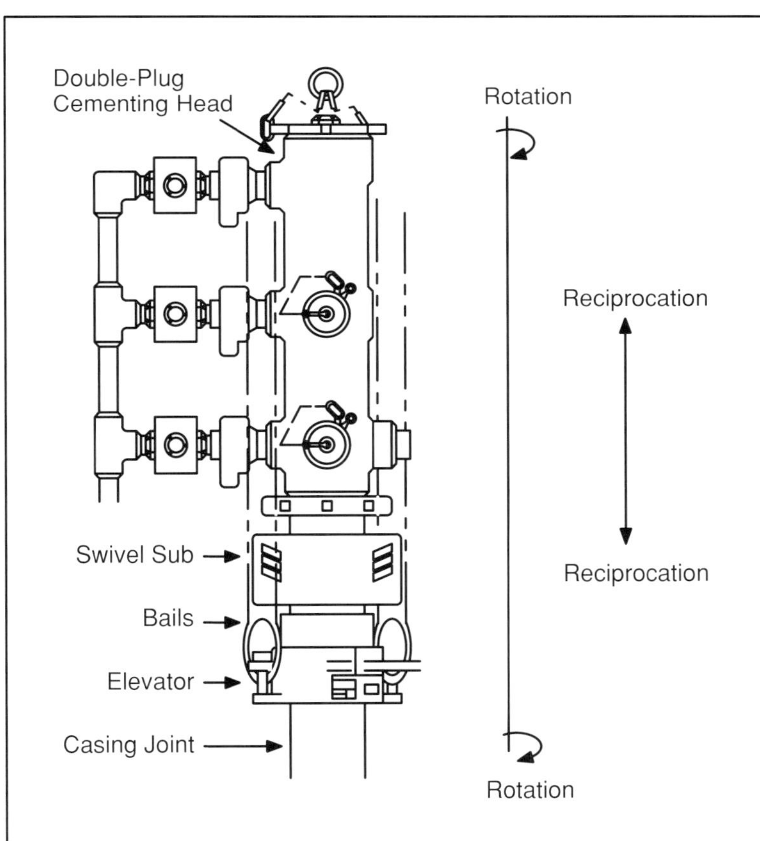

Figure 12–20—Rotating and reciprocating casing during cementing.

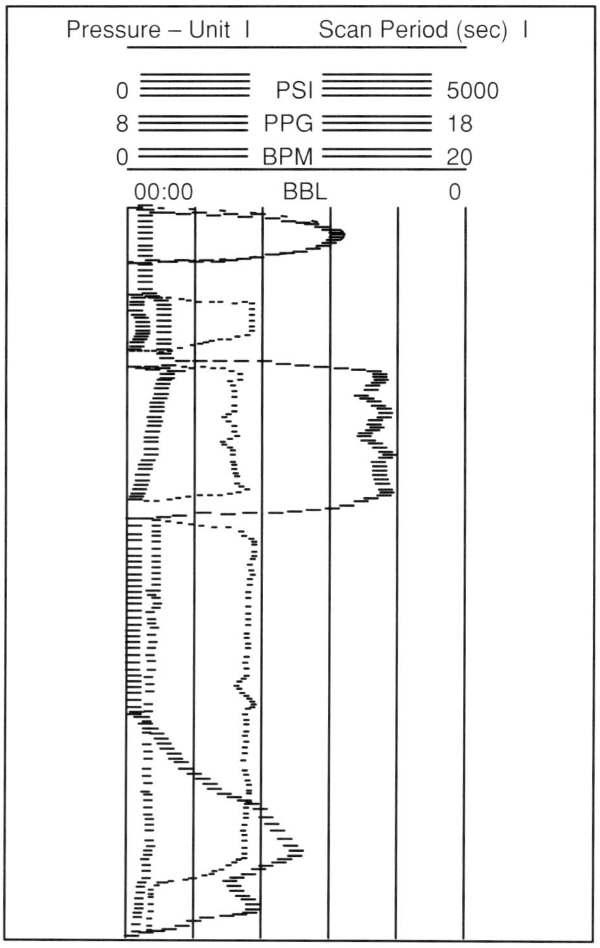

Figure 12–21—Recording device output for cement jobs.

the positive displacement pumps, or by counting the reciprocation of the pump plungers. The densities of the fluids are more accurately recorded with fully enclosed radioactive densitometers.

12–6.8 Casing Connections and Completions

When the conductor pipe has been driven to the desired depth, a mud return line or flowline is welded underneath the rig floor so that mud can return to the pits. The hole is then drilled to the depth required for the surface casing. After it has been run and cemented (always to surface), it is cut off underneath the rig floor at the desired height. The casing head (which will enable the next size of casing to be hung) is then welded to the surface casing, inside and outside. Some are available which screw onto a casing thread.

The BOPs with the connections for the kill line and choke line are flanged onto the casing head. Only one annular preventer may be attached, at this point, or a full set including blind rams and pipe rams. Before drilling can continue, the BOPs must be tested to the desired pressure. A sealing plug is run into the casing on drillpipe, and the BOPs are closed one by one. Pressure is then applied through the kill line.

All the BOPs and the wellhead connections must hold pressure before drilling can continue (Fig. 12–22). Then, when the next size of hole has been drilled to the desired depth, the next size of casing is run in and cemented. Cement may or may not be required to reach the surface.

The cement is usually allowed to set up while the casing is hanging from the elevators. After it has set, while the casing is still hung from the elevators, the BOPs are unscrewed from the casing head and suspended from the substructure. Slips are then set between the casing and the casing head.

It is very important that the casing be set with the same weight hanging from the slips that were taken by the elevator, to avoid buckling of the casing downhole. The casing can then be cut off either level with the casinghead flange or one or two feet higher. A sealing mechanism is normally placed above the slips to seal the annulus between the two casing strings. Then a new casing head is flanged to the previous head, the BOPs are reattached (or replaced by BOPs with a higher pressure rating), and (after testing) drilling can recommence. This entire process is known as "nippling up."

In this way, each time a new string of casing is run, it is hung from a casing head that was attached to the previous casing head. The production casing will have a head from which to hang the tubing—the tubing head. Therefore,

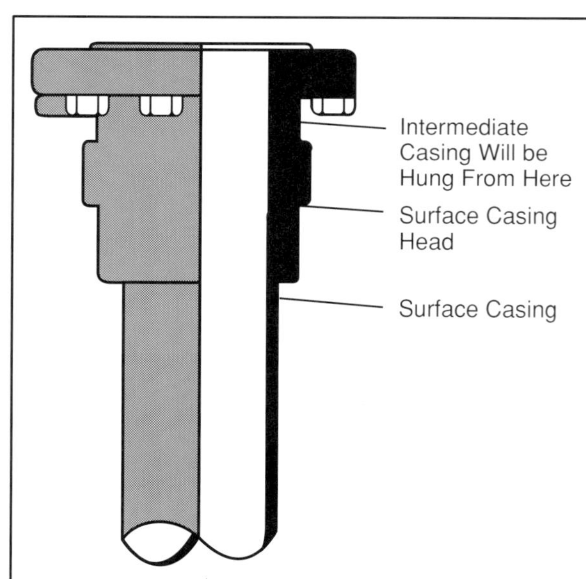

Figure 12–22—Connection of casing strings.

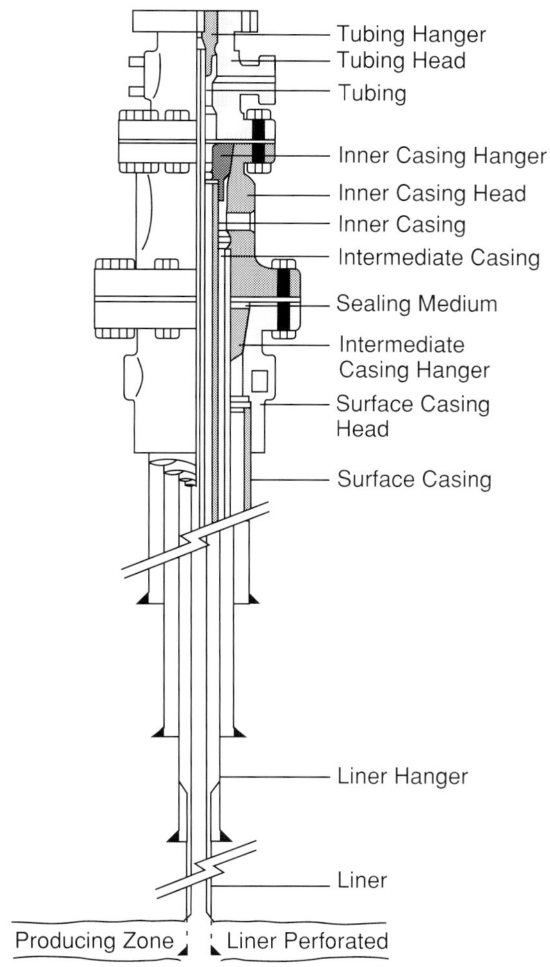

Figure 12–23—Typical wellhead assembly on a production well.

the weight of all the strings is partly supported by the surface casing (Fig. 12–23).

12–7 CONCLUSIONS

The basic mechanics of the more common primary cementing techniques has been presented in this chapter. When decoupled from the other related issues, such as fluid rheology, cement slurry design, cement slurry mixing procedures, annular gas migration, etc., these procedures may appear to be very simple. Such an impression is deceptive. It is essential that the engineer be intimately familiar with the procedures and devices which are used for primary cementing. In addition, it is critical that the engineer verify that all equipment is in proper working order before the job; otherwise, the long and meticulous planning process before each job may be wasted.

REFERENCES AND SUGGESTED READING

API Mid-Continent Dist. Study Committee on Casing Programs: "Casing String Design Factors," API paper 851-29-I (March 1955).

Beirute, R. M.: "A Circulating and Shut-in Well Temperatures Profile Simulator," paper SPE 17591, 1988.

Bowers, C. N.: "Design of Casing Strings," paper SPE 514-G, 1955.

Bowman, G.R. and Sherer, B.: "How to Run and Cement Liners," *World Oil* (July 1988) 84, 86–88.

Bulatov, A. I.: *Plugging Materials and the Cementing of Wells*, second edition, (1985).

Dowell Schlumberger, *Cementing Technology*: Nova Communications Ltd., London (1984).

Gerke, R. R., Logan, J. L., and Sabins, F. L.: "A Study of Bulk Cement Handling and Testing Procedures," paper SPE 14196, 1985.

Graves, K. S.: "Planning Would Boost Liner Cementing Success," *Oil & Gas J.* (April 15,1985) 47-52.

Jones, R. R.: "A Novel Economical Approach for Accurate Real-Time Measurement of Wellbore Temperatures," paper SPE 15577, 1986.

Lubinski, A.: "Influence of Tension and Compression on Straightness and Buckling of Tubular Goods in Oil Wells," *API Proc.*, Prod. Bull. 237 (1951) 31.

Mitchell, R. F. and Wedelich, H. F. III: "Prediction of Downhole Temperatures Can Be a Key for Optimal Wellbore Design," paper SPE 18900, 1989.

Oilwell Cementing Practices in the United States, American Petroleum Institute, Dallas (1959).

Smith, D. K.: *Cementing*, Henry L. Doherty Series, SPE, Richardson, TX (1987).

Smith, R. C.: "Checklist Aids Successful Primary Cementing," *Oil & Gas J.* (Nov. 1, 1982) 72–75.

Smith, T. R.: "Cementing Displacement Practices: Application in the Field," paper SPE/IADC 18167, 1989.

Suman, G.O. Jr. and Ellis, R.C.: *World Oil's Cementing Oil and Gas Wells Including Casing Handling Procedures*, Books on Demand, Ann Arbor, MI (1977).

13

Remedial Cementing

Christian Marca

Dowell Schlumberger

13–1 SQUEEZE CEMENTING—INTRODUCTION

Squeeze cementing has long been a common operation. Numerous squeeze jobs are performed daily under a wide variety of downhole conditions, and considerable experience has been accumulated over five decades of field practice. Although excellent literature describing this technology has been published and is readily available, misconceptions still exist and operating failures are not uncommon, resulting in increased drilling and completion costs. This chapter presents a review of squeeze cementing technology, with the aim of contributing to improvements in the design and execution of such treatments. The relevant equations which are useful for job design are found in Appendix C.

Squeeze cementing is defined as the process of forcing a cement slurry, under pressure, through holes or splits in the casing/wellbore annular space. When the slurry is forced against a permeable formation, the solid particles filter out on the formation face as the aqueous phase (cement filtrate) enters the formation matrix. A properly designed squeeze job causes the resulting cement filter cake to fill the opening(s) between the formation and the casing. Upon curing, the cake forms a nearly impenetrable solid (Suman and Ellis, 1977). In cases where the slurry is placed into a fractured interval, the cement solids must develop a filter cake on the fracture face and/or bridge the fracture.

Squeeze cementing has many applications during both the drilling and the completion phases. The most commonly cited applications are listed below.

- Repair a primary cement job that failed due to the cement bypassing the mud (channeling) or insufficient cement height in the annulus.
- Eliminate water intrusion from above, below, or within the hydrocarbon producing zone.
- Reduce the producing gas/oil ratio (GOR) by isolating the gas zones from adjacent oil intervals.
- Repair casing leaks due to corroded or split pipe.
- Abandon a nonproductive or depleted zone.
- Plug all, or part, or one or more zones in a multizone injection well so as to direct the injection into the desired intervals.
- Seal off lost-circulation zones.
- Protect against fluid migration into a producing zone.

These scenarios are discussed later in the chapter. First, a review of squeeze cementing theory, placement techniques, and slurry design is presented.

13–2 SQUEEZE CEMENTING—THEORY

Regardless of the technique used during a squeeze job, the cement slurry (a suspension of solids) is subject to a differential pressure against a filter of permeable rock. The resulting physical phenomena are filtration, filter-cake deposition and, in some cases, fracturing of the formation. The slurry, subject to a differential pressure, loses part of its water to the porous medium, and a cake of partially dehydrated cement is formed.

The cement cake, forming against a permeable formation, has a high initial permeability (Fig. 13–1). As the particles of cement accumulate, the cake thickness and hydraulic resistance increase; as a result, the filtration rate decreases, and the pressure required to dehydrate the cement slurry further increases. The rate of filter-cake buildup is a function of four parameters:

- permeability of the formation,
- differential pressure applied,
- time, and
- capacity of the slurry to lose fluid at downhole conditions.

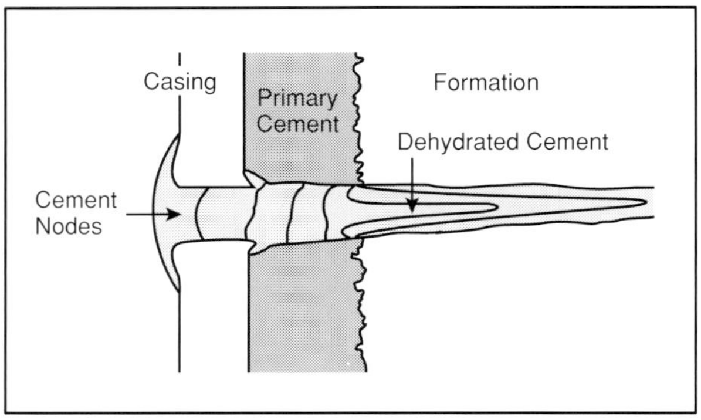

Figure 13–1—Cake permeability and dehydration rate of a slurry as a function of the fluid loss additive concentration (after Hook and Ernst, 1969).

When squeezed against a formation of given permeability, the rate at which slurry dehydration decreases is directly related to the fluid-loss rate (Fig. 13–2). When squeezed against low-permeability formations, slurries with low fluid-loss rates dehydrate slowly, and the duration of the operation may be excessive. Against a high-permeability formation, a slurry with a high fluid-loss rate dehydrates rapidly; consequently, the wellbore may become choked by filter cake, and channels which otherwise would have accepted cement would be bridged off. The ideal squeeze slurry should thus be tailored to control the rate of cake growth, and allow a uniform filter cake to build up over all permeable surfaces.

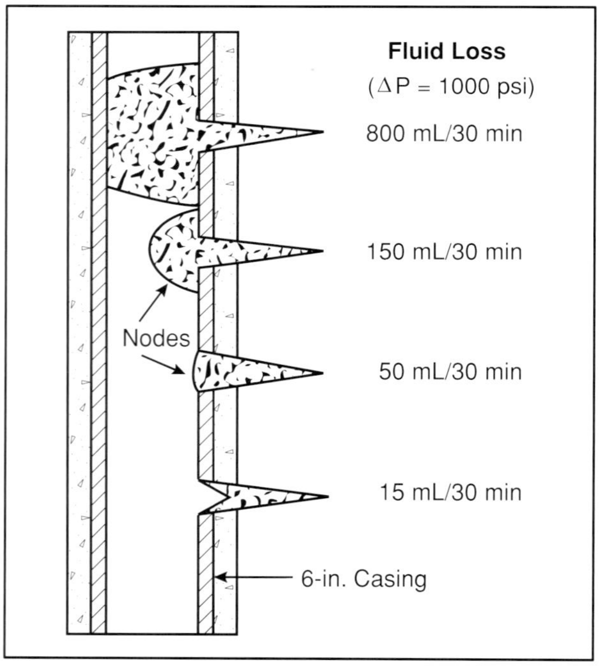

Figure 13–2—Node buildup after a 45-minute squeeze using slurries with different water loss (after Rike, 1973).

The basics of the theoretical and practical work regarding the fundamentals of filter-cake deposition in squeeze cementing can be found in the publications of Binkley, Dumbauld, and Collins (1958) and Hook and Ernst (1969).

13–2.1 Binkley, Dumbauld, and Collins Study

These authors developed the law of filter-cake formation for a suspension (such as a cement slurry). When a volume dQ of filtrate passes through a planar permeable surface of area A, a filter cake of thickness dS and of porosity ϕ is deposited. This relationship is illustrated in the following series of equations.

$$dS = \frac{f}{1 - f - \phi} \times \frac{dQ}{A}, \qquad (13\text{–}1)$$

where

f = fraction by volume of solids in the suspension, or

$$f = \frac{V_{solid}}{V_{solid} + V_{liquid}}. \qquad (13\text{–}2)$$

The ratio $w = f/(1-f-\phi)$ is called the "deposition constant."

The "law of filter-cake formation" may thus be written as

$$\frac{dS}{dt} = wq, \qquad (13\text{–}3)$$

where

q = flow rate of filtrate per unit area of surface, and

dS/dt = rate of growth of the filter-cake thickness.

Assuming that the pressure drop across the filtration surface is negligible, Binkley et al. (1958) applied Darcy's law to the flow of the filtrate through the cake, establishing the following equations.

1. *Growth of a filter cake as a function of filtration time.*

$$S = \sqrt{\frac{2kw\,\Delta P}{\mu}}\; t^{1/2}, \qquad (13\text{–}4)$$

where

k = permeability of the filter cake (constant),

μ = viscosity of the filtrate, and

ΔP = differential pressure.

2. *Cumulative volume of filtrate as a function of filtration time.*

$$Q = \sqrt{\frac{2kA^2\Delta P}{\mu w}}\; t^{1/2} \qquad (13\text{–}5)$$

3. *Filter-cake permeability.*

$$k = \frac{\mu QS}{2At\Delta P} \tag{13–6}$$

4. *Deposition constant.*

$$W = \frac{AS}{Q} \tag{13–7}$$

5. *Fill up of an unfractured perforation.*

Binkley et al. (1958) assumed the perforation geometry shown in Fig. 13–3. The depth of the perforation is considered large in relation to its diameter (at least four times greater). Assuming that the pressure drop in the formation is zero, Binkley et al. (1958) demonstrated that the time required to build a filter cake in close contact with the inside of the casing can be expressed with the following equation.

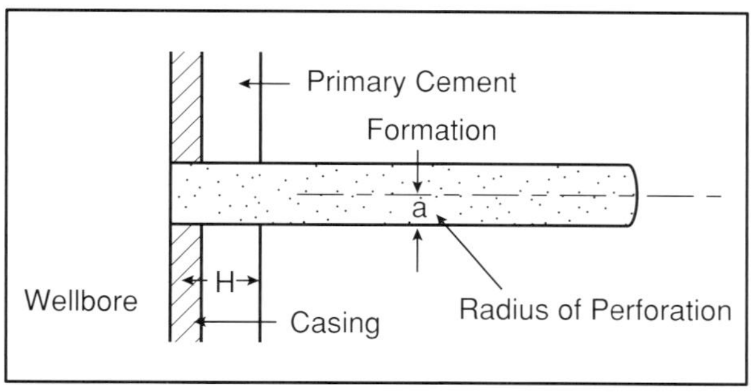

Figure 13–3—Geometry of perforation (after Binkley, Dumbauld, and Collins, 1958).

$$T = \frac{\mu}{kw\Delta P}\left(\frac{H^2}{2} + \frac{a^2}{4} + eaH\right) \tag{13–8}$$

Experiments determined e to be equal to 0.25. The ratio $\mu/kw\Delta P$ contains all the variables related to the deposition properties of the cement slurry, and is called the composition factor. It is interesting to note that, with the assumption regarding perforation depth vs its diameter, the depth of the perforation has a negligible effect on the deposition process.

By measuring the filter-cake thickness S and the cumulative volume of filtrate Q obtained when a cement slurry is in contact with a planar permeable surface of area A, and submitted to a differential pressure ΔP for a time T, it is possible to determine the filter-cake permeability k and the deposition constant w.

6. *Deposition of solids inside the casing following perforation fill-up.*

To simplify the calculation, Binkley et al. (1958) assumed that the node building up inside the casing has a spherical shape at every stage of the growth, and that the lateral growth occurs at the same rate as the vertical growth. The geometry is illustrated in Fig.

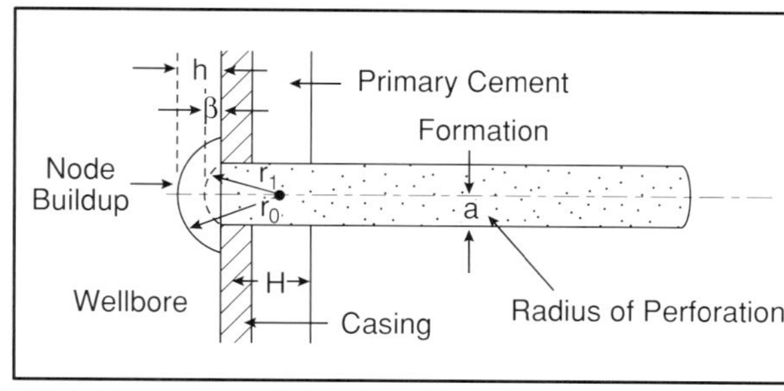

Figure 13–4—Geometry of the cement node (after Binkley, Dumbauld, and Collins, 1958).

13–4. If h is the height of the node building up inside the casing, then the equation below can be derived.

$$\frac{dh}{dt} = \frac{kw\Delta P}{\mu}\left[\frac{1}{(H + \beta + ea)\frac{r_0^2}{r_1^2} + \frac{r_0^2}{r_1^2} - r_0}\right], \tag{13–9}$$

where

H = combined thickness of the cement sheath and casing,

a = radius of the perforation,

$$\beta = r_1 - \sqrt{r_1^2 - a^2},$$

$$r_0 = \frac{h^2 + (a + h)^2}{2h}$$

$$r_1 = \sqrt{(r_0 - h)^2 + a^2}, \text{ and}$$

$$e = 0.25.$$

The results of the numerical integration of this complex equation have been plotted in Fig. 13–5, which represents the time required to fill a perforation and build a node, vs the ratio H/a, for different node heights.

13–2.2 Hook and Ernst Study

Hook and Ernst (1969) performed an experimental study of the effects of fluid-loss control additives, differential pressure, and formation permeability upon the rate of filter-cake buildup. Their conclusions are presented in Tables 13–1, 13–2, and 13–3.

Table 13–1 is a compilation of permeability measurements conducted on filter cakes which were formed with different concentrations of a fluid-loss additive. The permeability of a neat-cement filter cake was measured to be about 5 md—a value lower than that of many producing

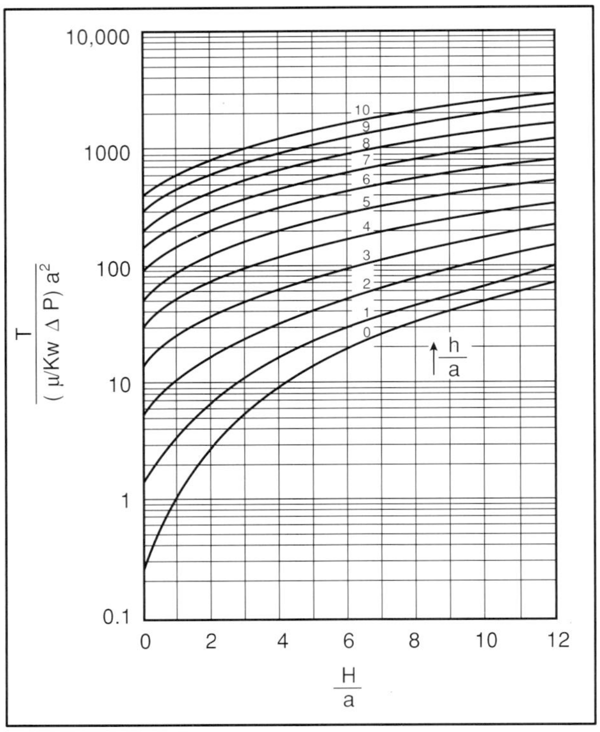

Figure 13–5—Relationship between slurry property, perforation geometry, and height of filter-cake node inside casing (after Binkley, Dumbauld, and Collins, 1958).

Slurry — API Class A Cement with Liquid Fluid-Loss Additive and 46% Water			
Concentration of Fluid-Loss Additive (gal/sk)	**API Fluid Loss at 1000 psi (cc/30 min)**	**Permeability of the Filter Cake Formed at 1000 psi (md)**	**Time to Form a 2-in. Thick Filter Cake (min)**
0.00	1200	5.00	0.2
0.07	600	1.60	0.8
0.13	300	0.54	3.4
0.17	150	0.19	14.0
0.19	100	0.09	30.0
0.22	50	0.009	100.0
0.24	25	0.006	300.0

Table 13–1—Cake permeability and dehydration rate of a slurry as a function of the fluid-loss additive concentration (after Hook and Ernst, 1969).

sandstone formations. When the slurry contains sufficient fluid-loss additive to reduce the API fluid-loss rate to 25mL/30 min, the resulting filter cake is approximately 1,000 times less permeable. This approaches the permeability of matrix formations that produce very slowly, and are difficult in which to pump.

Table 13–1 also shows that the rate of filter-cake growth is indirectly related to the permeability of the filter cake. Fluid-loss additives lower the permeability of the filter cake; consequently, the quantity of solids which can be filtered out from the slurry is also reduced.

The data in Table 13–2 demonstrate the influence of squeeze pressure upon the rate of filter-cake growth. First, it was shown that varying the squeeze pressure from 500 to 1,000 psi (3.5 to 6.9 MPa) does not influence the permeability of the resulting filter cake. Nevertheless, in keeping with Darcy's law, the data showed that the flow rate of fluid through the filter cake was directly proportional to the squeeze pressure.

Differential of Filter Cake Formation (psi)	Permeability of Filter Cake (md)	API Fluid Loss (cc/30 min)	Flow Rate Through Filter Cake (cc/min)
Slurry I			
500	5.8	1200	50
1000	6.0	1200	110
Slurry II			
500	1.9	600	17
1000	1.6	600	30
Slurry III			
500	0.53	300	4.7
1000	0.54	300	9.7

Slurry I — Class A Cement
46% Water

Slurry II — Class A Cement
0.5% Dispersant
0.07 gal/sk Liquid Fluid-Loss Additive
46% Water

Slurry III — Class A Cement
0.5% Dispersant
0.13 gal/sk Liquid Fluid-Loss Additive
45% Water

By increasing the pressure from 500 psi to 1000 psi, the filtration rate is increased by a factor close to 2.

Table 13–2—Effect of the differential pressure on the permeability of filter cakes and on the filtration rate (experimental) (after Hook and Ernst, 1969).

Table 13–3 shows the effect of formation permeability upon the rate of cement filter-cake growth. Against a 30-md formation, the time required to form a filter cake of given thickness is roughly twice that observed against a formation with a 300-md permeability. These results demonstrate the importance of knowing the formation permeability before designing the slurry; this point is discussed later in the chapter.

13–3 SQUEEZE CEMENTING—PLACEMENT TECHNIQUES

Normally, the slurry injection is performed through casing perforations. There are two fundamentally different squeeze job classifications.

- *Low-pressure squeeze:* The bottomhole treating pressure is maintained below the formation fracturing pressure.
- *High-pressure squeeze:* The bottomhole treating pressure exceeds the formation fracturing pressure.

Within these two classes, there are two basic techniques (the Bradenhead squeeze and the squeeze tool technique) and two pumping methods (the running squeeze and the hesitation squeeze). Each of these classifications and techniques is explained in this section.

13–3.1 Low-Pressure Squeeze

The aim of this operation is to fill the perforation cavities and interconnected voids with dehydrated cement. The volume of cement is usually small, because no slurry is actually pumped into the formation. Precise control of hydrostatic pressure of the cement column is essential, because excessive pressure could result in formation breakdown.

The following calculations are given as an example. A 500-psi (3.5 MPa) safety factor is taken, and friction losses are assumed negligible as a result of the very low rates at which the job is performed. The maximum column of cement the formation can withstand, X, is determined by the following equation.

$$X = \frac{[(FGh) - 500] - (0.052\,h\,\rho_c)}{0.052(\rho_s - \rho_c)} \qquad (13\text{–}10)$$

Therefore, the maximum volume of cement slurry is calculated by the equation below.

$$V_{max} = XV_t \qquad (13\text{–}11)$$

where

h = perforations depth (ft),

X = length of cement column (ft),

ρ_s = slurry density (lb/gal),

ρ_c = completion fluid density (lb/gal),

FG = formation fracture gradient (psi/ft),

500 = safety factor (psi),

0.052 = conversion factor (lb/gal to psi/ft),

V_t = tubing volume per unit of length (ft^3/ft), and

k = distance from packer to perforations.

Example

Perforations to be squeezed off at 6,250 ft through $2^1/_2$-in.- ID tubing inside 7-in. (30-lb/ft, 0.0325-ft^3/ft) casing. Packer to be set at 6,150 ft.

Estimated Fracture Gradient—0.7 psi/ft
Displacement Fluid—8.6 lb/gal brine
Cement Slurry—Class G + 0.2 gal/sk fluid-loss additive + 0.1 gal/sk dispersant

Time Required for the Formation of a $1^1/_4$-in. Long Filter Cake on a 1-in. Diameter Filtration Surface at 1000 psi			
	Bandera Sandstone 30 md	**Berea Sandstone 300 md**	**API 325-mesh Screen**
Slurry I	6 min	2.5 min	2.5 min
Slurry II	9 min	6.5 min	6.5 min
Slurry III	5 min	2.5 min	2.5 min

Slurry I — API Class A Cement
0.5% Dispersant
0.14 gal/sk Liquid Fluid-Loss Additive
46% Water

Slurry II — API Class A Cement
0.5% Dispersant
0.17 gal/sk Liquid Fluid-Loss Additive
46% Water

Slurry III — API Class A Cement
0.7% Solid Fluid-Loss Additive
46% Water

Table 13–3—Effect of formation permeability on the rate of the filter cake growth (after Hook and Ernst, 1969).

Density: 15.8 lb/gal
Yield: 1.15 ft^3/sk

Using Eq. 13–10, the following is obtained.

$$X = \frac{[(0.7 \times 6{,}250) - 500] - (0.052 \times 6{,}250 \times 8.6)}{0.052(15.8 - 8.6)}$$

$= 2{,}885$ ft

Therefore, using Eq. 13–11, $V_{max} = 2{,}885 \times 0.0325 = 93.75$ ft^3. Or, in terms of sacks of cement, 80 sk.

In low-pressure squeezes, it is essential that perforations and channels be clear of mud or other solids. If the well has been producing, such openings may already be free of obstructions; however, for newly completed wells, it may be necessary to clean the perforations before performing the squeeze job (Section 13–9.2)

A properly designed slurry will leave only a small node of cement filter cake inside the casing. Improperly designed systems can result in excessive development of cement filter cake. This can result in a complete bridging of the inside of the casing, with loss of pressure transmission to the formation, and insufficient contact of the cement filter cake with the formation.

Review of the literature shows that, according to most authors, a low-pressure squeeze should be run whenever possible, and that this technique has the highest success rate (Rike and Rike, 1981; Goodwin, 1984; Bradford and Reiners, 1985). The low-pressure squeeze requires only a small amount of cement slurry, while the high-pressure technique usually involves a larger volume of slurry.

13–3.2 High-Pressure Squeeze

In some cases, a low-pressure squeeze of the perforations will not accomplish the objective of the job. The channels behind the casing might not be directly connected to the perforations. Small cracks or microannuli that may allow flow of gas do not allow the passage of a cement slurry. In such cases, these channels must be enlarged to accept a viscous solids-carrying fluid. In addition, many low-pressure operations cannot be performed if it is impossible to remove plugging fluids, or debris, from ahead of the cement slurry or inside the perforations.

Placement of the cement slurry behind the casing is accomplished by breaking down the formation at or close to the perforations (Fig. 13–6). Fluids ahead of the slurry are displaced in the fractures, allowing the slurry to fill the desired spaces. Further application of pressure dehydrates the slurry against the formation walls, leaving all channels (from fractures to perforations) filled with cement cake.

However, during a high-pressure squeeze, the location and orientation of the created fracture cannot be controlled. Sedimentary rocks usually have an inherently low tensile strength, and are held together primarily by the weight or the compressive forces of overlying formations. These cohesive forces act in all directions to hold the rock together, but do not have the same magnitude in all directions. When sufficient hydraulic pressure is applied against a formation, the rock fractures along the plane perpendicular to the direction of the least principal stress (Fig. 13–7). A horizontal fracture is created if the fracturing pressure is greater than the overburden pressure. A vertical fracture occurs if overburden pressure is greater (Roegiers, 1987).

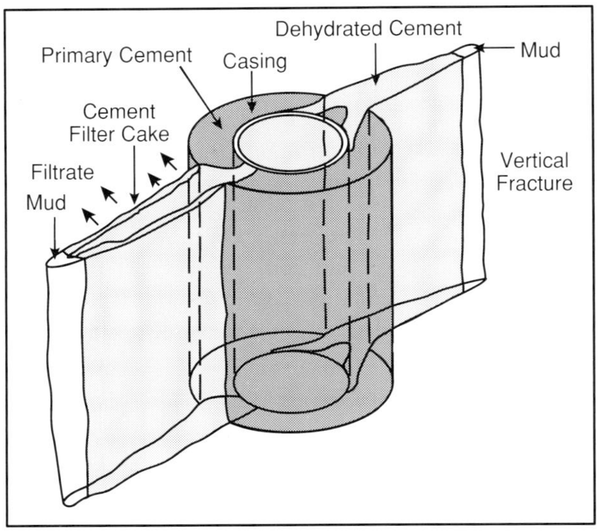

Figure 13–6—High-pressure technique: vertical fracture generated by high-pressure squeezing.

The extent of the induced fracture is a function of the pump rate applied after the fracture is initiated. The amount of slurry used depends on the way the operation is performed. High pump rates generate large fractures; thus, large volumes of cement are required to fill them. A properly performed, high-pressure squeeze should leave the cement as close to the wellbore as possible.

Drilling muds or other fluids with low fluid-loss rates should not be pumped ahead of the slurries. A wash with a high fluid-loss rate, such as water or a weak hydrochloric acid solution, not only opens smaller fractures but also cleans perforations and the cement path. The fracture initiation pressure is lower using this type of spearhead than using nonpenetrating fluids.

13–3.3 Bradenhead Placement Technique (No Packer)

This technique, illustrated in Fig. 13–8, is normally used when low-pressure squeezing is practiced, and when there are no doubts concerning the casing's capacity to withstand the squeeze pressure. No special tools are

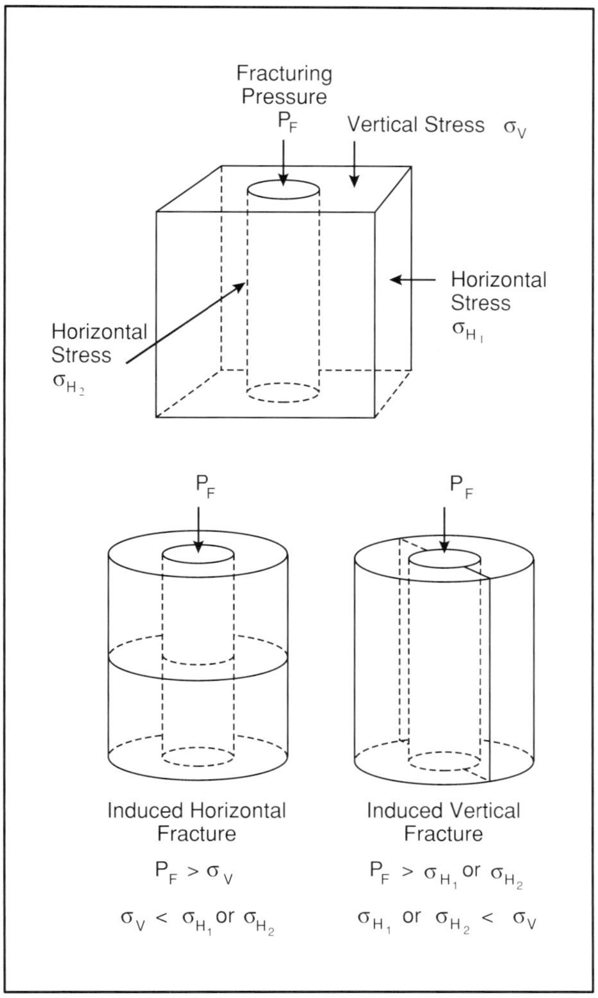

Figure 13–7—Effect of well depth and vertical/horizontal formation stresses on the orientation of hydraulic fracture induced by injected fluid. Horizontal fractures will not be created if fracture pressure is less than overburden pressure. This is usually the case at depths greater than 3,000 ft. (after Suman and Ellis, 1977).

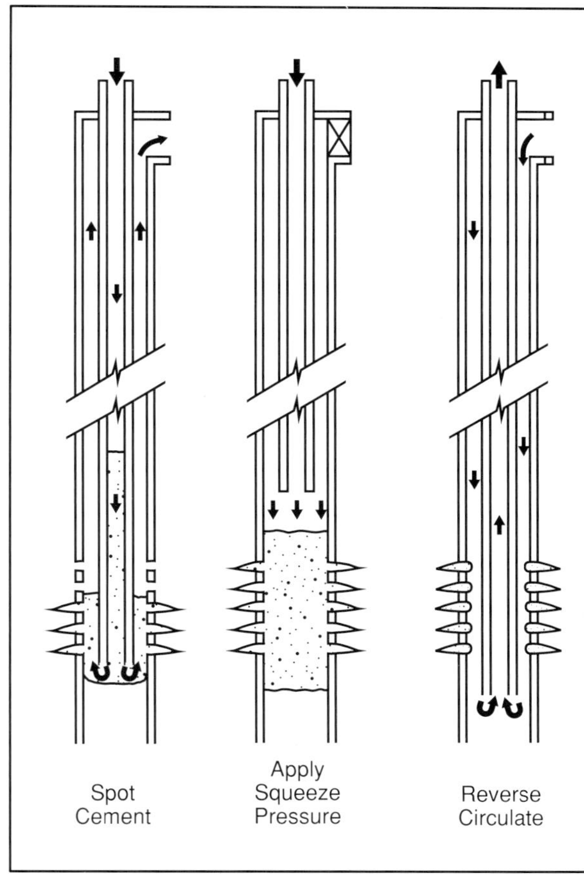

Figure 13–8—Bradenhead squeeze technique is normally used on low-pressure formations. Cement is circulated into place down drillpipe (left), then wellhead, or BOP, is closed (center) and squeeze pressure is applied. Reverse circulating (right) removes excess cement, or plug can be drilled out (after Suman and Ellis, 1977).

involved, although a bridge plug may be required to isolate other open perforations further downhole.

Open-ended tubing is run to the bottom of the zone to be cemented. Blowout preventer (BOP) rams are closed over the tubing, and the injection test is performed. The cement slurry is subsequently spotted in front of the perforations. Once the cement is in place, the tubing is pulled out to a point above the cement top, the BOPs are closed, and pressure is applied through the tubing. The Bradenhead squeeze is very popular because of its simplicity.

13–3.4 Squeeze Tool Placement Technique

This technique can be subdivided into two parts—the retrievable squeeze packer method, and the drillable cement retainer method. The main objective of using squeeze tools is to isolate the casing and wellhead while high pressure is applied downhole.

13–3.4.1 Retrievable Squeeze Packer Method

Retrievable packers with different design features are available (Chapter 10). Compression- or tension-set packers are used in squeeze cementing. As shown in Fig. 13–9, they have a bypass valve to allow the circulation of fluids while running in the hole, and once the packer is set. This feature allows the cleaning of the tools after the cement job, and the reversing out of excess slurry without excessive pressure; it also prevents a piston or swabbing effect while running in or out of the hole.

The principal advantage of the retrievable packer over the drillable retainer is its ability to set and release many times, thus allowing more flexibility. Retrievable bridge plugs can be run in one trip with the packer, and retrieved after the slurry has been reversed or drilled out. Most

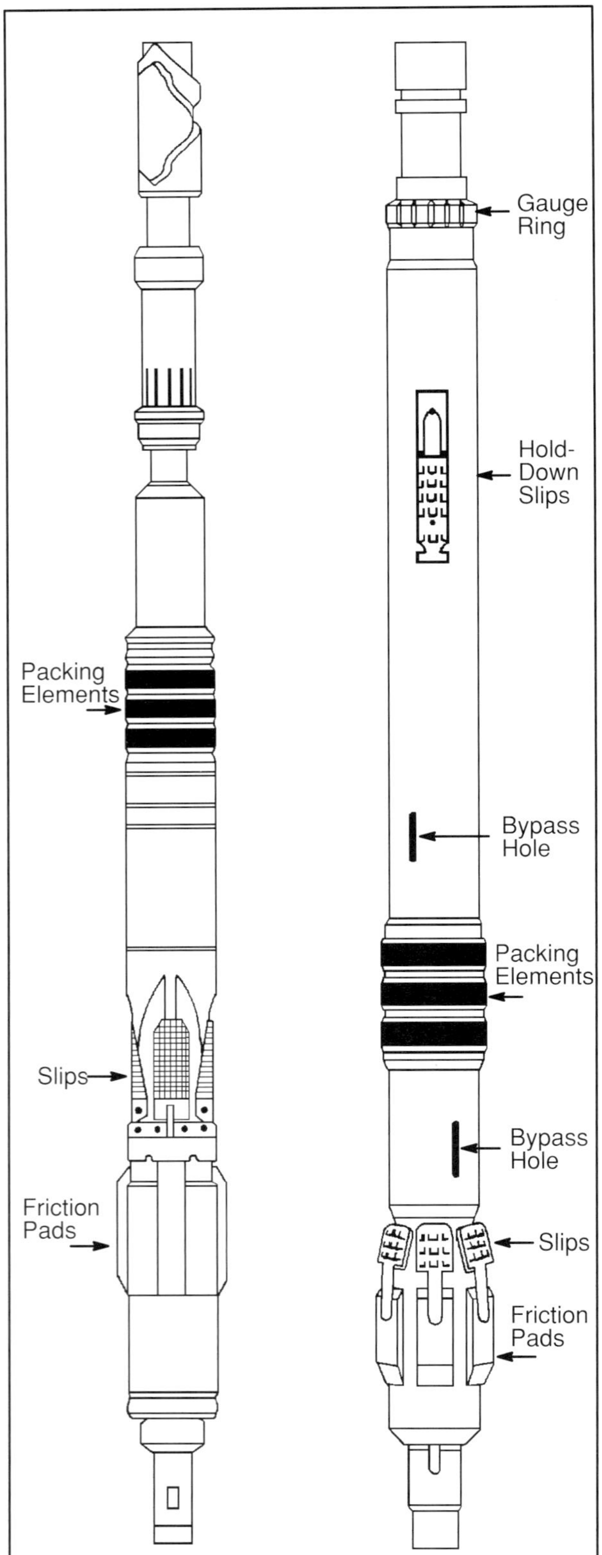

Figure 13–9—Bridge plug and squeeze packer.

operators drop one or two sacks of frac sand on top of the retrievable bridge plug before the job, to prevent the settling of cement over the releasing mechanism.

13–3.4.2 Drillable Cement Retainer

Cement retainers are used instead of packers to prevent backflow when no cement dehydration is expected, or when a high negative differential pressure may disturb the cement cake. In certain situations, potential communication with upper perforations makes the use of a packer a risky operation.

When cementing multiple zones, the cement retainer isolates the lower perforations, and subsequent zone squeezing can be performed without waiting for the slurry to set. Cement retainers are drillable packers provided with a valve which is operated by a stinger at the end of the work string (Fig. 13–10).

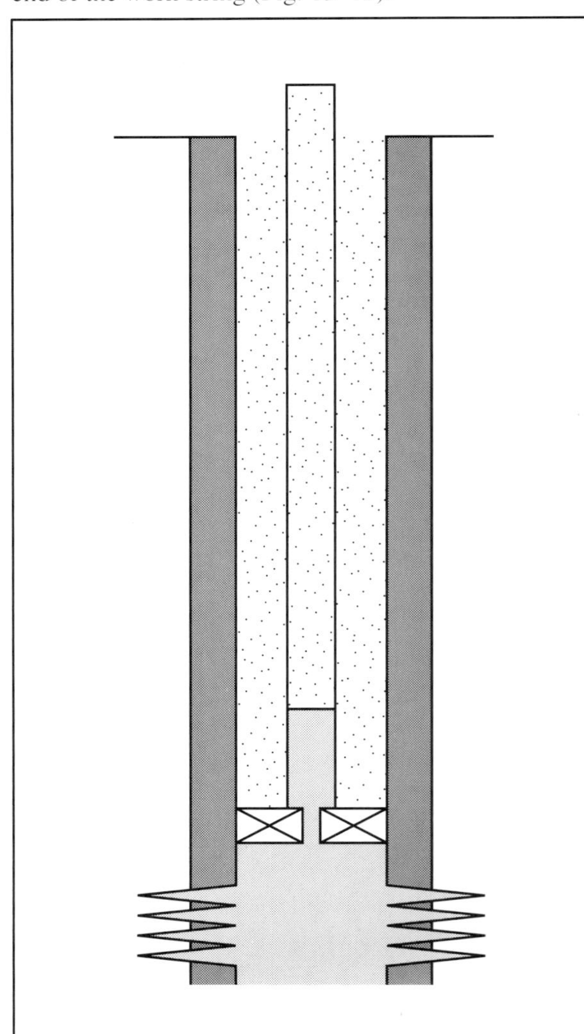

Figure 13–10—Squeeze with cement retainer.

A drillable retainer gives the operator more confidence in setting the packer closer to the perforations. This is also advantageous in that a lower volume of fluid below the packer is displaced through the perforations ahead of the cement slurry.

Drillable bridge plugs are normally used to isolate the casing below the zone to be treated. Their design is similar to that of cement retainers. They can be run with a wireline or with the work string. Bridge plugs do not allow flow through the tool.

13–3.5 Running Squeeze Pumping Method

During a running squeeze procedure, the cement slurry is pumped continuously until the final desired squeeze pressure (which may be above or below the fracture pressure) is attained. After pumping stops, the pressure is monitored and, if the pressure falls due to additional filtration at the cement/formation interface, more slurry is pumped to maintain the final surface squeeze pressure. This continues until the well maintains the squeeze pressure for several minutes without additional injection of cement slurry. The volume of slurry injected is usually large. Rike and Rike (1981) reported that volumes ranging from 10 to 100 barrels are commonly used.

13–3.6 Hesitation Squeeze Pumping Method

The relatively small amount of filtrate lost from the slurry makes impractical, if not impossible, continuous pumping at a rate slow enough to maintain a constant differential pressure. The minimum pump rate attainable with existing field equipment is excessive even for high-pressure squeezing where large fractures are created.

The only procedure which makes possible the dehydration of small quantities of cement into perforations of formation cavities is the hesitation squeeze pumping method. This procedure involves the intermittent application of pressure (at a rate of $^1/_4$ to $^1/_2$ bbl/min), separated by an interval of 10 to 20 minutes for pressure leakoff due to filtrate loss to the formation. The initial leakoff is normally fast because there is no filter cake. As the cake builds up, and the applied pressure increases, the filtration periods become longer and the difference between the initial and final pressures become smaller, until at the end of the job the pressure leakoff becomes negligible (Fig. 13–11). The volumes of slurry necessary for this technique are usually much less than those required for a running squeeze.

A loose formation normally requires a long hesitation period to begin building squeeze pressure. A first hesitation period of 30 minutes or more is not unreasonable. A much shorter initial hesitation period (possibly five minutes) is normally sufficient for tight formations (Grant and White, 1987).

13–4 INJECTION TEST

Prior to mixing and pumping the cement slurry, an "injection test" is performed. This procedure consists of pumping a fluid, typically water or a mud flush, into the well. The injection test is performed for several reasons:

- to ensure that the perforations are open and ready to accept fluids,
- to obtain an estimate of the proper cement slurry injection rate,
- to estimate the pressure at which the squeeze job will be performed, and
- to estimate the amount of slurry to be used.

Should the fluid fail to achieve the injection, acid should be injected under matrix conditions. Hydrochloric and hydrofluoric acids are commonly used.

13–5 DESIGN AND PREPARATION OF THE SLURRY

As discussed above, the properties of the cement slurry must be tailored according to the characteristics of the formation to be squeezed, and the technique to be used. It is generally agreed that a squeeze slurry should be designed to have the following general attributes:

- low viscosity—to allow the slurry to penetrate the small cracks,
- low gel strength—a gelling system restricts slurry movements and causes increases in surface pressure which are difficult to interpret,

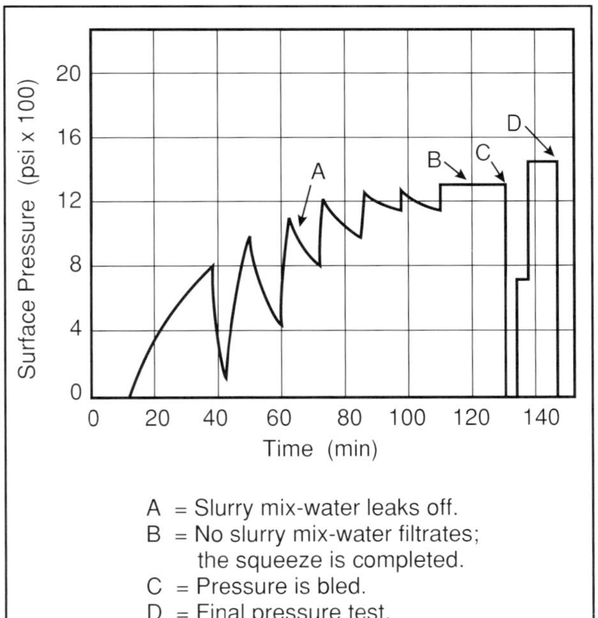

Figure 13–11—Hesitation squeeze pressure behavior.

- no free water,
- appropriate fluid-loss control—to assure optimum filling of the cracks or perforations, and
- proper thickening time—to safely meet the anticipated job time.

The specifications to test a squeeze slurry are found in API Spec 10 (1988) (Appendix B).

13–5.1 Fluid-Loss Control

As discussed earlier, fluid-loss control is particularly important when squeeze cementing across permeable formations. For low-pressure squeezing, a properly designed slurry should allow the complete filling of perforation cavities, leaving a minimum node buildup in the casing. The slurry fluid loss must be tailored to the formation type and the permeability (Young, 1967). The generally accepted API fluid-loss rates are listed below.

- Extremely Low-Permeability Formation: 200 mL/30 min
- Low-Permeability Formation: 100 to 200 mL/30 min
- High-Permeability Formation (>100 md): 35 to 100 mL/30 min

When squeezing fractured limestone or dolomite formations, the situation is different from that for sandstone, because the permeability consists of interconnected voids or fracture systems. All cement particles can enter these channels and, as the slurry slowly dehydrates, it will travel relatively long distances into the formation. Allowing this to happen may put the cement-free zone out of reach of the perforating gun. To confine the cement within a close range around the wellbore, the dehydration process must occur quickly. Cement systems with high fluid-loss rates (300 to 800 mL/30 min) are used to allow a fast cake build up. It is also useful to include lost-circulation prevention agents in the slurry (powdered coal, nut-shells, sand, etc.) to act as bridging agents across the cracks and voids.

Grant and White (1987) reported success with a two-slurry squeeze design for vugular zones. A lead slurry with a short pumping time and a fairly high fluid-loss rate (300 to 500 mL/30 min) is followed by a tail slurry with a longer pumping time and a lower fluid-loss rate. The tail slurry is used for hesitation.

In high-pressure squeezing, when overcoming the formation fracture pressure, the slurry is pumped into the induced fractures, and dehydrates against the fracture walls. If the formation permeability is sufficiently high, a medium- to high-fluid-loss slurry (200 to 500 mL/30 min) will usually permit the caking and subsequent diversion of slurry to smaller cracks.

13–5.2 Slurry Volume

The appropriate volume of cement slurry depends upon the length of the interval to be cemented, and the placement technique to be used. A low-pressure squeeze requires only enough slurry to build a cement filter cake in each perforation tunnel. In many cases, less than a barrel is sufficient. However, for job convenience a 5- to 15-bbl batch is normally prepared. Pumping rates as low as 0.25 bbl/min are common. Excess cement can be included according to experience in the area; however, it must be kept in mind that excess cement could be detrimental to the productivity of the formation being squeezed.

A high-pressure squeeze, in which the formation is fractured, requires a higher volume of slurry. The volume required is a function of the width and depth of the fractures created. It has been reported that in some instances of running squeeze, where the cracks generated were excessive, volumes of more than 100 barrels of slurry have been injected (Rike and Rike, 1981). The volume can be minimized by fracturing at a low pump rate, and maintaining the injection pressure below the propagation pressure of the fractures. If the squeezing is performed at high pressures and pump rates, the fractures will develop accordingly, resulting in large quantities of cement being pumped in the formation.

Smith (1987) cited several useful rules of thumb.

- The volume should not exceed the capacity of the run-in string.
- Two sacks of cement should be used per foot of perforated intervals.
- The minimum volume should be 100 sacks if an injection rate of 2 bbl/min can be achieved after breakdown; otherwise, it should be 50 sacks.
- The volume should not be so great as to form a column that cannot be reversed out.

The hydrostatic and surface pressures must be controlled during the job. A high cement column during the displacement could cause the breakdown of low-pressure or depleted formations. When large quantities of cement are necessary (natural fractures), the use of low-density slurries is recommended.

13–5.3 Thickening Time

As with primary cementing, the temperature and pressure are important factors which influence the placement time of a cement slurry. The temperatures encountered in squeezing can be higher than those on primary jobs, because fluid circulation before the job is usually less. For this reason, special API testing schedules exist for squeeze cement slurry design (Appendix B), and must be followed to prevent premature setting.

In a shallow well, the slurry can be designed for a fairly short pumping time (e.g., two hours). Accelerators are commonly used. However, a hesitation squeeze job may require a pumping time as long as six hours. Therefore, one must add sufficient retarder to assure slurry placement, and reversing out of the excess.

13–5.4 Slurry Viscosity

The ability of the slurry to flow into narrow channels is proportional to its fluidity. Thick slurries, although useful when cementing large voids, will not flow into small restrictions unless they are subjected to high differential pressures, which are limited by the formation fracture pressure. Therefore, low-viscosity slurries containing dispersants are commonly used.

13–5.5 Compressive Strength

High compressive strength, although desirable for withstanding shocks from subsequent tools and preventing cracking during the reperforation, is not a primary concern. A partially dehydrated cement cake of any normal cement slurry will develop sufficient compressive strength.

13–5.6 Spacers and Washes

There are two major concerns related to the success of cement placement.

- Cleaning of perforations and surrounding voids. Solids-carrying fluids or drilling mud must be removed from the perforation channels and the formation face to allow a proper dehydration process and complete fill-up.
- Avoiding contamination of the cement slurry. Slurry properties, such as fluid loss, thickening time, and viscosity, can be modified by cement contact with completion fluids. A small quantity of contaminated slurry, having a high fluid-loss rate or high viscosity, may readily block channels and prevent optimum slurry placement.

In low-pressure squeezing, treatments related to the first point are performed as a separate stage. Usually, cement slurry contamination is avoided by pumping a compatible water spacer ahead of and behind the cement. If the cement is not spotted, a chemical wash or weak acid solution may be squeezed ahead of the slurry, separated by a compatible fluid.

13–6 BASIC SQUEEZE-JOB PROCEDURES

Below is a list of the general sequence of events during a squeeze job.

1. The lower zones are isolated with a retrievable or drillable bridge plug.
2. The perforations are washed with a perforation washing tool, or are reopened with the back-surging technique (Section 13–9.2).
3. The perforation washing tool is retrieved and, if the packer method is chosen, is run in the hole with the work string, set at the desired depth and tested. An annular test pressure of 1,000 psi (6.9 MPa) is usually sufficient. If the cement is to be spotted in front of the perforations, a tail pipe, covering the length of the zone plus 10 or 15 ft (3 or 5 m), is run below the packer.
4. An injection test is performed using clean, solids-free water or brine. If a low-fluid-loss completion fluid is in the hole, it must be displaced from the perforations before starting the injection test. This test gives an idea of the permeability of the formation to the filtrate.
5. The spearhead fluid followed by the cement slurry is circulated downhole with the packer bypass open. This circulation is performed to avoid squeezing the damaging fluids ahead of the slurry into the formation. A small amount of backpressure is applied on the annulus to prevent slurry free-fall as a result of the "U" tube effect.

   If no tail pipe has been run, the packer bypass must be closed two to three barrels before the slurry reaches the packer. If the cement is to be spotted in front of the perforations, with the packer unset, circulation is stopped as soon as the cement covers the selected zone. The tail pipe is pulled out of the cement slurry, and the packer is set at the desired depth.

   The depth at which the packer is set must be carefully chosen. If a tail pipe is run, the minimum distance between the perforations and the packer is limited to the length of the tail pipe. The packer must not be set too close to the perforations, as pressure communication through the annulus above the packer may cause casing collapse. A safe setting depth must be decided upon after evaluation of the quality of the cement bond with the logs (Fig. 13–12).

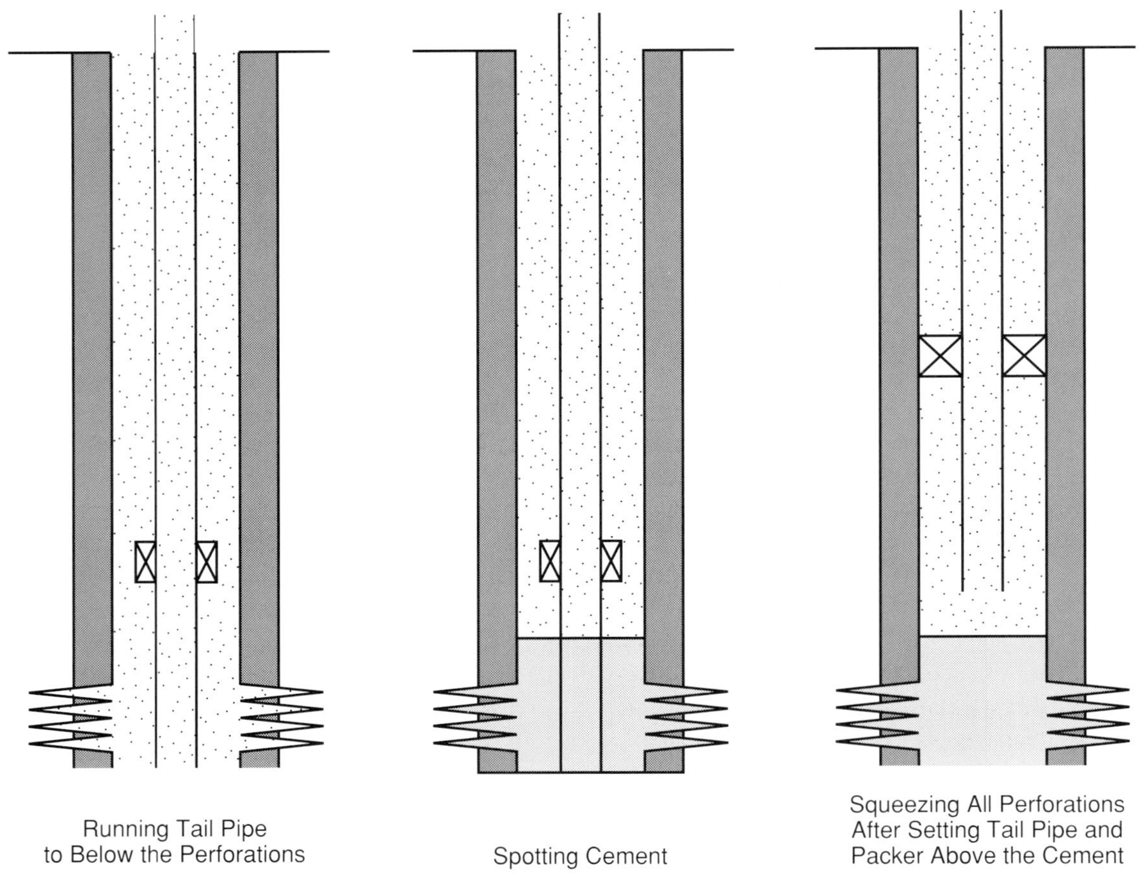

Figure 13–12—Squeeze with a retrievable packer and tail pipe.

Possible contamination of the squeeze cement by the fluid in the hole limits the maximum spacing between the packer and the treated zone. In Fig. 13–13, the packer is set too high, allowing cement slurry to be contaminated as it channels through the mud to reach the perforations. Shryock and Slagle (1968) recommended that the retrievable packer be set at no more than 25 ft (8 m) above the perforations.

6. Squeeze pressure is applied. If the hesitation method is used with the high-pressure squeeze technique, the formation is broken down, and the cement slurry is pumped into the fracture before hesitation pumping is applied. If the low-pressure squeeze technique is elected, the hesitation pumping is started as soon as the packer is set.
7. Pumping continues until no pressure leakoff is observed. A further pressure test of about 500 psi (3.5 MPa) over the final injection pressure indicates the end of the injection process. Usually, a well-cemented perforation accepts a pressure above the formation fracturing pressure, but the risk of fracturing exists if one attempts to verify such a condition.

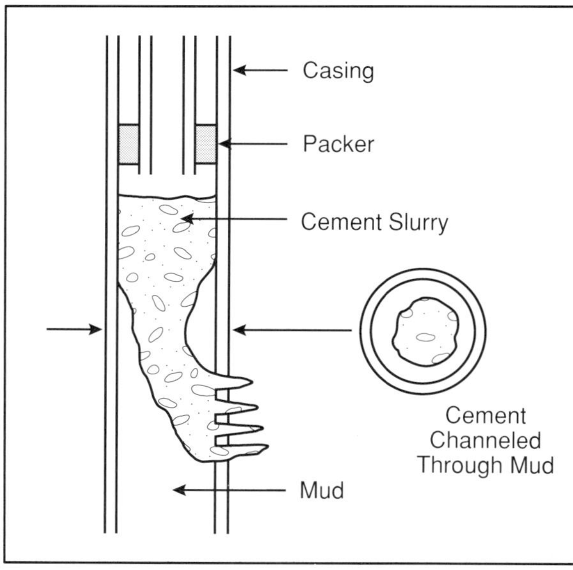

Figure 13–13—Cement slurry contamination (after Shryock and Slagle, 1968).

8. Pressure is bled off and returns are checked. The packer bypass is opened and excess cement is reversed out. Washing off cement in front of the perforations can be performed by releasing the packer, and slowly lowering the work string during the reversing; however, there is a danger of disturbing the unset cement filter cake.
9. Tools are retrieved, and the well is left undisturbed to allow the slurry to cure for the recommended waiting-on-cement (WOC) time.

When preparing the slurry, the use of a recirculating mixer or a batch tank is strongly recommended, because it ensures that the properties of the slurry pumped in the well are as close as possible to those of the slurry designed in the laboratory. On most squeeze jobs, the amount of slurry involved is quite small, but the requirements for its quality are high; therefore, special care must be taken in preparing it.

13–7 SQUEEZE CEMENTING—APPLICATIONS

13–7.1 Repairing a Deficient Primary Casing Job

Poor mud displacement during primary cementing causes the cement slurry to channel through the drilling mud. Consequently, pockets or channels of mud are left behind the casing (Fig. 13–14), and sufficient hydraulic isolation between the various permeable zones, which is the aim of the primary cementing job, is not achieved. Should such defects in zonal isolation not be corrected, serious problems may occur during the life of the well, such as those listed below.

- Stimulation treatments not meeting expected results because of poor control of the fluids placement.

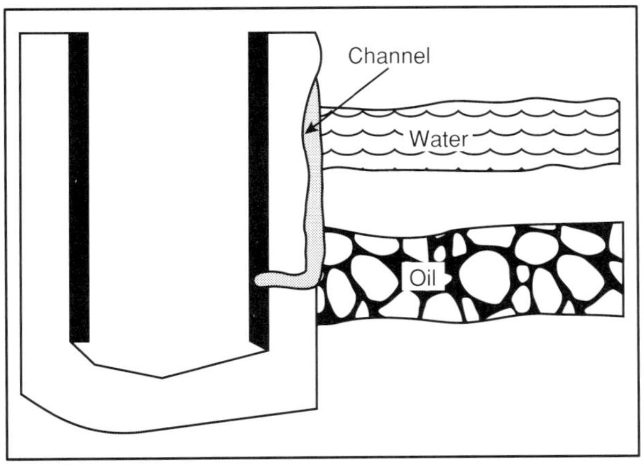

Figure 13–14—A defective primary cementing job.

- Inaccurate evaluation of the production potential of the well because of the parasitic effects of nearby flowing fluids.
- Poor well productivity as a result of water cut or high GOR.
- Failure of waterflooding project (Goolsby, 1969).

Electing to perforate a cemented casing to perform a squeeze is not an easy decision to make. Likewise, defining precisely where to perforate is also critical. A thorough analysis of the primary cementing job, based on an accurate record of all the parameters of the operation, and a careful interpretation of the logs (Chapter 16) are the key elements in aiding the decision process.

Two situations may exist behind the casing.

- The mud channel to be repaired is against a permeable formation. During the squeeze job, the cement filter cake builds and eventually fills the void.
- Circulation is established between two sets of perforations. A "circulation" or "channel" squeeze is performed to replace the mud in the channel by cement. Basically, this is a partial or total recementing of the interval of interest.

Both of these operations can be successful only if the downhole treating pressure remains below the formation fracturing pressure. Fractures created during the execution of the job would result in the opening of a preferential route through which a large quantity of the cement slurry will penetrate; as a result, damage to the permeable interval occurs, and the treatment objectives are not met.

The "circulation" squeeze, illustrated in Fig. 13–15, is often performed with a cement retainer in preference to a packer. Circulation is achieved with water or acid as a spearhead fluid. The interval is circulated with a wash fluid to ensure a good cleanup, and the cement slurry is then pumped and displaced. No pressure buildup occurs during the job, except for an increase due to the hydrostatic pressure of the column of cement as it flows up the annulus. Once the placement is completed, the stinger or packer is released. The excess cement circulating out of the upper perforations can be reversed out if desired.

There is a strong possibility that some of the cement slurry (the volume is not known, so an excess is always taken) may enter the casing, drillpipe, tubing, or the annulus above the squeeze tool during the job. Should this cement set, there is a risk that the drillpipe (or tubing) may become stuck in the hole. Thus, running a cement retainer instead of a packer is recommended to minimize

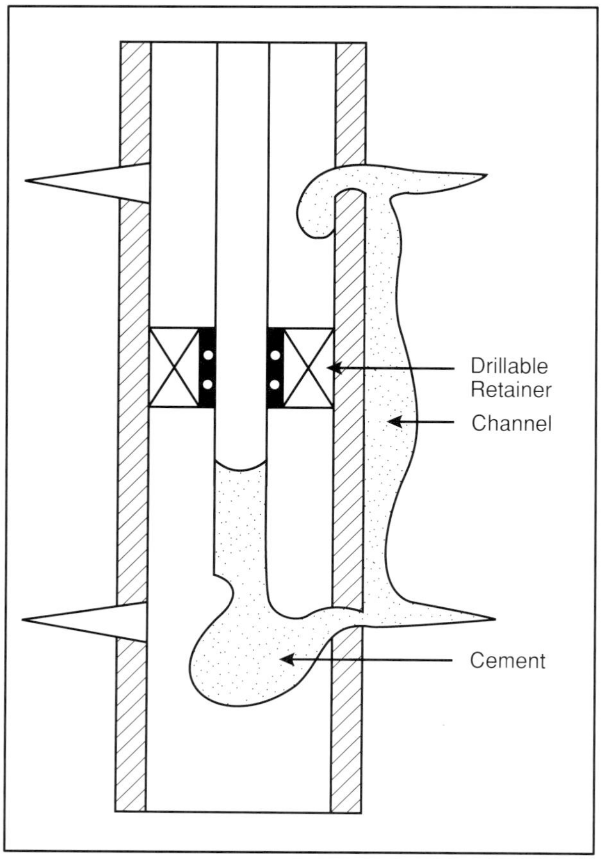

Figure 13–15—Circulating squeeze.

this risk. It is easier to remove the stinger assembly than the packer due to minimal casing clearance of the latter. Preferably, the retainer should be set as close as possible to the upper perforations to minimize the exposure of the drillpipe to cement which may enter the wellbore through the upper perforations.

13–7.2 Shutting Off Unwanted Water

Water intrusion (usually as a result of coning) may occur during the life of a well, resulting in an excessive water/oil ratio (WOR), and reduced well productivity. Remedial cementing is performed to seal off the unwanted water.

The coning water may be either bottomwater, or water which is migrating through annular channels. Such water production can only be stopped or altered if the water flows through natural or created fractures, or through channels in the primary cement sheath. Water flowing through the vertical permeability of the matrix is very difficult to stop because, during a squeeze treatment, only the cement filtrate penetrates the formation pores. The cement grains remain at the formation face.

Recently, nonparticle-laden and low-viscosity fluids, capable of forming strong gels under a wide range of downhole conditions, have been developed (Vidick et al., 1988). These fluids are able to penetrate deeper into the rock permeability, offering promising perspectives for solving water invasion problems.

Also, the use of coiled tubing units for such jobs has gained considerable popularity. Coiled tubing has proved to be a very economical method to accurately place the small volumes of cement slurry usually involved in these operations. Harrison and Blount (1986) reported that, in some instances, up to 85% savings on workover costs have been achieved when using these units. As illustrated in Fig. 13–16, the procedure can be divided into four stages.

- A supporting column of mud ("viscous pill") is injected until its level is just below the perforations to be squeezed. Some mud contamination may occur during placement, because of mixing with wellbore fluids; thus, the coiled tubing string nozzle is pulled up, and contaminated mud is circulate out. The wellbore above the mud is then loaded with water and oil or diesel.
- Cement is pumped with the nozzle located just above the mud/water interface. When the perforations are covered with cement slurry, squeeze pressure is applied. The nozzle must be kept below the water/cement interface.
- After the squeeze pressure has been attained, a contaminant fluid is injected to dilute the cement slurry.
- The contaminated cement and mud are reversed out, and the wellbore is flushed clean.

13–7.3 Reducing the GOR

During the life of a well the GOR may increase beyond the economic limit, necessitating remedial action. Such a situation is illustrated in Fig. 13–17. A common procedure is to squeeze off all the perforations in the oil and gas zones, and reperforate a selected interval (Goodwin, 1984).

13–7.4 Repairing a Casing Split or Leak

Squeeze cementing is also applied to repairing defects in the casings. However, when dealing with old and corroded casing, one should be aware that it will probably suffer more damage from the application of the treating pressure and packer-generated stresses. It may be advisable to pull out the old casing (if possible) and run a new string. Squeeze treatments performed on old wells with corroded casings often fail after a short period of time, because of the opening of new holes due to corrosion.

Casing leaks have also been spotted on new pipes, in which case a "patching" job can be designed. The

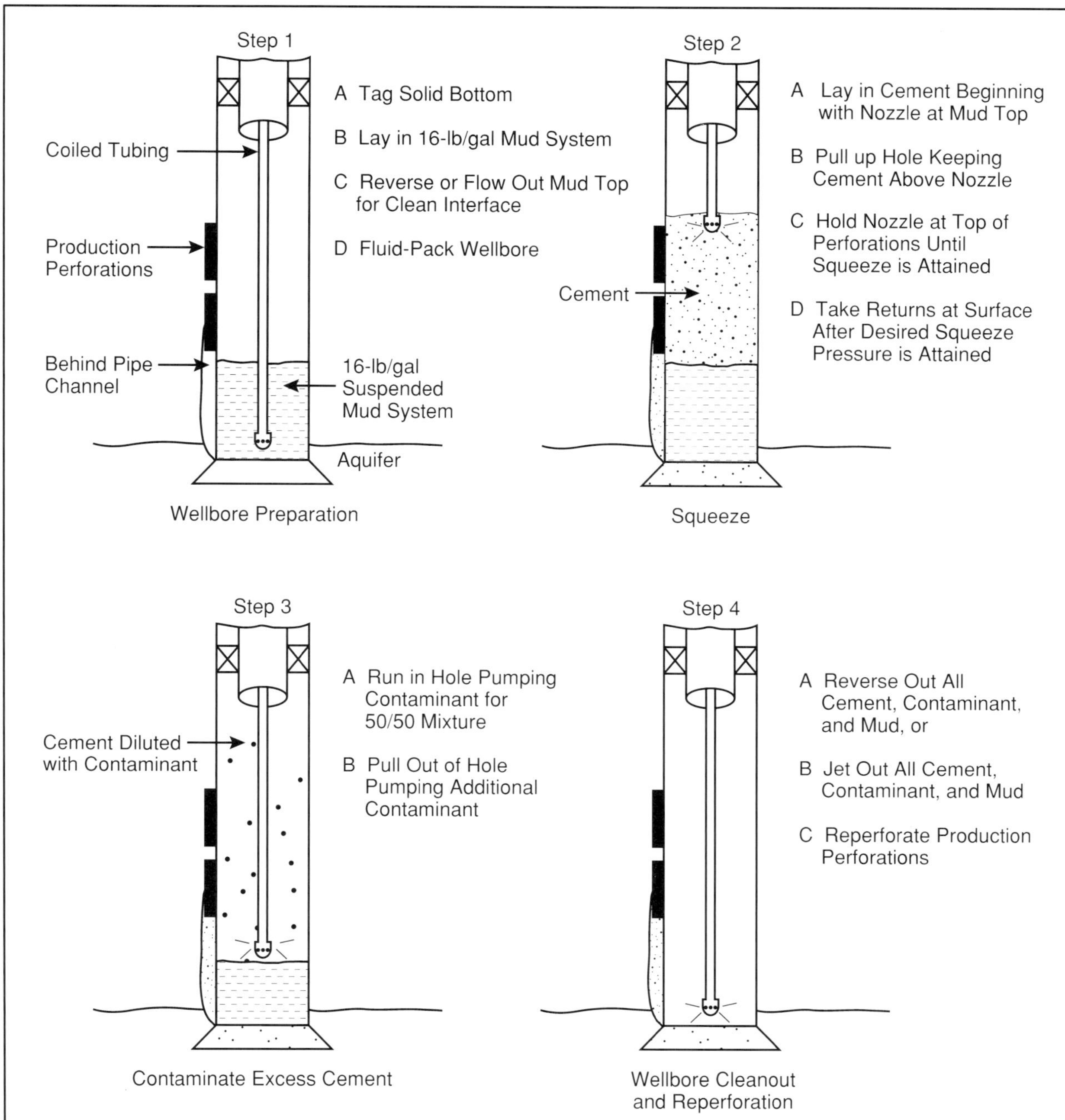

Figure 13–16—Coiled tubing squeeze (after Harrison and Blount, 1986).

squeeze is performed in the same way as one would set a cement plug, i.e, with an open-ended drillpipe (or tubing). The drillpipe or tubing is then pulled up above the cement, and squeeze pressure is applied while ensuring that the formation fracturing pressure is not reached.

13–7.5 Abandoning Nonproductive or Depleted Zones

Plugging off depleted zones is a commonly performed workover operation. The injection of the slurry is usually performed through a squeeze tool (packer or retainer),

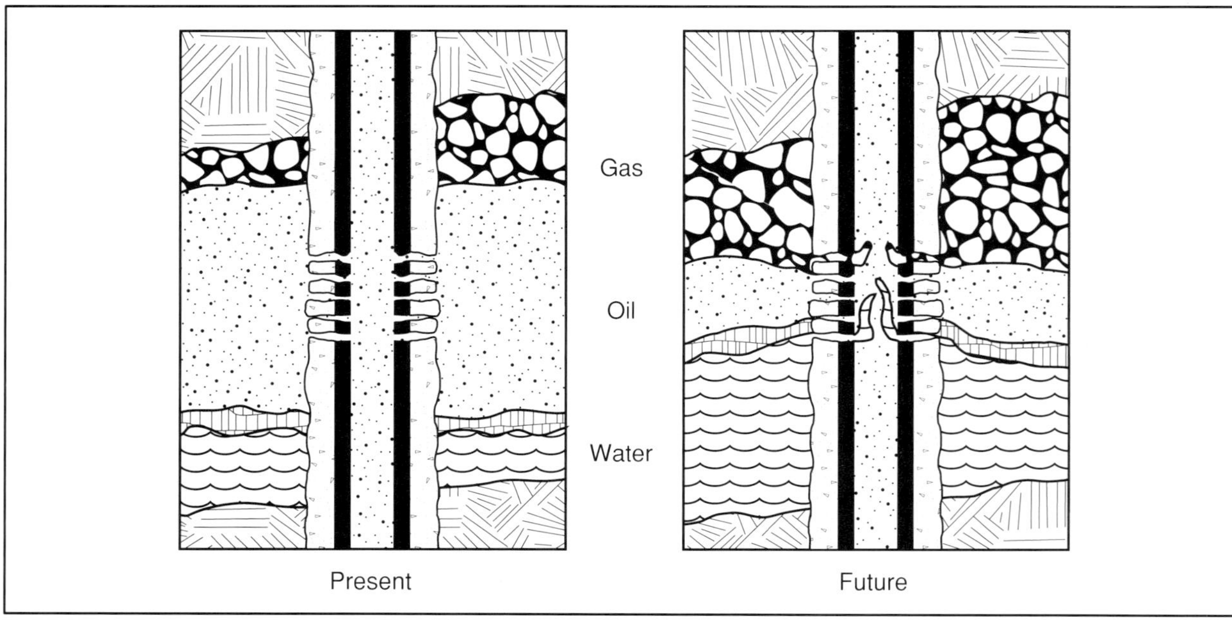

Figure 13–17—Water or gas invasion into oil zone (after Shryock and Slagle, 1968).

and at below formation fracturing pressure (a low-pressure squeeze). In recent years, however, operators have turned to coiled tubing units in preference to workover rigs to perform these operations economically.

13–7.6 Supplementing a Primary Cement Job

The height of the primary cement may be insufficient because of a number of factors:

- lost circulation occurring during the cement job,
- the hole volume is not accurately known, and
- the presence of large washouts, preventing proper measurements with a caliper tool.

The height of the cement column in the annulus can be extended by perforating the casing at the cement top, and pumping a cement slurry through the perforations to the desired depth. Bond logs and/or temperature surveys are available to locate the top of the cement accurately.

This operation is similar to a primary job and simply involves the circulation of a cement slurry. It is performed at below formation fracturing pressure.

13–7.7 Altering Injection Profiles

Squeeze cementing has been successfully used to alter unacceptable injection profiles in water injection wells. A practice reported by some authors consists of squeezing off all the perforations, and reperforating a new interval (Goolsby, 1969).

13–7.8 Block Squeeze

"Block squeezing" is a method to prevent fluid migration from either above or below a producing zone. The permeable section below the producing zone is perforated and squeezed; then, the procedure is repeated for the permeable section above the producing zone. Both plugs are then drilled out, and the productive zone is perforated for production. This method has been successful in many cases for the prevention of water coning.

13–7.9 Top of Liner

As discussed in Chapter 12, liner cementing is difficult because of the small annular clearance and often poor centralization. The amount of cement used during the primary job is small, and the wellbore geometry is often not conducive to good mud removal; consequently, mud contamination can occur. In addition, annular gas migration can be a complicating factor.

If the liner leaks at the overlap, remedial cementing is indicated, and the procedure is the same as squeezing a hole in the casing. If good coverage is not obtained, the casing must be perforated, and another squeeze job performed (Smith, 1987).

13–8 EVALUATION OF A SQUEEZE JOB

The need to evaluate the results of a squeeze job is determined by the requirements of the subsequent operations to be performed on the well.

13–8.1 Positive Pressure Test

After the WOC time has elapsed, it is common practice to test the plugged perforations. However, this must not be considered as a test of the ability of the cement to hold the formation fluid in place; rather, the test serves as a method to diagnose a gross failure of the squeeze treatment.

The pressure applied at the face of the perforations is predetermined at the job-design stage. It may be the reservoir pressure, but should not exceed the formation fracturing pressure (Crenshaw, 1985).

Mud filter cake has been known to withstand over 5,000 psi (34.5 MPa) of differential pressure when applied from the wellbore toward the formation. Yet the same filter cake cannot withstand a significant differential pressure when applied from the formation toward the wellbore (Rike and Rike, 1981).

13–8.2 Negative Pressure Test

The universally recognized technique for confirming whether the cement in place will hold the formation fluids under producing conditions consists of applying a negative differential pressure on the face of the plugged perforations. This is accomplished by the following steps:

- circulating a light fluid (i.e., through a concentric pipe),
- swabbing the well, and
- running a dry test (Fig. 13–18) (Chapter 16).

If the sealing achieved in the perforations is complete, no inflow should be recorded on the test pressure chart (Fig. 13–19).

13–8.3 Acoustic Log

When the objective of the squeeze is to repair a primary cementing job, the normal cement logs should be run to evaluate the effectiveness of the repair by comparing pre-squeeze and post-squeeze logs (Chapter 16).

13–8.4 Temperature Profile

Goolsby (1969) evaluated squeeze results on water injector wells by comparing pre- and post-squeeze temperature profiles. By logging the well temperature after a

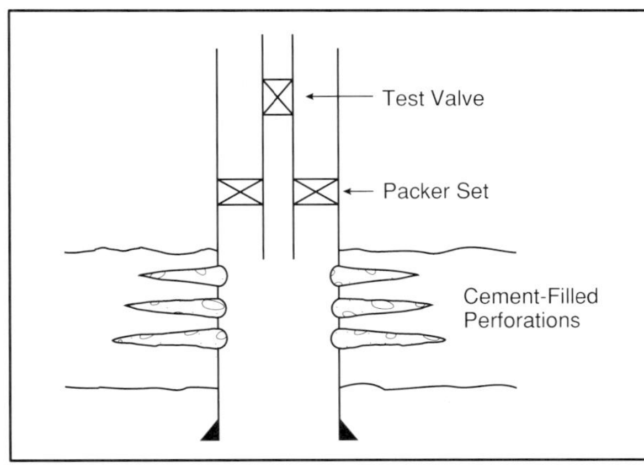

Figure 13–18—Running a dry test.

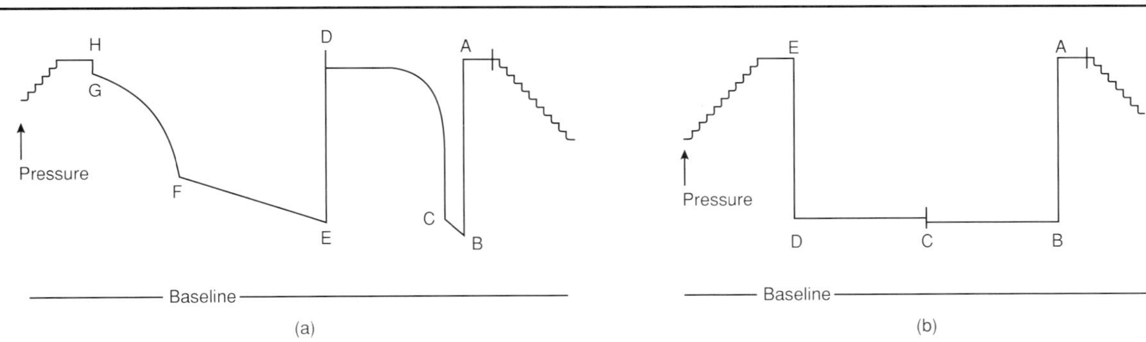

Typical DST chart of a perforated interval before cement squeezing.

A Initial hydrostatic pressure, packer set
B Initial flowing pressure (water cushion used) when test valve tool opened
B-C Flow period
C Test valve closed
C-D Pressure build up curve
D Initial shut-in pressure
E Flowing pressure after second opening of test valve tool
E-F Flow period
F Test valve closed
G Second (final) shut-in pressure
H Final hydrostatic pressure after unsetting packer

DST chart of the same interval after a successful squeeze job.

A Initial hydrostatic pressure, packer set
B "Flowing" pressure when test valve tool opened
C Pressure at the end of "flow period," test valve tool closed
D Pressure at the end of shut-in period
E Final hydrostatic pressure after unsetting packer

Figure 13–19—Dry test.

post-squeeze large injection test, he was able to demonstrate that the well was taking the injection water at the planned location.

13–8.5 Cement Hardness

Suman and Ellis (1977) reported that in squeeze jobs where cement was drilled out, a good indication of success was the nature of the cuttings. If the cement was hard throughout, the results were usually good. Soft spots or voids usually indicated a failure.

13–8.6 Radioactive Tracers

Radioactive materials may be added to the cement slurry, and subsequent tracer surveys can indicate whether the cement is in the desired interval. I^{131}, Ir^{192}, and Sc^{46} are appropriate because of their short half-lives—8 days, 75 days, and 85 days, respectively. The iridium and scandium radioisotopes are preferable, because iodine (present as iodide) is soluble, and may be squeezed out of the cement with the filtrate.

13–9 REASONS FOR SQUEEZE-CEMENTING FAILURES

Whenever a squeeze job has failed to meet the objectives, a thorough investigation must be conducted to analyze the job, understand why a failure occurred, and improve the design of subsequent treatments.

13–9.1 Misconceptions

- The cement slurry penetrates the pores of the rock.

  Only the mix-water and dissolved substances penetrate the pores, while the solids accumulate at the formation face and form the filter cake. It would require a permeability higher than 100 darcies for the cement grains to penetrate a sandstone matrix. The only way for a slurry to penetrate a formation is through fractures and large holes (vugs).
- High pressure is needed to obtain a good squeeze.

  If the formation fracturing pressure is exceeded, control of the placement of the slurry is lost, and the slurry enters unwanted areas. Pressure is of no help to place the slurry in all the desired locations. Once created, a fracture may extend across various zones, and open unwanted channels of communications between previously isolated zones.

13–9.2 Plugged Perforations

Another common misconception concerning squeeze cementing, which can lead to failure, is that all perforation holes are open and receptive to fluids (Rike and Rike, 1981). The mud filter cake, which is capable of withstanding a large differential pressure when applied from the wellbore toward the formation, easily cleans up when submitted to a differential pressure in the other direction. In addition to mud cake, debris, scale, paraffin, formation sand, pipe dope, rust, paint, etc., can accumulate in the perforations, and contribute to the plugging. Goodwin (1984) reported that, in a producing well, the upper perforations are usually open, while the plugged perforations are generally found in the lower zones. Squeezing under these well conditions results in the failure to fill all the perforations with cement, and the plugged perforations will allow the flow of formation fluids and indicate the failure of the squeeze.

Perforation washing before the squeeze job is a useful method for making all perforations receptive to the squeeze cement slurry. This can be done by mechanical or chemical means.

Mechanical perforation washing involves the use of a washing tool and back-surge techniques. The perforation washing tool (Fig. 13–20) isolates a small number of perforations at a time. A wash fluid is pumped down the tubing, forced into the perforations, then outside the casing and back through upper perforations into the annulus. The tool is slowly moved upward to cover the entire perforated interval. Common wash fluids are chemical washes containing surfactants, followed by weak acids when scales or drilling muds are to be removed. Solvents are used when paraffin deposits are present.

The surge tool (Fig. 13–21) is basically an air chamber between an upper and lower valve. The tool is run in the hole with a packer to isolate the desired interval. Once the packer is set, the lower valve is opened (annulus pressure operated), allowing fluids to enter the air chamber. The rapid depressurization of the borehole creates a high differential pressure across the perforations, and the subsequent cleanup of debris and other plugging materials from them. To establish circulation after surging, the upper valve is opened (this is accomplished by tubing movement, tubing pressure, or disk rupturing) and the debris is reverse circulated out of the hole.

The chemical perforation cleaning techniques involved the use of acids and solvents, pumped ahead of the squeeze slurry, as spearhead fluids to clean the perforations.

13–9.3 Improper Packer Location

Should the packer be set too high above the perforations, the cement slurry becomes contaminated as it channels through the mud or completion fluid. Slurry properties such as fluid loss, thickening time, and viscosity are adversely affected by contamination, and slurry placement results are altered.

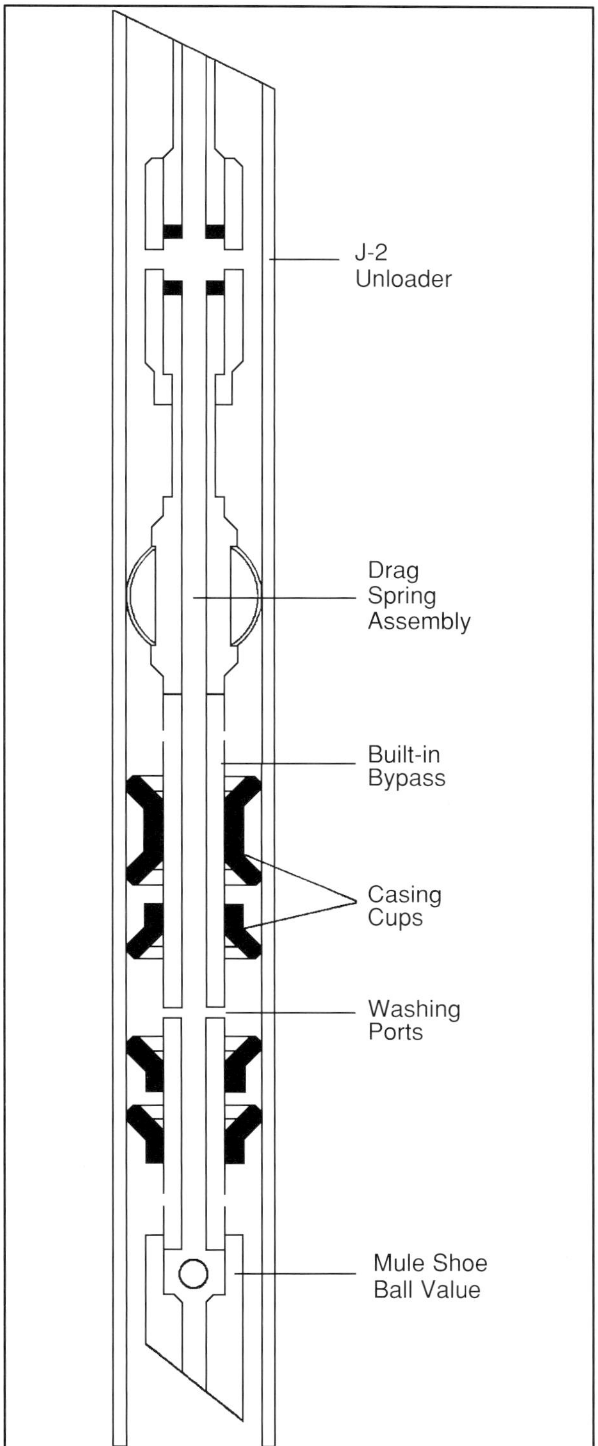

Figure 13–20—A perforation washing tool.

Shryock and Slagle (1968) recommended that the squeeze packer be set at no more than 75 ft (23 m) above the perforations. Suman and Ellis (1977) recommended that the packer be set between 30 and 60 ft (9 and 18 m) from the perforations. The use of compatible spacer fluids ahead and behind the cement slurry was also recommended.

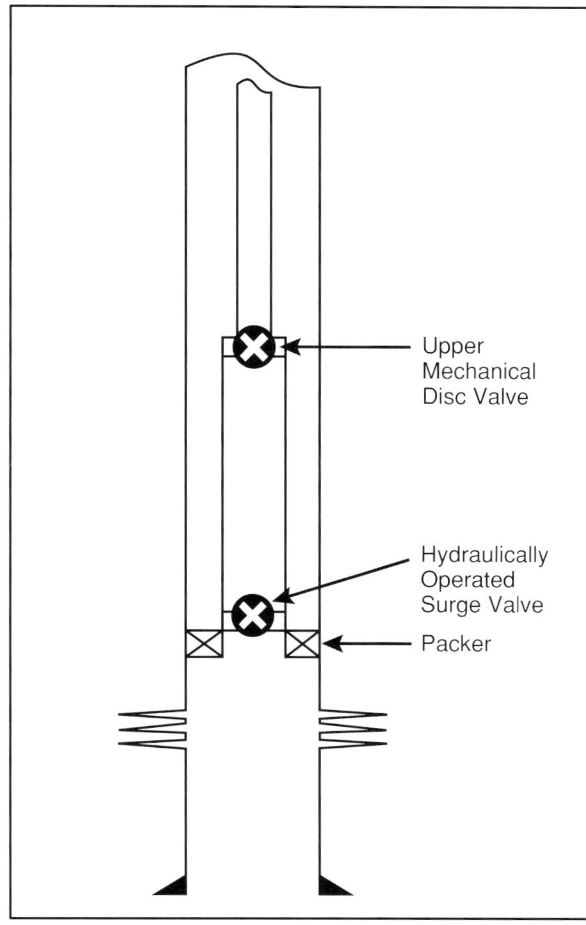

Figure 13–21—Surge tool.

13–9.4 High Final Squeeze Pressure

A high final pressure does not increase the chances of success; on the contrary, it increases the chances of fracturing the formation, and losing control of the cement slurry placement. It is important that a "think downhole" attitude be developed among all personnel (designer or operator) involved in this operation.

13–10 SQUEEZE CEMENTING—CONCLUSIONS

Successful squeeze cementing starts at the job-design stage. The following questions must all be answered before executing the operation.

- What is the problem?
- What are the objectives of the job?
- Which squeeze technique will be used?
- Which types of tools are to be used?
- At which depth should the packer be set?

- To which depth should the tail pipe be lowered?
- Which well preparation technique(s) is needed?
- Which type of fluid is in the hole?
- What will be the maximum job pressure?
- Which job procedure will be followed?
- What is the estimated job time?
- What are the composition and properties of the slurry?
- What quantity of slurry is necessary?
- How will the perforations be opened or cleaned?
- What is the formation breakdown pressure?
- Have other formation data (lithology, permeability, pressure, water/oil and gas/oil contact, bottomhole temperature) been considered?
- What is the WOC time?
- How will the job be evaluated?

One cannot be confident of job success, unless satisfactory answers to these questions have been received.

However, there is common thinking among authors that the most successful method is the low-pressure squeeze, with a packer or retainer as the isolation tool, using a low-fluid-loss slurry and the hesitation pumping technique. The high-pressure technique should be applied with extreme caution.

13–11 CEMENT PLUGS—INTRODUCTION

Setting a cement plug in a wall is a common oil-field operation. A cement plug involves a relatively small volume of cement slurry placed in the wellbore for various purposes:

- to sidetrack above a fish or to initiate directional drilling,
- to plug back a zone or plug back a well,
- to solve a lost-circulation problem during the drilling phase, and
- to provide an anchor for openhole tests.

The necessary equations for job design are presented in Appendix C.

13–11.1 Sidetrack and Directional Drilling (Whipstock Plug)

During directional drilling operations, it may be difficult to achieve the correct angle and direction when drilling through a soft formation (Fig. 13–22). It is a common practice to set a "whipstock plug" across the zone to achieve the desired course and target. Also, in some instances where fishing cannot be performed economically, the only remaining solution is to plug the hole with cement, and sidetrack the well above the fish.

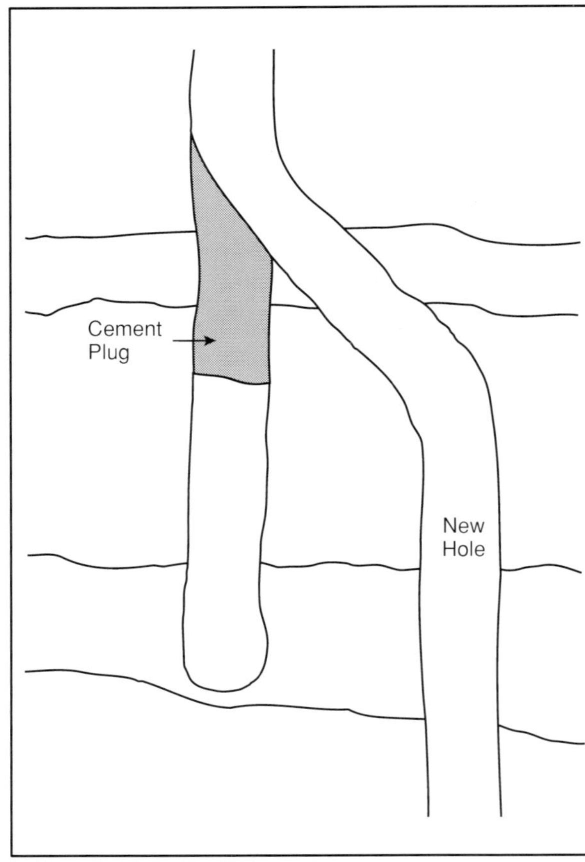

Figure 13–22—Sidetrack plug.

13–11.2 Plugback

Several cement plugs at various depths are set to abandon a well and prevent zonal communication or the migration of fluids which might pollute underground freshwater sources (Fig. 13–23). Depleted producers are also plugged with cement when they are abandoned (Fig. 13–24). In many countries, oil and gas well operators are compelled to precisely follow abandonment procedures which are dictated by government authorities.

13–11.3 Lost Circulation

Loss of drilling fluid can be stopped by setting a properly formulated cement slurry across the thief zone. Although the slurry may be lost to the thief zone, it will harden and consolidate the formation (Fig. 13–25). A cement plug can also be set on top of a zone, to protect it from being fractured under the hydrostatic pressure that might develop during the cementing of a casing string. Lost-

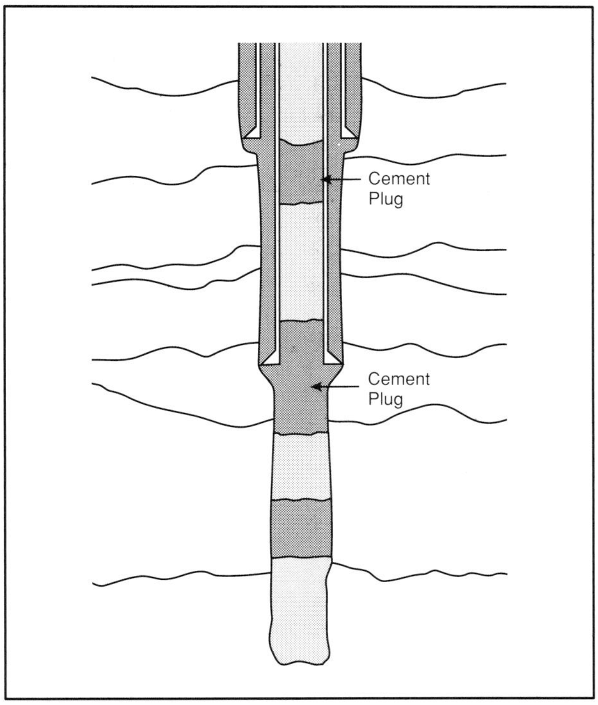

Figure 13–23—Well abandonment plugs.

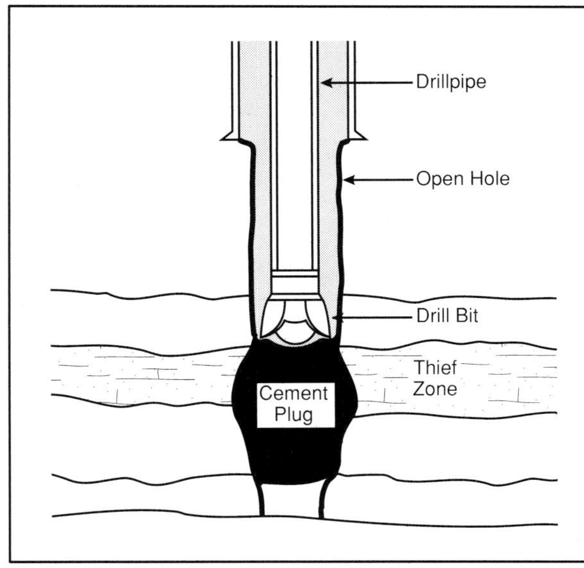

Figure 13–25—Lost circulation plug.

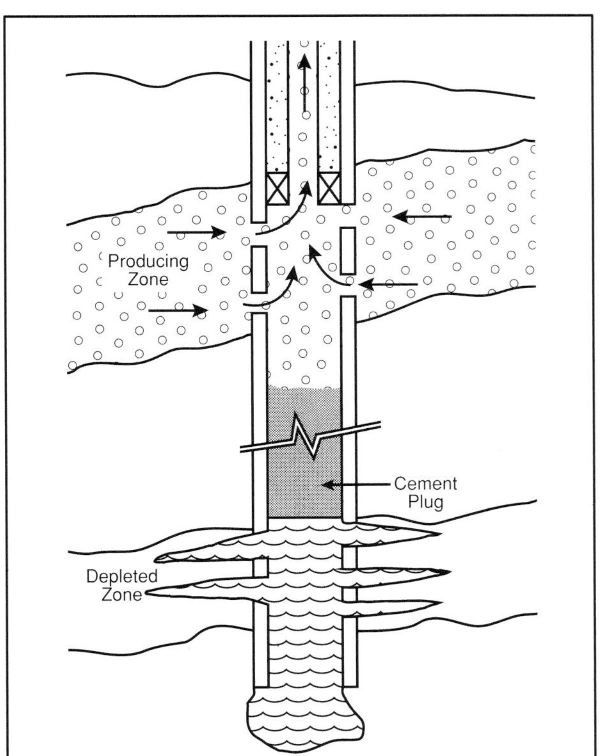

Figure 13–24—Plugging a depleted zone.

circulation additives are often added to the slurry to assure a successful job in this environment (Chapter 6).

13–11.4 Test Anchor

Cement plugs are set when a soft or weak formation exists in an open hole below the zone to be tested, and when it is impractical or impossible to set a sidewall anchor (Fig. 13–26).

13–12 PLUG PLACEMENT TECHNIQUES

There are three common techniques for placing cement plugs:

- balanced plug,
- dump bailer, and
- two-plug method.

13–12.1 Balanced Plug

The most common placement method is the balanced plug technique (Fig.13–27). Tubing or drillpipe is run in the hole to the desired depth for the plug base. An appropriate volume of spacer or chemical wash is pumped ahead and behind the slurry to avoid any detrimental contamination of the cement by the mud. The slurry is often batch mixed for better density and rheology control.

The volumes of spacer or wash are such that their heights in the annulus and in the drillpipe or tubing are the same. Displacement is completed to the depth of the calculated plug top in the pipe. It is common practice to slightly underdisplace the plug (usually by two or three barrels) to avoid mud flowback on the rig floor when

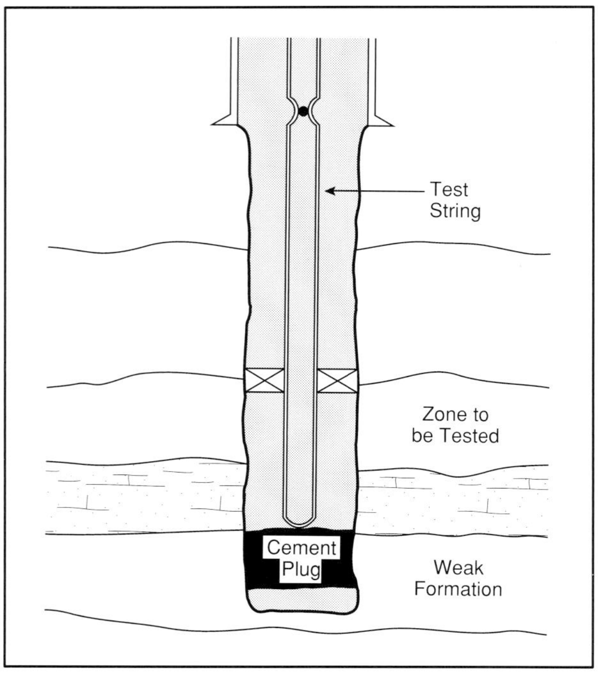

Figure 13–26—Plug set as an anchor for a test.

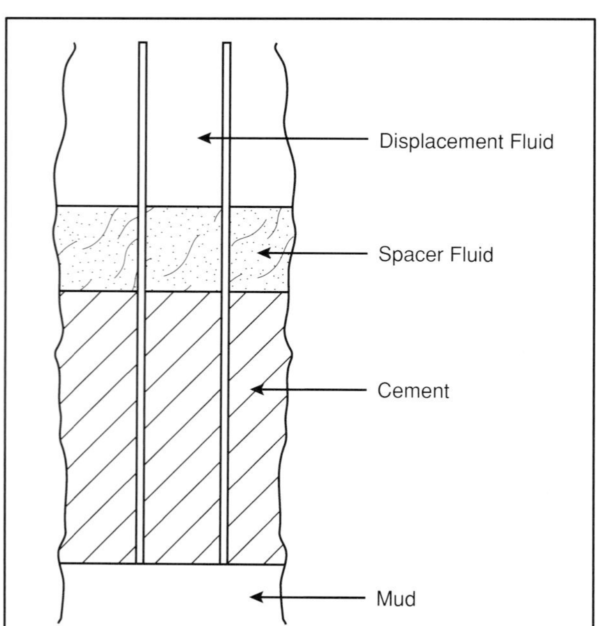

Figure 13–27—Balanced plug.

breaking the pipe after the placement, and allow the plug to reach hydrostatic balance. Once the plug is balanced, the pipe is slowly pulled out of the cement to a depth above the plug, and excess cement is reversed out.

13–12.2 Dump Bailer Method

The cement is placed by running a dump bailer, containing a measured quantity of cement slurry, on a wireline. The bailer is opened by touching a permanent bridge plug placed below the desired plug interval, and the cement is dumped on the plug by raising the bailer (Fig. 13–28). Usually employed for setting plugs at shallow depths, the dump bailer method can also be used at greater depths by using properly retarded cement systems. The advantages of this method are that the depth of the cement plug is easily controlled, and it is relatively inexpensive. The principal disadvantage is that the available quantity of cement is limited to the volume of the dump bailer.

13–12.3 Two-Plug Method

This method uses a special tool to set a cement plug in a well at a calculated depth, with a maximum of accuracy and a minimum of cement contamination. The tool essentially consists of a bottomhole sub installed at the lower end of the drillpipe, an aluminum tail pipe, a bottom wiper plug (which carries a dart), and a top wiper plug (Fig. 13–29).

The bottom plug is pumped ahead of the cement slurry to clean the drillpipe wall and isolate the cement from the mud. The shear pin connecting the dart to the plug is broken by increased pump pressure, and pumped down through the aluminum tail pipe. The top plug is pumped behind the cement slurry to isolate the cement from the displacement fluid. Increased surface pressure is observed when the plug arrives at its seat. The drillpipe is pulled up until the lower end of the tail pipe reaches the calculated depth for the top of the cement plug. The shear pin between the catcher sub body and the sleeve is then broken, allowing the sleeve to slide down and unmask the reverse circulating path. If in the course of the operation the aluminum tail pipe becomes stuck in cement, an increase in the pull will break the tail pipe and free the drillpipe.

13–13 JOB-DESIGN CONSIDERATIONS

The design of the job starts with the definition of the objective. Setting a plug for lost circulation is quite different from setting a plug to abandon a depleted zone or to plug back a well. Before each job the following questions need to be answered.

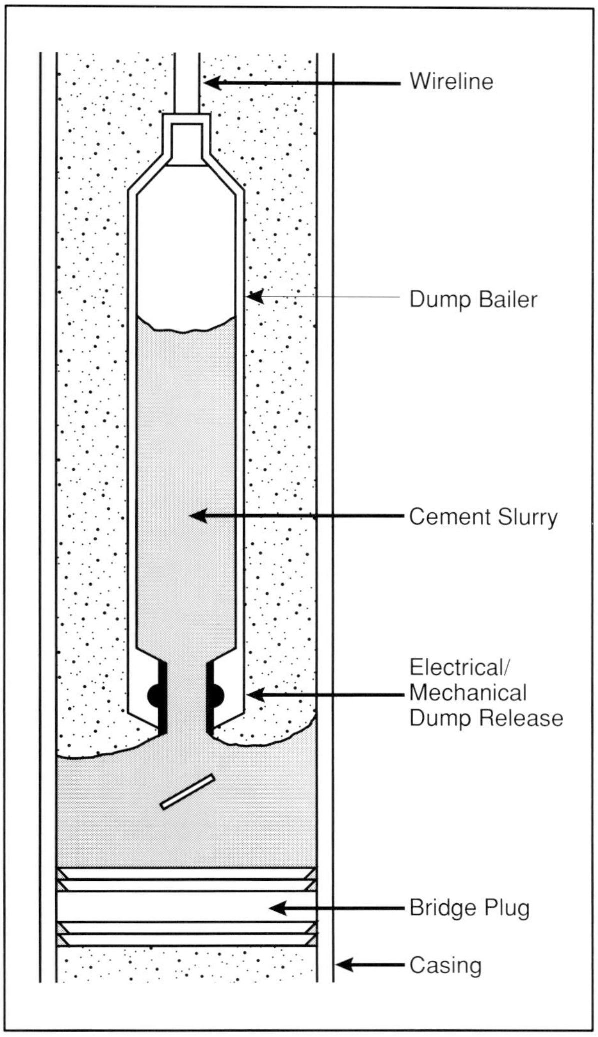

Figure 13–28—Dump bailer method.

- *At what depth will the plug be set?*

  The chances of success are greatly improved if the plug is set where the hole is near gauge. The logs should be consulted.

- *Across which formation is the plug going to be set?*

  A cement plug is best set in a competent hard rock. Shales should be avoided as they are often caved and out of gauge. However, if kicking off is the objective, the plug should not be set in an excessively hard formation. Ideally, the plug should extend from a soft shale down to a hard formation (Dees and Spradlin, 1982). In any case, the logs and the drilling rate record should be consulted when selecting a location to set a plug.

- *At what density should the slurry be mixed?*

  The higher the density differential between the slurry and the drilling fluid, the higher the chance that the slurry will migrate downward. Proven techniques that prevent this phenomenon are described below. On the other hand, a lighter slurry will result in a lower the compressive strength. On average, a 15.8-lb/gal (1.90-g/cm^3) slurry develops a final compressive strength of at least 5,000 psi (34.5 MPa). A reduced water slurry of 17.5 lb/gal (2.10 g/cm^3) develops a final compressive strength of at least 8,500 psi (58.6 MPa).

  For better control of the slurry density, the batch-mixing technique is preferable (Smith et al., 1983). Slurry densities usually range from 15.6 lb/gal (1.87 g/cm^3) to 17.5 lb/gal (2.10 g/cm^3) to ensure good compressive strength development.

- *What is the bottomhole temperature?*

  The API recommends that well simulation test procedures follow simulated Squeeze Cementing Schedules 12 through 21 (Appendix B).

- *What volume should be pumped?*

  The amount of cement depends upon the objective of the plug. The lengths and depths of abandonment plugs are usually dictated by government regulations. Whipstock plugs must be very long to provide for a gradual deviation of the bit. A caliper of the hole is very useful. The size of the cement plug should be 300 to 900 ft (91 to 274 m) of annular fill (Smith et al., 1983). Care must be taken to avoid excessive hydrostatic pressure on lower depleted or weak zones; otherwise, the plug may not be placed at the desired depth.

- *Is mud conditioning necessary prior to the operation?*

  A low-rheology mud is easier to displace.

- *What are the appropriate slurry properties?*

  Viscous slurries with high gel strength are needed for lost-circulation plugs, to restrict flow into voids or fractures. When the difference between cement density and hole fluid density is high, the cement will tend to fall through the lighter fluid. In this case, thixotropic slurries may solve the problem. Another approach is to place a viscous bentonite mud pill below the intended plug depth to act as a support medium for the cement (Fig. 13–30) (Smith et al., 1983). As shown in Fig. 13–31, a diverter tool is recommended. Such tools minimize the risk of the heavy slurry "telescoping" through the mud, but diverting the fluid flow through side ports at the bottom of the work string.

  High compressive strength is mandatory in whipstocking to have a sharp contrast between the plug and formation hardness. Since compressive strength is a function of the water/solids ratio, high-density (low water/solids ratio) slurries are best suited for such

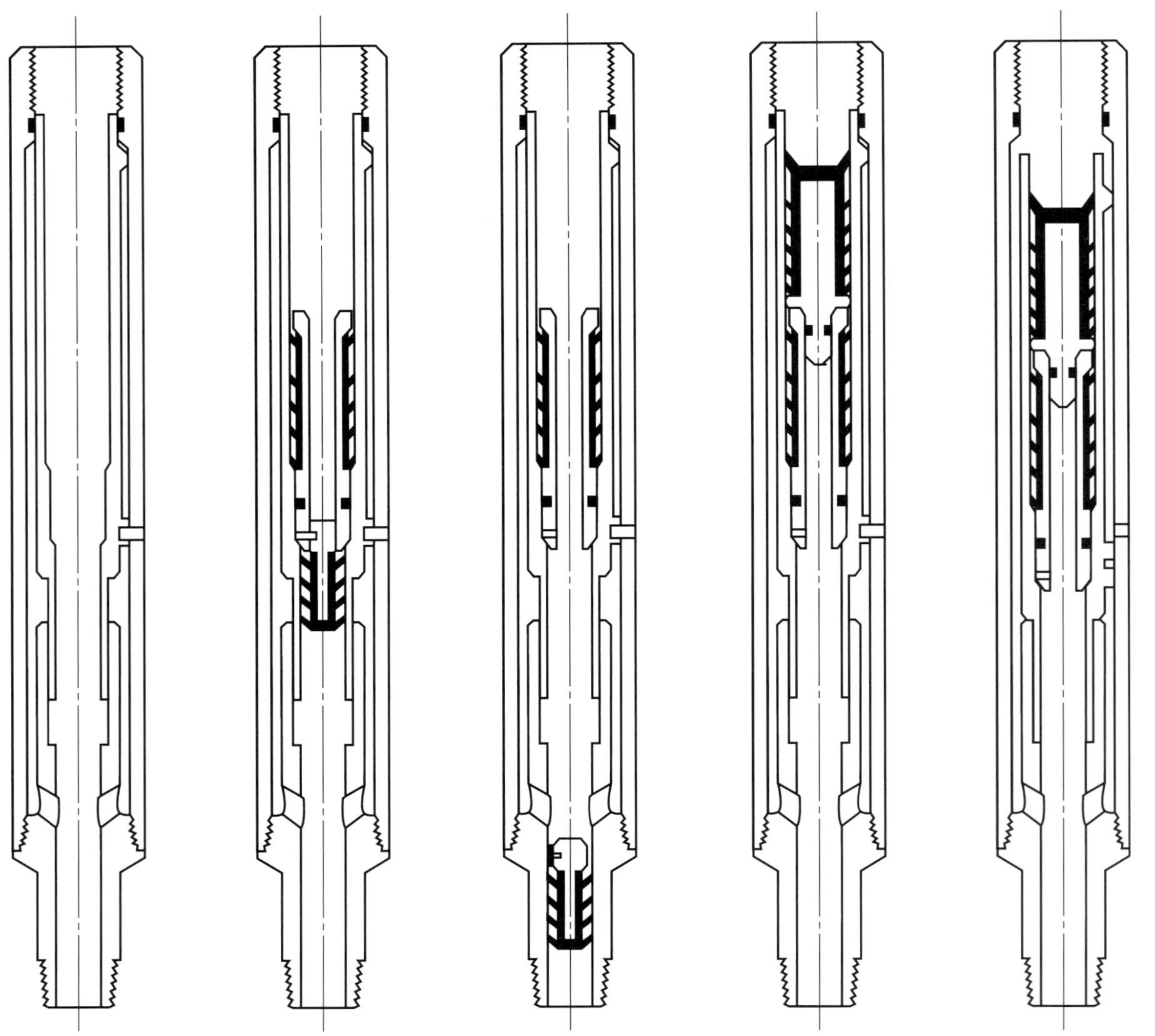

Figure 13–29—Two-plug method.

plugs. Addition of sand or weighting materials will not improve the compressive strength of a lower water content slurry. On the other hand, lost circulation or plugback jobs may require viscous low-density slurries to avoid losing the plug in the formation.

- *What is the appropriate thickening time?*

  Smith et al. (1983) recommended that the slurry pumping time be equal to the anticipated job time plus 30 minutes.

- *How does one ensure that the cement will not be contaminated with the mud?*

  The use of spacers and washes is a must, as most muds are incompatible with cement slurries. Bradford (1982) recommended that the spacer be 1 to 2 lb/gal heavier than the mud, to gain the effect of buoyancy for improved mud displacement. Smith et al. (1983) recommended that, whenever possible, spacers and washes be pumped in turbulent flow conditions. An annular height of 500 to 800 ft (152 to 244 m) is recommended. If turbulent flow is not feasible, plug flow spacers are perfectly acceptable. In addition, the use of densified cement slurries can help reduce the likelihood of mud contamination, as well as reduce the impact of the mud contamination should it occur.

- *Are pipe centralization and pipe rotation necessary?*

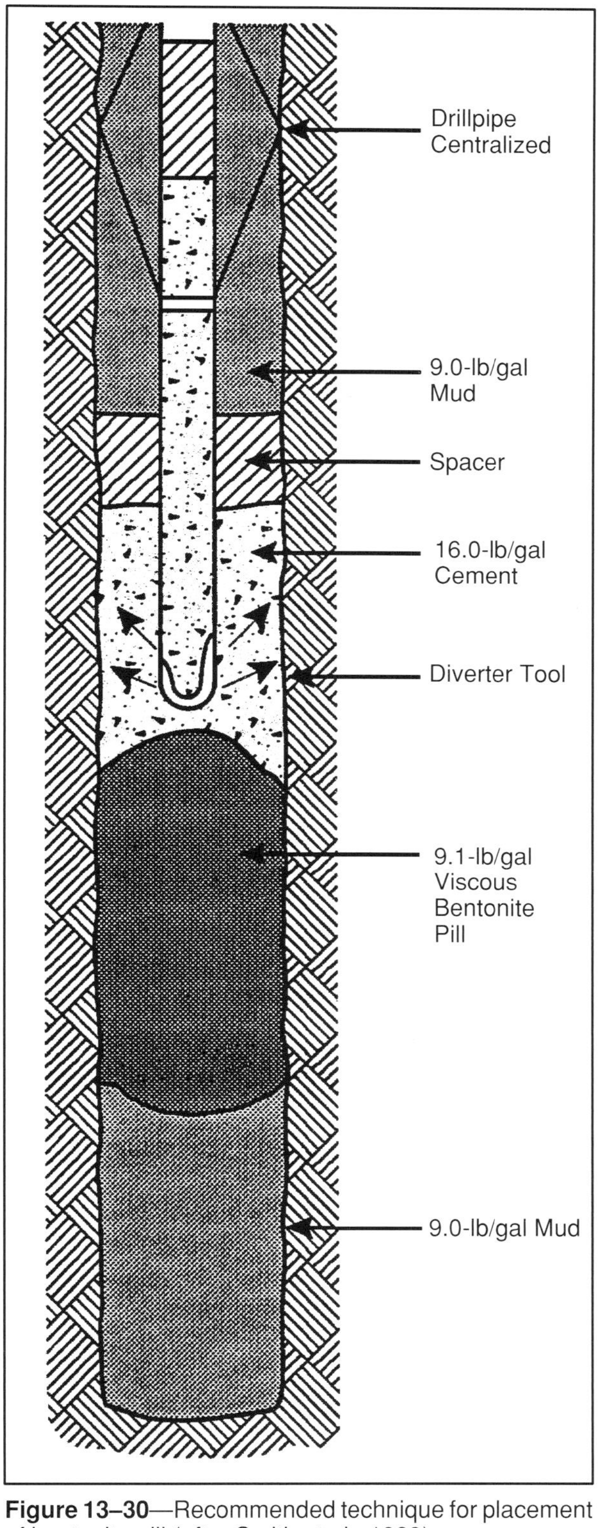

Figure 13–30—Recommended technique for placement of bentonite pill (after Smith et al., 1983).

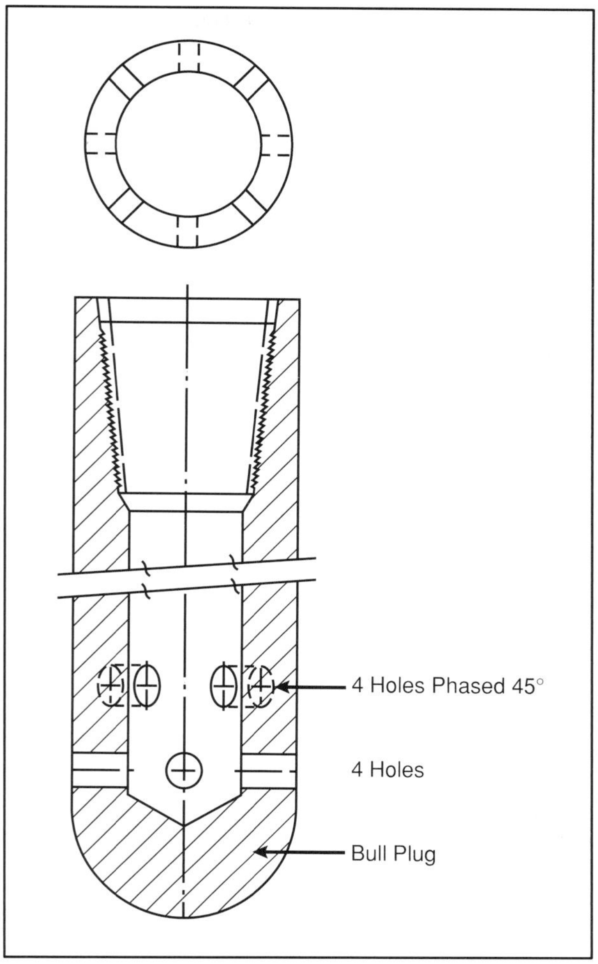

Figure 3–31—Flow diverter tool (after Smith et al., 1983).

Bradford (1982) recommended that the pipe be carefully centralized. This precaution can dramatically improve mud removal. Pipe rotation is also cited as an advisable practice.

- *Waiting on cement time?*

  Early compressive strength depends heavily on the thickening time. Rig time can be saved with a proper slurry design. The slurries must be designed for a thickening time in accordance with well conditions and job procedures, plus a reasonable safety factor. Smith et al. (1983) recommended that ample WOC time be allocated (12 to 24 hours). Since the well temperature for a cement plug job is difficult to know accurately, a common practice is to allow for longer WOC times. A minimum of 500-psi (3.5-MPa) compressive strength is normally recommended for drilling out cement.

13–14 EVALUATION OF THE JOB, REASONS FOR FAILURES

After the WOC time has elapsed, the job results are evaluated. This is normally done by tagging the cement. Depth of the top of the plug and hardness of the cement are the key indicators to measure success or failure. Whenever a cement plug has failed to meet the objectives of the job, the reason(s) for the failure should be carefully investigated to modify and improve the design of the repeated attempt and to be successful the next time. Some of the most common causes of failure are listed below.

- *Mud Contamination*

  Mud contamination is recognized as a major cause of cement plug failure (Bradford, 1982; Smith et al., 1983). Mud contamination dramatically affects the cement compressive strength (Table 13–4). This contamination may result from a poorly centralized pipe. If the tubing or drillpipe is not properly centralized, it will stand on one side and the slurry coming out of the bottom will follow the path of least resistance—the open side. Cement channels in the mud, and mixes with the mud when the pipe is pulled out of the hole.

- *Insufficient Cement Volume*

  The plug may have been set in a washout or a poorly calibrated section of hole; therefore, its height is not adequate. Also, in large hole sections, mud displacement is difficult, immobile gelled mud is present, and chances of contaminating the slurry are high. It is strongly recommended to place the plug in a well-calibrated hole section.

  Bradford (1982) and Dees and Spradlin (1982) recommended a minimum of 500 ft (152 m) for plug height. Smith et al. (1983) recommended that the plug size should be 300 to 900 ft (91 to 274 m). The extra cement required is very economical when compared to the costs associated with repeating the job, waiting again on cement, retesting the plug, etc.

- *Plug Slipped Downhole*

  Smith et al. (1983) recognized that a heavy cement resting on top of a lightweight mud forms an unstable interface; as a result, the cement channels downward and becomes diluted in the mud (Fig 13–32).. The common practice of using an open-ended drillpipe is a major contributor to plug failure, as the cement turning the shoe causes considerable disturbances to the interface. As discussed earlier, this problem can be remedied by the placement of a viscous mud pill with a diverter tool. As an alternative, a delayed-gel fluid could be used instead of the bentonite pill.

  Bour et al. (1986) recommended the placement of a reactive fluids system (RFS), which creates a rapid forming gel acting as a bridge upon which the cement slurry can rest. As a slurry, they recommended the use of an adequate gel strength cement (thixotropic) to counteract the density differential driving forces.

13–15 PLUG CEMENTING—CONCLUSIONS

The implementation of the simple techniques and guidelines described above have resulted in a significant success rate improvement. There are no requirements for large investment, and yet significant savings can be achieved. The following is a list of measures that should be taken (Smith et al., 1983).

- Place the plug in a competent formation (i.e., a hard formation).
- Use ample cement.
- Use a tail pipe through plugback intervals.
- Use scratchers or wipers and centralizers on the tail pipe where the hole is not excessively washed out.
- Use a drillpipe plug and a plug catcher.

Neat Class H Cement 16.5 lb/gal			Effect of Mud Contamination*		
Mud Contamination (% by volume)	Compressive Strength (psi at 170°F) 8 hr	16 hr	Mud Contamination (%)	Normal Slurry 15.6 lb/gal	Reduced Water Slurry** 17.5 lb/gal
0	4,647	5,862	0	4,082 psi	8,600 psi
5	3,512	5,300	10	2,950 psi	8,237 psi
10	2,619	4,538	40	2,426 psi	3,850 psi
20	2,378	2,331	60	593 psi	2,967 psi
50	245	471			

* Compressive strength in 18 hr at 230°F.
** Contains dispersant.

Table 13–4—Effect of mud contamination on set cement compressive strength.

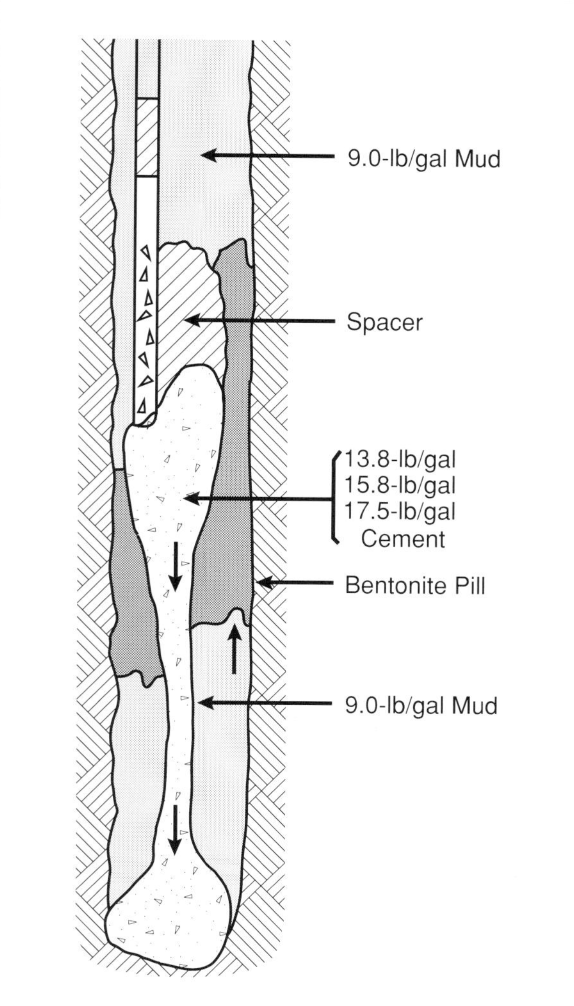

Figure 13–32—The unstable interface (after Smith et al., 1983).

- Circulate the hole sufficiently before running the job. Use a mud of low yield point and low plastic viscosity, but of sufficient weight to control the well.
- Ahead of the cement, run a flush and/or bentonite pill that is compatible with the mud and will prevent the cement from sliding down from the hole.
- Use densified cements with a dispersant to combat the effects of mud contamination. Spacers and washes are also useful.
- Allow ample time for the cement to set.

13–16 REFERENCES

American Petroleum Institute, *Specifications for Materials and Testing for Well Cements*, API Spec 10, third edition, API, Dallas (1986).

Beach, H. J., O'Brien, T. B., and Goins, W. C. Jr: "Controlled Filtration Rate Improves Cement Squeezing," *World Oil* (May 1961) 87–93.

Beirute, R. M: "Flow Behavior of an Unset Cement Plug in Place," paper SPE 7589, 1978.

Binkley, G. W., Dumbauld, G. K., and Collins, R. E.: "Factors Affecting the Rate of Deposition of Cement in Unfractured Perforations During Squeeze-Cementing Operation," *Trans.*, AIME (1958) **213**, 51–58.

Bour, D. L., Sutton, D. L., and Creel, P. G.: "Development of Effective Methods for Placing Competent Cement Plugs," paper SPE 15008, 1986.

Bradford, B. B.: "Setting Cement Plugs Requires Careful Planning and Execution for Successful Cementing Job," *Oil & Gas J.* (Dec. 13, 1982) 102–104.

Bradford, B. and Reiners, B.: "Analysis Gives Successful Cement Squeeze," *Oil & Gas J.* (April 1, 1985) 71–74.

Cementing Technology, Dowell Schlumberger, Nova Communications Ltd., London (1984).

Crenshaw, P. L.: "How to Avoid Myths of Squeeze Cementing," *Oil & Gas J.* (April 22, 1985) 93–95.

Dees, J. M. and Spradlin, W. N. Jr.: "Successful Deep Open-Hole Cement Plugs for the Anadarko Basin," paper SPE 10957, 1982.

Goodwin, K. J.: "Principles of Squeeze Cementing," paper SPE 12063, 1984.

Goolsby, J. L.: " A Proven Squeeze-Cementing Technique in a Dolomite Reservoir," paper SPE 2473, 1969.

Harrison, T. W. and Blount, C. G.: "Coiled Tubing Cement Squeeze Technique at Prudhoe Bay, Alaska," paper SPE 15104, 1986.

Hook, F. E. and Ernst, E. A.: "The Effects of Low-Water-Loss Additives, Squeeze Pressure, and Formation Permeability on the Dehydration Rate of a Squeeze Cementing Slurry," paper SPE 2455, 1969.

Messenger, J.: *Lost Circulation*, Pennwell Books, Tulsa, OK (1981).

Murphy, W. C.: "Squeeze Cementing Requires Careful Execution for Proper Remedial Work," *Oil & Gas J.* (Feb. 16, 1976) 87–88, 90, 93–94.

Rike, J. L.: "Obtaining Successful Squeeze-Cementing Results," paper SPE 4608, 1973.

Rike, J. L. and Rike, E.R.: "Squeeze Cementing: State of the Art," paper SPE 9755, 1981.

Roegiers, J-C.: "Elements of Rock Mechanics," *Reservoir Stimulation*, M. J. Economides and K. G. Nolte (eds.), second edition, Schlumberger Educational Services, Houston (1987) 2-1—2-22.

Shryock, S. H. and Slagle, K. A.: "Problems Related to Squeeze Cementing," *JPT* (Aug. 1968) 801–810.

Smith, R. C., Beirute, R. M., and Holman, G. B.: "Improved Methods of Setting Successful Whipstock Cement Plugs," paper SPE 11415, 1983.

Suman, G. O. Jr. and Ellis, R. E.: *World Oil's Cementing Oil and Gas Wells Including Casing Handling Procedures*, Books on Demand, Ann Arbor, MI (1977).

Toor, I. A.: "Problems in Squeeze Cementing," paper SPE 11499, 1983.

Vidick, B., Yearwood, J. A., and Perthuis, H.: "How to Solve Lost Circulation Problems," paper SPE 17811, 1988.

Young, V. R.: "Well Workover With Remedial Rig," Petroleum Engineer Refresher Course No. 4—Well Analysis, Los Angeles Basin Section of SPE–AIME (1967).

14

Foamed Cement

Jean de Rozières and Tom J. Griffin

Dowell Schlumberger

14-1 INTRODUCTION

Foamed cement can be a solution for cementing problems related to formations which are fractured (or have a low fracture gradient), highly permeable, vuggy, or cavernous. Fluids in wellbores exert hydrostatic pressure on points downhole, which is dependent upon the fluid column height and the density of the fluid. Some formations penetrated by wellbores may be very weak or contain holes or vugs, and are able to support the hydrostatic pressure from only very light fluids. Some formations will not even support a column of water. Conventional cements mixed with water must always have a density in excess of 8.33 lb/gal (1.00 g/cm^3); thus, it is frequently not possible to place such slurries in a wellbore. To allow cementing in such cases, ultralow-density cements have been developed. Such slurries can be obtained by mixing a conventional cement slurry with hollow microspheres (Chapter 3) or with a gas. For foamed cements, the resulting densities could vary between that of the base slurry, usually about 15 to 16 lb/gal (1.80 to 1.92 g/cm^3), and that of the gas, for all practical purposes zero lb/gal.

Although foamed cement was first used by the construction industry over 50 years ago, its first use in a well occurred in 1979, to seal an LPG storage cavern leached from salt. The cavern was communicating with an old, large-diameter mine shaft which had been backfilled with rubble and abandoned. The foamed cement was mixed at 3.5 to 4.2 lb/gal (0.42 to 0.50 g/cm^3). The first treatment consisted of 8,000 ft^3 (226 m^3) of foamed slurry. Two subsequent jobs were required to completely seal off the leak (Montman et al., 1982). Following this introduction, the use of foamed cement technology in oil and gas wells developed rapidly during the early 1980s.

The practical lower density limit for conventionally extended slurries is between 11.0 and 12.0 lb/gal (1.32 and 1.44 g/cm^3) (Benge et al., 1982; Bozich et al., 1984; Harms and Febus, 1985). Below this limit, the compressive strength is too low and the permeability is too high to provide adequate zonal isolation.

As described in Chapter 3, hollow ceramic or glass microspheres are also used to prepare ultralow-density cement slurries. They are expensive (4 to 10 times that of conventional slurries), and can be used at densities as low as 8.0 lb/gal (0.96 g/cm^3) (Smith et al., 1979; Root et al., 1982). Such slurries also require special handling techniques. The slurry rheology must be carefully controlled to prevent the spheres from floating. When attempting to prepare slurries at densities close to that of water, using a densitometer for correct proportioning of the water-to-cement ratio can be difficult, because large variations of the amount of mix water have only a small effect upon the actual density. Also, the strongest microspheres have a collapse pressure of about 7,250 psi (50 MPa); consequently, the depths to which such slurries are feasible is limited (Smith et al., 1984).

Foamed cements are less expensive than microsphere systems, and the slurries are easier to design (Edmondson and Benge, 1983). In addition, foamed cement can be mixed at lower densities and yet maintain better properties. Montman et al. (1982) reported useful properties to 3.5 lb/gal (0.42 g/cm^3). Of course, this depends on the application, with densities below 6.0 lb/gal (0.72 g/cm^3) having only limited applications. Slaton (1981) reported the use of foamed cement at densities as low as 5.0 lb/gal (0.60 g/cm^3) where strength was not critical.

Foamed cement has several advantages in addition to its low density. It has relatively high compressive strength (which is developed in a reasonable time), causes less damage to water-sensitive formations (Bozich et al., 1984), can reduce the chance of annular gas flow (Tinsley et al., 1980; Hartog et al. 1983), and allows cementing past zones experiencing total losses. Also, because the addition of gas has little effect on the cement's placement properties (e.g., thickening time), the density can be varied easily by changes in the gas concentration. Since the density is less, and cement losses to potential producing zones are less, increased well productivity may be a benefit (Colavecchio and Adamiak, 1987).

14–2 THEORY

14–2.1 Foam Stability

Foamed cement is a coarse dispersion of gas, usually nitrogen, in a cement slurry which contains a surfactant as a foaming agent and other chemical products to improve foam stability. Foams are characterized by their "quality," which is the ratio (expressed as a percentage) between the volume occupied by the gas and the total volume of the foam. Depending upon the quality, two extreme structural situations can be encountered. Concentrated foams are mostly gas phase, and consist of polyhedral gas cells separated by thin liquid films. Dilute foams consist of nearly spherical bubbles separated by rather thick films of somewhat viscous liquid. Foamed cement belongs to the second category, with a quality not exceeding 80% and usually less than 50%.

Foamed cement is a compressible fluid; consequently, the quality will change during its circulation in the well because of significant pressure variations. Typically, the quality of a foam generated at around 1,000 psi (6.9 MPa) pressure will decrease when flowing down the casing where pressures may exceed 10,000 psi (69.0 MPa). The quality will then increase again when the foam flows up the annulus. This variation in the quality can be predicted as a first approximation by considering the compressibility laws for nitrogen, and its solubility in the base slurry. While properties such as thickening time have been reported not to depend on the quality (McElfresh and Boncan, 1982), studies concerning foamed fracturing fluids (Harris, 1985) have shown that, for the same foam quality, higher pressures promote the formation of smaller bubbles.

In addition to pressure, many other parameters must be considered. Foamed cement is a three-phase system (liquid/gas/solid) with many phenomena occurring at the interfaces. This system is in constant evolution because of the reorganization of bubbles that may grow, shrink, or coalesce, and to chemical reactions which occur in the cement. Such systems are difficult to characterize, because foams are shear-history-dependent fluids whose texture is strongly affected by the mixing procedure. Foamed cements made under large-scale field conditions, where high shear rates and high pressure are used, have been found to be more stable than foamed cements made under laboratory conditions (Davies et al., 1981). Moreover, foams are not reproducible (Monsalve and Schechter, 1984), because it is nearly impossible to consistently produce two samples with the same initial bubble-size distribution (henceforth abbreviated BSD).

Close examination of cured foamed cement reveals a network of cement matrix and pore structures. If the foam quality is sufficiently high, each gas bubble is adjacent to several other gas bubbles (Figs. 14–1 and 14–2).

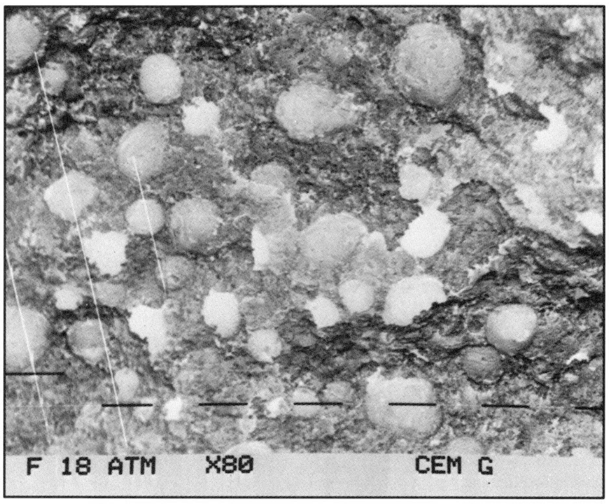

Figure 14–1—Microstructure of 18% quality foam.

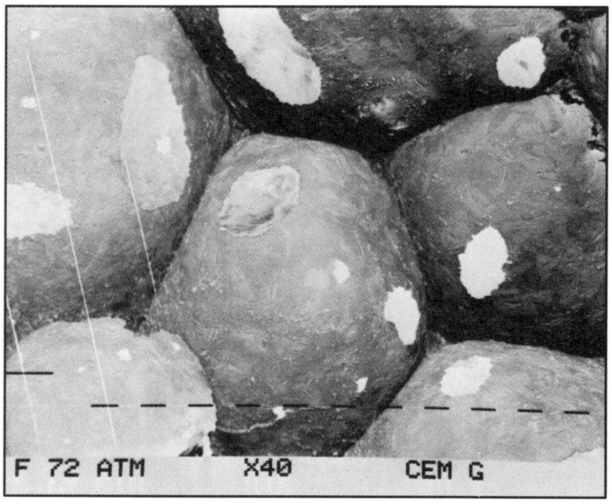

Figure 14–2—Microstructure of 72% quality foam.

The integrity of the structure depends on the maintenance of the interfaces between them. Eventually, outside forces such as dehydration will cause some interfaces to rupture; as a result, an interconnection occurs between two or more bubbles.

Unstable slurries result in a pore structure which is nonspherical and interconnected. This phenomenon occurs while the cement sets. It is caused by the rupture of unstable nitrogen bubble walls upon contact with other nitrogen bubbles, resulting in a coalescence and larger gas pockets. This results in a sponge-like structure with lower compressive strength, higher permeability, and inferior bonding properties. On the other hand, stable foam systems exhibit spherical, discrete, disconnected pore structures with a well-defined cement matrix.

Foam stability is of prime importance. In this section, the relevant factors which affect stability are discussed in detail.

14–2.1.1 Thermodynamic Properties

Stability, in a thermodynamic sense, is described by the following relationship (Ross, 1980).

$$\Delta F = \gamma \Delta A - \int P_{gf} \, . dV_{gf} + W_{desorption} \,, \qquad (14\text{–}1)$$

where

ΔF = the Helmholtz free energy of coalescence of bubbles at constant temperature and fixed amount of components,

γ = interfacial tension between the liquid and the gas phase,

A = total interfacial area,

P_{gf} = absolute pressure of the gas in the foam,

V_{gf} = volume occupied by the gas, and

W = work necessary to desorb a surface-active agent from the interface.

A decrease in free energy results from a loss of area, and from an expansion of the gas. Thus, a stable foam cannot be created from a pure liquid. An additional component must be present to retard the coalescence of bubbles. Surfactants, solute, finely divided solids, and liquid crystals have been reported to be effective (Ross, 1980). This surface-active component must be preferentially adsorbed at the interface to cause a positive work of desorption as described in Eq. 14–1.

From a thermodynamic point of view, interfacial tension may be thought of as being due to the tendency of a fluid to reduce its surface to a point of minimum surface potential energy. This condition produces the smallest area per unit volume, i.e., a sphere. Surface tension opposes forces that tend to distort the bubble shape. Hence, bubble deformation and the breakup of larger bubbles into smaller bubbles is favored by a lowering of the surface tension. Lowering the surface tension of a liquid permits a greater dispersion of the gas bubbles and the formation of a more stable foam.

An equation of state has been proposed (Ross, 1969) to describe the thermodynamic properties of a foam.

$$P_s V_{gf} + \frac{2}{3} \gamma A = nRT, \qquad (14\text{–}2)$$

where

P_s = absolute pressure of the surroundings.

Whatever may be the range of application of Eq. 14–2, it must be noted that a foam has sufficient surface area to contribute significantly to the total energy. When maintaining the interfacial area as high as possible, the energy is minimized by reducing the interfacial tension.

14–2.1.2 Physico-Chemistry

This competition between the decrease in the surface area and the stabilization by surface-active agents becomes more evident when one considers the microscopic mechanisms which affect bulk foam stability. An extensive study of these phenomena was performed by Sanchez (1987), and a brief list of the relevant effects is given in Table 14–1.

Stabilizing Phenomena	Destabilizing Phenomena
Laplace diffusion	Gravity drainage
Plateau border suction	Surfactant concentration gradient
Marangoni effect	Gibbs elasticity
Surface viscosity	

Table 14–1—Parameters which affect foam stability.

In a spherical bubble of radius R, P_{gf} exceeds the pressure in the fluid (P_s) by $2(\gamma/R)$. When two bubbles of different radius ($R > r$) are in close vicinity, the Laplace pressure difference between the smaller and larger bubble is

$$\Delta P_{gf} = 2\gamma \left(\frac{1}{r} - \frac{1}{R} \right). \qquad (14\text{–}3)$$

This pressure difference sets up a driving force for gas diffusion from the smaller bubble to the larger one (Fig. 14–3). Application of hydraulic pressure will reduce the differential pressure, and tend to stabilize the foam. Therefore, the higher the external pressure, the less

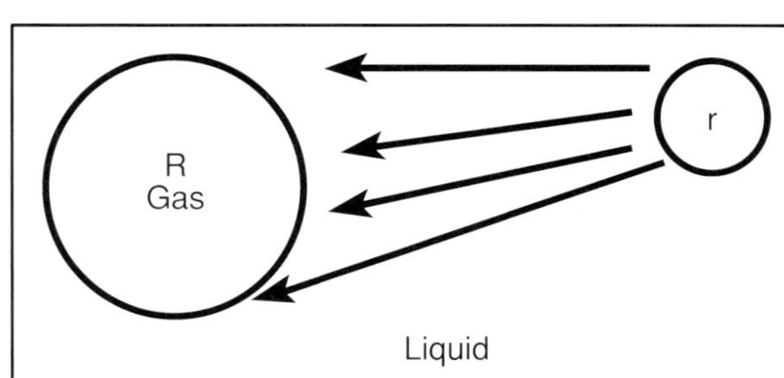

Figure 14–3—Laplace diffusion.

significant will be slight differences in bubble pressure (Davies and Rideal, 1963). Therefore, the BSD is a factor governing the stability of a foam. Foams having a uniform BSD are more stable. Stability also varies inversely with bubble size. Large bubbles are the result of unstable pressure conditions, causing the small bubbles to diffuse into the larger ones with a rate of diffusion directly proportional to the difference in bubble size. Thus, high pressures will stabilize foams which have been created at low pressures, because the BSD will tend to narrow with increasing pressure and diminishing compressibility. Therefore, the coalescence process is also controlled by dynamics as well as by equilibrium thermodynamics (Monsalve and Schechter, 1984). The authors are unaware of any study concerning the BSD in foamed cements. A log-normal distribution may be a good approximation to start with (Harris, 1985), because it is known from the central limit theorem that a random variable which depends in a multiplicative way on a large number of random variables will follow such a distribution.

Gravity drainage and Plateau border suction also contribute to the destabilization of a foam. Both mechanisms cause the thinning and subsequent breaking of the lamella because of gravitational forces in the vertical lamella (Fig. 14–4) or a pressure-driving force (Fig. 14–5) resulting from high interfacial curvature. Foamed cements are less affected by gravity drainage than so-called conventional aqueous surfactant foams, probably because of the higher lamellar viscosity. Because of these phenomena, without any surface-active agent, a foam could not be stable.

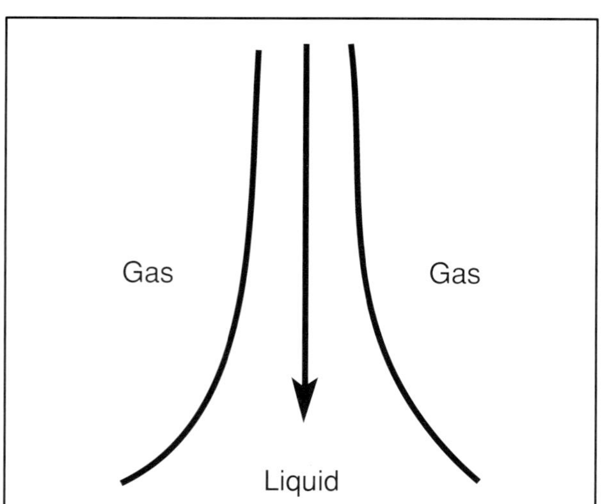

Figure 14–4—Gravity drainage.

The crucial effect of surfactants is that they allow a surface tension gradient to develop in the thin films between droplets. The first step in producing films thin

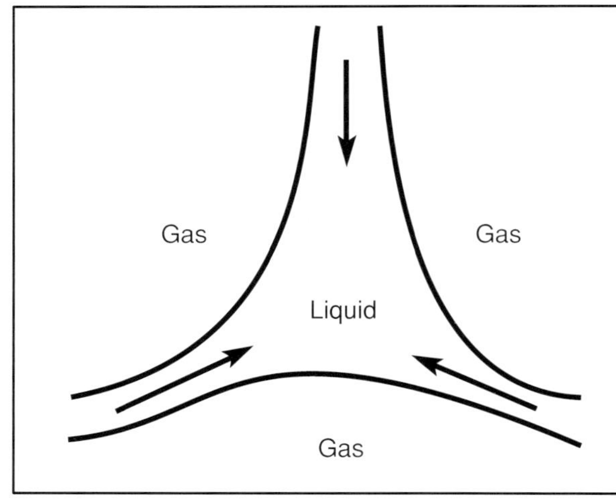

Figure 14–5—Plateau border suction.

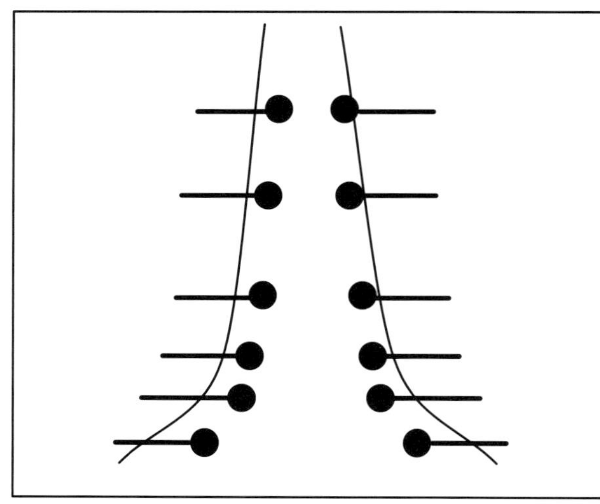

Figure 14–6—Surfactant concentration gradient.

enough to break down is the drainage (Gravity drainage or Plateau suction) of the continuous phase between two bubbles. At first, the surfactant molecules move with the continuous-phase fluid. Thus, the surfactant tends to pile up in the region where the film is thicker, producing a strong surface tension gradient (Fig. 14–6). This gradient opposes any further movement of surfactant molecules. The flow is strongly retarded and may even stop if the film is sufficiently thin. This is one of the strong mechanisms by which surface-active agents prevent coalescence (Manev et al., 1974).

The Marangoni effect is also an important factor for stabilizing thick films. Freshly formed films have a higher initial surface tension than those at equilibrium; consequently, thinning and rupture are minimized (Marangoni, 1878). Many factors are involved in the Marangoni effect. Among these are (1) the rate of diffusion of surfactant to the surface, (2) convective transport, (3) electrical repulsion of incoming molecules in the case of a charged molecule, and (4) possible steric hindrance discouraging the entrance of large molecules into the surface region.

In many cases, the breakdown of a thin film involves a fluctuation which leads to a thinning in a small region. However, these fluctuations usually also produce a local stretching of the interface. This stretching lowers the surface concentration of surfactant and leads to a locally high surface tension. The resistance to this stretching is often quantified by the Gibbs elasticity (Fig. 14–7). This effect depends on the fact that the surfactant is soluble in the continuous phase.

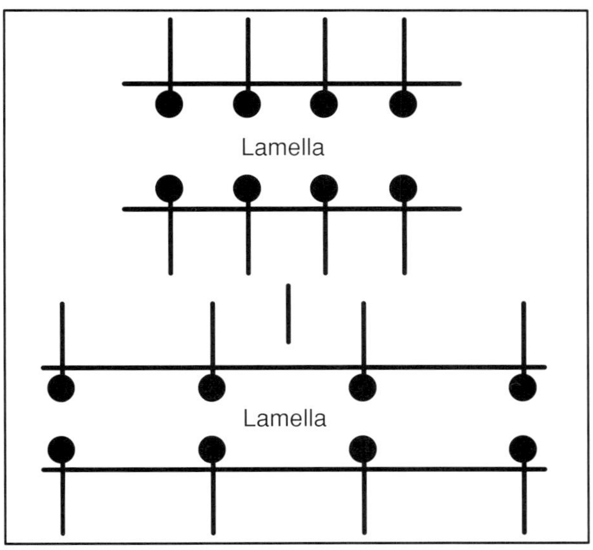

Figure 14–7—Gibbs elasticity.

Surface viscosity is another stabilizing factor brought by the surfactant because a high-viscosity lamella drains more slowly than one with a low viscosity (Ross, 1980). Inversely, the relative rigidity of the surface layer in a foamed cement, compared to a liquid foam, is detrimental to foam life. A viscous fluid having an elastic surface layer is optimum for foam stability.

14–2.2 Rheology

The rheological behavior of foams is unlike that of other fluids. The difference arises from many factors. Foams are compressible fluids, and also cannot be regarded as mathematical continua. They present anisotropy, heterogeneity in composition, and variable properties under shear. Foams are shear-history-dependent fluids in which the bubble structure may be continuously destroyed and rebuilt. They may also present dynamic instability. For this last reason, rotational viscometers with a fixed amount of sample are not suitable for rheological measurements of foams (Heller and Kuntamukkula, 1987). The sample undergoes modifications during shear which lead to a collapse or, in the best case, to a rearrangement of the network of foam bubbles. In any event, foam decays during the course of testing, and the torque measured is not truly indicative of the rheological properties of the foam. Therefore, continuous-flow tube viscometers are more suitable despite the fact that the foam is compressible and non-Newtonian, and the flow will never be steady state (Reidenbach et al., 1986). Viscosity, density, and flow rate will vary continuously as the system pressure changes. As a result, the equations to calculate wall shear stress, wall shear rate, etc., require corrections. Viscosity measurements of foamed cements at low pressure may also not be representative of field conditions. One way to eliminate these effects would be to work with a small pressure drop compared to the absolute pressure at which the test is performed, and at a pressure as close as possible to what could be encountered in the field.

To the best of the authors' knowledge, only one dissertation (Al-Mashat, 1976) has been devoted to the rheology of foamed cement. All other studies have dealt with aqueous foams. Al-Mashat reported flow experiments on foamed cement performed in two capillary tubes with diameters of 0.093 in. and 0.132 in. (2.3 mm and 3.4 mm), respectively. The foam qualities during the tests ranged from 30% to 65%, and the shear rate varied between 1,000 sec^{-1} and 10,000 sec^{-1}. His results indicated that the plastic viscosity of foamed cement increases with the foam quality, and decreases with an increasing shear rate to an almost constant value. This indicates a Newtonian behavior at high shear rates. The curves also showed that the measured viscosity to be higher when measured in the larger capillary tube. The scale of the fluid structure interferes with the definition and measurement of its viscosity. The apparent viscosity of foam depends upon the size of the bubbles relative to the flow boundary. The reported viscosities ranged between 10 cP and 60 cP. Unfortunately, very little information was given concerning the type of cement tested, the absolute pressure at which the experiments were conducted, or the rheological properties of the base slurry. Moreover, the range of shear rates in which the experiments were conducted did not correspond to that normally experienced by a foamed cement during a job operation—less than 1,000 sec^{-1} (Smith et al., 1983; Smith et al., 1984).

Literature concerning the flow of aqueous foams is more abundant. Many studies have been devoted to the rheological properties of foamed fracturing fluids, and the results are presented in terms of the classical rheological parameters such as yield stress and apparent viscosity. A complete description of foam flow behavior should include the interfacial rheological properties of foam lamellae. Their role is not understood despite some attempts to relate the yield stress to the interfacial tension and the bubble radius (Princen, 1983). After Einstein's flow theory (1906), the viscosity (μ) of a dilute suspension of non-deformable spheres varies linearly with Q provided no interaction between spheres occurs (in the range zero to five percent).

$$\mu = \mu_0 (1.0 + 2.5\,Q) \qquad (14\text{–}4)$$

The term μ_0 is the viscosity of the base fluid.

Taking into account interferences between deformable bubbles, Hatschek (1913) derived a flow theory from Stoke's Law which is valid in the quality range of 52% to 74%.

$$\mu = \mu_0 (1.0 + 4.5\,Q) \qquad (14\text{–}5)$$

Above 74%, and considering the deformation of the bubbles, Hatschek derived the following expression.

$$\mu = \frac{\mu_0}{1.0 - Q^{0.333}} (1.0 + 2.5\,Q) \qquad (14\text{–}6)$$

Flow-tube experiments performed on aqueous foams at 1,000 psi (6.9 MPa) (Mitchell, 1969) showed that, in the foam quality range of 0% to 54%, foam has a Newtonian behavior, and the viscosity follows the linear law shown below.

$$\mu = \mu_0 (1.0 + 3.6\,Q) \qquad (14\text{–}7)$$

This is not very different from the theories of Einstein and Hatschek. Above 54% quality, Mitchell reported foam to be plastic at low shear rates (<15,000 sec^{-1}), and Newtonian for higher shear rates.

As a general rule, foams are considered to be pseudoplastic, and more viscous than their gas or liquid fraction (Davies et al., 1981; Rickard, 1985). Apparent viscosity increases with quality and pressure (Harms and Febus, 1985), and decreases when the temperature and the shear rate increase (Cawiezel and Niles, 1987). Although a flowing foam behaves like a pseudoplastic fluid, static foams have a measurable gel strength which increases with foam quality (David and Marsden, 1969). The development of a "foam-structure" and the crowding of the bubbles are the most probable reasons for the development of a yield point before continuous shearing motion can occur. The yield point is very important for predicting foam fluid-flow behavior, as it governs the minimum pressure gradient required to initiate flow through a pipe. With aqueous foam, the existence of a yield stress above 54% quality has been reported (Blauer et al., 1974). The yield stress, as the apparent viscosity, varies inversely with bubble size, directly with quality, increases with pressure, and decreases with temperature (Heller and Kuntamukkula, 1987).

Visual observations of foams flowing in plastic tubes have shown that, at low flow rates, foams flow as a rigid plug, moving on a thin film of water next to the wall (Princen, 1982). There is an overwhelming influence of the boundary film thickness on the pressure drop required to make foam flow through a pipe or an annulus, because this film acts as a lubricant at the wall. The accepted rheological procedure by which slip effects at the tube wall are considered (Mooney, 1931) cannot be rigorously applied in data interpretation for a compressible fluid. Efforts to explain the variation of apparent foam viscosity with the tube radius, based upon slippage at the tube wall, were not entirely successful (David and Marsden, 1969) because the slip-corrected viscosity still retained a dependence on the tube diameter.

14–3 DESIGN

There are two principal aspects concerning the design of a foamed cement treatment:

- the laboratory design or the adjustment of the slurry properties to ensure placement at the right location, and adequate cement performance during the life of the well, and
- the prejob planning and engineering to achieve proper placement.

14–3.1 Laboratory Design

Early in the development of foamed cementing, the compositions of slurries to make foamed cement were limited. With today's surfactants and foaming techniques, virtually any base slurry can be foamed. This includes all classes or types of Portland cement, and other hydraulic cements. Although some slurry compositions may not have application in foamed cementing, slurries containing most cementing additives may be used. In addition, the laboratory testing of foamed cements has achieved a higher level of sophistication.

14–3.1.1 Cement, Foaming Agents, Stabilizers, and Additives

In practice, several conditions must be fulfilled to obtain sufficient foam stability. First, slurries should always be mixed at their optimum water ratio. A good rule of thumb for an initial design is to avoid using a water-to-cement ratio which would not be stable in an unfoamed slurry. The same mechanism that produces free water in common slurries will contribute to foam segregation in a foamed slurry.

The base slurry density can be selected according to the properties desired for the final foamed cement. Normal density base slurries can be selected to obtain higher compressive strength. Such slurries require larger volumes of gas to achieve a given foam density; thus, the resulting permeability will be higher. On the other hand, slightly lower density base slurries (prepared using lightweight filler materials) will allow foamed cement with lower permeabilities, but also with lower strength.

To select suitable foamers and stabilizers for cement, one should consider the following criteria: efficiency, stability, compatibility, effect on the cement strength and permeabilty, cost, safety, and any handling concerns. The chemicals used to generate and stabilize the foamed cement must be effective at elevated temperatures and pressures in the highly alkaline, calcium-containing environment of the water phase of a cement slurry. It is essential that the foam be stable for a period longer than the time required for the cement to set (Hengst and Tressler, 1982).

The choice of the foaming agents and chemical stabilizers is important. Since nitrogen is inert, most of the usual cement additives can be used. Foaming agents are usually anionic surfactants, while stabilizers may be surfactants, polymers, latices, or even solids.

Mixtures of surfactants have a minimum in surface tension and a maximum in surface viscosity. A minimum bubble size is obtained when the surface-active agents have the same chain length. Similar chain lengths allow for a tighter packing of surfactants at the gas liquid interface. Mixed surfactant foaming agents, in which an ionic surfactant is blended with a nonionic surfactant, have produced aqueous foams with good stability (Shah et al., 1978).

Polymers, latices, or long-chain alcohols can be used as foam stabilizers; however, their mechanism of action is not well understood. They probably viscosify or thicken the base slurry, preventing the bubbles from coalescing until the cement sets (Sita Ram Sarma et al., 1988).

The cement particles themselves play a significant role in the stabilization of the foam (Davies et al. 1981). When a solid particle is attached to a bubble, it reduces bubble coalescence and enhances foam stability (Sharma et al., 1982). Stability is related to the size of the particle and its wettability. The way in which solid particles are retained at a liquid/gas interface is analogous to the adsorption of solute molecules. In both cases, work is required to transfer the material out of the surface into the bulk solution. This work of desorption is the phenomenon that confers thermodynamic stability to the foam. In the case of finely divided solids, Ross (1980) reported that stability is favored by a reduction in the particle size. Also, particle size is significant as it affects the sedimentation rate. Foam stability is also related to the wettability of the particle. If the contact angle is too low, the particles are wetted by the liquid and sink. If the angle is too high, adhesion to the lamella is insufficient to stabilize the particle against the cohesion of the liquid. The optimum contact angle lies between 40° and 70°.

In addition to the base slurry, foaming agent, stabilizers, and additives, one must use a gas which is inert with respect to the cement properties (e.g., nitrogen), and a foam generator with sufficient energy and mixing action.

14–3.1.2 Laboratory Test Methods

Testing of foamed cement under downhole conditions is difficult. Because of the pressure and temperature dependence of the foam volume, curing a foamed cement at high pressure and temperature requires different equipment from used for conventional slurries. Early investigations have shown that there are few differences between the results at low and high pressure (Montman et al., 1982; Tanner and Harms, 1983). Therefore, most testing has been performed with foams generated and cured at atmospheric pressure. However, some new equipment has been developed which allows testing under downhole conditions. The key tests required to design a foamed cement slurry are described below (de Rozières and Ferrière, 1990).

Stability

The stability must be tested to ensure that the gas will not break out of the slurry. If the gas were to coalesce, increasing bubble size would cause it to rise to the surface or form gas pockets in the cement. A simple test for stability consists of slicing a column of foamed cement cured in the laboratory under appropriate conditions. The density of each slice is then measured, and the existence of a density gradient indicates insufficient stability. More sophisticated tests can be developed to study the stability of the foam under shear or during a pressure cycle. These tests are aimed at determining the behavior of the foam when flowing down the casing and up the annulus.

Compressive Strength and Permeability

Compressive strength and permeability are normally tested using slurries cured at atmospheric pressure, which contain the same volume percent of gas that the slurry would under downhole conditions. The data reported in Figs. 14–8 and 14–9 show that typical 7.0- to 12.0-lb/gal (0.84- to 1.44-g/cm^3) foamed cement slurries may have compressive strengths between 500 to 1,200 psi (3.45 to 8.27 MPa) and air permeabilities from 0.02 to 0.16 md.

When the density of conventional cement slurries is reduced by adding water or other extenders, the amount of cementitious material is diluted. Because of the large density difference between water and gas, much less gas is required to reduce the density by an equal amount. This results in less dilution and, therefore, less deterioration of cement properties. Consequently, the physical properties of foamed cements are similar to those of conventional lightweight cements which are 2.0 to 4.0 lb/gal (0.24 to 0.48 g/cm³) heavier (Tanner and Harms, 1983; Smith et al., 1984). The compressive strength and permeabilities of foamed cement compared to those of microsphere cement at 600°F (315°C) are shown in Figs. 14–10 and 14–11. Additional information is presented in Chapter 9.

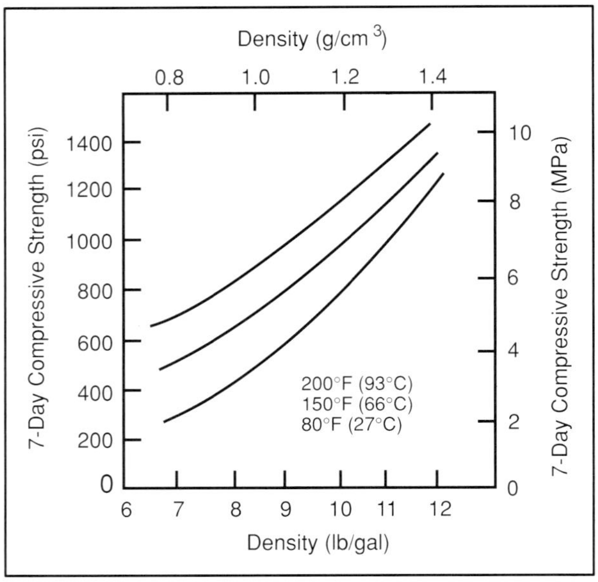

Figure 14–8—Foamed cement compressive strength (after Root, Barrett, and Spangle, 1982).

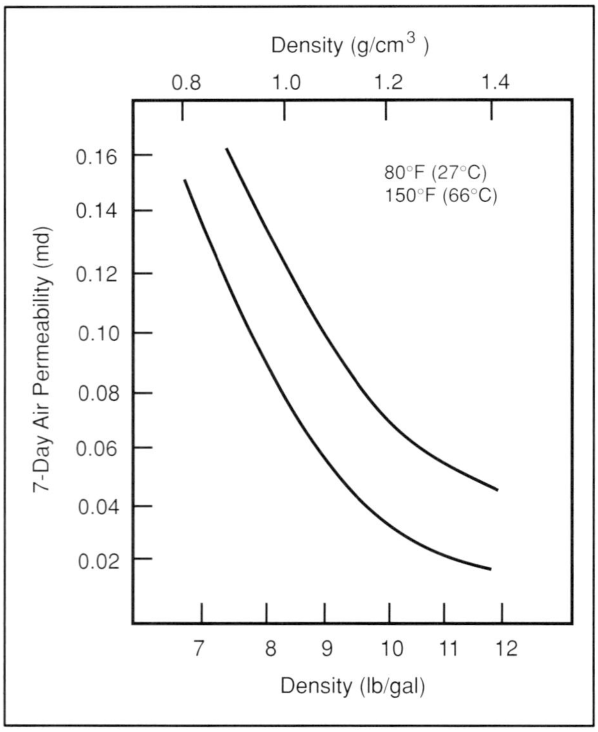

Figure 14–9—Permeability of foamed cement (after Root, Barrett, and Spangle, 1982).

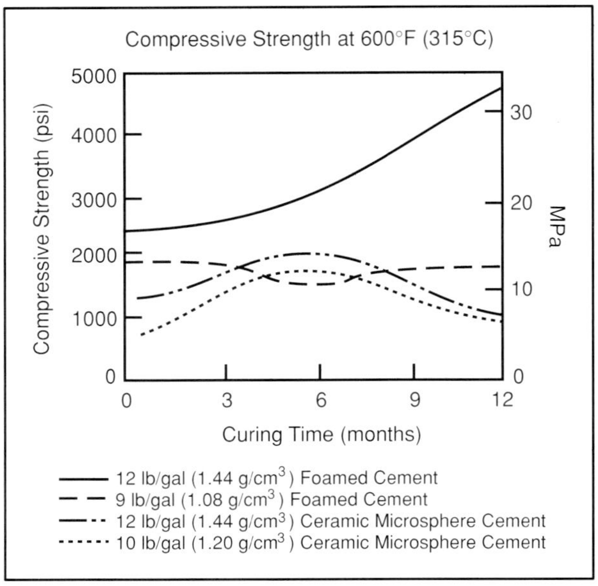

Figure 14–10—Foamed cement vs microspheres: compressive strength (after Nelson, 1986).

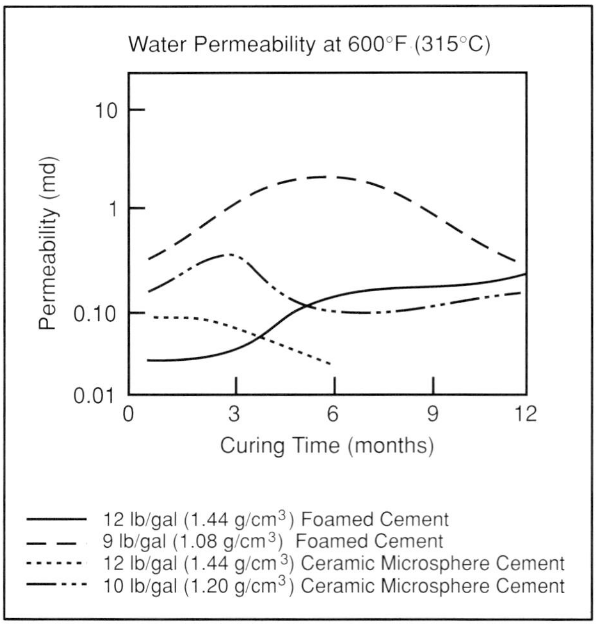

Figure 14–11—Foamed cement vs microspheres: permeability (after Nelson, 1986).

Foamed cement systems can more easily meet the regulatory requirements for compressive strength and permeability at low densities (Olanson, 1984). However, care must be taken for some applications. Chekiri (1978) reported that perforation of foamed cements with porosities above 40% tends to cause excessive fracturing. As a rule, the permeability will dramatically increase when the foam quality exceeds 40% (Table 14–2); however, this depends upon the additives, surfactants, and the curing conditions (Aldrich and Mitchell, 1975; Smith et al., 1984).

Density (lb/gal)	Density (g/cm³)	Percent Gas	Water Permeability (md)
7	0.84	51	117
8	0.96	45	0.5
9	1.08	37	< 0.001
10	1.20	29	< 0.001
11	1.32	23	< 0.001

Base Slurry—14.2 lb/gal (1.70 g/cm³)
Cured 7 days at 80°C.

Table 14–2—Permeability vs gas percent (after Moran, Spangle, and Evans, 1983).

Shear Bond

Although no data were developed, Davies et al. (1981) reported that foamed cement can undergo an expansion prior to the time of set. In some situations, this can result in improved bonding (Slaton, 1981). This hypothesis is supported by indirect evidence from the improved bond logs obtained on the foamed cemented wells, and can be explained as an effect of pressure maintenance by the compressed gas in the cement. As the cement loses hydrostatic pressure because of gelation, the pressure of the gas is available to maintain tight contact between the cement and the casing or formation. Smith et al., (1984) reported that foamed cement at 7.9 lb/gal (1.14 g/cm³) has higher shear bond strength than a 12% bentonite cement at 12.7 lb/gal (1.52 g/cm³). They also found that the ratio of shear bond to compressive strength is higher for foamed cements, and increases with nitrogen concentration. These data are presented in Table 14–3.

Thickening Time

Among the tests performed on foamed cement, measurement of thickening time has proved to be the most difficult to perform and the least conclusive. To be significant, this test should be performed under downhole conditions with a foam mixed in a manner comparable to what occurs on location. Thus, ideally, the slurry should be prepared in a pressurized mixer, and transferred under pressure to the pressurized consistometer. This is not an easy task. The thickening time test consists of measuring the evolution of a rheology-related property in a shear field (Appendix B). Because of the particular rheological behavior of a foam, the shear field is not uniform in a consistometer. A large part of the foam remains static while the small amount which is sheared is finally destabilized.

Instead of working with foamed systems, the common procedure today is simply to measure the thickening time of the base slurry containing the additives, surfactants, and stabilizers. This method gives a reasonable estimate of the available pumping time for the foamed slurry (Davies et al., 1981; McElfresh and Boncan, 1982).

Calorimetry experiments performed at atmospheric pressure under static conditions have shown that the time at which the maximum temperature is recorded does not depend on the foam quality (Fig. 14–12). This test, which measures the evolution of a property related to the chemical aspect of the hydration process, must not be considered as a true thickening time. However, these experiments show that the hydration process is not affected by the presence of nitrogen in the system.

Fluid Loss

Although few studies have been done, introduction of a gas phase causes a significant reduction in the flow of

Composition	Density (lb/gal)	Density (g/cm³)	Compressive Strength (psi)	Compressive Strength (MPa)	Shear Bond Strength (psi)	Shear Bond Strength (MPa)	Ratio of Shear to CS (%)
Class G	15.8	1.90	4200	29.0	403	2.8	9.6
Class G + 12% Bentonite	12.7	1.53	722	5.0	70	0.5	9.5
40% Gas	9.5	1.14	873	6.0	118	0.8	13.5
50% Gas	7.9	0.95	571	3.9	97	0.7	17.1

Cured 24 hr at 80°F (27°C), then 24 hr at 176°F (80°C).

Table 14–3—Compressive strength and shear bond strength (after Smith, Lukay and Delorey, 1984).

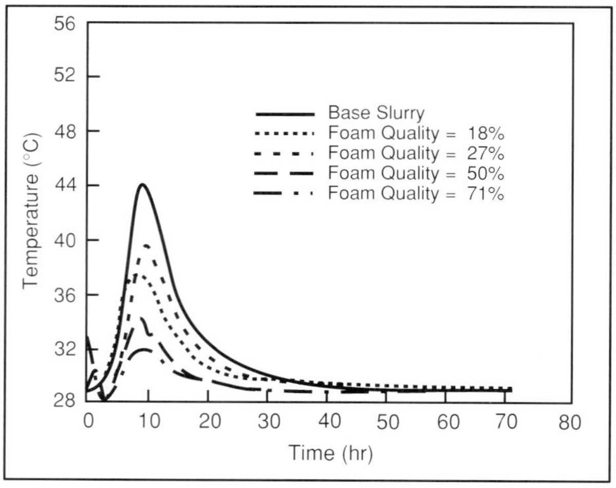

Figure 14–12—Effect of foam quality upon thermal behavior of cement slurries.

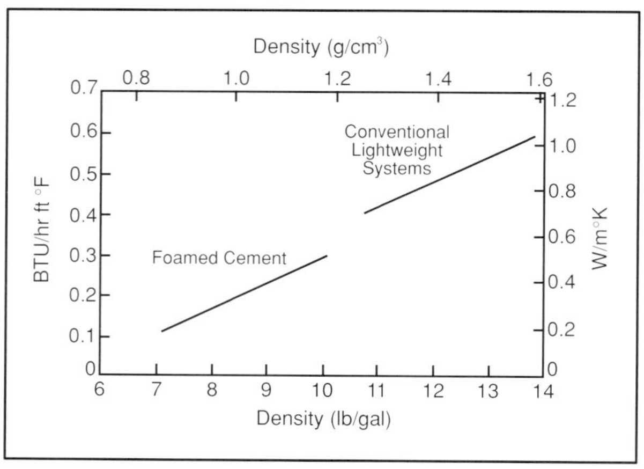

Figure 14–13—Thermal conductivity.

liquids in porous media (Anderson, 1975). McElfresh and Boncan (1982) reported a decrease in fluid loss with an increase in gas volume.

Thermal and Electrical Conductivity

Short et al. (1961) reported reductions in thermal conductivity because of the reduced amount of solids, and gas voids in the foamed cement. Nelson (1986) reported the thermal conductivity to be roughly proportional to the density of the slurry, regardless of the presence of gas. These data are presented in Fig. 14–13.

Studies of the resistivity of foamed cement indicate that the electrical conductivity is the same as for conventional cements (Loeffler, 1984) (Table 14–4).

Rheology

To the best of the authors' knowledge, no routine rheological measurement is made on foamed cement slurries. As discussed earlier, conventional rotational viscometers are unsuitable. To obtain truly representative results, a rheometer system should be pressurized and should not destabilize the foam. Until a practical method which satisfies these conditions is developed, rheology cannot be used as a laboratory design parameter.

Cement Type	Density (lb/gal/g/cm^3)	Specific Resistivity* (ohm–cm)
Conventional	15.7/1.88	1.25 x 10^4
Foamed	9.3/1.11	1.45 x 10^4

* ASTM D-257

Table 14–4—Electrical resistivity.

14–3.2 Engineering Design Parameters

In addition to the physical and chemical properties of cement, the location and placement technique must be designed for the conditions of the well to be cemented. Most often, foamed cements are used to cement casings in formations where the fracture pressure is low. Therefore, one of the problems encountered by field engineers is to assure that the wellbore pressure, during and after the placement of the foam, never exceeds the fracture pressure. Experience has shown that compression of the gas by the combination of friction and hydrostatic pressures may result in higher foamed cement density, and a lower top of cement than that calculated based on static conditions. Because of the difficulty modeling the rheological behavior of the cement, the determination of the proper amounts of cement, and the density to achieve the proper placement of the foamed cement in the wellbore must be based on practical experience. Wellbore pressure is very often the first design criterion, and the engineer must make decisions regarding the densities of all the wellbore fluids, and the rates at which they will be displaced. Unfortunately, because of the current lack of a method to determine the rheology of foamed cement slurries, these decisions must be made with incomplete information.

The wellbore fluids include the mud, a preflush, a cap cement slurry, the foamed cement, and a tail slurry. The nature of these fluids, and the design considerations concerning each, are discussed below.

14–3.2.1 Muds and Preflushes

Conventional drilling fluids are often used to complete wells with low fracture gradients; however, in some cases, the mud is foamed to reduce its density. Mud foaming can be advantageous in that it allows the use of higher density cement systems. It is frequently helpful to run chemically reactive preflushes to clean the wellbore and to promote better bonding of the cement to the casing and formation surfaces. Preflushes can also be foamed.

When the fracture gradient is very low, wells treated with foamed cement are often drilled with air. In such cases, a preflush is required to wet the formation before foamed cement is pumped; otherwise, severe dehydration of the cement may occur.

14–3.2.2 Cap Slurry

A cap slurry is usually placed above the foamed cement. Its purpose is to compress the gas in the upper foamed stage, and to prevent gas from breaking out of the foamed slurry and migrating to the surface. The cap is normally mixed at a higher density, and its hydrostatic pressure must be considered when determining the potential for wellbore breakdown or losses. If the foamed cement is circulated to the surface (or near the surface), the cap should be pumped from the surface down the annulus to recompress the foamed cement slurry. This assures circulation to the surface, and precise placement of the cap. In this case, a thixotropic (Copening, 1983) or a very rapid setting slurry should be used, and pressure should be maintained for at least four hours, or until the cement has had time to reach its initial set. When pumping down the annulus, to ensure equal distribution of the cap slurry around the pipe, swirl-type centralizers should be used. The BOPs should be cleared by pumping water behind the cement.

14–3.2.3 Foamed Cement Slurry

The following criteria impose boundary conditions on the design of the foamed cement column:

- density of the lead slurry,
- fracture pressure profile,
- pore pressure profile,
- formation permeability,
- safety factors,
- top of the foam and pressure at this depth, and
- bottom of the foam.

Generally, the slurry should be designed to have a permeability less than one-tenth that of the critical formations. The compressive strength should be above 100 psi (0.7 MPa) to support the casing (Smith, 1987), or above 500 psi (3.5 MPa) where required by regulations. The foamed cement must also be able to contain the formation pore pressure. All of these factors impose a lower limit upon the foamed cement density. The fracture gradient is the principal criterion which determines the upper limit of foamed cement density. Depending upon the boundary conditions, there are two methods by which the foamed cement column can be designed.

If it is possible to find a single nitrogen-to-base slurry ratio which satisfies the boundary conditions, a "constant nitrogen ratio" design can be followed. Operationally, this is the simplest method, because the nitrogen injection rate remains constant during the cement job. This method results in a variable foam quality throughout the cement column, with a low density at the top, and constantly increasing density with depth because of compression (Fig. 14–14). One disadvantage is that the upper cement has lower strength and higher permeability, which may allow fluid invasion and subsequent corrosion of the casing; however, if there are no flowing or corrosive zones in the upper portion of the well, the constant ratio method is appropriate. A bradenhead squeeze may be required to place the cap slurry if foamed cement is circulated to surface.

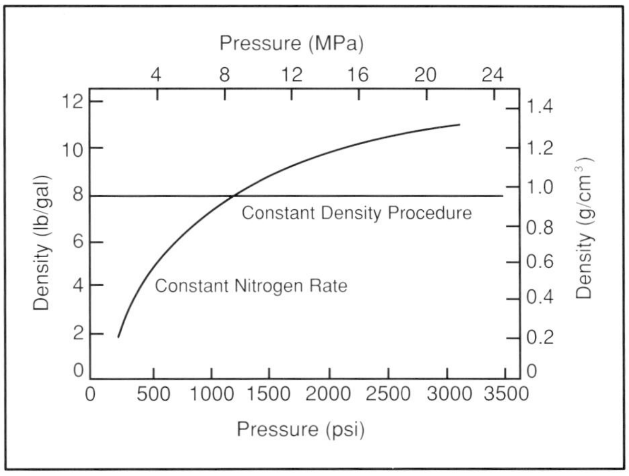

Figure 14–14—Comparison of constant nitrogen rate and constant density procedures.

If the constant ratio method is unacceptable, it may be preferable to follow the "constant density" procedure, where the foamed cement column is divided into several stages, each with a different nitrogen ratio. As depth increases, each successive stage contains more nitrogen; as a result, the density from top to bottom is relatively constant. Therefore, the cement has the same properties throughout the cemented interval. This is better for wells with multiple producing zones. However, many practical disadvantages appear with such designs. They are difficult to perform with small cement volumes, for it requires a complex pumping schedule with close coordination and control of the treatment on location. Additionally, a low-nitrogen-ratio slurry (more dense) in the first stages may break down weak formations. Finally, the location of slurries in the wellbore and the resultant hydrostatic pressure are sensitive to the reliability of hole geometry measurements. Cement excesses will also affect the placement position.

During some foamed cement treatments, a backpressure is maintained on the return line to compress the foamed column. The friction pressure of the fluid flowing in the return line is sometimes sufficient to ensure a sufficient backpressure. If not, the backpressure is controlled with a valve. Typical values range between 100 and 500 psi (1.4 and 3.5 MPa).

14–3.2.4 Tail Slurry

An unfoamed tail slurry is often pumped just after the foamed column. This cement should be strong enough to provide good support for the "shoe" of the casing and/or isolate a producing zone at the bottom of the well.

14–3.2.5 Calculations

Once the hydrostatic description of the wellbore column has been made, the end of the design consists of calculating the quantity and the rate of all the fluids to be pumped. A hole caliper is essential to calculate volumes, especially for jobs designed in several stages. The hole dimensions can dramatically affect the locations of the various stages.

To obtain an engineering design of the foamed cement portion of the column, one must determine the amount of nitrogen which must be injected at various times throughout the job. This number is expressed by the nitrogen ratio, which represents the amount of nitrogen in standard cubic feet per barrel of base slurry. The calculation is based on three simple physical parameters:

- hydrostatic pressure,
- compressibility laws for nitrogen, and
- nitrogen solubility in the base slurry.

Since the foam quality is constantly changing because of changes in hydrostatic pressure, the calculations are very tedious and time consuming if done by hand (Appendix C). They are best done by computer simulation (Chapter 11). It is important to point out that these calculations assume a static equilibrium situation; consequently, the final position of the top of the foamed cement column is often different from what had been planned. In fact, the error has been reported to be as high as 30% (Kulakofsky et al., 1986; Colavecchio and Adamiak, 1987).

With conventional slurries, it is possible to estimate both frictional and hydrostatic pressures separately and then total the overall pressure drop (Lord, 1981). This is not the case with compressible fluids, where frictional and hydrostatic pressure contributions influence each other through the pressure-dependent fluid density. Because of friction pressure and subsequent compression, the top of cement may be below the calculated point. Because of the thixotropic nature of foamed cement, this compression may not be relieved, preventing the foam from expanding to the calculated static conditions. This can result in greater hydrostatic pressure, and an inaccuracy in the position of the top of the column; therefore, it should be accounted for in the design. For the first use in an area, 25% over caliper should be used and adjustments made after the first one or two wells in the area.

Because of changing hydrostatic pressure as the slurry is pumped deeper into the well, the volume ratio of nitrogen changes, which results in changes in density, hydrostatic pressure, rheological parameters, flow rates, etc. Thus, free-fall, friction pressure, etc., are changing. In many foamed cement treatments, free-fall may not occur during pumping of the foamed cement, but during the pumping of the tail slurry and displacement. This effect will vary with the pipe and annular dimensions. Finally, because of the nature of these fluids, turbulent flow placement is unlikely (Harms and Febus, 1985).

14–4 EXECUTION AND EVALUATION

The procedure for a foamed cement job is more complex than that for a conventional one, because dry cement, water, additives, surfactants, and gas must all be mixed in the correct proportions. A foamed cement treatment also requires additional personnel and equipment. Planning is a very critical part of a successful foamed cement treatment. Table 14–5 presents a list of items of concern, and the party that is normally responsible for them.

A typical location layout and a hookup to the BOP are shown in Figs. 14–15 and 14–16. If air is used instead of nitrogen, high rate and pressure compressors are needed. When considering the use of air for foamed cement, several other factors must be considered:

- The compressibility of air is different from that of nitrogen,
- Oxygen is more soluble than nitrogen,
- The presence of oxygen and carbon dioxide in the air might result in excessive corrosion of the casing, and deterioration of cement properties.

The latter point has not been investigated sufficiently. Since air is often more convenient to use than nitrogen, particularly in remote areas, further study is indicated.

Operator		
Depth	Location of Weak Zones	Fracture Gradients
Pore Pressures	Acceptable Safety Margins	BHCT and BHST
Future Well Temperature	Hole Caliper	Desired Top of Cement
Service Company		
Cement Pump Trucks(s)	Bulk Cement Equipment	Batch Mixing Equipment
Cement Slurry Flow-Rate Meter	Pressurized Densitometer	Pressure Gauges
Nitrogen Unit(s)—tank, vaporizer, and pump		Nitrogen Flow-Rate Meter
Check Valves	Surfactant	Surfactant Tank
Surfactant Pump	Surfactant Flow-Rate Meter	Foam Generator
Radios	Kill Line for Top Job	Gas-Tight Cementing Head
Staked Treating Lines		
Drilling Contractor		
Fresh Water	Displacement Fluid	Layout
Annular Pressure Control (BOP)		Staked Return Lines
Disposal of Returns	Annular Choke	

Table 14–5—Factors to consider during prejob planning of a foamed cement job.

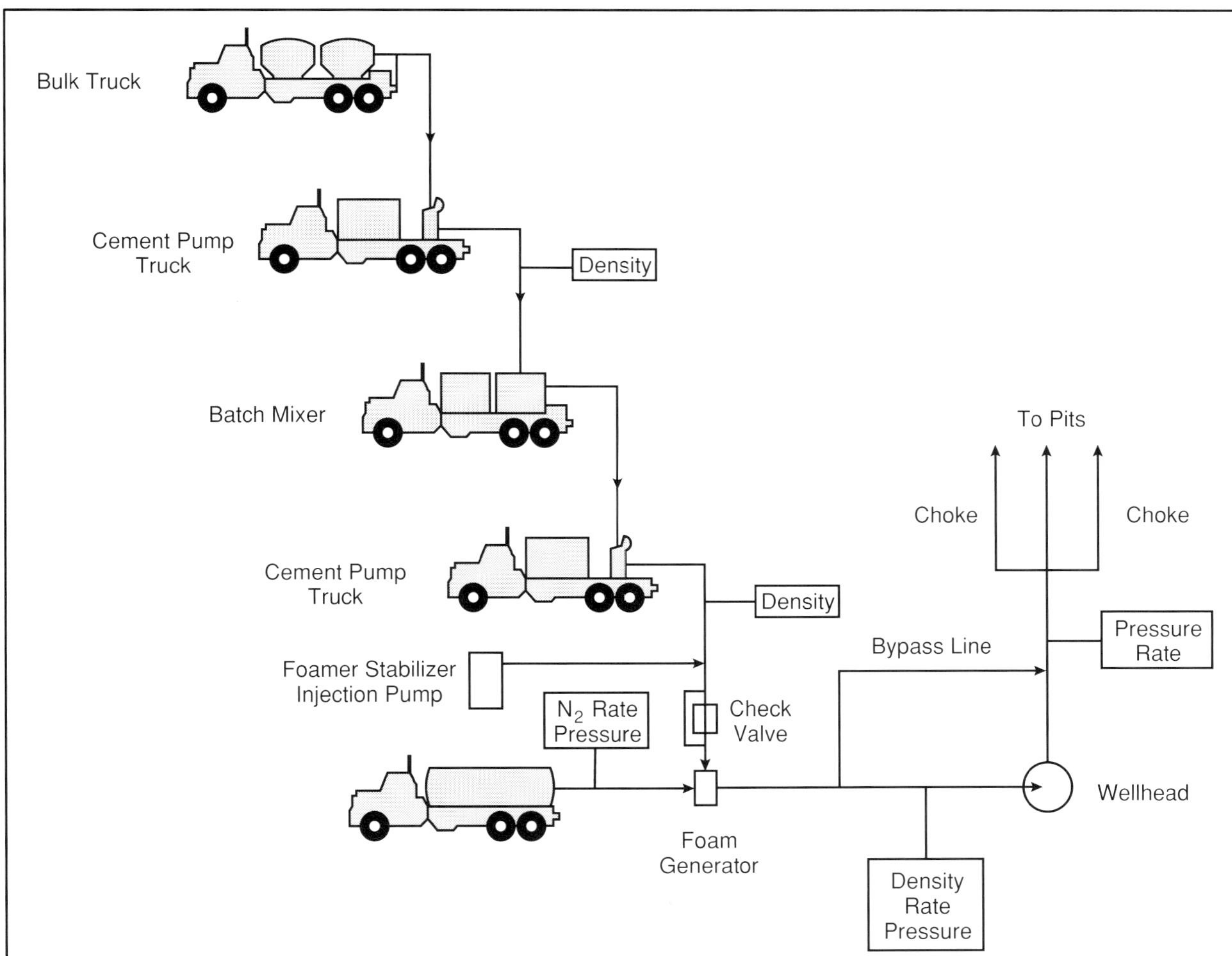

Figure 14–15—Location layout for foamed cementing.

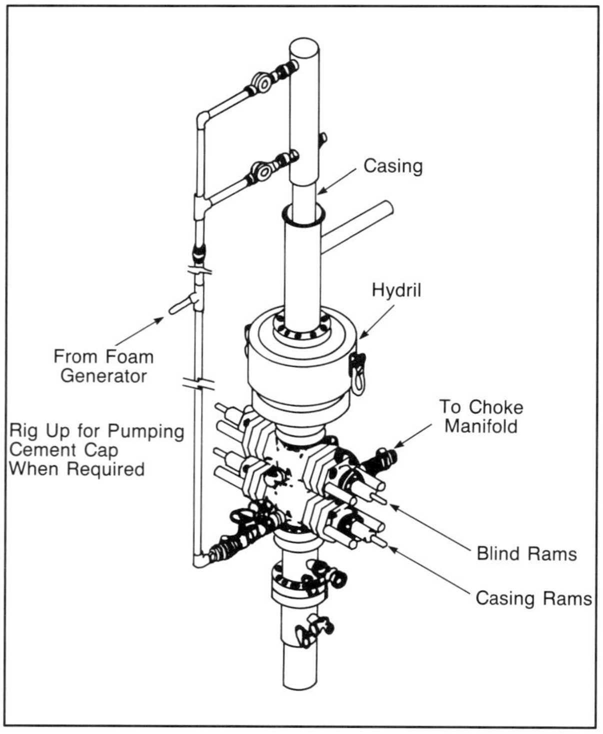

Figure 14–16—Typical BOP arrangement.

14–4.1 Operationally Critical Job Parameters

Because of the nature of foamed cement treatments, there are several factors which must be given very close consideration during the execution of a foamed cementing treatment.

14–4.1.1 Safety

Since compressed gases are used to produce foamed cement, additional safety precautions are required when performing foamed cement jobs. Also, due to the compressed gases, the fluids have much more potential energy than conventional slurries. This makes it necessary to stake all treating lines, so that they are unable to whip around in the event of a failure of the treating iron. The gasified cement, as it flows to surface, expands and can produce great force. This can cause it to blow across the pit and erode the pit walls. Flow lines to the pit must also be staked and pointed down into the pit. Due to this expansion, the volume of flow may be greater than the flow lines can handle at atmospheric pressure. Thus, the annular preventers must be closed to force the flow through the line to the pit. Casing should be landed at the floor to assist in making a gas-tight connection of the head to the casing.

14–4.1.2 Mixing

Because of the necessity of having multiple components mixed together in the right proportions, it is advisable to batch mix the base cement slurry. Batch mixing provides uniform slurry density, and allows better control of the rate while pumping downhole, thus allowing better control of the ratio of the various components as surfactants and gas are mixed with the cement slurry. To achieve the coordination required for an operation of this type, it is necessary that the supervisor and the operators of the cement, nitrogen, and surfactant pumps be able to communicate with one another. This is best done with radio headsets such as those more frequently found on large stimulation treatments. The supervisor should prepare a table showing the rates of the different components for variations in the slurry pump rate. Each of the operators should have a copy, and the supervisor should make sure they follow the prescribed design. If the job is to be constant density, the table should show the rates and volumes of each stage. The supervisor should constantly check the material volumes pumped, and those remaining, against his table of the job procedure. This backup of the measurements made by the sensors is necessary to ensure that the job is proceeding as scheduled, in case any of the sensors are out of adjustment.

14–4.1.3 Surfactants

Addition of the foamer and stabilizer is the most critical part of the job. This requires a reliable metering pump. The pump must be checked to ensure it will handle the required surfactant rate at all pressures anticipated during the treatment. It is best if the surfactant is measured with a flow-rate meter and backed up with a physical measurement. The treatment will be more trouble-free if all of the surfactant is contained in one tank rather than in drums. The foamer/stabilizer tank should be elevated to maintain a positive suction head.

14–4.1.4 Foam Generator

To generate a stable foamed cement, specially designed generators which provide a fine gas dispersion, and exert sufficient energy to ensure efficient mixing of the gas with the base slurry, are required. The resulting slurry contains uniformly dispersed, microscopic gas bubbles.

14–4.1.5 High-Pressure Densitometer

For control during the treatment, the density of the foamed cement should be measured under pressure. Because of the compressibility of the gas, the pressure must be measured simultaneously. Since the pressures change during a cementing treatment, a chart must be prepared to interpret this pressure/density data. This is the final control measure used by the supervisor.

14–4.1.6 Backpressure

If the foamed cement is circulated to the surface, some form of backpressure control is required to assure a competent cement in the upper part of the hole, and to control flow from the well. Normally, the treatment is performed with the annular blowout preventer closed. Otherwise, the rapidly expanding gasified cement would flow out the bell nipple and spray into the derrick. A choke is normally used to control flow through the line to the pit, and to maintain backpressure on the foamed cement at the top of the annulus. If the slurry at the top of the annulus were allowed to expand without backpressure, the density would decrease excessively, resulting in a set cement with low compressive strength and high permeability. Under most circumstances, 50 to 200 psi of backpressure is sufficient.

14–4.2 Evaluation

Normal methods of evaluating cement jobs must be modified when a well has been cemented using foamed cement (Chapter 16).

14–4.2.1 Temperature Surveys

Temperature surveys should be performed 8 to 24 hours after the treatment. Because of the dilution of the cement and the insulating properties of the gas, the magnitude of the thermal change may be less than that observed with conventional extended cements. The thermal gradient will be greater for the cap and tail cements, while the foamed cement may show a thermal gradient about the same or below the normal background (Montman et al., 1982; Tanner and Harms, 1983). Thus, the presence of a cap will aid in identifying the top of the foamed cement. Without the cap, it may not be possible to identify the foamed cement.

14–4.2.2 Cement Bond Log (CBL)

Because of the presence of gas in the slurry, special considerations must be made when running CBLs to account for the change in attenuation by the gas-filled slurry (Chapter 16). Bruckdorfer et al., (1983) and Jutten et al., (1987) developed interpretation curves and monographs (Chapter 16) for use in evaluating bond logs of wells cemented with foamed cement. Epps and Tello (1988) described the response of the Pulse Echo Tool[SM1] to foamed cement.

14–5 FIELD APPLICATIONS AND CASE HISTORIES

Because of its unique properties, foamed cement has numerous applications.

14–5.1 Prevention of Fracturing in Weak Formations

There are numerous methods which have been used over the years to prevent fracturing or loss to weak formations. These include reducing the density by using extenders mixed with additional water, or using lightweight filler materials such as ceramic or glass microspheres. This can also be done by reducing the length of the column, using external casing packers (ECPs), or stage cementing.

Stage cementing is very useful, but has disadvantages which can be overcome by foamed cement. Stage tools sometimes leak, require drillout with special bits, and occasionally fail to operate properly. Since the pumping is performed in two or more stages (sometimes several hours apart), additional rig time is necessary to complete the cement job and to drill out the stage tool. It is also necessary to make a decision concerning the placement of the tool, which is not always straightforward. Foamed cement is generally less expensive than stage cementing (Phipps and Krajeski, 1983; Davis, 1984; Bozich et al.,1984).

Smith et al., (1984) found foamed cement to be ideal for cementing in the deep waters off the east coast of Canada. The seabed there is characterized by an unconsolidated sand and silt formation with a very low fracturing gradient, 0.51 psi/ft (11.5 kPa/m). Good cement is needed for the conductor casing to support subsea equipment. This is compounded by the very low temperatures of the deep (4,920 ft or 1,500 m) water. Nitrogen was added to the slurry to produce foamed cement with a density of 8.6 lb/gal (1.03 g/cm^3) at the seafloor and 9.9 lb/gal (1.19 g/cm^3) at TD. With an average density of 9.4 lb/gal (1.12 g/cm^3), the pressure gradient of the cement was only 0.488 psi/ft (11.04 kPa/m).

The regulatory body (ERCB) of Alberta, Canada, requires 500 psi (3.5 MPa) compressive strength in 48 hours at the uppermost hydrocarbon-bearing formations. Formations in the southeast Alberta, Grande Prairie, and Lindbergh areas are too weak to withstand slurries extended with conventional fillers. Foamed cement was found to be the most economical solution (Olanson, 1984).

---

[1]Service mark of Gearhart Industries, Inc.

14–5.2 Thermal Wells

The properties of foamed cement make it useful for cementing many thermal wells (Chapter 9). Because geothermal, steamflood, and fireflood wells are frequently drilled in areas having weak formations, foamed cement is often the best material meeting both the strength and density requirements (Nelson and Eilers, 1985; Nelson, 1986).

Rickard (1985) described the cementing of geothermal wells in the Gunung Salak Project, West Java, Indonesia. A base slurry of Class G cement with silica flour, mixed at 15.6 lb/gal (1.87 g/cm^3), was used. The slurry was foamed to densities between 4.0 to 15.0 lb/gal (0.48 to 1.80 g/cm^3). The slurry was preceded by a reactive preflush, which was also foamed. Severe lost-circulation wells were treated with alternating stages of calcium chloride solution, reactive preflush, and ultralightweight foamed cement.

14–5.3 Wells Drilled With Air

As discussed earlier, many areas also use air for drilling. When the cement is placed in the wellbore, lost circulation is likely. Foamed cement is ideally suited for cementing such wells. Many are found in the Appalachian basin of the United States. The wells are drilled to about 5,000 ft (1,524 m) through formations having fracture gradients of less than 0.51 psi/ft (11.3 kPa/m) (Colavecchio and Adamiak, 1987). About 3,000 ft (914 m) of fill-up is necessary. Conventional slurries with densities of 13.0 to 14.0 lb/gal (1.56 to 1.68 g/cm^3) result in inconsistent fill-up. During a series of cement jobs, cement was foamed in five stages to produce a constant density of 9.0 lb/gal (1.08 g/cm^3). The fill-up on early treatments was less than calculated, because of the compression of the gas by friction pressure. This effect was not as severe for wells with larger annular dimensions. The change to foamed cementing in this area seemed to have an additional effect. Breakdown pressures during fracturing treatments have averaged about 825 psi (5.69 MPa) less than for wells cemented with the conventional slurries. This may be due to less loss of cement and, therefore, damage to the producing formations.

Wells in the Marmul field in Oman are drilled with water or, if water is not available, with air (Davies et al., 1981; Hough, 1982). The 17½-in. (44.5-cm) hole for the protective string is drilled with total losses. The 13⅜-in. (34-cm) casing is set using foamed cement, because of its higher compressive strength, thixotropy, and expansion while setting. The treatments are of constant gas ratio, the gas being air. Air is used because of the ready supply from the compressors for drilling with air. The ratio is adjusted to give a maximum porosity of 25%. For more critical wells, the air may be added to give a maximum porosity of 50%. During one series of treatments, a pozzolanic cement slurry was mixed at 13.7 lb/gal (1.64 g/cm^3), and foamed to 10.0 lb/gal (1.20 g/cm^3). Cement bond logs on these wells were far superior to those of wells cemented conventionally in this field.

14–5.4 Lost Circulation in Natural Fractures

Because of its thixotropic nature, foamed cement helps to reduce the penetration of vuggy or cavernous formations. As the cement is forced into these voids, it expands and its viscosity increases, further resisting flow, until it eventually stops (Turki and Mackay, 1983).

In the Mt. Poso field of Kern County, California, steam is injected to stimulate production (Davis, 1984). The field is characterized by zones of very high permeability, through which the wells are drilled with total losses. Returns are lost in the Olcese formation at 500 to 1,000 ft (152 to 305 m). It is necessary to circulate cement from about 1,850 ft (564 m) to the surface. When cemented using a 1:1 Class G:perlite lead slurry mixed at 12.3 lb/gal (1.47 g/cm^3), cement was circulated to the surface only about 25% of the time. The cement treatment was changed to a base slurry of Class G + 40% silica, foamed to 7.5 lb/gal (0.90 g/cm^3). The foam was preceded by a foamed, reactive preflush. The success rate using this technique was 89%, and the cost was significantly lower than that for a conventional job.

The density of foamed cement can also be matched to that of the fluid in the voids, thus preventing it from slumping to the bottom of the void. Since foamed cement can be designed to float on water, it can be used to seal the tops of liquid-filled caverns.

14–5.5 Improved Bonding Across Salt Formations

The foam matrix in the slurry results in reduced mobility of the interstitial water during the hydration of the cement. This reduces the dissolution of salt when placed across salt zones. Because of the reduction in leaching, microannulus formation is much less, and bonding and zonal isolation across salt formations are much improved. As discussed previously, the compressed nature of the gas in the matrix of the slurry will cause the slurry to expand. This is an additional factor preventing the formation of a microannulus prior to the time the cement has set.

In west Texas and southeastern New Mexico, the Salado formation produces a corrosive brine which caused cementing problems during the early development of fields in this area. About 2,000 ft (610 m) of open hole across this formation was left uncemented between the top of the production casing cement and the shoe of the surface casing. Corrosion was a major problem. In addition, there is also a weak zone (fracture gradient of 0.60 to 0.65 psi/ft [13.6 to 14.7 kPa/m]) about 500 ft (152 m) below the brine. Foamed reactive preflushes and foamed 50:50 gypsum:cement were used to squeeze through perforations to fill the annulus with competent cement. Sometimes, as many as three separate perforations followed by cement were required to fill the void completely to the surface. This method was far superior to repairs using conventional cements, which required up to 20 squeezes because of the contamination of the cement by the flowing brine (Garvin et al., 1984).

14–5.6 Thermal Insulation

The low conductivities of lightweight cements are quite useful in certain wells with extremes of temperature. In thermal enhanced recovery wells, the insulating nature of these cements can aid in the maximum delivery of heat to the formations into which the heated fluids are being injected. For thermal recovery wells, the insulation can maximize the amount of heat recovered, and improve the efficiency of the power-generating process. In wells through permafrost, the insulating properties protect the sensitive formations from thawing and causing environmental as well as integrity problems. This insulating property can also reduce the problem of scale or paraffin formation in the tubing during production.

14–5.7 Squeeze Cementing of Weak or Depleted Zones

Because of lower hydrostatic pressures, foamed cement can be used to squeeze into zones which normally could not be "squeezed" with conventional slurries (Kondratoff and Chmilowski, 1989). Conventional slurries would immediately break down these formations, be lost in them, and fail to cover the entire area to be squeezed. Because of the thixotropic nature of the foamed cement, it resists in-flow from the formation (Bour and Creel, 1987). The cost of these treatments compares favorably with those using perlite or microsphere cements. Major savings are achieved resulting from reductions in rig time, because the need for the setting of lost-circulation plugs is eliminated (Rickard 1985).

14–5.8 Gas Channeling

Foamed cement has been cited as a tool for the prevention of annular gas migration (Chapter 8). Because the compressed gas maintains pressure within the matrix of the cement, the pore pressure within the formation is "overbalanced." As a result, the gas in the formation is unable to penetrate the wellbore (Tinsley et al., 1980; Cooke et al., 1983; Stewart and Schouten, 1986). Additionally, because of its relatively large percentage of gas, cement volume losses due to fluid loss or volumetric shrinkage have a smaller impact than conventional slurries (Davies et al., 1981).

14–6 CONCLUSIONS

From the above discussion, one can see that foamed cement has practical applications in a variety of well situations. In some cases, it is advantageous economically. In others, it provides significant technical benefits. Although a more complete understanding of the rheological properties of foamed cement is badly needed, sufficient knowledge is available which, when coupled with experience, can provide satisfactory results.

Foamed cements are more difficult to design and place than conventional systems, but the risk is less than that associated with stage cementing. Foamed cement must be carefully evaluated for its merits on each type of cementing treatment. Often it will be the most desirable system, but it should not be considered as a panacea for all cementing problems.

REFERENCES

Al–Mashat, A. M.: "Rheology of Foamed Cement," PhD dissertation, Colorado School of Mines, Golden, CO (1976) 18.

Aldrich, C. H. and Mitchell, B. J.: "Strength, Permeabilities, and Porosity of Oilwell Foam Cement," paper ASME 75–Pet–10, 1975.

Anderson, G.: *Coring and Core Analysis Handbook*, PennWell Publishing Co., Tulsa, OK (1975).

Benge, O. G., Spangle, L. B., and Sauer, C. W. Jr.: "Foamed Cement—Solving Old Problems With a New Technique," paper SPE 11204, 1982.

Blauer, R. E., Mitchell, B. J., and Kohlhaas, C. A.: "Determination of Laminar, Turbulent and Transitional Foam Flow Losses in Pipes," paper SPE 4885, 1974.

Bour, D. and Creel, P.: "Foam Cement for Low-Pressure Squeeze Applications," *Proc.*, Southwestern Petroleum Short Course, Lubbock, TX (1987) 1–11.

Bozich, M. P., Montman, R. C., and Harms, W. M.: "Application of Foamed Portland Cement to Deep Well Conditions in West Texas," paper SPE 12612, 1984.

Bruckdorfer, R. A., Jacobs, W. R., and Masson, J-P.: "CBL Evaluation of Foam-Cemented and Synthetic-Cemented Casings," paper SPE 11980, 1983.

Cawiezel, K. E. and Niles, T. D.: "Rheological Properties of Foam Fracturing Fluids Flowing Under Downhole Conditions," paper SPE 16191, 1987.

Chekiri, K.: "Effect of Jet Perforation on Foam Cement," MS thesis, Colorado School of Mines, Golden, CO (1978) T–2069.

Colavecchio, G. P. and Adamiak, R.: "Foamed Cement Achieves Predictable Annular Fill in Appalachian Devonian Shale Wells," paper SPE 17040, 1987.

Cooke, C. E., Kluck, M. P., and Medrano, R.: "Field Measurements of Annular Pressure and Temperature During Primary Cementing," *JPT* (Aug. 1983) 1429–1438.

Copening, W. L.: "Lightweight Cement Slurry and Method of Use," U.S. Patent No. 4,415,366 (1983).

David, A. and Marsden, S. S.: "The Rheology of Foam," paper SPE 2544, 1969.

Davies, D. R., Hartog, J. J., and Cobbett, J. S.: "Foamed Cement—A Cement With Many Applications," paper SPE 9598, 1981.

Davies, J. T. and Rideal, E. K.: *Interfacial Phenomena*, Academic Press Inc., Orlando, FL (1963).

Davis, Robert N.: "Foam Cementing Program," *Drilling* (Sept. 1984) 70.

Davis, R.N.: "Shell Foam Cements to Surface in California's Mt. Poso Field," *Drilling* (Sept. 1984) 67–69.

de Rozières, J. and Ferrière, R.: " Foamed Cement Characterization under Downhole Conditions and its Impact on Job Design," paper IADC/SPE 19935 (1990).

Edmondson, T. D. and Benge, O. G.: "A Case Study of Ultralightweight Cementing Practices in the Northeastern United States," paper SPE 12318, 1983.

Einstein, A.: "Eine Neue Bestimmung der Molekuldimensionen," *Ann. Phys.* (1906) **19**, Ser. 4, 289.

Epps, D. S. and Tello, L. N.: "Improved Compressive Strength Evaluations in Foamed Cements Using the Pulse Echo Tool," *Proc.*, Southwestern Petroleum Short Course, Lubbock, TX (1988) 10–18.

Exerowa, D., Manev, E. D., and Scheludko, A.: "Effect of Surfactant Concentration on the Critical Thickness of Liquid Films," *Colloid & Polymer Sci.* (July–Aug. 1974) 252, 586–593.

Garvin, T. and Creel, P.: "Foamed Cement Restores Wellbore Integrity in Old Wells," *Oil & Gas J.* (Aug. 20, 1984) 124–127.

Harms, W. M. and Febus, J. S: "Cementing of Fragile–Formation Wells With Foamed Cement Slurries," paper SPE 12755, 1985.

Harris, P. C.: "Effects of Texture on Rheology of Foam Fracturing Fluids," paper SPE 14257, 1985.

Hartog, J. J., Davies, D. R., and Stewart, R. B., "An Integrated Approach for Successful Primary Cementations," *JPT* (Sept. 1983) 1600–1610.

Hatschek, E.: "The General Theory of Viscosity of Two-Phase Systems," *Trans.*, The Faraday Society (1913–1915) 9, 80.

Heller, J. P. and Kuntamukkula, M. S.: "Critical Review of the Foam Rheology Literature," *Ind. Eng. Chem. Res.* (1987) **26**, 318–325.

Hengst, R. R. and Tressler, R. E.: "Fracture of Foamed Portland Cements," *Cement & Concrete Res.* (1983) **13**, 127–134.

Hough, R. B.: "Drilling in Oman's Marmul Field," *Oil & Gas J.* (March 8, 1982) 139–144.

Jutten, J. J., Parcevaux, P. A., and Guillot, D. J.: "Relationship Between Cement Slurry Composition, Mechanical Properties, and Cement Bond Log Output," paper SPE 16652, 1987.

Kondratoff, L. and Chmilowski, W.: "Evaluation of Foamed Cement Squeeze Treatments for Low-Pressure Highly Permeable Formations," paper Petroleum Society of CIM 89–40–81, 1989.

Kulakofsky, D. S., Creel, P. G., and Kellum, D. L.: "Techniques for Planning and Execution to Improve Foam Cement Job Performance," paper SPE 15519, 1986.

Loeffler, N. R.: Unpublished Data, 1984.

Lord, D. L.: "Analysis of Dynamic and Static Foam Behavior," *JPT* (Jan. 1981) 39–45.

Manev, E. D., Scheludko, A. and Exerowa, D., "Effect of Surfactant Concentration on the Critical Thickness of Liquid Films," *Colloid and Polymer Sci.* (1974) **252**, 586.

Marangoni, C.: *Nuovo Cim.* (1871) **5**, 239.

Marangoni, C.: *Nuovo Cim.* (1878) **3**, 97.

McElfresh, P. M and Boncan, V. C. G.: "Applications of Foam Cement," paper SPE 11203, 1982.

Mitchell, B. J.: "Viscosity of Foam," PhD dissertation, Oklahoma U., Norman, OK (1969).

Monsalve, A. and Schechter, R. S.: "The Stability of Foams: Dependence of Observation on the Bubble Size Distribution," *J. Colloid & Interface Sci.* (Feb. 1984) **97**, No. 2, 327–335.

Montman, R., Sutton, D. L., Harms, W. M., and Mody, B. G.: "Low–Density Foamed Portland Cements Fill Variety of Needs," *Oil & Gas J.* (July 26, 1982) 209–216.

Montman, R., Sutton, D., Harms, W., Mody, B.: "Low Density Foam Cements Solve Many Oil Field Problems," *World Oil* (June 1982) 171–186.

Montman, R., Sutton, D. L., Harms, W. M., and Mody, B. G.: "Oil Field Application of Low Density Foamed Portland Cements," *Proc.*, Southwestern Petroleum Short Course, Lubbock, TX (1982) 13–36.

Mooney, M.: *J. Rheol.* 1931, **2**, 210.

Moran, L. K., Spangle, L. B., and Evans, J. R., "Foamed Cement—Theory and Practice," Proc., Southwestern Petroleum Ass. Short Course, Lubbock, TX (1983) 22–26.

Nelson, E. B., and Eilers, L. H.: "Cementing Steamflood and Fireflood Wells—Slurry Design," *J. Cdn. Petr. Tech.* (Sept.–Oct. 1985) 58–63.

Nelson, E.B.: "Improved Cement Slurry Design for Thermal EOR Wells," *Oil &Gas J.* (Dec. 1, 1986) 39–44.

Olanson, M. T., "Application of Foam Cements in Alberta," paper Petroleum Society of CIM 84–35–72, 1984.

Phipps, S. R., and Krajeski, B.: "One–stage Cementing Technique Saves Dollars," Western Oil Reporter (May 1983) 62–65.

Princen, H. M.: Rheology of Foams and Highly Concentrated Emulsions—1. Elastic Properties and Yield Stress of Cylindrical Model System, *J. Colloid & Interface Sci.* (1983) **91**, 160–175.

Reidenbach, V. G., Harris, P. C., Lee, Y. N. and Lord, D. L.: "Rheological Study of Foam Fracturing Fluids Using Nitrogen and Carbon Dioxide," *SPEPE* (Jan. 1986) 31–41.

Rickard, W. M.: "Foam Cement for Geothermal Wells," Trans., Geothermal Resources Council (Aug. 1985) 9, Part 1, 147–152.

Root, R. L., Barrett, N. D., and Spangle, L. B.: "Foamed Cement—A New Technique to Solve Old Problems," paper SPE 10879, 1982.

Ross, S., "Bubbles and Foam," *Ind. & Eng. Chem.* (1969) 61, 48.

Ross, S.: "Foams," *Kirk–Othmer Encyclopedia of Chemical Technology*, third edition, John Wiley & Sons, Inc., New York (1980) **11**, 127–145.

Sanchez, J. M.: "Surfactant Effects on the Two–Phase Flow of Steam/Water and Nitrogen/Water Through an Unconsolidated Permeable Medium", PhD dissertation, University of Texas, Austin (1987).

Shah, D. O., Djabbarah, N. F., and Wasan, D. T.: *Colloid & Polymer Sci.* (1978) **256**, 1002.

Sharma, M. K., Shah, D. O., and Brigham, W. E.: "The Chain Length Compatibility and Surface Properties of Foaming Solutions in Relation to Fluid Displacement Efficiency in Porous Media," paper SPE 10612, 1982.

Short, A., Struct, M. I., and Kinniburgh, W.: "The Structural Use of Aerated Concrete," *Structural Eng.* (Jan. 1961) 3.

Sita Ram Sarma, D. S. H., Pandit, J., and Khilar, K. C.: "Enhancement of Stability of Aqueous Foams by Addition of Water-Soluble Polymers—Measurements and Analysis," *J.Colloid & Interface Sci.* (July 1988) **124**, No. 1, 339–348.

Slaton, M.: "Foamed Cement—Ultra Light and More," *Drilling* (Dec. 1981) 15.

Smith, D. K.: *Cementing*, Monograph Series, No. 4, SPE, Richardson, TX (1987) 47.

Smith, R. C., Powers, C. A., and Dobkins, T. A.: "A New Ultra Lightweight Cement with Super Strength," paper SPE 8256, 1979.

Smith, T., Lukay, R., and Delorey, J.: "Foamed Cement Application in Canada," paper Petroleum Society of CIM 83–34–21, 1983.

Smith, T., Lukay, R., and Delorey, J.: "Light, Strong Foamed Cement: A New Tool for Problem Wells," *World Oil* (May 1984) 135–144.

Stewart, R. B. and Schouten, F. C.: "Gas Invasion and Migration in Cemented Annuli: Causes and Cures," paper SPE 14779, 1986.

Tanner, C. H. and Harms, W. M.: "Unique Ultra Light-Weight Cement Slurry Compositions for Use in Unique Well Conditions, Laboratory Evaluation, and Field Performance," paper SPE 11486, 1983.

Tinsley, J. M., Miller, E.C., Sabins, F. L., and Sutton, D. L., "Study of Factors Causing Annular Gas Flow Following Primary Cement Placement," paper SPE 8257, 1980.

Turki, W. H. and Mackay, A. S.: "Primary Cementing Across Massive Lost-Circulation Zones," paper SPE 11490, 1983.

15

Horizontal Well Cementing

Robert E. Cooper and George Birch

Dowell Schlumberger

15-1 INTRODUCTION

Horizontal boreholes to increase production were first drilled in the late 1920s in Texas. In the 1930s, several patents were registered with the U.S. Patent Office covering the development of the technique. From that auspicious start, development was slow. The 1940s and 1950s saw some progress, with the drilling of a significant number of short-length horizontal wells, usually less than 100 ft (30 m), to enhance production from low-pressure formations in the U.S.A. At the same time over 40 horizontal wells were drilled in the U.S.S.R. (Bosio and Reiss, 1988). In the mid-1960s, two horizontal wells were drilled in China. Although technically feasible, such wells were not considered economically viable; consequently, the technique was largely ignored.

Horizontal drilling technology began to make a comeback during the late 1970s, when several wells were drilled at Norman Wells in Canada, to produce from a reservoir below the MacKenzie River (Moore, 1987), and to put the Cold Lake tar sands into production (MacDonald, 1985). To date, several hundred horizontal wells have been drilled, and their commercial viability is no longer in doubt. The equivalent production from vertical wells would require almost six times as many wells. Most of these horizontal producers are located in North America (Mall and Fincher, 1986; Wilkirson et al., 1986), Indonesia (Clary and Stafford, 1987), and Europe. In Europe, horizontal wells have been drilled in France (Reiss et al., 1982), England (Hardman, 1986), Germany (Prevedel, 1985), Denmark (Howes, 1988), and especially Italy, with the full development of the Rospo Mare field (Dussert et al., 1988).

At present, most major oil companies are practicing high-angle drilling, and are beginning to develop extensive practical experience. In addition, most oil and service companies are presently engaged in research programs dedicated to one or more aspects of high-angle well technology. This chapter concentrates on work performed to improve the opportunity for success during the cementing phase. However, an introductory discussion is also presented that defines the various types of horizontal (or deviated) wells, reviews the applications to which horizontal well technology is suited, and describes some typical completion designs.

15–2 HORIZONTAL WELL CLASSIFICATION

Horizontal wells are those in which part of the wellbore is inclined 90° with respect to vertical, although less-than-horizontal, high-angle wells often receive this designation. The horizontal portion of the well is often called a "drainhole." Horizontal drilling techniques can be subdivided into three different groups, depending upon the angle build rate: long, medium, and short radius (Fig. 15–1). The principal characteristics of these well types are summarized in Table 15–1.

15–2.1 Long Radius

The long-radius system uses standard oilfield directional drilling technology. The buildup angle ranges from 3° to 8°/100 ft (30 m) and, depending on reach, requires the buildup to be performed in two or three sections.

The horizontal drainhole of a long-radius well can be relatively long, with a maximum length in excess of 4,000 ft (1,220 m) (Ackert et al., 1988). Highly deviated wells drilled conventionally may or may not be "extended reach" wells. As shown in Fig. 15–2, such wells generally have an initial build after the "kickoff point (KOP)" to 40°–50° deviation, followed by a long ramp displacement section, and may ultimately build to horizontal in the reservoir (Jourdan et al., 1988). With present technology, the drilling of horizontal drainholes with lengths exceeding 2,000 ft (610 m), diameters of 5½ to 9½ in. (14 to 24 cm), and vertical control within 16 ft (5 m) does not present major problems, even across problematic formations, and at vertical depths up to 10,000 ft

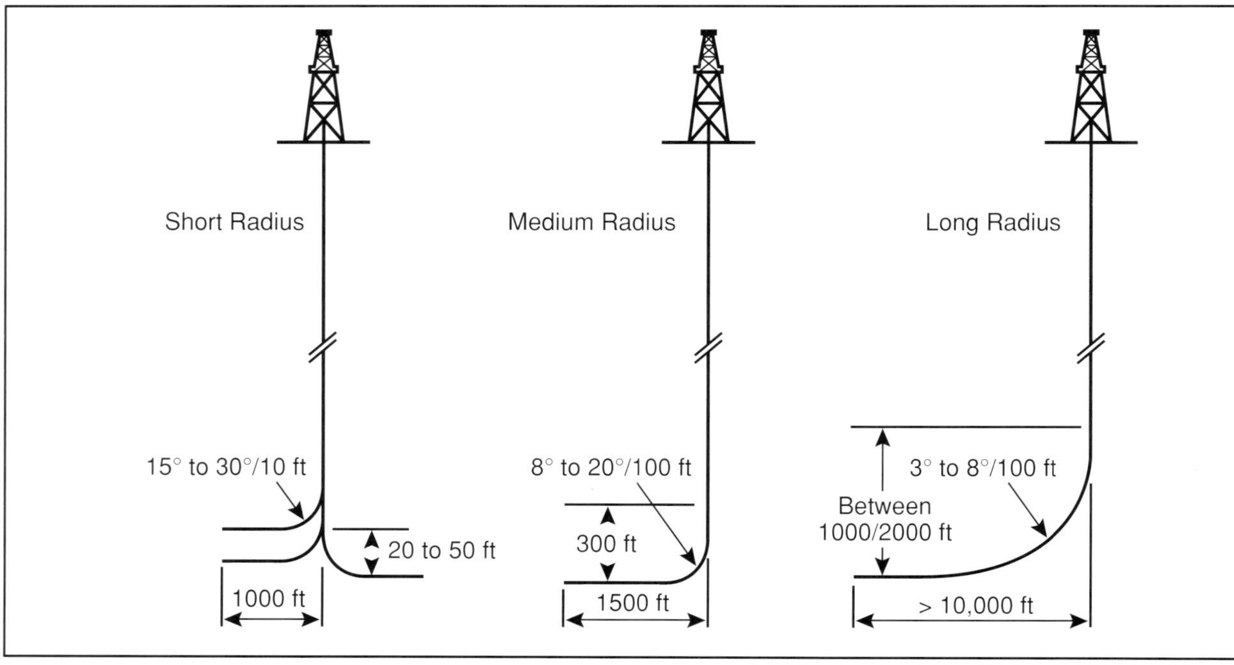

Figure 15–1—Horizontal well classification.

Type of Well	Applications	Drilling Method	Cost	Drainhole Size	Maximum Drainhole Length (to date)	Completion Method	Drainhole Logging Capability
Long Radius	Thin reservoirs; fractured reservoirs; coning reduction; large fields on land; offshore work; EOR methods such as steam or gas injection	Conventional directional drilling	~1.2 to 4X vertical hole	$8^1/_2$ in.	4,012 ft (1,223 m)	Slotted liner; cemented and perforated liner	Drillpipe- and coiled tubing-conveyed tools; MWD
Medium Radius	Same as above except suited for smaller fields	Rotary or down–hole motor below the kickoff level	~2X	4 to $8^1/_2$ in.	1,500 ft (460 m)	Slotted liner	Some tools with drillpipe and coiled tubing systems; MWD
Short Radius	Marginally producing wells; EOR methods; multiple drainholes from single vertical	Rotary side-track generally from existing vertical well	~1.4 to 3X	$4^1/_2$ to $6^3/_4$ in.	1,150 ft (350 m)	Open hole; slotted liner	None

Table 15–1—Comparison of horizontal drilling methods (from Ackert et al., 1988).

(3,048 m) (Armessen et al., 1988). One of the most ambitious extended reach well to date was drilled in the Bass Straits of Australia with a lateral displacement of 15,082 ft (4,597 m), a true vertical depth of 7,974 ft (2,430 m), and an angle of between 69° and 72° (Howes, 1988).

15–2.2 Medium Radius

Medium-radius drilling employs modified conventional drilling equipment, and produces buildup rates ranging from 8° to 20°/100 ft (30 m), although buildup rates as

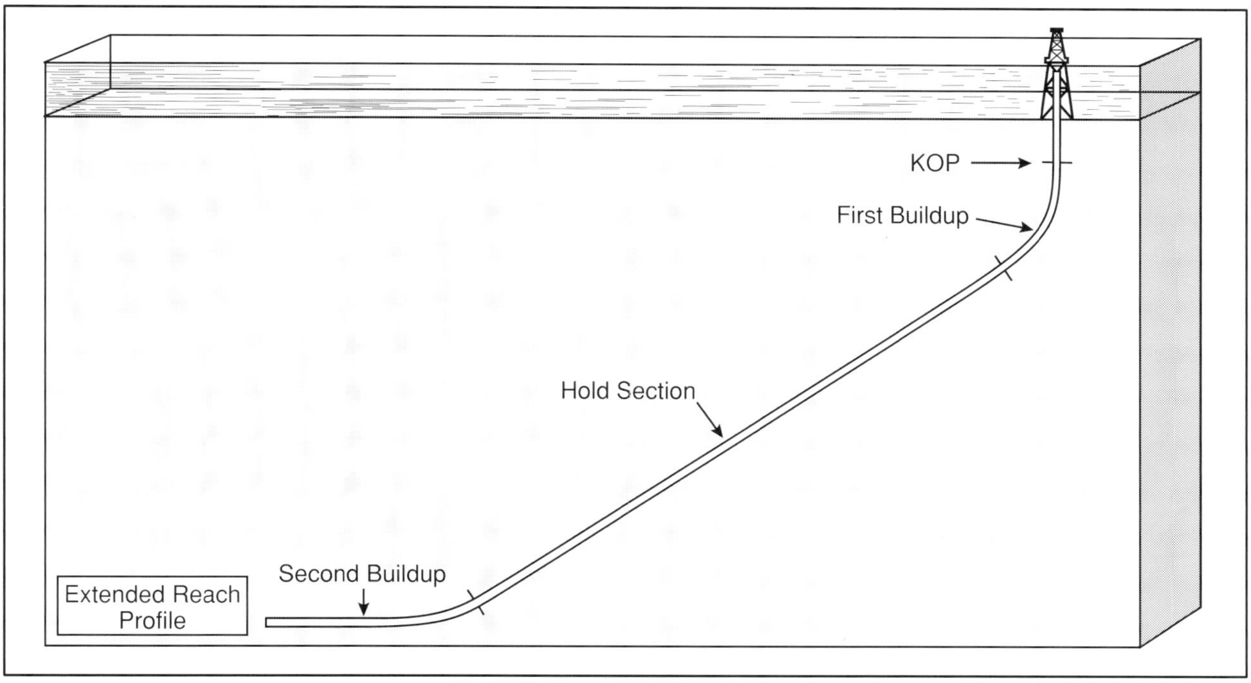

Figure 15–2—Extended reach well profile (from Jourdan et al., 1988).

high as 50°/100 ft are theoretically possible. The length of the horizontal section can be 3,000 ft (915 m) or more, and of the same diameter as long-radius wells.

15–3.3 Short Radius

The short-radius lateral drilling method produces rates of buildup between 1.5° to 3°/ft, allowing the well to deviate from vertical to horizontal in less than 100 ft (30 m). Lateral penetrations up to 900 ft (274 m) are typical. Very specialized equipment is required, combining rotary tools with jointed drill collars fashioned from articulated pipe (Moore, 1981; Parson and Fincher, 1986; Jourdan et al., 1988). Multiple drainholes are often drilled from the same vertical hole with this technique.

15–3.4 Ultrashort-Radius System

A fourth category, sometimes referred to as the ultrashort-radius method (Jourdan et al., 1988), uses the jetting action from a high-pressure nozzle mounted on the end of an oriented flexible pipe. Angle build rates as high as 90°/ft are possible; however, the length and diameter of such holes are generally limited from 100 to 200 ft (30 to 60 m) and 2 in. (5 cm), respectively. More than 10 small drainholes can be drilled in the same plane at right angles from the vertical hole, and are referred to as "star jet holes" (Fig. 15–3). Such wells are beyond the scope of this chapter.

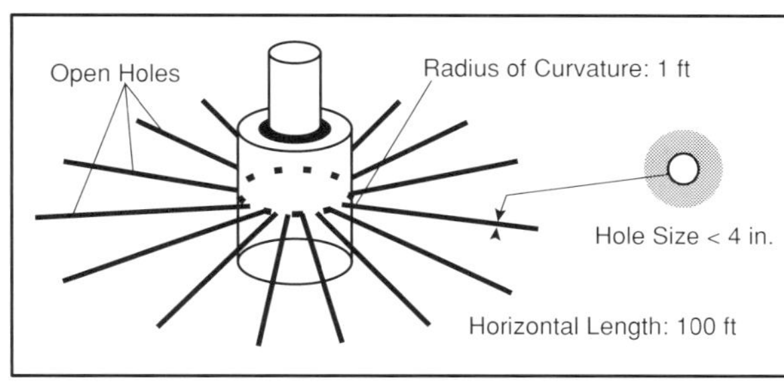

Figure 15–3—Star jet well diagram.

15–3 HORIZONTAL WELL APPLICATIONS

The potential applications of horizontal drilling are numerous. They can be related to the targeted location, reservoir characteristics, the nature and properties of formation fluids, or even to an overall field development plan. In general, long-radius wells are used when an extended reach from an existing surface location is required. Medium-radius wells are used to develop reservoirs whose depth and thickness require accurate well placement. The limited hole size of short-radius wells makes them applicable mainly to the lower permeability, naturally fractured reservoirs (Zaleski and Spatz, 1988).

The usual objective is to provide improved production relative to that obtained with a less-deviated well. Horizontal wells produce, on average, approximately four times more than vertical wells, primarily because of the greater productive formation surface area exposed to the wellbore; thus, not only the productivity but also the total production is increased. As shown in Fig. 15–4, vertical wells have a radial flow geometry and a concentrated pressure drop, while horizontal wells have a parallel flow and a uniform pressure drop.

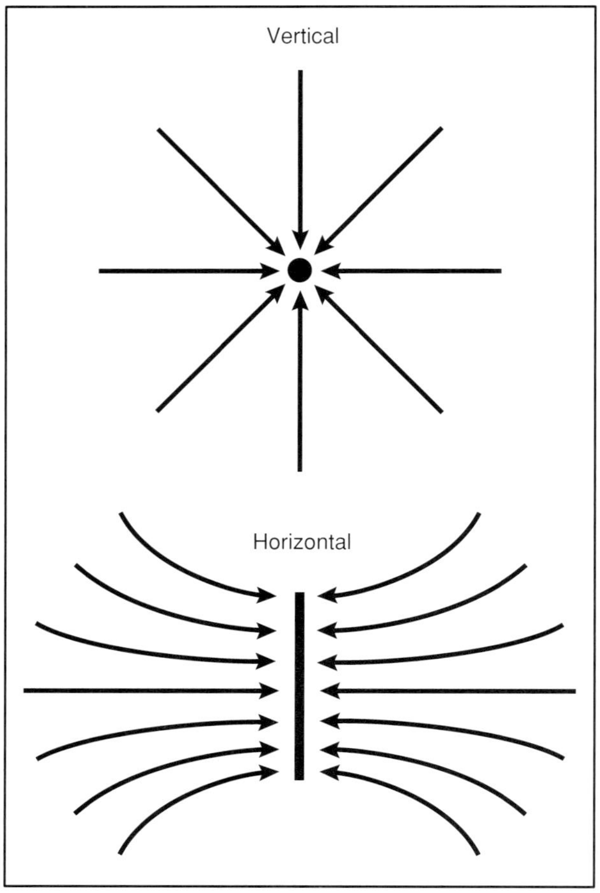

Figure 15–4—Comparison of flow geometries of vertical and horizontal wells (from Howes, 1988).

The advantage of horizontal wells can best be appreciated by referring to Fig. 15–5. This graph compares the productivity index ratio of horizontal and vertical wells with the horizontal length of the wellbore. This simplified production analysis approach is based on the following equations (Giger et al., 1984).

$$\frac{(PI)_H}{(PI)_V} = \frac{\ln\,(r_{eH}/r_w)}{\ln \dfrac{a + \left((a^2 - (L/2)^2\right)^{1/2}}{L/2} + \dfrac{\beta h}{L} \ln \dfrac{\beta h}{2\,r_w}}, \qquad (15\text{–}1)$$

where

h = net reservoir thickness,

L = horizontal length of the well,

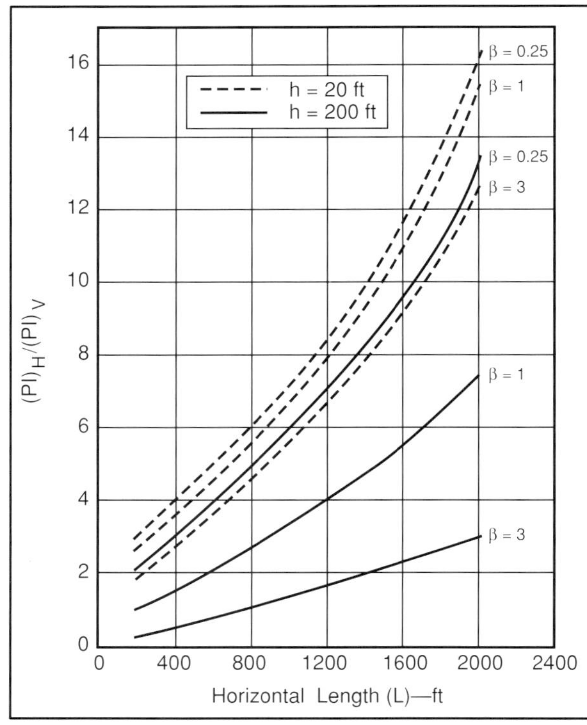

Figure 15–5—Performance comparison between fully completed vertical and horizontal wells (from Giger et al., 1984).

r_w = well radius,

r_{eH} = drainage radius,

$$\beta = \left(\frac{k_h}{k_v}\right)^{1/2} \qquad (15\text{–}2)$$

= permeability anisotropy, where k_h and k_v are the horizontal and vertical formation permeabilities, respectively,

and

$$a = L/2\left(0.5 + \left(0.25 + \left(\frac{r_{eH}}{L/2}\right)^4\right)^{1/2}\right)^{1/2}. \qquad (15\text{–}3)$$

The graph represents a comparison of productivity index ratios for three situations—common anisotropy ($\beta = 3$), complete isotropy ($\beta = 1$), and highly favorable anisotropy ($\beta = 0.25$). The comparison is for two net thicknesses—20 ft (6.1 m) and 200 ft (61.0 m). It is assumed that the drainage area is 40 acres ($r_{eH} = 745$ ft) and the well radius is 7⁷/₈ in. (20 cm).

The graph indicates that the value of β is crucial, and horizontal wells are comparatively more attractive for thinner sections. Assuming a productivity index ratio equal to two as the minimum requirement for drilling a horizontal well, it can be seen that a 1,400-ft (427-m) reservoir with common permeability anisotropy ($\beta = 3$) is not attractive when compared to a vertical well. On the other hand, any horizontal well over 200 ft (61.0 m) in a naturally fissured reservoir ($\beta = 0.25$) is more attractive than a vertical well. Thus, an extended drainhole (L = 2,000 ft or 610 m) can dramatically improve the productivity index ratio for $\beta < 1$.

Joshi (1986) performed a study to determine the productivity ratio between horizontal and vertical wells for different producing zone thicknesses and drainhole lengths. As shown in Fig. 15–6, the thicker the producing formation, the longer the drainhole must be to obtain a given increase in production compared with an equivalent vertical well. Giannesini and Bosio (1988) stated that the optimum drainhole length is 10 times the formation thickness. Thus, for a 100-ft (30-m) thick reservoir, the wellbore should be 1,000 ft (300 m) long. This roughly agrees with Joshi's results.

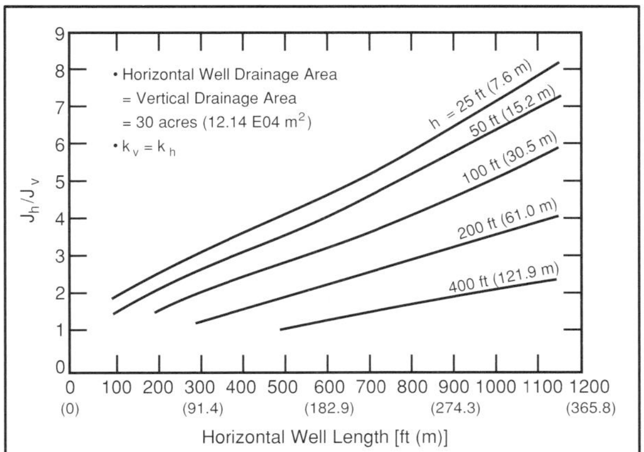

Figure 15–6—The influence of reservoir height on horizontal and vertical well productivity indices ratio (from Joshi, 1986).

Many applications exist where horizontal wells can achieve production more economically than vertical wells (Fig. 15–7). A few of them are discussed below.

15–3.1 Gas and Water Coning

This is a major application. Here, the longer horizontal drainhole increases the exposure of the pay zone, and allows a higher production rate at a lower drawdown pressure. In addition, as shown in Fig. 15–8, the water (or gas) "cresting" or "cylindering" of a horizontal well as opposed to the "coning" of a vertical well ensures increased total oil recovery prior to breakthrough. Wells in this category have been drilled in Indonesia, Prudhoe Bay, and Holland. Production increases of four- to eightfold are being realized in these areas.

15–3.2 Tight Reservoirs and Heavy Oil

In tight and heavy oil reservoirs, the increased exposure from drilling the drainhole horizontally significantly increases production. In tight reservoirs, further production increases can be attained by inducing multiple vertical fractures at right angles to the wellbore. Such a process has been performed in the Dan field of Denmark. Horizontal wells producing heavy oil have been drilled in the Kern River field in California.

15–3.3 Fractured Reservoirs

Another major application is producing pay zones that have sparsely distributed vertical fractures. Here, a vertical well has less chance of intersecting a fracture system compared to a horizontal well, and very significant production increases can be achieved. Vertical wells making a few barrels of oil per day (BOPD) in the Austin chalk (Texas) are being plugged, horizontally sidetracked, and converted to horizontal producers of 300 BOPD. Another example is the Rospo Mare field, where wells have production rates up to five times higher than an equivalent vertical well.

15–3.4 Edge-Water or Gas-Drive Reservoirs

Here the application is to convert "depleted" vertical wells to horizontal wells to produce oil from the edge of the reservoir. Similarly, this technique of horizontal wells on the reservoir edge could easily be applied in the initial development to exploit the reservoir more effectively.

15–3.5 Inaccessible Reservoirs

Horizontal wells can be used to develop inaccessible oil and gas reservoirs, such as under cities, water, and rugged terrain. An example is the previously mentioned project to develop a reservoir under the MacKenzie River in Canada. Environmental considerations are also an issue that horizontal drilling could address under these circumstances.

15–3.6 Enhanced Oil Recovery

Horizontal wells can improve the injectivity and the areal sweep efficiency of oil fields initially produced by vertical wells. The horizontal line drives can be more efficient than the conventional five-spot vertical patterns. Combining horizontal and vertical production in existing vertical well fields could also be used to further enhance production and ultimate recovery.

15–3.7 Others

Other possible applications include—

- reducing the number of offshore platforms and wells needed to develop a field,
- infill drilling from an existing platform to provide better coverage of the reservoir, and

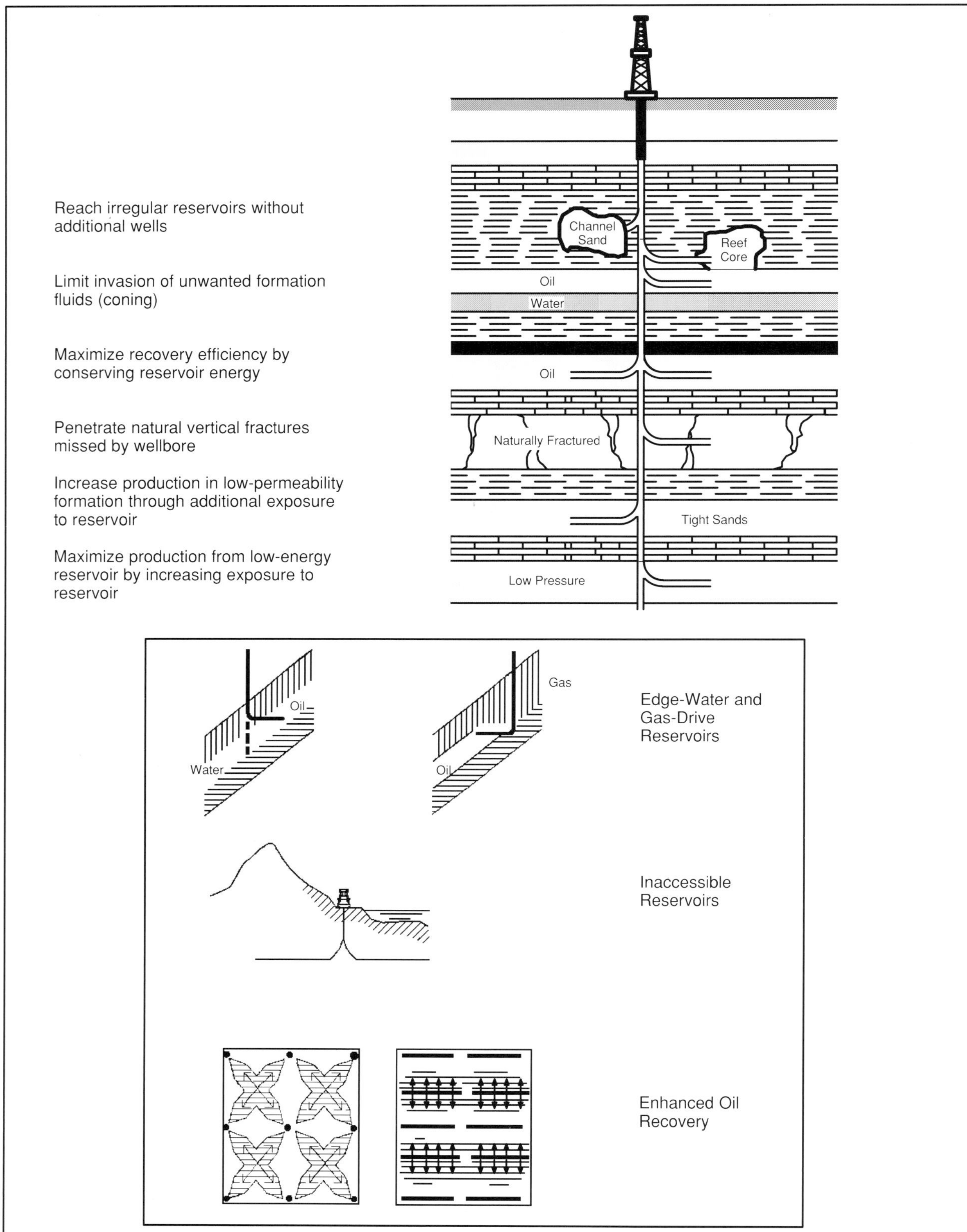

Figure 15–7—Horizontal well applications.

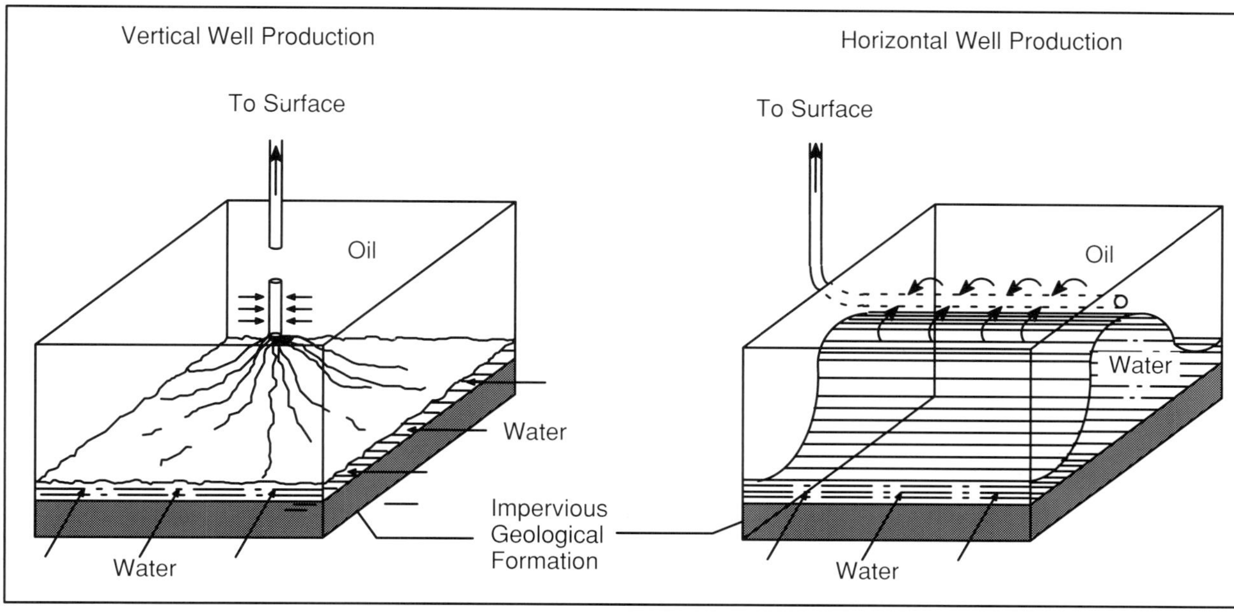

Figure 15–8—Comparison of water coring in vertical and horizontal wells (from Howes, 1985).

- evaluation wells, after initial discovery, to provide better information concerning the reservoir before the final development decision is made.

15–4 COMPLETION PROCEDURES

At present, most horizontal holes are completed without cementing. The horizontal section is often lined with a slotted liner, preperforated liner or, in some cases, wire-wrapped sand control liners. In such wells, the formation rock must have sufficient integrity to prevent collapse or sloughing, particularly when approaching depletion. Very rarely can horizontal wells be completed as an open hole without some method of lining.

The previous intermediate casing, which is frequently highly deviated, must have a good cement job. This is necessary to protect the intermediate string from produced fluids, and to provide isolation between the upper cased off zones and the lower producing intervals.

Often, however, there are horizontal well completion and production circumstances which dictate that casing must be run, and some form of isolation initiated. Some of these are listed below.

- When subsequent multi-interval stimulation treatments of the reservoir are planned.
- When "gas-coning" and "water-coning" control problems are foreseen due to the borehole penetrating or being too close to the gas cap or water table. This may result from loss of directional control causing the boreholes to meander, or simply penetration of the gas cap prior to entering the oil producing zone.
- When current producing intervals may require remedial cementing to prevent unwanted water or gas breakthrough.

Examples of typical horizontal completions and cementing methods are illustrated in Figs. 15–9 and 15–10. The first method is a totally cemented liner, and selected perforated intervals.

Another method is multizone isolation using external casing packers (ECPs) (Chapter 10) and selectively placed sections of slotted liners or subsequent selective perforations. Cement-filled ECPs appear to provide better long-term sealing than the mud-filled type. There is some argument that short (<6.5 ft [<2 m]) packers are easier to center, and provide better sealing than the conventional longer ones (Lessi and Spreux, 1988). However, several operators have reported successful zone isolation using conventional 20- or 40-ft (6- or 12-m)

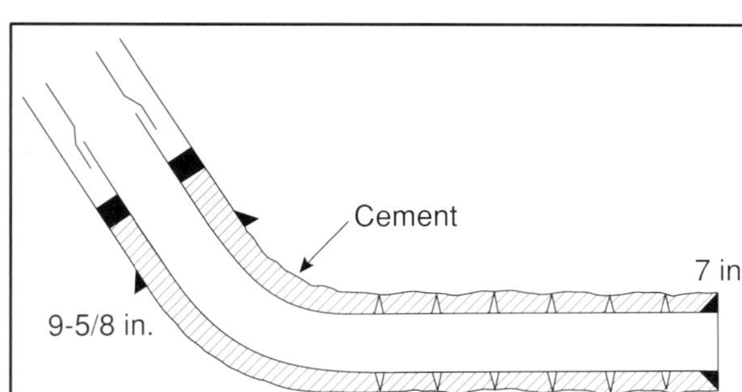

Figure 15–9—Horizontal completion with cemented and fully perforated liner.

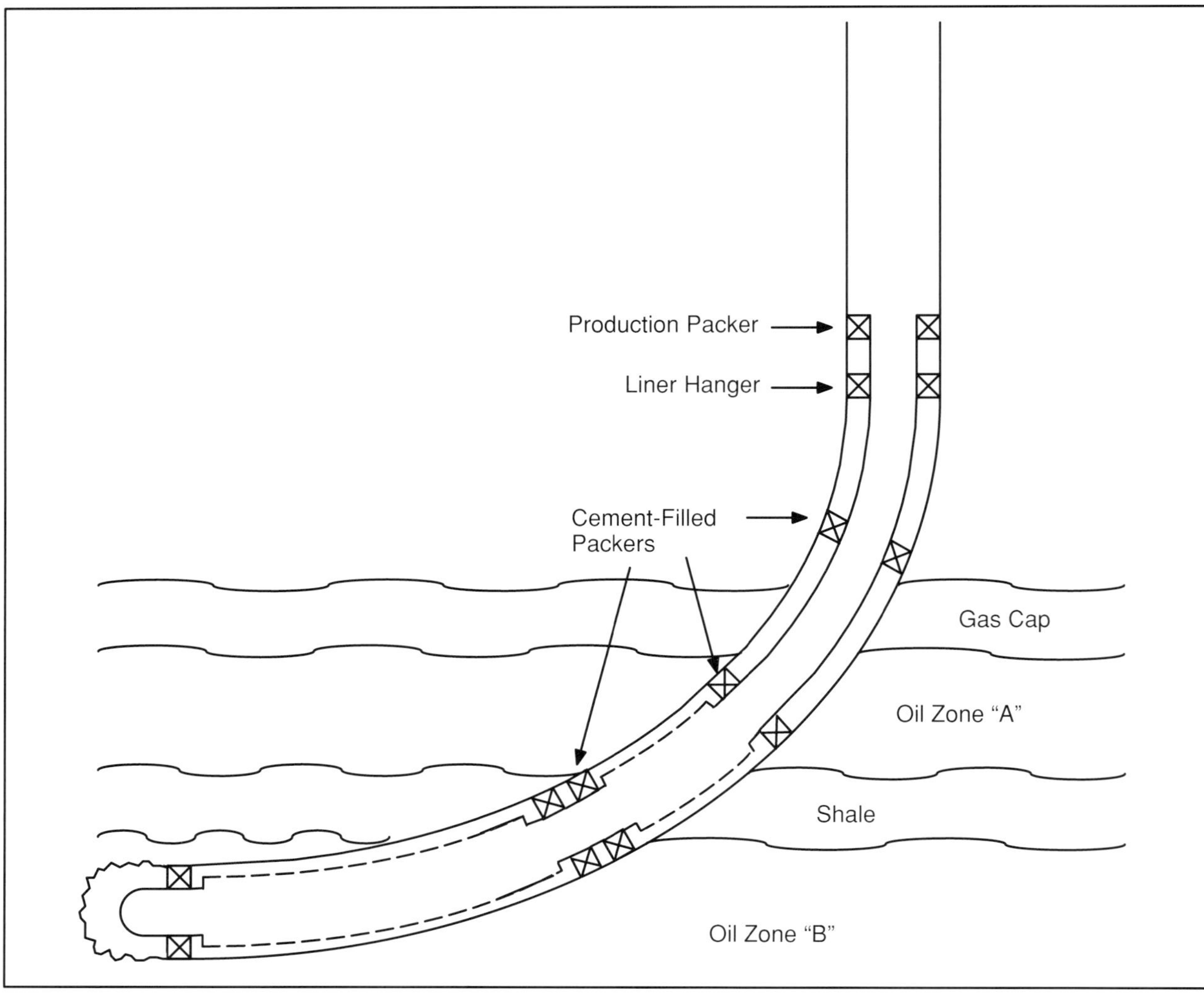

Figure 15–10—Horizontal completion with slotted liner and ECPs.

packers (Reiley et al., 1988). Some controversy exists concerning the efficiency of ECPs and cement-filled formation packers. Reiss (1987) reported case histories where inflation problems and poor sealing were experienced. ECPs are probably suited to complement a cement job, but not to replace it. Such devices may provide temporary sealing after a cement job, but the set cement should be designed to provide long-term zonal isolation by itself.

A third method uses uncemented predrilled liners with selected blank sections. This method assumes that the formation will collapse around the blank sections with time and depletions. This technique is controversial and needs to be proved in field practice (Cooper and Troncoso, 1988).

15–5 MUD REMOVAL

As in conventional cementing, mud displacement is absolutely essential to obtain a good primary cement job. The normal principles of effective mud removal, previously described in Chapter 5, apply in horizontal wellbores; however, there are some additional important factors.

15–5.1 Mud Properties

Hole cleaning during drilling has always been one of the major concerns of directional drilling (Dellinger et al., 1980). This is because cuttings tend to settle on the low side of the hole, and are difficult to remove (Iyoho, 1980). A series of large-scale laboratory tests was con-

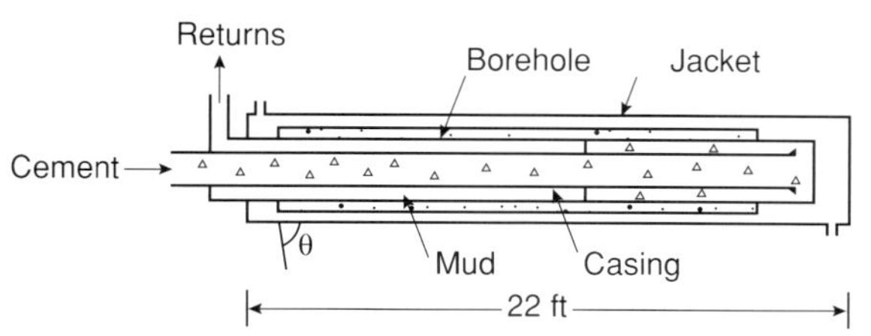

Figure 15–11—Schematic diagram of apparatus in Keller et al. study (from Keller et al., 1987).

ducted by Keller et al. (1987) to determine how solids settling from the drilling fluid affects mud displacement during cementing. They constructed an apparatus (Fig. 15–11) to simulate a full-scale section of a deviated oil or gas well. The model simulated a 5-in. (13-cm) casing in a $6^1/_2$-in. (17-cm) hole, and could be fitted with either a permeable or a nonpermeable "formation," made from man-made sandstone or steel, respectively. The model was operated at three inclinations—0°, 60°, and 85°.

Two water-base mud formulations and one cement composition (described in Table 15–2) were used.

Additives per 1 bbl for 16-lbm/gal mud	
Water, gal	29.80
Bentonite, lbm	15.00
Carboxymethylcellulose, lbm	0.25
Barite, lbm	409.00
Lignosulfonate, lbm	4.00
Sodium hydroxide	
Additives per 1 bbl for 12-lbm/gal mud	
Water, gal	35.80
Bentonite, lbm	19.50
Barite, lbm	186.00
Lignite, lbm	1.00
Sodium hydroxide	
Additives per 1 sack for 16.8-lbm/gal cement	
Class H cement, lbm	94.00
Water, gal	3.91
Retarder, %	0.50
Dispersant, %	0.50

Table 15–2—Fluid compositions in Keller et al. study (from Keller et al., 1987).

Spacer fluids were not used. The test procedure was designed to simulate an actual cement job. Mud circulation and conditioning were performed, and sufficient cement slurry was pumped through the model (>30 times the actual annular volume) to simulate the contact time the casing and formation would actually experience during a job. The cement was then allowed to cure; the apparatus was disassembled, and cut into segments. The displacement efficiencies were then determined based upon examination of all the segments for mud channels. The results are shown in Table 15–3.

Examination of the data in Table 15–3 suggests that for a given flow rate, the ability of a mud to prevent solids settling is related to its yield point and gel strength. For example, scrutiny of the results at 85° deviation (impermeable formation) reveals that the displacement efficiencies were generally higher for muds with higher yield points and gel strengths. In addition, Keller et al. (1987) found that the severity of the solids settling was significantly worse at 85° than at 60°.

Crook et al. (1987) investigated the yield-point issue further, using the same wellbore model described above. Mud-displacement tests similar to those performed by Keller et al. (1987) were performed at 60° and 85° deviation. The rheological properties of both the mud and the cement slurry were monitored, and the displacement efficiency was determined after dismantling the model when the cement had set. Their results (Table 15–4) led to two conclusions—(1) there appears to be a threshold value below which a continuous solids channel will appear, and (2) the yield-point value required to prevent the formation of a channel increases with an increase in the deviation angle. A minimum yield point of 20 lbf/100 ft^2 for 60° deviation, and 28 lbf/100 ft^2 for 85°, was recommended to prevent settling from the mud.

Laboratory testing and industry experience also indicate that turbulent flow is essential for removing the cuttings from a horizontal hole (Parcevaux, 1987). To promote turbulent flow, several major operators drill with a low apparent viscosity mud, while maintaining a high circulation rate up to 500 gal (1,900 L)/min ($8^1/_2$ in. or 22-cm hole size), and a yield point/plastic viscosity (τ_y/μ_p) ratio above 1.

Apart from keeping the hole clean, the mud must also avoid damaging the reservoir, avoid incompatibility with the reservoir fluids, ensure borehole stability in unconsolidated formations, and reduce torque and drag of the drillstring and casing. To satisfy these conditions, oil-base muds are often preferred.

15–5.2 Mud Circulation

Mud circulation prior to cementing is as important in horizontal wellbores as it is in conventional wells. Proper circulation at the highest allowable pump rate is necessary to break the gel strength of the mud, and facilitate its removal by the displacing fluids. As in conventional wells, circulation should be at least "bottoms up" (but

Angle (°)	Formation	Drilling Fluid: Density (lbm/gal)	Solids (%)	PV/Yield Point at 72°F (cp/lbf/100 ft²)	Gels* 10 sec/10 min (lbf/100 ft²)	Fluid Loss** LT/HT (cm³/30 min)	Cement: Density (lbm/gal)	PV/Yield Point at 180°F (cp/lbf/100 ft²)	Volume (bbl)	Rate (bbl/min)	Displacement Efficiency: Top/Bottom (%)
0	impermeable	12.2	20	34/9	2/2	13/34	16.8	16/32	10	1	98/98
0	impermeable	12.1	18	31/7	2/2	14/28	16.8	18/30	10	7	98/98
60	impermeable	16.0	30	36/4	3/5	10/13	16.8	23/29	20	4	99/77
60	impermeable	15.9	29	91/15	3/5	8/11	16.8	23/29	10	7	88/80
60	impermeable	16.2	30	83/24	6/10	7/20	16.8	21/33	20	7	93/87
60	impermeable	15.9	26	61/11	3/13	11/11	16.8	20/30	10	4	99/92
85	impermeable	15.8	22	29/2	2/2	12/25	16.8	16/30	10	1	96/31
85	permeable	15.7	29	53/6	2/3	11/13	16.8	24/29	10	7	95/45
85	permeable	15.8	27	55/8	2/4	9/15	16.8	25/30	10	4	74/48
85	impermeable	15.7	28	56/14	4/5	8/14	16.8	15/30	10	4	99/51
85	impermeable	15.8	28	55/11	4/4	9/26	16.8	17/29	10	4	99/55
85	impermeable	15.9	31	72/19	5/7	7/13	16.8	20/30	10	1	97/68
85	impermeable	15.8	26	56/19	5/9	6/13	16.8	18/29	10	7	98/80
85	impermeable	11.8	12	31/6	2/3	9/34	16.8	20/30	10	4	92/14
85	impermeable	12.3	13	34/4	2/3	9/12	16.8	21/29	10	4	99/23
85	impermeable	11.9	14	44/9	2/3	8/12	16.8	15/29	10	4	93/50
85	impermeable	11.5	20	65/23	9/10	6/11	16.8	19/31	10	4	99/59
85	impermeable	11.6	12	72/25	5/5	8/10	16.8	14/29	10	4	99/67
85	impermeable	12.0	16	78/47	19/26	5/10	16.8	20/28	10	4	99/99
85	impermeable	12.2	20	36/12	2/4	14/29	16.8	15/30	10	4	91/37

* At 72°F
** Low temperature at 72°F and 100 psi; high temperature at 180°F and 500 psi.

Table 15-3—Summary of displacement test results (from Keller et al., 1987).

preferably two hole volumes), and should be continued until a minimum of 95% of the circulatable mud is moving. One operator "tags" the mud frequently to determine when the maximum volume of mud is flowing. By measuring the time between the introduction of the tagging material (e.g., oats) and the time between returns of the individual tags, one can determine when the maximum mud volume is flowing. When these times are equal, the maximum is moving (Chapter 5). Turbulent flow should be maintained provided formation breakdown pressures are not exceeded. This can be verified with computer simulators (Chapter 11).

The use of a top-drive drilling system (TDS) is almost mandatory to provide adequate hole cleaning in highly deviated and horizontal wells (Stewart and Williamson, 1988). The TDS provides several advantages over a conventional rotary table/kelly drive system —

- the ability to reciprocate and rotate the drillstem during circulation to improve hole cleaning when conditioning the hole prior to pulling out of the hole (backreaming),
- the ability, on making connections, to backream each stand with the top drive at least once in gauge hole, and at least three times over washed out zones, and
- the ability to make frequent wiper trips back to the casing shoe with circulation for at least the first 10 stands off bottom.

15-5.3 Pipe Movement

Movement of the casing or liner is important to aid in breaking the gel strength of the mud, and to allow the displacing fluids to sweep away the mud (Chapter 5). Both rotation and reciprocation are preferred over either method alone. Rotation is preferred in gauge holes because the rotational forces on the fluid will cause it to be swept entirely around the annulus (Webster et al., 1987). Reciprocation is an acceptable alternative, and should be used in washed out holes. Rotation should be at 10 to 20 RPM, and reciprocation should be in 10- to 20-ft (3- to 6-m) strokes, with one to two strokes every one to two minutes. Movement should begin with the initial mud circulation, and continue until the final plug is bumped. Combinations of rotation and reciprocation have been used in horizontal wells, and can be used for either full strings or liners. It should be noted that pipe movement is much easier with oil-base mud than with water-base mud, because the wall friction is about one-half (Reiley et al., 1987).

	16.8-lbm/gal Cement at 180°F*			Drilling Fluid at 72°F**		Displacement Efficiency (%)		
Deviation Angle (°)	Volume (bbl)	PV (cp)	Yield Point (lbf/100 ft^2)	PV (cp)	Yield Point (lbf/100 ft^2)	Overall	Top Half	Bottom Half
85	10	21	29	34	4	65	99	23
85	10	20	30	31	6	51	92	14
85	10	15	29	31	9	80	93	50
85	20	32	22	63	17	92	100	79
85	20	25	27	63	23	86	100	58
85	10	19	31	65	23	80	99	59
85	10	14	29	72	25	84	99	67
85	20	35	33	85	28	99	97	99
85	20	27	34	104	36	100	100	100
85	10	20	28	78	47	99	99	99
60	20	25	31	73	15	83	94	70
60	20	63	37	47	17	95	100	90
60	20	18	37	61	20	100	100	99
60	20	55	42	48	24	99	100	97
60	20	22	30	62	29	100	100	100

* Displacement rate = 4 bbl/min.
** 12-lbm/gal water-base mud

Table 15–4—Variation of mud yield point under deviated conditions (from Keller et al., 1987).

		16.8-lbm/gal Cement* at 180°F			Preflush*		Drilling Fluid at 72°F**		Displacment Efficiency (%)			Mud Mobiity Factors	
Test	Type of Movement	Volume (bbl)	PV (cp)	Yield Point (lbf/100 ft^2)	Type	Volume (bbl)	PV (cp)	Yield Point (lbf/100ft^2)	Overall	Top 5 ft	Bottom 5 ft	10-min Gel Strength	Filtrate Volume (cm^3)
39	None	30	50	55	Water	20	76	38	65	67	63	14	15,900
40	Reciprocation	30	55	22	Water	20	63	29	71	70	72	3	10,350
41	Rotation	30	47	56	Water	20	64	29	62	81	43	6	5,100
42	Rotation and reciprocation	30	59	40	Water	20	68	34	78	75	80	10	7,900
43	Rotation, Wipers***	30	62	46	Water	20	77	41	78	69	86	8	16,200
44	Reciprocation, wipers***	30	44	41	Water	20	59	32	83	67	98	30	5,900
45	Reciprocation, rotation, and wipers***	30	60	48	Water	20	75	35	96	92	99	22	8,200

* Displacement rate = 4 bbl/min. **12-lbm/gal water-base mud. *** Across bottom 5 ft only.

Table 15–5—Effect of pipe movement on displacement of a nonsettling mud under 80° permeable, deviated conditions (from Crook et al., 1987).

15–5.4 Cable Wipers

Cable wipers can be an aid in breaking the gel strength of the mud, but they should be used in conjunction with pipe movement. Indeed, in a laboratory study, Crook et al. (1987) observed a dramatic improvement in mud removal when cable wipers were used in conjunction with pipe movement (Table 15–5).

15–5.5 Centralization

Centralization is essential to provide an annulus with open flow paths. If the casing is not centralized, the pipe will trap mud against the low side of the hole. Because of the differences in the flow path, there is no flow regime or practical flow rate which can remove the trapped mud. Field experience indicates that a minimum of 67% standoff is necessary to provide the best chance to remove the mud from the narrow side of the annulus. This was largely confirmed by Wilson and Sabins (1988) who, in a laboratory study, observed mud contamination and poor displacement efficiency when the API casing standoff was less than 60%, despite careful control of mud, preflush, and cement slurry properties.

The difficulty of maintaining turbulent flow around an eccentered casing is graphically shown in Fig. 15–12. The average critical Reynolds number increases 2.5 times when the standoff is reduced from 67% to 40%. Crook et al. (1987) observed that some turbulence can be induced by the centralizer bows, resulting in improved mud removal; however, the effect was localized to within a few feet of the centralizer.

Casing centralization is difficult when the angle of deviation is high, because of the increasing load on the centralizers (Reiss, 1987). To maintain optimum standoff, a rule of thumb is to keep the spacing between the centralizers below 20 ft (6.1 m). Rigid bar centralizers, like those described by Mikoljczyk (1986), Milam (1987), or Reiley et al. (1988), are recommended when cementing in near-gauge hole. Welded bow centralizers can be used in washed out sections. The centralizers should include a bearing sleeve which allows the pipe to be rotated and reciprocated without moving the centralizers (Chapter 10). The required number and positioning of centralizers can accurately be determined by computer simulation (Figs. 15–13 and 15–14).

Buoyancy effects and density differentials should also be considered in planning the centralizer program. Placement of a high-density cement when low-density mud is in the wellbore can result in poor centralization, because the heavier cement can cause the centralizers to collapse or embed. Proper centralizer specifications must be made as well. Because of the importance of centralization, some operators recommend a minimum annular clearance of 0.75 to 1.0 in. (2 to 3 cm) to achieve mud removal and proper cement placement. For an 8½-in. (22-cm) horizontal hole, 5½-in. (14-cm) casing is preferred from a cementing standpoint, but often this is not practical because production considerations may dictate the larger 6⅝- or 7-in. (17- or 18-cm) casing size.

15–5.6 Wedge Effect

At low flow rates (laminar flow), there is a possibility that the heavier cement can act like a wedge and channel

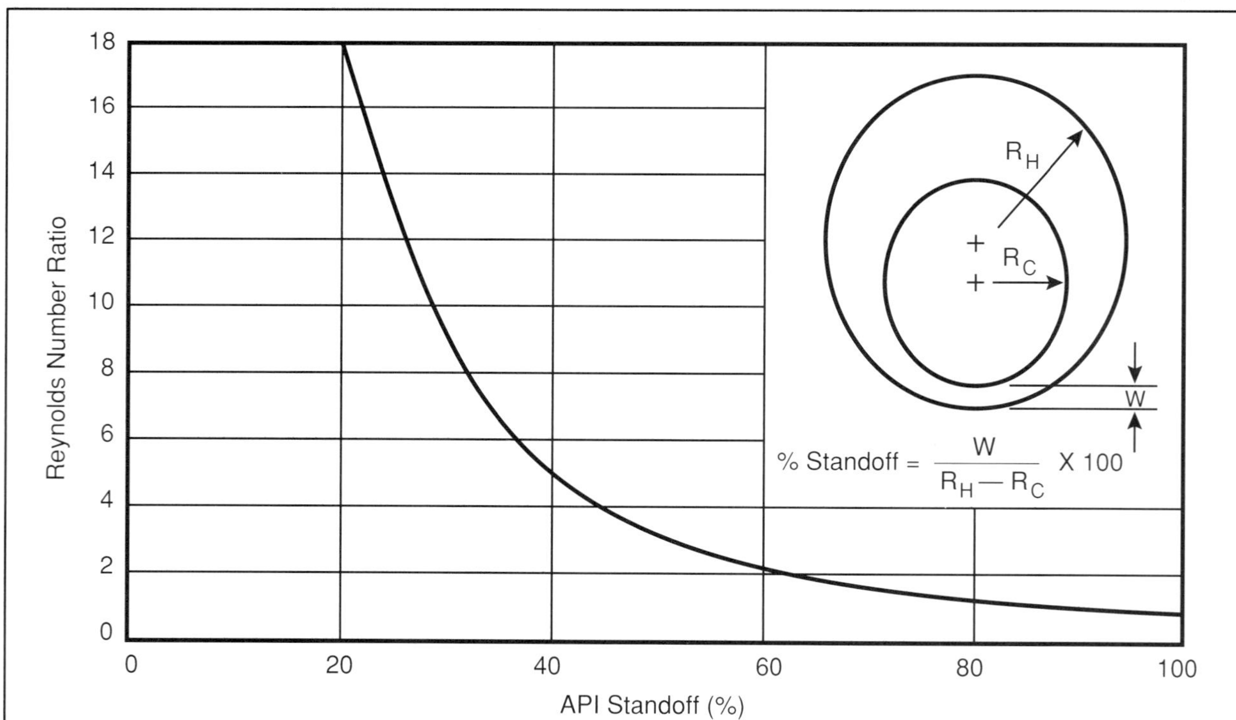

Figure 15–12—Relative variation of the average critical Reynolds number as a function of eccentration.

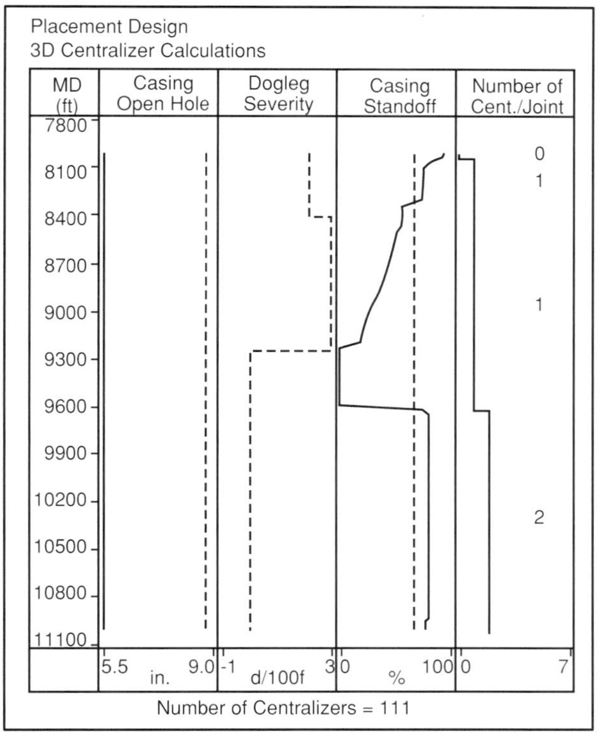

Figure 15–13—Computer simulation of inadequate centralizer placement.

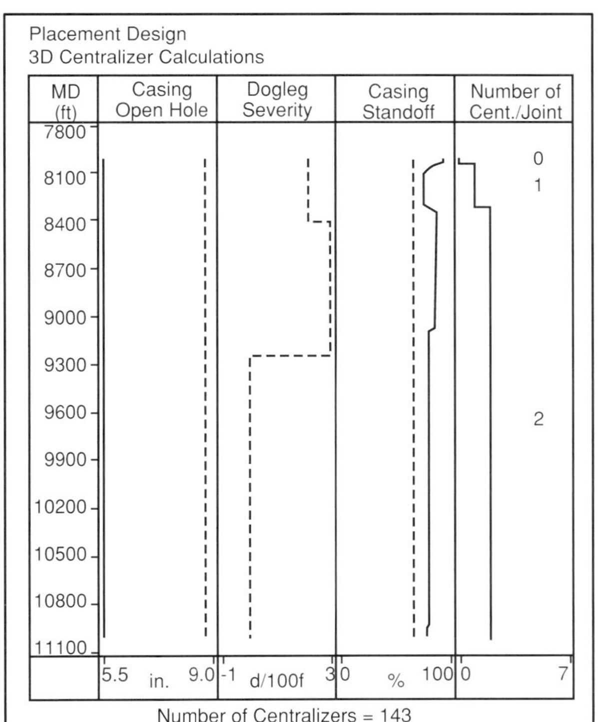

Figure 15–14—Computer simulation of proper centralizer placement.

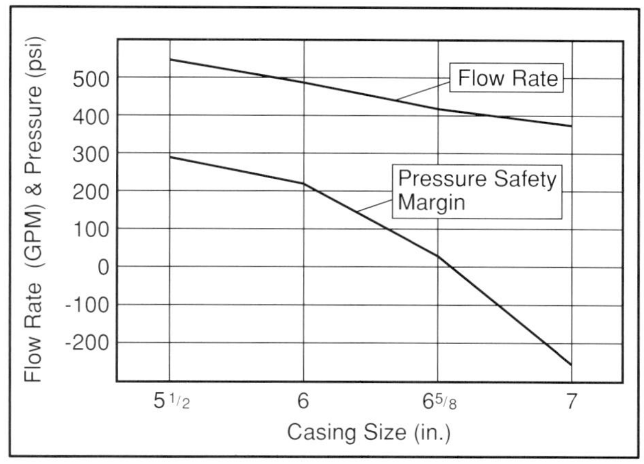

Figure 15–15—Fracture pressure safety margin with mud in turbulent flow.

under the mud. Nevertheless, this effect can also be offset by velocity differentials between the top and the low side of the annulus because of eccentered casing, or apparent eccentering because of solids settling from the drilling mud. Moreover, there appear to be no published theoretical or experimental studies concerning the effects of density differential and casing standoff; thus, recommendations on this point should rely mainly on field experience.

15–5.7 Preflushes and Spacer Fluids

Spacers and chemical washes should always precede a cement slurry. Ideally, all fluids should be in turbulent flow, including the mud. However, if the cement cannot be displaced in turbulent flow, then it must be preceded by a turbulent flow spacer or chemical wash. The cement slurry can then easily displace the thinner fluid, which has a low resistance to flow. For one set of cementing conditions, Fig. 15–15 indicates the rate to achieve turbulent flow of the mud in the annulus for different casing sizes in an 8$^1/_2$-in. (22-cm) wellbore, and the corresponding safety margin. It clearly shows that even with the higher flow rate needed to achieve turbulent flow in 5$^1/_2$-in. (14-cm) casing, a 300-psi (2-MPa) safety margin remains at the end of the displacement, whereas in the 7-in. (18-cm) case, the formation fracturing pressure has been exceeded by 200 psi (1.4 MPa).

15–6 CEMENT SLURRY PROPERTIES

Several cement slurry properties need to be considered for successful cementing (Chapter 5). Some of these properties are more critical in horizontal cementing than in less deviated cementing. Two of the most important properties are slurry stability and fluid loss.

15–6.1 Slurry Stability

Stability of the cement is always important, but even more so in a deviated wellbore. There are two properties that determine the stability of the slurry —free water and sedimentation. Free water is important because it can migrate to the high side of the hole and create an open channel through which well fluids can flow (Keller et al., 1987; Wilson and Sabins, 1988). Sedimentation can result in a low-strength, highly porous cement in the upper part of the wellbore. Loss of zonal isolation may occur, resulting in fluid migration and reduced reservoir control efficiency. Although free water and sedimentation can occur together, they are not necessarily connected. One can easily exist without the other; therefore, testing should be conducted to assure that neither occurs.

Free water should be maintained at zero. In the laboratory, the free water and settling should be measured at the anticipated maximum angle of deviation. Although a standard API method currently does not exist for horizontal slurry testing, many major operators and service companies have developed in-house procedures for free-water testing. Some methods involve the tilting of a slurry-filled graduated cylinder at the anticipated angle of deviation in the well (Webster and Eikerts, 1979; Parcevaux, 1987). Free water and sedimentation can be prevented by chemical means, such as the addition of thickening agents and/or metallic salts which form hydroxide complexes (Defossé, 1983; Baret, 1988) (Chapter 3).

15–6.2 Fluid Loss

Fluid-loss control is particularly important in horizontal wells, because slurry exposure to long, permeable sections is more extensive than in vertical wells. Low fluid-loss rates are necessary to preserve the carefully designed rheological properties of the slurry (Baret, 1988). The fluid-loss rate should always be less than 50 mL/30 min. One method to obtain very low fluid-loss rates without adversely affecting free water and viscosity is to use a properly designed, latex-modified cement system (Cooper and Troncoso, 1988) (Chapter 7).

15–6.3 Other Slurry Properties

Density control of the slurry and uniform additive concentration are particularly important to assure that the cement properties will be consistent throughout the cemented interval. Batch mixing of the slurry should be performed if possible (Cooper and Troncoso, 1988). If ultralow-density cement systems are required because of lost-circulation difficulties, microsphere-extended (Chapter 3) cements would be preferred over the conventional lightweight systems to obtain maximum compressive strength. The use of expansive cements (Chapter 7) could also be envisaged; however, the use of such systems in deviated wellbores has not been reported. Finally, it is important to assure that the cement is compatible with the formation. Special systems are required when traversing salt zones (Chapter 7) or gas zones (Chapter 8), and can be designed to provide all of the desired properties described above.

Once the cement slurry has been designed, the flow rates and pressures should be checked on a "U-tube simulator" (Chapter 11). It is important to verify that the fracture and pore pressures of the formation will not be exceeded during the job.

15–7 SUMMARY—KEYS TO CEMENTING HORIZONTAL WELLS

Based on field experience and laboratory investigations, the main keys to the successful cementing of horizontal wells can be summarized as follows.

- Prevent mud solids from settling.
- Optimize cement slurry properties.
- Maximize annular clearance.
- Centralize the casing.
- Circulate mud at least one hole volume.
- Reciprocate and rotate the casing.
- Batch mix the slurry.
- Run compatible preflushes.
- Design displacement rates for turbulent flow (within fracture- and pore-pressure limits).

Experience within the industry confirms that with good cementing practices and rigid attention to the special details of design and execution, horizontal wells can be cemented with successful results.

15–7 REFERENCES

Ackert, D. et al.: "Looking Sideways for Oil," *Schlum. Tech. Rev.* (1988) **36**, No. 1, 22–31.

Anderson, S. A., Hansen, S. A., and Fjedgaard, K.: "Horizontal Drilling and Completion," paper SPE 18349, 1988.

Armessen, P. et al.: "Horizontal Drilling Has Negative and Positive Factors," *Oil & Gas J.* (May 23, 1988) 37–40.

Baret, J. F.: "Why Cement Fluid-Loss Additives are Necessary," paper SPE 17630, 1988.

Bosio, J. and Reiss, L. H.: "Site Selection Remains Key to Success in Horizontal Well Operations," *Oil & Gas J.* (March 21, 1988) 71–76.

Carter, L. G. and Evans, G. W.: "A Study of Cement-Pipe Bonding," paper SPE 764, 1963.

Clary, M. M. and Stafford, T. W.: "MWD Performance and Economic Benefits in the ZU Horizontal Drilling Program," paper SPE/IADC 16171, 1987.

Cooper, R. E. and Troncoso, J. C.: "An Overview of Horizontal Well Completion Technology," paper SPE 17582, 1988.

Crook, R. J., Keller, S. R., and Wilson, M. A.: "Deviated Wellbore Cementing: Part 2—Solutions," *JPT* (Aug. 1987) 961–966.

Defossé, C. A.: "Composition de laitiers de ciment pour cimentation de puits pètroliers, permettant de controler l'eau libre, et le proce'de' de cimentation correspondant," French Patent No. 2,540,097 (1983).

Dellinger, T. B., Gravley, W., and Tolle, G. C.: "Directional Technology Will Extend Drilling Reach," *Oil & Gas J.* (Sept 15, 1980) 153–169.

Dussert, P., Santoro, G., and Soudet, H.: "A Decade of Drilling Developments Pays Off in Offshore Italian Oil Field," *Oil & Gas J.* (Feb. 29, 1988) 33–40.

Giannesini, J. F. and Bosio, J.: "Horizontal Wells Cut Offshore Production Costs," paper OSEA 88127, 1988.

Giger, F. M., Reiss, L. H., and Jourdan, A. P.: "The Reservoir Engineering Aspects of Horizontal Drilling," paper SPE 13024, 1984.

Hardman, P.: "Beckingham 36 Horizontal Well," paper SPE 15895, 1986.

Howes, J.: "Horizontal and Extended-Reach Drilling Come of Age," *The Oilman* (April 1988) 15–39.

Iyoho, A. W.: "Drilled-Cuttings Transport by Non-Newtonian Drilling Fluids Through Inclined, Eccentric Annuli," PhD dissertation Tulsa U., Tulsa, OK (1980).

Joshi, S. D.: "Augmentation of Well Productivity Using Slant and Horizontal Wells," paper SPE 15375, 1986.

Jourdan, A. P. et al.: "Elf Has Set Up Rules for Horizontal Drilling," *Oil & Gas J.* (May 9,1988) 33–40.

Keller, S. R., Crook, R. J., Haut, R. C, and Kulakofsky, D. S.: "Deviated Wellbore Cementing: Part 1—Problems," *JPT* (Aug. 1987) 955–960.

Lessi, J. and Spreux, A.: "Completion of Horizontal Drainholes," paper SPE 17572, 1988.

MacDonald, R. R.: "Drilling the Cold Lake Horizontal Well Pilot No. 2," paper SPE 14428, 1985.

Mall, T. and Fincher, R.: "Michigan Operator Salvages Well Using Lateral Drilling," *Oil & Gas J.* (June 9,1986) 33–38.

Mikoljczyk, R. F.: "Casing Centralizer," UK Patent No. GB 2,171,436A (1986).

Milam, J. J.: "Apparatus for a Tubular String and Method of Attaching the Same Thereto," U.S. Patent No. 4,658,896 (1987).

Moore, S. D.: "Horizontal Drilling a Success at Norman Wells," *Pet. Eng. Intl.* (Sept. 1987) 19–22.

Moore, W. D. III: "ARCO Drills Horizontal Drainhole for Better Reservoir Placement," *Oil & Gas J.* (Sept. 15, 1981) 130–148.

Parcevaux, P. A.: "Guides Emerge for Cementing Horizontal Strings," *Oil & Gas J.* (Oct. 19, 1987) 35–42.

Parson, R. S. and Fincher, R. W.: "Short Radius Lateral Drilling: A Completion Alternative," paper SPE 15943, 1986.

Prevedel, B.: "How One Operater Drills Horizontally Through a Salt Dome," *World Oil* (Dec. 1985) 69–80.

Reiley, R. H., Black, J. W., Stagg, T. O., Walters, D. A., and Atol, G. R.: "Cementing of Liners in Horizontal and High-Angle Wells at Prudhoe Bay, Alaska," paper SPE 16682, 1987.

Reiley, R. H., Black, J. W., Stagg, T. O., Walters, D. A., and Atol, G. R.: "Improving Liner Cementing in High Angle/Horizontal Wells," *World Oil* (July 1988) 69–80.

Reiss, H. et al.: "Le forage horizontal: premie'res realisations en Europe," *Pètr. et Tech.* (Dec. 1982) No. 294, 33–36.

Reiss, L. H.: "Production from Horizontal Wells After 5 Years," *JPT* (Nov. 1987) 1411–1425.

Stewart, C. D. and Williamson, D. R.: "Horizontal Drilling Aspects of the Helder Field Redevelopment," paper SPE 17886, 1988.

Webster, M. B., Ottot, G. E., and Rice, D. L.: "Cementing High-Angle Wells Using Cement Expanded Formation Packers and/or Casing Rotation," paper SPE/IADC 16136, 1987.

Webster, W. W. and Eikerts, J. V.: "Flow After Cementing—A Field and Laboratory Study," paper SPE 8259, 1979.

Wilkirson, J. P., Smith, J. H., Stagg, T. O., and Walters, D. A.: "Horizontal Drilling Techniques at Prudhoe Bay, Alaska," paper SPE 15372, 1986.

Wilson, M. A. and Sabins, F. L.: "A Laboratory Investigation of Cementing Horizontal Wells," *SPEDE* (Sept. 1988) 275–280.

Zaleski, T. E. Jr. and Spatz, E.: "Horizontal Completions Challenge for Industry," *Oil & Gas J.* (May 2, 1988) 58–70.

16

Cement Job Evaluation

Jacques Jutten
Dowell Schlumberger

Steven L. Morriss
The University of Texas at Austin

16–1 INTRODUCTION

Cement job evaluation consists of checking whether the objectives have been reached *after the job has been performed.* No evaluation of the cement job will be efficient if the objectives are not clear. In addition to the first obvious objective, which is to support the pipe, there are others depending upon the nature of each cement job. For a conductor casing, the essential intention of the cement job is to prevent erosion by stopping the circulation of drilling fluids outside the casing. Surface casings must be cemented to seal off and protect water formations, and to help support deeper casing strings. Intermediate strings are cemented to seal off abnormal pressure formations, isolate incompetent formations, and shut off lost-circulation zones. Production strings are cemented to prevent the migration of fluids in the annulus, and to ensure zonal isolation. Cement also provides some corrosion protection to all the casing strings. For remedial cementing, the objectives are to improve the quality of a primary cement job, seal off perforations, repair a casing leak, isolate productive layers, etc.

Before the development of cement bond logs, the evaluation of cement jobs was performed either by testing the hydraulic isolation or locating the top of the cement. The first type of test often requires casing perforation or additional drilling out. In cases where only pipe support is required, the location of the top of the cement may be sufficient. In other cases where interzonal isolation is desired, more sophisticated methods of evaluation must be used.

The method of evaluation must be selected according to the objective to be reached. The purpose of this chapter is to cover the techniques presently available to evaluate cement jobs—hydraulic testing (Section 16–22), nondestructive methods such as temperature, nuclear, or noise logging (Section 16–3), and acoustic cement logs (Section 16–4). The importance of cement job monitoring is also discussed in Section 16–4.

16–2 HYDRAULIC TESTING

Hydraulic testing primarily consists of testing the isolation provided by the cement. This can be after primary cement jobs, when water zones are located near the oil or gas zone to be produced. It can also be after remedial cementing, e.g., to test if perforations have been effectively sealed.

Several techniques exist which are used to evaluate the degree of isolation provided by the cement job. The most common techniques are pressure testing and dry testing. In some areas, the quality of the cement is established through a production test, or communication testing through perforations.

16–2.1 Pressure Testing

Pressure testing is without any doubt the most common method. It is generally performed after every surface or intermediate casing cement job, once the casing shoe has been drilled. The internal casing pressure is increased until the pressure at the casing shoe becomes larger than the expected pressure to be applied at this point during the next drilling phase. A casing shoe which does not hold pressure indicates a poor cement job, and remedial cementing is required.

16–2.2 Dry Testing

Dry testing is in fact a drillstem test (DST) specially applied to assess the isolation provided by the cement. Dry tests are particularly useful to test the effectiveness of a cement squeeze or a cement seal at the top of the liner. While the objective of a DST is to evaluate the formation parameters from an influx into the wellbore and pressure buildup, the objective of the dry test is to prove that when the pressure is reduced inside the casing, nothing is coming into the wellbore. A successful dry test shows no downhole pressure change during the opening of the downhole valve, or during the following shut-in period (Fig. 16–1) The dry test can also be used to test the

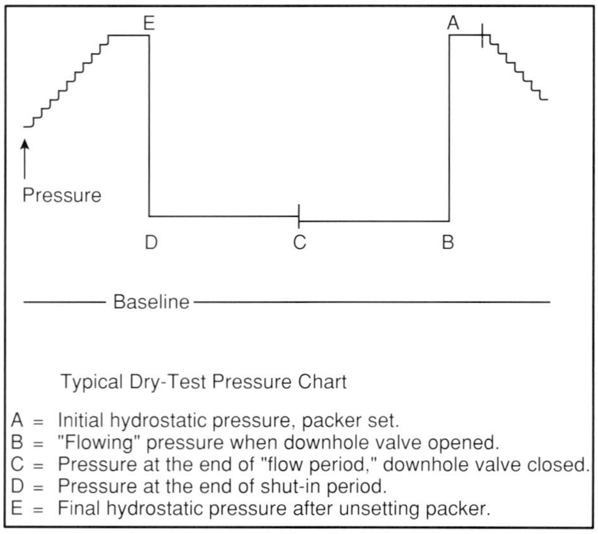

Figure 16–1—Typical dry-test pressure chart.

cement seal around the casing, once the casing has been perforated across an impermeable layer, or after drilling out of the casing shoe in an impermeable layer.

16–2.3 Tests Through Perforations

In some areas, especially when the production interval has a low permeability, the isolation provided by the cement is evaluated after perforating the interval(s) to be produced. The well then produces through the perforations and the production is analyzed. Water-cut production indicates annular communication, and indicates the need for remedial cementing.

In other cases, essentially when cement bond logs show poor results, or when effective isolation is required over short intervals, the casing is perforated in two different locations. A packer is set between both sets of perforations, and pressure is applied on the lower perforations; this is a communication test, as explained by Abdel-Mota'al (1983). If pressure transmission or annular transmission is observed, there is a definite lack of hydraulic isolation in the annulus, and remedial cementing is to be performed.

16–3 TEMPERATURE, NUCLEAR AND NOISE LOGGING MEASUREMENTS

16–3.1 Temperature Logging

Temperature logging is often used for cement evaluation. It is used for primary cementing evaluation, mainly in top-of-cement detection. Temperature surveys are also run to detect leaks or channeling.

16–3.1.1 Cement Hydration Detector

Temperature surveys are often used to detect cement in the annulus several hours after cement placement, because of the exothermic character of cement hydration (Chapter 2, Fig. 2–12). The heat generated by the cement raises the temperature of the wellbore, and induces a deviation from the normal temperature gradient. Figure 16–2 is a typical temperature survey performed after a primary cement job. Such measurements are particularly convenient, and accurately detect the top of the cement. When the volume loss is known, volumetric calculations can be performed to evaluate the displacement efficiency when combined with hole size.

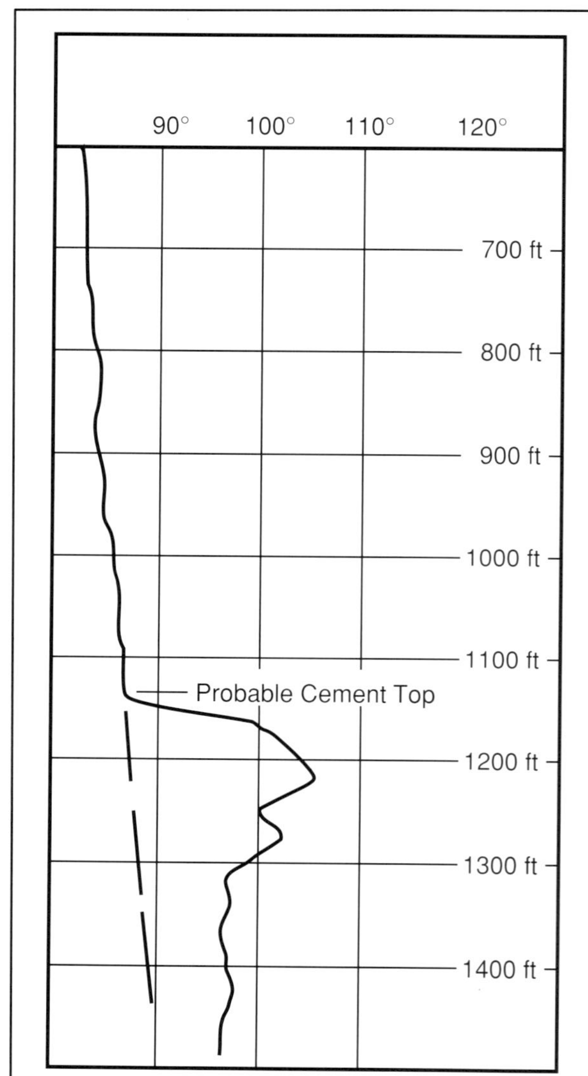

Figure 16–2—Typical temperature survey.

Field experience shows that the maximum temperature anomalies due to cement hydration range from 10° to 40°F (5.6° to 22.2°C), and strongly depend upon the annular thickness and the heat conductivity of the formation. The heat generated during cement hydration is also intimately related to the amount of cement and additives present in the slurry. When compared with standard cements, lightweight cements tend to have longer thickening times and generate far less heat per unit of volume. In this case, the temperature increase may not be sufficiently clear to be detected.

Well cooling by the circulation of fluids prior and during cementing will also have an influence on the kinetics of cement hydration—the longer the circulation, the lower the temperature. This leads to longer thickening times, and delayed and smaller heat increases in the well. When considering a long cement column, temperature logs may not be suitable to evaluate the cement job. This is due to the large temperature differential between the top and the bottom of the well, and the often extremely long cement set times at the top. The API is presently developing new thickening time schedules for long liners, with heating followed by cooling.

The influence of well temperature is illustrated in Fig. 16–3. For the same slurry, the heat peak is higher and sharper at a higher temperature. Consequently, the resulting wellbore temperature increase may be larger in hot or deep wells. In most cases, the peak temperature is attained 4 to 12 hours after cement placement in the wellbore, but the temperature remains elevated for more than 24 hours (Suman and Ellis, 1977). For best results, the temperature survey should be run within the first 12 to 24 hours, depending upon the cement thickening time and downhole conditions. However, it may be impossible to locate the cement if the temperature survey is run too early or too late.

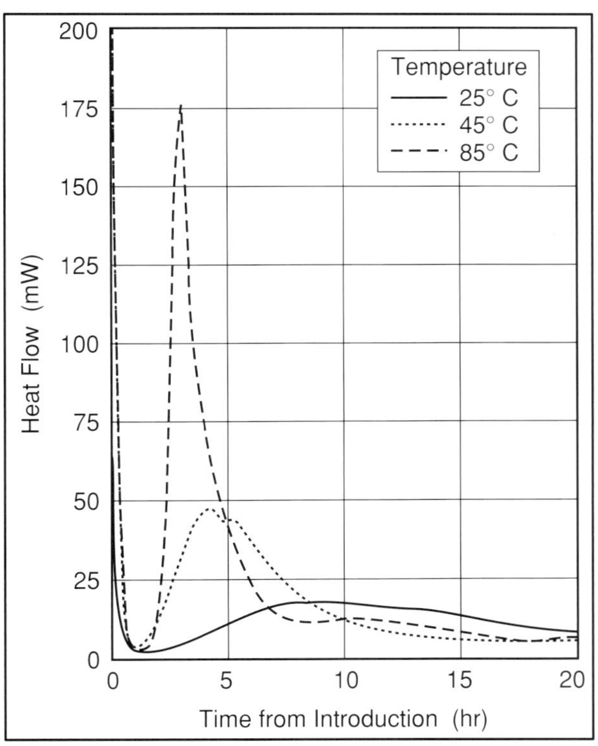

Figure 16–3—Effect of temperature on hydration kinetics of typical Class G cement system (15.8 lb/gal).

Since the amount of generated heat is related to the cement volume, the temperature increase may be higher in large annuli (e.g., in washouts) when compared to in-gauge holes across homogeneous formations. In some cases, a temperature increase in a smooth hole can be due to cement invasion in a fractured or fragile formation (Suman and Ellis, 1977). On the other hand, in narrow annuli, the heat generated during cement hydration may not be sufficient to significantly alter the wellbore temperature profile.

16–3.1.2 Communication Indicator

Once a well has been completed, and channeling behind the casing is suspected due to contamination of the production, temperature logs can be very helpful to better identify the problem. Figure 16–4 is a typical example. The first temperature survey shows the temperature profile prior to the injection of 80 barrels of diesel oil. The second temperature survey, run a short time after injection, shows a large temperature decrease above the perforations, and temperature variations down to the oil/water contact, indicating communication behind the casing. In such cases, remedial cementing must be performed to seal the annulus and reduce the water content.

16–3.2 Nuclear Logging

In the oil industry, it is common practice to add radioactive materials as tracers. This technique can be used to tag drilling mud and to estimate circulation times and volumes, by detecting the radioactive material in the returns. In stimulation, addition of a radioactive material to the treating fluids can help to estimate the extent of the treatment, by comparing gamma ray logs run before and after injection (Ahmed, 1987). This type of technique, essentially qualitative, is also used in cementing, mainly to locate the top of the cement (Fig. 16–5). Kilne and Smith (1986) proved that radioactive tracers are useful to perform a quantitative cement evaluation, using a uniform concentration of radioactive material in the cement, and spectral gamma ray and caliper log results.

Several radioactive tracers can be used in cementing. Soluble tracers (e.g., I^{131}) can be added to the mix water.

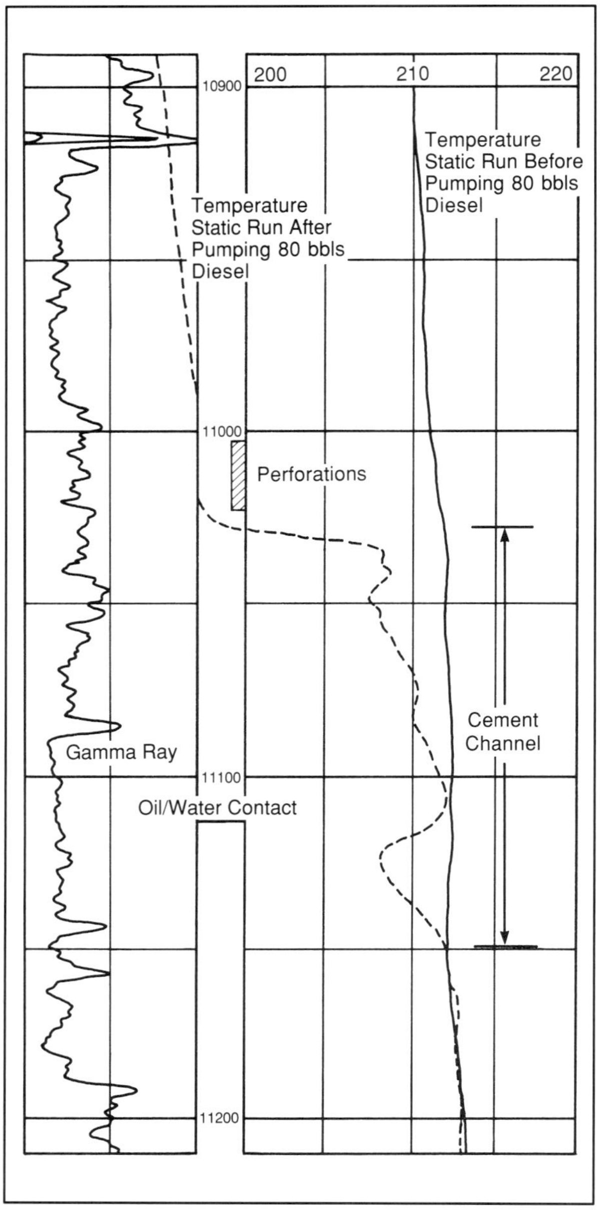

Figure 16–4—Temperature composite profile log before cement squeeze.

Sand or glass beads coated with a radioactive material (e.g., Ir^{192}) can also be used. The standard concentration of radioactive material is about three millicuries per cubic meter of mixing water.

The primary criterion for the selection of the radioactive tracer is its half-life. When long, the alteration to the original gamma ray log will be permanent. This may not be desired for well completion. When short, the influence of the radioactivity of the tracer will die within a few weeks or months, and the gamma ray log will return to its initial state.

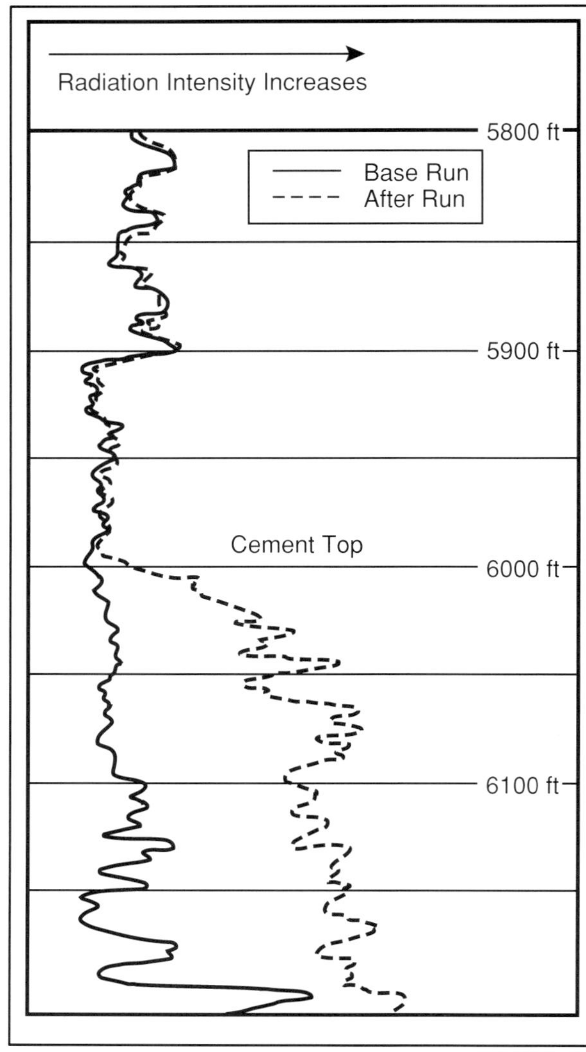

Figure 16–5—Typical radioactive tracer survey.

A second criterion of selection is the energy of the dominant gamma ray emitted by the tracer. When using a spectral gamma ray log, it is possible to selectively measure the radioactivity of the tracer, and the amount of radioactive matter can be significantly decreased. Table 16–1 is a list of the more common radioactive tracers, with information regarding their half-lives and gamma ray energies. It is important to stress that special safety and health precautions must be taken when handling and using radioactive materials, especially for long-half-life materials.

16–3.3 Noise Logging

Any flow, whether gas, water, or oil, produces a noise. Noise logging can be used to detect fluid flow behind the casing, or fluid/gas entry inside a wellbore. This

Element Name	Half-Life (days)	Main γ-ray Energy (MeV)
Cr^{51}	27.7	0.32
Fe^{59}	45	1.10
Br^{82}	1.5	0.77
I^{131}	8	0.36
Ir^{192}	74	0.32
Au^{198}	2.7	0.41

Table 16–1—Radioactive tracers used in nuclear logging.

identification of fluids flowing, and also gives some information concerning the magnitude of the problem. In 1973, McKinley et al. discussed cases where noise logs could identify interzonal communication more accurately than temperature logs. A noise log is a succession of static noise measurements, because it is difficult to detect formation-related sounds if the tool moves continuously. As a result, this technique is marginally used in the oil field.

16–4 ACOUSTIC LOGGING MEASUREMENTS

16–4.1 Introduction

Acoustic logs are without a doubt the most widely used and efficient method to evaluate cement jobs. Cement job evaluation through acoustic log interpretation seeks the relationship between the response of a tool and the quality of the cement job *after a given time* following cement placement. With the response of acoustic tools, *related to the acoustic properties of the surrounding environment* (casing, cement, and formation), it is possible to determine the quality of the acoustic coupling between the casing, cement, and formation. The present lack of a relationship between the acoustic coupling and the hydraulic isolation is a major limitation of the acoustic log interpretation. "Good Bond" indicates a good acoustic coupling, but does not necessarily mean good zonal isolation.

Nevertheless, acoustic log interpretation can still provide a large amount of valuable information concerning the cement job, provided the acoustic properties of the cement and formation are known. Since this is a fairly new concept in the oil industry, one part of this chapter is devoted to acoustic properties in general.

Cement is only one of many parameters that can affect the log response. The analysis of the log must be performed very carefully to determine the origin of the log response. Most of the time, detailed information regarding the well geometry, formation characteristics, and cement job is required.

A fair interpretation of an acoustic log can only be made when it is possible to anticipate the log response. A valid cement job evaluation is the result of the analysis of discrepancies between the expected and the actual log response. Today, it can qualify and possibly quantify the results of the cement job, mainly in terms of cement quality and cement coverage.

The following prerequisites must be met for a meaningful acoustic log interpretation (Fig. 16–6).

- Good quality-control procedure of the field log.
- Knowledge of the well and casing data.
- Knowledge of the cement job events.
- Knowledge or a good estimate of the relevant cement properties.
- Knowledge of pre- and postjob well history in some cases.

In this section, each of these points is covered in detail for CBL/CBT-VDL and CET* logs.

16–4.2 Quality Control

Before developing topics specific to each type of log, general quality-control procedures are discussed which apply to ***EVERY SINGLE LOG***, and without which a log loses all credibility.

16–4.2.1 Measurement Repeatability

All logs should have a *repeat section*. A repeat section is a short log pass, generally over about 200 ft (61 m) of

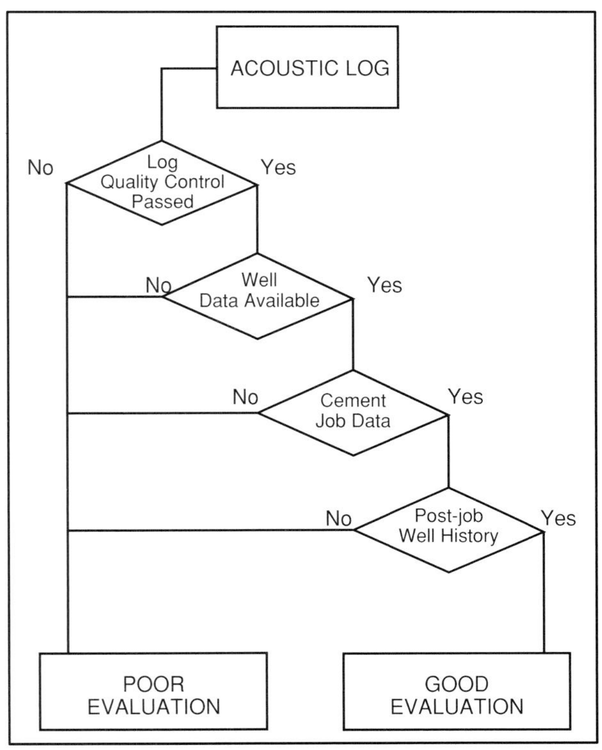

Figure 16–6—Acoustic log flowchart.

* Mark of Schlumberger

16–4.2.1 Measurement Repeatability

All logs should have a *repeat section*. A repeat section is a short log pass, generally over about 200 ft (61 m) of hole, recorded immediately before the main pass. The hole interval logged for the "repeat section" must also be part of the main pass, so that the two independent logs can be compared. Both logging passes must occur under conditions as identical as possible, including tool settings and hole conditions. The purpose is to verify that the logging tool produces the same reading repeatedly under the same conditions. It is a tool functionality check of fundamental importance. If the tool response repeats itself, it is no guarantee that there is no problem; however, if it does not repeat itself there is most assuredly a problem. This verification is very important for all measurements, but it takes on added significance in well logging because of the environmental extremes to which the logging tool is typically subjected, and the need to assure that the time spent logging will result in good data.

The standard of what constitutes good repeatability depends upon the type of tool, its principle of operation, and the design. For acoustic tools, as opposed to nuclear, there is no statistical fluctuation; thus, repeatability is primarily a function of tool design quality. The conventional bond log should exhibit a virtually identical response on the "repeat" and the "main pass."

A complication unique to cement evaluation logging results from the common practice of running the log while holding pressure in the hole from the surface. The sensitivity of the measurement to small changes in pressure requires that identical pressures be established and maintained throughout the repeat pass and the corresponding interval of the main pass. This may not be possible, depending upon the equipment used to hold the pressure and to maintain a wireline seal. The not uncommon practice of running the repeat with no additional surface pressure, then the main pass under pressure, clearly defeats the entire purpose of the repeat. There is no repeat of anything, because the well conditions are not identical for both passes. From the foregoing, it might be concluded that the only way to assure a valid repeat section is to run two passes with no surface pressure.

16–4.2.2 Calibration Summary

A true "calibration" is made against an accepted reference standard. This standard may be agreed upon internationally as with time, pressure, and length. It may be an industry-wide standard such as the API limestone "pit" for nuclear tools. Or, it may be a company-specific standard as is currently the case with bond logs. The latter case is perhaps better referred to as a "tool check" than as a calibration, because the reference is not a standard and the objective is to assure *consistency* in tool response over time among all the tools in use. To properly interpret a log, a calibration summary should be printed. It enables the log analyst to determine if the setup of the measurement was correct. Without a calibration summary, it is impossible to know exactly what was measured and how it was measured; thus, log results are doubtful and interpretation may lead to erroneous conclusions.

At present, there is no industry standard for cement log calibration, although the issue is being addressed by the API. Current proposals would extend the tool check concept to provide industry-wide consistency.

Some logging tools lack an explicit calibration procedure. For consistency, they depend on the extreme accuracy and stability of various electronic circuits, e.g., a quartz-crystal-base clock. A good example of this is the Borehole Compensated Sonic tool, whose basic measurement is of "time." The ultrasonic cement evaluation tools currently are on the edge of this category. They have the potential for an absolute calibration. Since the acoustic cement evaluation tools respond to the acoustic properties of the annular materials, an absolute calibration in the truest sense of the word is possible.

16–4.3 Acoustic Properties

16–4.3.1 Definitions

Acoustics deals with the characteristics of the propagation of sound waves (Chang, 1985). The propagation of sound is actually the periodic compression and rarefaction of molecules (in the case of a gas or liquid) or the squeezing and stretching of the grain fabric (in the case of a solid). When this motion occurs in the same direction as the traveling propagation, the phenomenon is called a *compressional wave*.

In a solid, a second type of wave (the *shear wave*) can propagate. It does not exist in fluids. When it passes through a solid, the grain fabric vibrates perpendicularly to the direction of wave propagation. The shear wave always travels more slowly than the compressional wave. Compressional and shear wave velocities are intimately related to the elastic properties of the material (Young's modulus, shear modulus, and Poisson's ratio), and are almost independent of the frequency. These elastic properties relate the stress to the strain in the material following Hooke's law. In well logging, sound waves are generally characterized by their slowness (ΔT), traditionally expressed in μs/ft or μs/m, which is the inverse of velocity.

Today, for cased hole log interpretation, one is mainly interested in the propagation velocity of compressional waves. The knowledge of the velocity of the compressional waves through a material enables us to determine the compressional acoustic impedance (Z) of this material, traditionally expressed in 10^6 kg/m^2 s, also called Megarayleigh (Mrayl).

$$Z = \rho V_c , \qquad (16\text{–}1)$$

where

ρ = density of material (kg/m^3), and
V_c = velocity of compressional wave (m/s).

While propagating through a material, a sound wave loses energy. This loss of energy, called *attenuation*, is characteristic of the material and increases with the frequency of the wave. No general relationship exists between attenuation and frequency. For a given frequency, the attenuation is normally expressed in decibels (dB) per unit of distance.

$$A = \frac{20}{L} \log_{10} \frac{P_X}{P_{X+L}} , \qquad (16\text{–}2)$$

where

A = signal attenuation (dB),
P_x = signal pressure (amplitude) in x,
P_{x+L} = signal pressure (amplitude) in $x+L$, and
L = distance between two measured points.

16–4.3.2 Acoustic Properties of Formations

Acoustic properties of the formation have an influence on acoustic logs. The familiar terms *fast formation* and *slow formation* refer to sound velocity. Traditionally, for cement evaluation purposes, a formation is called "fast" when sound travels through it faster than along the casing, i.e., less than 57 µs/ft. The typical characteristics of common formations and common fluids are given in Table 16–2.

The same table cannot be constructed for attenuation characteristics, because they are frequency dependent. Generally speaking, attenuation is high when slowness is high. Attenuation is very high in nonconsolidated materials, e.g., shales at shallow depth. Attenuation is negligible in strong consolidated rocks.

Nonporous Solids

Material Type	ΔT (µs ft^{-1})	Sound Velocity (ft s^{-1})	Sound Velocity (m s^{-1})	Acoustic Impedance (Mrayl)
Casing	57.0	17,500	5,334	41.60
Dolomite	43.5	23,000	7,010	20.19
Anhydrite	50.0	20,000	6,096	18.17
Limestone	47.6	21,000	6,400	17.34
Calcite	49.7	20,100	6,126	16.60
Quartz	52.9	18,900	5,760	15.21
Gypsum	52.6	19,000	5,791	13.61
Halite	66.6	15,000	4,572	9.33

Water-Saturated Porous Rock

Material Type	Porosity (%)	ΔT (µs ft^{-1})	Sound Velocity (ft s^{-1})	Sound Velocity (m s^{-1})	Acoustic Impedance (Mrayl)
Dolomite	5 to 20	50.0 to 66.6	20,000 to 15,000	6,096 to 4,572	16.95 to 11.52
Limestone	5 to 20	54.0 to 76.9	18,500 to 13,000	5,639 to 3,962	14.83 to 9.43
Sandstone	5 to 20	62.5 to 86.9	16,000 to 11,500	4,877 to 3,505	12.58 to 8.20
Sand	20 to 35	86.9 to 111.1	11,500 to 9,000	3,505 to 2,743	8.20 to 6.0
Shale		58.8 to 143.0	17,000 to 7,000	5,181 to 2,133	12.0 to 4.3

Fluids

Material Type	ΔT (µs ft^{-1})	Sound Velocity (ft s^{-1})	Sound Velocity (m s^{-1})	Acoustic Impedance (Mrayl)
Water	208	4800	1463	1.46
Water + 10% NaCl	192.3	5200	1585	1.66
Water + 20% NaCl	181.8	5500	1676	1.84
Sea Water	199	5020	1531	1.57
Kerosene	230	4340	1324	1.07
Air (15 psi, 32°F (0°C))	920	1088	331	0.0004
Air (3000, psi, 212°F (100°C))	780	1280	390	0.1

Table 16–2—Acoustic characteristics of common formations and fluids.

16–4.3.3 Acoustic Properties of Cements

Cased hole acoustic log response is primarily dependent upon the acoustic properties of the hard cement. The acoustic properties of rocks are well known; however, it is more difficult to know those of the cement because *they change with time*. This fundamental difference makes the log analysis critical in some cases.

- The log can change with time, because the physical properties of the cement are changing with time.
- The cement is not in the same physical state all along the casing string. This can produce a strong difference in the log response on longstrings where a large temperature difference exists between the bottom and top of the cement.

The acoustic properties of various cement formulations at ambient conditions are reported in Table 16–3 (Jutten et al., 1987). From these results, it appears that low-density slurries have a low acoustic impedance, that can change significantly after several days. The acoustic impedance of the denser slurries changes less than 20% between one and seven days. This can be critical with hollow silica microsphere-extended slurries which may have a low acoustic impedance. Foamed cement also has an extremely low acoustic impedance. When the quality (porosity) of the foam is high, it may be difficult to differentiate the cement from water.

Acoustic property fluctuations over days or even weeks have also been experienced in the field. It is evident by the improvement of acoustic logs with time after cement jobs. Such problems often originate from an overestimation of the bottomhole circulating temperature, and subsequent overretardation of the slurry. At higher temperatures, the kinetics of cement hydration are faster, and the acoustic properties stabilize more rapidly; thus, less time dependence is observed. Since the acoustic properties of the cement are not yet measured on a routine basis, the values reported in Table 16–3 can be used as a first approximation.

16–4.4 Cement Bond Log—CBL/VDL

16–4.4.1 History

Cement Bond Logging (CBL) was developed about 30 years ago, and is still the most widely used method to evaluate cement jobs. Its use can be traced to 1959, when Tixier et al. found that the transit-time curve on openhole sonic logs often "cycle skipped" when run inside the

Slurry Type	Density (lb/gal)	Time (days)	V_c (m s^{-1})	Z (Mrayl)	ΔZ (% of $Z_{1\,day}$)
Neat G	15.8	1	3,000	5.68	0
		2	3,250	6.16	8
		7	3,400	6.44	13
Latex + Hollow Silica Microspheres	11.2	1	1,650	2.21	0
		2	2,200	2.95	33
		7	2,500	3.36	52
Soluble Silicate	12.0	1	1,600	2.30	0
		2	1,750	2.52	9
		7	2,000	2.88	25
Hollow Silica Microspheres 4% $CaCl_2$	12.0	1	2,600	3.74	0
		2	2,800	4.03	8
		7	3,000	4.32	16
Soluble Silicate	13.3	1	1,750	2.79	0
		2	2,200	3.51	26
		7	2,500	3.99	43
Latex	15.8	1	2,900	5.49	0
		2	3,150	5.97	9
		7	3,350	6.35	16
18% NaCl	16.1	1	2,850	5.50	0
		2	3,200	6.18	12
		7	3,375	6.51	18
Hematite	19.0	1	3,300	7.59	0
		2	3,400	7.74	2
		7	3,530	8.04	6
36% Quality Foam	10.0	7	2,300	2.75	—

Table 16–3—Acoustic properties of various cement formulations.

casing, because of considerable attenuation of the acoustic signal.

In 1961, two major service companies published the results of laboratory and field testing of tools designed to measure the actual amplitude of the acoustic signal (Grosmangin et al., 1961; Anderson and Walkers, 1961). One company reported that more than 500 CBLs had been run between April 1959 and November 1960 in North and South America. Both observed many of the interpretation complexities still discussed today, such as cement sheath thickness, pipe eccentering, bond to pipe vs bond to formation, and "fast formation."

In 1963, Winn et al. reported the results of a laboratory study of the effect of various cement compositions on the attenuation of a sonic signal, and compared the acoustic measurement to hydraulic communication. In 1963, Bade reported a field study involving approximately 250 CBLs run in the Williston Basin of central North America. Four commercially available tools were used. He observed the need for proper centralization, as well as the need for narrow gate widths, to exclude the formation signals. In 1963, Pardue et al. provided cement-to-log measurement correlations that are still used in interpretation charts, along with a theoretical model employing a thin steel plate coupled to an infinite half-space. Also in 1963, Pickett argued for some form of full waveform presentation on both openhole and cased hole (CBL) logs. Using data from Bade's field study, Pickett showed the possible wide variation in amplitude measurements because of the various gate settings used, and suggested the need for some kind of waveform presentation to eliminate these ambiguities.

In 1964, Carter and Evans provided information from experiments on the pipe-to-cement bond, with consideration given to casing surface finishes, cement placement techniques, and timing of operations in the cemented casing. In a paper directed primarily at formation evaluation, Anderson and Riddle (1964) pointed out the efficacy of a direct attenuation measurement using two receivers in place of a single-receiver amplitude measurement. In 1968, Walker presented an overview of the CBL, and the economic considerations in the decision on whether to run one. Brown et al. (1970) showed the advantages to be gained from a shorter spacing between the transmitter and receiver for amplitude measurements than for full waveform measurements. They suggested 3 ft (0.9 m) and 5 ft (1.5 m), respectively, as good compromises.

A paper was later published in 1974 by Fertl et al., and was oriented toward operational considerations, including a checklist, along with a fairly complete description of tool design differences and the resulting effects on the log. The authors lauded the potential of the bond log measurement and lamented the lack of standardization and understanding.

A resurgence of CBL-related research occurred in the 1980s. McGhee and Vacca (1980) summarized the possible causes of a "microannulus," and emphasized the importance of dealing with it. In 1982, Gollwitzer and Masson presented the justification for a borehole-compensated CBL which directly measured attenuation by employing multiple receivers in the logging sonde. In 1983, Bruckdorfer et al. extended the available empirical correlations between cement compressive strength and the CBL attenuation rate to include foamed cements.

In 1984, Nayfeh et al. published the results of an experimental and numerical study of the effects of wellbore fluids on CBL amplitude measurements. Considerable differences in free-pipe amplitude were shown to occur for various brines, in particular. Pressure and temperature corrections were given for the transducers used in CBL tools. In 1985, Allen and Wood offered a solution to the operational problem of assuring that the amplitude measurement gate is correctly set, by proposing the simultaneous recording of sliding- and fixed gate CBL amplitudes. In 1985, Bigelow presented a comprehensive guide to CBL interpretation, with an emphasis on qualitative interpretation of the full waveform presentation based on understanding the basic acoustic processes involved and the cement job/well completion history. Bigelow emphasized the need for intelligent interpretation. Tubman et al. (1986) provided a theoretical foundation for the response of acoustic logs in cased boreholes.

In 1987, Leslie et al. analyzed the theoretical response of the borehole-compensated bond log, and separated a "coupling attenuation" (insensitive to a microannulus) from the "propagation attenuation," which requires shear coupling between the casing and the cement. Jutten and Parcevaux (1987) and Jutten et al. (1988) advanced the understanding of the effects of borehole geometrical parameters and cement slurry composition/mechanical properties on bond log response.

16–4.4.2 Description of the Conventional Bond Logging Tool

Figure 16–7 is shows a schematic diagram of a representative bond log tool, along with the cross section of a cased and cemented well. There is an acoustic transmitter which is usually made of a piezoelectric ceramic. There are two receivers, also of piezoelectric ceramic, in most tools. Some designs incorporate a single receiver. In the former case, the two receivers are generally located 3 ft (0.9 m) and 5 ft (1.5 m) from the transmitter. In the latter, the single receiver is 4 ft (1.2 m) from the transmitter. Some hostile-environment tools use magnetorestrictive transducers rather than those made of ceramics. This requires different pressure/temperature corrections (Nayfeh et al., 1984). Not shown in the figure, but always a required part of the tool, are a sufficient number of centralizers to ensure that the transmitter/receiver section of the tool remains absolutely centered in the pipe.

16–4.4.3 Acoustics of the Bond Log Measurement

The transmitter repeatedly emits short bursts of acoustical energy. The duration of each burst is about 50 μs, and the repetition rate is between 10 to 60 Hz, depending upon the particular tool design (and on the setting made by the logging engineer, in some cases). The frequency content of each burst is centered at about 20 kHz for larger diameter tools (larger than 3 in. or 8 cm) and about 30 kHz for smaller diameter tools (less than 2 in. or 5 cm). One company offers a tool with a center frequency of 12 kHz. In the time interval between transmitter bursts, the receiver picks up the signal and makes the bond log measurements. Most of the signal of interest arrives at the receiver within one to two milliseconds after the transmitted burst.

The transmitter burst creates an approximately spherical wave front expanding away from the tool in all directions. As the wave front strikes the inside wall of the casing, it is refracted according to Snell's law. There is one particular direction of travel of the wave front that will result in a refraction straight down the pipe. This is the "critical angle." It is about 16.5° with fresh water in the hole. The part of the wave front which is refracted straight down the pipe ultimately determines the "amplitude" and "transit-time" measurements which appear on the log. Some parts of the original wave front travel directly through the mud, and some parts are refracted into the annulus and formation. Part of the latter ultimately arrives at the receiver as a "formation signal," and the former shows up as "mud waves."

Figure 16–8 is a schematic representation of these various "paths" which the original burst can follow and still arrive at the receiver. The waveforms in the figure are meant to convey the relative times of arrival of the acoustic energy which has traveled along the various paths. The wave which is refracted directly down the casing wall usually arrives first because of the high velocity of sound in steel combined with the relatively short distance. A relatively low sound speed in fluids results in the mud wave arriving very late in spite of having the shortest distance to travel. The arrival time of the formation wave, both shear and compressional, is highly variable (Section 16–4.3). The signal from the receiver will be a mixture, or composite, of waves from all these paths. The interpretation of the actual bond log measurement (as opposed to a picture of the entire composite wave) depends upon the casing wave arriving before anything else. Since they are not used in bond log interpretation, the Stoneley and Rayleigh waves are not discussed here. If the annulus contains a fluid, so little energy arrives at the formation that the received signal consists almost exclusively of the casing signal and the mud waves.

The so-called casing wave is the portion of the original acoustic burst which propagates directly down the casing wall. It loses energy into the annulus and borehole as it propagates, because of the shear coupling with the adjacent materials. The greater the shear coupling, the greater the energy "lost" into the adjacent materials. The loss to the borehole is low and constant; thus, the loss to the annulus is the variable. The rate of this loss is reflected in the "amplitude" or "attenuation" appearing on the log. It should be expected that there will be little attenuation of the casing signal if there is a fluid in the annulus. In fact,

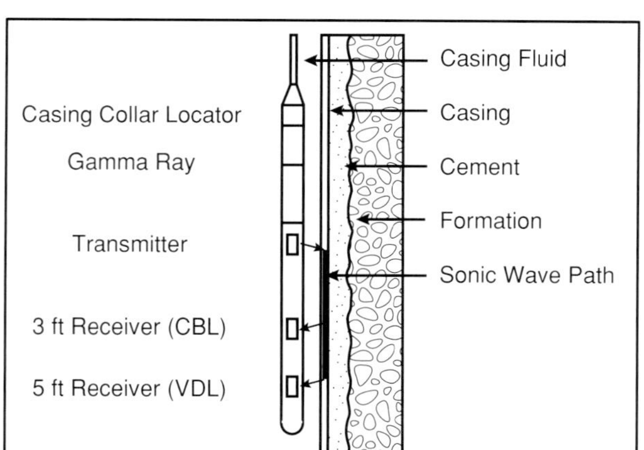

Figure 16–7—CBL-VDL tool configuration.

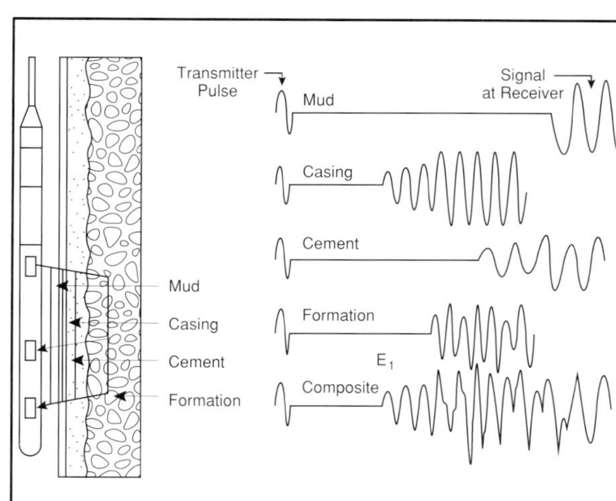

Figure 16–8—Sonic wave paths.

all fluids would be expected to "look" alike because there is no shear coupling for any fluid. This is also the reason why even a microscopic gap of a few thousandths of an inch between the pipe and a cement sheath, referred to as a "microannulus," has a strong effect on the signal (Section 16–4.4.13).

16–4.4.4 Description of the Full Acoustic Wave Display—VDL

The presentation of the full waveform provides some information about the cement job. The actual "composite" signal received is presented in Fig. 16–9. While its use in bond log interpretation is primarily qualitative at present, it contains all available information in "picture" form. There are two ways to present or display this signal on the log. One is to display an actual waveform. It has the disadvantage that it is not a continuous display with depth, but a discrete display with usually one waveform for every 2 to 4 ft. (0.6 to 1.2 m) of hole depth. The other presentation is the "variable intensity" display. Figure 16–9 is an illustration of how the "amplitude" information is transformed to the "intensity" information. An amplitude of zero is converted to the median intensity level. Positive amplitudes have higher intensities (darker) as they become more positive. Negative amplitudes have lower intensities (lighter shades) as they become more negative. Continuous and discrete (five-level) intensity scales are used in the industry. This display is continuous with depth and easy to read, but depends on intensity or darkness to convey the information. This places a great demand on log production (e.g, film development) and reproduction (blueprint) techniques.

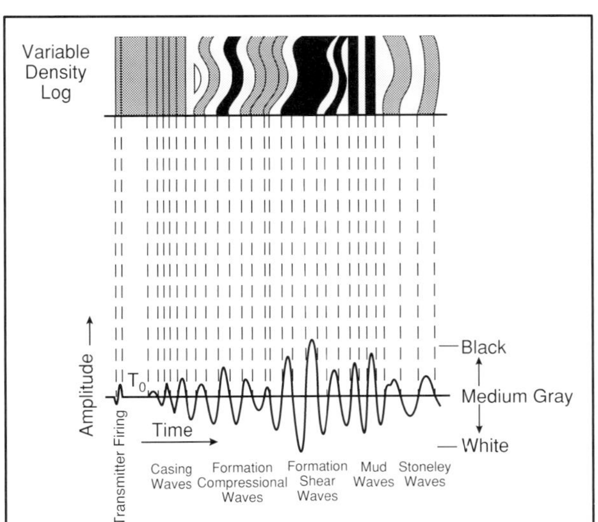

Figure 16–9—Presentation of complete waveform signal from CBL tool.

Furthermore, photocopying and telecopying use equipment (in actual oil-field practice) that produces black and white results, so all the intensity-conveyed information is lost. A "good" variable-intensity display contains the most information possible. A "bad" variable-intensity display contains the least. A waveform display discretizes the "depth" variable. A bad variable-intensity display discretizes the "amplitude" into a binary, black-and-white display. Examples of both displays of the full acoustic wave are shown in Fig. 16–10.

In multiple-receiver tools, the full wave display generally comes from the 5-ft (1.5-m) spaced receiver. Increasing the spacing between the transmitter and receiver has the advantage that the various constituents of the composite wave (casing, formation compressional, and formation shear) are spread apart from one another, because the effect of their velocity differences is more pronounced as the distance increases. However, increased spacing is problematic because the received wave is more attenuated; thus, a compromise is necessary. Since the full wave display is used qualitatively in most cases, the very high attenuation at 5 ft (1.5 m) is not a problem, because the qualitative characteristics are still distinct.

16–4.4.5 CBL-VDL—Qualitative Interpretation

The analysis of the full wave display gives only qualitative information about the cement job. If the cement is well bonded to the casing, most of the sonic energy will leave the casing and pass into the cement—casing waves will have an extremely low amplitude. If the cement is well bonded to the formation, the energy will go through the cement into the formation. The sonic waves will then propagate (compressional and shear) and attenuate through the formation. Since formations are never perfectly homogeneous, their acoustic properties change with depth. Wavy patterns on the received waveforms are the perfect illustration of this, a qualitative indication of good acoustic coupling between cement and formation, and also between casing and cement. Several special cases should be taken into consideration.

Unconsolidated formations—Occurring at shallow depth, this type of formation strongly attenuates the sound. The VDL does not show any formation waves because their amplitude is too low.

Fast formations are those through which sound travels faster than along the casing. The wavy pattern on the VDL is seen earlier than the casing arrivals. Obviously, both the casing-to-cement bond and the cement-to-formation bond are generally good.

Salt formations, which are highly plastic, have little heterogeneity. Across such a zone, the VDL is very

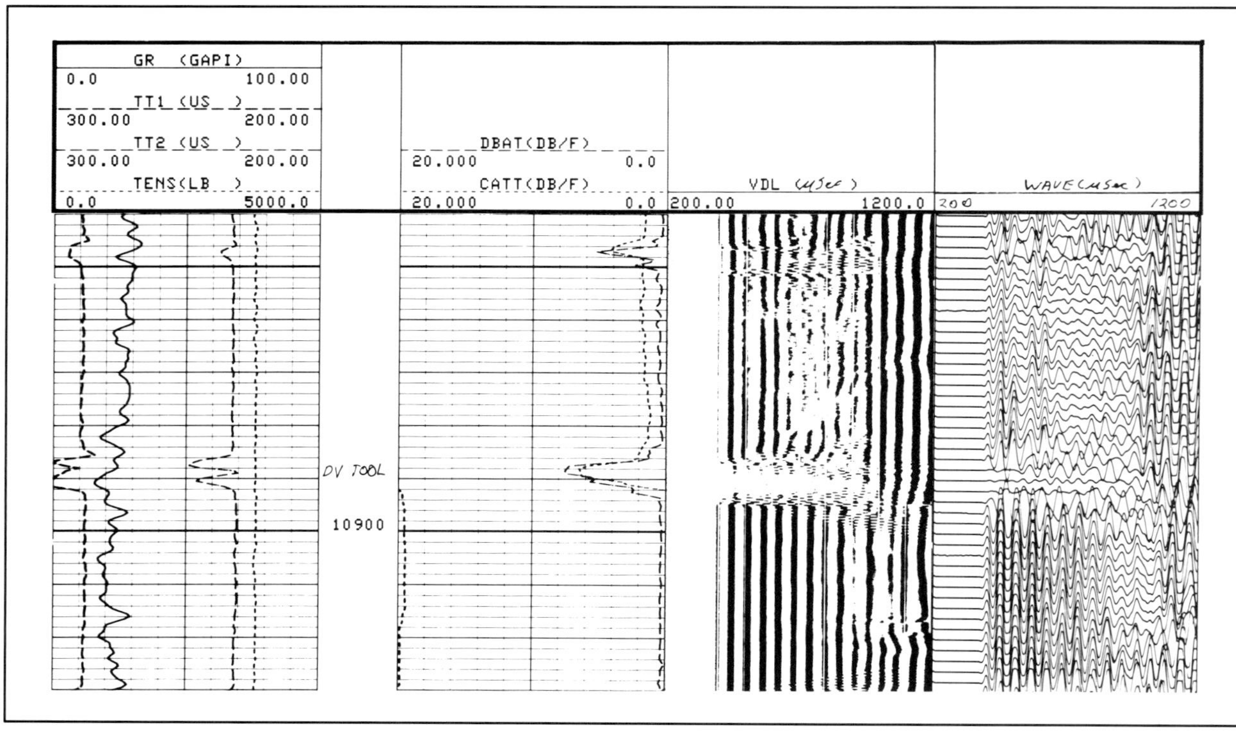

Figure 16–10—Displays of the full acoustic wave.

regular most of the time, sometimes appearing similar to free pipe.

Concentric casings—If the inner casing is well cemented, the VDL will show the quality of the previous cement job, but often shows parallel stripes as in free pipe. The chevron patterns at the casing collars are visible. When the annular gap between both casings is well cemented, the casing collars from the previous casing can also be seen on the VDL. Often in this type of situation, an apparent frequency increase is seen on the signal, which is visualized on the VDL by a larger number of thinner black-and-white stripes (Jutten, 1988).

Intimate contact between the casing and formation—When the casing is not centralized, even when not well cemented, formation arrivals may be seen on the VDL. The casing signals are also very strong.

16–4.4.6 Quantitative Data Taken From the Acoustic Wave

The first few cycles of the received waveform are shown in Fig. 16–11 (note that this is stylized—amplitudes of all peaks are not the same in actuality). The traditional convention in acoustic logging is to label the half-cycles as shown, E_1, E_2, E_3,..., with odd numbers referring to positive peaks and even numbers to negative peaks. The first cycles will be from the casing signal, ideally. In "free" pipe (i.e.,a fluid-filled annulus) this is true, but in cemented pipe the picture is not so clear, depending on the formation properties and the thickness of the cement sheath (Section 16–4.4.11—"Cement Thickness" and "Fast Formations").

The underlying premise in quantitative bond logging is that the strength of the casing signal is a function of the annular material adjacent to the pipe; thus, it is necessary to measure the strength of the casing signal. From this premise, and the considerations in the preceding paragraph, it follows that "the sooner, the better." In other words, the earliest significant part of the wave (i.e., the peak E_1) should be measured, but this peak is considerably smaller than the following ones. Hence, every conceivable scheme has been implemented to quantify the "strength of the casing signal," including measuring the

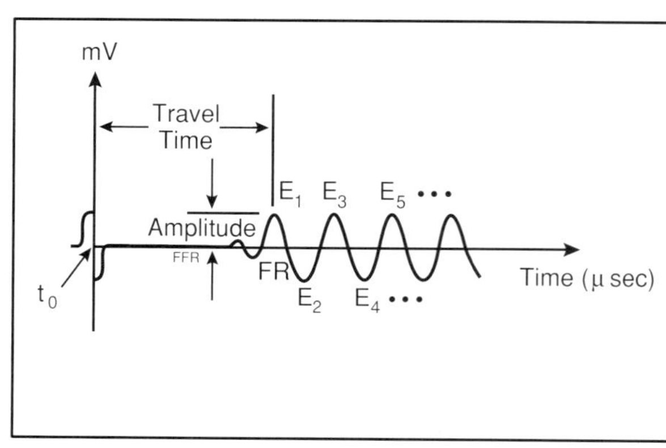

Figure 16–11—Sonic waveform nomenclature.

peak amplitudes of E_1, E_2, and E_3, the area under the half-cycles, and the area under multiple half-cycles (Bade, 1963; Pickett, 1963). For this reason, the log curves generated by different tool designs in the same well at the same time can look completely different. Since there is no analytical description of the received waveform in terms of the geometry, physical parameters, etc., involved, there is no definitively "correct" measurement scheme to be applied to the waveform. However, recent laboratory work (Jutten and Parcevaux, 1987; Jutten, 1988) indicated that even the tail end of the E_1 half-cycle may be affected by the typical geometry of actual wells. Thus, the recent trend toward measuring the amplitude of E_1 is not only reasonable, but necessary.

The first quantitative measurement performed on the full wave is the time from transmitter firing to the arrival of the first part of the wave exceeding a preset amplitude threshold (Fig. 16–12). It is known as the *transit time*. The actual measurement value then depends on (1) the threshold level, and (2) the use of positive or negative peaks.

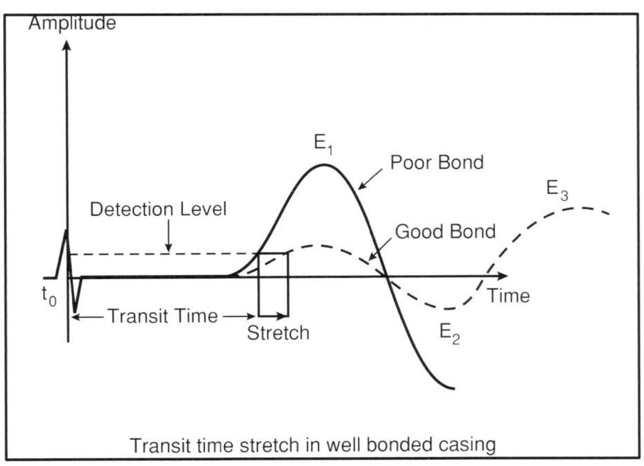

Figure 16–12—Transit-time stretch in well-bonded casing.

The time measured also depends on the size of the inside diameter (ID) of the casing and the outside diameter (OD) of the tool, as well as the speed of sound in the borehole fluid. Because of the threshold (necessary to avoid noise), the measured time increases slightly as the amplitude of the relevant half-cycle (E_1 in Fig. 16–12) drops close to the threshold level ("detection level" or "bias level"). When the amplitude drops below the threshold, the arrival time of the next half-cycle is measured, e.g., E_3. Thus, the transit-time measurement responds to changes in amplitude in a crude way. It is sometimes assumed (mistakenly) that the transit-time curve is there to say something about the cement. In fact, its most valuable (and essential) function is one of quality control (Section 16–4.4.10). As long as the amplitude is well above the threshold level (on the order of five millivolts, commonly), the transit-time curve is a highly sensitive measure of tool centering in the pipe.

The second quantitative measurement made on the wave is the amplitude measurement from which a quantitative cement evaluation can be made. Since the arrival time of the peaks is related to the geometry of the CBL tool and casing, and wellbore fluid properties, the amplitude measurement can be made in different ways: either using a "fixed gate" or a "sliding gate." The technique employed commonly today is to position a fixed gate over the time interval of the desired peak(s), preferably an E_1 half-cycle. For example, the time interval, during which the amplitude of the full wave is measured, is fixed or held constant relative to the time of the firing of the transmitter. To ensure precise placement of this "time window" over the desired part of the wave, the gate is set in the well to be logged using the "freest" pipe to be found. Note, however, that the gate must be repositioned for changes in pipe size and even "fine tuned" for changes in pipe weight. The positioning of the fixed gate is shown in Fig. 16–13. The correct setting of the gate is a crucial step in obtaining a valid bond log amplitude curve. In fact, it should be positioned on E_1.

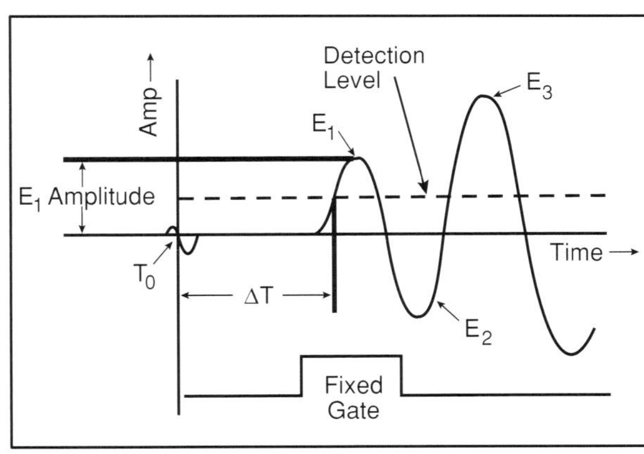

Figure 16–13—Fixed amplitude gate.

When the sliding gate is used, the CBL curve shows the maximum amplitude measured within a time interval positioned immediately after the detection of a signal larger than a preset detection level. The measurement is quantitatively useful only when the first half-cycle is measured. However, this is not the case when the amplitude of the actual casing signal falls below the bias setting (detection level) of the sliding gate. Since this technique appears to be obsolete as a primary means of measuring amplitude, and although its use was recently proposed (Allen and Wood, 1985) to verify the positioning of the fixed gate, it will not be considered further here.

16–4.4.7 CBL Attenuation Rate

It has been proved that many parameters have an influence on the amplitude measurement—calibration of the logging tool, centering of the tool, pressure, temperature, wellbore fluid, casing size and thickness, cement thickness, microannulus, spacing between transmitter and receiver, etc. To reduce the measurement sensitivity and quantify the results as a function of the cement, it is necessary to speak in terms of the attenuation rate.

A theoretical expression of the CBL attenuation rate was first published by Pardue et al. in 1963. It showed that for a thin, flat plate surrounded by a solid or a liquid, the attenuation rate can be well approximated by

$$\alpha = \frac{52.2\left(\frac{\rho_2}{\rho_1}\right)}{t\left(\left((C_p/C_c)^2-1\right)^{-0.5}+\left((C_p/C_s)^2-1\right)^{0.5}\right)}, \qquad (16\text{–}3)$$

where

ρ_2 = density of annular material,
ρ_1 = density of plate ,
C_p = wave velocity in plate ,
C_c = compressional wave velocity in annular material,
C_s = shear wave velocity in annular material,
t = thickness of plate (in.), and
α = attenuation rate (dB/ft).

This analysis shows the following.

- The primary cement variable affecting attenuation is the wave velocity.
- The attenuation is inversely proportional to the casing wall thickness and proportional to the cement density.
- The attenuation is independent of the frequency.
- The shear coupling is required at the casing/cement interface to produce full attenuation.

The analysis ignores the effects of plate curvature, the fluid within the pipe, and the thickness of the cement sheath.

The CBL attenuation rate can also be calculated from two amplitude measurements of the same transmitted signal made at two different receivers, according to the following formulas.

$$A = -\frac{20}{L}\log_{10}\frac{E_{1_2}}{E_{1_1}}, \qquad (16\text{–}4)$$

where

E_{1_1} = amplitude of E_1 at Receiver 1,
E_{1_2} = amplitude of E_1 at Receiver 2, and
L = distance (spacing) between Receivers 1 and 2 (ft).

In the case of tools with a single receiver, an approximate attenuation rate can still be computed using

$$A = -\frac{20}{L}\log_{10}\frac{E_1}{E_{1_{FP}}}, \qquad (16\text{–}5)$$

where

E_1 = amplitude of E_1,
$E_{1_{FP}}$ = amplitude of E_1 in free pipe, and
L = spacing between transmitter and receiver (ft).

This last expression, although still sensitive to tool calibration and wellbore parameters (it requires the free-pipe reading in mV), can be used.

To improve the accuracy of the attenuation-rate determination, special CBL sondes have been developed with two transmitters and two or three receivers, i.e., the Cement Bond Tool (CBT) (Fig. 16–14). The general principle is the same but, because of its conception (two transmitters and three receivers), it can compute a bottomhole-compensated attenuation rate on which many environmental effects (pressure, temperature, wellbore fluid characteristics, etc.) have almost no effect (Gollwitzer and Masson, 1982). However, it needs four amplitude measurements. The upper and lower transmitters are separated by 5.8 ft (1.8 m), and the three receivers are placed at 0.8 (short), 2.4 (near), and 3.4 ft (0.2, 0.7, and 1.0 m, respectively) from the upper transmitter.

The measurements made at 2.4- and 3.4-ft spacings are used to compute the Bottomhole-Compensated (BHC) attenuation rate, *which is quasi-independent of temperature and pressure.*

$$BHC\ Ratio = \frac{-10}{d_2 - d_1}\log_{10}\frac{A_{12}A_{21}}{A_{11}A_{22}} = \alpha, \qquad (16\text{–}6)$$

where

$A_{ij} = P_i\, S_j\, 10^{-(\alpha\, dj/20)}$,
P_i = acoustic pressure at transmitter i,
S_j = sensitivity of receiver j,
d_2 = 3.4 ft,
d_1 = 2.4 ft, and
α = attenuation rate,
with P_i and S_j temperature and pressure-dependent.

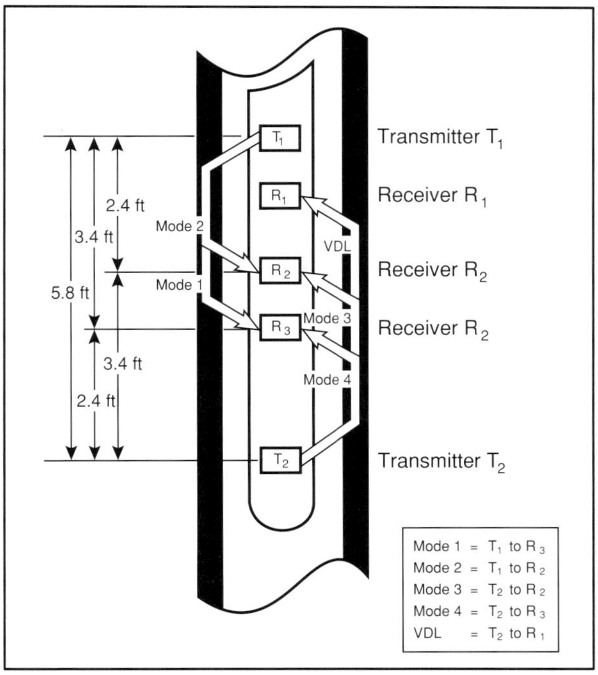

Figure 16–14—Cement bond tool.

16–4.4.8 CBL Quantitative Interpretation

Experiments proved the attenuation rate to be linearly related to the percentage of the circumference of the casing bonded by the cement (Fig. 16–15), from which the concept of Bond Index (BI) was derived (Pardue et al., 1963). Its validity was later extended (Jutten and Parcevaux, 1987) to the percentage of cemented area, regardless of the shape and fluid of the noncemented area. The Bond Index is the only quantitative information which can be derived from a CBL. Its calculation

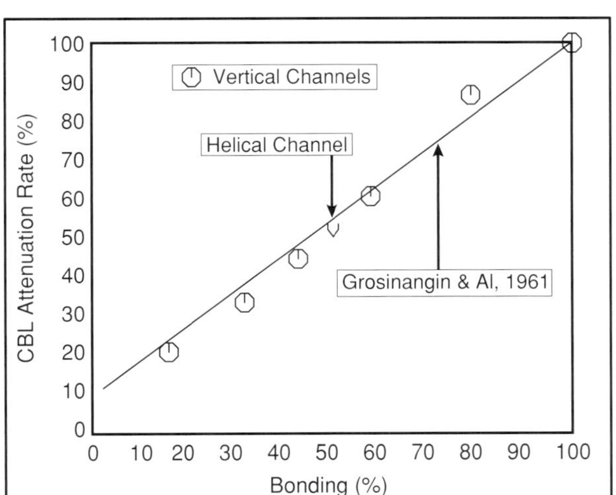

Figure 16–15—Relationship between percent bonding and CBL attenuation rate.

requires the knowledge of the log response in the well-cemented section, which is used as a reference for the computation of $A(100\%\ \text{cemented})$—

$$BI\,(x) = \frac{A(x)}{A\ (100\%\ \text{cemented})}, \qquad (16\text{–}7)$$

where A is the CBL attenuation rate. The Bond Index equation can also be solved graphically using semilog paper once the 100% cemented pipe amplitude is known (Fitzgerald et al., 1983).

Figure 16–16 is a synthetic view of the CBL interpretation in the form of a flowchart. Attenuation rate variations as a function of cement, casing size, and thickness were studied, leading to the construction of a famous nomograph known as the "CBL interpretation chart" (Pardue et al., 1963). This nomograph, which gives a correspondence between the CBL signal and the cement compressive strength, was later modified for lightweight cements (Bruckdorfer et al., 1983). More recent work performed over a large variety of cement slurry formulations proved the CBL attenuation rate to be *related to the acoustic impedance of the cement and not to compressive strength* (Jutten and Parcevaux, 1987). These experimental results, also valid for foamed cements, could be used to modify the CBL interpretation chart which now shows (Fig. 16–17) a relationship between Cement Acoustic Impedance (Z) and the CBL signal.

Unfortunately, a CBL-VDL interpretation does not give a direct access to hydraulic isolation. Empirical

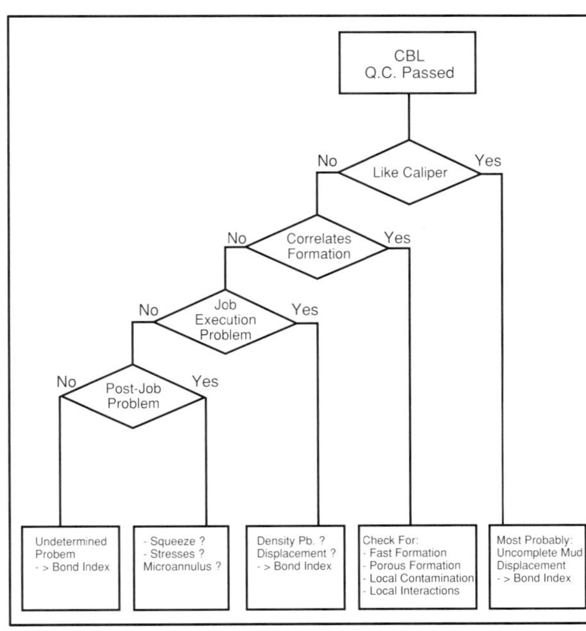

Figure 16–16—CBL interpretation—general flowchart.

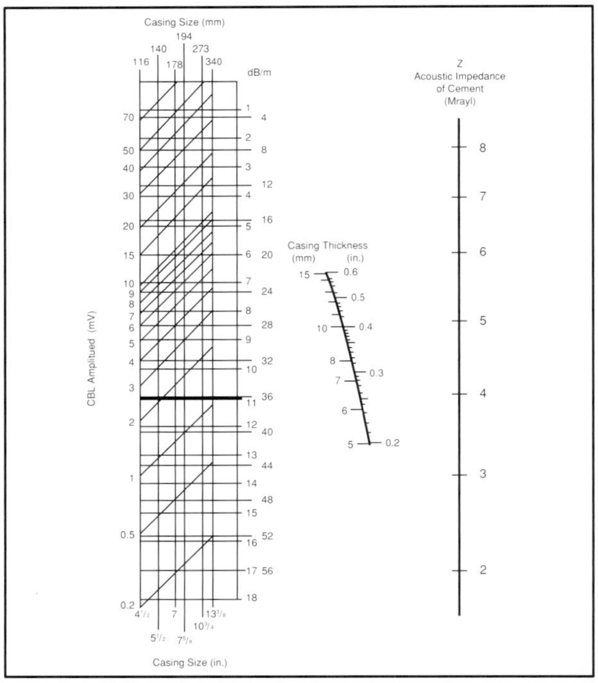

Figure 16–17—CBL interpretation chart (3-ft spacing).

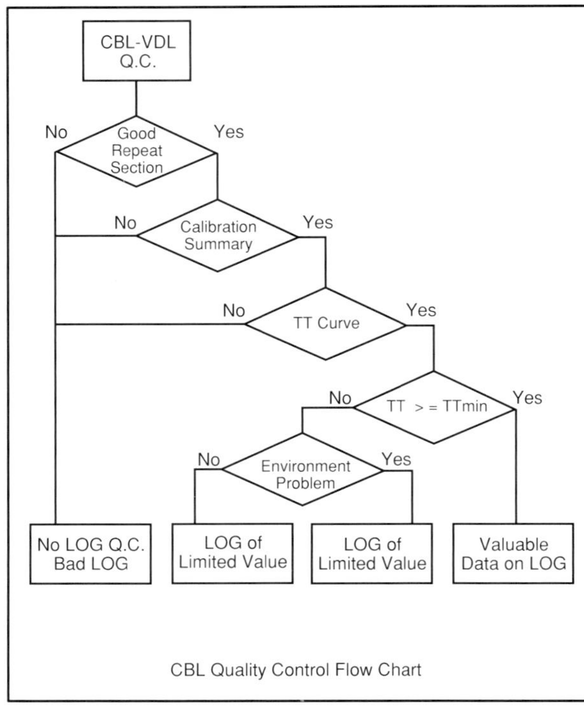

Figure 16–19—CBL quality control flowchart.

charts were developed which give the minimum length required for zonal isolation as a function of the Bond Index and the casing size. These charts, established without any theoretical or experimental background, are not valid and should not be used.

The CBL/CBT-VDL interpretation is thus restricted to an evaluation of the cement placement in relation to the quality of the cement. Well parameters, cement job events, and pre- and postjob well histories are required to make the best possible evaluation from a CBL/CBT. Well parameters and cement job events are necessary to compute the expected log response, while the cement job recording and pre- and postjob events are required most of the time to understand if the discrepancies between the expected response and the actual log, if any, are due to cement, incomplete mud displacement, a microannulus, etc.

16–4.4.9 Bond Log Presentation Formats

Bond logs are presented on a standard three-track log format with the depth track between data Tracks 1 and 2 (Fig. 16–18). Track 3 contains the full wave display, either as the waveforms or a variable intensity display. The common scale is 200 to 1,200 μs, although other scales are available from many companies for special cases (such as extra large holes or extra slow formation sound speeds).

Track 2 contains the amplitude and/or attenuation rate curves. The attenuation rate curve is usually presented on a scale of 20 to 0 dB/ft. Amplitude curve scales are not standardized, although 0 to 100 or 0 to 50 mV is very common with an amplified curve presented on a 0- to 20- or 0- to 10-mV scale. The double scale is very important because free-pipe readings can approach 100 mV (or more), while the very fine resolution of perhaps 1 mV or less may be required at very low amplitudes. Since the attenuation rate is inherently logarithmic in terms of amplitude, only one scale is needed.

Track 1 traditionally contains the transit-time measurement (or possibly a derivative thereof for the borehole-compensated tools), as well as a correlation curve (gamma ray or neutron). Casing collars are usually found here, but also may appear in the depth track or Track 2. A common scale for the conventional 3-ft transit time is 200 to 400 μs. This has the advantage of a single scale working for almost all casing sizes. However, the small changes in time which correspond to major eccentering (4 to 5 μs are recommended limits) require a more sensitive scale (100-μs width).

16–4.4.10 CBL-VDL Quality Control

CBL/CBT-VDL quality control can easily be divided into a step-by-step procedure (Fig. 16–19). If the first two steps apply for every log, the CBL has a special curve for

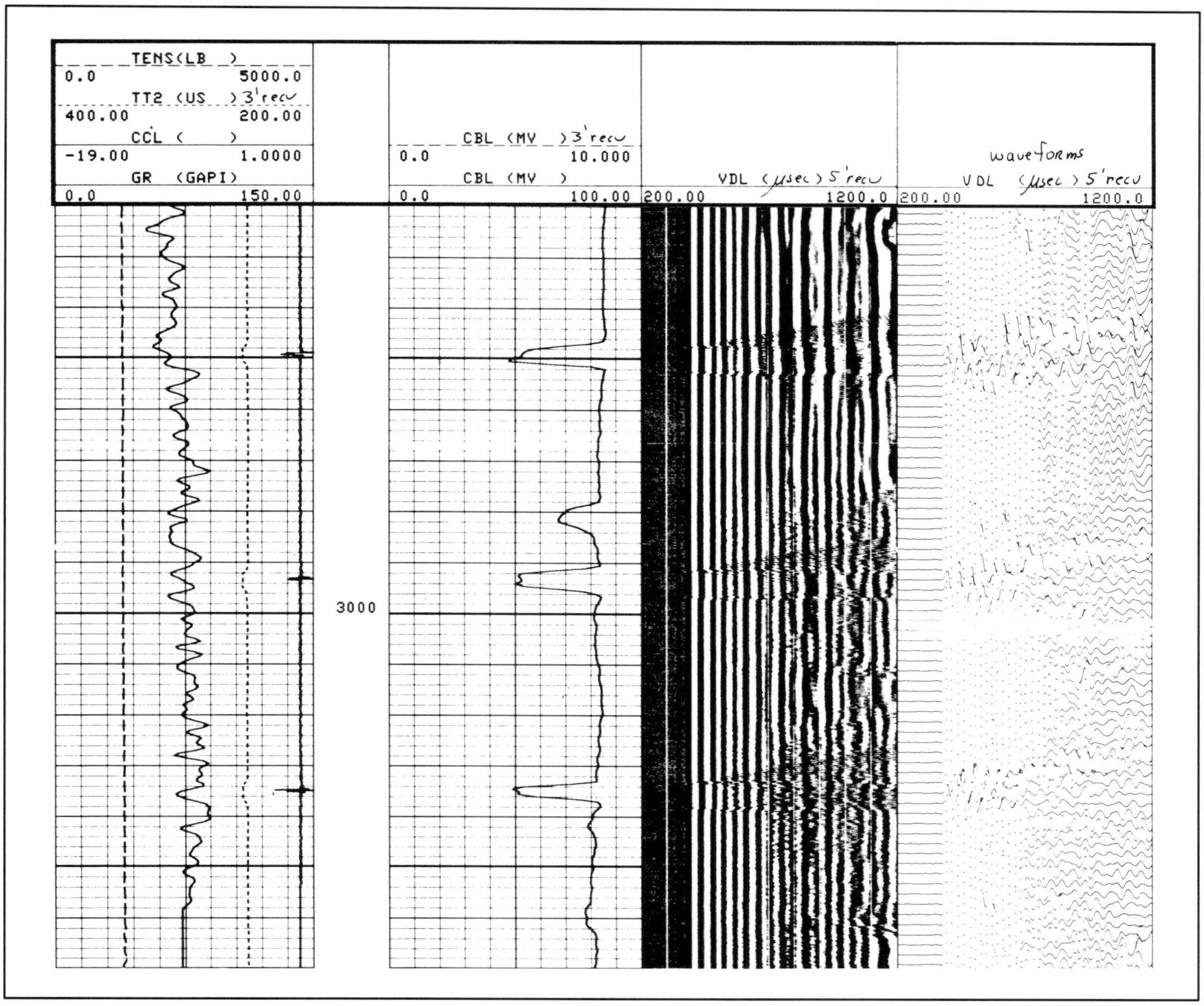

Figure 16–18—Standard three-track bond log format.

quality control—the transit-time curve (Fitzgerald et al., 1983; Bigelow, 1985). *If no transit-time curve is present on the log, no quality control of the log is possible, and the evaluation will be very restricted.* In the next part of this chapter, a transit-time curve is assumed to be present on the log.

By comparing the measured transit time with the expected transit time (time required for the sound wave to travel from the transmitter to the receiver through the mud and along the casing, which is normally the shortest path), the following conclusions can be drawn.

Shorter transit time is an indication of either poor centralization of the sonde or a fast formation. On the subject of eccentering, there have been recommendations that no more than a 4-μs decrease in transit time be considered acceptable. This corresponds to an eccentering of about ⅛ in. (0.32 cm) in fresh water. Since this amount of eccentering reduces the amplitude by more than 25% (Fig. 16–20), this recommendation is reasonable. There are several ramifications of this value of 4 μs.

- The measurement resolution of the logging instrumentation must be of the order of 1 μs or better.
- The log display must allow for ease of readability down to 1- or 2-μs visual resolution, leading to the recommendation of using a scale for the transit time of 100 μs across the track (Fertl et al., 1974; Fitzgerald et al., 1983). In the case of sonde eccentering, it is impossible to quantify the results of the cement job with the CBL. The influence of a fast formation, often also seen as a decrease in transit time, is discussed in Section 16–4.4.11.

Slightly longer transit time (stretching) is generally an indication of a good bond, and should correspond to reasonably low amplitudes. The Bond Index concept is applicable. If the CBL amplitude is still high, check for

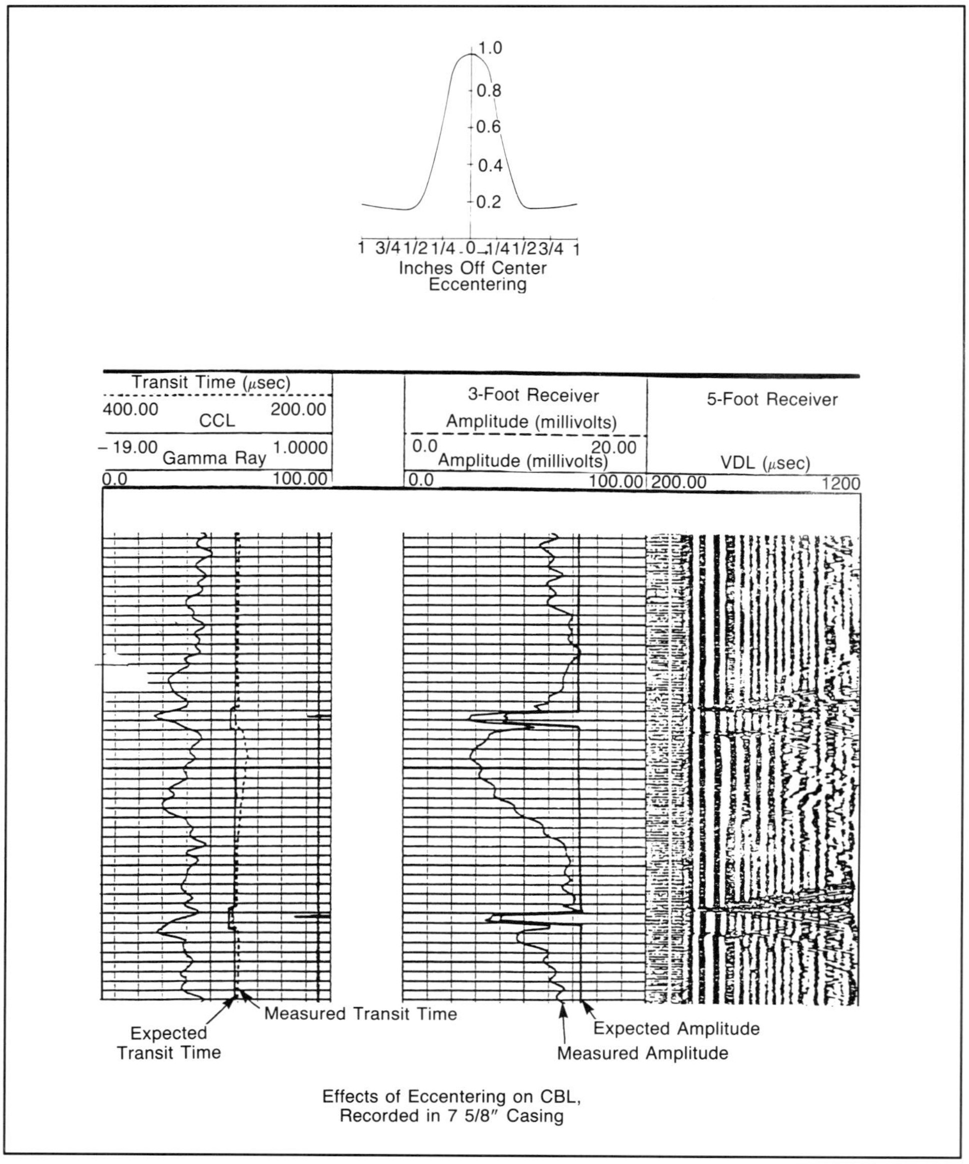

Figure 16–20—Effects of eccentering on CBL (recorded in $7^5/_8$-in. casing).

any interference because of the small cement thickness and high acoustic impedance contrast at the cement external interface.

Longer transit times (DTT > 15 μs), are called *skips*. In this case, E_1 is normally too small to be detected; thus, a good bond exists between the cement and casing. A cycle skip refers to a cycle of the original wave (50 μs for a 20-kHz signal). In this case, in fixed gate mode, the CBL amplitude *MUST* be below the detection level and the Bond Index concept applies. However, it is fairly common to have stable skips of more than 20 μs but less than 50 μs. This is due to energy reflections at the cement external interface, enhanced by large acoustic-impedance contrasts as in concentric strings (Jutten, 1988). If applied in this situation, the Bond Index concept will lead to erroneous conclusions, because the amplitude measured was not E_1.

16–4.4.11 Influence of Well Parameters on CBL

Temperature and pressure—The deeper a well, the higher the temperature and pressure. As for all the materials, velocity and attenuation of sound inside the wellbore fluid will be affected by downhole conditions. The response of the transducers also will vary. In 1984, Nayfeh et al. published a paper where pressure and temperature corrections were given for the transducers used in CBL tools.

Wellbore fluid properties—Wellbore fluid properties have an effect on both the transit time and the CBL amplitude. Experimental and numerical work has been performed to study the effects of wellbore fluids on CBL amplitude measurements (Nayfeh et al., 1984). In particular, considerable differences in free-pipe amplitude were shown to occur for various brines.

Casing size and thickness—The larger the casing size, the longer the path through the wellbore fluid where some attenuation occurs. It leads to decreasing free-pipe amplitudes for increasing casing sizes. In cemented pipe, experience also shows that for the same cement, the CBL amplitude is higher in larger casing. This can easily be explained by the increase of steel thickness, providing a smaller attenuation rate, and also by the reduction in effective spacing (effective shortest path along the casing wall), leading to a smaller total cemented pipe attenuation.

Cement thickness—When cement thickness is too small, reflections of energy at the cement external interface can interfere with the casing signal. These interferences are seen mainly in concentric strings, or cylindrical holes with a small annular clearance and well-centralized pipe (Jutten and Parcevaux, 1987). To determine if reflections interfere with E_1, it is necessary to accurately measure the openhole size *and the acoustic properties of the cement when the log is run.* Typical cement thicknesses which do not interfere with E_1 (20-kHz signal) can vary from 1 in. (2.54 cm) to 3 in. (7.6 cm), as a function of the velocity of the compressional waves through the cement.[1]

In the special case of concentric strings (e.g., top of the liner), the resonance of the external casing induces strong signal perturbations, leading to an apparent frequency increase of the first few arches of the waveform. Recent experiments confirmed with field logs (Jutten, 1988) proved that high CBL amplitudes obtained in concentric casings are often an artifact because of an excellent cement job between both casings, combined with a measuring gate of too large an amplitude. This problem can easily be solved by shortening the width of the measuring gate.

Fast formations—The well-known "fast-formation effect" is a decrease in transit time. Because of the energy path through the formation, it is not possible to quantify the results with the Bond Index, but qualitative evaluation is possible. If there is sufficient sound energy propagating through the formation to interfere on the early part of the waveform, it indicates that a good acoustic link exists between the casing and the formation. Most of the time this implies a good bond. However, if shorter transit time implies a fast-formation for a well-centered tool, it does not mean that the converse must also be true.

As is known from openhole sonic logging, the amplitude of the formation signals varies considerably, and the amplitude of the first positive peak (E_1) is especially small relative to subsequent peaks. This is one reason why the E_2 transit-time detection is used rather than the E_1 in openhole BHC sonic logging. This also explains the frequent observation that, in zones known to have a high sound speed (at much less than the 57 μs/ft value for casing), there is an apparent contradiction between the increase in transit time and the presence of a fast formation indicated by all other available information. This includes the fact that transit time is clearly tracking formation changes with depth; at the 5-ft (1.5-m) receiver the formation signal has clearly overtaken the casing signal. There could be a "fast-formation" phenomenon at the 5-ft receiver, but not at the 3-ft (0.9-m) receiver. It depends on the annular thickness, the speed of sound in the cement, and the difference in the speed of sound between the rock and the casing. It is important to be aware of this subtlety, because the formation signal can be present in the fixed gate, and thereby increase the amplitude considerably while the transit time is reading a larger value than in free pipe. An erroneous interpretation will result.

For example, given a bias level threshold of 5 mV and an amplitude of 1 mV in a "slow formation," both quite realistic values, the earliest formation arrival in an adjacent fast formation could drive the amplitude reading on the log up to as much as 5 mV without causing a decrease in transit time. However, standard interpretation techniques (Bond Index) would indicate a channel covering 35% of the circumference of the pipe. The bottom line is this—*do not make the common mistake of assuming that the amplitude is uncontaminated by a fast formation because the transit time has not decreased.*

On some compensated bond tools, an additional spacing of 0.8 ft (0.2 m) was chosen to minimize the effect of fast formations in casing sizes smaller than 7 in. (18 cm). When the formation has no influence on the measured peaks, the attenuation rate should be constant regardless of the spacing. In the presence of a fast formation, the measured attenuation rate decreases with increasing spacing, because of the increasing part of the sound energy arriving in the early portion of the waveform. In this case, the 0.8-ft (0.2-m) spacing attenuation rate is

[1] *Remark:* If the measuring gate is much larger than one-half of the period of the original signal, interference may induce erroneous amplitude measurements for even larger cement thicknesses.

larger than other attenuation rates (or the amplitude is lower).

16–4.4.12 Influence of Cement Job Parameters on CBL

The most common cause of cement-job failure is poor mud removal, and poor mud removal will never produce a good CBL (Chapter 5). Some cases are obvious—casing not centralized, slurry lighter than mud displaced at low flow rate, thin slurry pumped behind viscous fluids at low flow rates, etc. This can be determined by a prior log analysis, providing one has an *actual* job recording (flow rate, pressure and density) *during the entire job*, of openhole size and casing centralization. The evaluation of slurry placement is beyond the scope of this chapter; nevertheless, log evaluation still requires actual job data, especially for slurry density and volumes pumped.

Since slurries of different densities normally have different acoustic properties, it should be easy to detect the transition between the lead and tail slurries on the log. This concept is of prime importance for cement coverage estimation, because 100% bonding across the lead section corresponds to a much lower CBL attenuation rate than 100% bonding across the tail section. A Bond Index log should be computed section by section, without forgetting that the minimum cement thickness required to apply these rules also depends on the acoustic properties of the cement.

Sometimes it is possible to estimate mud removal by comparing the expected top of the cement with the one computed from the hole geometry and volumes pumped. However, such an estimation must be done quite carefully, because many parameters are involved (the accuracy of the caliper and flowmeters, and volume changes because of fluid loss and lost circulation).

When mixed lighter than the designed density, a slurry will often exhibit higher free water and sedimentation, longer thickening times, and lower acoustic impedance. This can be seen as well on the log.

16–4.4.13 Influence of PostJob Events on the CBL

Several postjob events can influence the CBL results. Any pressure and temperature change applied inside the casing will induce casing deformations that modify the stresses in the cement and at the cement-to-formation and cement-to-casing interfaces, and will possibly break these bonds, leading to the creation of a microannulus.

Recent work (Leslie et al., 1987) showed that the amplitude reduction of the sonic signal depends not only on the attenuation along the casing, but also on the efficiency of the acoustic coupling between the transducers and the casing wave. In the presence of a microannulus, the shear coupling is lost and the attenuation along the casing is negligible; however, the coupling is not lost when a fluid is in the microannulus. Using multiple-receiver tools, it is possible in theory to separate the coupling and attenuation rate, detect a microannulus, and even quantify the cement coverage behind the pipe.

In the total absence of experimental work on the subject, everything is based on field experience, rules of thumb, and "know-how." In a recent paper (Pilkington, 1988), the origin of microannuli is described in great detail, with the effect produced on the CBL. Guidelines are also given to run the CBL in the presence of a microannulus. However, for cement job evaluation purposes, a CBL performed under pressure may be contraindicated because of the potential detrimental effects on hydraulic isolation. One should first try to analyze the origin of the potential microannulus. Several cases need to be taken into consideration.

Thermal expansion or retraction—During the setting of the cement, heat is generated which increases the temperature in the wellbore. As explained at the beginning of this chapter, this heat is sometimes used to detect the top of the cement. It will also produce expansion of the tubular goods inside the wellbore. In particular, the casing diameter will expand. An approximate value is given by the following formula.

$$\Delta D = 6.9 \times 10^{-6} D \Delta T \quad , \qquad (16\text{–}8)$$

where

ΔD = diameter change (in.),

C = casing circumference (in.), and

ΔT = temperature change (°F).

During the life of the well, the production of hot fluids or the injection of cold or hot fluids can also produce expansion or retraction of the tubular goods. The above formula can be used to estimate the magnitude of the geometrical change induced.

Mechanical expansion or retraction—Such effects are mainly due to internal casing pressure applied during pressure tests, remedial cementing, or stimulation jobs. Sometimes the casing is kept under pressure while the cement sets, because of a leak at the float equipment. After cementing a production string, it is also fairly common to replace the drilling mud with a lighter completion fluid. The downhole hydrostatic pressure reduction can produce a significant retraction of the casing, and induce a microannulus if the bond between the casing and cement is not sufficiently strong. The diameter expansion of unsupported pipe because of an increase in internal pressure was detailed by Carter and Evans (1964). Often,

1,000-psi differential pressure is sufficient to create a fairly large microannulus, especially for large casing sizes.

Mechanical fatigue—In deviated wells and on intermediate strings, drilling can produce a great amount of vibration and mechanical stress, concentrated in special places (e.g., kickoff points). It can damage the quality of the bond between the casing and cement.

For all of the cases mentioned above, when the cement is strong enough to withstand the deformation, nothing will happen and the bond will not be affected. If the cement is still plastic when the stress is applied, the annular geometry will change. If the cement is not set when the stress is released, the bond should not be affected. However, if the cement hardens while the casing is significantly expanded, it may not follow the casing back when the stress is released, and can lead to the formation of a microannulus.

16–4.4.14 CBL-VDL Examples

Well cemented section—A 7-in. (18-cm) casing (23 lbm/ft) was cemented at shallow depth. The average hole size was between 12 and 17 in. (30 and 43 cm) for a bit size of 9⅞ in. (23.7 cm). The casing was cemented using two different slurries—a lead mixed at 10.6 lb/gal (1,270 kg/m^3) extended with hollow silica microspheres, and a tail mixed at 15.8 lb/gal (1,890 kg/m^3). Both formulations contained 35% silica flour BWOC. The job was pumped in "plug flow," at a maximum annular flow rate of 2 BPM for the slurries, achieving good mud removal.

The CBL was run several weeks after the job. The selected section shows the transition between the tail and the lead slurry (Fig. 16–21). The CBL amplitude is about 1 mV between 370 and 420 ft (113 and 128 m) across the tail, and between 8 and 14 mV across the lead. At that time, the estimated compressive strengths were in excess of 5,000 psi for the tail and about 1,000 psi for the lead cement. Using the standard "CBL interpretation chart," the CBL amplitudes were expected to be less than 1 mV for the tail and about 4 mV for the lead, which would give a pessimistic Bond Index of 65% on the section showing a 12-mV CBL amplitude. When using the modified CBL chart, with measured acoustic impedances of 6.0 Mrayl for the tail and 3.2 Mrayl for the lead, attenuation rates were computed and extrapolated to be about 1 mV for the tail and 8 mV for the lead. The CBL interpretation is similar for the tail cement across the bottom section. However, the discrepancy becomes critical for the lead, because the relationship between the CBL attenuation rate and cement acoustic impedance enables us to compute a more realistic Bond Index of 85%.

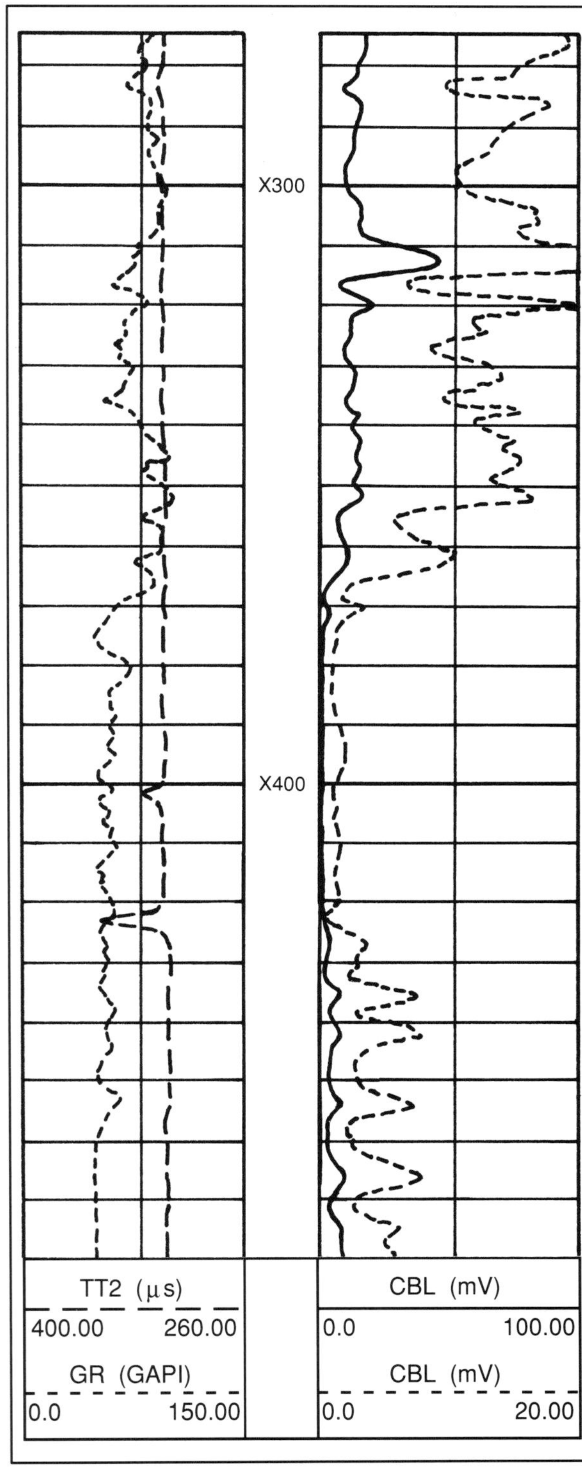

Figure 16–21—CBL-VDL section across slurry change —10.6 lb/gal slurry above 370 ft, 15.8 lb/gal slurry below 370 ft.

Fast formation—The log example in Fig. 16–22 shows a short interval of free pipe (A) from uphole and a second example of log across the pay zone (B,C,D, and E). Look first at the VDL display which contains much information. The free-pipe VDL character is distinct—very straight parallel bands and chevron-shaped diffraction patterns at the collars. In contrast, the rest has wavy bands. The pattern of the bands corresponds to changes in the rock, as indicated by comparing the VDL data to the gamma ray curve. These are formation signals. There is no evidence of a pipe signal. This indicates that the pipe is acoustically coupled to the rock; thus, the cement is

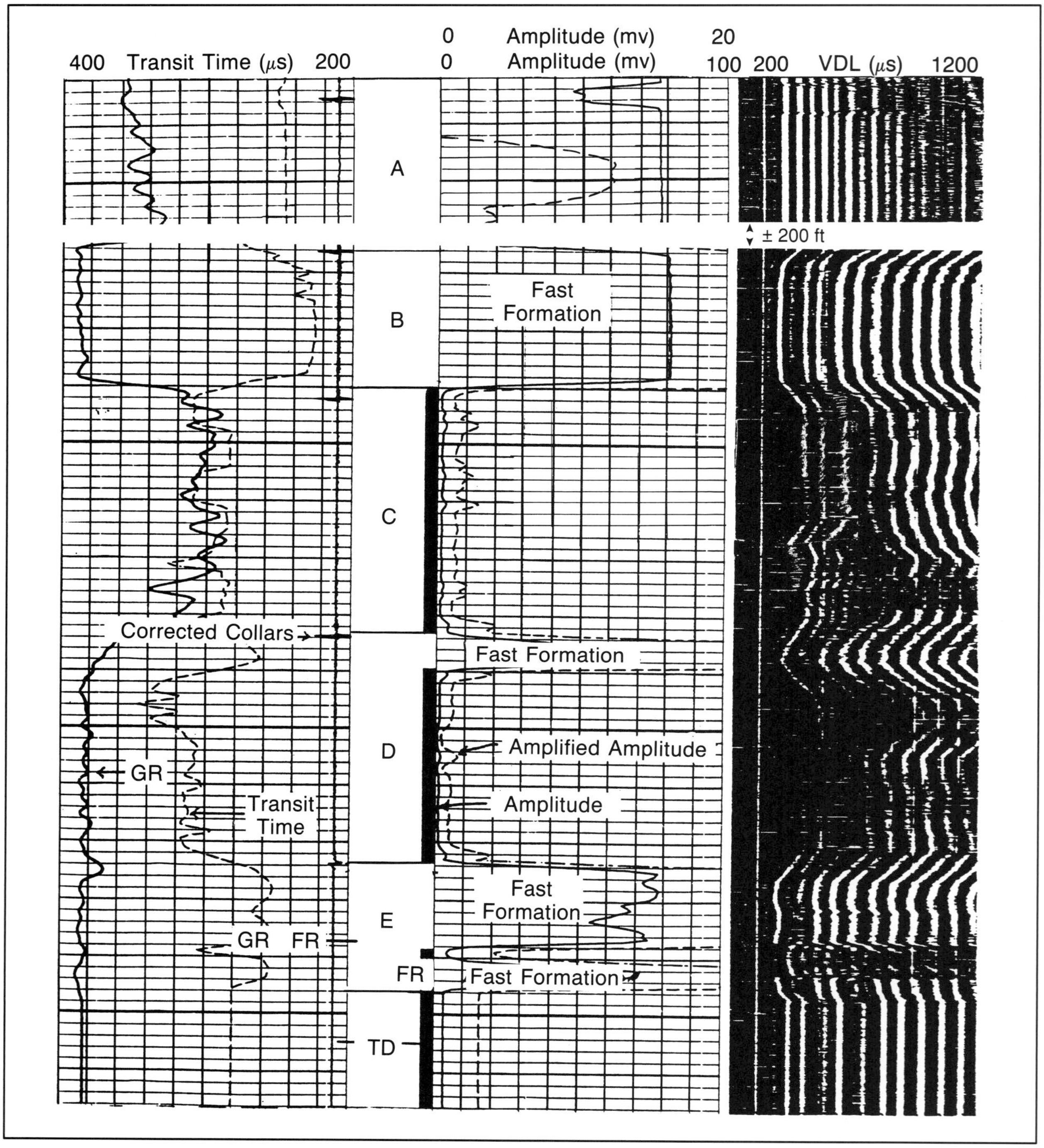

Figure 16–22—CBL-VDL example showing effect of fast formation.

filling the annulus and is bonded to the pipe and formation.

Now look at the amplitude curve. In the short free-pipe interval at the top (A), the amplitude is high and steady at about 76 mV, with a "kick" to the left at the collar. In the main interval either very low amplitudes, about 1 mV (C,D), or very high amplitudes in the 60- to 80-mV range (B,E) are seen. The high amplitudes are caused by a fast formation, not by a lack of cement. To identify the fast formation, look along the VDL display. Notice how the first white band in the main interval (B or E) is further left than the first white band in the free pipe (A). The first formation arrival occurs sooner than the first pipe arrival in the fast formation.

The important fact is that the amplitude measurement in a fast formation is meaningless. The amplitude is some part of the formation signal, not the first positive peak E_1 of the free-pipe signal. Notice the 25-ft zone at the top of the main interval (B), where the gamma ray reading is extremely low. The amplitude curve reads a steady 80 mV through this zone, which is higher than the free pipe. The VDL bands are fairly straight within the zone, looking like free pipe. In this case, the transit time, which is about 10 μs shorter than in the free pipe, clearly indicates the fast formation, as does the lack of a chevron pattern in the VDL at the collar.

Change in pipe weight—In the log shown in Fig. 16–23, there is an abrupt change in pipe weight at 6,953 ft. The $5^{1}/_{2}$-in. (14-cm) casing is 17 lbm/ft above the change and 23 lbm/ft below. Note the difference in CBL amplitude between the two weights—2 mV in the lightest pipe and 5 mV in the thicker one. If a Bond Index is to be run in this well, the reference point for 100% cemented must be changed for each pipe weight.

Microannulus—The two log sections shown in Fig. 16–24 demonstrate the effect of a microannulus on a CBL. The first section was logged without additional pressure at the surface.

- The pipe signals are visible in the VDL display as straight parallel bands at the earliest time.
- "Chevron" patterns are visible at the casing collars.
- The formation signals appear later in the VDL as wavy bands that can be correlated to the gamma ray curve.
- The amplitude is erratic at moderate values.

The second log is a repeat pass over the same interval with 1,200 psi applied at the surface (Fig. 16–25). The pipe signals have disappeared from the VDL display, and the amplitude has decreased to much lower values. In this size and weight of pipe, the expansion caused by the 1,200-psi increase is about 0.001 in. of radius.

Limitations—Unfortunately, with traditional cement bond logs, high amplitude over a cemented section can be due either to channeling or to a microannulus. In both cases, the VDL will show strong casing signals (parallel stripes) and weak formation arrivals. The only way to differentiate both cases is to run a CBL under internal casing pressure. If it is a microannulus, there will be a significant amplitude reduction. If the CBL does not improve when pressure is applied, it can be either a large microannulus or a channel: zonal isolation is probably not achieved.

The alternative is to use recent ultrasonic tools similar to the CET to evaluate the cement job. These tools, which are described in the next section, can differentiate bettween channeling and a microannulus without internal casing pressure.

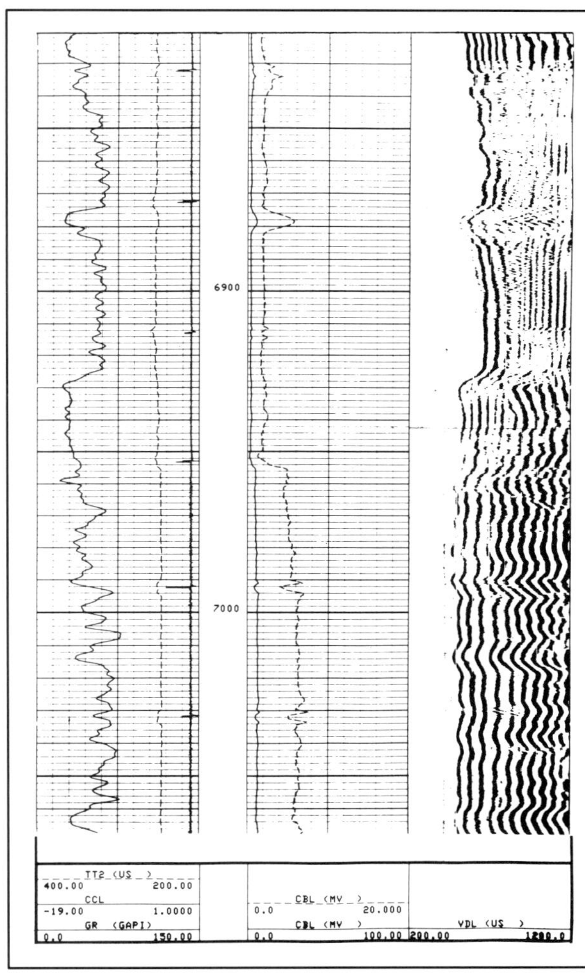

Figure 16–23—CBL-VDL example showing abrupt change in pipe weight.

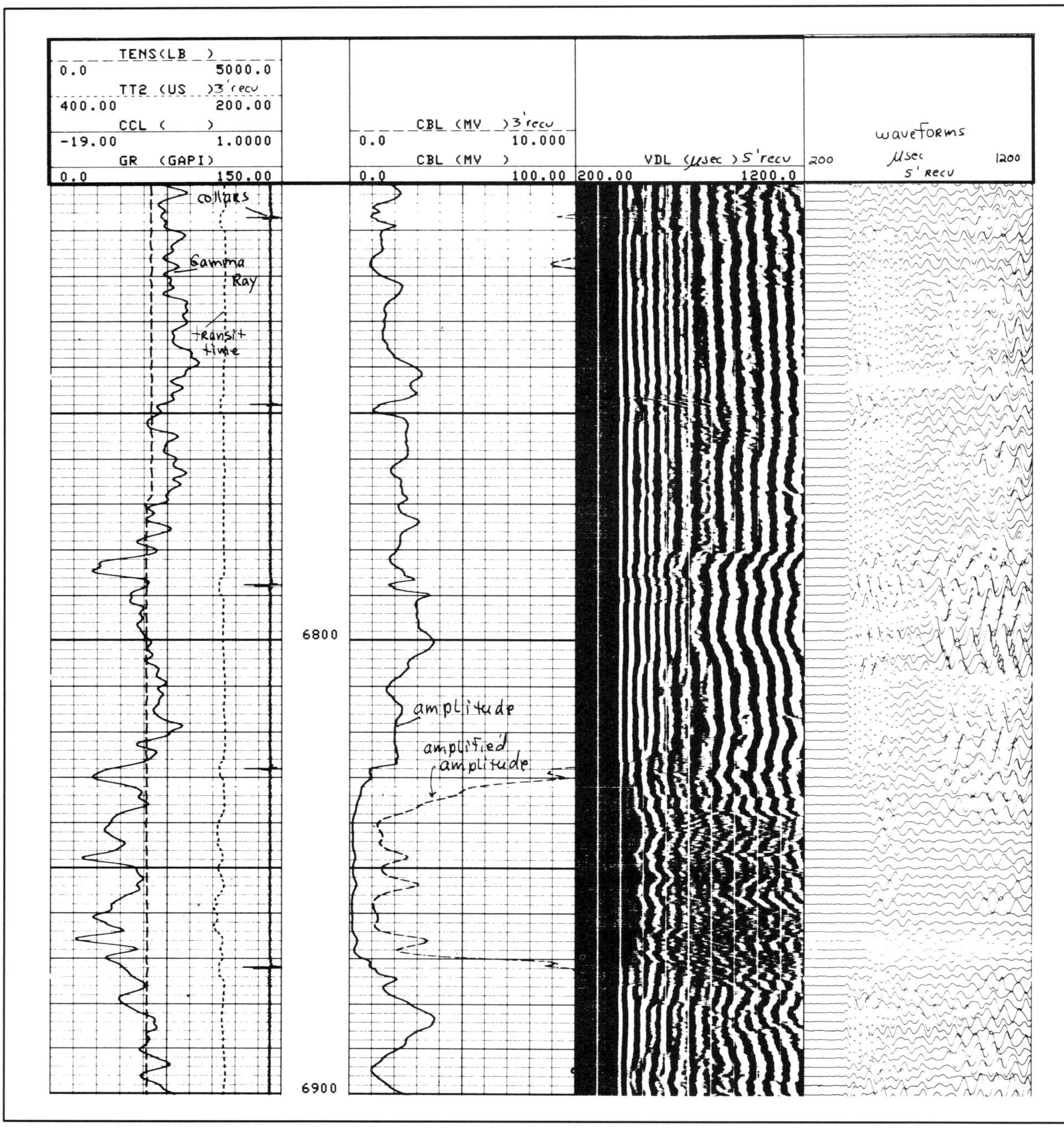

Figure 16–24—CBL/VDL example, showing effect of microannulus.

16–4.5 Ultrasonic Pulse Echo Cement Evaluation

16–4.5.1 History

The use of ultrasonic pulse echo technology in well logging dates back at least to the mid-1960s, when Mobil Oil Corporation developed a tool called the Borehole Televiewer (Zemanek and Caldwell, 1969). It was designed for imaging the borehole wall. The technology has also been applied extensively in the medical and materials testing industries. Application of ultrasonic pulse-echo technology to cement evaluation behind pipe was investigated by Schlumberger in the mid- to late-1970s, and presented by Havira in 1979. In 1981, Froelich et al. presented the results of the field testing of the tool based on this investigation designated the "Cement Evaluation Tool" (CET).

Havira described the theory of the measurement in greater detail in 1982. Modifications to measure casing wall thickness were presented in 1984 by Dumont et al.

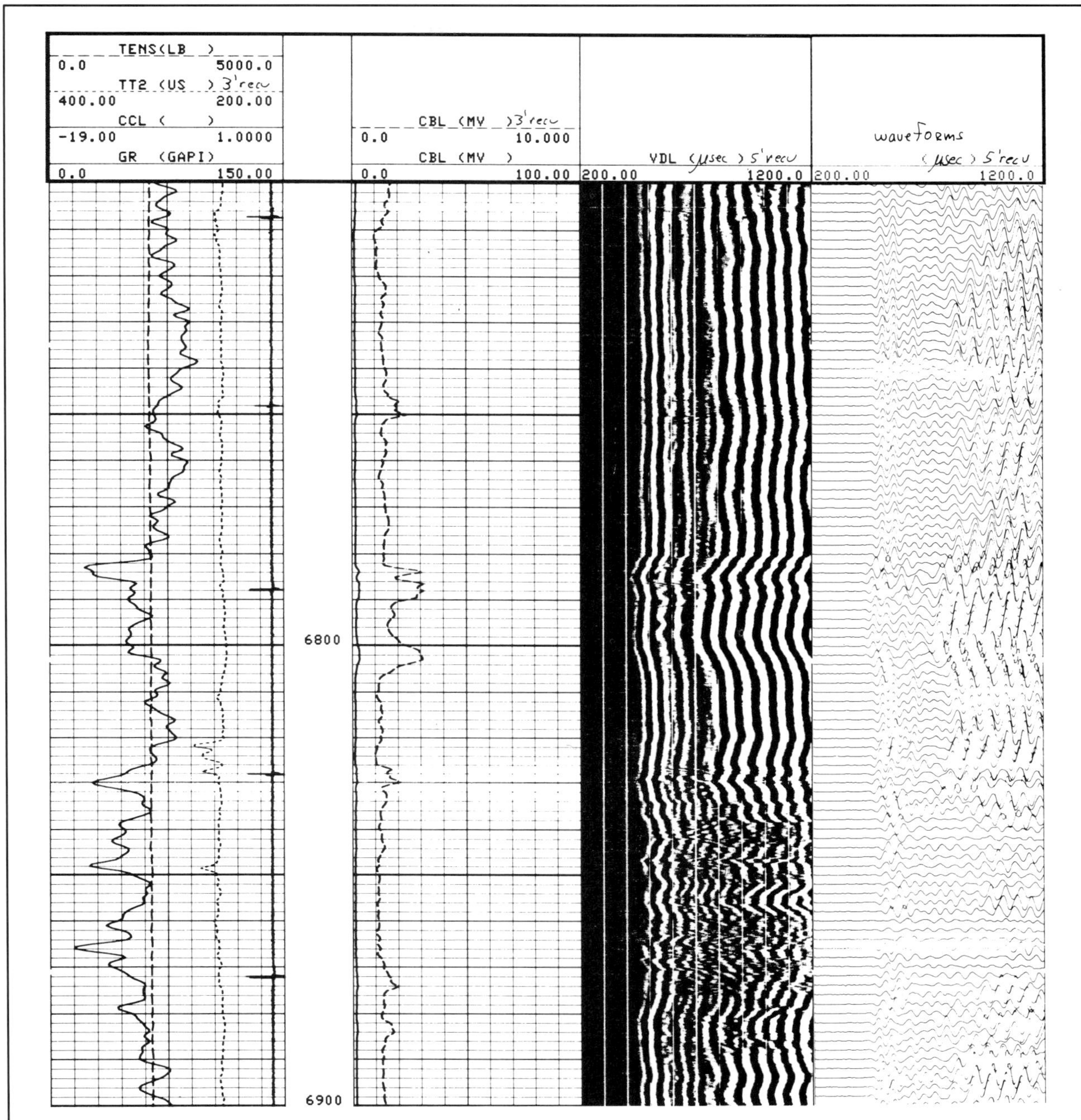

Figure 16–25—CBL-VDL example showing effect of pressure applied at surface.

The same year, Leigh et al. described the results of a field test, along with results from an interpretation model incorporating gas bubbles in the annulus. Later, Gearhart Industries introduced an ultrasonic pulse-echo cement evaluation tool, designated the "Pulse Echo Tool" (PET). A comparison between the PET and conventional and compensated bond logs, based on runs in a test well constructed by the U.S. Environmental Protection Agency, was presented by Albert et al. in 1987.

16–4.5.2 General Description

Ultrasonic tools induce casing resonance by transmitting a broad band pulse (300 to 600 kHz) normal to the casing wall. Two major advantages of this technique are:

- good spatial resolution—about one square inch of cross-sectional area, and
- a lack of need for perfect shear coupling between the pipe and cement.

The disadvantages, for the purpose of cement evaluation, have to do mainly with the combination of high frequency and short wavelength.

- Extreme surface roughness on the pipe (more than 0.1 in) can prevent cement measurement, and
- suspended particles in the borehole fluid can cause high attenuation of the acoustic wave.

16–4.5.3 Tool Configuration

Eight ultrasonic transducers are arranged in a helical spiral around the sonde, each facing outward (Fig. 16–26). They are evenly spaced around the circumference of the tool body—one transducer every 45°. A ninth transducer, facing downward in an opening in the tool body just above the bottom centralizer, measures the speed of sound in the borehole fluid. An integral reflector (located a precisely known distance below the transducer) is used.

Three general measurements are possible with each transducer-cement (or annular material) properties, internal casing radius, and pipe wall thickness. The ultrasonic beams are cylindrical, and are about 1 in. (2.54 cm) in diameter; thus, the measurements made by a transducer represent an average value over an area of roughly one square inch for the cement and wall thickness measurements, which are based upon the energy and frequency content of the reflected signal, respectively. The radius measurement represents a minimum radius over this area, because it is based on the round-trip time

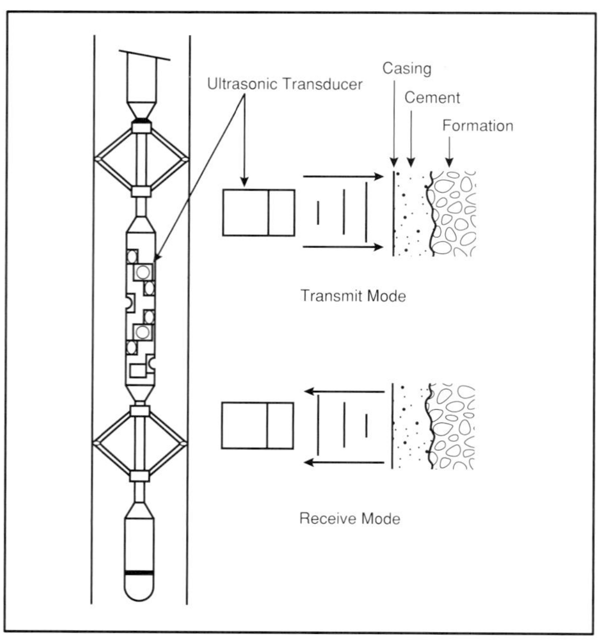

Figure 16–26—Cement evaluation tool.

of the first arrival in the echoed signal. Note that each transducer is a transmitter and receiver.

16–4.5.4 Acoustics of the Measurements

Figure 16–27 is a schematic of the path taken by a pulse

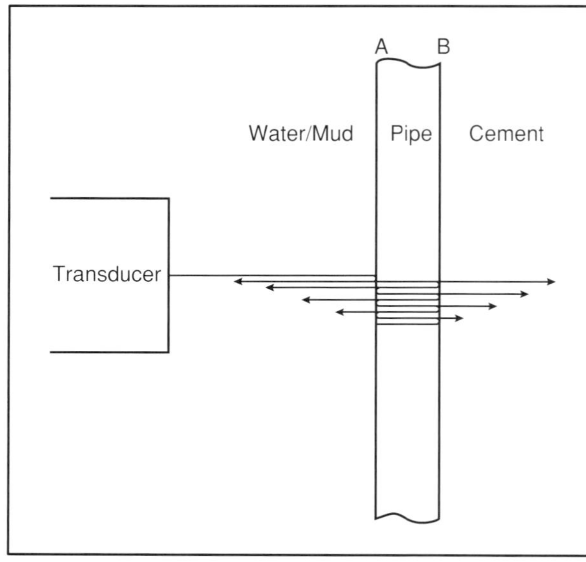

Figure 16–27—Schematic of path taken by ultrasonic energy through various media.

of ultrasonic sound energy transmitted by one of the eight transducers. When reaching the pipe wall, the energy coming from the transducer (the incident energy) is divided. At the boundary between the pipe and the borehole fluid, some fraction is reflected, and the balance is transmitted into the pipe wall. The relative fractions of the acoustic pressure of the incident wave are described by the following formulas.

$$R = \frac{Z_2 - Z_1}{Z_2 + Z_1} \tag{16–9}$$

and

$$T = 1 - R\,. \tag{16–10}$$

R is the reflection coefficient at the boundary between two materials of acoustic impedance Z_1 and Z_2, and T is the transmission coefficient at the same boundary. These coefficients are plane wave relations, and are not strictly valid for the curved pipe wall. However, they are good approximations when the pipe radius of curvature is large compared to the width of the acoustic beam.

The first reflection at the pipe wall returns to the transducer, and provides a measurement of the radius of the pipe. The energy transmitted into the pipe wall propagates to the outside of the pipe where the energy again divides, with some reflection back toward the inner surface, and some transmission into the annulus. This process continues with the effect that an acoustic pulse bounces back and forth inside the pipe wall, losing energy into the annulus and into the hole at every bounce.

The energy returning to the tool is related to the reflection coefficients at the two boundaries. Since the acoustic impedances of the borehole fluid and pipe are known or readily determinable, the only unknown variable is the acoustic impedance of the material in the annulus (when the energy emitted is known).

The two graphs in Fig. 16–28 show the signal transmitted by a transducer. The top graph shows the amplitude over time. Notice that the "pulse" is only about 10 μs in duration. The bottom graph shows the frequency spectrum of the signal, i.e., the magnitude of the Fourier transform. The frequency domain representation of the transmitted pulse shows that the pulse contains high-frequency sound energy in the range of 300 to 650 kHz. This is the range corresponding to the inherent thickness mode resonance frequencies of most oil-field casings. For example, 5½-in. (14-cm) (15.5 lbm/ft) casing has wall thickness of 0.275 in. (0.7 cm), which results in a natural resonance frequency of about 440 kHz. Casing with a wall thickness greater than 0.4 in. (1.0 cm) will resonate at a harmonic, because the fundamental resonance frequency is below 300 kHz. Casing with a wall thickness less than 0.18 in. (0.5 cm) will not be excited into

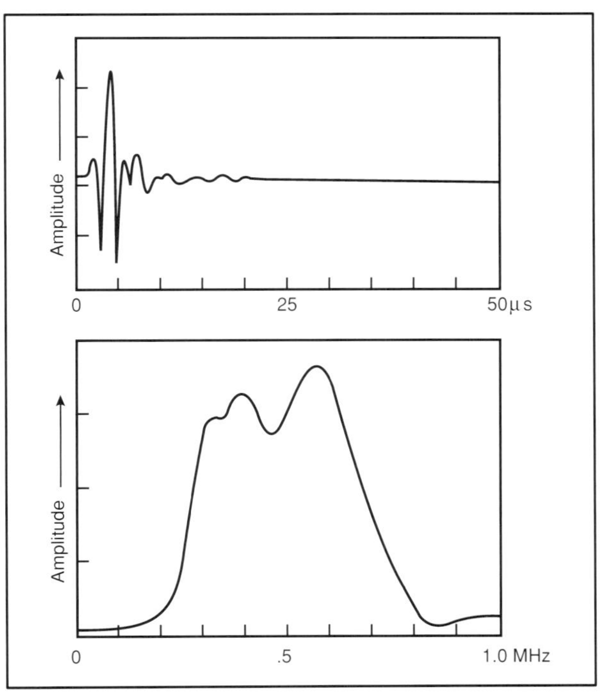

Figure 16–28—Characteristics of signal transmitted by ultrasonic transducer.

resonance, because a frequency greater than 650 kHz would be required.

Figure 16–29 is a representative echo as received by a transducer. First to return is a series of three large peaks looking like a misshapen "W," a direct reflection of the transmitted pulse. The remainder of the received signal is the resonance energy. The elapsed time from transducer firing to the first arrival, labeled DT, is used to calculate a radius. The sum of the measured radii from two transducers facing opposite directions and the tool outside diameter is a pipe diameter. Eight transducers give four diameters oriented 45° apart. The average of the four diameters gives an average inside diameter, and the difference between the largest and smallest is a measure of ovality. The distance that the tool is eccentered in the pipe is calculated from the radii measurements. The difference between opposite radii gives access to the eccentering of the tool.

The magnitude of the peak labeled W1 (Window 1) is measured. It is used as a first-order correction for mud attenuation and the surface condition of the pipe, because it is from a part of the signal which reflected off the inside of the pipe and, therefore, only traveled in the mud.

The resonance signal is measured by rectifying it and integrating it over time intervals—W2 and W3 for the CET and RW (resonance window) for the PET. W2 (or RW) energy is the primary measurement from which the acoustic impedance of the material in the annulus is determined. W3 is used to perform a check on W2. On digital tools, the frequency content of the resonance signal is used to calculate the wall thickness of the pipe.

16–4.5.5 Physical Relationships Used in Interpretation

The relationship between the energy measurement W2 and the acoustic impedance of the material in the annulus is shown in Fig. 16–30 (Tables 16–2 and 16–3). Notice that the relationship depends upon the casing wall thickness. Values for the wall thickness can come from the specified value for the size and weight of the pipe, from

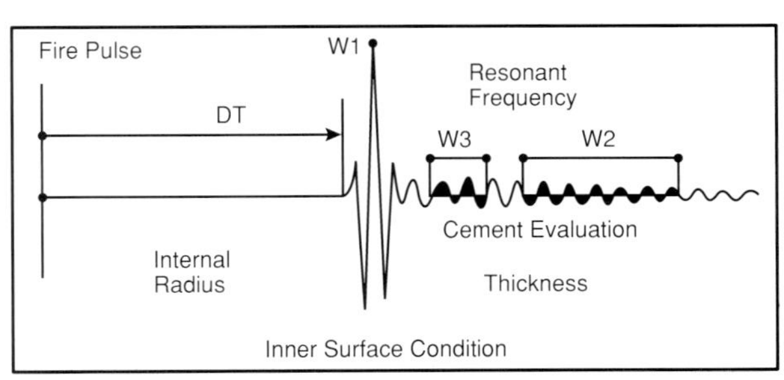

Figure 16–29—Characteristics of echo received by transducer.

the difference between the ultrasonic caliper— measured ID and nominal OD, or from the analysis of the frequency spectrum of the echo. The normalization of W2 is a crucial calculation step. A first-order correction for the borehole fluid is made by dividing W2 and W3 by W1. Second-order mud corrections and pipe curvature corrections are accounted for by free-pipe constants used to normalize the ratios W2/W1 and W3/W1. These constants, called W2FP and W3FP, are simply the values by which W2/W1 and W3/W1 must be divided to obtain a value of 1.0 with fresh water outside the pipe. Tables of values of W2FP and W3FP for most casings have been compiled from test-well studies, laboratory work, and theoretical models. A crossplot technique can be used in situations where table values are not available, or further analysis is warranted. It will be illustrated later in this chapter. Only the normalization process makes it possible to do quantitative analysis with the log.

At the time these tools were developed, the industry traditionally described cement in terms of compressive strength. Laboratory testing performed on different formulations has shown some correlation between the acoustic impedance and compressive strength, on which a straight line was fitted. It must be emphasized that this is an empirical correlation with extremely limited validity, and that only the measured value (acoustic impedance) should be used for interpretation.

If the W2 and W3 measurements from some interval in a well are crossplotted, they will ideally (in theory) define a curve as shown in Fig. 16–31. Each point on the plot represents the values of the two measurements at a particular depth. The data plotted can be from any one transducer or some combination of the eight. The upper right end of the curve corresponds to a very low acoustic impedance. The lower left end corresponds to a high acoustic impedance. Cements, liquid, and gas tend to plot in certain intervals along the curve as shown in the figure. Fresh water will plot at the coordinates (1,1) *if the normalization is accurate*. This type of plot is used to interpret the log. The plot is a compact presentation of considerable data. It is a picture of sorts, and certain patterns that correspond to significant downhole situations can be identified. These patterns are described later in this chapter.

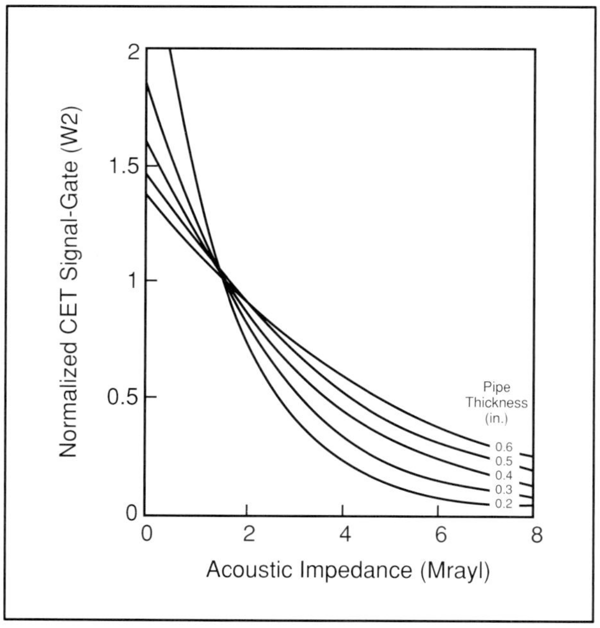

Figure 16–30—Relationship between energy measurement W2 and acoustic impedance.

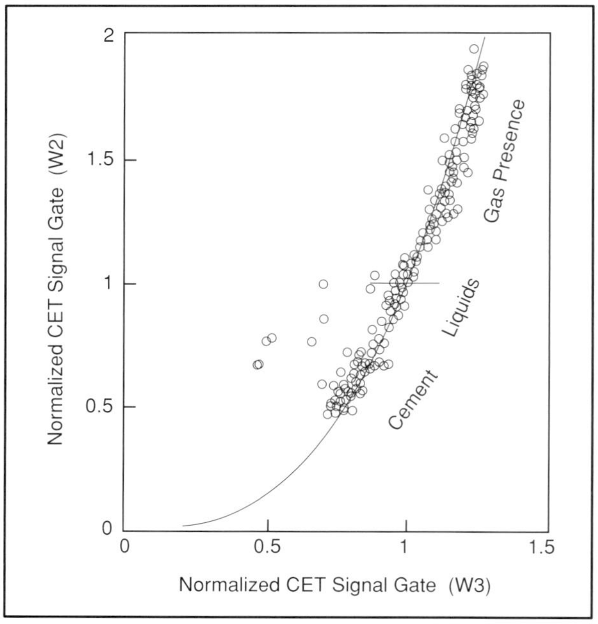

Figure 16–31—Crossplot of W2 and W3 measurements through various media.

16–4.5.6 Complicating Factors

In this section some situations are studied which complicate the interpretation of ultrasonic cement logs. It is important to recognize them, because their effect on the log can range from slightly biasing it (optimistically or pessimistically) to rendering it useless.

Secondary Reflections—The interpretation of ultrasonic pulse-echo tools is made from the model where only mud, pipe, and annulus material are present. However, the annulus material has a finite thickness, and the reflection of energy occurs at the annulus/formation boundary. If a sufficient number of factors exist simultaneously, they may be sufficient to significantly alter the echo and complicate the analysis. These factors include the following.

- "thin" cement sheath,
- pipe centered in the hole,
- gauge hole,
- smooth surface on the wall of the hole, and
- large acoustic impedance contrast between the cement and formation.

Secondary reflections should not be confused with fast formations. The latter is a phenomenon that relates to bond logs, whereas the former relates to ultrasonic measurements made perpendicular to the surface of the pipe.

The cement sheath may be considered to be too thin when the energy reflected at the formation overlaps in W2 with the original casing echo. Calculations made using standard values for pulse velocity through the cement and W2 gate position show that 2.4 in. (6 cm) can often be considered as thin. Using the same velocity value, W3 remains unaffected unless the cement sheath is less than 0.6 in. (1.5 cm) thick.

Note that when the reflections at the annulus/formation boundary interfere with the original echo, the energy can either increase or decrease, depending upon the phase difference between both signals.

Warning flags are placed on the log to alert the interpreter when the influence of a secondary reflection increases the W2 reading. The event is detected by a comparison of the normalized W2 and W3 readings (Fig. 16–32). Since secondary reflections affect the W2 reading but generally not the W3 reading, contaminated data will plot above the theoretical line. In fact, the flags will convey the nonexponential nature of the decay of the waveform, which does happen in well-cemented pipe when the energy reflected at the formation increases the W2 energy. The flags can also be triggered when the signal-to-noise ratio is low, a common situation when the surface condition of the pipe is bad, or when the borehole fluid has a high attenuation. In any case, all log data computed from W2 are invalid.

Gas in the annulus—When gas is present behind the pipe, normalized W2N values will be larger than in free pipe (one with water in the annulus). Gas detection is based on W2N exceeding the value W2GS set by the logging engineer. The determination of this value must be based on the wall thickness of the pipe. As seen earlier, gas has a very low acoustic impedance. The expected W2N reading for gas can be read from the plot of W2N vs acoustic impedance (Fig. 16–30). Whenever the W2N is higher than the gas threshold, but only when no secondary reflections are detected, warnings called "gas flags"

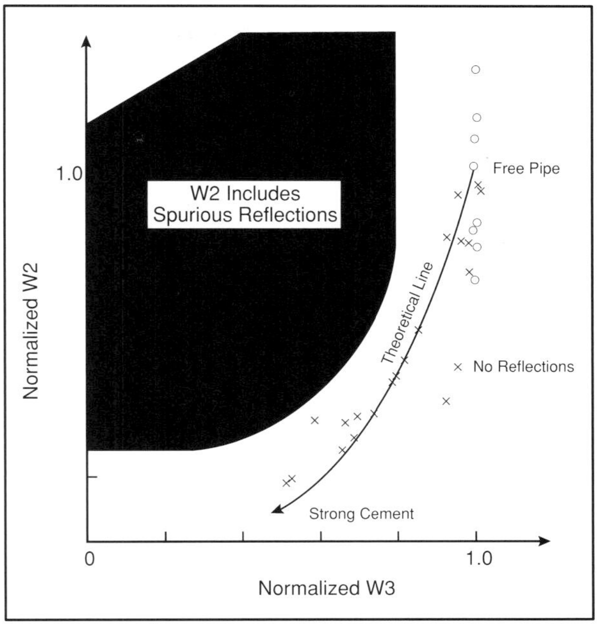

Figure 16–32—Comparison of normalized W2 and W3 readings.

will be displayed on the log. Gas flags are often seen across gas zones, even when well cemented.

Microannulus—The effect of a microannulus on the pulse echo signal is a function of the material present in the microannulus. The graph, presented in Fig. 16–33, shows how the transmis.sion of the signal is affected by the microannulus (Havira, 1982). As long as the microannulus size is small when compared to the wavelength of the signal in the gap, the effect is marginal. It means also that the sensitivity to a gas-filled microannulus will be higher than that for a liquid- or fluid-filled microannulus. Experiments and theory show that for a water-filled microannulus, the measurement is little af-

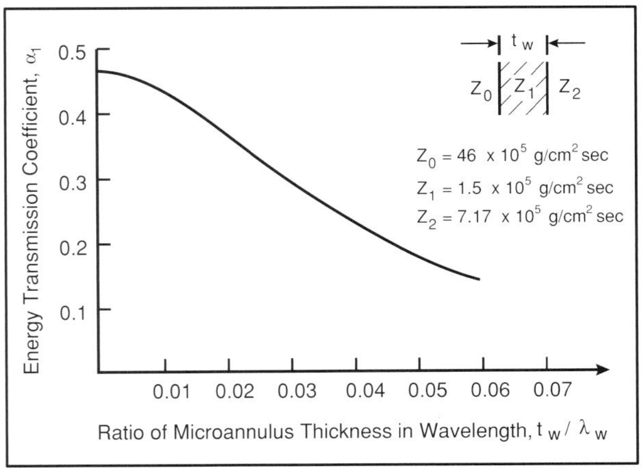

Figure 16–33—Transmission coefficient vs microannulus thickness.

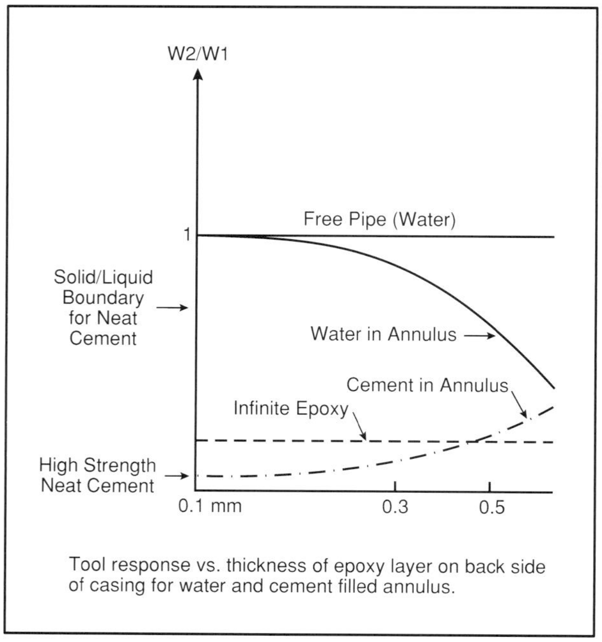

Figure 16–34—Tool response vs thickness of epoxy layer on back side of casing (for water- and cement-filled annuli).

fected for a gap up to about 0.005 in. (0.1 mm). However, if the microannulus is filled with gas, the values are much smaller, because the wavelength through gas is much shorter. In a study published in 1984, Leigh et al. stated that a gas-filled microannulus would start to affect the CET signal at gap values as small as 0.5 µm. The effect of the microannulus, related to the wavelength of the signal, will be directly related to the casing thickness. Thinner casings resonate at higher frequencies (shorter wavelengths), and will be affected more by smaller microannuli than thicker casings.

Thin Pipe Coatings—A complication closely related to the microannulus is that of thin coatings on the outside of the pipe; e.g., mill varnish and epoxy/sand coatings. Figure 16–34 shows the effect on W2N as a function of epoxy thickness for two cases—water in the annulus and cement in the annulus.

Cement Case: Starting at the left end of the "cement annulus" line, the W2N is the appropriate value for the cement. As the epoxy layer reaches about 0.01 in. (0.25 mm) in thickness, the reading begins to rise toward that of the epoxy. The epoxy value is reached at about 0.02 in. (0.5 mm) The calculated cement acoustic impedance at this point would be pessimistic. So the measurement will be quantitatively correct up to about 0.01 in. of epoxy and qualitatively useful up to about 0.02 in. of epoxy.

Water Case: Starting at the left end again, but with the "water in annulus" curve, a W2N water reading of 1.0 is observed. The reading begins to fall at about 0.005 in. (0.13 mm), but liquid in the annulus will be indicated on the log until W2N is below about 0.74 (for the neat cement correlation). This occurs at an epoxy thickness of about 0.016 in. (0.4 mm). For a greater thickness, cement will be indicated on the log.

Overall an epoxy thickness of up to 0.02 in. (0.4 mm) can be tolerated with little or no effect on the log.

Corroded casings—Corroded casings can produce a drastic scattering of the ultrasonic waves. So little energy is reflected back to the receiver that W1, W2, and W3 are all small. The results are unpredictable. Corroded casings are well evidenced by "cloudy" crossplots.

Heavy and oil-base muds—These produce strong ultrasonic attenuations, which lead to very low levels of received energy. The measurements will have very poor accuracy which is evidenced by fairly "cloudy" crossplots.

Casing thickness—The presence of more than one casing weight will spread the data on a W2/W3 crossplot, and indicates the need for zoned free-pipe parameters. In other words, the W2FP and W3FP values must be adjusted each time the casing weight changes. Also, casing thicknesses around 10 to 11 mm are problematic because the casing resonates not only with its natural frequency, but also with the first harmonic, both included in the transducer bandwidth. If the calibration only takes the fundamental into account, the acoustic impedance will be underestimated. Small variations in wall thickness result in large changes in free-pipe constants. The present API casing specifications (API Spec 5A) allow considerable variation in pipe thickness; thus, the accuracy of pulse-echo measurements will always be problematic.

16–4.5.7 Log Presentation

Log presentations vary slightly from region to region and between service companies. Mnemonics used for labeling curves also can change. For these reasons curves will be related (whenever possible) to the basic measurements made downhole.

The ultrasonic cement tools make many measurements downhole due largely to the multiple transducers. Some of these data are combined in the process of directly computing useful information such as acoustic impedance. Even so, there are many curves on a typical log. The log is often assembled in two parts spliced end-to-end to accommodate as many useful data as possible. For introductory purposes, the data are divided into four functional groups (Fig. 16–35).

Cement/annular data—The primary purpose of the tools is to determine what is in the annulus. For ease of interpretation, a cement map is derived from the measured acoustic impedance of each transducer, and presented on the right track. Since there are eight radially-oriented transducers, the display consists of eight segments. The adjacent segments can optionally be averaged at their boundaries to create a smooth appearance. This is, in effect, a circumferential interpolation between each of the eight readings at 45° intervals. The display is normally based on annular acoustic impedance: from totally white to totally black according to whether the computed acoustic impedance values are below or above two values (IMAL and IMAU, expressed in Mrayl). If the hole is deviated from vertical, the display can be adjusted to present the image from the low side of the hole in the middle of the track and the high side at the edges.

On the right side of the display are eight "flag tracks," one per transducer. The secondary reflections are indicated by a wide mark at the depth of occurrence. The gas flags are indicated by a narrow mark in the appropriate flag track.

The cement compressive strength is presented on the center track, derived from the acoustic impedance Z using an empirical relationship of restricted validity. This relationship is based on two parameters (CSCO and CSCG), which are compressive strength of cement offset and gain ($CS = Z \times CSCG - CSCO$). Most of the time, a minimum curve and a maximum curve (average of the three lowest and the three highest values calculated from the height transducers) are presented.

An average of the eight normalized W2N readings is also usually presented. This curve allows for log quality checks and more detailed interpretation (Section 16–4.4.8).

Caliper/casing data—It was seen earlier that the round-trip travel times from the eight transducers could be converted to radii and then to pipe inside diameters. The average of all four diameters is usually presented on the log. In most cases, the curve is usually labeled "mean diameter." Often, the individual diameters are presented on a separate log, or attached to the bottom of the cement presentation as a detailed caliper log. Also presented is a curve labeled "eccentering." This is the tool eccentering relative to the pipe center, and is calculated by comparing opposite radii differences. It is used for log quality control.

Another caliper-derived curve often presented is called "ovality." This is the difference between the largest and smallest of the four diameters. It is useful for locating pipe problems. Like eccentering, it will ideally read a value of zero. In practice, both curves will normally read a few hundredths of an inch.

Borehole fluid data—The ninth transducer continuously measures the speed of sound in the borehole fluid. This information is required for the distance calculations, and may also be displayed on the log. It is presented as slowness, the inverse of velocity. This curve reflects changes in density with high sensitivity. Occasionally, it proves very useful in locating fluid changes in the well (e.g., oil on top of water and salt influx).

Inclinometer data—There is a mechanical inclinometer built into the tool. It provides two useful measurements when the deviation is larger than 5°-hole deviation and tool rotation ("Relative Bearing"). The Relative Bearing is presented as a curve. It also provides the ability to shift the cement map image to display the measurements from the low side of the hole in the center of the display. Hole deviation is not used directly for any computations, but is extremely valuable in analyzing the results of a primary cement job.

Miscellaneous data—Collars may be detected with a conventional magnetic collar locater, or from the ultrasonic caliper measurements. A gamma ray or neutron curve is usually presented for correlation purposes.

16–4.5.8 CEL Quality Control

A step-by-step procedure should be followed to control the quality of the cement evaluation log (Fig. 16–36). The first two steps are common to every log.

- *Repeat section* and main log must *roughly* look the same. Some minor discrepancies are expected because of the partial casing coverage in most cases. Hence, each logging pass will likely investigate different points and yield slightly different results. Thus, it is important to compare a data curve representing a mathematical average of all the readings, e.g.,"WWM." See Fig. 16–37 for a typical example, where the curves are very similar. More to the point, identical interpretation would result from either pass.
- *Calibration summary/parameters* must be written on the log. Carefully check IMAL, IMAU, CSCG, and CSCO. The cement's acoustic impedance should be close to the expected impedance. The wrong setting of the imaging parameters can turn the cement map either totally white or totally black.

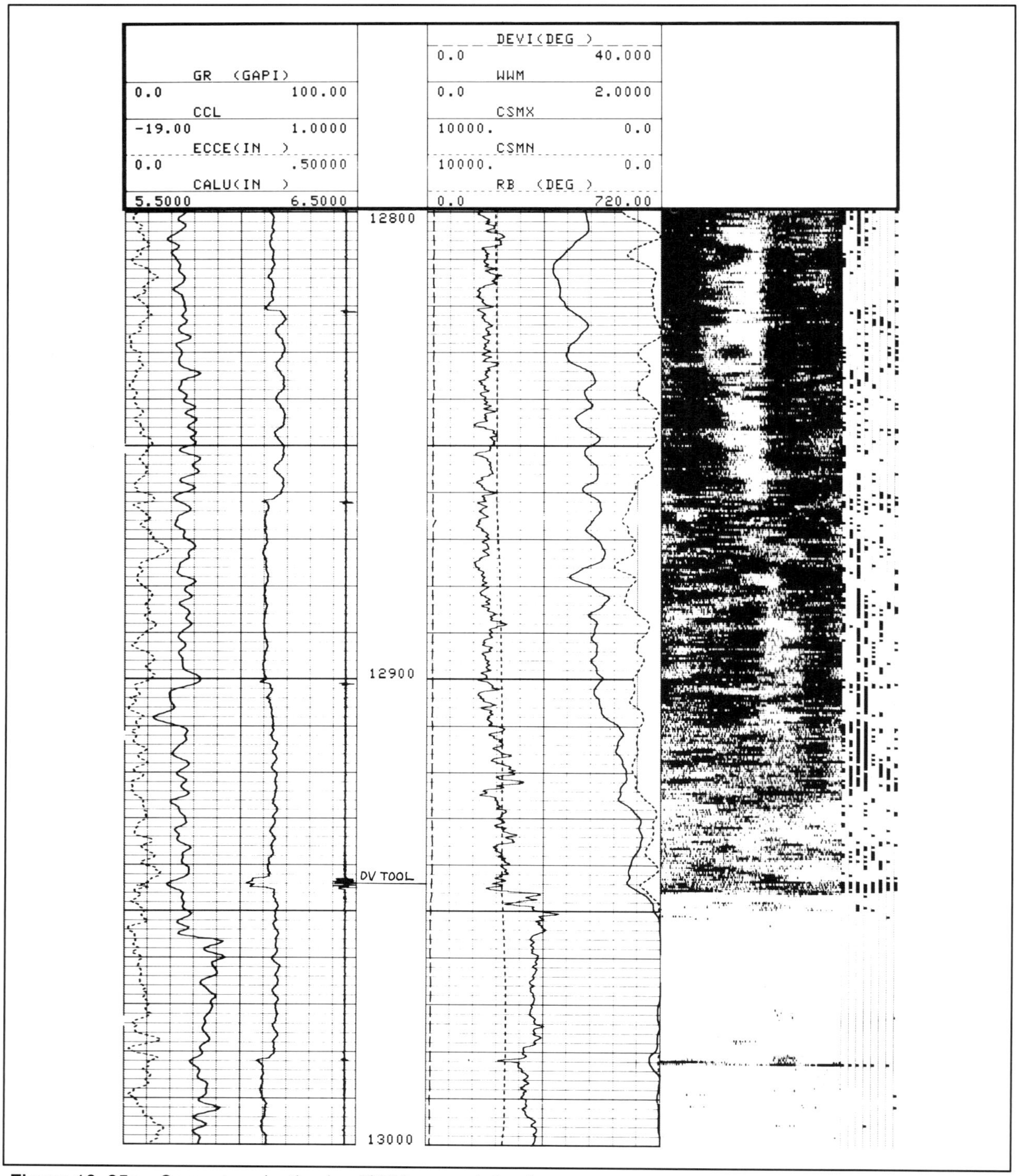

Figure 16–35a—Cement evaluation log (CEL) presentation format.

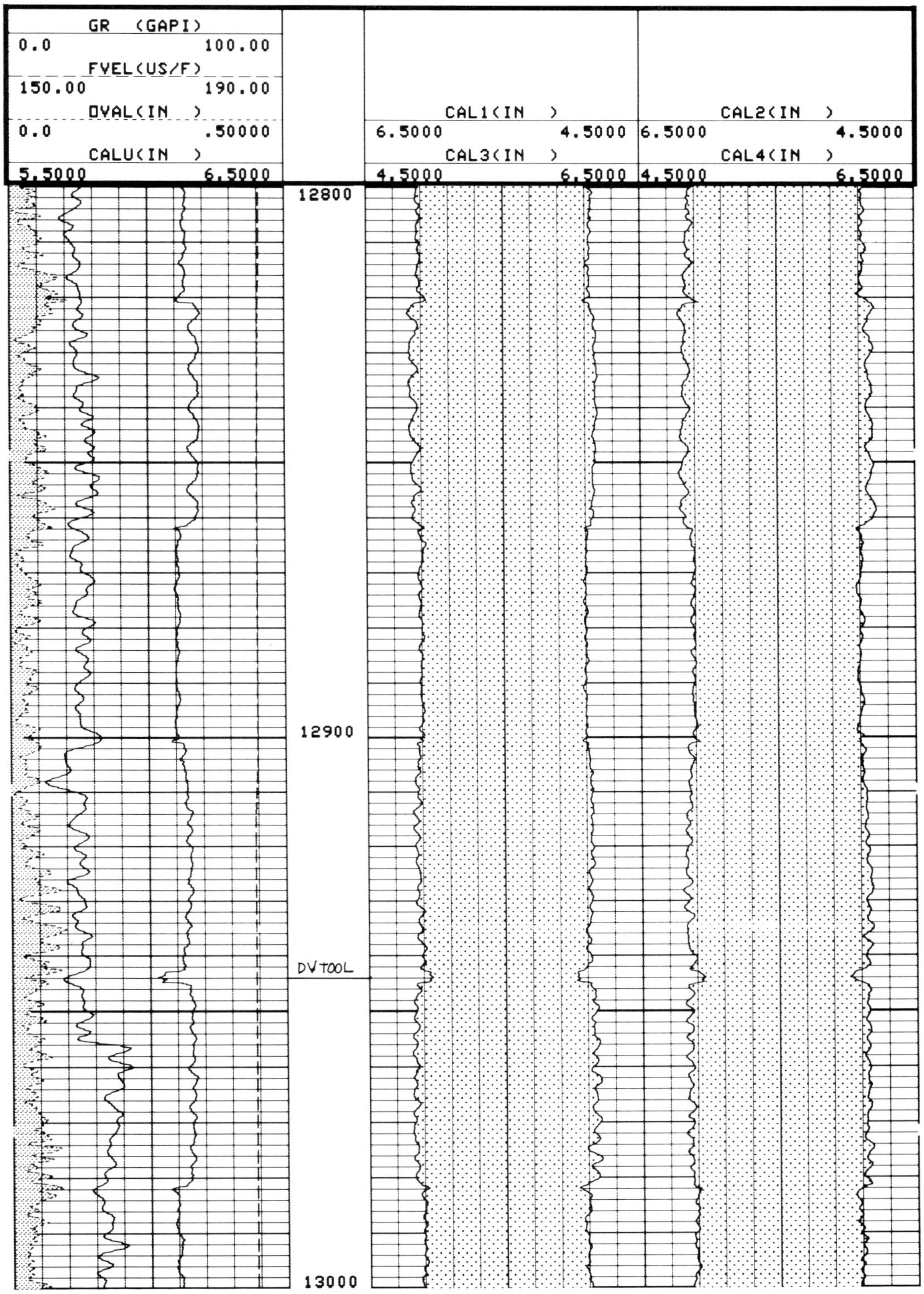

Figure 16–35b—Cement evaluation log (CEL) presentation format.

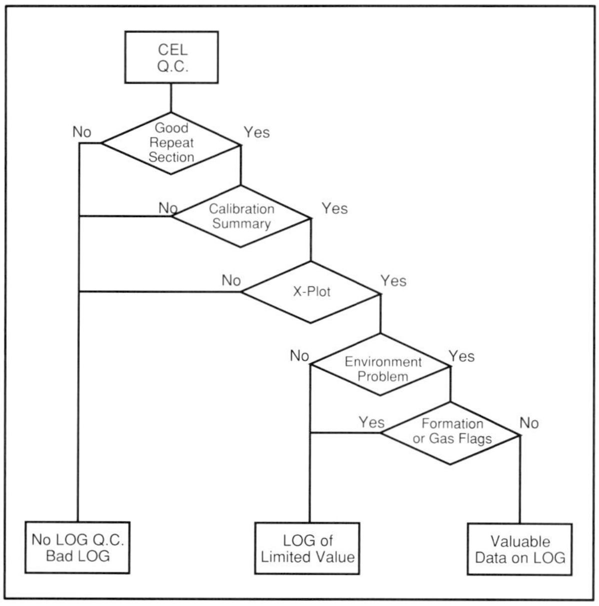

Figure 16–36—CEL quality control flowchart.

- *Crossplots* have two main purposes—to check both the calibration and the data integrity (Fig. 16–31). The first objective of the crossplot is the calibration of the tool: the choice of calibration parameters W2FP and W3FP should be such that the normalized CET gate values W2 and W3 are equal to 1-in. free pipe. Furthermore, to be valid, the data points should fall around a single curve in agreement with the model. Observe also that incorrect normalization of W3 can have serious effects. If W3FP is too high, the data cluster will shift left, and the data points in the fluid region will fall inside the secondary reflection region. Conversely, if W3FP is too small, the data cluster will be shifted out, and W2N values still considered as valid would incorrectly indicate lower acoustic impedance. In the absence of an crossplot, the quality check of the data is limited to the "regularity" of the tool response.
- *WWM* is the mean value of the eight W2N calculations representing the eight transducers. With fresh water behind the pipe, it will read 1.0; with gas at atmospheric pressure it will read higher (about 1.5 for heavy walled pipe, and 2.5 for thin walled pipe). With cement, it will read less than 1.

Tool eccentering—If the tool travels far out of center, the signals will strike the pipe wall at an angle, and will not return directly to the transducers. This will distort the energy measurement, resulting in an incorrect cement measurement. The admitted acceptable value for sonde eccentering is 4 mm in 7-in. casing (5 mm in 9-$^5/_8$-in. casing). Since the ultrasonic sondes are short, stiff, and light, eccentering is rarely a problem even in highly deviated wells.

16–4.5.9 CEL Interpretation —Examples

When a CEL has passed the quality-control step, the interpretation should be fairly straightforward. It gives the cement acoustic impedance and the distribution around the casing. This also means that for a proper setting of the display parameters, the cement acoustic impedance must be known at the time of the log.

Channeling example—On the log section presented in Fig. 16–38, mud channels are well evidenced on the cement map. Note as well that the orientation of the mud channel follows the rotation of the tool. In this case, the image was not corrected to present the low side of the pipe in the middle of the cement map. However, do not forget that the azimuthal coverage is complete only in 4.5-in. (11-cm) casing.

CEL across gas formation—The pulse echo measurement is very sensitive to the presence of gas in the annulus, especially at the interface between the casing and the cement. Quite often gas enters into the cement across the gas zone (Chapter 8), leading to a decrease in the cement acoustic impedance. Figure 16–39 is a section across a gas zone. At the level of the gas zone, the cement map is white and gas flags indicate the presence of gas in the annulus. Below and above the gas zone, the cement map is dark, showing a larger cement acoustic impedance.

If the log is showing uniformly poor results, check for slurry overretardation, incomplete mud removal leaving a film of mud on the casing wall, gas in the cement, free gas, liquid-filled large microannulus, etc.

16–4.6 Combined CBL/CET Interpretation

Comparing CBL and CET results presents the advantage of making a coherent interpretation, to enhance microannulus effects and to back up CET results in adverse or limiting conditions (corrosion and heavy muds).

16–4.6.1 Combined Interpretation Examples

As explained earlier, CBL response is similar in the presence of channeling and a microannulus. With only a CBL, it is not possible to identify one from another, unless another logging pass is run under pressure. The following examples show the benefits of having both logs.

Microannulus Identification—

- Fluid-Filled Microannulus: Figure 16–40 is a CBL with an amplitude between 10 and 50 mV. The transit-time curve is regular, without stretch and skip, clearly showing the casing collars. The VDL shows strong casing signals, and also continuous weak formation

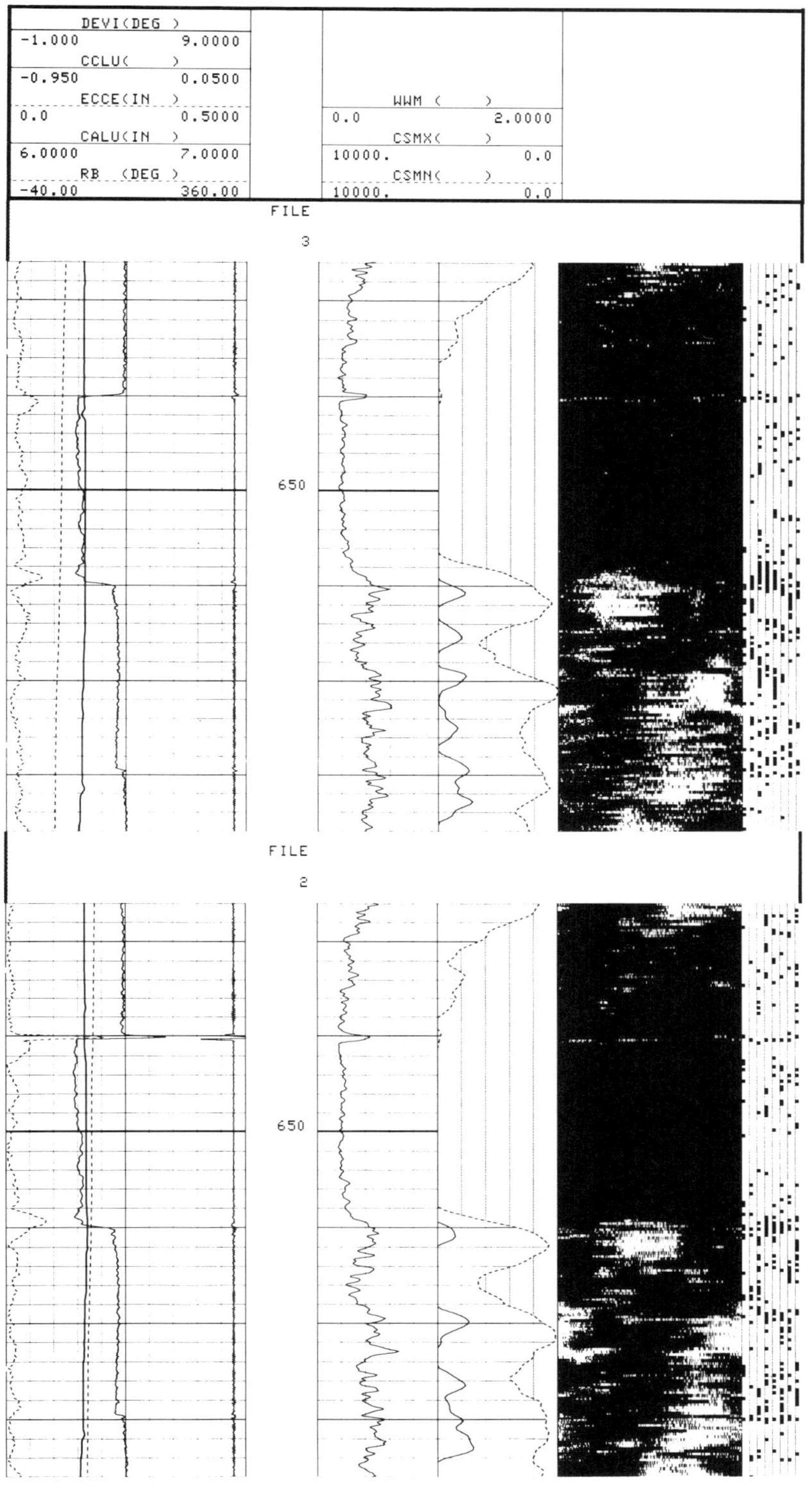

Figure 16–37—CEL example showing repeat section.

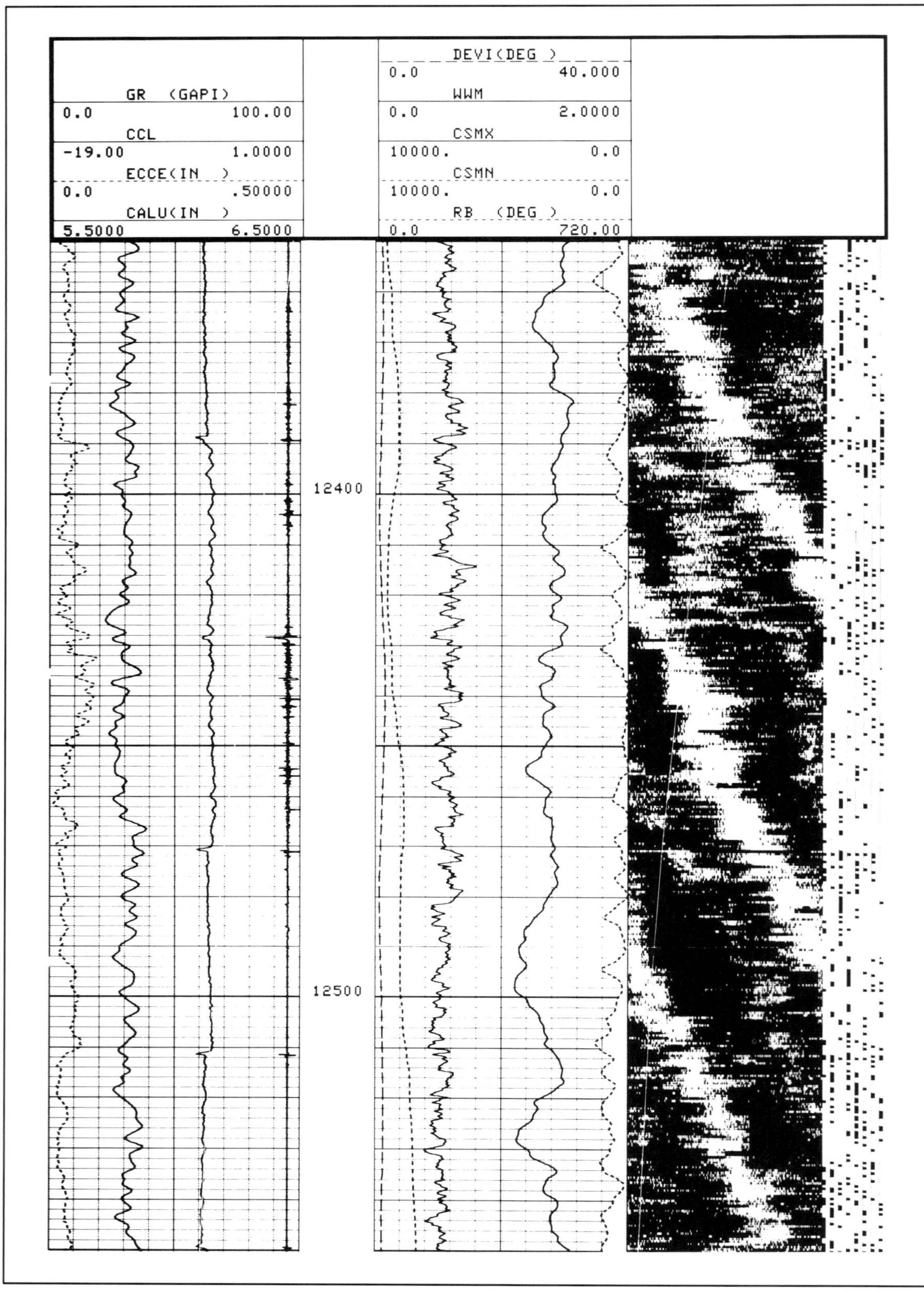

Figure 16–38—CEL example showing mud channels.

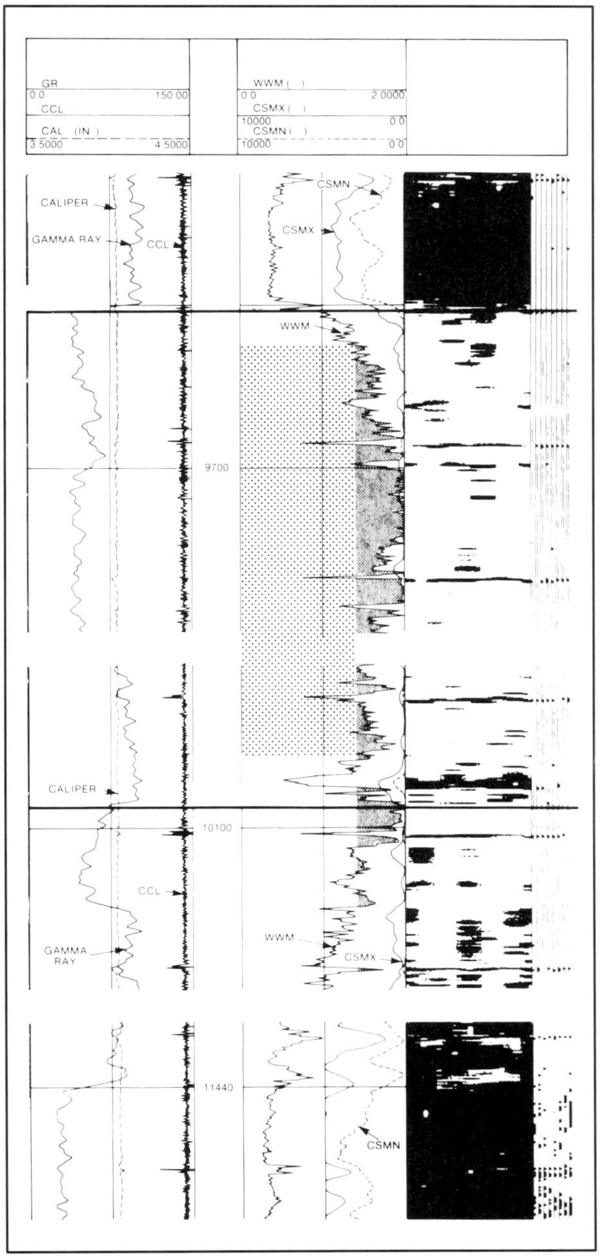

Figure 16–39—CEL example showing gas zone.

arrivals. This could be interpreted either as channeling or a microannulus. The CET was run on the same well (Fig. 16–41). The cement map makes the interpretation very clear. There is no indication of a channel; thus, there is a fluid-filled microannulus, which can be indicated without internal casing pressure. With 2,000 psi applied at the surface, the amplitude fell to the range of 5 to 15 mV over the interval in the figures; the CET was unchanged.

- Gas-Filled Microannulus: Figure 16–42 is a CET log run in 5½-in. (14-cm) casing. Because of the gas flags, a uniformly poor cement map, and a value of WWM between 0.8 and 1.6, one could conclude that gas contaminated cement exists and predict zonal isolation problems. The CBL run across the same section (Fig. 16–43) shows a relatively low amplitude, an indication of shear coupling between the casing and cement. The VDL, which shows mostly formation signal, confirms this interpretation. A small gas gap between the pipe and cement is indicated. It has a small effect on the CBL, but a large effect on the pulse-echo measurement. Such a gap is not detrimental to isolation.

Channeling identification—The CBL presented in Fig. 16–44 indicates extremely good bonding in the interval shown—1,622 ft to 1,675 ft. The attenuation is at least 30 dB/m through most of this interval. The VDL is dominated by a strong formation signal, and there is little casing or collar signal present. But there is, in fact, a large mud channel behind the pipe as can be seen clearly in the CET cement map (Fig. 16–45). Note that the value of WWM is larger where the channel size increases. The orientation of the channel, which is following the relative bearing curve, is always on the same side of the pipe.

Logs across a fast formation—This log interval is in the middle of a massive limestone formation that is generally tight, but highly fractured. The amplitude curve on the bond log (Fig. 16–46) is strongly influenced by fast formation effects and is unusable. All CBL interpretation must be done qualitatively from the VDL data. Note also that the transit-time curve is very irregular, with the lowest values in zones of low amplitude. The VDL display does not show strong casing signals, but fuzzy patterns which could be due to the fractures in the formation or in the cement.

Since ultrasonic cement logs are not influenced by a fast formation, they provide a clearer indication of the cement quality in this type of environment. The CET (Fig. 16–47) shows a large amount of cement, but with a channel from the perforated interval downward about 50 ft (15 m). A comparison of the two logs shows that the diffraction patterns on the VDL display correspond largely to problem areas in the cement sheath.

16–4.6.2 Enhancements

In 1984, Catala et al. proposed to merge the interpretations of the CET and CBL/CBT, so individual limitations of each log could be overcome. Comparing the response of both tools allows better identification of the quality of the cement job—mud or liquid, hard cement, gaseous cement, and free gas at the casing interface can be identified and visualized on the log.

Figure 16–48 shows how the four classes of materials can be identified from a CBL/CET crossplot.

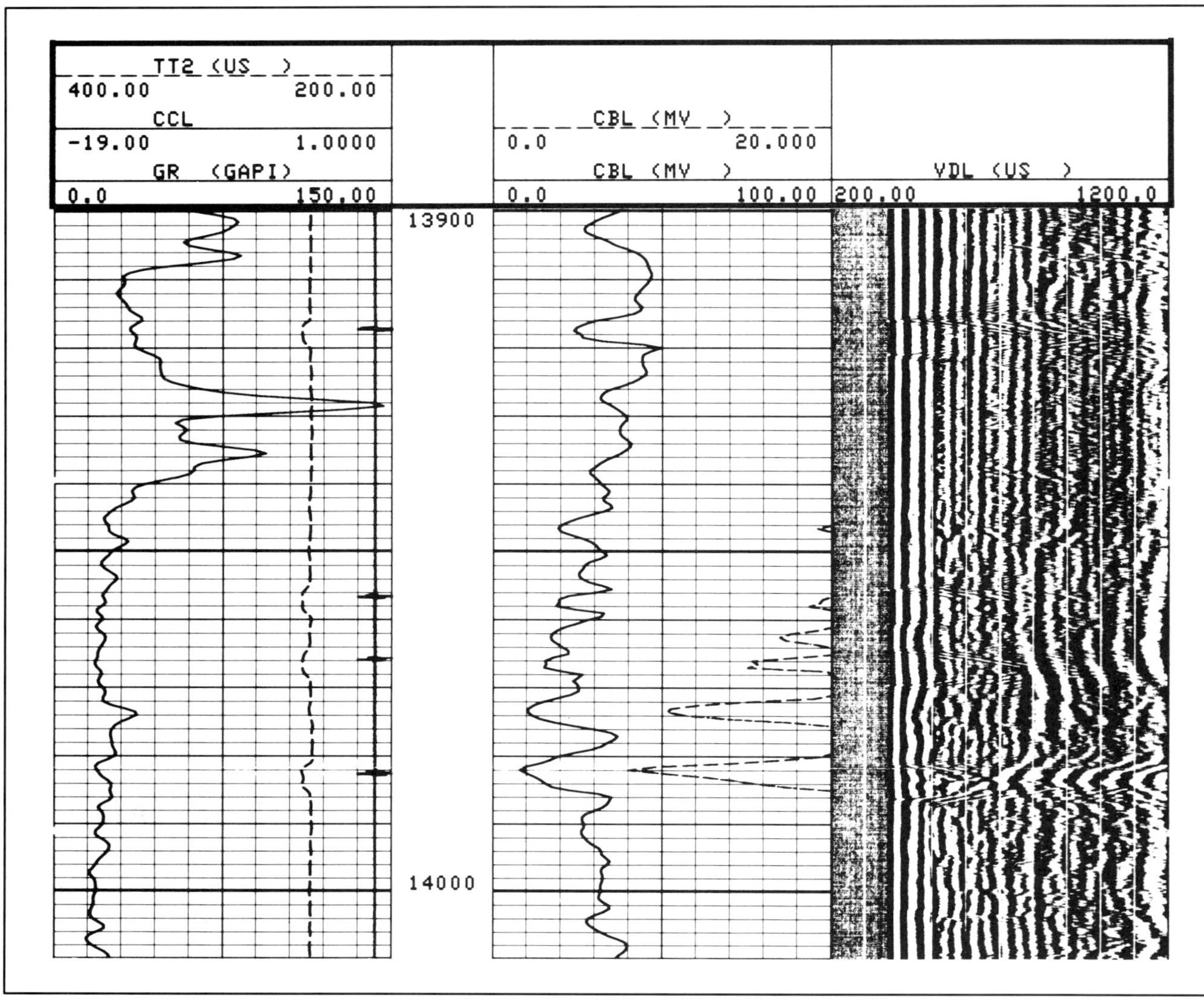

Figure 16–40—CBL example showing microannulus.

- *Mud or liquid*—In the presence of mud or liquid in the annulus, both the CBL and CET will read free pipe.
- *Good cement* produces low CBL amplitudes, and has high acoustic impedance; W2 and W3 energies are low.
- *Free gas*—The CET is much more sensitive to gas than the CBL. In the presence of free gas, the CBL response is about the same as in the presence of liquid: free-pipe amplitude will be indicated. However, the CET will measure the gas acoustic impedance, which is much smaller than the mud or liquid; W2 and W3 energy are higher than in free pipe with fluid.
- *Gaseous cement* is identified by a low CBL amplitude, but the CET is affected by the presence of gas; W2 and W3 energies are high.

These four classes of materials can be identified automatically, based on two distinct computations. A CBL amplitude measurement must be completely free of microannulus effects to be incorporated in this calculation.

- *Acoustic impedance from CET*—The following classes of materials are identified: gas or gaseous cement when acoustic impedance is low, cement when the acoustic impedance is high, and liquid in between.
- *Bond Index from CET and CBL*—Both curves should approximately superimpose. Large discrepancies allow the identification of either a fluid-filled microannulus when the Bond Index from the CBL is worse, or *"gaseous cement,"* when the Bond Index from the CBL is better (although it could also be a gas-filled

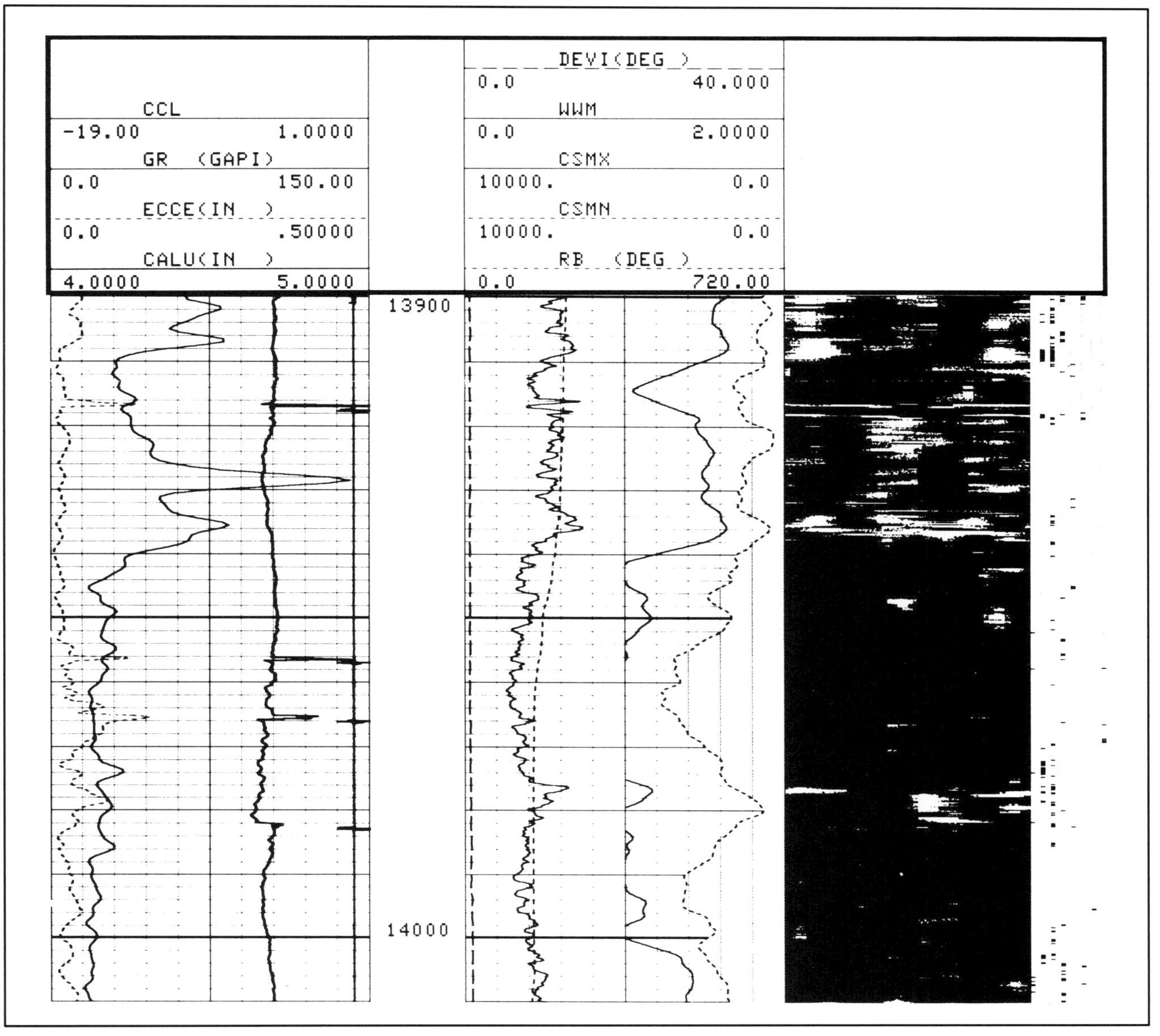

Figure 16–41—CEL example showing microannulua (for comparison with Fig. 16–40).

small microannulus). Thus, the distinction can be made between gas and gaseous cement from the CET/CBL combination.

16–4.7 Conclusions

Acoustic logs are a record of an electrical signal which is subject to caution, especially when the signal has been processed. The acoustic impedance of the cement is only one of many parameters which influences the acoustic logs. The combination of CBL and ultrasonic cement logs provides more information about the quality of the cement job, but the knowledge of well data, cement job events, and pre- and postjob well histories is still often determinant to the quality of the evaluation.

NOMENCLATURE

Sonic Related

CBL Cement Bond Log (E_1 amplitude)

CBLG E_1 Amplitude Direction Window Length (μs)

CCL Casing Collar Locator

TT Transit Time

VDL Variable Density Log

CBT Cement Bond Tool

Ultrasonic Related

CALU Ultrasonic Caliper (mean value)

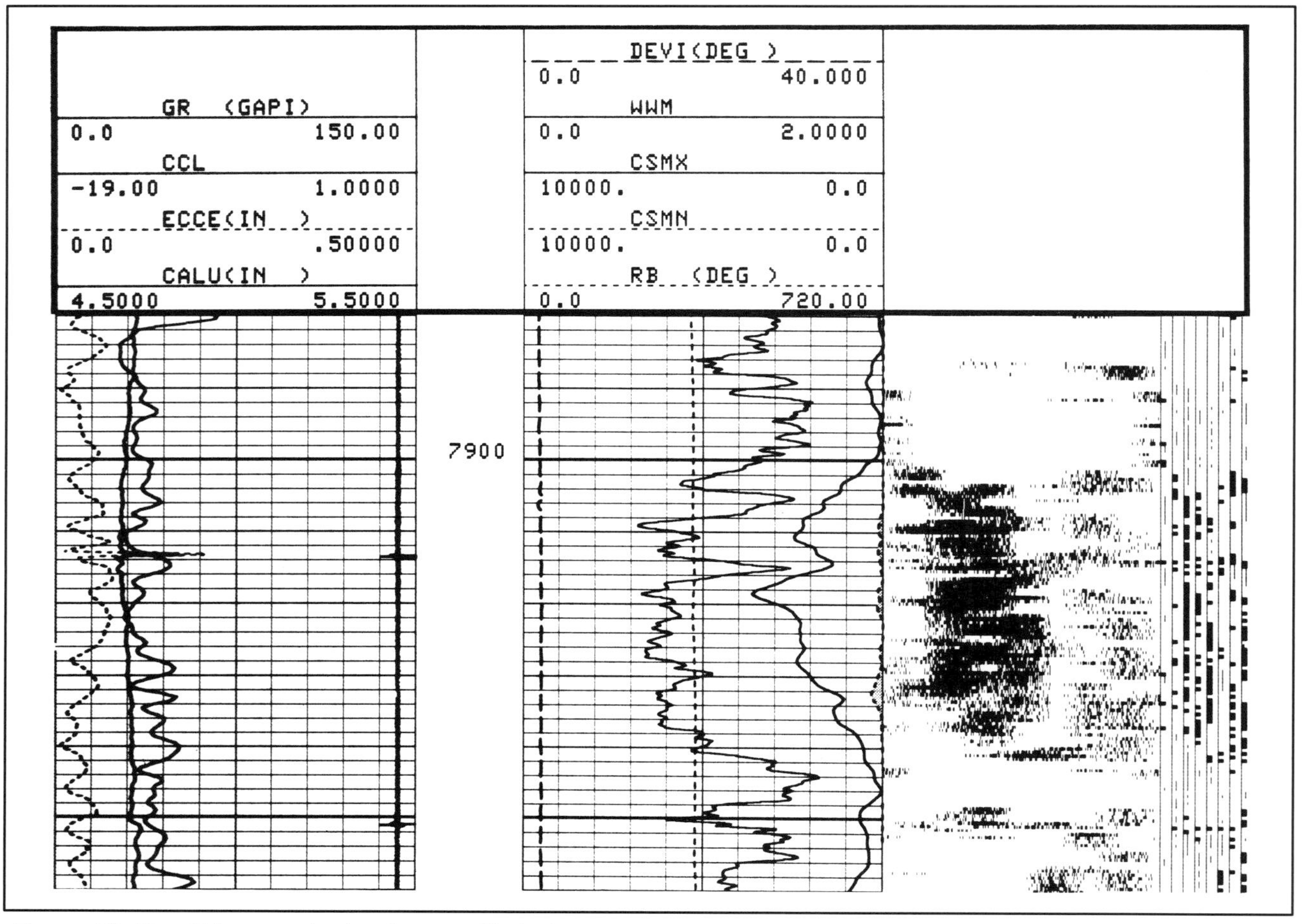

Figure 16–42—CEL example showing gas-filled microannulus.

CCLU	Ultrasonic Casing Collar Locator
CET	Cement Evaluation Tool (Mark of Schlumberger)
CSCG	Compressive Strength Conversion Gain
CSCO	Compressive Strength Conversion Offset
CSMN	Minimum Compressive Strength
CSMX	Maximum Compressive Strength
DEVI	Deviation
ECCE	Eccentering
FVEL	Velocity in Mud
IMAL	Image Lower Limit
IMAR	Image Rotation
IMAU	Image Upper Limit
OVAL	Ovalization
PET	Pulse Echo Tool (Trademark of Gearhart Industries)
RB	Relative Bearing
W1Ti	W1 Transducer i
W2N	Normalized Value of W2
W2FP	Free-Pipe Value of W2
W2Ti	W2 Transducer i
W3N	Normalized Value of W3
W3FP	Free-Pipe Value of W3
W3Ti	W3 Transducer i
WWM	Mean Ratio W2/W1 over 360°

Other

Z	Acoustic Impedance
BHC	Borehole Compensated

REFERENCES

Abdel-Mota'al, A. A.: "Detection and Remedy of Behind-Casing Communication During Well Completion," paper SPE 11498, 1983.

Ahmed, U.: "Post-Treatment Measurements," *Reservoir Stimulation*, M. J. Economides and K. H. Nolte, eds., Prentice Hall, New York, 1987.

Albert, L. E., Standley, T. E., Tello, L. N., and Alford, G. T.: "A Comparison of CBL, RBT and PET Logs in a Test Well with Induced Channels," paper SPE 16817, 1987.

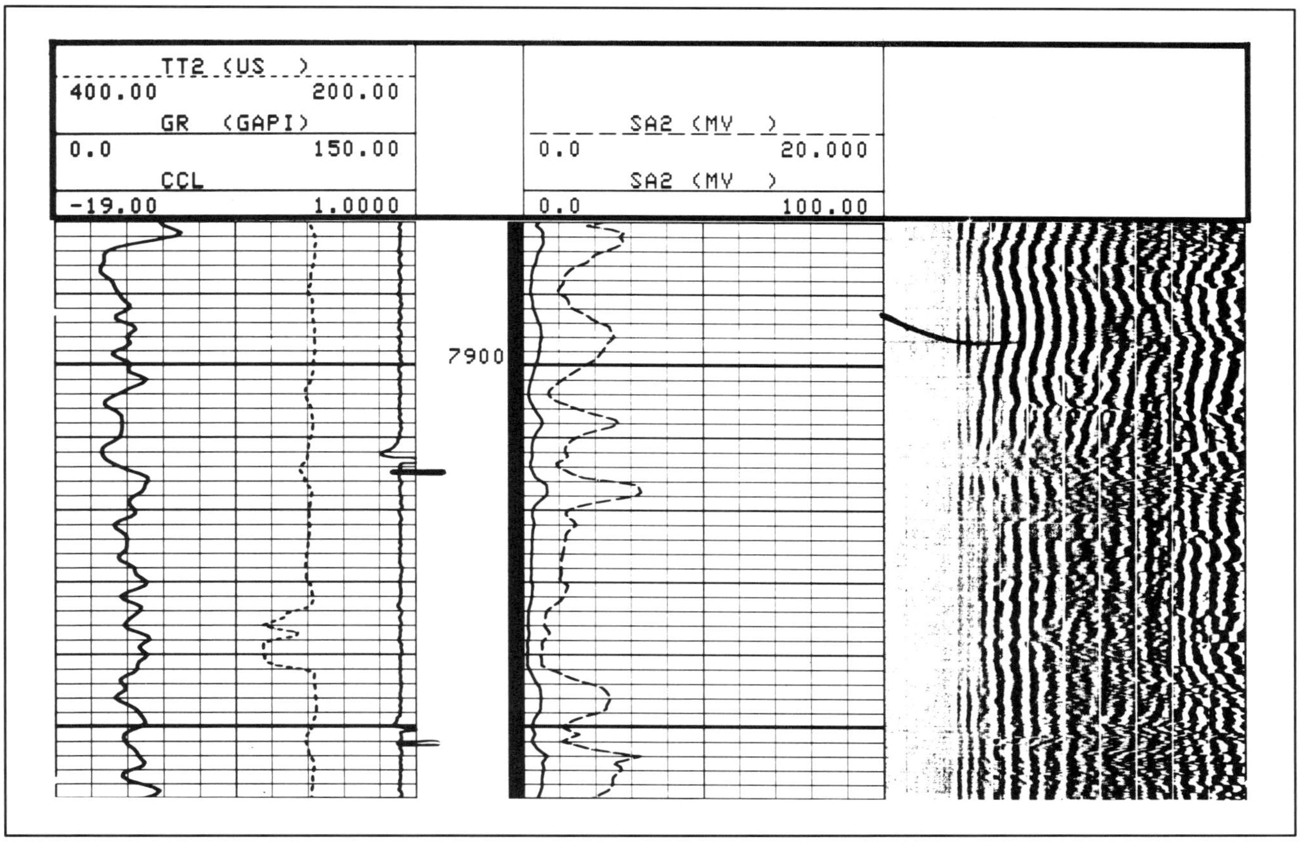

Figure 16–43—CEL example showing gas-filled microannulus (for comparison with Fig. 16–42).

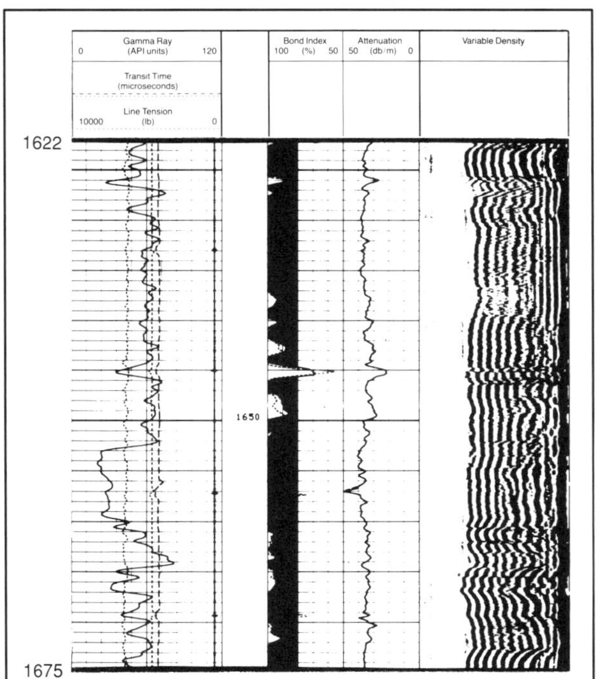

Figure 16–44—CBL example showing channel.

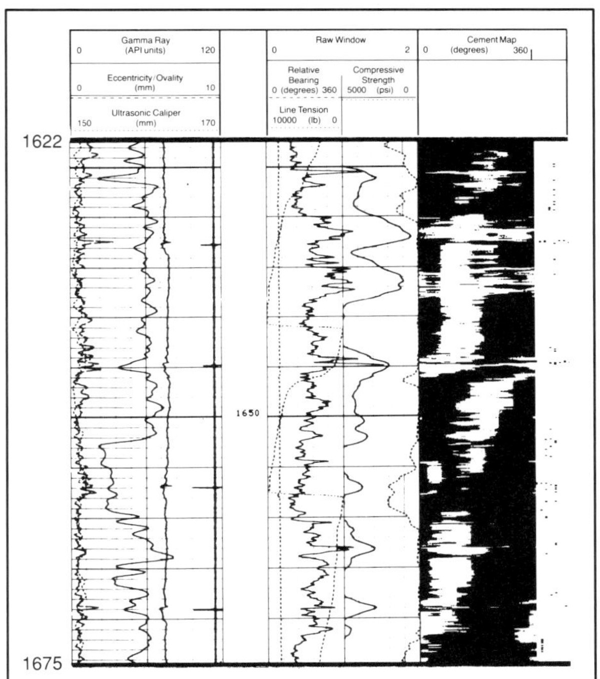

Figure 16–45—CBL example showing channel (for comparison with Fig. 16–44.

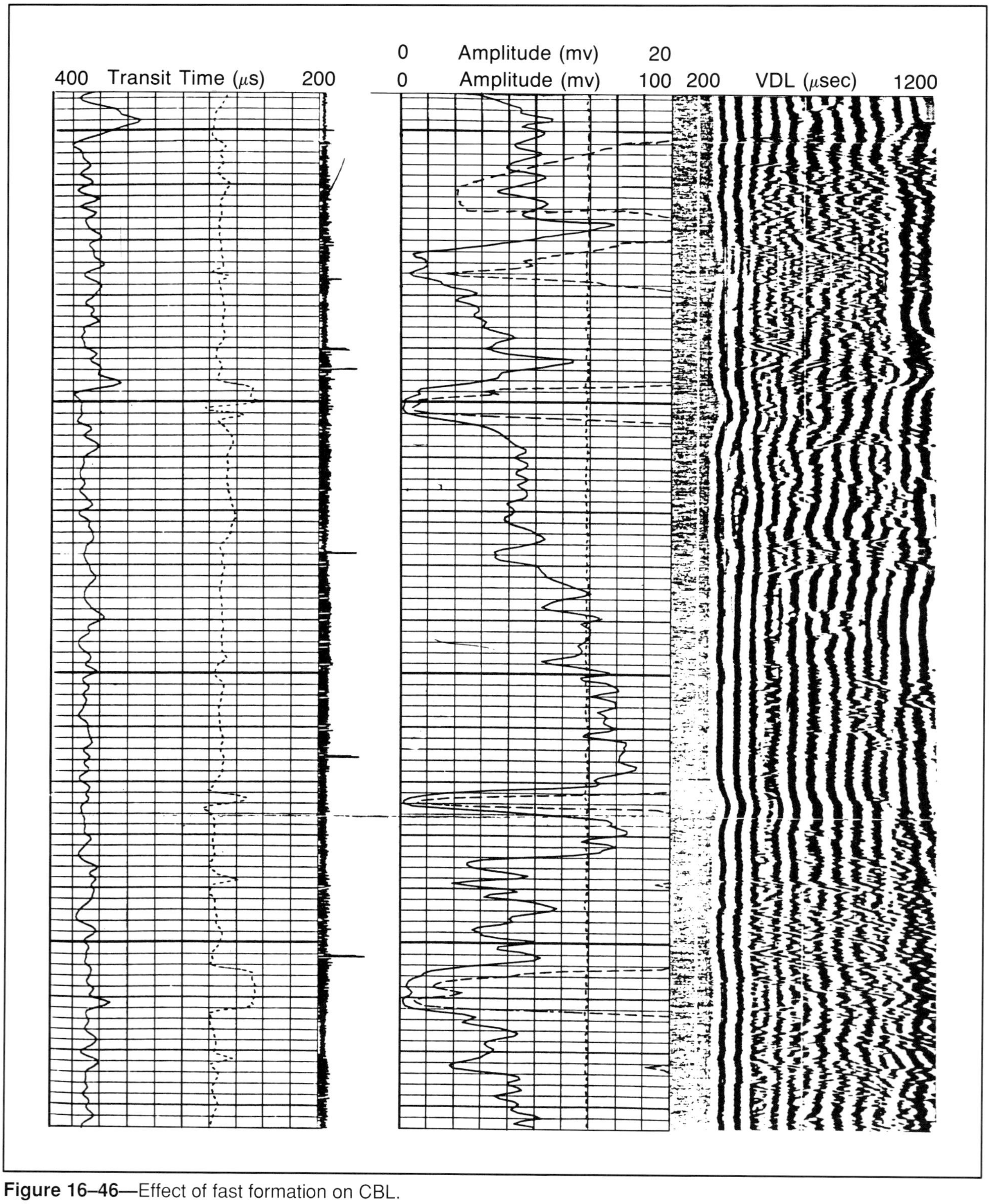

Figure 16–46—Effect of fast formation on CBL.

Allen, S.L. and Wood, M.W.: "Cement Bond Log Quality Control Through Simultaneous Recording of Fixed Gate and Sliding Gate Amplitudes With Transit Times," *Proc.*, SPWLA 26th Annual Logging Symposium (June 1985).

Anderson, W.L. and Walker, T.: "Research Predicts Improved Cement Bond Evaluations with Acoustic Logs," *JPT* (Nov. 1961).

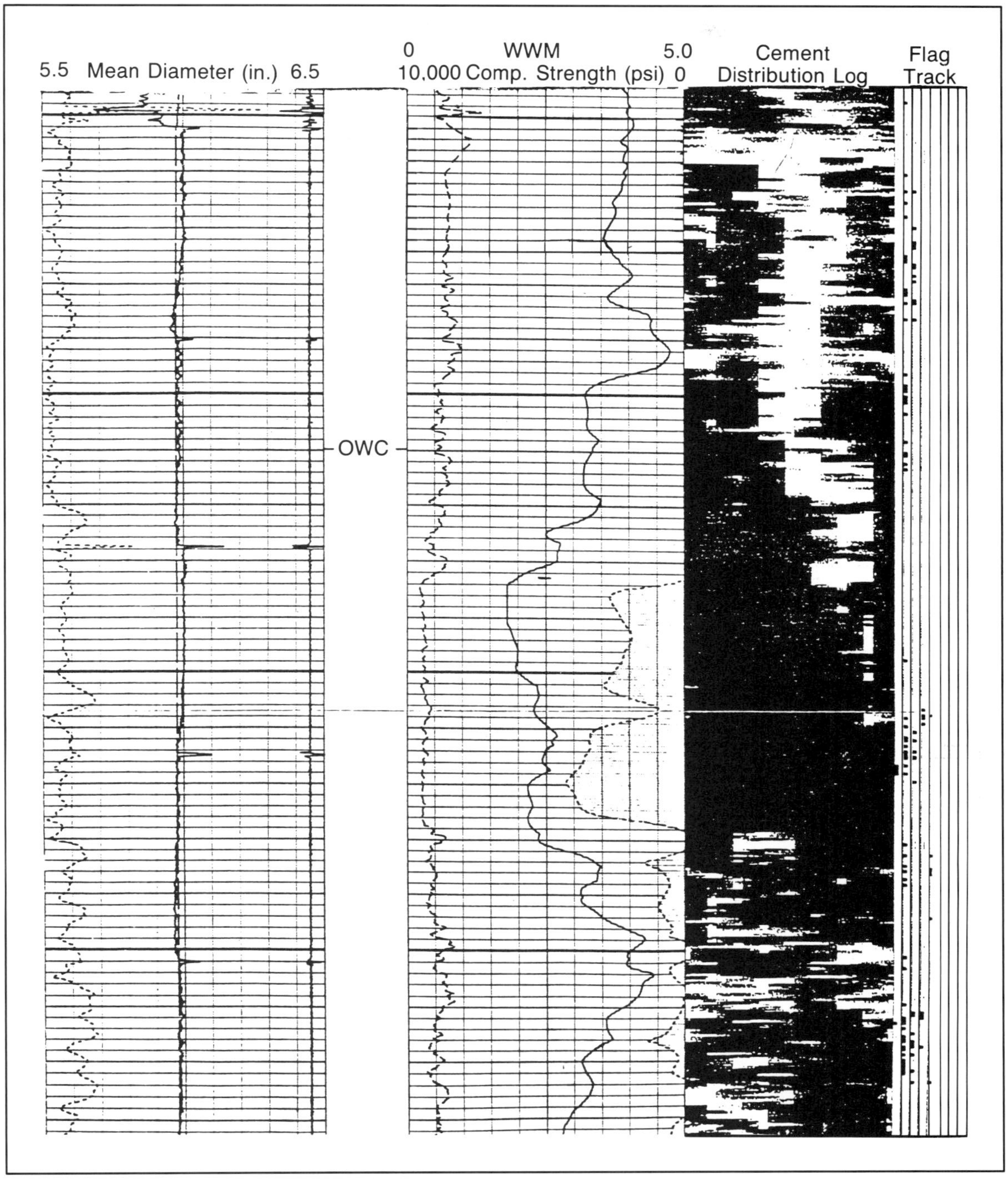

Figure 16–47—Effect of fast formation on CEL.

Anderson, W. L. and Riddle, G. A.: "Acoustic Amplitude Ratio Logging," *JPT* (Nov. 1964).

Bade, J. F.: "Cement Bond Logging Techniques—How They Compare and Some Variables Affecting Interpretation," *JPT* (Jan. 1963).

Bigelow, E. L.: " A Practical Approach to the Interpretation of Cement Bond Logs," *JPT* (July 1985).

Brown, H. D., Grijalva, V. E., and Raymer, L. L.: "New Developments in Sonic Wave Train Display and Analysis in Cased Holes," *Proc.*, SPWLA 11th Annual Logging Symposium (1970).

Bruckdorfer, R. A., Jacobs, W. R., and Masson, J. P.: "CBL Evaluation of Foam-Cemented and Synthetic-Cemented Casings," paper SPE 11980, 1983.

Carter, L. G., and Evans, G. W.: "A Study of Cement-Pipe Bonding," *JPT* (Feb. 1964).

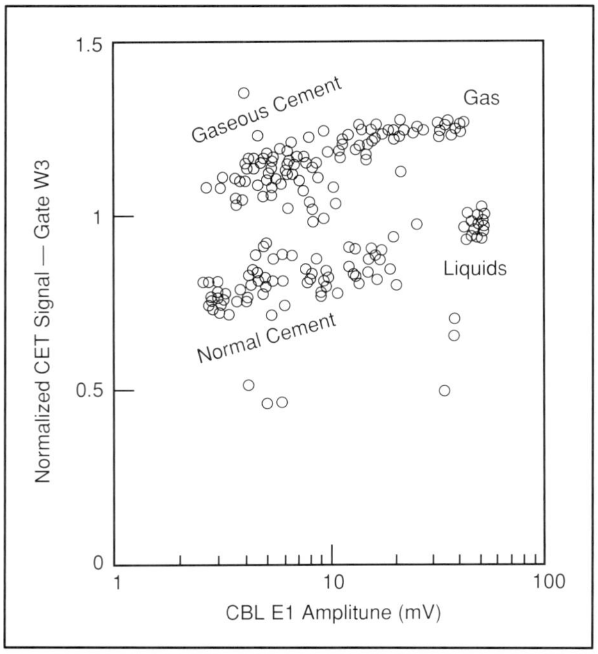

Figure 16–48—Comparison of CET/CBT response.

Catala, G. N., Stowe, I. D., and Henry, D. J.: "A Combination of Acoustic Measurements to Evaluate Cementations," paper SPE 16139, 1984.

Chang, S.: "Waves, Elasticity, and Velocities," *Schlum. Tech. Rev.* (1985) **33,** No. 1.

Dumont, A., Patin, J. B., and Le Floch, G.: "A Single Tool for Corrosion and Cement Evaluation," paper SPE 13140, 1984.

Fertl, W. H., Pilkington, P. E., and Scott, J. B.: "A Look at Cement Bond Logs," *JPT* (June 1974).

Fitzgerald, D. D., McGhee, B. F., and McGuire, J. A.: "Guidelines for 90% Accuracy in Zone Isolation Decision," paper SPE 12141, 1983.

Folmar, L. W.: "Methods of Detecting Top of Cement Behind Casing," *Oil-Well Cementing Practices in the United States*, API, Dallas (1959).

Froelich, B., Pittman, D., and Seeman, B. : "Cement Evaluation Tool—A New Approach to Cement Evaluation," paper SPE 10207, 1981.

Froelich, B., Dumont, A., Pittman, B., Seeman, D., and Havira, M.: "Ultrasonic Imaging of Material Mechanical Properties Through Steel Pipes," *Proc.*, IEEE Ultrasonic Symposium (Oct. 1981).

Gollwitzer, L. H. and Masson, J. P.: "The Cement Bond Tool," *Proc.*, SPWLA 23rd Annual Logging Symposium (July 1982).

Grosmangin, M., Kotesh, F. P., and Majani, P. "A Sonic Method for Analyzing the Quality of Cementation of Borehole Casings," *JPT* (Feb. 1961).

Havira, M.: "Ultrasonic Bond Evaluation in Multilayered Media" *J. Acoust. Soc. Amer.* (Fall 1979) **66**, Suppl. 1.

Havira, M.: "Ultrasonic Cement Bond Evaluation," *Trans.*, SPWLA (July 1982).

Jutten, J.: "Studies with Narrow Cement Thicknesses Lead to Improved CBL in Concentric Casings," paper SPE 18028, 1988.

Jutten, J. and Parcevaux:, P.: "Relationship Between Cement Bond Log Output and Borehole Geometrical Parameters," paper SPE/IADC 16139, 1987.

Jutten, J., Parcevaux, P., and Guillot, D.: "Relationship Between Cement Slurry Composition, Mechanical Properties and Cement Bond Log Output," paper SPE 16652, 1987.

Kilne, W. E. and Smith, W. E.: "Evaluation of Cementing Practices by Quantitative Radiotracer Measurements," paper SPE/IADC 14778, 1986.

Leigh, C. A., Finlayson, C. G., van der Kolk, C., and Staal, J.: "Results of Field Testing the Cement Evaluation Tool," *Proc.*, SPWLA 25th Annual Logging Symposium (June 1984).

Leslie, H. D., de Selliers, J., and Pittman, D. J.: "Coupling and Attenuation: A New Pair in Cement Bond Logging," paper SPE 16207, 1987.

McGhee, B. F. and Vacca, H. L.: "Guidelines for Improved Monitoring of Cementing Operations," *Proc.*, SPWLA 21st Annual Logging Symposium (July 1980).

McKinley, R. M., Bower, F. M., and Rumble, R. C.: "The Structure and Interpretation of Noise from Flow Behind Cemented Casing," paper SPE 3999, 1973.

Nayfeh, T. H., Wheelis, W. B. Jr., and Leslie, H. D.: "The Fluid-Compensated Cement Bond Log," paper SPE 13044, 1984.

Pardue, G. H. et al.: "Cement Bond Log—A Study of Cement and Casing Variables," *JPT* (May 1963) 545–554.

Pickett, G. R.: "Acoustic Character Logs and Their Applications in Formation Evaluation," *JPT* (June 1963).

Pilkington, P. E.: "Pressure Needed to Reduce Microannulus Effect on CBL," *Oil & Gas J.* (May 1988).

Suman G. O. Jr., and Ellis, R. C.: *World Oil's Cementing Handbook*, Petroleum Publishing Co., Houston (1977).

Tixier, M. P., Alger, R. P., and Doh, C. A.: "Sonic Logging," *Trans.*, AIME (1959).

Tubman, K. M., Cheng, C. H., Coles, S. P., and Toksoz, M. N.: "Synthetic Full-Waveform Acoustic Logs in Cased Boreholes, II—Poorly Bonded Casing," *Geophysics* (1986) **151**, No. 4.

Winn, K. H., Anderson, T. O., and Carter, L. G.: "A Preliminary Study of Factors Influencing Cement Bond Logs," *JPT* (Jan. 1963).

Walker, T.: "A Full-Wave Display of Acoustic Signals in a Cased Hole," *JPT* (Aug. 1968).

Zemanek, J. and Caldwell, R. L.: "The Borehole Televiewer—A New Logging Concept for Fracture Location and Other Types of Borehole Inspection," *JPT* (June 1969) 762–774.

A

Digest of Rheological Equations

Dominique Guillot

Dowell Schlumberger

FLOW CALCULATIONS FOR ALL FLUIDS – COHERENT UNIT SYSTEM

	Pipe Flow	Annular Flow Slot Approximation
Shear stress at the wall	$\tau_W = \frac{D}{4}\frac{dp}{dz}$	$\tau_W = \frac{(D_o - D_i)}{4}\frac{dp}{dz}$
Fanning friction factor defined such that: where:	$fr = \frac{\frac{D}{4} \times \frac{dp}{dz}}{\frac{1}{2}\rho V^2}$ $V = \frac{4 \times Q}{\pi D^2}$	$fr = \frac{\frac{(D_o - D_i)}{4} \times \frac{dp}{dz}}{\frac{1}{2}\rho V^2}$ $V = \frac{4 \times Q}{\pi (D_o^2 - D_i^2)}$
Reynolds number defined such as that in laminar flow:	$fr = \frac{16}{Re_{MR}}$	$fr = \frac{24}{Re_{AN}}$

Table A–1—Definition of the main parameters.

	Pipe Flow	Annular Flow Slot Approximation
Newtonian shear rate at the wall	$\dot{\gamma}_{NW} = \frac{8V}{D}$	$\dot{\gamma}_{NW} = \frac{12V}{D_o - D_i}$
Shear rate at the wall	$\dot{\gamma}_W = \frac{3n' + 1}{4n'} \dot{\gamma}_{NW}$	$\dot{\gamma}_W = \frac{2n' + 1}{3n'} \dot{\gamma}_{NW}$
where:	$n' = \frac{d \log (\tau_w)}{d \log (\dot{\gamma}_{NW})}$	$n' = \frac{d \log (\tau_w)}{d \log (\dot{\gamma}_{NW})}$
Velocity profile	$v(x) = \frac{D}{2\tau_w} \int_{\tau_x}^{\tau_w} \dot{\gamma}(\tau)\, d\tau$	$v(x) = \frac{D_o - D_i}{4\tau_w} \int_{\tau_x}^{\tau_w} \dot{\gamma}(\tau)\, d\tau$
Newtonian shear rate shear stress relationship	$\dot{\gamma}_{NW}(\tau_w) = \frac{4}{\tau_w^3} \int_0^{\tau_w} \dot{\gamma}(\tau)\, \tau^2 d\tau$	$\dot{\gamma}_{NW}(\tau_w) = \frac{3}{\tau_w^2} \int_0^{\tau_w} \dot{\gamma}(\tau)\, \tau\, d\tau$

Table A–2—Main equations for laminar flow.

	Pipe Flow	**Annular Flow Slot Approximation**
Reynolds Number	$Re = \frac{\rho V D}{\eta}$	$Re_{AN} = \frac{\rho V (D_o - D_i)}{\eta}$
Laminar flow		
Reynolds number value for end of laminar flow	$Re_1 = 2100$	$Re_1 = 2100$
Fanning friction factor	$fr = \frac{16}{Re}$	$fr = \frac{24}{Re_{AN}}$
Frictional pressure drop	$\frac{dp}{dz} = \frac{128 \eta Q}{\pi D^4}$	$\frac{dp}{dz} = \frac{192 \eta Q}{\pi (D_o + D_i)(D_o - D_i)^3}$
Normalized velocity profile	$\frac{v(x)}{V} = 2 \times (1 - x^2)$	$\frac{v(x)}{V} = 1.5 \times (1 - x^2)$
Turbulent flow		
Critical Reynolds number for turbulent flow	$Re_2 = 3000$	$Re_2 = 3000$
Critical flow rate for turbulent flow	$Q_C = 2356 \frac{\eta D}{\rho}$	$Q_c = 2356 \frac{\eta (D_o + D_i)}{\rho}$
Fanning friction factor	$\frac{1}{\sqrt{fr}} = 4.0 \log\left[Re \sqrt{fr}\right] - 0.4$	$\frac{1}{\sqrt{fr}} = 4.0 \log\left[\frac{2}{3} Re_{AN} \sqrt{fr}\right] - 0.4$
Power law approximation for $Re_2 < Re < 10^5$	Blasius approximation $fr = 0.0792 \times Re^{-0.25}$	$fr = 0.0992 \times Re_{AN}^{-0.26}$
	$\frac{dp}{dz} = \frac{0.241 \rho^{0.75} \eta^{0.25} Q^{1.75}}{D^{4.75}}$	$\frac{dp}{dz} = \frac{0.302 \rho^{0.74} \eta^{0.26} Q^{1.74}}{(D_o + D_i)^{1.74} (D_o - D_i)^{3.0}}$

Table A–3—Newtonian fluids.

	Pipe Flow	Annular Flow Slot Approximation
Reynolds Number	$Re_{MR} = \dfrac{\rho V^{2-n} D^n}{8^{n-1}\left[(3n+1)/4n\right]^n \times k}$	$Re_{AN} = \dfrac{\rho V^{2-n}(D_o - D_i)^n}{12^{n-1}\left[(2n+1)/3n\right]^n \times k}$
Laminar flow		
Reynolds number value for end of laminar flow	$Re_1 = 3150 - 1150n$	$Re_1 = 3150 - 1150n$
Fanning friction factor	$fr = \dfrac{16}{Re_{MR}}$	$fr = \dfrac{24}{Re_{AN}}$
Frictional pressure drop	$\dfrac{dp}{dz} = \dfrac{2^{3n+2}k}{D^{3n+1}}\left[\dfrac{(3n+1)Q}{\pi n}\right]^n$	$\dfrac{dp}{dz} = \dfrac{2^{2n+2}k}{(D_o - D_i)^{2n+1}}\left[\dfrac{4(2n+1)Q}{\pi n(D_o + D_i)}\right]^n$
Normalized velocity profile	$\dfrac{v(x)}{V} = \dfrac{3n+1}{n+1} \times \left(1 - x^{1+1/n}\right)$	$\dfrac{v(x)}{V} = \dfrac{2n+1}{n+1} \times \left(1 - x^{1+1/n}\right)$
Turbulent flow		
Critical Reynolds number for turbulent flow	$Re_2 = 4150 - 1150n$	$Re_2 = 4150 - 1150n$
Critical flow rate for turbulent flow	$Q_c = \dfrac{\pi}{4}(4150 - 1150n)^{1/(2-n)} \times \left[8^{n-1}\left(\dfrac{3n+1}{4n}\right)^n \dfrac{k}{\rho} D^{4-3n}\right]^{1/(2-n)}$	$Q_c = \dfrac{\pi(D_o + D_i)}{4}(4150 - 1150n)^{1/(2-n)} \times \left[12^{n-1}\left(\dfrac{2n+1}{3n}\right)^n \dfrac{k}{\rho}(D_o - D_i)^{2(1-n)}\right]^{1/(2-n)}$
Fanning friction factor	$\dfrac{1}{\sqrt{fr}} = \dfrac{4.0}{n^{0.75}} \log\left[Re_{MR} fr^{1-n/2}\right] - \dfrac{0.4}{n^{1.2}}$	$\dfrac{1}{\sqrt{fr}} = \dfrac{4.0}{n^{0.75}} \log\left[\dfrac{2}{3} Re_{AN} fr^{1-n/2}\right] - \dfrac{0.4}{n^{1.2}}$
Power law approximation for $Re_2 < Re_{MR} < 10^5$ and $Re_2 < Re_{AN} < 10^5$	$fr = a(n) \times Re_{MR}^{b(n)}$	$fr = a(n) \times \left[\dfrac{2}{3} Re_{AN}\right]^{b(n)}$
where:	$a(n) = 0.0792 + 0.0207 \log(n)$ $b(n) = -0.251 + 0.141 \log(n)$	$a(n) = 0.0893 + 0.0246 \log(n)$ $b(n) = -0.263 + 0.138 \log(n)$
Frictional pressure drop	$\dfrac{dp}{dz} = 2a(n)\left[\dfrac{4Q}{\pi}\right]^{2+(2-n)b(n)} \times \dfrac{\rho^{1+b(n)}}{\left[8^{n-1}\left(\dfrac{3n+1}{4n}\right)^n k\right]^{b(n)} D^{5+4b(n)-3nb(n)}}$	$\dfrac{dp}{dz} = 2a(n)\left[\dfrac{4Q}{\pi(D_o + D_i)}\right]^{2+(2-n)b(n)} \times \dfrac{\rho^{1+b(n)}}{\left[12^{n-1}\left(\dfrac{2n+1}{3n}\right)^n k\right]^{b(n)} (D_o - D_i)^{3+2(1-n)b(n)}}$

Table A–4—Power law fluids.

	Pipe Flow	Annular Flow Slot Approximation
Bingham Reynolds number	$Re_{Bg} = \frac{\rho V D}{\mu_p}$	$Re_{Bg} = \frac{\rho V (D_o - D_i)}{\mu_p}$
Hedstrom number	$He = \frac{\rho D^2 \tau_y}{\mu_p^2}$	$He = \frac{\rho (D_o - D_i)^2 \tau_y}{\mu_p^2}$
Metzner and Reed Reynolds number where:	$Re_{MR} = \frac{\rho V^{2-n'} D^{n'}}{8^{n'-1} k'}$ $n' = \frac{d\log(\tau_w)}{d\log(\dot{\gamma}_{NW})}$ $k' = \tau_w \dot{\gamma}_{NW}^{-n'}$	$Re_{AN} = \frac{\rho V^{2-n'} (D_o - D_i)^{n'}}{12^{n'-1} k'}$ $n' = \frac{d\log(\tau_w)}{d\log(\dot{\gamma}_{NW})}$ $k' = \tau_w \dot{\gamma}_{NW}^{-n'}$
Laminar flow		
Expressions for *Re_{AN} $n'k'$, dp/dz, v(x) V* as a function of *Q* are:	implicit	implicit
Local power law	$n' = \frac{(1-\psi)\left(1 + \frac{2}{3}\psi + \frac{1}{3}\psi^2\right)}{(1+\psi)(1+\psi^2)}$	$n' = \frac{(1-\psi)(1+\psi/2)}{(1+\psi+\psi^2)}$
and consistency index	$k' = \left[\frac{\tau_y}{\psi}\right]^{1-n'} \left[\frac{\mu_p}{1 - \frac{4}{3}\psi + \frac{1}{3}\psi^4}\right]^{n'}$	$k' = \left[\frac{\tau_y}{\psi}\right]^{1-n'} \left[\frac{\mu_p}{\left(1 - \frac{3}{2}\psi + \frac{1}{2}\psi^3\right)}\right]^{n'}$
Reynolds number value for end of laminar flow	$Re_1 = 3150 - 1150n'$	$Re_1 = 3150 - 1150n'$
Fanning friction factor	$fr = \frac{16}{Re_{MR}}$	$fr = \frac{24}{Re_{AN}}$
Frictional pressure drop can be obtained from:	$\dot{\gamma}_{NW} = \frac{\tau_y}{\mu_p}\frac{1}{\psi}\left[1 - \frac{4}{3}\psi + \frac{1}{3}\psi^4\right]$	$\dot{\gamma}_{NW} = \frac{\tau_y}{\mu_p}\frac{1}{\psi}\left[1 - \frac{3}{2}\psi + \frac{1}{2}\psi^3\right]$
or *fr (Re_{Bg}, He)*	$fr = \frac{16}{Re_{Bg}} + \frac{8He}{3Re_{Bg}^2} - \frac{16He^4}{3fr^3 Re_{Bg}^8}$	$fr = \frac{24}{Re_{Bg}} + \frac{3He}{Re_{Bg}^2} - \frac{4He^3}{fr^2 Re_{Bg}^6}$
Normalized velocity profile When: $x > \xi$ or $\tau > \tau_y$	$\frac{v(x)}{V} = 2\left[\frac{(1-\psi)^2 - (x-\psi)^2}{1 - \frac{4}{3}\psi + \frac{1}{3}\psi^4}\right]$	$\frac{v(x)}{V} = 1.5\left[\frac{(1-\psi)^2 - (x-\psi)^2}{1 - \frac{3}{2}\psi + \frac{1}{2}\psi^3}\right]$
When: $x < \xi$ or $\tau < \tau_y$	$\frac{v(x)}{V} = \frac{2}{1 + \frac{2}{3}\psi + \frac{1}{3}\psi^2}$	$\frac{v(x)}{V} = \frac{1.5}{1 + \frac{\psi}{2}}$
where the dimensionless shear stress is	$\psi = \frac{\tau_y}{\tau_w}$	$\psi = \frac{\tau_y}{\tau_w}$

Table A–5—Bingham plastic fluids.

	Pipe Flow	Annular Flow Slot Approximation
Turbulent flow		
Expressions for Re_{MR}, n', k', dp/dz, as a function of Q are:	implicit	implicit
For frictional pressure drop: Start with a guess value for dp/dz, Determine n' and k' Solve the (fr, Re_{MR}) or (fr, Re_{AN}) equations power law fluids with: Iterate till convergence is obtained	$n = n'$ $k = \left[\frac{4n'}{3n'+1}\right]^{n'} k'$	$n = n'$ $k = \left[\frac{3n'}{2n'+1}\right]^{n'} k'$

Table A–5, continued—Bingham plastic fluids.

	Pipe Flow	Annular Flow Slot Approximation
Metzner and Reed Reynolds Number	$Re_{MR} = \dfrac{\rho V D}{\mu_p \left[1 + (D\tau_y)/(6V\mu_p)\right]}$	$Re_{AN} = \dfrac{\rho V (D_o - D_i)}{\mu_p \left[1 + \dfrac{(D_o - D_i)\tau_y}{8V\mu_p}\right]}$
Laminar flow		
Reynolds number value for end of laminar flow	$3150 - 1150n'$	$3150 - 1150n'$
Local power law	$n' = 1 - \dfrac{1}{1 + \dfrac{6V\mu_p}{D\tau_y}}$	$n' = 1 - \dfrac{1}{1 + (8V\mu_p)/(D_o - D_i)\tau_y}$
and consistency index	$k' = \mu_p \left[\dfrac{8V}{D}\right]^{1-n'} + \dfrac{4\tau_y}{3}\left[\dfrac{8V}{D}\right]^{-n'}$	$k' = \mu_p \left[\dfrac{12V}{D_o - D_i}\right]^{1-n'} + \dfrac{3\tau_y}{2}\left[\dfrac{12V}{D_o - D_i}\right]^{-n'}$
Frictional pressure drop	$\dfrac{dp}{dz} = \dfrac{128\mu_p Q}{\pi D^4} + \dfrac{16\tau_y}{3D}$	$\dfrac{dp}{dz} = \dfrac{192\mu_p Q}{\pi(D_o + D_i)(D_o - D_i)^3} + \dfrac{6\tau_y}{(D_o - D_i)}$
Turbulent flow		
Expressions for Re_{MR}, n', k', dp/dz as a funstion of Q are:	implicit	implicit
Approximation for the critical rate for turbulent flow	$Q_c = \dfrac{1500\,\pi}{4} \dfrac{\mu_p D}{\rho} \times \left[1 + \sqrt{1 + 0.000307\,He}\right]$	$Q_c = \dfrac{1500\,\pi}{4} \dfrac{\mu_p (D_o + D_i)}{\rho} \times \left[1 + \sqrt{1 + 0.000231\,He}\right]$
For frictional pressure drop: Start with a guess value for *dp/dz* Determine n' and k' Solve the *(fr, Re_{MR}) or (fr, Re_{AN})* equation for power law fluids with: Iterate until convergence is obtained	$n = n'$ $k = \left[\dfrac{4n'}{3n' + 1}\right]^{n'} k'$	$n = n'$ $k = \left[\dfrac{3n'}{2n' + 1}\right]^{n'} k'$

Table A–6—Bingham plastic fluids with $\psi << 1$.

NOMENCLATURE

D	m	pipe inner diameter
D_0, D_i	m	pipe outer diameter, open hole diameter
dp/dz	Pa m^{-1}	frictional pressure drop
fr	—	Fanning friction factor
He	—	Hedstrom number
k	Pa.s	Consistency Index for a power law fluid
k'	Pa.s	local Consistency Index
n	—	Power Law Index for a power law fluid
n'	—	local Power Law Index
Q	m^3 s^{-1}	flow rate
Q_c	m^3 s^{-1}	critical flow rate for turbulent flow
Re	—	Reynolds number
Re_{AN}	—	annular generalized Reynolds number
Re_{BG}	—	Bingham Reynolds number
Re_{MR}	—	Metzner and Reed Reynolds number
Re_1	—	critical Reynolds number for end of laminar flow
Re_2	—	critical Reynolds number for beginning of turbulent flow
r	m	distance from pipe axis or from plane symmetry of the slot
v	m s^{-1}	velocity at a given distance from pipe axis or from plane of symmetry of the slot
V	m s^{-1}	average velocity
x	—	normalized distance from pipe axis $2r/D$ or from the plan of symmetry of the slot $2d/(D_0-D_i)$
$\dot{\gamma}$	s^{-1}	shear rate
$\dot{\gamma}_w$	s^{-1}	shear rate at the wall
$\dot{\gamma}_{NW}$	s^{-1}	Newtonian shear rate at the wall
μ	Pa s	plastic viscosity of a Bingham plastic fluid
ρ	kg m^{-3}	fluid density
τ	Pa	shear stress
τ_x	Pa	shear stress at a normalized distance x from pipe axis or from plane of symmetry of the slot
τ_w	Pa	shear stress at the wall
τ_y	Pa	yield stress of a Bingham plastic fluid
ψ	—	dimensionless shear stress

B

Laboratory Testing, Evaluation, and Analysis of Well Cements

David R. Bell and Erik B. Nelson

Dowell Schlumberger

B–1 INTRODUCTION

Laboratory testing of cements and cementing materials is an essential part of the entire cementing process. Testing begins at the manufacturing sites of the cement and additives to monitor product quality, and it continues through the slurry design stages at the pumping service company or operating company laboratories as a specific formulation is developed. Evaluation is frequently conducted on samples obtained from the bulk plant as the blend is prepared, and on samples taken from storage silos as the material is placed on location. Field blend samples of both the dry-blended cements and the wet slurry can be obtained during mixing for subsequent evaluation, either in the laboratory or on location using portable laboratory equipment. Laboratory examination of samples obtained from the field can sometimes be used as an aid in post treatment investigations.

In general, there are two types of laboratory testing of cements and cementing materials: performance evaluation and chemical characterization. The typical oilfield laboratory is engaged primarily in the performance evaluation of cements, through physical measurement of specific slurry properties under simulated downhole conditions. This type of evaluation is used mainly in the slurry design stages of a cementing treatment, and in the execution stages to monitor the preparation of the blended material. Chemical characterization typically involves quantitative or qualitative analysis of the slurry components prior to mixing, to ensure their suitability for use. Analytical techniques are used for quality-control purposes at the point of manufacture, to determine that components of a dry-blended cement system are present in the desired quantities, and are blended sufficiently at the bulk plant. Such techniques are also used to monitor the quality of the mix water on location. Correct application of a wide variety of laboratory testing methods is necessary to achieve a successful cementing treatment.

This appendix presents a broad overview of laboratory testing procedures and equipment, and is not proposed as a manual for cement laboratory workers. The reader must instead consult the official publications of organizations such as the American Petroleum Institute (API) or the American Society for Testing and Materials (ASTM) before commencing actual work.

B–2 SAMPLE PREPARATION

A meaningful laboratory evaluation of a material is not possible unless the testing is performed using a representative sample of that material. Statistical considerations regarding the choice of sample size may be found in the ASTM Standard C183. Sampling and handling procedures for sacked and bulk cement are described in Section 3 of API Specification 10 (Spec 10). The use of proper storage procedures is particularly important to avoid exposure of the cement to moisture and/or carbon dioxide in the air. Several studies regarding the sampling of blended cements have been reported (Pace et al., 1984; Cobb and Pace, 1985; Gerke et al., 1985; Kunze, 1986; Bell et al., 1988).

The best overall sampling device for blended cements is a diverted flow sampler (Fig. B–1), which permits sampling from a complete cross section of a flowing stream of material. Prior to testing in the laboratory, field blend samples may be split using a mechanical splitter as described in ASTM Specification C 702, because segregation of blended components may occur during shipping to the laboratory. Similarly, a mechanical splitter may be appropriate for use in obtaining laboratory samples from bulk or sack quantities of multicomponent additives in the field.

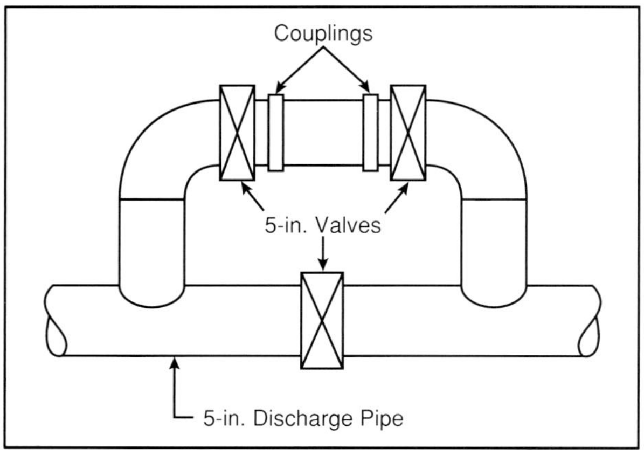

Figure B–1—Diverted flow sampler—top view.

B–3 PERFORMANCE EVALUATION OF CONVENTIONAL CEMENT SLURRIES

The API Committee on Standardization of Well Cements (Committee 10) acts as a governing body regarding the development of standardized test procedures for the performance evaluation of well cement slurries. The procedures are published by the American Petroleum Institute as API Spec 10. This publication contains specification tests for neat cement slurries, as well as operational testing procedures designed to encompass all conventional slurries. As of this writing, no API testing procedures exist for foamed cements. The procedures are designed to simulate downhole conditions for performance testing in a reasonably equipped field laboratory, and are based on a compromise between realistic wellbore conditions and the practical limitations of the laboratory environment.

B–3.1 Slurry Preparation

The equipment specification and operational procedures for the preparation of oil-well cement slurries in the laboratory are contained in Section 5 and Appendix A of API Spec 10. The mixing device is a two-speed, propeller-type mixer, shown in Fig. B–2. Specifications are given for the propeller speeds, mixer blade wear, batch size, and mixing time. Normally, 600 mL of slurry are prepared. The mixer is operated at 4,000 RPM for 15 seconds (during which all of the cement solids should be added to the mix water), followed by 35 seconds at 12,000 RPM. Cement slurries are very abrasive; as a result, careful monitoring of the mixer blade condition is essential.

Since this method was developed for neat cement slurries, specific procedures for the addition of additives (both liquid and solid) are not provided. Variations in mixing procedures can alter the resultant slurry properties significantly (Roy and Asaga, 1979); therefore, individual laboratories should establish and adhere to supplemental procedures covering items not specifically addressed by Spec 10. Variations in field mixing procedures can produce similar differences in slurry properties at the well site, and attempts have been made to define and improve the correlation between field mixing procedures and those used in the laboratory.

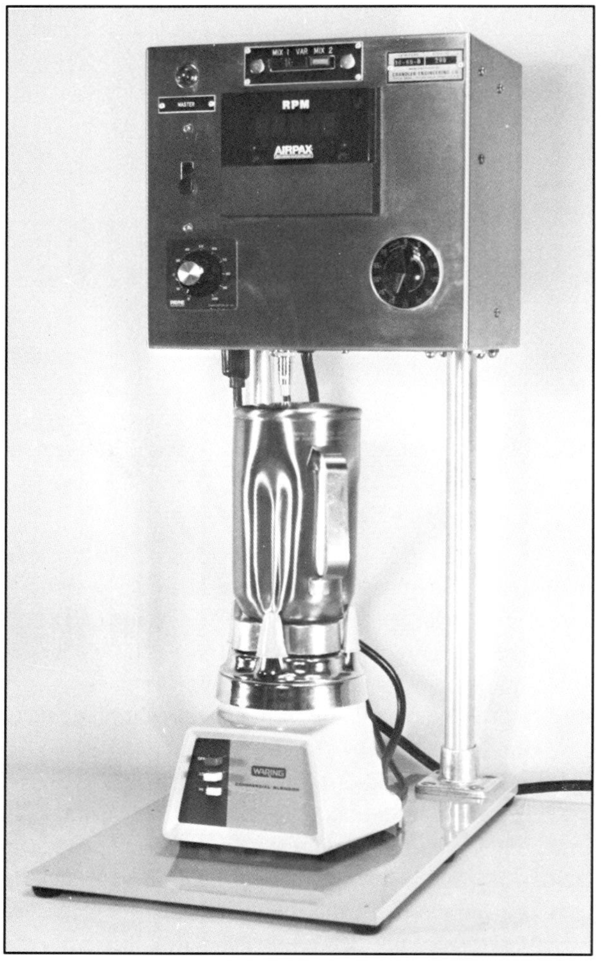

Figure B–2—Propeller-type mixing device commonly used to prepare well cement slurries (photo courtesy E G & G Chandler Engineering).

The slurry mixing procedure specified by the API is not suitable for the recently developed ultralow-density systems containing microspheres or nitrogen as an extender (Chapter 3). As of this writing, no standardized procedure for preparing such systems in the laboratory has been approved. Hollow microspheres are easily broken under high shear; consequently, the mixer must be operated at a low propeller speed. A kitchen-type mixer, which operates in the fashion of an egg beater, is also appropriate.

Foamed cements are routinely prepared in a propeller-type mixer. A base cement slurry containing surfactant is foamed at a high propeller speed until it completely fills the mixer bowl. The foamed cement density is varied by adjusting the amount of base slurry added to the mixer bowl. The disadvantage of this procedure is that the foam is not prepared under simulated high-pressure field conditions. However, some pressurized testing methods were recently described by de Rozières and Ferrière (1990).

B–3.2 Thickening Time

Thickening time tests are designed to determine the length of time which a cement slurry remains in a pumpable fluid state under simulated wellbore conditions of temperature and pressure. The specifications and operational procedures for determining slurry thickening time are contained in Section 8 and Appendix E of Spec 10.

The test slurry is evaluated in a pressurized consistometer, shown in Fig. B–3, which measures the consistency of the test slurry contained in a rotating cup while under simulated wellbore conditions. Most apparatuses are capable of exposing cement slurries to a maximum temperature and pressure of 400°F and 25,000 psi (204°C and 175 MPa); however, special units capable of 600°F and 40,000 psi (315°C and 280 MPa) are available for deep-well applications. A smaller portable consistometer, using a rotating paddle in a stationary cup, has been introduced recently (Fig. B–4). Thickening time results are reportedly consistent with those obtained with the conventional apparatus, and the API is considering a modification of Spec 10 procedures to include this configuration. A nonpressurized, or "atmospheric," consistometer (Fig. B–5) can be used to obtain a thickening time for low-temperature cement systems; however, today it is most commonly used for conditioning of slurries (according to various API procedures) before rheology, fluid-loss, or free-water tests.

The pumpability or consistency of the slurry is measured in Bearden units (Bc), a dimensionless quantity with no direct conversion factor to more common units of viscosity such as the poise. The end of a thickening time test is defined when the cement slurry reaches a consistency of 100 Bc; however, 70 Bc is generally considered to be the maximum pumpable consistency. Figure B–6 shows the output from a typical thickening time test. No provision for slurry fluid loss is made in the design of a consistometer slurry cup; therefore, the thickening time for a slurry in the wellbore may be different from that for the same slurry in the laboratory, particularly if the design specifies little or no fluid-loss control.

Appendix E of API Spec 10 contains schedules which specify the rate at which temperature and pressure are increased, as well as the final temperature and pressure, during the thickening time test. Such schedules have been derived from field data on wells with different depths and temperature gradients, and are designed to more accurately simulate the conditions to which the cement slurry would be exposed in a well. Schedules are given for casing, liner, and squeeze cementing treatments. A typical example is shown in Fig. B–7. Appendix K of API Spec 10 contains special methods for testing the thickening time of cement systems to be placed in an Arctic environment.

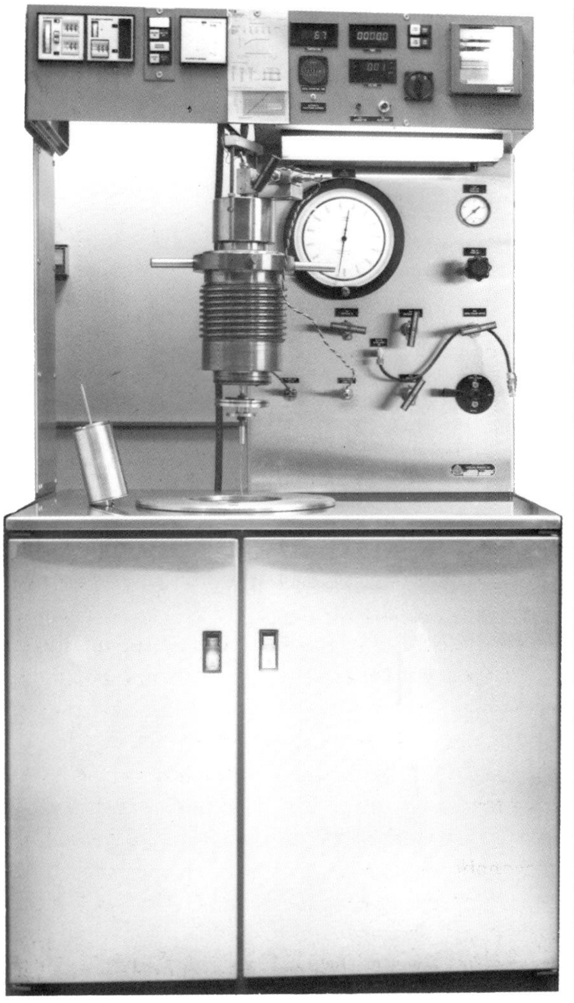

Figure B–3—Pressurized consistometer.

During the job execution, a flowing cement slurry is exposed to continuously changing pressure and temperature; consequently, measurement of the circulating temperature and pressure profile in the wellbore is very

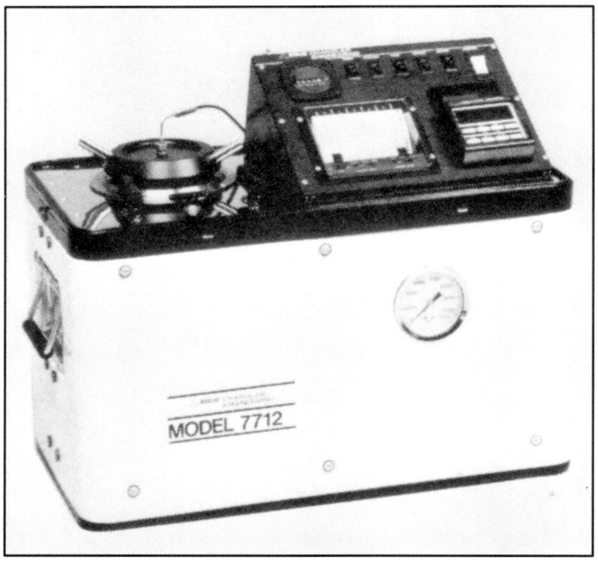

Figure B–4—Portable consistometer.

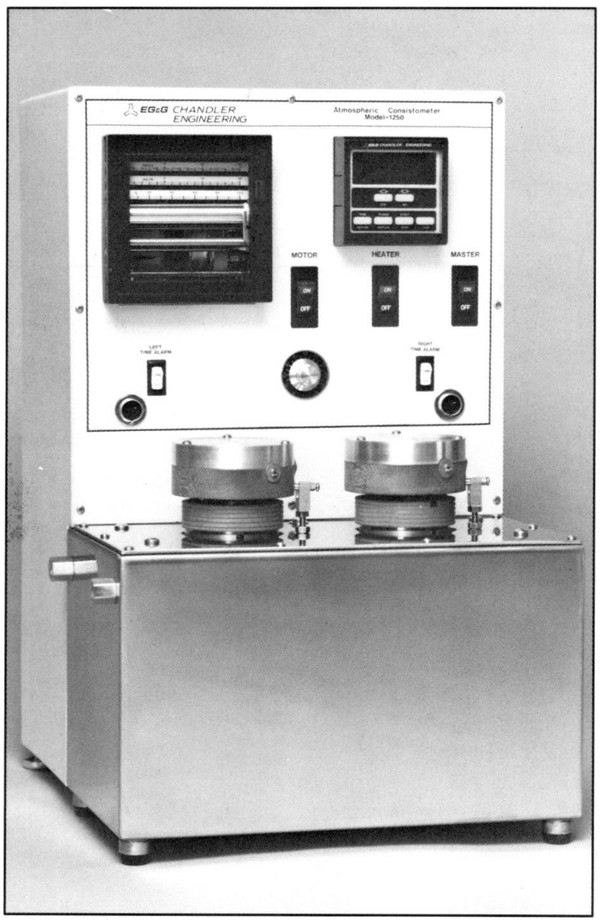

Figure B–5—Atmospheric consistometer (photo courtesy E G & G Chandler Engineering).

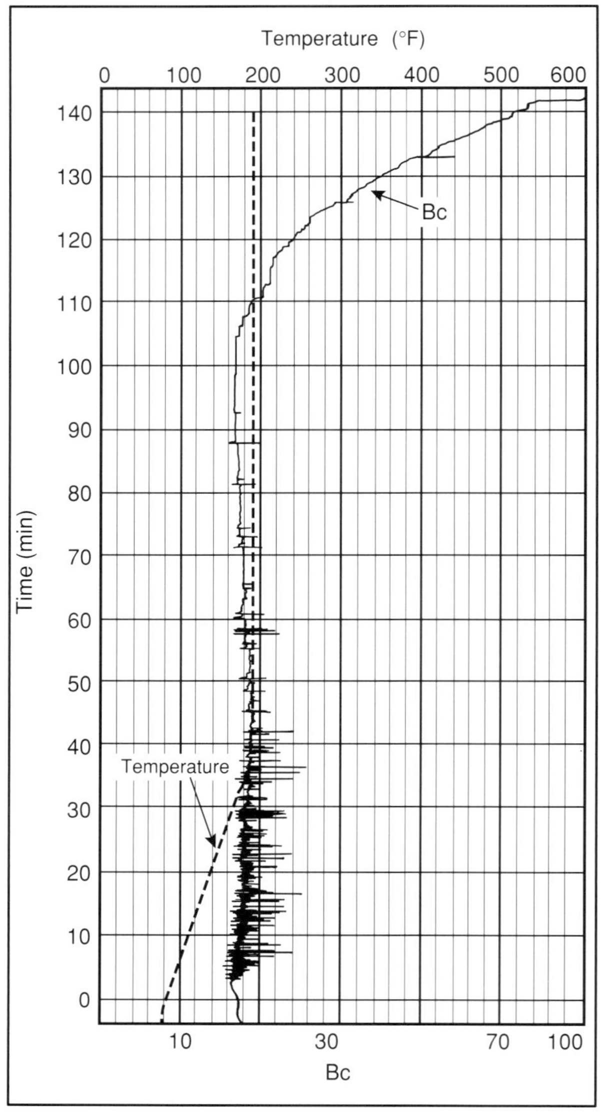

Figure B–6—Typical thickening time test output.

difficult. The highest temperature and highest pressure to which a slurry is exposed may not occur at the same point in the wellbore; consequently, they may not occur at the same time. Also, a slurry placed near the top of a long cement column almost certainly will be exposed to higher temperatures and pressures during circulation at deeper depths. At present, none of these factors is considered in the official API schedules.

Temperature and pressure can have a pronounced effect on measured thickening time. Although simulating the dynamic wellbore environment in the laboratory is difficult, refinements in procedures and improvements in equipment are helping make possible more realistic labo-

1	2	3	4	5	6	7	8
		Temperature Gradient, °F/100 ft depth (°C/100 m depth)					
		Temperature, °F (°C)					
	Pressure,	0.9 (1.6)	1.1 (2.0)	1.3 (2.4)	1.5 (2.7)	1.7 (3.1)	1.9 (3.5)
Time (min)	psi (kPa)	°F (°C)	°F (°C)	°F (°C)	°F (°C)	°F (°C)	°F (°C)
SCHEDULE 5g							
Depth: 8000 ft (2440 m)					**Mud Density: 10 lb/gal (1.2 kg/L)**		
0	1000 (6900)	80 (27)	80 (27)	80 (27)	80 (27)	80 (27)	80 (27)
2	1300 (9000)	83 (28)	83 (28)	83 (28)	84 (29)	85 (29)	86 (30)
4	1600 (11000)	86 (30)	87 (31)	87 (31)	89 (32)	90 (32)	92 (33)
6	1900 (13100)	90 (32)	90 (32)	91 (33)	93 (34)	94 (34)	97 (36)
8	2200 (15200)	93 (34)	94 (34)	95 (35)	97 (36)	99 (37)	103 (39)
10	2500 (17200)	96 (36)	97 (36)	99 (37)	101 (38)	104 (40)	109 (43)
12	2800 (19300)	100 (38)	101 (38)	103 (39)	106 (41)	108 (42)	115 (46)
14	3100 (21400)	103 (39)	104 (40)	107 (42)	110 (43)	113 (45)	120 (49)
16	3400 (23400)	106 (41)	108 (42)	111 (44)	114 (46)	118 (48)	126 (52)
18	3700 (25500)	110 (43)	111 (44)	115 (46)	119 (48)	122 (50)	132 (56)
20	4000 (27600)	113 (45)	115 (46)	119 (48)	123 (51)	127 (53)	138 (59)
22	4300 (29600)	116 (47)	118 (48)	123 (51)	127 (53)	132 (56)	143 (62)
24	4600 (31700)	120 (49)	122 (50)	127 (53)	131 (55)	136 (58)	149 (65)
26	4900 (33800)	123 (51)	125 (52)	131 (55)	136 (58)	141 (61)	155 (68)
28	5200 (35900)	126 (52)	129 (54)	135 (57)	140 (60)	146 (63)	160 (71)
Heating Rate, °F/min (°C/min)		1.64 (0.91)	1.75 (0.97)	1.96 (1.09)	2.14 (1.19)	2.36 (1.31)	2.86 (1.59)
SCHEDULE 6g							
Depth: 10,000 ft (3050 m)					**Mud Density: 12 lb/gal (1.4 kg/L)**		
0	1250 (8600)	80 (27)	80 (27)	80 (27)	80 (27)	80 (27)	80 (27)
2	1600 (11000)	83 (28)	84 (29)	84 (29)	85 (29)	85 (29)	86 (30)
4	1900 (13100)	87 (31)	87 (31)	89 (32)	90 (32)	91 (33)	93 (34)
6	2300 (15900)	90 (32)	91 (33)	93 (34)	95 (35)	96 (36)	100 (38)
8	2600 (17900)	94 (34)	95 (35)	98 (37)	99 (37)	102 (39)	107 (42)
10	3000 (20700)	97 (36)	98 (37)	102 (39)	104 (40)	107 (42)	113 (45)
12	3300 (22800)	101 (38)	102 (39)	107 (42)	109 (43)	113 (45)	120 (49)
14	3700 (25500)	104 (40)	106 (41)	111 (44)	114 (46)	118 (48)	127 (53)
16	4000 (27600)	108 (42)	109 (43)	116 (47)	119 (48)	124 (51)	134 (57)
18	4400 (30300)	111 (44)	113 (45)	120 (49)	124 (51)	129 (54)	140 (60)
20	4700 (32400)	115 (46)	117 (47)	124 (51)	129 (54)	135 (57)	147 (64)
22	5100 (35200)	118 (48)	120 (49)	129 (54)	133 (56)	140 (60)	154 (68)
24	5400 (37200)	122 (50)	124 (51)	133 (56)	138 (59)	146 (63)	161 (72)
26	5700 (39300)	125 (52)	128 (53)	138 (59)	143 (62)	151 (66)	167 (75)
28	6100 (42100)	129 (54)	131 (55)	142 (61)	148 (64)	157 (69)	174 (79)
30	6400 (44100)	132 (56)	135 (57)	146 (63)	153 (67)	162 (72)	180 (82)
32	6800 (46900)	135 (57)	139 (59)	150 (66)	158 (70)	168 (76)	187 (86)
34	7100 (49000)	138 (59)	142 (61)	154 (68)	163 (73)	174 (79)	193 (89)
36	7500 (51700)	141 (61)	146 (63)	158 (70)	167 (75)	180 (82)	200 (93)
Heating Rate, °F/min (°C/min)		1.69 (0.94)	1.83 (1.00)	2.17 (1.21)	2.42 (1.34)	2.78 (1.54)	3.33 (1.85)

Figure B–7—Example of API thickening time schedule.

ratory evaluations of slurry performance. As a result, the API schedules are presently being revised, based on new industry surveys made in 1987–1988. New schedules will be derived from and published with formulas for calculation of appropriate test conditions for laboratory evaluation, including provisions for reducing test pressure and temperature to simulate slurry placement at the top of a long cement column. Computer simulators for the calculation of test parameters from well data are being developed, and new instrumentation is available which provides precise control of temperature and pressure during a thickening time test.

B–3.3 Fluid Loss

Fluid-loss tests are designed to measure the slurry dehydration during and immediately following the completion of the placement phase of a cementing treatment. Operational test procedures for the determination of the fluid-loss rate are contained in Appendix F of API Spec 10. After being subjected to simulated wellbore conditions in a consistometer, the test slurry is placed in a heated filter press cell, shown in Fig. B–8, and the filtrate loss at either 100 psi or 1,000 psi differential pressure is measured across a standard filtration medium (325 mesh screen supported on a 60 mesh screen). The duration of the test is 30 minutes, and the filtrate volume (F_{30}) is noted. If all of the filtrate passes through the screen in less than 30 minutes, the following equation is used to calculate a hypothetical F_{30}.

$$F_{30} = F_t \frac{5.477}{\sqrt{t}} \qquad \text{(B–1)}$$

F_t is equal to the volume of filtrate (mL) collected at time t (min).

The prescribed test evaluates slurry fluid loss under static conditions (immediately following placement). No provision is made in this procedure for the measurement of fluid loss during placement, although results of fluid-loss determinations under dynamic conditions have been reported (Bannister, 1978).

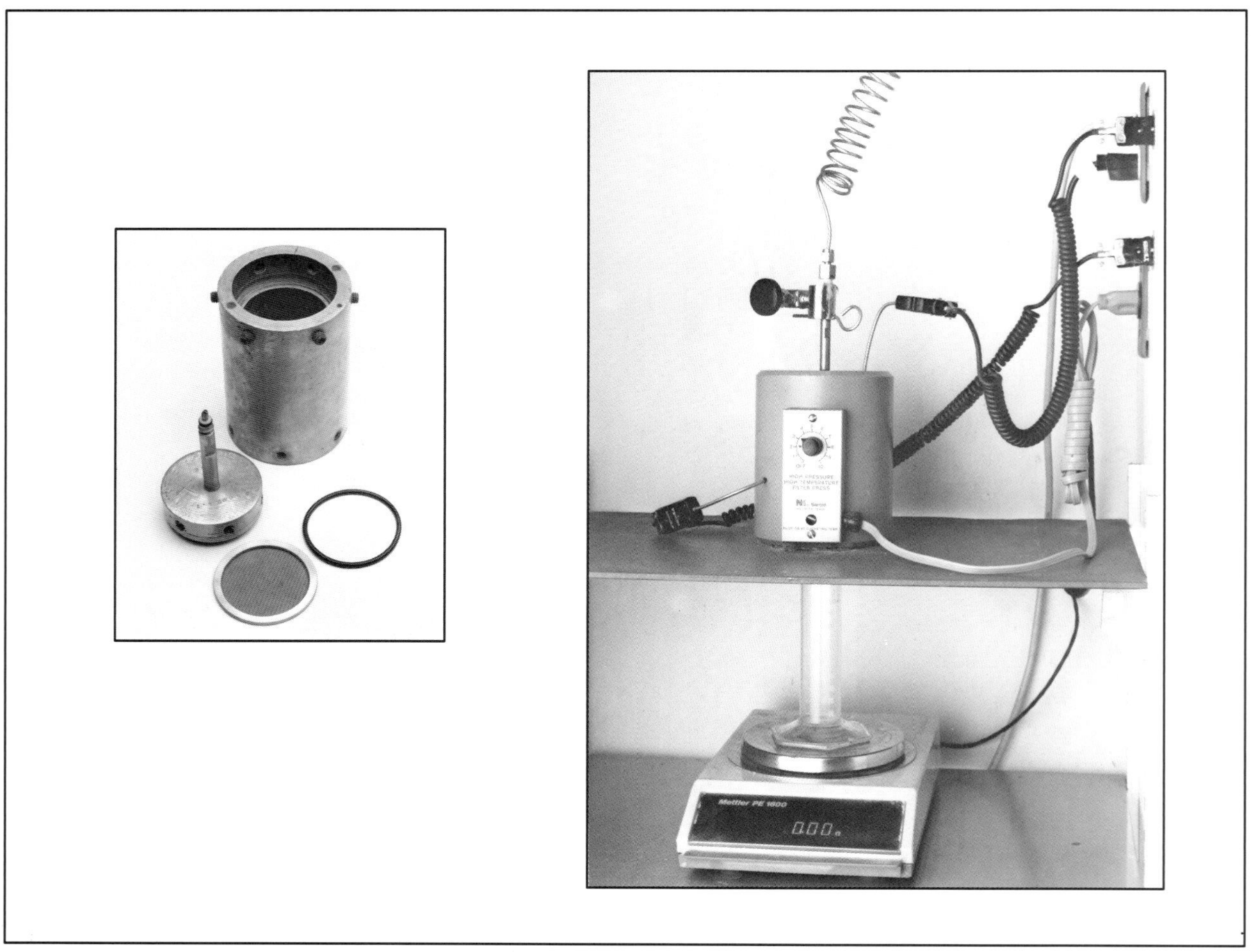

Figure B–8—Filter press cell and assembly for API fluid-loss tests.

Figure B–9—Filter press assembly for high-temperature fluid-loss tests.

The maximum temperature for which there is a specified API test procedure is 250°F (121°C). Higher circulating temperatures are routinely encountered in geothermal and deep oil and gas wells, and a special filter press assembly is available for fluid-loss testing at temperatures as high as 400°F (204°C) (Fig. B–9); however, for safety reasons, the differential pressure is limited to 500 psi.

A major logistical disadvantage of the present fluid-loss testing procedure is the necessity to transfer a heated slurry from the consistometer to the heated filter press cell. The process is awkward, and presents a safety hazard when high circulating temperatures are involved. To ameliorate this situation, several instrument manufacturers have recently developed stirring fluid-loss test equipment, which provides the capability of determining a slurry's fluid-loss rate without the necessity of transferring a heated slurry from one vessel to another (Fig. B–10). The API is presently working to revise the approved procedures, and to include the new equipment in the specifications.

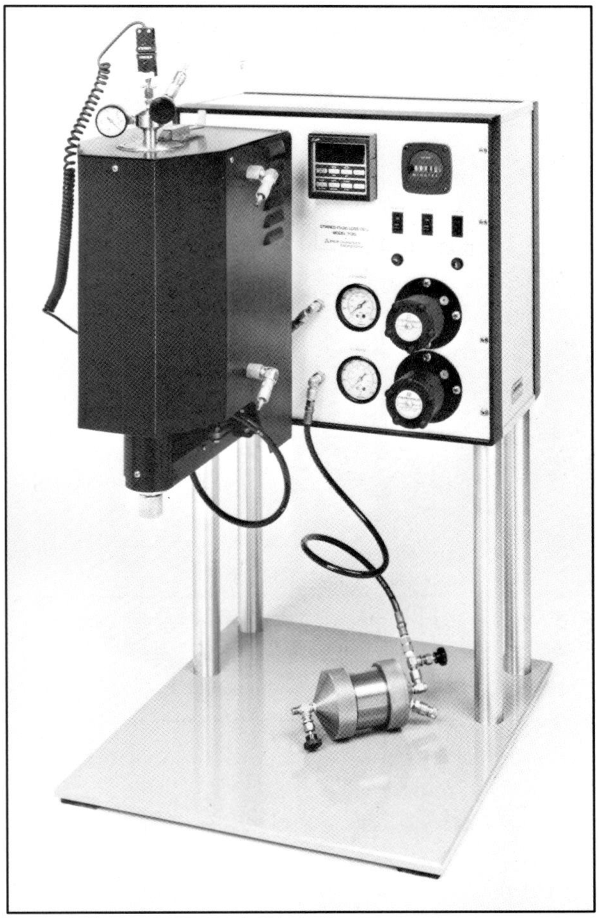

Figure B–10—Stirring fluid-loss testing apparatus (photo courtesy of E G & G Chandler Engineering).

B–3.4 Compressive Strength

The API specifications and procedures for the determination of compressive strength are described in Section 7 and Appendix D of API Spec 10. Test cement slurries are prepared according to the API mixing procedure, poured into two-inch cube molds, and cured for various time periods at specific temperatures and pressures. The set cement cubes are removed from the molds, and placed in a hydraulic press where increasing uniaxial pressure is exerted on each until failure. The compressive strength is then calculated by dividing the pressure at which failure occurred by the cross-sectional area of the specimen.

Figure B–11 shows a typical curing mold which makes two test specimens. API procedures are given for curing at pressures from atmospheric to 3,000 psi. For atmospheric tests, curing can be performed in a water bath, or a cooling bath to simulate cold weather or permafrost

Figure B–11—2-in. curing molds for compressive strength testing.

conditions. Pressurized curing chambers, such as the device shown in Fig. B–12, are available in various sizes and with varying performance capabilities. One can presently obtain units which hold up to 32 specimens, and with maximum operating conditions of 600°F (315°C) and 20,000 psi. Appendix D of API Spec 10 contains the prescribed heat-up and pressurization schedules which, like the thickening time schedules, are derived from field data and the anticipated wellbore temperature gradient. For Arctic cement systems, special curing methods are given in Appendix K. In addition, a method for evaluating a cement system's resistance to freeze/thaw cycling is given. When the specimens are placed in the hydraulic press for strength measurement (Fig. B–13), the rate of loading is regulated depending upon the anticipated strength of the specimen.

A relatively recent development is the estimation of compressive strength from the ultrasonic velocity (Rao et al., 1982). The Ultrasonic Cement Analyzer (UCA), shown in Fig. B–14 , measures the sonic travel time of ultrasonic energy through a cement sample as it cures under simulated wellbore conditions of temperature and pressure. The ultrasonic measurement is nondestructive, and may be made continuously as the cement sample cures at high pressure and elevated temperature. The ultrasonic velocity directly measures the bulk compressibility of the sample, but this is found to be well-correlated with compressive strength. The compressive strength estimate can be output directly via a preprogrammed microprocessor.

Compressive strength measurements are designed to furnish some indication of the ability of a set cement to

Figure B–12—Pressurized curing chamber for compressive-strength tests (photo courtesy E G & G Chandler Engineering).

provide zonal isolation, and to protect and support the pipe. However, the compressive strength values obtained using either the API crush test or the UCA are indicative of the integrity of the cement under uniaxial loading (no lateral restraint). In the wellbore, the cement is subject to complex triaxial loading, and the failure stresses may be substantially different from those observed in the standard compressive strength test (Neville, 1981). Furthermore, the compressive strength measurement provides no guide to the shear strength of the casing/cement or the casing/formation bond (Parcevaux and Sault, 1984).

B–3.5 Free Water and Slurry Sedimentation

When a slurry is allowed to stand for a period of time prior to the set, water may separate from the slurry, migrate upward, and accumulate either in pockets or at the

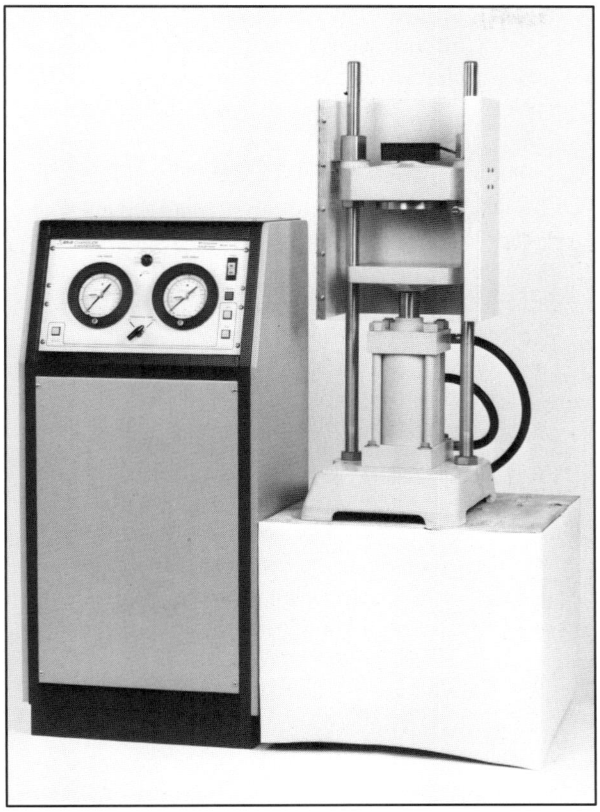

Figure B–13—Hydraulic press for measurement of compressive strength (photo courtesy of E G & G Chandler Engineering).

top of the column. This separation can result in incomplete zonal isolation, particularly in a highly deviated wellbore (Chapter 15). The free-water test is designed to measure this separation tendency in the laboratory, using a 250-mL graduated cylinder as a simulated wellbore. The duration of the test is two hours. The specification and operational test procedures are contained in API Spec 10, Section 6, and Appendix M.

The operational procedure permits preparation of the slurry at elevated temperatures and pressures. As with other tests, no provision is made for fluid loss. As interest increases in cementing deviated wellbores, many operators are evaluating free-water development by orienting the graduated cylinder at the angle of deviation. Typically, an increase in free water is observed in these situations; however, there is no clear understanding regarding how free water depends upon column height.

As with free-water development, when a slurry is allowed to stand for a period of time prior to development of a set, the suspended solids may tend to separate from the slurry and settle toward the bottom of the cement column. This separation is particularly evident in slurries containing weighting agents. Such "sedimentation" can produce a change in slurry density, leading to annular invasion and possible loss of well control (Chapter 8).

There is no published procedure specifically for the determination of the degree of settling in a cement slurry. Most field laboratories use the free-water tests described

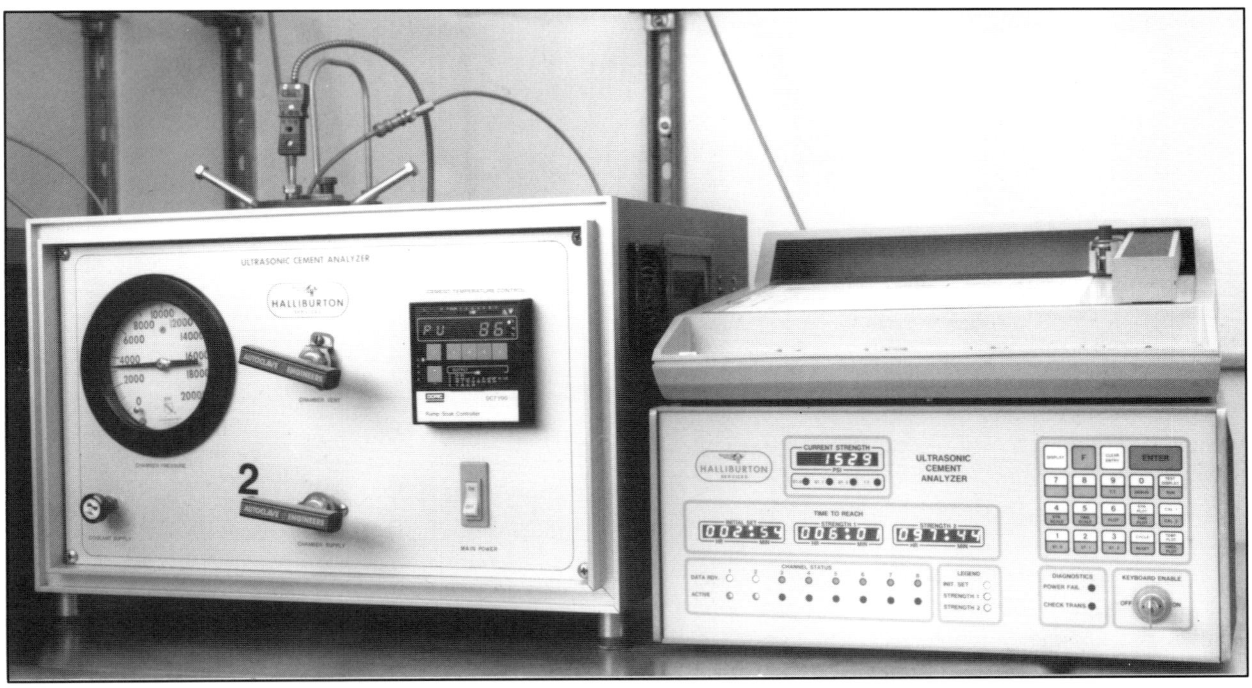

Figure B–14—Ultrasonic Cement Analyzer (UCA) (available from Halliburton Services).

in Spec 10, making a visual observation of any distinct settling which may be present. Settling can also be determined by sectioning a column of set cement, and comparing the density of the individual segments.

B–3.6 Permeability

The permeability of the cement sheath is a vital parameter with regard to zonal isolation. An operational procedure for determining the water permeability of set cement is contained in API Spec 10, Appendix G. The apparatus and procedure were developed in the early 1950s (Morgan and Dumbauld, 1952). Water at a differential pressure of 20 to 200 psi (100 to 1,400 kPa) is forced through a sample of set cement. Water is flushed through the specimen for a maximum of 15 minutes, or until one milliliter has been accumulated in the measuring tube. Darcy's law (Eq. B–2) is used to calculate the permeability.

$$K = 14{,}700 \frac{Q \mu L}{AP} \qquad \text{(B–2)}$$

where

K = permeability (md),
Q = flow rate (mL/sec),
μ = water viscosity (cp),
L = sample length (cm),
A = sample cross-sectional area (cm^2), and
P = differential pressure (psi).

Many laboratories today use a Hassler sleeve-type holder prescribed in an alternative test preparation (Fig. B–15), and measurement of permeability to air, methane, or other gases is fairly routine using the newer instrumentation (API, 1960). This procedure is also presented in Appendix G of API Spec 10.

B–3.7 Rheological Measurements

A detailed discussion of cement rheological properties and their significance is found in Chapter 4, and a summary of the pertinent relationships is presented in Appendix A. The use of these relationships for accurate prediction of friction pressure drops and slurry flow properties depends upon reliable laboratory measurements of the rheological parameters.

Two basic types of apparatuses are used today for rheological measurements: capillary pipe rheometers and coaxial cylinder rotational viscometers. As shown in Fig. B–16, rotational viscometers are designed either with a rotating outer cylinder (Couette type, covered by API Spec. 10, Appendix H) or a rotating inner cylinder (Searl type). Although the intrinsic design of the capillary pipe rheometer (Fig. B–17) makes it the preferred technique for use with non-Newtonian fluids (Bannister, 1980; Bannister and Benge, 1981) (Chapter 4), it is difficult and time consuming to use routinely. Conversely, the coaxial cylinder rotational viscometers are quick and easy to use.

In addition to friction pressure and flow regime calculations, laboratory measurements of rheological parameters can provide information about other slurry characteristics. A decrease in values for shear stress when measured in order of decreasing shear rate compared to values measured in order of increasing shear rate may indicate that the slurry is thixotropic. A determination of gel strength, using the procedure described in API RP 13B, Section 2, will provide an indication of a slurry gelation tendency. The procedure in API Spec 10 prescribes a conditioning period of 20 minutes in an atmospheric consistometer prior to measurement of the rheological parameters; however, measurements taken immediately after mixing may provide some indication of slurry mixability, particularly in a batch mixer.

B–3.8 Expansion

Set cement expansion can be measured using the procedure described in ASTM Specification C 151. This pro-

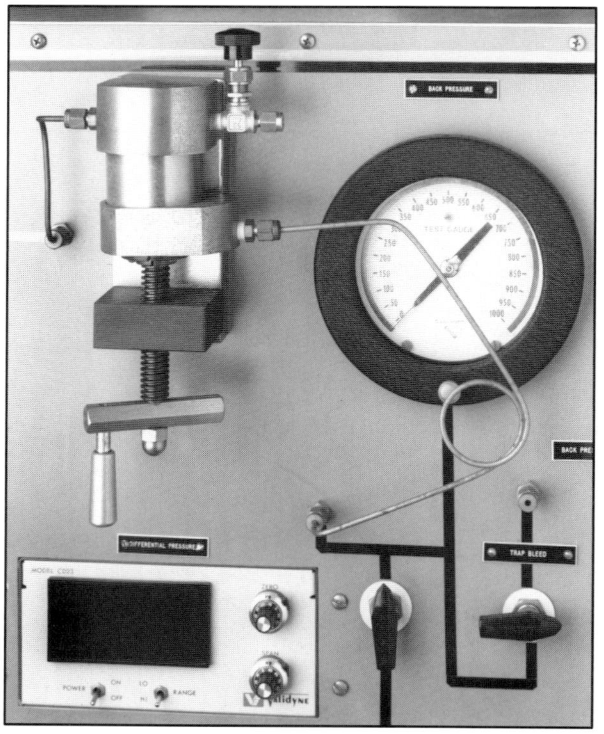

Figure B–15—Apparatus to measure water permeability.

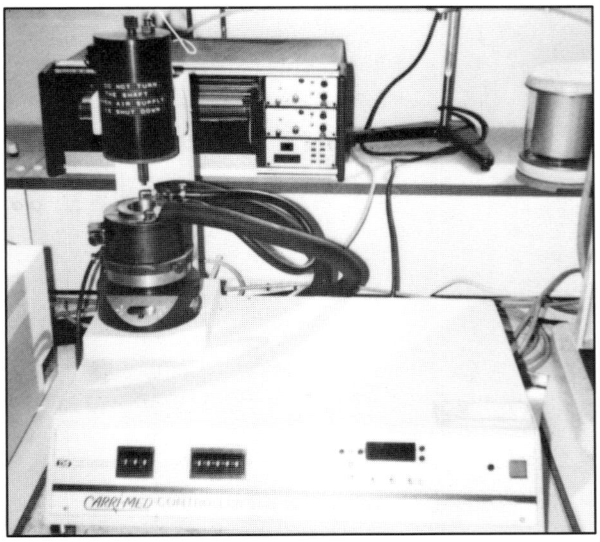

Figure B–16—Couette-type and Searl-type rotational viscometers.

cedure, developed for the concrete industry, involves placing the cement slurry into a bar-type mold and curing under water at atmospheric pressure. The cement bar is removed from the mold when it is sufficiently strong, the

Figure B–17—Pipe-flow rheometer.

length is carefully measured, and the bar is returned to the water bath for further curing. Periodically, during the curing period, the bar is removed for subsequent length measurements. The ASTM method has two major drawbacks: (1) since the cement must develop some strength before a length is measured, it is impossible to obtain a reliable "zero reading," and (2) there is no provision for studying the effect of pressure.

In 1983, Spangle invented an apparatus for measuring cement expansion which allows the simulation of the wellbore environment, and does not require the removal of the cement from a mold for an initial measurement. As shown in Fig. B–18, the device consists of a cylindrical sleeve which has a vertical slit therein. The sleeve is designed to fit inside a standard two-inch cube mold (Section B–3.4). Mounted on the outside of the sleeve are three sets of two pins, each located on opposite sides of the slit, and secured with a coil spring. The sleeve is in a closed position when the slurry is first introduced, and a zero reading of the distance between the three sets of pins is taken with a micrometer. If the cement composition expands during the curing period, the circumference of the sleeve increases and the longitudinal slit opens. After

Figure B–18—Curing sleeves for measurement of cement expansion.

the mold is removed from the curing chamber, the distances between the pins are remeasured, and the degree of expansion is easily calculated according to the following equation.

$$\% \text{ expansion} = \left(\frac{r}{R}\left(\frac{C_2}{C_1} - 1\right)\right), \qquad \text{(B–3)}$$

where

C_1 = distance between the pins when the sleeve is empty,

C_2 = distance between the pins when the sleeve is expanded,

R = inside radius of sleeve, and

r = R plus the length of the pin.

B–3.9 Slurry Density

An operational procedure for determination of slurry density is found in API Spec 10, Appendix C. The procedure uses a pressurized mud balance, shown in Fig. B–19. In addition to laboratory use, this simple device is frequently used on location to check the accuracy of automated in-line devices. The slurry pressurization in this device compresses entrained air, yielding a more accurate determination of slurry density than with a nonpressurized balance.

B–3.10 Static Gel Strength

The static gel strength of a cement slurry is routinely determined by a method developed for drilling fluids (API RP 13B, Section 2). A couette-type rotational viscometer is used.

Procedures specifically developed for the measurement of the static gel strength of a cement slurry have been reported (Sabins et al., 1980). These procedures use a consistometer-type apparatus equipped with a low-friction magnetic drive and torque measuring system coupled to the paddle. The slow movement of the paddle apparently does not inhibit gel-strength development while permitting measurements of torque.

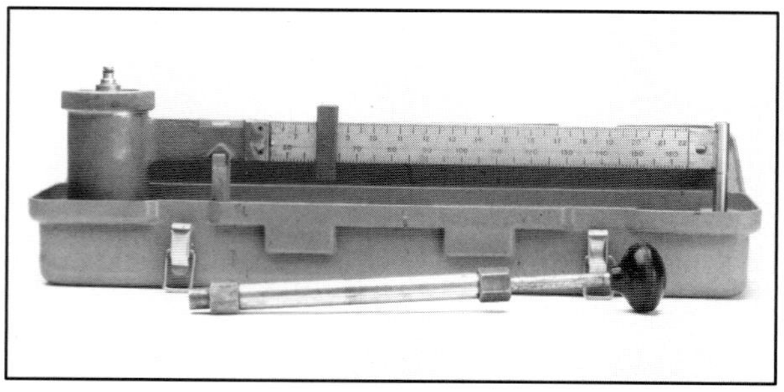

Figure B–19—Pressurized mud balance (available from Halliburton Services).

B–4 PERFORMANCE EVALUATION OF SPACERS AND CHEMICAL WASHES

As discussed in Chapter 5, spacers and washes perform two important functions in the cementing operation: to clean and remove the drilling fluid from the wellbore, and to minimize cement contamination by the drilling fluid. To accomplish these objectives, the spacer and wash must impart some degree of cleaning effect in the wellbore, and must be compatible with the drilling fluid being displaced from the hole and with the cement slurry being placed in the hole. Laboratory testing of these materials must include procedures to evaluate the cleaning effect and the compatibility.

Operational procedures to evaluate compatibility are contained in API Spec 10, Appendix P. The effects of a spacer or chemical wash upon the thickening time, compressive strength, fluid-loss control, and rheological characteristics of the cement system are investigated. The effects of the preflush upon the mud are also tested. Various volumetric ratios of mud, cement, and spacer are specified for the tests.

No specific criteria are provided to aid in the interpretation of the test results. No such criteria appear to have been published, and the user must employ sound engineering judgment in evaluating and interpreting test results to determine fluid compatibility.

Standard procedures to evaluate the mud removal capability of preflushes have not been published. A qualitative observation of the cleaning ability can be obtained by soaking a glazed tile in the drilling fluid, clamping the tile on the end of a rod, attaching the rod to a stirring motor, and rotating the tile in a container of the spacer/wash being evaluated for a period of time equal to the designed contact time of the spacer/wash across a specified section of the wellbore. Results of this observation may indicate a need to modify the spacer/wash formulation and/or the contact time.

B–5 CEMENT CHARACTERIZATION AND ANALYSIS

Characterization of cement and cementing materials in the laboratory involves the application of appropriate analytical techniques to provide a chemical and/or physical description of the sample as a whole or of the chemical constituents of the sample. This physico-chemical identification may include a qualitative determination of the chemical species present, a quantitative measurement of the amount of one or more of these species present, and a determination of physical properties of one or more of these species or of the sample as a whole. Four types of samples typically are examined in an oilfield cementing laboratory: neat cement powder, dry-blended cement, set cement, and mix water. Occasionally, if a failure to obtain a set cement in the wellbore is suspected, a sample of fluid returned from the wellbore may be examined to determine the presence or absence of cementitious material.

A wide variety of analytical techniques is available to characterize cement and cementing materials in the laboratory. Obtaining an accurate analysis of a sample may require the use of several of these techniques. An excellent review of techniques applicable to the chemical characterization of neat and dry-blended cements was published by Simpson (1988). Many techniques applicable to the physical characterization of these materials are contained in the ASTM Standards (1988). A general description of the most common applications for these techniques is provided here, organized by sample type and by characterization classification.

B–5.1 Chemical Characterization of Portland Cement

Chemical analysis of Portland cement powder is typically performed to ascertain the relative amounts of the four principal phases present (tricalcium silicate, dicalcium silicate, tricalcium aluminate, and tetracalcium aluminoferrite), gypsum, and certain minor oxides (Chapter 2). X-ray diffraction (XRD) is commonly used in the laboratory to determine the phases qualitatively, although reliable quantitative analyses are possible only with careful attention to sample preparation and the use of consistent standards (Aldridge, 1982). A more accurate quantitative analysis of the principal cement phases is currently obtained by first performing a complete chemical analysis. Wet chemical methods, or atomic excitation spectroscopic methods such as atomic absorption (AA) or plasma emission (ICP/DCP), and X-ray fluorescence (XRF), are normally used. The oxide composition of the cement is calculated, and finally the "potential phase composition" is calculated using a system of equations developed by Bogue (1929). Bogue's method is based upon various equilibrium relationships between the clinker phases. Taylor (1989) recently proposed a modification of the Bogue equations to more closely reflect current cement manufacturing practice.

Selective chemical extraction and complexometric techniques have also been used for the separation and subsequent determination of individual phases (Simpson, 1988). Some of these techniques can be applied specifically to determine the reactive phases (those located on the surface of the cement particle).

Specific identification of surface phases can also be performed using scanning electron microscopy (SEM) and light microscopy. As with XRD, considerable operator skill is required for accurate quantitative analysis. In addition, it is difficult to assure that one is looking at a representative sample of the material. Thermal methods such as thermogravimetric analysis (TGA) and differential thermal analysis (DTA) can be used to determine gypsum, hemihydrate, and free lime in neat cement.

B–5.2 Physical Characterization of Neat Cement and Cementing Materials

Physical characterization of neat cement powder in the laboratory usually includes measurement of the particle-size distribution, surface area, and specific gravity. The determination of specific surface area is frequently performed using a Blaine permeameter (ASTM C 204), shown in Fig. B–20. The apparatus is used to measure air permeability through a sample, and the result is then used to calculate a value for specific surface area. The accuracy is highly dependent on operator skill. Another technique which can be used to determine the specific surface area is turbidimetry, in which the change in intensity of a beam of light passing through a suspension of particles can be related to the size of the particles. A Wagner turbidimeter is commonly used to determine particle size by this principle (ASTM C 115); in addition, a value for

Figure B–20—Blaine permeameter.

Wagner fineness is included in the specifications for some classes of API cements (Chapter 2). One limitation of this technique is that particles are assumed to be spherical. A measurement of specific surface area which is independent of particle shape can be obtained using a gas adsorption technique, the Brunauer/Emmett/Teller (BET) method (1938). The instrumentation for this measurement is considerably more sophisticated and expensive than a Blaine permeameter or a Wagner turbidimeter, and BET surface area measurements are not obtained routinely in the field laboratory.

The surface area measurements provide a determination of average particle size/surface area within a sample, but they provide no information about the range of particle sizes within that sample. Using more sophisticated instrumentation based on a light-scattering principle, a particle-size distribution can be determined (Fig. B–21) over a wide range of particle sizes, typically from 200 μm to 0.1 μm (Wertheimer and Wilcock, 1976). These distribution profiles can be much more useful in explaining and predicting performance variations between cement samples than simple surface area measurements.

The specific gravity of neat cement powder is assumed to be 3.15 for purposes of calculating the slurry density and other properties, and the specific gravity of common extenders can be determined easily. These values are considered to be relatively constant from one sample to another; however, some differences have been observed, with significant variations in slurry density noted. The standard method for determination of specific gravity of these materials in the laboratory uses the Le

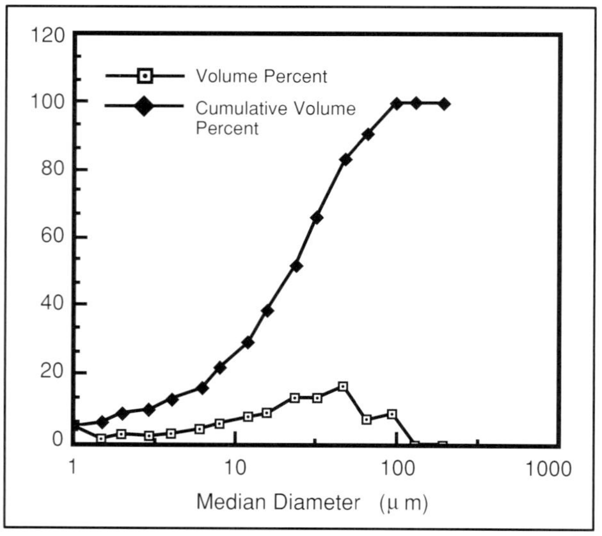

Figure B–21—Typical Portland cement particle-size distribution.

Chatelier flask (ASTM C 188). The procedure is simple in concept, but considerable skill and time are required for accurate results. Use of a pycnometer is preferred (ASTM C 128), because the procedure can be performed quickly and with accuracy comparable to that obtained with the Le Chatelier flask.

B–5.3 Chemical Analysis of Dry-Blended Cements

Chemical analysis of dry-blended cements can provide a more accurate indication of blend homogeneity than performance testing, and the determination of additive content also can be useful in explaining performance variations from one blend to another. Many analyses involve some type of separation technique to isolate the material of interest.

Most retarders and dispersants are chemically structured so that they absorb ultraviolet radiation; hence, these materials can be selectively dissolved or extracted and determined by UV-absorption spectrophotometry (ASTM C 114). Extraction techniques have also been applied to fluid-loss additives, although the diverse chemical nature of these materials has necessitated use of a wide variety of analytical determinations for the separated species. Lost-circulation materials and other additives with large particle size (such as $CaCl_2$ flakes) frequently are physically separated from a cement blend by sieving, followed by identification and quantification using standard techniques. Determination of salts, extenders, weighting agents, and silica can be performed using a combination of XRD and XRF techniques without separation of these materials from the blend (Simpson, 1988). Accurate quantitative results require considerable care, and may not be possible to obtain when small quantities of the additive are present.

Optical microscopy has been shown to be a useful tool for the qualitative analysis of cement blends (Reeves et al., 1983). With the use of a microscope equipped with polarization and fluorescence, and simple laboratory techniques, the presence of various cement additives can be confirmed.

B–5.4 Chemical Characterization of Set Cement

Many of the same techniques used to characterize neat cement powders and dry-blended cements can be used to examine set cement, although obtaining meaningful descriptions of material requires skillful sample preparation and interpretation of analytical results. Accurate quantitative analysis frequently is possible only with considerable knowledge of the properties of the components of the material under investigation. A combination of XRD and XRF techniques has been used with some success (Simpson, 1988).

B–5.5 Analysis of Cement Mix Water

Procedures for the analysis of cement mix water are contained in API RP 45. The reagents and procedures for determination of common species of interest are distributed in a portable field test kit, which is widely used for the analysis in the field laboratory and on location. Such kits are available from several suppliers. Techniques such as AA and ICP are used in laboratories with large sample throughput for efficiency, but these techniques offer no increase in accuracy if good laboratory practices are followed in all analyses. Ion chromatography can also be used for rapid determination of many anionic and cationic species in water.

B–6 SUMMARY

The technology currently available for the testing of well cements is sophisticated; however, there is much room for improvement. Consequently, most oil companies and service companies, and many instrument manufacturers, are engaged in research to improve existing techniques or to invent new procedures and equipment which simulate downhole conditions more accurately. Many methods or devices exist as "in-house" technology, where use is often limited to the company where the invention was made. Some of these are gaining acceptance throughout the industry, and may eventually appear as standard API tests.

Table B–1 is a summary of the procedures which are presently available to all industry laboratories.

REFERENCES

Aldridge, L. P.: "Accuracy and Precision of Phase Analysis in Portland Cement by Bogue, Microscopic and X-ray Diffraction Methods," *Cement & Concrete Res.* (1982) **12,** No. 3, 381–398.

American Petroleum Institute: *Specifications for Materials and Testing for Well Cements*, fourth edition, API Spec 10, API, Dallas (1988).

American Petroleum Institute: *Recommended Practice for Core-Analysis Procedures*, first edition, API Pub. RP40, API, Dallas (1960).

American Society for Testing and Materials: *Annual Book of ASTM Standards*, ASTM, Philadelphia (1989) **4.01** and **4.02**.

Bannister, C. E.: "Evaluation of Dynamic Fluid-Loss Behavior of Cement Slurries," paper SPE 7592, 1978.

Bannister, C. E.: "Rheological Evaluation of Cement Slurries: Methods and Models," paper SPE 9284, 1980.

Bannister, C. E. and Benge, O. G.: "Pipe Flow Rheometry: Rheological Analysis of a Turbulent Flow System Used for Cement Placement," paper SPE 10216, 1981.

Bell, D. R., Dammel, T. C. and Nahm, J. J.: "Evaluation and Procedures for Preparing and Sampling Dry-Blended Cements," paper SPE 18525, 1988.

Bogue, R. H.: "Calculation of the Compounds in Portland Cement," *Ind. Eng. Chem. Anal.* (1929) **1**, No. 4, 192–197.

Brunauer, S. *et al.*: "Adsorption of Gases in Multi-molecular Layers," *J. Amer. Chem. Soc.*, (1938) **60**, 309–319 .

Cobb, J. A. and Pace, R. S.: "Elements Affecting Thickening Time of a Cement Blend," paper SPE 14195, 1985.

El-Jazairi, B. and Illston, J. M.: "A Simultaneous Semi-Isothermal Method of Thermogravimetry and Derivative Thermogravimetry, and its Application to Cement Pastes," *Cem. Concr. Res.*, (1977) **7**, 247–257.

Gerke, R. R., Logan, J. L., Sabins, F. L. and Simon, J. M.: "A Study of Bulk Cement Handling and Testing Procedures," paper SPE 14196, 1985.

Kunze, K. R.: "Obtaining and Verifying Quality Cement Blends," paper SPE 15576, 1986.

Midgley, H. G.: "The Determination of Calcium Hydroxide in Set Portland Cements," *Cement & Concrete Res.* (1979), **9**, 77–82.

Morgan, B. E. and Dumbauld, G. K.: "Measurement of the Permeability of Set Cement," *JPT* (June 1952) **4**, 16.

Neville, A. M: *Properties of Concrete*, Pitman Publishing Ltd., London (1981) 281–285.

Orban, J. and Parcevaux, P.: "Viscometers Evaluated for Accurate Determination of Cement Slurry Rheology," *Oil & Gas J.* (June 30, 1986) 94–100.

Pace, R. S, McElfresh, P. M., Cobb, J. A., Smith, C. L. and Olsberg, M. A.: "Improved Bulk Blending Techniques for Accurate and Uniform Cement Blends," paper SPE 13041, 1984.

Parcevaux, P. A. and Sault, P. H.: "Cement Shrinkage and Elasticity: A New Approach for a Good Zonal Isolation," paper SPE 13176, 1984.

Rao, P. P., Sutton, D. L., Childs, J. D., and Cunningham, W.C.: "An Ultrasonic Device for Nondestructive Testing of Oilwell Cements at Elevated Temperatures and Pressures," *JPT* (Nov. 1982) 2611–2616.

Reeves, N. K., Bailey, D. E., and Caveny, W. J.: "Microscopic Analysis of Dry Cement Blends," paper SPE 11820, 1983.

Roy, D. M and Asaga, K.: "Rheological Properties of Cement Mixes: III. The Effects of Mixing Procedures on Viscometric Properties of Mixes Containing Superplasticizers," *Cement & Concrete Res.* (1979) **9**, 731–739.

de Rozières, J. and Ferrière, R. : "Foamed Cement Characterization under Downhole Conditions and its Impact on Job Design," paper SPE/IADC 19935 (1990).

Sabins, F. L., Tinsley, J. M., and Sutton, D. L.: "Transition Time of Cement Slurries Between the Fluid and Set State," paper SPE 9295, 1980.

Simpson, B. E.: "Analytical Chemistry of Portland Cement and its Oil-Field Admixtures," *SPEPE* (1988) 158–166.

Spangle, L. B.: "Apparatus and Method for Measuring the Expansion Properties of a Cement Composition," U.S. Patent No. 4,408,489, (1983).

Waechtler, H.J., Ilgnar, R., and Feldrappe, D.: "Thermoanalytical Studies in Cement Chemistry. 6. Differential Calorimetric Characterization of Portland Cements," *Cement & Concrete Res.*, (1984) **14**, 407–412.

Wertheimer, A. L. and Wilcock, W. L.: "Light Scattering Measurements of Particle Distributions," *Applied Optics* (1976), 1616.

Test Category	Equipment	Procedure Reference
sampling	mechanical splitter tube sampler diverted flow sampler	ASTM C 702 API Spec 10 (Section 3) Gerke et al., 1985
slurry preparation	two-speed propeller mixer	API Spec 10 (Section 5, Appx. A)
thickening time	atmospheric consistometer pressurized consistometer	API Spec 10 (Section 8, Appx. E)
fluid loss	high pressure fluid-loss cell using 325 mesh screen	API Spec 10 (Appx. F)
compressive strength	2 in. x 2 in. curing mold, placed in a water bath, or in a pressurized autoclave—strength measured with hydraulic press ultrasonic cement analyzer	API Spec 10 (Section 7, Appx. D) Rao et al., 1982
free water	250 mL graduated cylinder	API Spec 10 (Section 6, Appx. M)
permeability	water permeameter	API Spec 10 (Appx. G)
rheology	rotational viscometer (couette type) rotational viscometer (Searl type) pipe flow rheometer	API Spec 10 (Appx. H) Orban and Parcevaux, 1986 Bannister, 1978
static gel strength	rotational viscometer (couette type)	API RP13B (Section 2)
expansion	bar mold cylindrical sleeves	ASTM C 151 Spangle, 1983
spacer/wash/cement compatibility	rotational viscometer (couette type) with pressurized consistometer, fluid-loss cell, pressurized autoclave, and hydraulic press	API Spec 10 (Appx. P)
Portland cement phase analysis	wet chemical methods, XRD, XRF, AA, or ICP—phases calculated by Bogue equations	Aldridge, 1982 Bogue, 1929
determination of gypsum, hemihydrate, and free lime in Portland cement	wet chemical methods thermogravimetry (TGA) differential thermal analysis (DTA)	Simpson, 1988 El-Jazairi and Illston, 1977 Waechtler et al., 1984 Midgley, 1979
particle analysis of Portland cement	Blaine permeameter Wagner turbidimeter BET Laser light scatterin.g	ASTM C 204 ASTM C 115 Brunauer, et al., 1938 Wertheimer and Wilcock, 1976
specific gravity of Portland cement	Le Chatelier flask pycnometer	ASTM C 188 ASTM C 114
chemical analysis of dry blends	UV–absorption spectrophotometry XRD, XRF optical microscopy	ASTM C 114 Simpson, 1988 Reeves et al., 1983
chemical analysis of mix water	wet chemical method AA, ICP	API RP 45 Simpson, 1988

Table B–1—Summary of test procedures for well cements.

C

Cementing Calculations

Tom J. Griffin

Dowell Schlumberger

C–1 INTRODUCTION

The performance of calculations is an integral part of a cement job design. Calculations are necessary to determine the properties of a cement system (density, yield, volume of mix water, and proportions of additives). In addition, depending upon the type of cement job, calculations are necessary to determine the volume requirements, pressures, etc. In this appendix, five categories of calculations are discussed.

- Cement Slurry Properties
- Primary Cement Job Design
- Squeeze Cement Job Design
- Cement Plug Design
- Foamed Cement Job Design

C–2 CEMENT SLURRY PROPERTIES

The API Spec 10 (1988) specifies an amount of water to be mixed with neat cements. These API *water concentrations*, as well as the corresponding slurry densities (assuming a specific gravity of 3.14 g/cm^3 for Portland cement), are dependent upon the class of cement (Table C–1), and are mainly a function of the cements' surface areas. However, when additives are present in the system, the appropriate water concentration (to obtain the desired cement slurry properties) may change. As discussed earlier in this book, the important properties include density (for well control and avoiding lost circulation), free water, sedimentation, rheology, compressive strength, fluid-loss control, and permeability. All are a direct function of the relative quantities of cement, water, and additives. In this appendix, the methods of calculating the proportions of materials for the various types of cement systems are presented.

Class	Mix Water (% BWOC)	Slurry Density (lb/gal)	Slurry Density (g/cm^3)	Yield (ft^3/sk)
A	46	15.6	1.87	1.18
B	46	15.6	1.87	1.18
C	56	14.8	1.77	1.32
D	38	16.45	1.97	1.05
E	38	16.45	1.97	1.05
F	38	16.45	1.97	1.05
G	44	15.8	1.89	1.15
H	38	16.45	1.97	1.05

Table C–1—API mix water, resultant slurry density, and yield for oilwell cements.

C–2.1 Specific Gravity of Portland Cement

The specific gravity of Portland cement varies between about 3.10 and 3.25, depending on the raw materials used in its manufacture. For the calculations to be precise, the specific gravity of each cement should be measured (Appendix B). For the calculations in this appendix, a specific gravity of 3.14 will be assumed.

C–2.2 Absolute and Bulk Volumes

The absolute volume of a material is the volume occupied only by the material itself (not including the volume occupied by the air surrounding its particles). The volume occupied by the dry material, plus the air surrounding it, is its *bulk volume*. Portland cement normally has a bulk volume of 1 cubic foot for 94 lb, which is commonly referred to as a "sack." The absolute volume occupied by a 94-lb sack of cement is 3.59 U.S. gal or 0.48 ft^3. Other cements (e.g., commercial lightweight formulations or calcium aluminate cement) have different absolute and bulk volumes. Table C–2 is a listing of the bulk and absolute volumes of several cements, presented in English and SI units. In this appendix, the calculations will be presented in English units.

The absolute and bulk volumes of cement additives are available from literature published by the major cementing companies. Table C–3 is a listing of such information for some commonly used materials.[1]

	Sack Weight (lb)	Bulk Volume (ft^3/sk)	Absolute Volume (gal/lb)	(m^3/T)
API Classes A through H	94	1.0	0.0382	0.317
Class J	94	1.0	0.0409	0.341
Trinity Lite-Wate™ 1	75	1.0	0.0409	0.375
TXI Lightweight	75	1.0	0.0425	0.355
Ciment Fondu™ 2	87.5	1.0	0.0373	0.312
Lumnite™ 3	94	1.0	0.0380	0.317

1 Trademark of Lafarge Corporation
2 Trademark of Lafarge Fondu International
3 Trademark of Lehigh Portland Cement Company.

Table C–2—Cement volume factors.

Materials that dissolve in the water do not occupy as much space as their dry absolute volumes. For soluble additives like retarders, dispersants, and fluid-loss additives, which are added in relatively small amounts, the difference is negligible. However, salt (NaCl) is usually added in much larger concentrations; consequently, the difference must be taken into account. This point is discussed later.

C–2.3 Concentrations of Additives

The concentrations of most solid cement additives are expressed as a percentage by weight of cement (BWOC) or cementitious material. This method is also used for water. For example, if 35% (BWOC) silica sand is used in a cement blend, the amount for each sack of cement is $94\ \text{lb} \times 0.35 = 32.9$ lb of silica sand. This results in $94 + 32.9 = 126.9$ lb of total mix. The true percent of silica sand in the mix is $32.9 / 126.9 = 25.9\%$. Salt is a special exception. It is added by weight of mix water (BWOW). In addition, weighting materials such as barite are often added on a "pounds per sack (lb/sk)" basis. This is done for convenience, as it eliminates the need to convert from percent BWOC to pounds in the bulk plant.

Liquid additive concentrations are most commonly expressed in gallons per sack of cement or cementitious material. For example, according to Table C–3, liquid sodium silicate has an absolute volume of 0.0859 gal/lb. If a concentration of 0.4 gal/sk sodium silicate is prescribed, the weight of the material is $(0.4\ \text{gal/sk}) / (0.0859\ \text{gal/lb}) = 4.66$ lb/sk.

C–2.4 Slurry Density and Yield

The slurry density is calculated by adding the masses of the components of the cement slurry and dividing by the total of the absolute volumes occupied. In other words, to determine the density in lb/gal, divide the total pounds by the total gallons.

[1]This listing should not be used for slurry design purposes. Cementing companies may obtain materials from different sources; consequently, the absolute and bulk volumes may vary.

Material	Absolute Volume (gal/lb)	(m^3/T)	Specific Gravity
Barite	0.0278	0.231	4.33
Bentonite	0.0454	0.377	2.65
Coal (ground)	0.0925	0.769	1.30
Gilsonite	0.1123	0.935	1.06
Hematite	0.0244	0.202	4.95
Ilmenite	0.0270	0.225	4.44
Silica Sand	0.0454	0.377	2.65
NaCl (above saturation)	0,.0556	0.463	2.15
Fresh Water	0.1202	1.000	1.00

Table C–3—Absolute volumes of common cementing materials.

$$\rho_{slurry}\ (\text{lb/gal}) = \frac{\text{lb}_{cement} + \text{lb}_{water} + \text{lb}_{additives}}{\text{gal}_{cement} + \text{gal}_{water} + \text{gal}_{additives}} \quad \text{(C–1)}$$

The yield of a cement system is the volume occupied by one unit of the cement plus all of the additives and mix water. For cement measured in sacks, the yield is expressed in cubic feet per sack (ft^3/sk). This value is then used to calculate the number of sacks required to achieve the desired fill-up in the annulus. Most slurry density calculations are performed on the basis of one sack of cement (94 lb). This simplifies the calculation of the slurry yield.

Example Calculation

Consider a slurry composed of Class G cement plus 44% water (94 lb × 0.44 = 41.36 lb water).

Component	Weight (lb)	Absolute Volume (gal/lb)	Volume (gal)
Cement	94	0.0382	3.59
Water	41.36	0.1202	4.97
TOTAL	135.36		8.56

$$\rho_{slurry} = \frac{135.36\ \text{lb}}{8.56\ \text{gal}} = 15.8\ \text{lb/gal}$$

The yield is now determined by dividing the volume of the total slurry per 94-lb sack of cement (8.56 gal) by the conversion factor of 7.48 gal/ft^3.

$$\textit{Slurry Yield} = \frac{8.56\ \text{gal/sk}}{7.48\ \text{gal/ft}^3} = 1.14\ \text{ft}^3\text{/sk}$$

Another important calculation is the amount of mix water required. This is necessary to ensure that enough water is available for the cementing operation. It is simply the gallons calculated above (4.97) multiplied by the number of sacks of cement to be mixed.

Most additives are handled in the same manner as shown above. Often, when the calculations are performed by hand, the additives present in minor amounts (less than 1%) are ignored. Today, most laboratories use computers to calculate the slurry mixes and the density, and to determine the amounts of additives to use in the laboratory mix. All additives are taken into consideration.

Example Calculation

Consider a slurry composed of Class G cement + 35% silica flour + 1% solid cellulosic fluid-loss additive + 0.2 gal/sk liquid PNS dispersant + 44% water.

Component	Weight (lb)	Absolute Volume (gal/lb)	Volume (gal)
Cement	94	0.0382	3.59
Silica Flour	32.9	0.0454	1.49
Cellulosic Fluid-Loss Additive	0.94	0.0932	0.088
Liquid PNS Dispersant	1.97	0.1014	0.20
Water	41.36	0.1202	4.97
TOTAL	171.17		10.34

$$\rho_{slurry} = \frac{171.17 \text{ lb}}{10.34 \text{ gal}} = 16.55 \text{ lb/gal}$$

$$Slurry\ Yield = \frac{10.34 \text{ gal/sk}}{7.48 \text{ gal/ft}^3} = 1.38 \text{ ft}^3\text{/sk}$$

C–2.4.1 Special Additives

Salt

As mentioned earlier, salt concentration is expressed as a percentage by weight of water (BWOW). The absolute volume of NaCl when mixed with water is less than it is dry; since it is usually added at a high concentration, this must be reflected in the density and yield calculations. The absolute volume of salt is dependent upon its concentration in the water. Table C–4 is a listing of the absolute volumes that should be used for various salt concentrations.

Concentration (% BWOW)	Absolute Volume in Water (gal/lb)	(m^3/T)
2	0.0371	0.310
4	0.0378	0.316
6	0.0384	0.321
8	0.0390	0.326
10	0.0394	0.329
12	0.0399	0.333
14	0.0403	0.336
16	0.0407	0.340
18	0.0412	0.344
20	0.0416	0.347
22	0.0420	0.351
24	0.0424	0.354
26	0.0428	0.357
28	0.0430	0.359
30	0.0433	0.361
34	0.0439	0.366
37.2 (saturated)	0.0442	0.369

Table C–4—Absolute volume of salt in water.

Example Calculation

Consider a slurry composed of Class G cement + 37.2% NaCl (BWOW) + 44% water. What are the slurry density and the yield?

94 lb cement × 0.44 = 41.36 lb water
41.36 lb water × 0.372 = 15.39 lb NaCl

Reading from Table C–4, the absolute volume of NaCl at a concentration of 37.2% BWOW is 0.0442 gal/lb. Thus, the calculation can be completed as follows.

Component	Weight (lb)	Absolute Volume (gal/lb)	Volume (gal)
Cement	94	0.0382	3.59
NaCl	15.39	0.0442	0.68
Water	41.36	0.1202	4.97
TOTAL	150.75		9.24

$$\rho_{slurry} = \frac{150.75 \text{ lb}}{9.24 \text{ gal}} = 16.31 \text{ lb/gal}$$

$$Yield = \frac{9.24 \text{ gal/sk}}{7.48 \text{ gal/ft}^3} = 1.24 \text{ ft}^3\text{/sk}$$

Fly Ash

As discussed in Chapter 3, fly ash is a pozzolanic extender which is often used to replace part of the cement. A special convention is used to describe fly ash/cement blends. These mixtures are normally written as ratios, with the ratio indicating the absolute volume contribution of the two components. A ratio of 35:65 refers to 35% fly ash and 65% cement (the first number always represents the fly ash and the second the cement). Other common ratios are 50:50 and 75:25. The quantity of fly ash necessary to prepare 3.59 gallons of blend may be calculated from the following formula.

$$Weight_{fly\ ash} = Weight\ of\ Cement\ Replaced \times \frac{SG_{fly\ ash}}{SG_{cement}} \quad (C\text{–}2)$$

The weight of the cement replaced is 94 lb minus the amount of cement remaining. For a 35:65 fly ash:cement blend, this is 94 lb – (0.65 × 94 lb) = 32.9 lb. Thus, for a 35:65 fly ash:cement blend, using a cement with a specific gravity of 3.14 and a fly ash with a specific gravity of 2.46, the required weight of fly ash is

$$Weight_{fly\ ash} = 32.9\ \text{lb} \times \frac{2.46}{3.14} = 25.8\ \text{lb}.$$

The weight of cement is

$$Weight_{cement} = 0.65 \times 94\ \text{lb} = 61.1\ \text{lb}.$$

The mixture of 25.8 lb of fly ash with 61.1 lb of cement weighs 86.9 lb. This mixture is referred to as the *equivalent sack*, because the absolute volume of the blend is 3.59 gal. For such systems, additive concentrations are calculated as a function of the equivalent sack, *not* a 94-lb sack of Portland cement. Those calculated BWOC are calculated based on the 86.9-lb equivalent sack.

There is much variation in the specific gravity of fly ashes, and it must be determined for each. The water requirements for the different fly ashes will vary according to the desired performance properties, fineness, and chemical composition.

Example Calculation

Consider a slurry consisting of a 50:50 fly ash:Class G blend plus 2% bentonite and 54% water ($SG_{fly\ ash}$ = 2.48; therefore, Absolute Volume = 0.0483).

$$\text{Weight of Cement Replaced} = 94\ \text{lb} - (0.50 \times 94\ \text{lb}) = \text{lb}) = 47.0\ \text{lb}$$

$$Weight_{fly\ ash} = 47.0\ \text{lb} \times \frac{2.48}{3.14} = 37.12\ \text{lb}$$

$$Weight_{cement} = 0.50 \times 94\ \text{lb} = 47.0\ \text{lb}$$

$$Weight_{water} = 0.54 \times (37.12\ \text{lb} + 47.0\ \text{lb}) = 0.54 \times 84.12 = 45.47\ \text{lb}$$

Component	Weight (lb)	Absolute Volume (gal/lb)	Volume (gal)
Cement	47	0.0382	1.795
Fly Ash	37.12	0.0483	1.795
Bentonite	1.88	0.0454	0.085
Water	45.47	0.1202	5.47
TOTAL	131.47		9.145

$$\rho_{slurry} = \frac{131.47\ \text{lb}}{9.145\ \text{gal}} = 14.38\ \text{lb/gal}\text{ , and}$$

$$Yield = \frac{9.145\ \text{gal}}{7.48\ \text{gal/ft}^3} = 1.22\ \text{ft}^3\text{/equiv sk}.$$

Note that in this case, the yield is expressed in cubic feet per equivalent sack (ft³/equiv sk) to show that it is based on both the cement and the fly ash.

Bentonite

As discussed in Chapter 3, bentonite and attapulgite are clays which are used to develop gel strength and allow the addition of extra water. The resulting slurries are less dense, and more economical. In editions prior to 1989, Section 10 of API Spec 10 recommended the addition of 5.3% water for each percent of bentonite (BWOC). This is only a general guide, because the efficiency of bentonite varies from batch to batch. The actual amount of water should be based on the desired performance properties of the cement (e.g., free water, gel strength, etc.) as determined in the laboratory.

Bentonite can be dry blended with the cement or prehydrated in the mix water. About 30 minutes is required to completely hydrate bentonite in fresh water, and the efficiency of the material as an extender is improved up to four times. It is commonly stated that bentonite "yields." This is an oilfield term to describe the ability of bentonite to absorb water. Thus, it is often said that prehydrated bentonite yields four times as efficiently as dry blended. Unfortunately, this has resulted in some confusion concerning calculations involving prehydrated bentonite. The improved efficiency of prehydrated bentonite has no bearing on the density or the yield of a sack of cement. Bentonite calculations are based on the actual amounts of materials that are present in the blend.

Example Calculation

For a slurry consisting of Class G cement + 2% pre-hydrated bentonite + 86.4% water (0.864 × 94 lb = 81.22 lb), the calculation is performed as follows.

Component	Weight (lb)	Absolute Volume (gal/lb)	Volume (gal)
Cement	94	0.0382	3.59
Bentonite	1.88	0.0454	0.085
Water	81.22	0.1202	9.76
TOTAL	177.10		13.435

$$\rho_{slurry} = \frac{177.10 \text{ lb}}{13.435 \text{ gal}} = 13.2 \text{ lb/gal}$$

$$Yield = \frac{13.435 \text{ gal}}{7.48 \text{ gal/ft}^3} = 1.80 \text{ ft}^3\text{/equiv sk}$$

For 8% dry-blended bentonite at the same density (90.6% water), the calculation is shown below.

Component	Weight (lb)	Absolute Volume (gal/lb)	Volume (gal)
Cement	94	0.0382	3.59
Bentonite	7.52	0.0454	0.34
Water	85.15	0.1202	10.21
TOTAL	186.67		14.14

$$\rho_{slurry} = \frac{186.67 \text{ lb}}{14.14 \text{ gal}} = 13.2 \text{ lb/gal}$$

$$Yield = \frac{14.14 \text{ gal}}{7.48 \text{ gal/ft}^3} = 1.89 \text{ ft}^3\text{/equiv sk}$$

Thus, by comparison, the yield and mix water for a 13.2 lb/gal slurry is shown below.

Bentonite	Mix Water (gas/sk)	Yield (ft^3/sk)
dry blend	10.2	1.89
pre-hydrated	9.76	1.80

Weighting Agents

Certain materials are used in cement slurries to increase the density of the slurry for well-control purposes. When these materials are used, a specified density is required, and an unknown amount of additive must be used to achieve the necessary density. The following example illustrates the procedure for calculating the required amount of a weighting agent.

Example Calculation

For a slurry composed of Class G cement plus 44% water, how much hematite (absolute volume: 0.0244 gal/lb) is required to prepare a slurry with a density of 18.5 lb/gal?

Component	Weight (lb)	Absolute Volume (gal/lb)	Volume (gal)
Cement	94	0.0382	3.59
Hematite	x	0.0244	0.2444x
Water	41.36	0.1202	4.97
TOTAL	135.36 + x		8.56 + 0.0244 x

$$\rho_{slurry} = 18.5 \text{ lb/gal} = \frac{(135.36 + x) \text{ lb}}{(8.56 + 0.0244x) \text{ gal}}$$

$$x = 41.92 \text{ lb}$$

Thus, since the calculation was based on one sack of cement, 41.92 lb of hematite is required per sack to increase the density of the API Class G slurry to 18.5 lb/gal. This may be expressed in lb/sk (41.92 lb/sk) or in % (41.92 lb/94 lb × 100 = 44.6% BWOC). In reality, this increase might be accomplished by reducing the amount of water with the aid of dispersants, with the addition of a smaller amount of weighting agent. The properties of the slurry are also carefully measured to ensure that the slurry will perform as desired. Particularly important are the slurry's rheology, free water, and tendency for the weighting agent to settle.

Water Requirement

It is sometimes necessary to calculate the volume of water necessary to achieve a desired density, to ensure that a certain fracturing pressure is not exceeded. This is done with an unknown, similar to the calculation of a weighting agent as illustrated above. A variation is that it is convenient to set the unknown equal to the volume of water.

Example Calculation

For a slurry consisting of Class G cement plus 8% bentonite, how much water is necessary for a density of 13.0 lb/gal?

Component	Weight (lb)	Absolute Volume (gal/lb)	Volume (gal)
Cement	94	0.0382	3.59
Bentontite	7.52	0.0454	0.341
Water	8.34*x*	0.1202	*x*
TOTAL	101.52 + 8.34*x*		3.93 + *x*

$$\rho_{slurry} = 13.0 \text{ lb/gal} = \frac{(101.52 + 8.34x) \text{ lb}}{(3.93 + x) \text{ gal}}$$

$$x = 10.82 \text{ gal}$$

Thus, 10.82 gal of water is required for each sack of cement.

C–3 PRIMARY CEMENTING CALCULATIONS

Calculations are used for primary cementing to determine the following.

- Cement Volume (annular volume)
- Water to Mix Cement
- Cement Density and Yield
- Displacement to Land Plug
- Pump Pressure to Land Plug
- Hydrostatic Pressure on Formation (fracture and pore pressures)
- Pressure to Lift Casing

C–3.1 Annular Volumes

Annular volumes are calculated to determine the amount of cement required to obtain the desired fill-up. Often, during the design stage, these calculations are performed based on the bit size plus an excess volume determined from experience in a field. This allows the service company to determine the total time required to mix and pump the cement, and displace it into the annulus.

After the casing point is reached, caliper logs should be run, and the cement volume should be adjusted based upon the actual hole size. Even with the caliper, it is common practice to use an excess volume to ensure fill-up by cement across all critical zones. The type of caliper can affect the amount of cement computed, and the resulting fill by the cement. Two- and three-arm calipers, with arms that operate together, may underestimate (or overestimate in the case of the two-arm caliper) the size of the hole. This is especially true for deviated wells in which the holes tend to be oval (Fig. C–1). For these situations, four-arm calipers (with independently operating arms) are preferred. Normally, the interval being cemented is divided into increments, and the average hole diameters are estimated for each increment (Fig. C–2). Table C–5 is a listing of the annular volume calculations from caliper measurements.

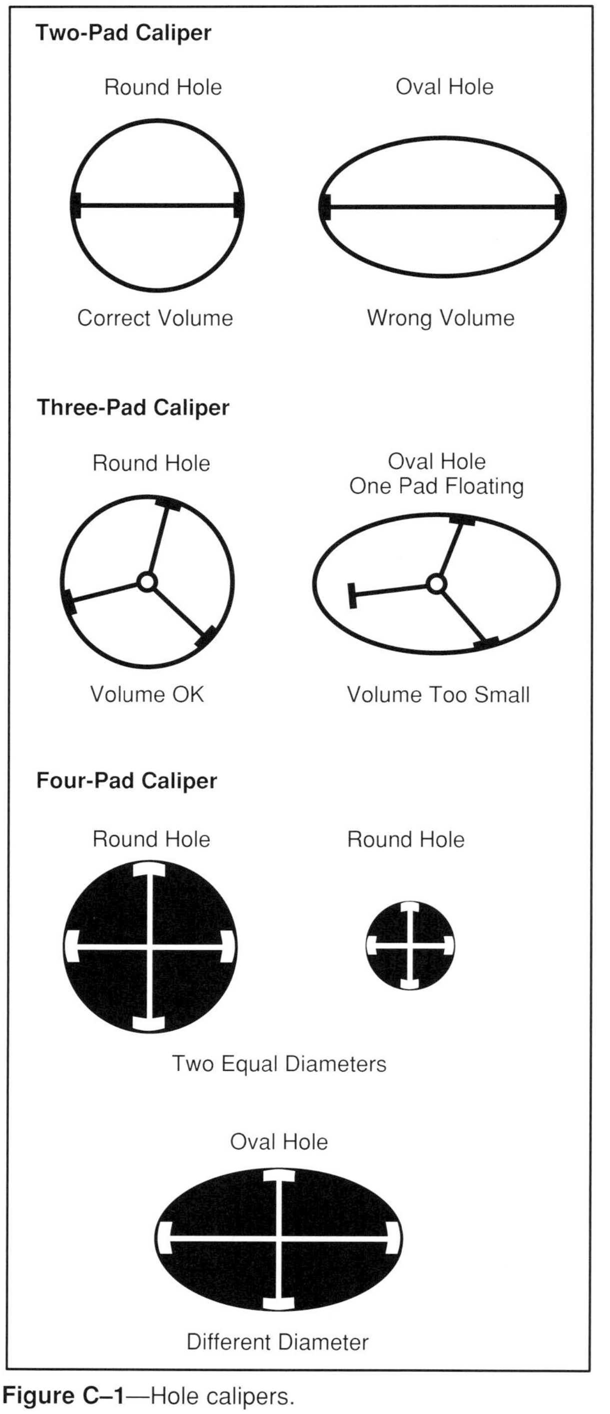

Figure C–1—Hole calipers.

9	9½	10	10½	11	11½	12	12½	13	13½	14	14½	15	15½	16
	10	10	10						10				10	
	10	10												
	10	10												
		10												
Totals	**30**	**40**	**10**						**10**				**10**	

for 7 in. in 10 in. hole	=	0.2782 $ft^3/ft \times 30$ ft	=	8.35 ft^3
for 7 in. in 10½ in. hole	=	0.3341 $ft^3/ft \times 30$ ft	=	13.36 ft^3
for 7 in. in 11 in. hole	=	0.3927 $ft^3/ft \times 30$ ft	=	3.93 ft^3
for 7 in. in 13½ in. hole	=	0.7267 $ft^3/ft \times 30$ ft	=	10.43 ft^3
				Total = 43.34 ft^3

Table C–5—Annular volume calculations from caliper measurements.

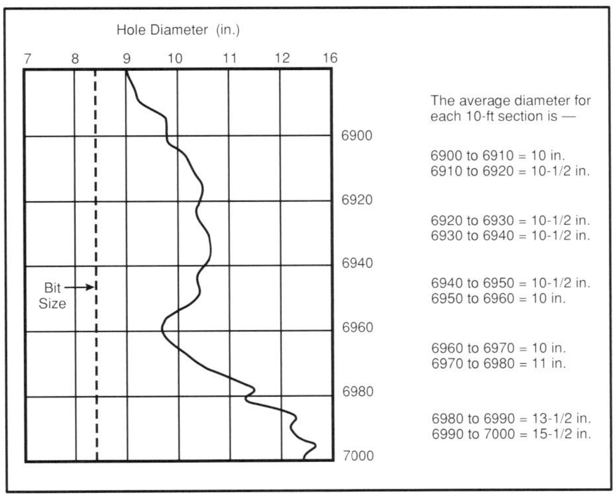

Figure C–2—Microcaliper log.

Having calculated the annular volume, an excess is added (normally 10% to 15%, based on experience in the field) and then the number of sacks required to fill this volume is computed based on the "yield" of the cement. For example, in this small section of hole (Table C–5),

$$\textit{Cement Volume} = 43.34 \text{ ft}^3 \times 1.10\ (\textit{excess factor}) = 47.7 \text{ ft}^3.$$

For a yield of 1.18 ft^3/sk, the amount of cement required is

$$\frac{47.7 \text{ ft}^3}{1.18 \text{ ft}^3/\text{sk}} = 40.4 \text{ sk.}$$

Most logging companies also offer computerized annular volume calculations which are presented on the log. A log with these calculations is shown in Fig. C–3. Tick marks on the depth track represent the total hole volume (left) and the annular volume between the casing and the open hole (right) in 10-ft^3 increments. The total volume of the hole (VHOL) and cemented annulus (VCEM) are also shown in the scale header.

Excess factors must be based on experience, whether excess over caliper, or excess over bit size. Normally, the excess is calculated only for the openhole portion being cemented. This excess is to account for cement which may be lost into the formation, for enlarged hole, or for fluid which may be lost from the cement into permeable zones. When returns to the surface are desired or required, excess volumes may be used to assure that this is achieved.

Care must be taken in using excess. If the well has weak formations which are close to being fractured, excess cement (which will raise the top of the cement) may cause the formation to be fractured because of increased hydrostatic and friction pressures.

The final step in calculating the amount of cement to be used is that which will remain in the shoe joints, between the float collar and the shoe. This is simply the casing volume between those two points. This volume should be added to that required based on the hole size and excess.

C–3.2 Density, Yield, and Mix Water

The cement mix water, density, and yield are determined as described in Section C–2.4. It is common for the cementing service company to calculate the volume of mix water required. This is the sum of the water required for mixing cement, spacers and preflushes, filling the

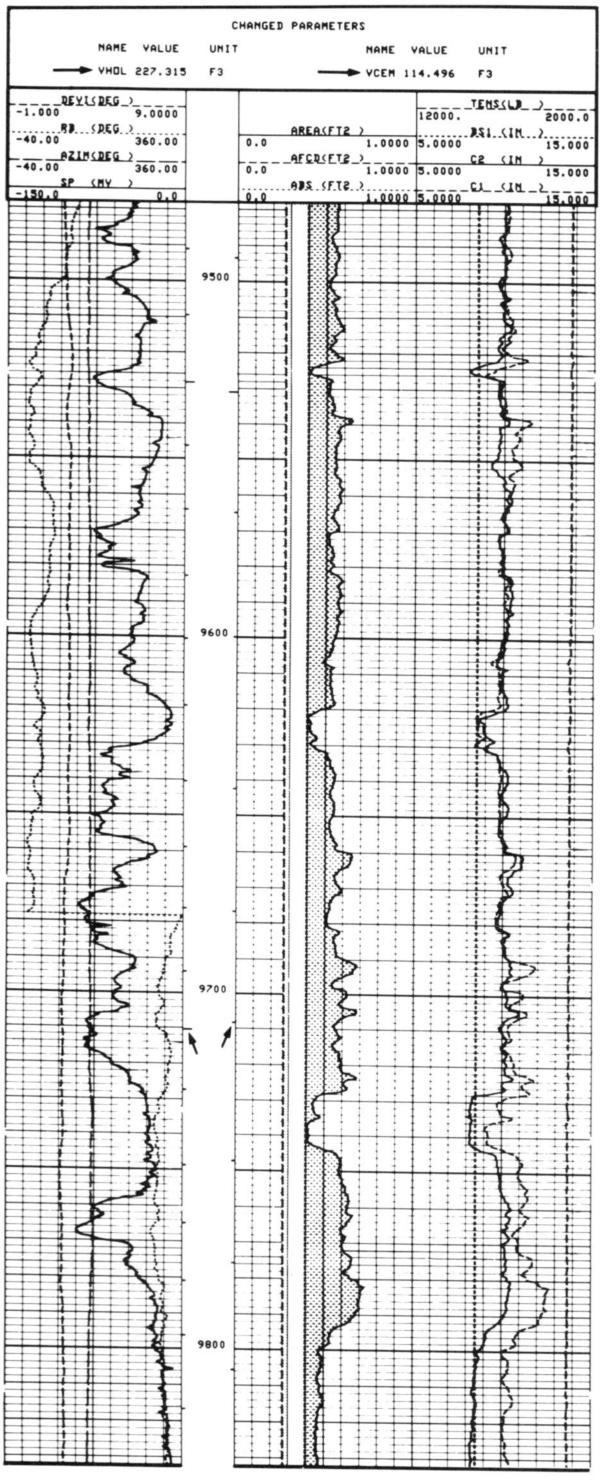

Figure C–3—Borehole Geometry Log.

dead volume of tanks, and for displacement. A safety factor or contingency factor volume should also be added.

C–3.3 Displacement Volume to Land Plug

The displacement volume to land the plug is simply a calculation of the capacity of the pipe. This is normally done by multiplying the length of the pipe (or segments of pipe if the entire string is not the same size or weight) times the capacity of that pipe. The capacity of the pipe is normally determined by looking in standard tables. The volume should be that between the pumps and the landing collar. A small excess volume may be pumped to allow for the compression of any air which may be entrained in the displacing fluid, and to account for pump inefficiency.

C–3.4 Pump Pressure to Land Plug

The pump pressure to land the plug is the difference in hydrostatic pressure of the fluids in the annulus and the pipe. Depending on the pump rate, additional pressure may be required to overcome friction pressure. This friction pressure is best determined by so-called "U-tube" or placement simulators (Chapter 11). These pressures are calculated to determine the type of pump required, to ensure that the cementing head is adequate and that there is no danger of bursting the casing.

C–3.5 Hydrostatic Pressure on the Formation (Fracture and Pore Pressure)

To ensure the safety of the well, it is necessary to determine if it is likely that the well will flow or be fractured during or after the cement treatment. This is done by calculating the hydrostatics at the critical points in the wellbore. Such calculations are a good first approximation, but friction pressure is particularly important when ascertaining the possibility of fracturing weak formations. Since the hydrostatic and friction pressures are constantly changing during a primary cement job, the only truly appropriate method is to use computerized placement simulators.

To determine the hydrostatic pressure component, the following equation is used.

$$P_h = 0.052 \times \rho \times H\,, \qquad \text{(C–3)}$$

where

P_h = hydrostatic pressure (psi),

ρ = density of the fluid (lb/gal), and

H = height of column having density ρ (ft).

If there are several fluids in the wellbore, then this calculation must be made for each, and the P_h for all the fluids totalled. This value should then be compared with the pressures of the critical formations to ensure that it lies between their pore pressures and fracturing pressures. Note that the hydrostatic pressure exerted by a column of cement slurry decays with time (Chapter 8).

C–3.6 Example Well Calculations

The following calculations would be performed for the well depicted in Fig. C–4. Additional well information is given below.

Well Information

Surface Casing: 13⅜ in.(54.50 lb/ft) to 1,700 ft

Open Hole: 12¼ in. to 4,950 ft

Casing: 9⅝ in. (36.00 lb/ft)

Excess Required: 25%

Shoe Joint : 42 ft

Top of Cement: 300 ft inside 13³/₈-in. casing

Top of Tail: 4,450 ft

Lead System: 13.0 lb/gal (yield: 1.50 ft³/sk)

Tail System: 16.4 lb/gal (yield: 1.05 ft³/sk)

Spacer: 12.5 lb/gal (volume: 40 bbl)

Displacement Fluid: 11.5 lb/gal mud

Weak Formation: 3,215-psi fracture pressure at 4,320 ft

Highest Pore Pressure: 3,150 psi at 4,800 ft

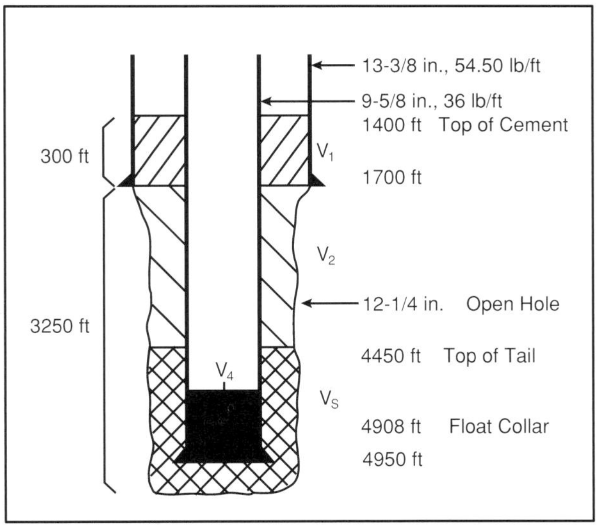

Figure C–4—Example well for primary cementing calculations.

[2]The volume factors (ft³/ft and bbl/ft) are obtained from standard tables published by most cementing companies.

Cement Volume Calculations[2]

Lead Cement Inside Surface Casings: volume between 9⅝-in. and 13⅜-in. casings

$V_1 = 300 \text{ ft} \times 0.3627 \text{ ft}^3/\text{ft} = 108.8 \text{ ft}^3$

Open Hole: volume between 9⅝-in. casing and 12¹/₄-in. open hole, with 25% excess

$V_2 = (4{,}450 - 1{,}700) \text{ ft} \times 0.3131 \text{ ft}^3/\text{ft} \times 1.25 = 1{,}076.3 \text{ ft}^3$

Total Lead: $V_L = V_1 + V_2 = 108.8 + 1{,}076.3 = 1{,}185.1 \text{ ft}^3$

Tail Cement Open Hole: volume between 4,450 ft and 4,950 ft

$V_3 = 500 \text{ ft} \times 0.3131 \text{ ft}^3/\text{ft} \times 1.25 = 195.7 \text{ ft}^3$

Shoe Joint: $V_4 = 42 \text{ ft} \times 0.4341 \text{ ft}^3/\text{ft} = 18.2 \text{ ft}^3$

Total Tail: $V_T = V_3 + V_4 = 195.7 + 18.2 = 213.9 \text{ ft}^3$

Displacement Volume

$V_D = (4{,}950 - 42) \text{ ft} \times 0.0773 \text{ bbl/ft} = 379.4 \text{ bbl}$

Pump Pressure to Land the Plug

Hydrostatic Pressure Inside Casing (P_{HI}):

$P_{HM} = 0.052 \times 11.5 \text{ lb/gal} \times (4{,}950 - 42) \text{ft} = 2{,}935.0 \text{ psi}$

$P_{HT} = 0.052 \times 16.4 \text{ lb/gal} \times 42 \text{ ft} = 35.8 \text{ psi}$

$P_{HI} = 2{,}935.0 \text{ psi} + 35.8 \text{ psi} = 2{,}970.8 \text{ psi}$

Hydrostatic Pressure Outside Casing (P_{HO}): The sum of hydrostatic pressures of all annular fluids must be calculated.

Mud: The height of the mud is 1,400 ft less the length of the spacer. The length of the spacer is 40 bbl × 15.48 ft/bbl (from standard tables) = 619.2 ft. Thus, the mud height is 1,400 – 619.2 = 780.2 ft.

$P_{HM} = 0.052 \times 11.5 \text{ lb/gal} \times 780.2 \text{ ft} = 466.9 \text{ psi}$

Spacer: $P_{HS} = 0.052 \times 12.5 \text{ lb/gal} \times 619.2 \text{ ft} = 402.5 \text{ psi}$

Lead Cement: $P_{HL} = 0.052 \times 13.0 \text{ lb/gal} \times (4{,}450 - 1{,}400) \text{ ft} = 2{,}061.8 \text{ psi}$

Tail Cement: $P_{HT} = 0.052 \times 16.4 \text{ lb/gal} \times (4{,}950 - 4{,}450 \text{ ft}) = 426.4 \text{ psi}$

Total: $P_{HO} = P_{HM} + P_{HS} + P_{HL} + P_{HT} = 466.9 + 402.5 + 2061.8 + 426.4 = 3{,}357.6 \text{ psi}$

Pressure to Land Plug (excluding friction pressure):

$P_{LP} = P_{HO} - P_{HI} = 3{,}357.6 - 2{,}970.8 = 386.8 \text{ psi}$

Hydrostatic Pressure on Formations

Fracture Pressure: 3,215 psi at 4,320 ft

P_H at 4,320 ft = P_{HM} + P_{HS} + P_{HL} (at 4,320 ft)

P_{HL} (at 4,320 ft) = 0.052 × 13.0 lb/gal x (4,320 – 1,400) ft = 1,973.9 psi

= 466.9 + 402.5 + 1,973.9 = 2,843.3 psi

Therefore, the hydrostatic pressure is 371.7 psi below the fracture pressure.

Pore Pressure: 3,150 psi at 4,800 ft

P_H at 4,800 ft = P_{HM} + P_{HS} + P_{HL} + P_{HT} (at 4,800 ft)

P_{HT} (at 4,800 ft) = 0.052 × 16.4 lb/gal × (4,800 – 4,450) ft = 298.5 psi

= 466.9 + 402.5 + 2,061.8 + 298.5

= 3,229.7 psi

Therefore, the hydrostatic pressure is 79.7 psi above the pore pressure.

C–3.7 Pressure to Lift the Casing

During some cementing treatments, there is a danger that the casing may be "pumped" out of the well. Conditions which favor such an occurrence are

1. lightweight pipe,
2. short pipe length,
3. large-diameter pipe,
4. high-density cement slurries,
5. low-density displacement fluids,
6. high annular friction pressures,
7. bridging in the annulus, and
8. backpressure.

Conditions 2 through 5 are all met when cementing surface or conductor casings. The lifting force is calculated as shown below, using the wellbore diagram shown in Fig. C–5.

Under static conditions,

$$\Delta F = (P_h \times A) - (W_c + W_d), \qquad \text{(C–4)}$$

where

P_h = hydrostatic pressure of well fluid (psi),

A = cross-sectional area (in.²),

W_c = casing weight (lb), and

W_d = weight of fluid inside casing (lb).

When pumping, the pump pressure acting on the cross-sectional area ID must be added to the above equation.

$$\Delta F = [(P_h \times A) + (P_p \times a)] - (W_c + W_d) \qquad \text{(C–5)}$$

ΔF is the differential force. If its value is positive, the casing may come out of the well. Working this problem backward, the value of P_p giving a ΔF value of zero is the critical pump pressure above which the casing may be "pumped from the well." The service crew should ensure that the pump pressure during the treatment does not exceed this value unless the casing is restrained.

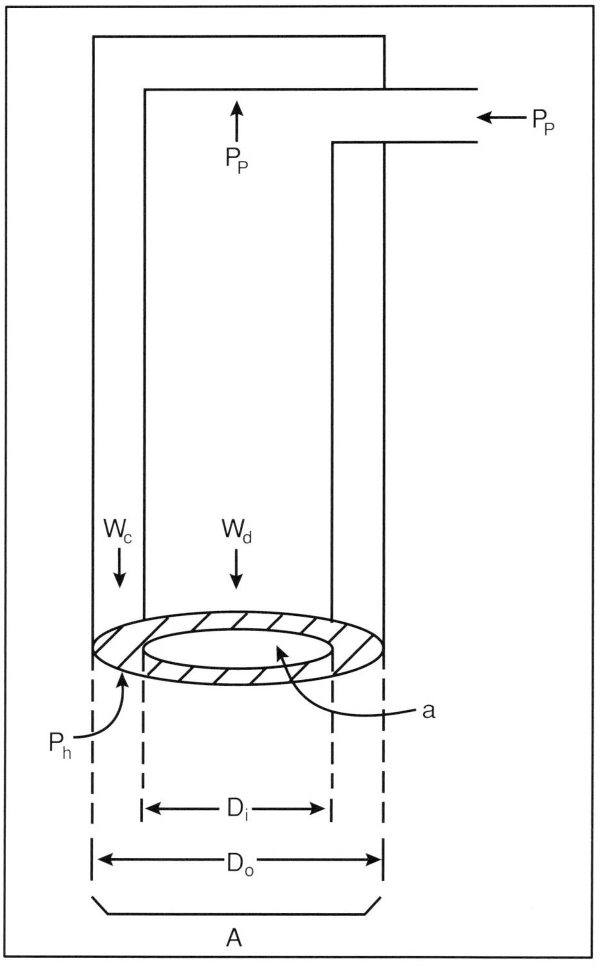

Figure C–5—Wellbore diagram for calculation of pressure to lift casing.

Example Calculation

Consider a 13³/₈-in., 61-lb/ft casing set at 800 ft with 14.8 lb/gal cement and 8.33 lb/gal water for displacement. Is there any danger of floating the casing out of the well?

Under static conditions,

$$\Delta F = [0.052 \times 14.8 \text{ lb/gal} \times 800 \text{ ft} \times (13.375)^2 \text{ in.} \times \tfrac{\pi}{4}] -$$

$$[(800 \text{ ft} \times 61 \text{ lb/ft}) + (800 \text{ ft} \times 0.052 \times 8.33 \text{ lb/gal} \times (12.515)^2 \text{ in.} \times \tfrac{\pi}{4})]$$

$$= 86{,}503 - 91{,}428 = -4{,}926 \text{ lbf}.$$

The negative force indicates that there is insufficient buoyancy to float the casing under static conditions. The pump pressure to bump the plug is

$$P_p = (14.8 \text{ lb/gal} - 8.33 \text{ lb/gal}) \times 0.052 \times 800 \text{ ft} = 269.2 \text{ psi}.$$

With the pump pressure acting on the inside of the casing, the additional force due to pump pressure F_p is

$$F_p = 269.2 \times (12.515 \text{ in.}^2 \times \frac{\pi}{4}) = 33{,}115 \text{ lbf}.$$

Therefore, the total force (F_T) acting on the casing is

$$F_T = -4{,}926 + 33{,}115 = 28{,}189 \text{ lbf}.$$

In this case, the hydrostatics are not sufficient to lift the casing from the well; however, with the pump pressure added, an unrestrained casing would be lifted out of the well.

This example uses single fluids in the casing and in the annulus. In practice, there may be several fluids in the annulus at the end of the job (tail, lead cement, spacer, etc.). In this case, the example is worked with the appropriate contributions from the different fluids.

C–4 PLUG BALANCING

Cement plugs, especially sidetrack or "kickoff" plugs, are normally "balanced" in the borehole. This means that the hydrostatic pressure in the annulus and in the work string are equal at the time of placement. This is done to prevent "U-tubing" after placement of the cement, and helps to prevent contamination and produce a strong plug. In practice, the cement is frequently slightly under-displaced from the balance point. This allows cement to fall while the pipe is being pulled, filling up the space that was occupied by the pipe. It also allows the pipe to be pulled "dry."

As stated above, when balancing a plug, the hydrostatic pressures in the pipe and in the annulus must be equal. To do this, the fluids used to displace the cement are the same fluids that are ahead of the cement, but in the reverse order. The heights of each of the fluids in the pipe and the annulus must be equal (Fig. C–6).

To ensure that the top of the plug is placed at the desired location, sufficient cement is normally run so that the top is well above the desired top. This excess may be left to set, and "dressed off" with the bit before it has developed its full strength. In some cases, the excess cement may be reversed out so that the top of the cement plug is where desired (Fig. C–7).

C–4.1 Equations

Volume of Cement

$$V_{cmt} = L \times C_h, \tag{C–6}$$

where

L = length of column of cement in open hole (ft),

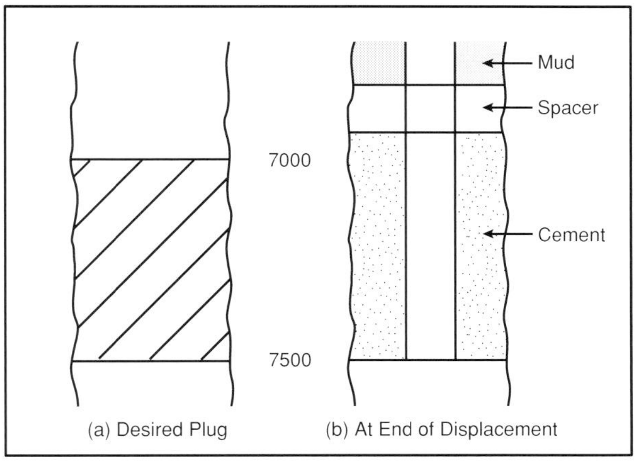

Figure C–6—Wellbore diagram for plug cementing calculations.

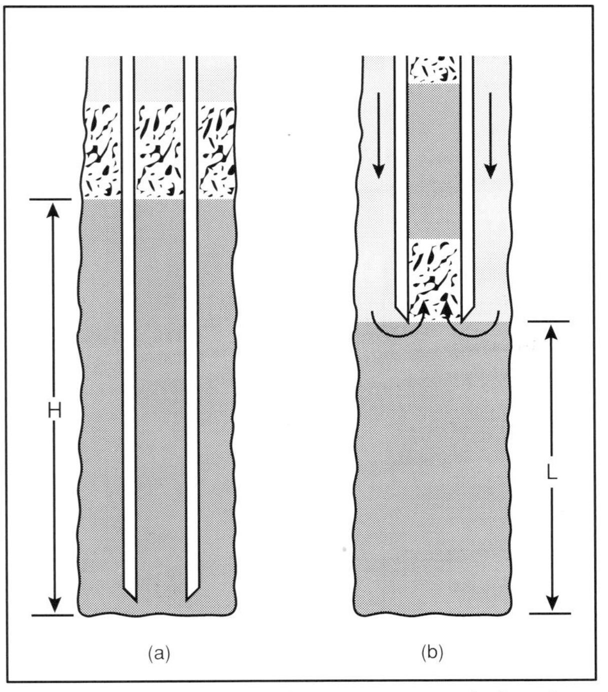

Figure C–7—Reversing of excess cement during plug cement.

and

C_h = capacity of open hole from standard tables (ft^3/ft).

Length of Balanced Plug (with work string in place)

$$L_{cmt} = \frac{V_{cmt}}{C_{an} + C_{tbg}}, \tag{C–7}$$

where

C_{an} = capacity of annulus between tubing or drillpipe and open hole (ft^3/ft), and

C_{tbg} = capacity of tubing or drillpipe (ft^3/ft).

Volume of Spacer Behind the Cement

$$V_{sp2} = \frac{V_{sp1}}{C_{an}} \times C_{tbg}, \qquad \text{(C–8)}$$

where

V_{sp1} = volume of spacer ahead of the cement.

Displacement Volume

$$V_d = C_{tbg} \times [D - (L_{cmt} + L_{sp2})] , \qquad \text{(C–9)}$$

where

D = depth of work string (bottom of cement plug) (ft), and

L_{sp2} = length of spacer behind (ft) = V_{sp2}/C_{tbg}.

C–4.2 Example Calculations

The well in question is shown in Fig. C–6.

Well Data

Open Hole: 8½-in.
Hole Capacity (C_h) = 0.3941 ft³/ft
Drillpipe: 4 in., 14.0 lb/ft
Pipe Capacity (C_{tbg}) = 0.01084 bbl/ft or 0.06084 ft³/ft
Annular Capacity (C_{an}) = 0.0546 bbl/ft or 0.3068 ft³/ft
Spacer Ahead of Cement (V_{sp1}): 10 bbl

Volume of Cement

$$V_{cmt} = L \times C_h = 500 \text{ ft} \times 0.3941 \text{ ft}^3/\text{ft} = 197.1 \text{ ft}^3$$

Length of Balanced Plug

$$L_{cmt} = \frac{V_{cmt}}{C_{an} + C_{tbg}} = \frac{197.1 \text{ ft}^3}{0.3068 + 0.06084 \text{ ft}^3/\text{ft}} = 536.1 \text{ ft}$$

Volume of Spacer Behind the Cement

$$V_{sp2} = \frac{V_{sp1}}{C_{an}} \times C_{tbg} = \frac{10\ bbl}{0.0546\ bbl/ft} \times 0.01084\ bbl/ft$$
$$= 2.0 \text{ bbl}$$

Displacement Volume

$$V_d = C_{tbg} \times [D - (L_{cmt} + L_{sp2})]$$
$$= 0.01084 \text{ bbl/ft} \times [7{,}500 \text{ ft} - (536.1 \text{ ft} + 2.0 \text{ bbl}/0.01084 \text{ bbl/ft})]$$
$$= 0.01084 \text{ bbl/ft} \times 6{,}779 \text{ ft} = 73.5 \text{ bbl}$$

Note that the use of inaccurate excess volumes may result in the plug being improperly balanced.

C–5 SQUEEZE CEMENTING

Calculations for squeeze cementing involve two separate sets of calculations; calculation of the volumes during the treatment and the pressures at various points in the wellbore at various stages of the treatment.

The volumes calculated for a squeeze treatment are:

- Cement slurry volume (from presumed void to be filled and/or experience),
- Volume to end of work string,
- Casing volume below the work string,
- Volume to spot water or cement to the perforations, and
- Volume to spot cement to the tool.

These are simple pipe volume calculations, taking into account the various fluids that are in the pipe.

Pressure calculations are necessary to ensure the safety of the well and to determine the anticipated pressure to squeeze the well. These include—

- Pressure to kill the well,
- Pressure to inject into the void—to avoid fracturing (maximum pressure limit) or to fracture if desired,
- Bottomhole pressure (at various stages of the treatment)—the sum of the hydrostatic pressure and pump pressure, less the friction pressure (because of low pump rates, the friction pressure is usually negligible during a squeeze treatment),
- Squeeze pressure—an established increment over injection pressure,
- Maximum surface pressure safely applied to the annulus,
- Forces on the casing at the packer,
- Maximum allowable squeeze pressure, and
- Maximum allowable pressure to reverse out .

Ideally, fracturing is not required and the final squeeze is the only calculation required. The final squeeze pressure is calculated based on the injection pressure. Normally, it is 500 or 1,000 psi over the injection pressure (Chapter 13). In some cases, because of well operations required prior to the time of cement setting, it is necessary to apply a higher squeeze pressure. A typical example is in reversing out excess cement, such as is done in coiled tubing squeezes. In this case, the hydrostatic pressure (because of the cement column in the coiled tubing) is greater than the desired final squeeze pressure. If so, the squeezed perforations have to be tested to the pressure necessary to reverse out.

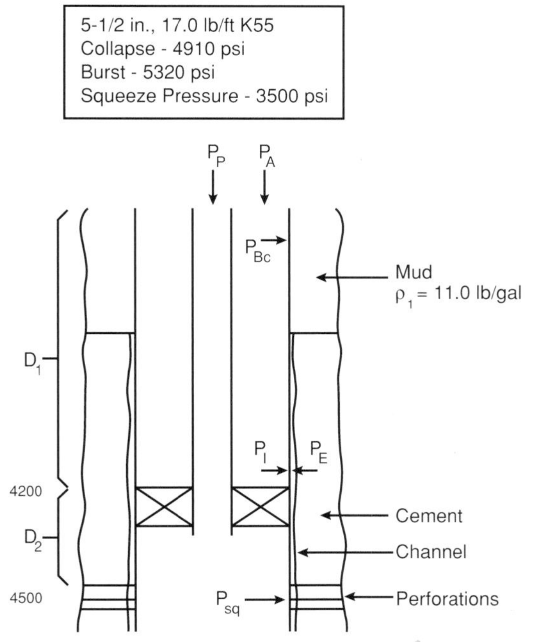

D_1 = depth from surface to the packer,
D_2 = distance from the perforations to the packer,
P_A = pressure applied to the annulus,
P_{Bc} = pressure tending to burst the casing,
P_E = external pressure to which the casing is exposed at the packer,
P_I = internal pressure on the casing at the packer,
P_P = pump pressure during the squeeze,
P_{Sq} = squeeze pressure experienced by the formation,
P_{CC} = collapse pressure rating of the casing, and
P_{CB} = burst pressure rating of the casing.

Figure C–8—Wellbore diagram for squeezing cement.

C–5.1 Example Calculations

Most squeeze cementing calculations involve simple volumetrics or hydrostatics; however, several of the calculations require further explanation, and are demonstrated with an example. The example wellbore situation is illustrated in Fig. C–8.

C–5.1.1 Forces on the Casing at the Packer

There are pressures on the casing at the packer from two directions. The greatest pressure is on the outside of the casing, P_E, tending to collapse it (Fig. C–8). This pressure is a combination of the hydrostatic pressure plus the squeeze pressure. To offset this pressure, it is sometimes necessary to apply pump pressure, P_A, to the annulus between the casing and the work string. This pressure must be sufficient to prevent the collapse of the casing, considering its condition. The external pressure is calculated by

$$P_E = P_{Sq} + (0.052 \times D_1 \times \rho_1) - (0.052 \times D_2 \times \rho_2)\ , \qquad \text{(C–10)}$$

where

ρ_1 = density of fluid in the casing/hole annulus,
ρ_2 = density of lightest fluid pumped during the treatment,
D_1 = depth to the packer, and
D_2 = distance from packer to squeeze perforations.

Of course, P_I is the sum of the pressure applied to the annulus and the hydrostatic pressure of the fluid in the work string/casing annulus. This pressure must be at least as large as the difference between the adjusted collapse pressure of the casing and the external pressure, P_E.

In this example, for a squeeze treatment requiring a final squeeze pressure of 3,900 psi and a completion fluid of 8.5-lb/gal brine (annular fluid and preflush),

$$P_E = P_{Sq} + (0.052 \times D_1 \times \rho_1) - (0.052 \times D_2 \times \rho_2)$$

$$P_E = 3{,}900 + (0.052 \times 4{,}200 \times 11.0) - (0.052 \times 300 \times 8.5), \text{ and}$$

$$P_E = 6{,}160 \text{ psi.}$$

With a collapse pressure (P_{cc}) of 4,910 psi, there must be at least 1,259 psi on the inside of the casing (P_I) at the packer to prevent collapse (assuming new casing). The hydrostatic pressure of the 8.5-lb/gal brine is 1,856 psi, so P_I is exceeded by 597 psi and should be safe if the casing is new. Depending on the condition of the casing, the operator may desire to have a larger safety margin, and require pressure held on the annulus. In addition, to make the appropriate calculation, the extent to which the collapse pressure has deteriorated must be determined.

C–5.1.2 Maximum Surface Pressure Safely Applied to the Annulus

As discussed above, in some squeeze treatments, it is necessary to apply pressure to the annulus between the work string and the casing to avoid collapsing the casing at the packer. This pressure can result in bursting the casing. The burst rating of the casing must be evaluated (and downgraded for old or used casing) to ensure that the casing will not be damaged.

C–5.1.3 Maximum Allowable Squeeze Pressure

Prior to the squeeze treatment, the maximum allowable pressure to be applied must be evaluated. The bottomhole squeeze pressure must be determined, considering the placement technique selected for the treatment and the integrity of all the pipe. If the placement technique involves fracturing the formation to be able to place the cement, then the integrity of the pipe is the main concern. If the treatment is to be done without fracturing, then the fracturing pressure must be evaluated to ensure that it is not exceeded. Of course, the pressure to which the pipe and the formation are exposed is the total of the hydrostatic pressure and the pump pressure. Once the maximum allowable bottomhole pressure has been determined, the hydrostatic pressure of the displacing fluids (and any cement column in the pipe) must be subtracted to determine the maximum allowable surface pressure during the squeeze.

C–5.1.4 Maximum Allowable Pressure When Reversing Out

The final pressure calculations required are those to which the well will be exposed while reversing out. In some cases, the only fluid reversed out will be the completion fluid. In others, the cement may be reversed out. The critical areas to be considered during this process are casing at the surface (burst), collapse of the tubing, and the squeezed interval itself. If reversing cement, it is best to assume that no cement was injected into the formation and that all the cement is in the tubing. The calculations are a simple matter of calculating the fill of the pipe and the hydrostatics of the fluids in the pipe and in the annulus. The difference in the two is the pump pressure required, and is the pressure to which the casing will be exposed. This pressure must be compared to the burst rating of the casing (appropriate to its condition). The total hydrostatic pressure in the work string (all cement in the work string) plus any between the work string and the perforations must be considered. If this pressure is greater than the final squeeze pressure, consideration should be given to a higher squeeze pressure. This is not necessary for the squeeze itself, but is necessary to ensure that the squeeze will not be broken down by the reversing process. The friction pressure may be significant, especially in the case of small-diameter work strings. If this is the case, it should also be considered. This is best done by placement simulators, but can be done by hand (Chapter 4). At any rate, a limit should be set on the pump pressure to reverse. This limit must protect the casing.

C–6 CALCULATIONS FOR FOAMED CEMENT JOBS

Most cementing companies use computer programs to design foamed cement jobs; however, it is useful to know how to perform the calculations manually. As discussed in Chapter 14, foamed cement jobs can be divided into two types (depending on the method of scheduling the nitrogen): *constant nitrogen (or air) ratio*, and *constant density*. During a constant-ratio job, the nitrogen is added to the base cement slurry at a constant rate (scf N_2/bbl of slurry). Because of compression, this method results in a cement with variable density—lightest at the top, heaviest at the bottom. A constant density job is actually one in which several stages of foamed cement, each with a constant ratio of nitrogen, are used. The nitrogen ratios are calculated so that each stage has the same average density at its final position in the annulus; however, the density varies within each stage because of differences in hydrostatic pressure. In this section, the design calculations for the constant-density method are presented. The same calculations may be used to design a constant-rate job. In that case, the entire interval is considered as one stage. The following items are calculated to design a foamed cement job:

- Minimum fracture pressure (less a safety factor),
- Hydrostatic pressure of the fluids above the weak zone (if tail slurry is above the weak zone, it must be included),
- Allowable average density of the foamed cement,
- Number of stages,
- Hydrostatic pressure at the midpoint of each stage,
- Nitrogen requirement for each stage based on the midpoint hydrostatic pressure,
- Foam quality for each stage,
- Yield of the foamed cement for each stage,
- Volume of each stage (from length of stage and caliper),
- Hydrostatic pressure as each successive stage of foamed cement enters the annulus (this is necessary as the quality of the first stages of foamed cement is low at greater depths because of compression of the gas; thus, their density as they pass the shoe will be higher), and
- Pump schedule, including:
  - base slurry volume,
  - nitrogen ratio,
  - nitrogen volume,
  - nitrogen pump rate, and
  - foamer pump rate.

Once these calculations are made, a series of tables and graphs should be constructed to show the parameters for each stage. The following example illustrates this procedure (Fig. C–9).

Design Data

TD:	9,000 ft
Lost Circulation:	5,000 to 7,700 ft
Fracture Gradient:	0.512 psi/ft at 7,700 ft

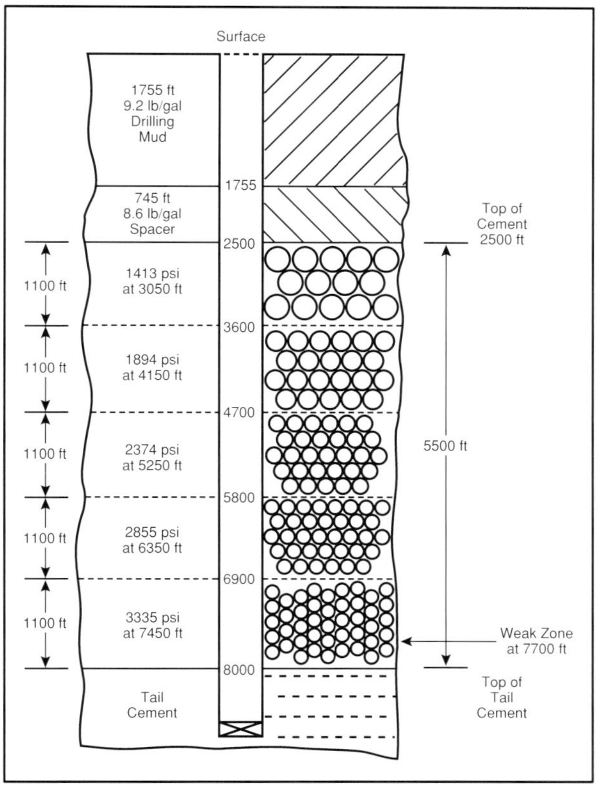

Figure C–9—Wellbore diagram for staged foamed cementing.

Full Circulation:	9.2 lb/gal mud
Casing:	5½ in., 17.0 lb/ft
Hole Caliper:	9¼ in.
Static Temperature:	180°F at 9,000 ft
	165°F at 7,700 ft
Circulating Temperature:	160°F at 7,450 ft
	150°F at 6,350 ft
	140°F at 5,250 ft
	130°F at 4,150 ft
	120°F at 3,050 ft
Spacer:	745 ft at 8.6 lb/gal
Top of Tail Cement:	8,000 ft
Top of Cement:	2,500 ft
Base Cement Density:	14.2 lb/gal
Base Cement Yield:	1.29 ft³/sk

Minimum Fracture Pressure

$$P_{Fmin} = (0.512 \text{ psi/ft} \times 7{,}700 \text{ ft}) - 500 \text{ psi (safety factor)} = 3{,}442 \text{ psi}$$

Hydrostatic Pressure of the Fluids Ahead of the Foamed Cement [3]

$$P_h = 0.052 \times [(1{,}755 \text{ ft} \times 9.2 \text{ lb/gal}) + (745 \text{ ft} \times 8.6 \text{ lb/gal})] = 1{,}173 \text{ psi}$$

Allowable Average Density of the Foamed Cement

$$P_f = \frac{P_{f\,min} - P_h \text{ Above Foamed Cement}}{0.052 \times \text{Height of Foamed Cement (above weak formation)}} \qquad \text{(C–11)}$$

$$= \frac{3{,}442 \text{ psi} - 1{,}173 \text{ psi}}{0.052 \times (7{,}700 \text{ ft} - 2{,}500 \text{ ft})}$$

$$= 8.4 \text{ lb/gal}$$

Number of Stages

Divide into intervals of 1,000 to 1,500 ft.

Total interval is 8,000 – 2,500 ft = 5,500 ft.

Divide into 5 stages of 1,100 ft each.

Hydrostatic Pressure at the Midpoint of Each Stage

$P_h = P_h$ above stage + P_h to midpoint of stage

Stage No. 1 (top)

$$\begin{aligned} P_{h1} &= 1{,}173 \text{ psi} + [(\text{Stage \#} - 0.5) \times 1{,}100 \text{ ft} \times \rho f \times 0.052] \\ &= 1{,}173 \text{ psi} + (0.5 \times 1{,}100 \text{ ft} \times 8.4 \text{ lb/gal} \times 0.052) \\ &= 1{,}413 \text{ psi} \end{aligned}$$

Stage No. 2

$$\begin{aligned} P_{h2} &= 1{,}173 \text{ psi} + (1.5 \times 1{,}100 \text{ ft} \times 8.4 \text{ lb/gal} \times 0.052) \\ &= 1{,}894 \text{ psi} \end{aligned}$$

[3] 0.052 psi• gal / lb• ft is a conversion factor.

Stage No. 3

P_{h3} = 1,173 psi + (2.5 × 1,100 ft × 8.4 lb/gal × 0.052)
= 2,374 psi

Stage No. 4

P_{h4} = 1,173 psi + (3.5 x 1,100 ft × 8.4 lb/gal × 0.052)
= 2,855 psi

Stage No. 5 (bottom)

P_{h5} = 1,173 psi + (4.5 × 1,100 ft × 8.4 lb/gal × 0.052)
= 3,335 psi

Nitrogen Requirement for Each Stage Based on the Midpoint

$$\text{Nitrogen Density } (\rho_{N2}) = (1.724 \times 10^{-3}) \times \text{Nitrogen Volume Factor}^4 \text{ (scf/bbl)} \quad \text{(C–12)}$$

$$\text{Foamed Cement Quality } (Q) = -\frac{\rho_f - \rho_{N2}}{\rho_{bs} - \rho_{N2}} \quad \text{(C–13)}$$

where ρ_f = foamed cement density,

ρ_{N2} = nitrogen density, and

ρ_{bs} = base slurry density.

$$\text{Foamed Cement Quality}(Q) = -\frac{\rho_f - \rho_{N2}}{\rho_{bs} - \rho_{N2}} \quad \text{(C–14)}$$

$$\text{Annular Volume} = \text{Length} \times \text{Annular Capacity} \quad \text{(C–15)}$$

$$\text{Cement Requirement (sk)} = \frac{\text{Annular Volume}}{\text{Foamed Cement Yield}} \quad \text{(C–16)}$$

$$\text{Nitrogen Requirement} = \text{Annular Volume} \times Q \quad \text{(C–17)}$$

The Nitrogen Requirement refers to the volume required at circulating temperature and pressure. For job-design purposes, this value must be converted to the equivalent volume of nitrogen in standard cubic feet (at STP).

$$\text{Nitrogen Volume (STP) (scf)} = \text{Nitrogen Requirement} \times \text{Nitrogen Volume Factor} \quad \text{(C–18)}$$

Stage No. 1, P_h = 1,413 psi

ρ_{N2} = 1.724 × 10^{-3} × Nitrogen Volume Factor (scf/bbl)
= 1.724 × 10^{-3} × 476 scf/bbl
= 0.8206 lb/gal

[4]The Nitrogen Volume Factor can be calculated based upon pressure and BHCT, or more easily looked up in standard tables published by most cementing companies.

$$Q = \frac{1 - 8.4 - 0.8206}{14.2 - 0.8206} = 0.4335$$

$$\text{Foamed Cement Yield} = \frac{1.29 \text{ ft}^3/\text{sk}}{1 - 0.4335} = 2.28 \text{ ft}^3/\text{sk}$$

$$\text{Annular Volume} = 1{,}100 \text{ ft} \times 0.3017 \text{ ft}^3/\text{ft} = 331.9 \text{ ft}^3$$

$$\text{Cement Requirement (sk)} = \frac{331.9 \text{ ft}^3}{2.28 \text{ ft}^3/\text{sk}} = 145.6 \text{ sk}$$

$$\text{Nitrogen Requirement} = 331.9 \text{ ft}^3 \times 0.4335 = 143.9 \text{ ft}^3$$

$$\text{Nitrogen Volume}_{(STP)} = 143.9 \text{ ft}^3 \times 0.178 \text{ bbl/ft}^3 \times 476 \text{ scf/bbl} = 12{,}192 \text{ scf}$$

Similarly, the requirements for the other stages are calculated, and tables (such as that shown in Table C–6) can be built.

Hydrostatic Pressure as Each Stage of Foamed Cement Enters the Annulus

Actually, this should be calculated for the position of each stage immediately above the weak formation(s). Since a job designed for constant density uses a slurry with the lowest concentration of nitrogen in the first stages, these stages will be significantly more dense when they pass the weaker formations below. Therefore, to ensure the integrity of the well, the hydrostatic pressure exerted on the weak formations must be calculated, and compared to the fracturing pressure of these formations. To do this, the following steps must be followed.

1. Determine the volume occupied by each stage at the weak zone.
2. Calculate its length based on the annular capacity.
3. Calculate the hydrostatic pressure of the fluids in the annulus above.
4. Calculate the hydrostatic pressure of the foamed cement stage(s).
5. Add the two to get the total hydrostatic pressure.
6. Compare this value to the fracturing pressure of the weak formation.

Stage	1	2	3	4	5
P_h (psi)	1,413	1,894	2,374	2,855	3,335
ρ_{N2} (lb/gal)	0.8206	1.0689	1.2930	1.4960	1.6810
Quality	0.4335	0.4417	0.4494	0.4565	0.4633
Foamed Cement Yield (ft^3/sk)	2.28	2.31	2.34	2.37	2.40
Annular Volume (ft^3)	331.9	331.9	331.9	331.9	331.9
Cement Requirement (sk)	145.6	143.6	141.8	140.0	138.3
Nitrogen Requirement (scf)	12,192	16,179	19,912	23,409	26,714
TOTAL CEMENT REQUIREMENT: 709.3 sk					
TOTAL NITROGEN REQUIREMENT: 98,406 scf					

Table C–6—Calculation of nitrogen requirement for each stage.

7. If the formation will be fractured, consider a constant nitrogen rate or a hybrid job (hybrid is several stages of foamed cement with different designed densities).

From the preceding example, this calculation is performed as shown below.

1. *Determine the volume occupied by each stage at the weak zone.*

   For Stage No. 1, 145.6 sk or (145.6 × 1.29 ft³/sk × 0.178 bbl/ft³) = 33.4 bbl of cement slurry are required. The nitrogen requirement is 12,192 scf. The weak zone is at 7,700 ft. The fluids ahead of the cement are 9.2-lb/gal mud (0.4784 psi/ft), and 745 ft of spacer at 8.6-lb/gal (0.4472 psi/ft). Assuming the foamed cement occupies 850 ft (because of compression), the length of the mud column is 7,700 – 850 (cement) – 745 (spacer) = 6,105 ft.

   The hydrostatic pressure from the mud is 6,105 ft × 0.4784 psi/ft = 2,920 psi, and the hydrostatic pressure from the spacer is 745 ft × 0.4472 psi/ft = 333 psi; thus, the total hydrostatic pressure is 2,920 + 333 = 3,253 psi. Reading from the nitrogen tables, the volume occupied by 12,192 scf of nitrogen is 12.7 bbl at 160°F. Thus, the volume of the foamed cement slurry is 12.7 + 33.4 = 46.1 bbl.

2. *Calculate its length based on the annular capacity.*

   With an annular capacity of 0.3017 ft³/ft and 46.1 bbl of slurry, the fill-up is

$$\frac{46.1 \text{ bbl}}{0.3017 \text{ ft}^3/\text{ft} \times 0.178 \text{ bbl/ft}^3} = 858 \text{ ft.}$$

If this result had not been close to the assumed foamed cement length of 850 ft (Step No. 1), the calculation would be repeated using an adjusted length.

3. *Calculate the hydrostatic pressure of the fluids in the annulus above.*

   The height of the mud column is 7,700 – 745 – 858 = 6,097 ft. Therefore, the hydrostatic pressure is 6,097 ft × 0.4784 psi/ft = 2,916.8 psi. For the spacer, the hydrostatic pressure is 745 ft × 0.4472 psi/ft = 333.1 psi.

4. *Calculate the hydrostatic pressure of the foamed cement stage(s)*

$$\begin{aligned}\rho_{N2} &= 1.724 \times 10^{-3} \times \text{Volume Factor (scf/bbl)}\\ &= 1.724 \times 10^{-3} \times 950 \text{ scf/bbl}\\ &= 1.638 \text{ lb/gal.}\end{aligned}$$

$$Q = \frac{12.7 \text{ bbl}}{12.7 \text{ bbl} + 33.4 \text{ bbl}} = 0.2755$$

From the previous calculation of foamed cement quality,

$$\begin{aligned}\rho_f &= (1 - Q)(\rho_{bs} - \rho_{N2}) + \rho_{N2}\\ &= (1 - 0.2755)(14.2 - 1.638) + 1.638\\ &= 10.74 \text{ lb/gal.}\end{aligned}$$

Therefore, the hydrostatic pressure from foamed cement is 10.74 lb/gal × 0.052 psi/ft/lb/gal × 858 ft = 479.2 psi.

5. *Add the three hydrostatic pressures to obtain the total hydrostatic pressure.*

   2,916.8 psi + 333.1 psi + 479.2 psi = 3,729.1 psi

6. *Compare this value to the fracturing pressure of the weak formation.*

   The fracturing pressure of the weak formation is 3,942 psi.

Note that the pressure calculated in the previous step does not exceed the fracturing pressure of the weak formation, but does exceed the safety margin set in the first step. This is what would be expected based on the method for calculating the foamed cement density, based on the fracturing pressure less the safety margin. This calculation should be repeated as each stage passes the weak formation.

Job Execution Tables

It is helpful for control of the job to construct tables of the pumping schedule containing the following information.

- Base Slurry Volume
- Nitrogen Ratio
- Nitrogen Volume
- Nitrogen Pump Rate
- Foamer Pump Rate

For the preceding example, the table would be as shown in Table C–7.

SUGGESTED READING

American Petroleum Institute: "API Spec 10", Specification for Materials and Testing for Well Cements, fourth edition, API, Dallas, (1988).

Dowell Schlumberger: *Dowell Schlumberger Field Data Handbook*, Pub. No. TSL 0538, 1985.

Job Schedule per Stage

Stage	1	2	3	4	5
Base Slurry Volume (bbl)	33.4	32.9	32.6	32.1	31.8
Nitrogen Ratio (scl/bbl base slurry)	365	492	611	729	840
Nitrogen Volume (scf)	12,192	16,179	19,912	23,409	26.714

Nitrogen and Foamer Rate

Base Slurry Rate (BPM)	3	4	5	6	7
N_2 Rate (scf/min)					
Stage No. 1	1,095	1,460	1,825	2,190	2,555
Stage No. 2	1,475	1,967	2,459	2,951	3,442
Stage No. 3	1,832	2,443	3,054	3,664	4,276
Stage No. 4	2,188	2,917	3,646	4,376	5,105
Stage No. 5	2,520	3,360	4,200	5,040	5,880
Foamer Rate (gal/min):					
All Stages	1.31	1.74	2.18	2.61	3.05

Table C–7—Foamed cement job execution table.

INDEX